ANNUAL REVIEW OF PHYSIOLOGY

EDITORIAL COMMITTEE (1981)

T. E. ANDREOLI
R. M. BERNE
I. S. EDELMAN
A. P. FISHMAN
J. G. FORTE
J. GERGELY
J. E. HEATH
J. F. HOFFMAN
D. E. KOSHLAND, JR.
D. T. KRIEGER
P. H. PATTERSON

Responsible for the organization of Volume 44
(Editorial Committee, 1980)

T. E. ANDROELI
R. M. BERNE
W. R. DAWSON
I. S. EDELMAN
A. P. FISHMAN
J. GERGELY
J. E. HEATH
D. E. KOSHLAND, JR.
D. T. KRIEGER
S. G. SCHULTZ

Production Editor	R. L. BURKE
Indexing Coordinator	M. A. GLASS
Subject Indexer	T. GREENBERG

ANNUAL REVIEW OF PHYSIOLOGY

VOLUME 44, 1982

I. S. EDELMAN, *Editor*
Columbia University College of Physicians and Surgeons

ROBERT M. BERNE, *Associate Editor*
University of Virginia Medical School

ANNUAL REVIEWS INC.   4139 El Camino Way   PALO ALTO, CALIFORNIA 94306 USA

ANNUAL REVIEWS INC.
Palo Alto, California, USA

COPYRIGHT © 1982 BY ANNUAL REVIEWS INC., PALO ALTO, CALIFORNIA, USA. ALL RIGHTS RESERVED. The appearance of the code at the bottom of the first page of an article in this serial indicates the copyright owner's consent that copies of the article may be made for personal or internal use, or for the personal or internal use of specific clients. This consent is given on the condition, however, that the copier pay the stated per-copy fee of $2.00 per article through the Copyright Clearance Center, Inc. (21 Congress Street, Salem, MA 01970) for copying beyond that permitted by Sections 107 or 108 of the US Copyright Law. The per-copy fee of $2.00 per article also applies to the copying, under the stated conditions, of articles published in any *Annual Review* serial before January 1, 1978. Individual readers, and nonprofit libraries acting for them, are permitted to make a single copy of an article without charge for use in research or teaching. This consent does not extend to other kinds of copying, such as copying for general distribution, for advertising or promotional purposes, for creating new collective works, or for resale. For such uses, written permission is required. Write to Permissions Dept., Annual Reviews Inc., 4139 El Camino Way, Palo Alto, CA 94306 USA.

International Standard Serial Number: 0066-4278
International Standard Book Number: 0-8243-0344-X
Library of Congress Catalog Card Number: 39-15404

Annual Reviews Inc. and the Editors of its publications assume no responsibility for the statements expressed by the contributors to this *Review*.

PRINTED AND BOUND IN THE UNITED STATES OF AMERICA

PREFACE

With Volume 44, I bring to a close ten years of association with the *Annual Review of Physiology* as a member of the Editorial Committee, an Associate Editor, and Editor. Julius H. Comroe, Jr., then Editor of the *ARP,* persuaded me to join the Editorial Board in 1971; his leadership and foresight played a key role in the evolution of *ARP* during the last decade. As in all of our other associations it was a deeply rewarding experience to have served with him. I also enjoyed a rich intellectual and personal harvest in working as Associate Editor during Ernst Knobil's tenure as Editor.

Volume 44 continues the presentation of reviews in sections representative of the various subspecialties of physiology, the format first introduced with Volume 41. The inexorable growth of the body of physiological information and its attendant demands for specialization dictated first the spin-off of neurophysiology into the *Annual Review of Neuroscience* (now in its fifth year) and then the organization of *ARP* into subspecialties. It was the hope of the Editorial Committee that despite sectionalization, the broad range of topics would provide the reader with access to almost all branches of physiology and serve to offset the trend towards tunnel vision in physiological scholarship.

I wish to express my deep gratitude to my colleagues, the Section Editors, who served so ably during my tenure as Editor and are inscribed in my personal honor roll: Thomas E. Andreoli, Robert M. Berne, William R. Dawson, Alfred P. Fishman, John G. Forte, John Gergely, James E. Heath, Joseph F. Hoffman, Samuel M. McCann, and Stanley G. Schultz. As Section Editors they carried the full burden of identifying timely topics, recruiting authors, and editing the chapters. Special thanks are also due to the Editors of the Special Topics Sections, Gerald Fishbach, Arthur Karlin, Daniel E. Koshland, Jr., and Paul H. Patterson.

Of course, the success of *ARP* ultimately depends on the high quality of the reviews. The authors of these reviews are too numerous to be individually acknowledged, but their contributions have certainly earned the gratitude of all of us.

To bring the *ARP* to the physiological public requires the efforts of many dedicated and skilled technical and professional personnel, namely the staff at Annual Reviews Inc: Dick Burke, Production Editor of *ARP,* and William Kaufmann, Editor-in-Chief of Annual Reviews Inc., were especially supportive throughout my tenure. I shall always be grateful for their help and friendship. It has been a privilege (and not too onerous) for me to have served a periodical that has contributed so much to scholarship in physiology for so long—in fact, for 44 years. I am especially pleased that the task of Editor of *ARP* will now be taken up by Robert M. Berne, a distinguished physiological researcher and scholar.

I. S. Edelman
Editor

ANNUAL REVIEWS INC. is a nonprofit corporation established to promote the advancement of the sciences. Beginning in 1932 with the *Annual Review of Biochemistry*, the Company has pursued as its principal function the publication of high quality, reasonably priced *Annual Review* volumes. The volumes are organized by Editors and Editorial Committees who invite qualified authors to contribute critical articles reviewing significant developments within each major discipline. Annual Reviews Inc. is administered by a Board of Directors, whose members serve without compensation.

1982 Board of Directors, Annual Reviews Inc.

Dr. J. Murray Luck, Founder and Director Emeritus of Annual Reviews Inc.
Professor Emeritus of Chemistry, Stanford University
Dr. Joshua Lederberg, President of Annual Reviews Inc.
President, The Rockefeller University
Dr. James E. Howell, Vice President of Annual Reviews Inc.
Professor of Economics, Stanford University
Dr. William O. Baker, *Retired Chairman of the Board, Bell Laboratories*
Dr. Robert W. Berliner, *Dean, Yale University School of Medicine*
Dr. Sidney D. Drell, *Deputy Director, Stanford Linear Accelerator Center*
Dr. Eugene Garfield, *President, Institute for Scientific Information*
Dr. Conyers Herring, *Professor of Applied Physics, Stanford University*
Dr. D. E. Koshland, Jr., *Professor of Biochemistry, University of California, Berkeley*
Dr. William F. Miller, *President, SRI International*
Dr. William D. McElroy, *Chancellor, University of California, San Diego*
Dr. Esmond E. Snell, *Professor of Microbiology and Chemistry,*
University of Texas, Austin
Dr. Harriet A. Zuckerman, *Professor of Sociology, Columbia University*

Management of Annual Reviews Inc.

John S. McNeil, Publisher and Secretary-Treasurer
Alister Brass, M.D., Editor-in-Chief
Sharon E. Hawkes, Production Manager
Mickey G. Hamilton, Promotion Manager
Donald S. Svedeman, Business Manager

Publications

PLANNED FOR 1983: Annual Review of Immunology

ANNUAL REVIEWS OF
Anthropology
Astronomy and Astrophysics
Biochemistry
Biophysics and Bioengineering
Earth and Planetary Sciences
Ecology and Systematics
Energy
Entomology
Fluid Mechanics
Genetics
Materials Science
Medicine

Microbiology
Neuroscience
Nuclear and Particle Science
Nutrition
Pharmacology and Toxicology
Physical Chemistry
Physiology
Phytopathology
Plant Physiology
Psychology
Public Health
Sociology

SPECIAL PUBLICATIONS
Annual Reviews Reprints:
  Cell Membranes, 1975–1977
  Cell Membranes, 1978–1980
  Immunology, 1977–1979
Excitement and Fascination
  of Science, Vols. 1 and 2
History of Entomology
Intelligence and Affectivity,
  by Jean Piaget
Telescopes for the 1980s

For the convenience of readers, a detachable order form/envelope is bound into the back of this volume.

Annual Review of Physiology
Volume 44, 1982

CONTENTS

GASTROINTESTINAL PHYSIOLOGY

Introduction, *Eugene D. Jacobson, Section Organizer*   1

Stomach Blood Flow and Acid Secretion, *Paul H. Guth*   3

Local Control of Intestinal Oxygenation and Blood Flow, *A. P. Shepherd*   13

Relationship Between Intestinal Blood Flow and Motility, *C. C. Chou*   29

Relationships Between Intestinal Absorption and Hemodynamics, *David Mailman*   43

Physiological Regulation of the Hepatic Circulation, *Peter D. I. Richardson and Peter G. Withrington*   57

Measurement of Gastrointestinal Blood Flow, *Barry L. Tepperman and Eugene D. Jacobson*   71

COMPARATIVE AND INTEGRATIVE PHYSIOLOGY

Introduction, *James E. Heath, Section Editor*   83

Brain Cooling in Endotherms in Heat and Exercise, *M. A. Baker*   85

Energetics and Mechanics of Terrestrial Locomotion, *C. Richard Taylor and Norman C. Heglund*   97

Avian Flight Energetics, *J. M. V. Rayner*   109

Energetics of Locomotion in Warm-Bodied Fish, *E. Don Stevens and Andrew E. Dizon*   121

Energetics of Locomotion in Endothermic Insects, *James Edward Heath and Maxine S. Heath*   133

RENAL AND ELECTROLYTE PHYSIOLOGY

Introduction, *Thomas E. Andreoli, Section Editor*   145

Recent Advances in Renal Morphology, *Ruth Ellen Bulger and Dennis C. Dobyan*   147

Heterogeneity of Tubular Transport Processes in the Nephron, *Christine A. Berry*   181

Biological Importance of Nephron Heterogeneity, *Larry A. Walker and Heinz Valtin*   203

Respiratory Physiology

Introduction, *Alfred P. Fishman, Section Editor*	221
Structural Bases for Metabolic Activity, *Una S. Ryan*	223
Processing of Endogenous Polypeptides by the Lungs, *James W. Ryan*	241
Pulmonary Metabolism of Prostaglandins and Vasoactive Peptides, *Sami I. Said*	257
The Fate of Circulating Amines within the Pulmonary Circulation, *C. N. Gillis and B. R. Pitt*	269
Chemotactic Mediators in Neutrophil-Dependent Lung Injury, *Joseph C. Fantone, Steven L. Kunkel, and Peter A. Ward*	283

Cell and Membrane Physiology

Introduction, *John Gergely, Section Editor*	295
Structure and Function of the Calcium Pump Protein of Sarcoplasmic Reticulum, *Noriaki Ikemoto*	297
Turnover of Acetylcholine Receptors in Skeletal Muscle, *D. W. Pumplin and D. M. Fambrough*	319
The Development of Sarcoplasmic Reticulum Membranes, *Anthony Martonosi*	337
Ionic Channels in Skeletal Muscle, *E. Stefani and D. J. Chiarandini*	357

Cardiovascular Physiology

Introduction, *Nick Sperelakis, Guest Editor, and Robert M. Berne, Section Editor*	373
Myocardial Membranes: Regulation and Function of the Sodium Pump, *Tai Akera and Theodore M. Brody*	375
Electrogenic Na Pumping in the Heart, *H. G. Glitsch*	389
Phosphorylation of the Sarcoplasmic Reticulum and Sarcolemma, *Michihiko Tada and Arnold M. Katz*	401
The Slow Inward Calcium Current in the Heart, *Terence F. McDonald*	425
Sodium-Calcium Exchange in the Heart, *G. A. Langer*	435
Calcium Release from Cardiac Sarcoplasmic Reticulum, *Saul Winegrad*	451
The Action of Cardiotoxins on Cardiac Plasma Membranes, *M. Lazdunski and J. F. Renaud*	463

Adrenergic Receptors in the Heart, *Brian B. Hoffman and Robert J. Lefkowitz*	475
Mechanisms of Cardiac Arrhythmias, *Joseph F. Spear and E. Neil Moore*	485

SPECIAL TOPIC: CHEMOTAXIS AND MOTILITY

Introduction, *D. E. Koshland, Jr., Section Editor*	499
Bacterial Chemotaxis, *Alan Boyd and Melvin Simon*	501
The Physiological Basis of Taxes in *Paramecium*, *Ching Kung and Yoshiro Saimi*	519
Chemotaxis in *Dictyostelium*, *Günther Gerisch*	535
Leukocyte Chemotaxis, *Elliott Shiffmann*	553

ENDOCRINOLOGY AND METABOLISM

Introduction, *Dorothy T. Krieger, Section Editor*	569
Behavioral Effects of Neuropeptides: Endorphins and Vasopressin, *George F. Koob and Floyd E. Bloom*	571
The Role of the Central Nervous System in the Control of Ovarian Function in Higher Primates, *C. R. Pohl and E. Knobil*	583
Neuroendocrine Control Mechanisms and the Onset of Puberty, *Edward O. Reiter and Melvin M. Grumbach*	595
The Evolution of Peptide Hormones, *Hugh D. Niall*	615
Post-Translational Proteolysis in Polypeptide Hormone Biosynthesis, *Kevin Docherty and Donald F. Steiner*	625
Receptors for Peptide Hormones: Alterations in Diseases of Humans, *Jesse Roth and Simeon I. Taylor*	639
Polypeptide and Amine Hormone Regulation of Adenylate Cyclase, *G. D. Aurbach*	653
Calmodulin in Endocrine Cells, *Anthony R. Means and James G. Chafouleas*	667

INDEXES

Author Index	683
Subject Index	713
Cumulative Index of Contributing Authors, Volumes 40-44	735
Cumulative Index of Chapter Titles, Volumes 40-44	738

OTHER REVIEWS OF INTEREST TO PHYSIOLOGISTS

From the *Annual Review of Neuroscience,* Volume 5 (1982)

Morphology of Cutaneous Receptors, A. Iggo and K. H. Andres

Electroreception, Theodore Holmes Bullock

Signaling of Kinesthetic Information by Peripheral Sensory Receptors, P. R. Burgess, Jen Yu Wei, F. J. Clark, and J. Simon

Where Does Sherrington's "Muscular Sense" Originate? Muscles, Joint, Corollary Discharges?, P. B. C. Matthews

The Synapse: From Electrical to Chemical Transmission, John C. Eccles

Developmental Neurobiology and the Natural History of Nerve Growth Factor, Rita Levi-Montalcini

From the *Annual Review of Pharmacology and Toxicology,* Volume 21 (1981)

Genetic Investigation of Adenylate Cyclase: Mutations in Mouse and Man, Zvi Farfel, Michael R. Salomon, and Henry R. Bourne

Mechanism of Action of Barbiturates, I. K. Ho and R. Adron Harris

Genetic Mechanisms Controlling the Induction of Polysubstrate Monooxygenase (P-450) Activities, Daniel W. Nebert, Howard K. Eisen, Masahiko Negishi, Matti A. Lang, Leonard M. Hjelmeland, and Allan B. Okey

Victor E. Hall

1901–1981

The Editorial Committee dedicates this volume of the *Annual Review of Physiology* to Victor Hall, Associate Editor of the *Review* from its beginning in 1938 until 1947 when he became Editor, a post he held until retirement in 1971. Under his guidance the *Review* came to fulfill the purpose which he formulated as "to sift the excellent from the merely good, to recognize and evaluate new trends in physiological thought, to warn against blind fashion, and to keep the enduring questions concerning the function of living matter firmly before us." The *Review,* as it continues to fulfill its role for the international community of physiologists, is a legacy of Victor Hall's imaginative approach and constant striving for excellence.

GASTROINTESTINAL PHYSIOLOGY

Introduction, Eugene D. Jacobson, *Section Organizer*

The gastrointestinal circulation has received comparatively little attention from either circulatory or gastrointestinal physiologists. The splanchnic blood vessels are relatively inaccessible and their organization is complex. The numerous organs they serve are functionally disparate and are composed of heterogenous tissues.

These morphological characteristics necessitate precise methods that measure not only the total blood flow to a gastrointestinal organ but also the distribution of that flow to the different tissues of the organ. Furthermore, such technics must not alter organ function, if the investigator hopes to explore relationships between blood flow and the physiology of the organ. In addition, methods are needed to determine tissue oxygenation during states of increased cellular activity. Unfortunately, available methods for assessing hemodynamics and metabolism of individual tissues are not well-developed.

Gastric mucosal blood flow appears to be regulated by changes in resistance vessels of the submucosa, which are arranged in series with mucosal vessels, and by the degree of secretory activity in the gastric glands of the mucosa. Presumably, the increased mucosal metabolism associated with secretion of acid gastric juice prompts a metabolic hyperemia.

Within the circulation of the gut two mechanisms regulate blood flow at the local level. With enhanced metabolism of parenchymal cells of the villi, oxygen is more avidly consumed, thereby lowering tissue pO_2 and relaxing nearby vascular smooth muscle of arterioles and precapillary sphincters. The result is an overall increase in mucosal blood flow and a recruitment of more capillaries to carry blood through the nutrient circulation and provide more oxygen to meet the enhanced metabolic demand. There are also intrinsic myogenic mechanisms in vascular smooth muscle that can autoregulate blood flow in the face of fluctuating blood pressures.

Mechanical activity of visceral smooth muscle in the gut may prompt opposing effects on the circulation—namely, a hyperemia reflecting increased metabolism of the muscle, or an increased resistance to blood flow reflecting extravascular compression by the contracting muscle. In addition, stimuli of intestinal motility may also exhibit direct vasoactive effects. Therefore, the relationship between intestinal motor activity and blood flow does not conform to a simple description and will require introduction of methods that can dissect tissue blood flow from total flow to the gut and correlate hemodynamic and electromyographic events.

Another primary intestinal function, absorption of nutrients, also involves a discrete component of the total intestinal circulation. Absorptive-site blood flow and the absorptive process interrelate through several local phenomena, such as countercurrent exchange, washout of absorbed substances, mucosal perfusion, oxygen delivery, and passive effects of Starling forces on absorption. Again, available technology has not permitted sufficiently precise estimations of concomitant occurrences.

In the liver, a dual blood supply is so regulated as to produce partial reciprocity, thereby vasoregulating a steady hepatic blood flow. Vasoactive agents present in either the hepatic artery or portal vein can alter resistance to blood flow in both circuits. Some of the gastrointestinal hormones may serve a vasoregulatory function in the liver.

The preceding brush strokes only suggest the contents of the six chapters in this section on the splanchnic circulation. In each chapter the well-accepted information, expressed succinctly, forms the framework on which each author then hangs the bright ornaments of exciting recent findings. Such findings are rapidly changing this field of physiology.

STOMACH BLOOD FLOW AND ACID SECRETION

Paul H. Guth

Wadsworth V.A. Hospital, CURE, and UCLA School of Medicine, Los Angeles, California 90073

INTRODUCTION

A decade ago the gastrointestinal system was thought to possess only a few interesting hormones. Today this tract has been proven to contain one of the major endocrine systems of the body. Prostaglandins, originally isolated from seminal fluid, are now known to be synthesized by gastrointestinal and vascular tissues and probably play an important role in the regulation of gastrointestinal functions, including blood flow. The development of histamine H_2-receptor antagonists has permitted determination of the role of H_1 and H_2 receptors in various parts of the gastrointestinal tract. There have also been significant advances in methods for the study of gastric blood flow, including the aminopyrine-clearance and radiolabelled-microsphere techniques. With the new knowledge of gastrointestinal hormones, prostaglandins, and histamine receptors we have also gained an improved understanding of factors controlling the blood flow to splanchnic organs.

This paper reviews our knowledge of the factors controlling gastric blood flow—neural, biogenic amines, hormones, and prostaglandins—and then addresses the role of blood flow in acid secretion.

FACTORS CONTROLLING GASTRIC BLOOD FLOW

Neural

CENTRAL NERVOUS SYSTEM "In ... predisposition, from whatever cause ... fear, anger, or whatever depresses or disturbs the nervous system —the villous coat [of the stomach] becomes sometimes red and dry, at other times pale and moist, and loses its smooth and healthy appearance...."

This description by Beaumont (2) was based on observations of the exposed gastric mucosa of his patient, Alexis St. Martin, in the early 19th century and is probably the first evidence that the central nervous system plays a role in gastric function, including blood flow. Observations of this nature were extended by Wolf in the study of the gastrostomy of his patient, Tom (38). He described two opposing patterns of gastric reaction to emotional conflicts. The arousal pattern was characterized by vascular engorgement of the mucosa, increased secretion of HCl, and increased motor activity. The other pattern was marked by a general attitude of withdrawal and was characterized by pallor of the mucosa with decreased acid secretion and motor activity.

Stimulation of the anterior hypothalamus in dogs resulted in a parasympathetic vascular and secretory pattern—i.e. increased blood flow (measured with a flowmeter)—as well as augmented secretion of acid (27). Stimulation of the posterior hypothalamus resulted in a sympathetic vascular acid and secretory pattern—i.e. marked decrease in gastric blood flow and inhibition of histamine-stimulated acid secretion. Unfortunately, results of this study can be accepted only with reservation pending confirmation by other investigators. There have been many more studies on the effect of central nervous stimulation on gastric acid secretion than on blood flow, and the results are highly discrepant (7, 27).

AUTONOMIC NERVOUS SYSTEM The sympathetic nerve supply to the stomach is derived from the sixth to the tenth spinal nerves, which terminate in the celiac ganglia. From here postganglionic fibers reach the stomach as discrete nerves, nerves mixed with vagal fibers, or as nerves accompanying the arteries. Electrical stimulation of the sympathetic fibers to the stomach decreased total venous outflow (22), celiac artery inflow (15), and gastric mucosal blood flow (34). When stimulation of sympathetic fibers was prolonged, blood flow, following the initial fall, increased and reached a new steady-state flow level within 3–4 min (22). This phenomenon has been termed autoregulatory escape from vasoconstrictor influence. Originally the escape was attributed to a flow redistribution in the stomach wall due to the opening of submucosal arteriovenous anastomoses during sympathetic stimulation. However, in vivo microscopic observations (11, 13) indicated that escape was due to relaxation of initially constricted vessels, not to the opening of shunts.

The in vivo microscopic studies also demonstrated the important role of constriction and dilation of submucosal arterioles in regulating gastric mucosal blood flow in the rat (13) and cat (15). Stimulation of the left splanchnic nerve for 3 min caused an initial constriction of submucosal arterioles followed by partial escape. No arteriovenous anastomoses or shunts were seen either in the resting state or during nerve stimulation. In vivo microscopic observations of the superficial mucosal blood flow during splanchnic nerve stimulation revealed a slowing of flow with progressively fewer red blood cells present in the capillaries and in the collecting veins into which they feed. Finally there was a complete cessation of flow with no red blood cells seen in the majority of capillaries, and the mucosa appeared blanched. Subsequently, despite continued splanchnic stimulation, escape occurred with a partial return of flow. There was no statistically significant difference between the mean time to maximum constriction of submucosal arterioles and the mean time to maximum blanching of the mucosa, nor between time to escape in the two areas. Thus the submucosal arterioles appear to be the vascular segment controlling gastric mucosal blood flow, constriction of these arterioles decreasing, and dilation increasing, mucosal blood flow. No evidence was found to support the widely held concept that submucosal arteriovenous anastomoses or shunts play a role in the control of mucosal blood flow.

The parasympathetic supply to the stomach is via the vagus nerves. In the cat these nerves may be divided into low and high threshold groups of fibers based upon a duration of electrical stimulus of less than or greater than 0.5 msec (31). Stimulation of high-threshold fibers induced an atropine-resistant, nonadrenergic relaxation of the stomach. Stimulation of the relaxatory fibers augmented both gastric blood flow (venous outflow measurement) and secretion. Atropine reduced the blood flow increase and completely abolished the secretory response, supporting the concept that the blood flow increase to the stomach during vagal stimulation is largely secondary to an augmented secretion. However, careful scrutiny of this work reveals a very prompt (within 1 min) increase in blood flow with the onset of vagal stimulation, which suggests a direct dilator effect of vagal stimulation preceding acid secretion (whose onset requires several min) as well as the increase in blood flow prompted by the increase in acid secretion.

With an in vivo microscopic technique in the rat, gastric submucosal arteriolar diameter was measured immediately before and after vagal nerve stimulation (8 V, 2 msec duration, 6 i.p.s.) (13). In 45-51 stimulation episodes in 11 rats the arterioles dilated within 10 sec of beginning stimulation, and constricted within 10 sec of cessation of stimulation. Similar results were obtained in the cat (15). The findings of nearly immediate submucosal arteriolar dilation with vagal stimulation and of constriction on cessation of stimulation are compatible with a direct dilator effect of vagal stimulation on the vascular smooth muscle of these vessels, although a local intermediate cannot be entirely excluded. Furthermore, the early dilator response to vagal stimulation does not preclude the important role of other factors secondary to augmented acid secretion in enhancing subsequent blood flow to the stomach. However, the immediate "on-off" response observed in the in vivo microscopic study is too rapid to be explained by factors secondary to acid secretion.

Vagotomy was reported to cause a prompt decrease in gastric mucosal blood flow measured by ^{85}Kr clearance (3) or aminopyrine clearance (33) techniques. There is disagreement over the long-term duration of the vagotomy-induced reduction in gastric mucosal blood flow, however (4, 6, 33).

Biogenic Amines

In reviewing the earlier literature concerning the effect of intravenous catecholamines on gastric blood flow one is struck by the markedly conflicting findings. The direct effects of these agents on the gastric circulation are confounded by their effects on cardiac output and arterial pressure. In vitro studies of the effect of agents on arterial strips, in vivo microscopic study of the topical application of catecholamines, and the close intraarterial infusion of vasoactive agents permit study of the direct gastric vascular effect of the drugs.

In vitro studies on isolated helical strips of canine gastric arteries showed contractile responses to increasing concentrations (10^{-5}–10^{-9} M) of both epinephrine and norepinephrine (36). The adrenergic blocking agent, phenoxybenzamine, caused dose-dependent inhibition of the response to both epinephrine and norepinephrine. The direct constricting effect of both catecholamines was also demonstrated in the intact animal in microscopic studies in vivo in the rat (12, 10). Application of 10^{-6} M epinephrine (12) and 10^{-5}–10^{-7} norepinephrine (10) to the intact, exposed gastric submucosal vascular bed produced prompt constriction of the arterioles.

Close intraarterial injection of adrenergic agents in the gastric circulation of anesthetized dogs induced constriction followed by dilation—i.e. escape from adrenergic constriction (39). The constrictor components were attenuated or abolished by alpha-adrenergic blockade with phenoxybenzamine, and the dilator components were attenuated by beta-adrenergic blockage with propranolol. The beta-adrenergic agonist isoproterenol produced vasodilation of both right and left gastric circulations, and this effect was attenuated by beta-adrenergic blockade. It is of interest that with epinephrine, 0.5 μg kg^{-1} min^{-1}, the dilator response following the initial constriction in the left gastric artery resulted in a blood flow that was significantly greater than control flow within 3 min. This might explain the dilator effect of epinephrine infusions reported by several investigators. The effect of close intraarterial injection of epinephrine, 0.5 μg kg^{-1}, has also been studied in the baboon (40).

The above in vitro and in vivo microscopic and close intraarterial injections studies indicate that both norepinephrine and epinephrine initially constrict gastric arteries and decrease gastric blood flow. In the dog but not in the baboon, the close intraarterial studies suggest that both catecholamines are "mixed" adrenergic agonists with an alpha-adrenergic constricting effect and a beta-adrenergic dilating effect. The beta-adrenergic dilating effect is much more pronounced with epinephrine. Isoproterenol is a "pure" beta-adrenergic agonist and increases gastric blood flow unless the agent is used in antisecretory doses in the secreting stomach, in which case it reduces mucosal blood flow (21).

HISTAMINE   The importance of considering route of administration and dosage in studying the effects of agents on the gastric circulation is clearly demonstrated by histamine. Histamine dilates the gastric circulation and increases gastric blood flow when it is infused locally under circumstances in which it has no effect on systemic arterial pressure, or when it is infused systemically in small doses (18). In contrast, administration of large amounts of histamine by a systemic route can cause abrupt hypotension and presumably a sympathetic discharge resulting in constriction and a decreased blood flow to the stomach (32).

Histamine acts via two types of receptors; first those blocked by the older antihistamine agents such as mepyramine—for example, the receptors mediating bronchial constriction—termed H_1; and, second, those not blocked by mepyramine—for example, the receptors mediating gastric acid secretion—termed H_2. The recently developed H_2 antagonists block the latter receptors. When acid secretion was inhibited by the histamine H_2 receptor antagonists burimamide and metiamide, histamine still increased gastric mucosal blood flow, measured by the clearance of ^{14}C-aniline (30). The selective histamine H_1 receptor agonist, 2-pyridyl ethylamine, had no effect on acid output but increased resting mucosal blood flow. These results suggested that histamine H_2-receptors, primarily concerned with acid secretion, and H_1-receptors, concerned with vasodilation, are both present in the gastric mucosa. In the submucosal arterioles of the corpus and antrum of the cat and rat stomachs, changes in the diameter of submucosal arterioles were determined with an image-splitting technique in response to superfusion of the exposed submucosal vascular bed with histamine before and after administering either the H_1 antagonist, mepyramine, or the H_2 antagonist, metiamide (16). H_1 and H_2 histamine receptors mediating vasodilation were demonstrated in both areas in both species. A similar vasodilation of gastric submucosal arterioles was observed in response to superfusion with H_1 agonists, 2-pyridyl ethylamine and 2-thiazole ethylamine, and the H_2 agonist, dimaprit. These studies indicate that histamine increases gastric blood flow by stimulating both H_1 and H_2 histamine receptors on the gastric submucosal arterioles.

Gastrointestinal Hormones

It has been clearly demonstrated in a number of investigations that both gastrin and pentagastrin are potent stimulants of gastric acid scretion; secondary to the increase in acid secretion there is an increase in gastric mucosal blood flow. This was found in all species studied, dog (21), cat (17), rat (30), and human (9). The evidence that the increase in blood flow is secondary to the increase in acid secretion was based on the finding that the ratio of gastric mucosal blood flow to acid secretion decreases as both increase in response to increasing doses of pentagastrin. When histamine and gastrin effects were compared (using aminopyrine clearance) the ratio was greater for histamine (19). This suggested that the increased mucosal flow due to histamine, in contrast to that due to gastrin, represented both a direct dilating property and an indirect metabolic effect secondary to secretion. This was confirmed by in vivo microscopic observations in the cat in which close intraarterial infusion of histamine, but not pentagastrin, in doses approximating 3% of the i.v. D_{50} dose for stimulation of acid secretion, caused prompt dilation of gastric submucosal arterioles (14).

Using the aminopyrine clearance technique in the dog, Konturek and his colleagues found that VIP (vasoactive intestinal peptide) and secretin (25), nor-leucine motilin (24), and somatostatin (26) decreased pentagastrin-stimulated acid secretion, an effect accompanied by a fall in gastric mucosal blood flow. The ratio of blood flow to acid secretion remained unchanged, indicating that the fall in blood flow with all these hormones was secondary to the inhibition of acid secretion.

Vasopressin decreased gastric blood flow in the baboon (40) and mucosal blood flow in the dog (21). In human angiographic studies, marked constriction of the gastric arteries occurred in response to the close intraarterial infusion of vasopressin (1). In the studies in the baboon and man, evidence for escape from vasopressin vasoconstriction was sought without success.

In in vivo microscopic studies (14) the close intraarterial infusion of glucagon was without effect on cat gastric submucosal arterioles. Whereas natural secretin (GIH) caused dilation, synthetic secretin did not. These findings with secretin were suggestive of a contaminating vasoactive agent in the GIH preparation, since both secretin preparations studied were potent stimulants of canine pancreatic secretion. The octapeptide of cholecystokinin, in a dose that could be considered physiological, 3% of the i.v. D_{50} for pancreatic secretion, caused vasodilation.

Prostaglandins

Indomethacin, in doses sufficient to inhibit prostaglandin formation, significantly reduced resting gastric mucosal blood flow in both rats and conscious dogs (23, 29). Indomethacin also decreased the caliber of gastric submucosal arterioles in the rat (10). These findings suggest that endogenous prostaglandins may modulate basal gastric mucosal blood flow.

In the rat, prostacyclin (PGI_2) increased resting mucosal blood flow, and, during PGI_2 inhibition of pentagastrin-induced acid secretion, the ratio of mucosal blood flow to acid output increased (37). The breakdown product of PGI_2, 6-oxo-$PGF_{1\alpha}$, had no significant effect on gastric acid secretion or mucosal blood flow. Prostaglandins of the E and A series, which inhibit gastric acid secretion, also have a direct vasodilator action on the rat gastric mucosa. Superfusion of the exposed submucosal vascular bed with PGE_2 resulted in a dose-related dilation of the submucosal arterioles (10). Intraarterial infusion of PGE_1 increased total gastric and mucosal blood flows (as measured by venous outflow and radiolabeled microspheres) and decreased acid secretion in an exteriorized, chambered preparation of canine fundic stomach (5).

GASTRIC BLOOD FLOW AND ACID SECRETION

In conscious dogs both histamine and gastrin stimulated acid secretion and gastric mucosal blood flow and decreased the ratio of gastric mucosal blood flow to acid secretion; this ratio fell from high values (70–100) at low secretory rates to values of 30–40 at maximal secretory rates. Vasopressin infusion during background stimulation with histamine inhibited acid secretion and was associated with a parallel decline in clearance. Infusion of isoproterenol at a low dose, 1.2 mg hr^{-1}, did not affect histamine-stimulated secretion but did markedly increase aminopyrine clearance. These findings indicate that when the stomach is stimulated to secrete, mucosal blood flow increases. If blood flow to the stomach is reduced sufficiently, secretion decreases. However, the converse is not true. Increasing blood flow, as by a low dose of isoproterenol, does not increase secretion. Thus mucosal blood flow plays a supportive and permissive role in gastric secretion.

Similar increases in gastric mucosal blood flow with increases in acid secretion in response to gastrin, histamine, or pentagastrin have been observed in the cat (17), rat (30), and human (9). In studies in the conscious dog, vagal stimulation (insulin or 2-deoxy-D-glucose) as well as histamine and gastrin-stimulated acid secretion were accompanied by an increase in mucosal blood flow (20). These data demonstrate that the increase in acid secretion by gastrin, histamine or cholinergic stimuli is accompanied by an increase in mucosal blood flow. Since gastrin, histamine, and acetylcholine stimulate acid secretion in the isolated parietal cell (35), blood flow is not a prerequisite for initiation of gastric secretion. However, it would appear that blood flow does become a rate-limiting factor at high rates of secretion.

While gastric secretion and blood flow generally change in parallel, there are situations where there is a clear separation between them. Thus increasing mucosal blood flow, as with isoproterenol (21), is not accompanied by an increase in acid secretion. The findings with prostaglandins are even more marked. In the rat, prostaglandins of the E and A series, which inhibit acid secretion, have a direct vasodilator action (28). In the anesthetized dog, the intraarterial infusion of PGE_1 increased gastric blood flow while completely inhibiting histamine-stimulated secretion (5).

The potent effect of prostaglandins on gastric acid secretion and gastric blood flow and the demonstration of prostaglandin generation by gastric as well as vascular tissue suggest that endogenous prostaglandins may mediate blood flow changes accompanying changes in acid secretion (8). Aspirin, 100 mg kg^{-1} i.v., and indomethacin, 8 mg kg^{-1} i.v., each decreased (a) resting left gastric artery blood flow, (b) the increase in blood flow stimulated by sodium arachidonate, and (c) the increase in flow associated with pentagastrin-stimulated acid secretion. Both the increase in flow accompanying pentagastrin infusion and the decrease in flow with indomethacin were entirely due to reduction in mucosal flow, the muscle layer flow remaining unaltered. Both aspirin and indomethacin potentiated pentagastrin-stimulated acid output while decreasing gastric blood flow. These data suggest that pentagastrin increases the output of vasodilatory and acid-inhibiting prostaglandins.

Literature Cited

1. Athanasoulis, C. A., Baum, S., Waltman, A. C., Ring, E. J., Imbembo, A., Vandersalm, T. J. 1974. Control of acute gastric mucosal hemorrhage. Intra-arterial infusion of posterior pituitary extract. *N. Engl. J. Med.* 290:603–610
2. Beaumont, W. 1833. *Experiments and Observations on the Gastric Juice and the Physiology of Digestion,* p. 107. NY: Dover. (Facsimile ed., 1959)
3. Bell, P. R. F., Battersby, C. 1968. Effect of vagotomy on gastric mucosal blood flow. *Gastroenterology* 54:1032–37
4. Bell, P. R. F., Shelley, T. 1968. Gastric mucosal blood flow and acid secretion in conscious animals measured by heat clearance. *Am. J. Dig. Dis.* 13:685–96
5. Cheung, L. Y., Lowry, S. F. 1978. Effects of intra-arterial infusion of prostaglandin E_1 on gastric secretion and blood flow. *Surgery,* 83:699–704
6. Delaney, J. P. 1967. Chronic alterations in gastrointestinal blood flow induced by vagotomy. *Surgery* 62:155–58
7. Feldman, S., Birnbaum, D., Behar, A. J. 1961. Gastric secretion and acute gastroduodenal lesions following hypothalamic and preoptic stimulation. *J. Neurosurg.* 18:661–70
8. Gerkins, J. F., Shand, D. G., Flexner, C., Nies, A. S., Oates, J. A., Data, J. L. 1977. Effect of indomethacin and aspirin on gastric blood flow and acid secre-

tion. *J. Pharmacol. Exp. Ther.* 203: 646–52
9. Guth, P. H., Baumann, H., Grossman, M. I., Aures, D., Elashoff, J. 1978. Measurement of gastric mucosal blood flow in man. *Gastroenterolgy* 74:831–34
10. Guth, P. H., Moler, T. L. 1979. Endogenous prostaglandins in the regulation of the rat gastric microcirculation. *Microvasc. Res.* 17:S15.
11. Guth, P. H., Ross, G., Smith, E. 1976. Changes in intestinal vascular diameter during norepinephrine vasoconstrictor escape. *Am. J. Physiol.* 230:1466–68
12. Guth, P. H., Smith, E. 1974. Vasoactive agents and the gastric microcirculation. *Microvasc. Res.* 8:125–31
13. Guth, P. H., Smith, E. 1975. Neural control of gastric mucosal blood flow in the rat. *Gastroenterology* 69:935–40
14. Guth, P. H., Smith, E. 1976. The effect of gastrointestinal hormones on the gastric microcirculation. *Gastroenterolgy* 71:435–38
15. Guth, P. H., Smith, E. 1977. Nervous regulation of the gastric microcirculation. In *Nerves and the Gut,* ed. F. P. Brooks, P. W. Evers, pp. 365–73. NY: Charles B. Slack
16. Guth, P. H., Smith, E., Moler, T. 1980. H_1 and H_2 receptors in rat gastric submucosal arterioles. *Microvasc. Res.* 19:320–28
17. Harper, A. A., Reed, J. D., Smy, J. R. 1968. Gastric blood flow in anesthetized cats. *J. Physiol. London* 194:795–807
18. Jacobson, E. D. 1963. Effects of histamine, acetylcholine and norepinephrine on gastric vascular resistance. *Am. J. Physiol.* 204:1013–17
19. Jacobson, E. D., Chang, A. C. K. 1969. Comparison of gastrin and histamine on gastric mucosal blood flow. *Proc. Soc. Exp. Biol. Med.* 130:484–86
20. Jacobson, E. D., Eisenberg, M. M., Swan, K. G. 1966. Effects of histamine on gastric blood flow in conscious dogs. *Gastroenterology* 51:466–72
21. Jacobson, E. D., Linford, R. H., Grossman, M. I. 1966. Gastric secretion in relation to mucosal blood flow studied by a clearance technique. *J. Clin. Invest.* 45:1–13
22. Jansson, G., Kampp, M., Lundgren, O., Martinson, J. 1966. Studies on the circulation of the stomach. *Acta Physiol. Scand.* 68 (Suppl. 277): 91
23. Kauffman, G. L., Aures, D., Grossman, M. I. 1980. Intravenous indomethacin and aspirin reduce basal gastric mucosal blood flow in dogs. *Am. J. Physiol.* 238:G131–34
24. Konturek, S. J., Dembinski, A., Krol, R., Wunsch, E. 1977. Effect of 13-Nle-motilin on gastric secretion, serum gastrin level and mucosal blood flow in dogs. *J. Physiol. London* 264:665–72
25. Konturek, S. J., Dembinski, A., Thor, P., Krol, R. 1976. Comparison of vasoactive intestinal peptide (VIP) and secretin in gastric secretion and mucosal blood flow. *Pflügers Arch.* 361:175–81
26. Konturek, S. J., Tasler, J., Cieskowski, M., Coy, D. H., Schally, A. V. 1976. Effect of growth hormone release-inhibiting hormone on gastric secretion, mucosal blood flow, and serum gastrin. *Gastroenterology* 70:737–41
27. Leonard, A. S., Long, D., French, L. A., Peter, E. T., Wangensteen, O. H. 1964. Pendular pattern in gastric secretion and blood flow following hypothalamic stimulation—origin of stress ulcer? *Surgery* 56:109–20
28. Main, I. H. M., Whittle, B. J. R. 1973. The effects of E and A prostaglandins on gastric mucosal blood flow and acid secretion in the rat. *Br. J. Pharmacol.* 49:428–36
29. Main, I. H. M., Whittle, B. J. R. 1975. Investigation of the vasodilator and antisecretory role of prostaglandins in the rat gastric mucosa by use of non-steroidal anti-inflammatory drugs. *Br. J. Pharmacol.* 53:217–24
30. Main, I. H. M., Whittle, B. J. R. 1976. A study of the vascular and acid-secretory responses of the rat gastric mucosa to histamine. *J. Physiol. London* 257:407–18
31. Martinson, J. 1965. The effect of graded vagal stimulation on gastric motility, secretion and blood flow in the cat. *Acta Physiol. Scand.* 65:300–9
32. Menguy, R. 1962. Effects of histamine on gastric blood flow. *Am. J. Dig. Dis.* 7:383–93
33. Nakamura, K., Ishi, K., Kusano, M., Hayashi, S. 1974. Acute and long-term effects of vagotomy on gastric mucosal blood flow. In *Vagotomy. Latest Advances with Special References to Gastric and Duodenal Ulcer,* ed. F. Halle, S. Andersson, pp. 109–111. NY: Springer.
34. Reed, J. D., Sanders, D. J., Thorpe, V. 1971. The effect of splanchnic nerve stimulation on gastric acid secretion and mucosal blood flow in the anesthetized cat. *J. Physiol. London* 214:1–13
35. Soll, A. H., Walsh, J. H. 1979. Regulation of gastric acid secretion. *Ann. Rev. Physiol.* 41:35–53

36. Van Hee, R. H., Vanhoutte, P. M. 1978. Cholinergic inhibition of adrenergic neurotransmission in the canine gastric artery. *Gastroenterology* 74:1266–70
37. Whittle, B. J. R., Boughton-Smith, N. K., Moncada, S., Vane, J. R. 1978. Actions of prostacyclin (PGI_2) and its product 6-oxo-PGF_1 on the rat gastric mucosa in vivo and in vitro. *Prostaglandins* 15:955–68
38. Wolf, S. 1965. *The Stomach,* pp. 179–93. NY: Oxford Univ. Press
39. Zinner, M. J., Kerr, J. C., Reynolds, D. G. 1975. Adrenergic mechanisms in canine gastric circulation. *Am. J. Physiol.* 229:977–82
40. Zinner, M. J., Kerr, J. C., Reynolds, D. G. 1976. Distribution and arteriovenous shunting of gastric blood flow in the baboon: Effect of epinephrine and vasopressin infusions. *Gastroenterology* 71:299–302

LOCAL CONTROL OF INTESTINAL OXYGENATION AND BLOOD FLOW[1]

A. P. Shepherd

Department of Physiology, University of Texas Health Science Center, San Antonio, Texas 78284

INTRODUCTION

For more than a century circulatory physiologists have struggled with the concept that some organs might possess intrinsic mechanisms through which they could regulate their own perfusion independently of nervous or humoral influences. The teleological appeal of this concept is obvious since each organ could sense its nutrient requirements and adjust its perfusion to achieve an optimal exchange of oxygen and other substrates between its tissues and the bloodstream. In the first half of this century little progress was made in applying the concept of local control to the intestinal circulation. Indeed, if pressure-flow autoregulation is defined as "the intrinsic ability of an organ to maintain a relatively constant blood flow despite imposed changes in perfusion pressure," it was not until 1949 that autoregulation was first demonstrated in the intestine (4). In the next two decades additional investigators addressed the question of intestinal autoregulation and publications began to appear reporting the presence (12–15) or absence (11) of autoregulation in the circulation of the small intestine. These studies served primarily to demonstrate that the gut was potentially capable of controlling its own perfusion. Pressure-flow autoregulation was shown to be dependent on local mechanisms since it was unaffected by adrenergic blockade or chronic denervation (13). These studies also indicated that autoregulation in the gut may not be as pronounced as in other organs, such as brain or kidney. In fact, pressure-flow autoregulation in the intestine was

[1]This work was supported by Grant HL-23435 from the National Institute of Health.

an evanescent phenomenon, appearing in some experiments but not in others. Although extensive external perfusion systems, pumps, and arterial cannulation (41) were known to reduce an investigator's chances of seeing the ghost of autoregulation past, investigators were unaware of the necessary experimental conditions required to elicit active autoregulatory behavior during perfusion pressure manipulations, rather than passive responses.

It is also a fair characterization of these early studies to say, in retrospect, that perhaps undue emphasis was placed upon blood flow per se, and that the other definition of autoregulation—"the intrinsic ability of an organ to regulate blood flow in accordance with its metabolic demands"—was underemphasized by investigators of the intestinal circulation. Metabolic rate or oxygen uptake, with some exceptions (12), was usually not measured. Even when pressure-flow autoregulation was demonstrable, the autoregulatory ability of the gut was often not quantitated. Finally, because the vascular response to elevated venous pressure (15, 18), namely vasoconstriction, was clearly consistent with the myogenic theory of autoregulation and opposite to the predictions of the metabolic hypothesis, the best evidence available in the late 1960s supported a myogenic basis for intestinal autoregulation whereas the possibility of metabolic control of the intestinal circulation was relatively unexplored.

In this review I survey the literature on intestinal autoregulation since 1970 and attempt to analyze the experimental evidence for local control within the theoretical framework of a metabolic model of autoregulation. I also discuss the role of pressure-flow autoregulation in oxygen transport, the relative contributions of blood flow and oxygen extraction to oxygen homeostasis, quantitative aspects of the ability of the gut to regulate blood flow, and the likely candidates for the vasodilator that might link the metabolic requirements of the gut to its perfusion. Although there are many relatively unexplored aspects of the intestinal circulation, I hope to emphasize particularly those now impeding the growth of our knowledge of intestinal autoregulation.

METABOLIC MODEL OF LOCAL CIRCULATORY CONTROL

Figure 1 is a schematic diagram of the oxygen delivery version of the metabolic theory of local circulatory control. In this scheme the oxygen flux to the parenchymal cells of the intestine is proposed to be the regulated variable, rather than blood flow per se. Thus two discrete microvascular effectors are available to regulate the rate at which oxygen is delivered to intestinal tissue: (a) Arterioles regulate blood flow and thus govern the convective flux of oxygen into the capillary beds. (b) Precapillary sphincters control the number of capillaries perfused at a given time. By thus determining the diffusion parameters, capillary surface area and diffusion distance, the precapillary sphincters regulate the diffusive flux of oxygen out of the capillary bed and into the tissue. In this model of local control we assume that in response to inadequate oxygenation, parenchymal tissue can produce a feedback signal to dilate resistance vessels and open precapillary sphincters, thereby increasing blood flow and capillary density. The increase in blood flow tends to increase or to maintain capillary pO_2. Because the diffusive flux of oxygen is a function of both the capillary density and the capillary-to-cell pO_2 difference, the two vascular responses restore the tissue oxygen level toward its normal value. In this scheme the term "precapillary sphincter" is used in a functional rather than in an anatomical sense. Physiological measurements of capillary density such as the capillary filtration coefficient (K_f) or the capillary transport coefficient (PS) invariably show that capillary density in the intestine changes 3- to 4-fold with alterations in vasomotor tone (15, 18, 25, 27, 30, 33, 37, 38, 40). In addition, some evidence supports the existence of precapillary sphincters as anatomical entities in the mesenteric circulation. Precapillary sphincters have been observed in vivo in the rat mesoappendix (1) and in the mucosa (2). More than 75% of the capillaries in the rat villus are encircled by a single contractile cell that appears to be a sphincter; however, few of these are observed in the muscular layer of the gut (2).

A computer model (36) published in 1973 was based on this classic capillary recruitment concept of August Krogh (20). The mathematical model simulated many of the experiments carried out in the gut and in other

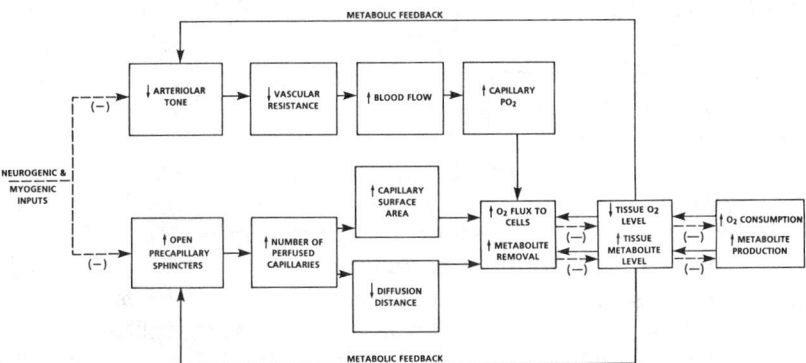

Figure 1 Two-component metabolic model of autoregulation. Oxygen flux to cells is regulated by metabolic feedback to resistance vessels and precapillary sphincters. Feedback signal could be vasodilator metabolites or interstitial hypoxia. The negative signs indicate that an increase in an input produces a change opposite to that shown by the arrow in the box. Sympathetic and myogenic inputs interact with the metabolic control system because they also affect resistance and exchange vessels.

perfused organs. It also generated a number of testable hypotheses. For example, in the computer model we could defeat one of the two feedback loops (e.g. resistance control) and study the efficacy of the other loop in maintaining tissue pO_2 in response to various perturbations. Such simulations made it obvious that capillary recruitment was quantitatively as significant as modulations of blood flow in regulating tissue oxygenation. Clearly, such an important variable can hardly be ignored, yet in most of the studies then available capillary density was not determined; it was also the uncontrolled variable in most experiments. It is relatively easy to "open the feedback loop" controlling blood flow simply by perfusing the organ at a constant blood flow with a pump, but it is uncertain that the "constant capillary density" experiment can actually be performed. A maximal pharmacological vasodilation might achieve this result; however, pharmacological vasodilation often depresses the transcapillary extraction of indicators such as ^{86}Rb and thus reduces the capillary exchange capacity from the optimal level that it would reach if the local feedback mechanisms were not altered by the exogenous vasodilator (28).

In the computer model of local control, capillary density and blood flow are almost completely independent variables because capillary recruitment has so slight an effect on total vascular resistance (36). I emphasize the independence of the vascular controls for blood flow and capillary density to point out the danger in making assumptions about the behavior of capillary density. If the oxygen demands of the tissue are constant and if blood flow is changed by altering perfusion pressure, capillary density and blood flow seem to be inversely related. In this situation, increasing blood flow above its resting level depresses capillary density to a minimal value. Similarly, reducing blood flow below its normal value induces a marked capillary recruitment to rectify the ensuing cellular hypoxia. By contrast, if the oxygen demands of the tissue increase and if both feedback loops are free to operate, blood flow and capillary density appear to be directly related. In this case, they increase concomitantly to enhance the oxygen flux and meet the increased metabolic demands of the tissue (10). A final point should be made about the necessity of considering capillary density changes in the interpretation of perfused organ data. Even in constant flow experiments the "washout" of a vasodilator metabolite would be enhanced by capillary recruitment regardless of whether the tissue-to-capillary concentration difference increased or not. Indeed, the familiar relationship (28) between capillary clearance and blood flow consists of a family of curves, each curve resulting from a particular capillary density (PS).

Experiments to test the predictions of this model of the intestinal circulation have proceeded in several stages. The first stage was to determine whether resistance and exchange vessels in the gut behaved qualitatively during various experimental perturbations (e.g. arterial hypoxia, increased metabolic rate, etc) as the metabolic model predicted. The second stage was to quantitate the ability of the resistance control loop to regulate flow during perfusion pressure manipulations and to determine the relative contributions of blood flow and capillary density to oxygen homeostasis. The third step will be to identify the specific substances that could act as the metabolic feedback signal.

EVIDENCE FOR METABOLIC CONTROL

Arterial Hypoxia

Arterial hypoxia can be used to study the ability of the local circulatory control mechanism to maintain tissue oxygenation. Any reduction in the oxygen availability-to-demand ratio should evoke vasodilation and capillary recruitment. The responses of isolated gut loops to arterial hypoxia are qualitatively consistent with this prediction. In feline intestine (40) perfused at constant arterial pressure, blood flow increased 48% when the inspired pO_2 was reduced to 55 mm Hg. Capillary filtration coefficients (K_f) increased 60%. In dog intestine an arterial pO_2 of 46 mm Hg increased blood flow 46% during constant pressure perfusion, while the same level of arterial hypoxia increased the capillary exchange capacity (PS for ^{86}Rb) 60% above control during constant flow perfusion (30, 31). Thus both resistance and exchange vessels respond to arterial hypoxia in a manner consistent with predictions of the purely metabolic model. The result of vasodilation and capillary recruitment should be to maintain tissue oxygenation despite the reduced oxygen supply. When the resistance feedback loop was opened by perfusing dog intestine at constant flow, the oxygen consumption rate fell to 52% of control during arterial hypoxia ($pO_2 = 46$ mm Hg); however, when both blood flow and capillary density were free to increase, the oxygen uptake rate remained within 26% of control despite the severe hypoxia (30). Skeletal muscle reportedly maintains its oxygen consumption within 15% of control (10) at a comparable arterial pO_2. Thus the intestine may be somewhat less effective than skeletal muscle in controlling the oxygenation of its tissues during hypoxia.

Although the observed responses to arterial hypoxia—vasodilation, capillary recruitment, and the tendency of oxygen uptake to be maintained —all support the metabolic model, these studies should be considered preliminary. The complete relationship between arterial pO_2 and capillary exchange capacity has not yet been determined. Neither has the dependence of oxygen uptake on arterial pO_2 been fully characterized. Finally, it should be noted that oxygen extraction can be increased either by capillary recruitment or by a widening of the capillary-to-cell pO_2 difference. The extent to which these two mechanisms participate in maintaining intestinal oxygen consumption has not yet been determined. Doing so would require that we obtain an "average tissue pO_2" value in the intestine—another problem awaiting solution.

Reactive Hyperemia

The overshoot in blood flow that follows arterial occlusion can be simulated reasonably well by purely metabolic models (10) although myogenic mechanisms and refilling of the vasculature may contribute to the response. In addition to the reactive hyperemia the metabolic model predicts that in response to an arterial occlusion the vasodilation and the capillary recruitment should cause an overshoot in both tissue pO_2 and arteriovenous oxygen difference upon the release of an occlusion (10). These predictions have generally been confirmed in skeletal muscle (10), but they have not been as thoroughly studied in the intestine. In skeletal muscle (10) the post-occlusive capillary recruitment has been detected by measurements of both PS and K_f. In addition, pO_2 microelectrode studies in muscle confirm the overshoot in "tissue pO_2" upon the release of the occlusion (10). In the intestine (10), reactive hyperemia contributes to the repayment of the "oxygen debt" incurred during the occlusion. However, upon the release of the occlusion, the intestinal arteriovenous oxygen difference is reduced throughout the hyperemia, a finding that indicates a relative overperfusion. Because capillary density values during reactive hyperemia have not yet been determined in the gut, the capillary recruitment remains hypothetical. If the recruitment occurs, the reduced oxygen extraction would not be the result of limitations in diffusion. In fact, if the overshoot in blood flow is prevented by perfusing the gut at constant blood flow, the oxygen debt in the intestine is still repaid promptly, but in this case by a marked increase in oxygen extraction that occurs when perfusion is restored (10, 23). Although these data from the gut generally support the two-component metabolic model of tissue oxygenation (Figure 1), the evidence would be greatly strengthened if capillary density data were available.

Functional Hyperemia

The increased blood flow that follows the ingestion of a meal has been designated by the descriptive term "postprandial hyperemia." The term is useful, since the postprandial hyperemia attending the ingestion of mixed meals by man or experimental animals is probably a complex response involving not only local mechanisms but also nervous and circulating humoral factors (6, 33). By contrast, evidence from bioassay and cross-perfusion studies (6) indicates that a "functional hyperemia" due solely to local mechanisms occurs in isolated gut loops luminally perfused with simple, nonlipid solutes such as glucose. Because the Na-coupled, active transport of glucose requires metabolic energy, installation of glucose solutions in the lumen of isolated gut loops is an expedient means to study the vascular reaction to an increased demand for oxygen by the mucosal tissues. The prediction of the metabolic model to such an increased oxygen demand is, of course, vasodilation and capillary recruitment. This prediction is confirmed by the finding in canine small bowel (33, 34) that the increased oxygen consumption rate is made possible by increased blood flow and increased oxygen extraction. During the increased oxygen extraction, capillary density increased as shown by K_f and PS measurements (27, 33). Similarly, if such gut loops are perfused at constant blood flow, the glucose-induced increase in oxygen uptake occurs solely through augmented oxygen extraction and it is associated with increased PS values (^{86}Rb). In functional hyperemia experiments it should be possible to demonstrate a complete relationship between capillary density and blood flow; however, at this writing it has only been possible to demonstrate that both blood flow and capillary density increase above control levels when the oxygen demands of intestinal tissues are stimulated.

Other studies have shown that the vascular responses to glucose are due to its being actively transported and not simply because of its presence, a hypertonicity of some of the glucose solutions employed or its catabolism. Both mannitol and 2-deoxyglucose, hexoses that are not actively transported, failed to increase blood flow (6, 27), whereas both glucose and 3-0-methyl glucose (a form of glucose that is actively transported but not metabolized) increased oxygen uptake and blood flow (6). Furthermore, there is a significant correlation between the oxygen uptake and the rate of fluid absorption accompanying the solute transport. This relationship varies from solute to solute—i.e. the oxygen cost of a given volume of absorbed fluid depends upon the specific solute transported (42).

The vascular responses to the hypermetabolic state induced by luminal glucose seem to be solely due to local mechanisms since the venous blood from these isolated gut loops, when recirculated, does not cause vasodilation in adjacent gut loops (6). Furthermore, if the data from microsphere studies are reliable, the glucose-induced hyperemia may even be confined to the mucosa-submucosal layer of the bowel wall (6), but contradictory results have also been reported (27). A functional hyperemia quite analogous to exercise hyperemia in skeletal muscle has also been reported (3) during increased motility; however, the oxygen costs of intestinal motility have not been quantitated and are poorly understood.

Effect of Metabolic Rate on Pressure-Flow Autoregulation

Early studies on skeletal and cardiac muscle suggested a relationship between pressure-flow autoregulation and metabolic rate (10). In some of these reports, the effectiveness of the flow control system in response to perturbations in perfusion pressure seemed to depend on the prevailing venous oxygen content. The physiological basis for such a relationship between pressure-flow autoregulation and venous oxygen content could be that precapillary sphincters are more sensitive than resistance vessels to changes in tissue oxygenation. If such an assumption is included in the computer model (10, 36), it predicts a more powerful autoregulatory response to reduced arterial pressure if (*a*) the resting blood flow level is low, (*b*) arterial oxygen saturation is below normal, or (*c*) the metabolic rate is high. The reason for this is that in response to a small reduction in the oxygen availability-to-demand ratio the more sensitive sphincters maintain tissue oxygenation by increasing oxygen extraction and the contribution by the feedback mechanism controlling blood flow is minor. However, when the venous oxygen reserve is depleted, the contribution of the flow control loop becomes more important while the contribution of oxygen extraction diminishes.

These predictions have been confirmed in a recent study (34) that showed pressure-flow autoregulation was enhanced by high metabolic rate. In isolated gut loops pressure-flow autoregulation was totally absent in the control situation. When arterial pressure was reduced, increases in oxygen extraction were the only means employed to maintain oxygen uptake. Indeed, passive vascular recoil predominated in the flow responses—i.e. flow fell proportionately more than perfusion pressure. In contrast, when the metabolic rate was elevated with intraluminal solutes and oxygen extraction was already high, each of the previously passive gut loops actively autoregulated its blood flow in response to the same change in perfusion pressure. This finding not only supports the metabolic hypothesis of local control; it also explains why pressure-flow autoregulation occurred in some early experiments but not in others. The enhancement of pressure-flow autoregulation when the metabolic rate is high has also been observed in the superior mesenteric arteries of intact dogs; however, in such preparations autoregulation-like behavior cannot be attributed with certainty to local mechanisms because nervous and humoral factors were not eliminated (8, 26). Nevertheless, the apparent autoregulation in these animals also confirms the metabolic model's prediction of enhanced flow-control in the hypermetabolic state. However, the behavior of flow differed in two respects from that observed in the isolated loops. First, in control or fasted animals blood flow was better regulated than in the control loops. Second, "super-regulation" was occasionally observed—i.e. flow actually rose sometimes following a reduction in perfusion pressure. This "super-regulation" is seldom seen in isolated gut loops.

In the metabolic model the oxygen flux to parenchymal cells, rather than blood flow, is the regulated variable. Besides the effects of metabolic rate on pressure-flow autoregulation two other lines of evidence indicate that the gut autoregulates its oxygenation rather than blood flow. First, if the oxygen demands are elevated by transportable solutes, the functional hyperemia in isolated gut loops is not nearly as pronounced as in other organs—e.g. skeletal or cardiac muscle. In fact, blood flow declined in some reports [for a brief review, see (33)]. Such variability is to be expected in a system regulating oxygen uptake rather than blood flow. Indeed, the most consistent response to intraluminal solutes is an increase in oxygen uptake brought about predominantly by increased oxygen extraction regardless of the blood flow response (27, 33, 34, 42).

The other evidence that indicates the oxygenation of intestinal tissues is "autoregulated" is the relationship between oxygen uptake and blood flow when blood flow is the independent variable. However, conflicting reports (21, 33) have been published regarding the relationship between blood flow and oxygen uptake in the intestine. In both studies blood flow was changed in isolated gut loops by altering perfusion pressure rather than by vasodilating or vasoconstricting the preparations pharmacologically. This is an important consideration if the two studies are to be meaningfully compared since vasoconstrictors affect precapillary sphincters as well as resistance vessels (32, 37, 38). The data from Lutz et al (21) indicate that the oxygen consumption rate of feline intestine is essentially independent of blood flow until critically low blood values are reached. These data and others (8) indicate that within a wide blood flow range capillary recruitment and a widening of the capillary-to-cell pO_2 difference allow the gut to "autoregulate" its oxygen uptake. Other data from canine small bowel (34) contradict this view. Such data show a slight but definite dependence of oxygen uptake on blood flow. Although contradictory results have been reported in skeletal muscle, it would be highly desirable to explain the differences in these studies.

WHAT IS THE FEEDBACK SIGNAL?

Although the predictions of the metabolic model agree with the overall behavior of resistance and exchange vessels in isolated gut loops during reactive hyperemia, functional hyperemia, hypoxic vasodilation, and pressure-flow autoregulations, the nature of the link between the metabolism of the gut and its microvascular regulation remains unknown. The vasodilatory effects of numerous substances (6) have been studied in the intestinal circulation: carbon dioxide, hydrogen ion, potassium, prostaglandins, locally released gastrointestinal hormones, hyperosmolality, serotonin, and others. Yet, none of these substances is now a serious candidate for "the metabolic vasodilator" in the gut. CO_2 levels, for example, change in a manner consistent with the metabolic model. However, hypercapnia is only a mild vasodilator; pCO_2 must be changed beyond the physiologically observed range to cause sufficient vasodilation to account for the commonly observed autoregulatory phenomena. Morever, the exchange vessel response to CO_2 is opposite to that required of the metabolic feedback signal. Hypercapnia causes a capillary de-recruitment as shown by K_f measurements (40).

More recently, adenosine and interstitial hypoxia have been studied. Adenosine, when administered into the superior mesenteric artery, is a potent vasodilator (22). However, the adenosine-blocker, theophylline, did not alter the control blood flow and did not affect pressure-flow autoregulation in empty gut loops. Theophylline also did not block the functional hyperemia induced by placing pre-digested dog food in the intestinal lumen. Moreover, dipyridamole, which inhibits the re-uptake of adenosine by parenchymal cells, did not vasodilate the empty intestine (9). Therefore, at the "resting" metabolic rate adenosine does not contribute to vascular tone. Nor does adenosine participate in the response to moderate changes in the oxygen availability-to-demand ratio. However, with more severe stress adenosine may be involved in the response. Adenosine levels in venous blood increase 4-fold following a 60-sec arterial occlusion, and theophylline shortens the duration of reactive hyperemia by 50% (24). The enhancement of pressure-flow autoregulation usually seen in the hypermetabolic state is attenuated by theophylline. Finally, although dipyridamole is inactive in the empty gut, it adds to the functional hyperemia by causing a 25% increase in blood flow (9). Adenosine concentrations, therefore, do not appear to reach the vasoactive range until the oxygen delivery control system is markedly challenged. Hence, adenosine alone could not be the metabolic feedback signal during mild perturbations in the oxygen availability-to-demand ratio. An additional finding inconsistent with adenosine's being the vasodilator metabolite is that the effect of exogenous adenosine on capillary density is opposite to that required of the feedback signal. Intra-arterial adenosine depresses PS (unpublished observation) and reduces K_f (7).

Interstitial hypoxia could possibly act as the feedback signal because the effects of hypoxia on both resistance vessels and on precapillary sphincters are consistent with the metabolic model (30, 40). Measurements of tissue pO_2 with microelectrodes in rat intestine indicate that a fall in tissue pO_2 occurs during the absorption of glucose. The time course of absorption is consistent with its causing the observed fall in the pO_2 measured in intestinal villi (2). Moreover, the fall in pO_2 from 15 to 7 mm Hg is in the range in which hypoxia causes the relaxation of vascular smooth muscle. This observation supports a role for interstitial hypoxia as the metabolic feedback signal. However, the perivascular pO_2 near upstream submucosal arterioles rises as they dilate during the glucose-induced hyperemia. The latter observation indicates that the submucosal vasodilation not only cannot be explained by a local hypoxia but is more consistent with a mechanism of "ascending vasodilation" along the arterioles themselves or with the release of a vasodilator metabolite.

THE MYOGENIC THEORY OF AUTOREGULATION

Myogenic Responses of Resistance and Exchange Vessels

The myogenic theory of autoregulation is based upon the assumption that a passive increase in vascular wall tension, caused by raising transmural pressure, will induce active contraction of vascular smooth muscle (5, 13, 14). Evidence to support this theory has been obtained from perfused organs (12, 15), from intra-vital microscopy of single arterioles (16, 17, 19), and from isolated vascular smooth muscle (39). The strongest evidence that myogenic mechanisms regulate intestinal blood flow is that the observed response to elevated venous pressure, namely vasoconstriction, is predicted by the myogenic theory but is inconsistent with the metabolic theory. Numerous explanations have been offered for the intestinal vasoconstriction elicited by increasing venous pressure: accumulation of interstitial fluid, direct encroachment of capacitance upon resistance vessels, extrinsic nerve activity, and local nervous reflexes. These possibilities have all been discredited and the vasoconstriction evoked by elevating venous pressure is now attributed to "a myogenic response."

In addition to contracting resistance vessels, elevated venous pressure in the gut apparently evokes a myogenic closure of precapillary sphincters and thus a derecruitment of capillaries. This effect has been detected by direct observation of precapillary sphincters in mesoappendix (1) and by measurements of reduced PS (29) and K_f (15, 18) in perfused gut loops. As Figure 1 shows, the effects of a myogenic response to elevated transmural pressure could potentially depress the oxygen flux to cells either by reducing blood flow or by making the oxygen flux diffusion-limited because of the capillary de-recruitment. Indeed, sympathetic stimulation and exogenous vasoconstrictors depress K_f, PS, and oxygen uptake in isolated gut loops (6, 32), even those perfused at constant flow (37, 38). Thus one might reasonably expect a stimulus of the myogenic mechanism to compete with the metabolic feedback at both the resistance vessels and at the precapillary sphincters (31).

Competition Between Metabolic and Myogenic Mechanisms

In well-perfused gut loops the vasoconstriction and the depressed capillary density produced by the myogenic response to elevated venous pressure do not reduce intestinal oxygen uptake (29). This can be explained because mitochondrial oxygen uptake is independent of intracellular pO_2 until some critically low pO_2 is reached and because oxygen uptake is also independent of capillary density until a critically low capillary density is reached (10). That the myogenic response reduces PS (29) and K_f (15, 18), yet oxygen extraction increases, means that both capillary density and intracellular pO_2 are above their critical values in the well-perfused gut. Thus, in well-perfused intestine a myogenic response can occur without impairing tissue oxygenation and possibly without causing a competition between the myogenic and the metabolic mechanisms. By contrast, if a gut loop that has just displayed a myogenic response is deliberately underperfused, the myogenic response is abolished. Raising venous pressure then causes a fall in vascular resistance, an increase in capillary density, and a rise in oxygen uptake (29). The increased oxygen uptake could be due to increased motility or secretion caused by the elevation of venous pressure. An alternate explanation for the increased oxygen consumption is an improvement in the diffusion parameters caused by the capillary recruitment. This explanation requires the assumption that oxygen uptake in the underperfused gut loop was restricted by a combination of diffusion- and perfusion-limitations. Although the reason for the increased oxygen uptake remains obscure, it is clear that myogenic responses to elevated venous pressure do not impair the oxygenation of the intestinal tissue either when it is well perfused or when it is deliberately underperfused.

Myogenic resistance responses are observed more frequently and they have a greater magnitude if arterial pressure is pulsatile rather than steady (31). The explanation for this observation is that vascular smooth muscle (39) not only responds to the steady-state level of passive stretch to which it is subjected (static component) but also is sensitive to the rate of stretch (dynamic component). Thus elevating venous pressure in the presence of arterial pressure pulsations elicits both the static and the dynamic components of the myogenic response; hence a greater response. As mentioned previously, an increased metabolic rate enhances pressure-flow autoregulation while arterial pressure pulses enhance the myogenic response. Thus by comparing the effects of metabolic rate and pulsatile pressure on intestinal pressure-flow curves, it should be possible to determine the relative contributions that myogenic and metabolic mechanisms make to autoregulation. In such a study (35), changing the amplitude of arterial pulse pressure had no systematic effect on the ability of the gut to regulate blood flow in response to reductions in mean arterial pressure. By contrast, increasing the metabolic rate with transportable solutes caused each of the previously passive gut loops actively to autoregulate blood flow in response to the same pressure step. This occurred regardless of the amplitude of the arterial pulse pressure. Therefore, these data support the view that pressure-flow autoregulation is more dependent on metabolic than on myogenic mechanisms.

One can also study the possible interaction of myogenic and metabolic mechanisms by recording myogenic responses to elevated venous pressure while altering either oxygen availability or oxygen demand. Unfortunately, such studies have yielded conflicting results. For example, arterial hypoxia has been reported to attenuate (40) or to have no effect (31) on the resistance response to elevated venous pressure. Because baseline resistance is different in normoxia and in hypoxia, such data are particularly difficult to evaluate. Reports also conflict regarding the effect of increased metabolic rate on myogenic responses. In one study (8) with innervated preparations, increasing metabolic rate attenuated the vasoconstriction in response to elevated venous pressure; however, in another study (34) with denervated intestinal loops, myogenic responses were absent in controls but appeared in the hypermetabolic state. A possible explanation of the latter finding is that the mild vasodilation during the functional hyperemia might increase microvascular diameters, increase intramural tension through the Law of Laplace, and thus enhance the myogenic sensitivity of resistance vessels.

CONCLUSIONS

From the foregoing discussion it is apparent that although myogenic-metabolic interactions are poorly understood, considerable evidence supports the existence of both types of mechanisms. Thus a dual or unified theory of autoregulation has been advanced (5, 16). The dual theory proposes that both metabolic and myogenic mechanisms participate in local circulatory control. Future studies, therefore, must not only examine the myogenic nature of the microvascular regulators and their relationship with the metabolism of parenchymal cells, but must also delineate the contributions that myogenic and metabolic mechanisms make to homeostasis of the intestinal circulation.

Literature Cited

1. Baez, S., Laidlaw, Z., Orkin, L. R. 1974. Localization and measurement of microvascular and microcirculatory responses to venous pressure elevation in the rat. *Blood Vessels* 11:260 76
2. Bohlen, H. G. 1980. Intestinal tissue PO_2 and microvascular responses during glucose exposure. *Am. J. Physiol.* 238:H164–71
3. Chou, C. C., Grassmick, B. 1978. Motility and blood flow distribution within the wall of the gastrointestinal tract. *Am. J. Physiol.* 235:H34–39
4. Folkow, B. 1949. Intravascular pressure as a factor regulating the tone of the small vessels. *Acta Physiol. Scand.* 17:289–310
5. Folkow, B. 1964. Description of the myogenic hypothesis. *Circ. Res.* 14(Suppl. 1):279–85
6. Granger, D. N., Richardson, P. D. I., Kvietys, P. R., Mortillaro, N. A. 1980. Intestinal Blood Flow. *Gastroenterology* 78:837–63
7. Granger, D. N., Valleau, J. D., Parker, R. E., Lane, R. S., Taylor, A. E. 1978.

Effects of adenosine on intestinal hemodynamics, oxygen delivery, and capillary fluid exchange. *Am. J. Physiol.* 235:H707–19

8. Granger, H. J., Norris, C. P. 1980. Intrinsic regulation of intestinal oxygenation in the anesthetized dog. *Am. J. Physiol.* 238:H836–843

9. Granger, H. J., Norris, C. P. 1980. Role of adenosine in local control of intestinal circulation in the dog. *Circ. Res.* 46:764–70

10. Granger, H. J., Shepherd, A. P. 1979. Dynamics and control of the microcirculation. In *Advances in Biomedical Engineering,* ed. J. H. U. Brown, 7:1–63. NY: Academic Press, Inc.

11. Hinshaw, L. B. 1962. Arterial and venous pressure-resistance relationships in perfused leg and intestine. *Am. J. Physiol.* 203:271–74

12. Johnson, P. C. 1960. Autoregulation of intestinal blood flow. *Am. J. Physiol.* 199:311–18

13. Johnson, P. C. 1964. Origin, localization, and homeostatic significance of autoregulation in the intestine. *Circ. Res.* 14(Suppl. I):225–32

14. Johnson, P. C. 1964. Review of previous studies and current theories of autoregulation. *Circ. Res.* 15(Suppl. I):2–9

15. Johnson, P. C. 1965. Effect of venous pressure on mean capillary pressure and vascular resistance in the intestine. *Circ. Res.* 16:294–300

16. Johnson, P. C. 1967. Autoregulation of blood flow in the intestine. *Gastroenterology* 52:435–41

17. Johnson, P. C. 1968. Autoregulatory responses of cat mesenteric arterioles measured in vivo. *Circ. Res.* 22:199–212

18. Johnson, P. C., Hanson, K. M. 1966. Capillary filtration in the small intestine of the dog. *Circ. Res.* 19:766–73

19. Johnson, P. C., Intaglietta, M. 1976. Contributions of pressure and flow sensitivity to autoregulation in mesenteric arterioles. *Am. J. Physiol.* 231:1686–96

20. Krogh, A. 1918. The number and distribution of capillaries in muscles with calculations of the oxygen pressure head necessary for supplying the tissue. *J. Physiol. London* 52:409–515

21. Lutz, J., Henrich, H., Bauereisen, E. 1975. Oxygen supply and uptake in the liver and the intestine. *Pflügers Arch* 360:7–15

22. Mailman, D., Pawlik, W., Shepherd, A. P., Tague, L. L., Jacobson, E. D. 1977. Cyclic nucleotide metabolism and vasodilation in canine mesenteric artery. *Am. J. Physiol.* 232:H191–96

23. Mortillaro, N. A., Granger, H. J. 1977. Reactive hyperemia and oxygen extraction in the feline small intestine. *Circ. Res.* 41:859–65

24. Mortillaro, N. A., Mustafa, S. J. 1978. Possible role of adenosine in the development of intestinal post-occlusion reactive hyperemia. *Fed. Proc.* 37:874 (Abstr.)

25. Mortillaro, N. A., Taylor, A. E. 1976. Interaction of capillary and tissue forces in the cat small intestine. *Circ. Res.* 39:348–58

26. Norris, C. P., Barnes, G. E., Smith, E. E., Granger, H. J. 1979. Autoregulation of superior mesenteric flow in fasted and fed dogs. *Am. J. Physiol.* 237:H174–77

27. Pawlik, W. W., Fondacaro, J. D., Jacobson, E. D. 1980. Metabolic hyperemia in canine gut. *Am. J. Physiol.* 239:G12–17

28. Renkin, E. M. 1968. Exchange of substances through capillary walls. In *Symposium on Circulatory and Respiratory Mass Transport,* ed. G. W. Wolstenholme, J. Knight, pp. 50–66. Boston: Little, Brown & Company

29. Shepherd, A. P. 1977. Myogenic responses of intestinal resistance and exchange vessels. *Am. J. Physiol.* 233:H547–54

30. Shepherd, A. P. 1978. Intestinal oxygen consumption and ^{86}Rb extraction during arterial hypoxia. *Am. J. Physiol.* 234:E248–51

31. Shepherd, A. P. 1978. Effect of arterial pulse pressure and hypoxia on myogenic responses in the gut. *Am. J. Physiol.* 235:H157–61

32. Shepherd, A. P. 1979. Intestinal O_2 uptake during sympathetic stimulation and partial arterial occlusion. *Am. J. Physiol.* 236:H731–35

33. Shepherd, A. P. 1979. Intestinal capillary blood flow during metabolic hyperemia. *Am. J. Physiol.* 237:E548–54

34. Shepherd, A. P. 1980. Intestinal blood flow autoregulation during foodstuff absorption. *Am. J. Physiol.* 239:H156–62

35. Shepherd, A. P. 1980. Effects of pulse pressure and metabolic rate on intestinal autoregulation. *Blood Vessels* 17:161 (Abstr.)

36. Shepherd, A. P., Granger, H. J. 1973. Autoregulatory escape in the gut: a systems analysis. *Gastroenterology* 65:77–91

37. Shepherd, A. P., Mailman, D., Burks, T. F., Granger, H. J. 1973. Effects of norepinephrine and sympathetic stimu-

lation on extraction of oxygen and ^{86}Rb in perfused canine small bowel. *Circ. Res.* 33:166–74
38. Shepherd, A. P., Pawlik, W., Mailman, D., Burks, T. F., Jacobson, E. D. 1976. Effects of vasoconstrictors on intestinal vascular resistance and oxygen extraction. *Am. J. Physiol.* 230:298–305
39. Sigurdsson, S. B., Johansson, B., Mellander, S. 1977. Rate-dependent myogenic response of vascular smooth muscle during imposed changes in length and force. *Acta Physiol. Scand.* 99:183–89
40. Svanvik, J., Tyllstrom, J., Wallentin, I. 1968. The effects of hypercapnia and hypoxia on the distrubution of capillary blood flow in the denervated intestinal vascular bed. *Acta Physiol. Scand.* 74:543–51
41. Texter, E. C. Jr., Merrill, S., Schwartz, M., Van Derstrappen, G., Haddy, F. S. 1962. Relationship of blood flow to pressure in the intestinal bed of the dog. *Am. J. Physiol.* 202:253–56
42. Valleau, J. D., Granger, D. N., Taylor, A. E. 1979. Effect of solute-coupled volume absorption on oxygen consumption in the cat ileum. *Am. J. Physiol.* 236:E198–203

RELATIONSHIP BETWEEN INTESTINAL BLOOD FLOW AND MOTILITY[1]

C. C. Chou

Departments of Physiology and Medicine, Michigan State University, East Lansing, Michigan 48824

INTRODUCTION

The relationship between intestinal motility and blood flow was reviewed briefly 11 years ago and has been reviewed only infrequently since (8, 19, 41). Here two aspects of this relationship are discussed: the influence of motility (rhythmic and tonic contractions, drug- or nerve-induced contractions, and luminal distension) on blood flow, and the effects of changes in blood flow on motility. References to studies published before 1970 may be found in (8, 19, 41).

Muscle contractions can affect blood flow by two mechanisms, namely by evoking a metabolic hyperemia and by causing an extravascular compression. These produce opposite effects on local blood flow, and their interplay in the regulation of blood flow in the heart and skeletal muscle has been well demonstrated. The effects of muscle contractions on blood flow in the gut, however, are different from effects in the heart and skeletal muscle, since intestinal contractions are random in time and space, and vary in rhythm, strength, and duration. Segmental contractions, the most common type, involve only 1–2 cm of intestine and last 3–8 seconds. Therefore, unless blood flow is measured at the site of contraction, the compressive effect of the contraction may be offset by relaxation of adjacent areas. Since the blood vessels of relaxed regions freely communicate with those at an adjacent site of contraction, the overall effect is no net change in blood flow measured in a sizable intestinal artery. Even when the entire gut segment

[1]Supported by Research Grant HL-15231 from the National Institutes of Health

is exposed to a rise in luminal pressure, the blood flow measured in its supplying artery may or may not manifest the influence of the pressure rise, owing to the complex anatomy and physiology of the gut wall and its vasculature.

The intestinal wall contains three main tissue layers. The muscle mass comprises 22–45% by weight (9), and it receives 5–30% of the total gut wall blood flow. When calculated as flow per unit weight, muscularis blood flow is only one half to one tenth of mucosal flow, depending on the segment of the gastrointestinal tract and the condition of the gut preparation (4, 9, 13, 34). The intestinal tissue pO_2 at the muscle-submucosal interface is about 27 mm Hg while that at the villus apex is only 13 mm Hg (3). Oxygen consumption of the muscularis is one fourth to one third mucosal oxygen consumption (41). Since the functions of the mucosa and muscularis are different, a stimulus may induce a functional hyperemia in only one layer. Also, the potency of a given vasoactive stimulus may be different on mucosal and muscularis vessels. Since these two sets of vessels are arranged in parallel, the change in blood flow measured at an intestinal artery or vein is the algebraic sum of flow changes in the two tissue layers. The effects of muscle contractions are likely to be confined only to the muscularis whereas the effect of an increased pressure in the lumen is likely to be more severe on the mucosal vessels; therefore, the flow changes that occur at the intestinal artery may not reflect the changes occurring in each layer. The most significant technical achievement in recent years has probably been the measurement of compartmental blood flow within the gut wall, which provides information about separate responses in at least two functionally dissimilar tissues.

Finally, the physiological feedback systems can further complicate the relationship between motility and blood flow. The feedback systems are activated by changes in muscle tone, vascular transmural pressure, and cellular metabolic activities. They include such mechanisms as receptive relaxation, autoregulation, active hyperemia, stimulation or inhibition of local nerves, and local chemicals (8).

The vascular network in the submucosa may be thought of as a vascular manifold for the distribution and drainage of blood to and from the mucosa and muscularis (14). The pressures of the submucosal arterioles and venules are about 30 and 12 mm Hg, respectively (14). The tissue pressure of the gastric submucosa, as measured by Guyton's interstitial pressure capsule, changes by exactly the same amount as lumen pressure in response to distension or infusion of cholinergic agents (1). In another study it was reported that at lumen pressures between 0 and 25 mm Hg, approximately 92% of lumen pressure is transmitted to the lymphatics and, presumably, to the interstitium of the gut wall (15). Thus, an increase in lumen pressure above 15–30 mm Hg ought to decrease flow to the entire wall. The fact that blood flow is not seriously affected at pressures much higher than 30 mm Hg indicates that the effect of motility on blood flow is not simply a matter of mechanical compression.

RHYTHMIC AND TONIC CONTRACTIONS

Rhythmic contractions, if not strong, produce no change in mean blood flow but are accompanied by cyclic changes in instantaneous blood flow (8, 19, 31). During contractions, arterial inflow decreases, and vascular resistance and venous outflow increase, whereas the opposite occurs during relaxation (8, 19, 20, 25, 31, 36). When rhythmic contractions are accompanied by an increase in basal pressure in the lumen (an indicator of gut wall tension) mean blood flow decreases (8, 31). Rhythmic contractions can also be accompanied by an increase in blood flow (8, 9). The mechanism for the increased flow is unclear but may be related to an active hyperemia, since the increased flow is accompanied by an increase in oxygen consumption (8). In contrast to the variable response to rhythmic contractions, most studies show that spontaneous or drug-induced tonic contractions decrease blood flow and increase vascular resistance (8, 19, 20, 31).

A reciprocal relationship between motility and blood flow is better demonstrated in studies utilizing short segments of the gut. In two recent investigations a jejunal 12–20 cm segment supplied by a single artery was prepared; in one study the constant flow technique was used (20) while in the other (31) the free flow technique was utilized. With constant flow, spontaneous transient (1–3 min) tonic contractions were accompanied by corresponding rises in perfusion pressure (indicating an increase in vascular resistance). There was a significant correlation between the increases in lumen pressure and the corresponding rises in perfusion pressure (change in perfusion pressure = 0.84 × change in lumen pressure − 0.23). During spontaneous rhythmic contractions, each increase in lumen pressure was accompanied by a corresponding rise in venous outflow, and there was a significant linear correlation between the pressure and the flow (change in flow = 0.52 × change in pressure + 1.66). Since the study was performed under constant flow conditions, it can be estimated that about 0.5 ml/min is expelled from the gut wall for each one mm Hg increment in lumen pressure. This massage effect of rhythmic contractions, which promoted venous outflow, was not observed in other reports (19). In one report (36), mild rhythmic contractions (4–8 mm Hg) produced an increase in venous outflow and a decrease in arterial flow, but strong contractions (13–19 mm Hg) produced a decrease in both arterial and venous flows.

Spontaneous or drug-induced (pilocarpine, 5–10 μg, and acetylcholine, 2.5–1.0 μg) rhythmic contractions were accompanied by cyclic changes in instantaneous blood flow, as measured by an electromagnetic flowmeter, in both conscious and anesthetized dogs (31). Mean blood flow, however, was unchanged. Administration of 20–70 μg pilocarpine or 10–70 μg acetylcholine produced a sustained tonic contraction and a decrease in blood flow proportional to the amplitude and duration of the contraction. Absorption of L-phenylalanine and L-serine decreased during tonic contractions but was not changed during isolated increases in rhythmic contractions. The motility index was linearly and negatively correlated with both the blood flow and amino acid absorption. Acetylcholine-, pilocarpine-, and mechanically induced stimulations of motility produced the same significant correlations, which suggests that the decrease in amino acid absorption is due to the decreased flow per se. Thus increased motility, particularly tonic contractions, can mechanically decrease flow and amino acid absorption, although the physiological significance of the effect is unclear.

When motility was decreased (both lumen pressure and rhythmic contractions) by either electrical stimulation of the serosa or luminal distension of an adjacent gut segment, blood flow decreased (37). Thus a 35–40% decrease in lumen pressure was associated with a 10–20% decrease in both arterial and venous blood flows. The result is opposite to that expected from a mechanical point of view. However, since the relaxation in this study was induced by a neural reflex, the decreased flow is probably due to a direct vasoconstrictor effect of the same nerves that inhibit motility (i.e. adrenergic nerves).

The use of microspheres can distinguish blood flow to the muscularis from that of the mucosa-submucosa but is unable to separate mucosal from submucosal flows [see (9) for references]. Also, mechanical handling of the gut during surgery can redistribute flow away from the mucosa-submucosa (9, 34). In both conscious and anesthetized dogs, 87–96% of the total wall flow is distributed to the mucosa-submucosa in those cases where the gut has not been manipulated (9, 13, 34), but only 51–80% of the total flow is distributed to the mucosa-submucosa when measured shortly after gastrointestinal surgery (9, 34). This shift in the distribution of blood flow within the gut wall is not due to either anesthesia or laparotomy (4, 34).

With the microsphere method, rhythmic contractions (12 cycles/min and 4–6 mm Hg or 15–25 mm Hg in amplitude) induced by mechanical manipulation or a mild degree of luminal distension increased total wall blood flow as measured by both venous outflow and microspheres (9). This increase in flow was confined to the muscularis; the mucosal-submucosal flow was unchanged. Physostigmine-induced (0.15–0.9 μg/kg/ min) tonic contractions decreased total wall flow of the entire gastrointestinal tract, but the decreased flow was confined only to the mucosa-submucosa; the muscularis flow was unchanged. Thus the vasculature of the muscularis escaped the effects of mechanical compression from tonic contractions. The percentage of total wall flow perfusing the muscularis was increased whether the total flow was increased or decreased. It is possible that mechanical manipulation may dilate only the muscular vasculature whereas physostigmine constricts only the mucosal-submucosal vasculature; however, the most likely explanation for these findings resides in the interplay of the two effects of muscle contractions. The mechanically induced rhythmic contractions are probably not strong enough to impede mucosal flow whereas the tonic contractions are sufficient to impede both mucosal and muscular flows. The muscularis exhibits active hyperemia during both rhythmic and tonic contractions, but in the former case the hyperemia is unopposed by mechanical compression whereas in the latter case the hyperemia is offset by the mechanical effect and flow to the muscularis is unchanged.

Other recent studies have also shown evidence for a contraction-induced hyperemia in the muscularis. Rhythmic contractions occur following 120-second complete arterial occlusion (27) and during elevation of venous pressure to 20–30 mm Hg (15, 16). With arterial occlusion (reactive hyperemia) the augmented flow to the muscularis is significantly greater than is the increase in the flow to mucosa-submucosa (27). With elevated venous pressure (venoarteriolar response leading to constriction) mucosal-submucosal flow decreases whereas muscularis flow increases (16).

EFFECTS OF CHEMICALS AND NERVE STIMULATION

Many chemicals and neurotransmitters that act on both intestinal and vascular smooth muscles are shown in Table 1. Some of these agents contract or relax both muscles whereas others contract intestinal but relax vascular smooth muscle and vice versa. In addition, some chemicals are relatively inactive with respect to one type of smooth muscle and yet exert a strong effect on the other—e.g. morphine. From the mechanical point of view, the effects of a stimulus, for example the potassium ion, that has the same effect on both types of smooth muscle should have an additive effect on blood flow. However, this additive effect has not been clearly demonstrated. When the effects are opposite in direction, as in the case of acetylcholine, the net result will be the algebraic sum of the two oppsing effects —i.e. the magnitude and type (rhythmic or tonic) of the change in motility and their potency as vasoactive agents. Thus administration of vasodilators such as acetylcholine, CCK, pentagastrin, and bradykinin produces an increase in blood flow when motility is only slightly increased, and a decrease in blood flow when greater lumen pressures in the lumen are reached, (8, 19, 31, 35, 40). Similarly, epinephrine can reduce vascular resistance while simultaneously decreasing motility in a hypermotile segment when administered at a dose that normally produces vasoconstriction in the quiescent gut (8).

Although strong contractions can mechanically alter the direct vascular effect of a stimulus, the vascular and motor actions of most stimuli are independent of each other. Intraarterial infusions of pentagastrin (0.1–0.5 μg/kg/min) or the synthetic octapeptide of CCK (0.01–0.1 μg/kg/min, i.a.) increased intestinal blood flow, motility, and oxygen consumption (28). These effects were blocked by atropine, which suggests that the effects may be mediated by cholinergic nerves. Secretin (0.03–0.3 μg/kg/min, i.a.) had no vascular effect but decreased intestinal motility. Intravenous bolus injections of pentagastrin (1–4096 ng/kg) or CCK (1–4096 mU/kg) produced variable vascular and motor responses in the stomcah, small intestine, and colon (35) (Table 1). Vasopressin, however, decreased flow to all three

Table 1 The concomitant effects of various chemicals on intestinal motility and total blood flow[a]

Effect on blood flow	Effect on motility	Substance and (Reference)
Increase	Increase	Cholinergics (8, 12, 19), Bradykinin (8, 11, 19), Ca^{2+} [high doses (38)], CCK (28, 35), Gastrin (28) [stomach (35)], Histamine (19, 34, 39), Met-enkephalin (30, 40), PGD_2 (40), Serotonin (8), Substance P [high doses (33)], VIP [colon (10)].
	Decrease	CO_2 (8), Glucagon (41), K^+ [low doses (8)], Mg^{2+} (8), Secretin (8), VIP [stomach (10)].
	No change	Substance P [low doses (33)], Adenosine (11)
Decrease	Increase	Angiotensin II (19, 40), CCK [stomach (35)], K^+ [high doses (8)], $F_{2\alpha}$ (40)
	Decrease	Adrenergics (8), Ca^{2+} [low doses (8)], Somatostatin (41)
	No change	Gastrin [colon (35)], Vasopressin (35)
No change	Increase	Morphine (40), Metacholine (41), Met-enkephalin (40)
	Decrease	Secretin [low doses (28)]
	No change	Gastrin (35), CCK [colon (35)]

[a] Variation in results is due to doses, administration routes, organ specificity, and techniques. All studies were done in the small intestine except those specified in brackets.

organs without affecting motility. Based on ED_{50}, the same study (35) has estimated that the gastric vascular and visceral smooth muscles are as sensitive to pentagastrin as gastric parietal cells, and the small intestinal vasculature is more sensitive to CCK than the gallbladder smooth muscle. Another study has also shown that CCK is a much more potent vasodilator in the duodenum and jejunum than in non-digestive organs, and the vasodilation can be induced at phsyiological blood concentrations (7).

Some vasoactive chemicals exert different effects on mucosal and muscularis blood flows, and the difference is in part due to their effects on motility. Histamine is a vasodilator that stimulates motility and selectively increases flow to the muscularis (34, 39). While both H_1 and H_2 receptors are responsible for the vasodilation, the H_1 receptors are responsible for the increase in motility and the redistribution of flow to the muscularis (29, 39). Methionine-enkephalin has little direct vascular action (40); but at doses of 0.03–1.0 μg/kg/min., i.a., it produced dose-dependent increases in intestinal blood flow, oxygen consumption, and mean motility index (30). At 0.5 μg/kg/min, total intestinal flow increased 23% while flow to the muscularis increased 50% and motility index increased 107%. The association of an increased motility and oxygen consumption and a preferential increase in muscle flow suggests that the increased flow is related to an active hyperemia in the muscle layer.

The effects of acetylcholine, prostaglandin D_2, methionine-enkephalin, morphine, prostaglandin F_{2a}, and angiotension II were examined on total wall flow as well as flow to the muscularis, motility, P-S product and VO_2 in the canine ileum (40). All six agents increased both the basal lumen pressure and rhythmic contractions. Acetylcholine increased total wall flow, with much of the increase in flow g⁻ing to the muscularis. Prostaglandin D_2 was a weak vasodilator, and methionine-enkephalin appeared to have no direct vascular effect. However, stimulation of motility by these two agents led to an increase in oxygen consumption and total wall flow that was predominantly due to an increase in flow to the muscularis. Prostaglandin F_{2a}, a mild vasoconstrictor, decreased total wall flow but increased oxygen consumption and did not significantly affect muscularis flow. The decreased total wall flow, therefore, was due to a decreased mucosal flow. Thus vasodilators, such as acetylcholine and prostaglandin D_2, seem to preferentially increase muscularis flow; a vasoinactive agent, methionine-enkephalin, increased only muscularis flow; and a mild vasoconstrictor, prostaglandin F_{2a}, decreased only mucosal blood flow. These findings could best be explained by the contraction-induced active hyperemia in the muscularis. Angiotension II, a potent vasoconstrictor, however, equally decreased both the mucosal and muscularis flows, while morphine, which increased motility, did not alter total flow or oxygen consumption but decreased flow to the muscle layer.

EFFECTS OF LUMINAL DISTENSION

Distension of the gut lumen to pressures below 30 mm Hg may increase (8, 9, 18) or decrease (15, 18) blood flow, but distension to higher pressures (40–70 mm Hg) decreases blood flow (8, 25). In most cases, flow tends to return toward control level even when distension is maintained (8, 18, 25, 32). At least four mechanisms are involved in the recovery of flow. The gut responds to distension with receptive relaxation; as a result the lumen pressure tends to return toward the control level with the recovery of lumen pressure being accompanied by recovery of flow (8). Also, prevention of intestinal expansion and movement by encasing the gut in plaster of Paris prevented the recovery of flow during distension. The influence of distension-induced rhythmic contractions (9) on blood flow has been discussed in the preceding section. Since papaverine, a muscle relaxant, not only abolished the autoregulation of intestinal blood flow but also increased hemodynamic effects of distension, the recovery of flow may be due partly to the myogenic autoregulatory response to a decrease in vascular transmural pressure (18). Neural reflexes may also be involved since mucosal anesthesia prevented flow from returning toward control level during distension (8).

For each one mm Hg rise in lumen pressure, the rise in vascular resistance produced by sustained tonic contractions was 1.61 times that produced by luminal distension (20). The lesser rise in vascular resistance during distension may have been due to the factors described above. Under constant lumen pressures, stepwise increases in lumen pressure to 100 mm Hg produced stepwise decreases in flow (8, 25). However, further increases in lumen pressure to 180 mm Hg did not significantly decrease flow further; the flow was about 20–35% of control (8). At these high pressures venous pO_2 almost equalled arterial pO_2, indicating that the residual flow was perfusing low- or nonnutritional vessels (8, 25). In most studies, deflation of the distended gut produced an increase in flow and a decrease in resistance. A hyperemia also occurs following cessation of tonic contractions (31, 36) and is probably related to a metabolically and myogenically induced reactive hyperemia.

Distension exerts different effects on mucosal and muscular blood flow. Effects of stepwise increases in lumen pressure of piglet gut to 15, 30, 45, and 60 mm Hg were studied by inflating 30 cm long small- and large-intestinal segments with air while blood flow was measured with microspheres after these pressures were maintained for 15 minutes (32). The stepwise increases in pressure to 45 mm Hg produced 24–55% and 55–72% decreases in total wall and mucosal flows in the small intestine and 13–35% and 55–69% decreases in the colon, respectively. Flows to the muscle layer of the small and large intestine, however, increased 118, 83, and 22%, and 161, 149, and 65% at lumen pressures of 15, 30, and 45 mm Hg, respectively. At 60 mm Hg, total wall flow decreased to 25% and mucosal flow declined to about 20% of control, but muscle flow was reduced only to 75–80% of control in both small and large intestine.

Luminal distension also affected local oxygen consumption, capillary filtration coefficient (CFC), lymphatic pressure, lymph flow, and transmucosal fluid absorption, but the effects differed with the degree of distension. At distending pressures below 20 mm Hg, oxygen consumption is either increased (26) or unchanged (20). As distension pressure was increased from 30 to 100 mm Hg, oxygen consumption decreased progressively as a function of the increase in pressure (8, 25). CFC increased or did not change at pressures of 20–40 mm Hg, but further distension to 100 mm Hg produced a progressive fall in CFC (25).

A mild luminal distension increased lymphatic pressure, lymph flow, and the rate of transmucosal fluid absorption (23). Since fluid absorption per se can increase lymph flow, a nonabsorbable silicone solution was used in a recent study to distend the gut lumen (15). At a venous pressure of 0 mm Hg, luminal distension to a pressure of 20 mm Hg resulted in a progressive reduction in both the oncotic pressure of lymph and the lymph-to-plasma protein concentration ratio (L/P), while lymphatic pressure and lymph flow progressively increased. When lumen pressure was further increased above 25 mm Hg, lymphatic pressure and lymph flow decreased progressively. The authors also compared the effects of venous pressure elevation at 0 and 20 mm Hg lumen pressure. At 0 mm Hg lumen pressure, elevation of venous pressure to 30 mm Hg resulted in a progressive increase in lymph flow and decrease in blood flow and L/P. However, at a lumen pressure of 20 mm Hg, venous pressure elevation had no effect on intestinal lymph flow, L/P, or blood flow until venous pressure exceeded the lumen pressure.

Approximately 92% of the lumen pressure was found to be transmitted to the lymphatics and, presumably, to the interstitium. Thus the effects of venous pressure elevation were not revealed until venous pressure exceeded tissue pressure (15). The investigators surmised that when tissue pressure exceeds venous pressure, venules collapse and the blood flow and capillary hydrostatic pressure in this situation ("waterfall" condition) would not be determined by venous outflow pressure; instead, they would be influenced by the collapse pressure—i.e. tissue pressure. The distension-induced reduction in vascular transmural pressure would dilate arterioles and precapillary sphincter (according to the myogenic theory of blood flow regulation), thereby causing both the capillary hydrostatic pressure and CFC to increase. Collapse of venules would amplify the effect of arteriolar relaxation on capillary hydrostatic pressure. Granger et al have also found that for a given increase in lymph flow, luminal distension produces a smaller reduction in interstitial oncotic pressure than that produced by either venous pressure elevation or intestinal fluid absorption. This suggests that luminal distension not only mechanically increases capillary filtration but also increases capillary permeability.

The highest lumen pressure that has been observed in pathological states is about 50 mm Hg (8), while that produced by intestinal obstruction is usually below 20 mm Hg (25). Thus, even in pathological states the effects of distension on gut blood flow, oxygen consumption, and CFC are probably minimal. In fact, a mild and transient distension that occurs when chyme enters a gut segment benefits intestinal function. Rhythmic contractions induced by distension promote mixing and transport of chyme. The increased metabolic demand for contractions is adequately met by an increase in muscularis flow. A mild distension also increases CFC, capillary permeability, and lymph flow, all of which tend to enhance fluid and nutrient absorption.

EFFECTS OF BLOOD FLOW ON MOTILITY

A 40–60% reduction in total blood flow to the feline small intestine by partial occlusion of the superior mesenteric artery produced 48–70% and 31–50% reductions in muscular and mucosal-submucosal flow, respectively (6). In spite of the greater ischemia in the muscularis, the study also showed that histologically the muscularis looked normal while total destruction of the villi and various degrees of mucosal damage were observed. Probably the intestinal muscle is more resistant to ischemic hypoxia than is the mucosa because of its lower metabolic rate. A recent in vitro (2) study has demonstrated that exposure of the feline colonic muscle strip to a physiological solution equilibrated with gas mixtures containing 75, 55, or 35% O_2 did not alter the myogenic electrical activity. Exposure to 0% O_2, however, decreased the slow wave frequency from 5 to 3 cpm and the time occupied by the migrating spike bursts from 17 to 6%. While the maximum changes occurred within 45 min of the exposure, the two electrical events persisted for 3 hr. Furthermore, the two events returned to control states within 15 min upon restoration of 95% O_2. The migrating spike bursts in the colon are related to prolonged contractions that occur independently of slow waves. Thus anoxia but not hypoxia can reduce the frequency of colonic slow waves and contractions. However, the persistence of the two events under anoxia for 3 hr indicates that these events are not exclusively dependent on oxidative metabolism.

Although doubling the blood flow does not significantly alter motility (8), intestinal ischemia and hypoxia can alter electrical contractile activities of the stomach and intestine depending on the duration and severity of the ischemia and hypoxia. Ischemia and hypoxia produced a biphasic change in motility—i.e., an initial transient increase followed by a prolonged paralysis. Thus occlusion of the artery perfusing a gut segment (5, 8, 17), thrombin-induced mesentric arterial and venous thrombosis (24), or a 75% reduction in inhaled oxygen (24) produced an immediate but transient increase in spike potentials and contractions lasting for 1–5 min. The increased contractions appeared to be mediated by intrinsic nerves since local administration of tetrodotoxin abolished this response (8). This increase in motility progressively diminished and the gut became quiescent 1–15 min after ischemia and hypoxia with the disappearance of spike potentials. Although the slow waves were present even after 18 hr of ischemia (5), their frequency progressively decreased as the ischemia was prolonged (5, 22, 24). Partial ischemia or a 50% reduction in inhaled oxygen produced only an initial increase in motility, which returned to near control levels in 5 min; the slow waves were not affected (24).

The duration of ischemia determines whether or not the normal motor activity recovers after revascularization. When the circulation was restored within 1–3 hr, the slow waves, spike potentials and spontaneous contractions returned to normal 1–13 min following revascularization (5), and the gut segment responded to luminal HCl with normal rhythmic contractions (17). When complete ischemia persisted more than 4 hr, revascularization did not restore the spontaneous contractions and spike potentials (5, 17). Furthermore, the gut segments did not respond to mucosal mechanical (irritation or distension) and chemical (HCl, or bethanechol) stimulation with contractions (5, 17).

Long-term effects of anoxia were assessed in jejunal segments subjected to anoxia by intravascular perfusion of Ringer-Tyrode's solution for 4 hr (8). Normal circulation was restored and the motor activities were observed in the conscious state. Throughout the 30–90 day observation period the slow wave frequency was subnormal, the contractions in the transverse muscle were not coordinated with longitudinal muscle, and the contractions in the hypoxic segment were not coordinated with adjacent normal segments. The changes may have been due to hypoxic insult on the myenteric plexuses (8).

Alternatively, it may be that both muscle and nerves were affected by hypoxia (22). The responses to intraarterial infusions of metacholine, nicotine, DMPP, and phenyldiguanide were reduced or absent shortly after 4 hr of anoxia. This indicates that ischemic anoxia affects both nerves and smooth muscle. The smooth muscle Na^+, Ca^{2+}, Cl^-, and water concentrations increased while K^+ and Mg^{2+} concentrations decreased. Muscle strips obtained from the ischemic segments showed no spontaneous rhythmic contractions and a reduced contractile response to methacholine but exhibited normal responses to stimuli that act on nerves (22). These acute in vitro studies seem to suggest that ischemic anoxia affects muscle more adversely than nerves. Similar pharmacological and chemical studies performed on the ischemic segments 10–15 days after revascularization indicated that muscle electrolyte concentrations were normal and the contractile responses to the stimuli which act on nerves were also normal. The motility responses to luminal HCl or distension, however, were absent. Some intrinsic nerves may still have been functioning, and their histological examination revealed that 25% of the ganglion cells were normal, 50% had varying degree of reversible changes, and 25% had irreversible changes.

Ischemia produced similar effects on the electrical and mechanical activities of the stomach (21). Normal activities could be restored when the duration of the ischemia was shorter than 2 hr, and gastrin plays a role in the restoration.

CONCLUDING REMARKS

Muscle contractions can affect blood flow by two mechanisms: the active hyperemia due to muscle contractions, and a decrease in vascular transmural pressures due to extravascular compression. The former increases flow in the muscularis while the latter tends to decrease flow to all tissue layers of the gut wall. The effect of compression, however, is modified by receptive relaxation of the intestinal muscle and by metabolically and myogenically induced feedback mechanisms. Furthermore, the stimuli that alter motility often exert their direct effect on vascular smooth muscle. The effect of motility on blood flow, therefore, varies with the nature of the motility, the initiating stimulus, and the feedback mechanisms. Although the muscularis is relatively resistant to hypoxia, severe hypoxia and ischemia for several hours can permanently alter motility. Since the contractile state of the intestine normally varies with time and space, the ideal method to elucidate the relationship between blood flow and motility would simultaneously and continuously measure electrical and mechanical events associated with contractions and local tissue blood flow in multiple areas of the intestine in conscious animals.

Literature Cited

1. Altamirano, M., Requena, M., Perez, T. C. 1975. Interstitial fluid pressure in canine gastric mucosa. *Am. J. Physiol.* 229:1414–20
2. Anuras, S., Chien, S. M., Christensen, J. 1980. Metabolic dependence of the electromyogram of the cat colon. *Am. J. Physiol.* 239:G173–76
3. Bohlen, H. G. 1980. Intestinal tissue Po_2 and microvascular responses during glucose exposure. *Am. J. Physiol.* 238:H164–71
4. Bond, J. H., Prentiss, R. A., Levitt, M. D. 1980. The effect of anesthesia and laparotomy on blood flow of the stomach, small bowel, and the colon of the dog. *Surgery* 87:313–18
5. Cabot, R. M., Kohatsu, S. 1978. The effects of ischemia on the electrical and contractile activities of the canine small intestine. *Am. J. Surg.* 136:242–46
6. Cassuto, J., Cedgard, S., Haglund, U., Redfors, S., Lundgren, O. 1979. Intramural blood flows and flow distribution in the feline small intestine during arterial hypotension. *Acta Physiol. Scand.* 106:335–42
7. Chou, C. C., Hsieh, C. P., Dabney, J. M. 1977. Comparison of vascular effects of gastrointestinal hormones on various organs. *Am. J. Physiol.* 232:H103–9
8. Chou, C. C., Gallavan, R. H. 1981. Blood flow and intestinal motility. *Fed. Proc.* To be published
9. Chou, C. C., Grassmick, B. 1978. Motility and blood flow distribution within the wall of the gastrointestinal tract. *Am. J. Physiol.* 235:H34–39
10. Eklund, S., Jodal, M., Lundgren, O., Sjogvist, A. 1979. Effects of vasoactive intestinal polypeptide on blood flow, motility and fluid transport in the gastrointestinal tract of the cat. *Acta Physiol. Scand.* 105:461–68
11. Fasth, S., Hulten, L., Johnson, B. J., Nordgren, S., Zeithin, I. J. 1978. Mobilization of colonic kallikrein following pelvic nerve stimulation in the atropinized cat. *J. Physiol. London* 285:471–78
12. Fasth, S., Hulten, L., Nordgren, S. 1980. Evidence for a dual pelvic nerve influence on large bowel motility in the cat. *J. Physiol. London* 298:159–69
13. Gallavan, R. H., Jr., Chou, C. C., Kvietys, P. R., Sit, S. P. 1980. Regional blood flow during digestion in the conscious dog. *Am. J. Physiol.* 238:H220–25
14. Gore, R. W., Bohlen, H. G. 1977. Microvascular pressures in rat intestinal muscle and mucosal villi. *Am. J. Physiol.* 233:H685–93
15. Granger, D. N., Kvietys, P. R., Mortillaro, N. A., Taylor, A. E. 1980. Effect of luminal distension on intestinal transcapillary fluid exchange. *Am. J. Physiol.* 239:G516–23
16. Granger, D. N., Richardson, P. D. I., Taylor, A. E. 1979. Volumetric assessment of the capillary filtration coefficient in the cat small intestine. *Pflügers Arch.* 381:25–33
17. Guisan, Y. J., Hreno, A., Gurd, F. N. 1975. Effect of acute ischemia on the motility of the small bowel in the awake dog. *Eur. Surg. Res.* 7:23–33
18. Hanson, K. 1973. Hemodynamic effects of distension of the dog small intestine. *Am. J. Physiol.* 225:456–60
19. Jacobson, E. D., Brobmann, G. F., Brecker, G. A. 1970. Intestinal motor activity and blood flow. *Gastroenterology* 58:575–79
20. Kachelhoffer, J., Pousse, A., Marescaux, J., Iturizaga, M., Grenier, J. F. 1978. Effects of motility and luminal distension on dog small intestine hemodynamics. *Eug. Surg. Res.* 10:184–93
21. Kowalewski, K., Zajac, S., Kolodej, A. 1976. Effects of ischemic anoxia on electrical and mechanical activity of the totally isolated procine stomach. *Eug. Surg. Res.* 8:12–25
22. Kyi, J. K. J., Daniel, E. E. 1970. The effects of ischemia on intestinal nerves and electrical slow waves. *Am. J. Dig. Dis.* 15:959–81
23. Lifson, N. 1979. Fluid secretion and hydrostatic pressure relationships in the small intestine. In *Mechanisms of Intestinal Secretion,* ed. H. J. Binder, pp. 249–61, NY: Liss
24. Meissner, A., Bowes, K. L., Sarna, S. K. 1976. Effects of ambient and stagnant hypoxia on the mechanical and electrical activity of the canine upper jejunum. *Can. J. Surg.* 19:316–21
25. Ohman, U. 1975. Studies on small intestinal obstruction. Blood circulation in obstructed and artificially distended small intestine in the cat. *Acta Chir. Scand. Suppl.* 452
26. Ohman, U. 1976. Blood flow and oxygen consumption in the feline small intestine: Responses to artificial distension and intestinal obstruction. *Acta Chir. Scand.* 142:329–33
27. Parker, R. E., Granger, D. N. 1979. Effect of graded arterial occlusion on ileal blood flow distribution. *Proc. Soc. Exp. Biol. Med.* 162:146–49

28. Pawlik, W., Bowen, J. C., Jacobson, E. D. 1977. Vasoactive and metabolic effects of gastrointestinal hormones in the intestine. *Mat. Med. Pol.* 31:151–54
29. Pawlik, W., Tague, L. L., Tepperman, B. L., Miller, T. A., Jacobson, E. D. 1977. Histamine H_1 and H_2 receptor vasodilation of canine intestinal circulation. *Am. J. Physiol.* 233:E219–24
30. Pawlik, W. W., Walus, K. M., Fondacaro, J. D. 1980. Effects of methionine-enkephalin on intestinal circulation and oxygen consumption. *Proc. Soc. Exp. Biol. Med.* 165:26–31
31. Pytkowsk, B., Michalowski, J. 1977. Motility- and blood flow-dependent absorption of amino acids in canine small intestine. *Eur. J. Clin. Invest.* 7:79–86
32. Ruf, W., Suehiro, G. T., Suehiro, A., Pressler, V., McNamara, J. J. 1980. Intestinal blood flow at various intraluminal pressure in the piglet with closed abdomen. *Ann. Surg.* 191:157–63
33. Schrauwen, E., Houvenaghel, A. 1979. Influence of substance P on mesenteric hemodymanics in the pig. *Acta Int. Pharmacodyn.* 242:315–17
34. Schwaiger, M., Fondacaro, J. D., Jacobson, E. D. 1979. Effects of glucagon, histamine and perhexiline on the ischemic canine mesenteric circulation. *Gastroenterology* 77:730–35
35. Schuurkes, J. A. J., Charbon, G. A. 1978. Motility and hemodynamics of the canine gastrointestinal tract. Stimulation by pentagastrin, cholecystokinin and vasopressin. *Arch. Int. Pharmacodyn. Ther.* 236:214–27
36. Semba, T., Kazumoto, F., Fujii, Y. 1971. The influence of rhythmic and tonic contraction of the small intestine on blood flow through the intestinal segment. *Jpn. J. Physiol.* 21:1–14
37. Semba, T., Mizonishi, T., Ikeda, Y., Nagao, Y. 1977. Influence of intestinal inhibitory reflex on mesenteric blood flow through an intestinal segment of the dog. *Jpn. J. Physiol.* 27:439–50
38. Walus, K. M., Fondacaro, J. D., Jacobson, E. D. 1981. Effects of calcium and its antagonists on the canine mesenteric circulation. *Circ. Res.* 48:692–700
39. Walus, K. M., Fondacaro, J. D., Jacobson, E. D. 1981. A further characterization of histamine H_1 and H_2 effects and blockade in canine intestinal circulation. *Digest. Dis. Sci.* 80:1542–49
40. Walus, K. M., Fondacaro, J. D., Jacobson, E. D. 1981. Local vascular responses to stimulation of intestinal motility. *Digest. Dis. Sci.* In press
41. Walus, K. M., Jacobson, E. D. 1981. Relationship between small intestinal motility and circulation. *Am. J. Physiol.* 241:G1–15

RELATIONSHIPS BETWEEN INTESTINAL ABSORPTION AND HEMODYNAMICS

David Mailman

Biology Department, University of Houston, Houston, Texas 77004

INTRODUCTION

The primary function of the intestine is the absorption of nutrients, salts, and water. These molecules move across the epithelium by several carrier-mediated or passive transport mechanisms. The substances then enter the interstitial space and are carried away by the blood flow near the absorptive site. Both the transport processes and blood flow have been well studied separately, but relatively few studies have focused on the relationship between these functions, for two reasons. First, there is no generally accepted technique for measuring blood flow at the absorptive site; second, interactions between absorption and blood flow are complex.

VILLOUS MICROVASCULATURE AND BLOOD FLOW

Perfusion of the gut is distributed in four parallel blood flows to the muscularis, submucosa, crypts, and villi. The parallel arrangement allows total blood flow and villous blood flow to vary independently of one another.

The villous microvasculature is composed of a dense network of fenestrated capillaries that have an extremely large surface area for exchange of small solutes but are relatively impermeable to colloids. The protein reflection coefficient (12) indicates that 90% of the theoretical colloid osmotic pressure can be exerted across the intestinal capillary.

The villous capillaries are arranged in two configurations. In most species the capillaries arise near the tip from one or two villous arterioles and descend, fountain-like, to near the villous base. In the dog the capillaries arise, ladder-like, along the length of the villous arteriole (11). Because only the upper third of the villous is in effective contact with the luminal contents, the anatomical blood flow to villi with a ladder-like arrangement will be greater than the effective villous blood flow perfusing the absorptive site. With the fountain-like arrangement of capillaries there is a countercurrent configuration and the effective blood flow could be less than the anatomical blood flow because of diffusion between the arterioles and capillaries.

Effective mucosal or absorptive-site blood flow has been estimated by several techniques including microspheres, the clearance of highly permeable substances such as inert gasses or ^3H_2O from either the blood or lumen, compartment flow analysis of blood labeled with low-energy radioisotopes measured with a luminal detector, carbon monoxide uptake from the gut, and optical techniques. Estimates of effective mucosal blood flow with the various techniques ranged from 50–100 $\mu l/g \cdot gut \cdot min$ in the cat and dog, 80–200 $\mu l/g \cdot gut \cdot min$ in the rabbit, and 200–250 $\mu l/g \cdot gut \cdot min$ in the rat (29). However, a large discrepancy between microsphere (250 $\mu l/g \cdot min$) and inert gas clearance (40 $\mu l/g \cdot min$) values was found in the dog and was attributed to countercurrent exchange, although in the cat, which has comparable villi, effective mucosal blood flow measured with plasma labeled with low energy radioisotopes gave values similar to those measured by ^{85}Kr clearance. Microspheres may bypass more basal capillaries in the canine villus and be trapped in the most apical group of capillaries, thereby overestimating blood flow to the villous absorptive site.

The volume absorbed from the lumen is a significant fraction of the effective mucosal blood flow. In the dog, volume absorption under control conditions is about 20% of absorptive site blood flow (27). Thus entry of this volume of colloid free fluid into the local blood flow can significantly alter hemodynamics and influence the driving forces for tissue uptake of absorbed fluid.

ABSORPTIVE PATHWAYS AND BLOOD FLOW

Epithelial transport can be divided into facilitated, carrier-mediated transcellular transport through the membranes of the cell and passive paracellular transport through the tight junction and lateral space. A molecule absorbed by transcellular transport can be subject to the same forces exerted on passively transported substances when it is in the lumen or after it has crossed the cell into the interstitial space. At luminal concentrations above the Km of the carrier-mediated transport systems even substances that are actively absorbed may have significant absorption through the passive pathway. When transported substances reach the basal area of the cell, they must cross the interstitial space and capillary wall before being removed by blood flow. In general, blood flow acts on carrier-mediated transcellular transport through the delivery of oxygen to support cell metabolism. Paracellular passive transport is affected primarily through physical mechanisms such as Starling forces.

Winne (39) has reviewed a number of studies showing a positive correlation between blood flow and the absorption of passively and actively transported substances. Donowitz et al (8) studied the relationship between mucosal-submucosal blood flow, measured with microspheres, and absorption or secretion stimulated by several agents such as glucocorticoids or cholera toxin. The authors concluded that increased absorption was associated with increased blood flow, but increased secretion was not associated with significant decreases in blood flow. However, if all of their data are considered (28), the correlation of blood flow with absorption is just short of being significant ($r = 0.744$, $5\%r = 0.754$). Therefore, there is evidence that a general cause/effect correlation exists between blood flow and absorption, although which is cause and which effect may vary under different experimental conditions.

OXYGEN DELIVERY

Intestinal blood flow can be reduced by 40% without decreasing O_2 consumption. This suggests that the delivery of O_2 may not be rate-limiting under nonabsorbing conditions. Luminal nutrients promptly increase blood flow to the gut, which reaches a plateau 20–100% above control in about 20 min. Lipid digestion products, amino acids, and absorbed sugars, particularly in the presence of bile (21), increase blood flow to the segment of intestine in which the nutrients are placed. Glucose absorption is associated with larger increases in blood flow and O_2 consumption than occurs with the nonmetabolizable but actively transported 3-0-methylglucose (36). This suggests that both active transport and local metabolism increase local O_2 consumption, and in turn blood flow, through metabolic feedback. Feeding increases autoregulatory responses, presumably because blood flow must supply O_2 when local metabolism has already increased O_2 extraction. Blood flow and O_2 consumption may also be partly controlled through other mechanisms. Corn oil placed in a gut segment caused an increase in blood flow to the entire gut, an effect attributed to the release of CCK.

Under some experimental conditions luminal glucose can increase O_2 consumption by increasing O_2 extraction with little effect on total gut blood flow (35). In these experiments, the capillary filtration coefficient increased, suggesting that local nutrient blood flow in some portion of the gut increased. This is consistent with local metabolic control of blood flow during absorption. Absorption need not always be associated with increased O_2 consumption. Increased volume absorption due to luminal glucose caused a proportional increase in O_2 consumption, but taurocholate-induced volume absorption caused a slight decrease in O_2 consumption (37).

Results from experiments in which local blood flow was measured by optical techniques while local pO_2 was monitored by O_2 microelectrodes suggested that local metabolic control over blood flow may not be the major factor in the hyperemia in response to luminal nutrients (2). In these experiments, mucosal pO_2 decreased in response to low concentrations of glucose in the lumen, and mucosal blood flow increased; this is consistent with local metabolic control of blood flow. However, submucosal and muscularis blood flow also increased despite a slight increase in pO_2 in these areas. Results from experiments in which mucosal pO_2 was changed by luminal O_2 plus glucose suggested that only about one third of the increased mucosal blood flow was responsive to pO_2, with the remainder perhaps sensitive to oxygenation or such other factors as metabolites or paracrine agents (2).

STARLING FORCES

Direct observation of the villous blood flow and simultaneous measurements of absorptive rate in the dog showed that net absorption was associated with low, transient villous perfusion through the central arteriole and a few longitudinally oriented blood vessels. Secretion was associated with blood-engorged villi in which transverse capillaries appeared between the central arteriole and the longitudinal blood vessels (38). Experiments in which isolated gut loops were mechanically perfused through their vascular supply also indicated that hemodynamic factors were important contributors to absorption. In order to prevent secretion and allow absorption, colloids had to be present in the vascular perfusate, the blood pressure had to be kept relatively low, and vasoactive responses had to be modified.

Volume absorption from the lumen causes proportional changes in tissue hydrostatic pressure, reflected in lymph flow, and tissue colloid osmotic pressure, reflected in lymph colloid osmotic pressure (16). These changes are independent of the solute used to stimulate absorption and O_2 consumption. Increasing tissue hydrostatic pressure and reducing colloid osmotic pressure when caused by a primary increase in fluid absorption will facilitate fluid entry from the interstitial space into the blood. Thus, fluid absorption promotes its uptake by the blood through altered Starling forces.

Altered Starling forces caused by primary changes in local hemodynamics alter gut absorption through physical effects. Hydrostatic pressure applied to the mucosal surface in vitro does not affect absorption, but serosal pressures greater than a threshold value of about 4 cm H_2O and less than 20 cm H_2O reduce absorption or increase secretion proportionally. Both the passive secretory driving forces and tissue conductance are increased. A larger conductance increase occurs at a serosal pressure above 20 cm H_2O because secretion increases markedly and the gut cannot maintain a greater hydrostatic pressure gradient.

Increasing mesenteric venous pressures between 0–30 cm H_2O did not affect mucosal hydration, lymph flow, or secretion; but above a threshold level of about 35 cm H_2O there were linear proportional increases in these three parameters relative to mesenteric venous pressure (40). These findings suggest that below a threshold level of venous pressure there are mechanisms for maintaining Starling force balance constant. It is possible that small increased volumes of capillary ultrafiltrate from increased capillary hydrostatic pressure dilute tissue colloids, thereby decreasing the colloid osmotic pressure gradient between plasma and tissue. This effect would reduce the pull of fluid from the plasma. Above threshold mesenteric venous pressure, the compensation for ultrafiltration is no longer sufficient and significant fluid begins to enter the interstitial space. Thus tissue pressure, mucosal hydration and consequently conductance increase, and tissue colloid osmotic pressure is further reduced. The net effect is to increase the driving forces for passive secretion and the ease of fluid movement into the lumen. The compensation for maintaining a constant balance among Starling forces is restricted to the mucosa because nonmucosal hydration increases proportionally to venous pressure without a threshold.

The balance among Starling forces can be upset by volume expansion with saline, which results in net intestinal secretion due primarily to an increase in the unidirectional secretory fluxes of sodium and H_2O (31). Total intestinal blood flow during saline infusion increased markedly, but absorptive-site blood flow (measured as venous drainage by ^3H_2O clearance) did not change. The secretion was passive due to an increase in capillary pressure caused mainly by an increase in venous pressure and/or by a decrease in plasma colloids with consequent entry of fluid into the interstitial space, increased interstitial fluid pressure, and subsequent movement of fluid into the lumen. The entry of fluid into the lumen was associated with increased mucosal conductance as suggested by increased movement of inulin or mannitol across the mucosa (9).

There may be a problem in the interpretation of changes in lymph flow or colloid osmotic pressure as reflecting changes in tissue pressure or colloid osmotic pressure at the absorptive site following changes in absorption or hemodynamics. During moderate and higher rates of absorption, lymph flow increases about 20-fold and equals about 20% of the total volume absorbed. The assumption is made that the lymph must, therefore, come from the absorptive site. However, if labelled H_2O is placed in the lumen, only about 2% of that absorbed is found in the lymph [Table 18 in (39)]. This finding raises the possibility that the increased lymph flow may come from areas other than the gut absorptive site and may reflect an indirect stimulation of ultrafiltration.

Direct measurement of lymphatic pressure has shown a correlation between lymphatic pressure and absorptive rate. However, the lymphatic pressures appear to be low relative to those required to physically affect absorption in vitro (greater than about 4 cm H_2O). Lymphatic pressures were no greater than 2.5 cm H_2O during absorption from glucose-containing Krebs-Ringer solution. Maximum average lymph pressure was only about 6.7 cm H_2O during rapid absorption from hypotonic solutions with occluded lymphatics (23). It is possible that interstitial resistance to fluid flow causes lymphatic pressure to be less than the average interstitial hydrostatic pressure.

CONDUCTANCE CHANGES

Changes in Starling forces or absorption would be relatively ineffective in moving fluid between the lumen and blood if the hydraulic conductance of the gut were not increased from its low value during normal nonabsorbing states (13, 40). The lateral intercellular spaces open as fluid enters during absorption or pressure driven secretion (25), thus increasing epithelial conductance. If plasma volume expansion is superimposed on an absorbing gut, then the lateral spaces expand further and "blisters" appear within the tight junctions that remain intact above and below the disruption (25). Both ultrastructural changes would likely increase epithelial conductance and are associated with increased movement of Na and H_2O from blood to lumen.

Increased absorption (and presumably increased capillary ultrafiltration also) is associated with an increased volume of interstitial space and a decreased fraction of interstitial space from which albumin is excluded (13). This suggests that increasing hydration of the interstitial matrix expands the meshwork of interstitial fibers and increases conductance as indicated by the increased entry of albumin into the interstitial space. Smaller solutes and water would also have proportionally greater increases in their conductance through the interstitial space.

The increased lymph flow that occurs following volume expansion or mesenteric venous occlusion is only transient and decreases about the time that intestinal secretion increases. This suggests that an increase in epithelial conductance may drain off interstitial fluid and pressure, thus relieving the stimulus for increased lymph flow (14).

REGULATION OF BLOOD FLOW AND ABSORPTION

Glucagon

Glucagon has variable effects on absorption depending upon its cardiovascular effects, species of animal and dosage used, and the route of administration in the in vivo preparation (12, 24, 34). Under in vitro conditions glucagon does not affect absorption in the presence of glucose, although absorption is decreased in the absence of glucose. Glucagon has been shown to raise total intestinal blood and lymph flow (12). Mucosal flow increased in proportion to total flow in dogs as determined by 7–10 micron microspheres (4) but was preferentially increased in dogs as determined by ^3H_2O clearance (24) and in rats as observed by optical measurements (17). Glucagon increased capillary pressure and capillary permeability with a resultant pressure-driven secretion (14, 24). Glucagon may exert different effects on absorption because of an interplay between its direct intestinal hemodynamic effects and indirect effecs exerted through central sympathetic activity and catecholamine release (24, 34). Glucagon can decrease intestinal absorption despite an increase in total blood flow and oxygen consumption potentiated by atropine. Hence, glucagon may also act through a cholinergic mechanism.

Somatostatin

Somatostatin (SRIF) also has variable effects on absorption. SRIF increased net Na and Cl absorption across rat ileum (but no other part of the entire gut) in vitro by increasing the absorptive flux in selected segments that had a low rate of basal NaCl absorption. SRIF increased NaCl absorption up to, but not greater than, the rate seen in segments that, owing to their lower secretory flux, were absorbing NaCl more rapidly in the basal state. In rabbit ileum in vitro, SRIF increased net NaCl absorption due primarily to an increase in the absorptive flux but did not affect 3-O-methylglucose absorption. SRIF, infused i.v. did not affect basal Na, H_2O, glucose, or urea absorption across rat jejunum but inhibited PGE_1 or theophylline-induced secretion (7). SRIF decreased jejunal glucose and amino acid absorption in humans (19), without affecting net Na and H_2O absorption, despite decreasing both unidirectional H_2O fluxes. These results were attributed partly to a decrease in functional mucosal surface area (19). SRIF decreased total gut blood flow and glucose absorption in diabetics. SRIF may exert its effects through modulation of the release of another hormone or paracrine agent, although hemodynamic effects of SRIF are suggested by the decrease in total blood flow and by the observation that both the absorptive and secretory H_2O fluxes are decreased (19). The latter findings are consistent with a reduction in effective mucosal blood flow and pressure (24, 26, 29, 31) as well as effective mucosal surface area.

Opiates

Opiates are similar to somatostatin in that they do not necessarily increase absorption over basal conditions in some species but can reduce secretion. Intestinal absorption of NaCl and H_2O in fed dogs, but not in fasted dogs, was increased by intravenous morphine owing to an increase in the absorptive fluxes and was associated with an increase in absorptive-site blood flow, although total blood flow decreased (27). Estimated absorptive-site capillary pressure was increased by morphine, but the secretory fluxes in fed dogs were decreased relative to those in fasted dogs for any given capillary pressure. It was suggested that the increased absorptive-site blood flow increased the washout of absorbed salt and water, but an increase in active absorption could also have contributed to the increased absorption.

Vasoactive Intestinal Polypeptide

Vasoactive intestinal polypeptide (VIP) increases active Cl and possibly Na secretion in vitro through stimulation of a cyclic AMP mediated process and thus affects transcellular transport. In vivo, VIP reduced net Na and H_2O absorption or increased secretion from the small intestine and colon when infused i.v. or i.a. in humans, rats, or dogs, effects due primarily to reduced absorptive fluxes (10, 20, 26). VIP infused i.v. either decreased or did not change total or absorptive-site blood flow as measured by ^3H_2O clearance (26) or optically (17). These hemodynamic responses to VIP were probably secondary to arterial hypotension, since the vascular effects of intravenous infusion of VIP could be partially blocked by both atropine and guanethidine (26), but the effects of intraarterial infusion could not be blocked by atropine (10). Perhaps VIP acts in part through systemic autonomic stimulation and also through local hemodynamic responses in the gut (26).

Bacterial Toxins

Both cholera and *E. coli* heat-labile toxins have effects that are also mediated through cyclic AMP stimulation. Cholera-induced secretion does not exhibit passive filtration characteristics because the lateral spaces are closed during secretion, there is no protein secreted into the lumen, lymph flow is not increased, and mechanically reducing intestinal blood flow and pressure does not reduce secretion (14). Increased or unchanged mucosal flow have been measured after cholera toxin administration, and decreased absorptive-site blood flow was observed following luminal administration of the very similar *E. coli* heat-labile toxin (6, 30). A reduction in blood flow by hemorrhage superimposed upon choleraic secretion increased the volume-secretion/mucosal-blood-flow ratio from 13% to 31%, suggesting that there is little effect of blood flow on choleraic secretion.

Neural Regulation

Both cholinergic stimuli and reduction of adrenergic stimuli reduce intestinal absorption or increase secretion, and adrenergic stimuli or reduction of cholinergic stimuli increase absorption (18, 32). Beta-adrenergic stimuli increase total gut blood flow but do not alter net water absorption; alpha-adrenergic stimuli and sympathetic nerve stimulation decrease total blood flow but increase absorption (5). However, villous blood flow increased during sympathetic nerve stimulation even though total blood flow decreased.

Direct visual observation following vagotomy also demonstrated vasoconstriction and pale villi, similar to villi in their absorptive state (38); engorged, vasodilated villi were observed following sympathectomy (33). The capillary filtration coefficient was decreased following alpha-adrenergic drugs. Sympathetic stimulation reduces mucosal arteriolar pressure and reduces blood flow by constricting precapillary sphincters and (to a lesser degree) venules (3). This raises the possibility that altered Starling forces may favor more rapid uptake of absorbed fluid by passive mechanisms. Alternatively, vasoconstriction in the crypt region (which may be a net secretory area) during sympathetic stimulation may alter the unidirectional fluxes across the gut to favor net absorption (5). However, atropine did not block the effect of luminal bile-oleic acid to increase total gut blood flow or O_2 consumption. It did block the secretory effect, although the disrupted villi reflecting previously increased tissue pressure still remained (22). The finding of reduced secretion with disrupted villi after bile-oleic acid plus atropine makes unclear the function of the absorptive cells. Could local hemodynamic factors be major forces for net fluid absorption?

Prostaglandins

Prostaglandins are released by mechanical stimulation of the mucosa and may be the direct stimulus for the functional hyperemia following food ingestion (1). Both PGE_1 and $PGF_{2\alpha}$, intraluminally, decreased absorption in rat jejunum and decreased absorptive fluxes of H_2O at doses that did not affect total blood flow (1). Both the absorptive and secretory fluxes of ^3H_2O and total blood flow were linearly increased by intraluminal concentrations of PGE_1 that were greater than the minimal concentration causing reduced absorption. $PGF_{2\alpha}$ increased the secretory flux of ^3H_2O but decreased the absorptive flux. Indomethacin decreased total blood flow and the secretory ^3H_2O flux but increased the absorptive ^3H_2O flux. The effects of prostaglandins on the unidirectional H_2O fluxes may reflect changes in transcellular transport or changes in local hemodynamics. Increased secretory fluxes may be an effect of increased capillary pressure, and decreased absorptive fluxes may be due to decreased effective mucosal blood flow (24, 26, 29, 31).

Prostaglandins may increase intestinal secretion because of stimulation of cyclic AMP and consequent transcellular secretion. However, recent work has suggested that the primary effect of prostaglandins to increase secretion is cardiovascular and exerted through increasing hydrostatic capillary pressure (15). This finding raises the possibility that other agents that increase mucosal cyclic AMP and cause active secretion in vitro may have additional cardiovascular effects in vivo.

The observations mentioned above suggest several problems complicating interpretation of control mechanisms relating to intestinal absorption, blood flow, and their interdependence. These complexities include (*a*) possible release of other agents that may cause the observed changes, and (*b*) direct effects of interventions on transcellular transport and/or on local hemodynamics independent of or interactive with any change in the other process. Results from in vitro experiments that suggest direct effects of a stimulus on transcellular transport may be inapplicable to in vivo experiments in which simultaneous cardiovascular changes can effect the magnitude of net transport. Also, the transcellular transport changes may modify the epithelium to make it more or less sensitive to hemodynamic factors; thus an increase in transcellular absorption that enlarges the lateral spaces would also increase the conductance of the epithelium, which could make it more sensitive to hemodynamic secretory forces in vivo.

BLOOD FLOW EFFECTS

Blood flow per se, independent of concurrent effects on hydrostatic pressure or conductance changes, can influence absorption through a washout effect.

Increasing blood flow increases the rate of absorption of certain substances (39). The absorption of highly permeable substances is more sensitive to blood flow than is the absorption of poorly absorbed compounds. Highly permeable compounds, such as ^3H_2O or inert gasses, approach blood flow limited absorption where their rate of absorption is nearly a linear function of blood flow. Very impermeable compounds have diffusion-limited absorption where their absorption rate is determined primarily by their slow rate of absorption across the epithelium. Blood flow influences absorption by changing the concentration of substance in the interstitial space and thus the rate of diffusion across the epithelium.

COUNTERCURRENT EXCHANGE

The arrangement of vessels in the villus permits countercurrent blood flows through the central arteriole and capillaries or venules. In theory, permeable substances should move from a higher concentration in one limb to a lower concentration in the other limb. Thus the diffusion of physically dissolved oxygen from arteriole to venule at the base of the villus would tend to deprive the absorptive cells at the tip of the villus of some oxygen. Absorbed materials can diffuse from venule to arteriole and slow the rate of their absorption. The efficiency of exchange is greater as blood flow decreases (39). In addition, countercurrent exchange may allow the creation of a hypertonic NaCl solution at the villous tip, thus promoting fluid absorption. The hypertonic NaCl concentration at the tip of the villus seems very great (1,000 mosm/kg) when compared to the similar concentration at the tip of the renal papilla. The latter contains more efficient countercurrent mechanisms because of their much greater length, more parallel arrangements, closer approximation, and the countercurrent multiplier action of the water-impermeable segments of the loops of Henle.

Similarly, it is theoretically possible that the countercurrent exchange may trap CO_2 at the tip of the villus, leading to an increased concentration of CO_2–HCO_3 at the absorptive site that could modify transcellular transport, presumably by some HCO_3/Cl exchange mechanism (25). This may explain the observation that cholera toxin causes NaCl secretion in vitro but more $NaHCO_3$ secretion in vivo.

SUMMARY

There are many interactions between intestinal blood flow and absorption. These include the support of transcellular transport by O_2 delivery; the passive effects of Starling forces on absorption, secretion, and mucosal conductance; the washout of absorbed materials; and countercurrent exchange. Blood flow, hemodynamics, and absorption interact. Regulatory mechanisms add a tertiary influence to these interactions. Further progress in this area would be greatly helped by techniques that could measure blood flow and hemodynamic factors at the absorptive and secretory sites more precisely.

Literature Cited

1. Beubler, E., Juan, H. 1978. PGE-release, blood flow and transmucosal water movement after mechanical stimulation of the rat jejunal mucosa. *Naunyn-Schmied. Arch. Pharmacol.* 305:91–95
2. Bohlen, H. G. 1980. Intestinal mucosal oxygenation influences absorptive hyperemia. *Am. J. Physiol.* 239:H489–93
3. Bohlen, H. G., Henrich, H., Gore, R. W., Johnson, P. C. 1978. Intestinal muscle and mucosal blood flow during direct sympathetic stimulation. *Am. J. Physiol.* 235:H40–45
4. Bond, J. H., Levitt, M. D. 1980. Effect of glucagon on gastrointestinal blood flow of dogs in hypovolemic shock. *Am. J. Physiol.* 238:G434–39
5. Brunsson, J., Eklund, S., Jodal, M., Lundgren, O., Sjovall, H. 1979. The effect of vasodilation and sympathetic nerve activation on net water absorption in the cat small intestine. *Acta. Physiol. Scand.* 106:61–68
6. Cedgard, S., Hallback, D., Jodal, M., Lundgren, O., Redfors, S. 1978. The effects of cholera toxin on intramural blood flow distribution and capillary hydraulic conductivity in the cat small intestine. *Acta. Physiol. Scand.* 102:148–58
7. Dharmsathaphorn, K., Sherwin, R. S., Dobbins, J. W. 1980. Somatostatin inhibits fluid secretion in the rat jejunum. *Gastroenterology* 78:1554–58
8. Donowitz, M., Wicklein, D., Reynolds, D. G., Hynes, R. A., Charney, A. N., Zinner, M. J. 1979. Effect of altered intestinal water transport on rabbit ileal blood flow. *Am. J. Physiol.* 236:E482–87
9. Duffy, P. A., Granger, D. N., Taylor, A. E. 1978. Intestinal secretion induced by volume expansion in the dog. *Gastroenterology* 75:413–18
10. Eklund, S., Jodal, M., Lundgren, O., Sjoqvist, A. 1979. Effects of vasoactive intestinal polypeptide on blood flow, motility and fluid transport in the gastrointestinal tract of the cat. *Acta. Physiol. Scand.* 105:461–68
11. Gannon, B. J. 1979. A comparison of the microvascular architecture of small bowel villi in laboratory mammals and man. *Microvasc. Res.* 17:S13 (Abstr.)
12. Granger, D. N., Kvietys, P. R., Wilborn, W. H., Mortillaro, N. A., Taylor, A. E. 1980. Mechanism of glucagon-induced intestinal secretion. *Am. J. Physiol.* 239:G30–38
13. Granger, D. N., Mortillaro, N. A., Kvietys, P. R., Rutili, G., Parker, J. C., Taylor, A. E. 1980. Role of the interstitial matrix during intestinal volume absorption. *Am. J. Physiol.* 238:G183–89
14. Granger, D. N., Mortillaro, N. A., Taylor, A. E. 1977. Interactions of intestinal lymph flow and secretion. *Am. J. Physiol.* 232:E13–18
15. Granger, D. N., Schackelford, J. S., Taylor, A. E. 1979. PGE_1-induced intestinal secretion: mechanism of enhanced transmucosal protein efflux. *Am. J. Physiol.* 236:E788–96
16. Granger, D. N., Taylor, A. E. 1978. Effects of solute-coupled transport on lymph flow and oncotic pressures in cat ileum. *Am. J. Physiol.* 235:E429–36
17. Holliger, C., Radzyner, M., Villiger, A., Knoblauch, M. 1979. Effects of glucagon, vasoactive intestinal peptide (VIP) and lysine-vasopressin on villous microcirculation and superior mesenteric artery blood flow of the rat. *Bibl. Anat.* 18:129–31
18. Hubel, K. A. 1978. The effects of electrical field stimulation and tetrodotoxin on ion transport by the isolated rabbit ileum. *J. Clin. Invest.* 62:1039–47
19. Krejs, G. J., Browne, R., Raskin, P. 1980. Effect of intravenous somatostatin on jejunal absorption of glucose, amino acids, water and electrolytes. *Gastroenterology* 78:26–31
20. Krejs, G., Fordtran, J. S., Bloom, S. R., Fahrenkrug, J., Schaffalitzky de Muckadell, O. B., Fischer, J. E., Humphrey, C. S., O'Dorisio, T. M., Said, S. I., Walsh, J. H., Schulkes, A. A. 1980. Effect of VIP infusion on water and ion transport in the human jenunum. *Gastroenterology.* 78:722–27

21. Kvietys, P. R., Gallavan, R. H., Chou, C. C. 1980. Contribution of bile to postprandial intestinal hyperemia. *Am. J. Physiol.* 238:G284–88
22. Kvietys, P. R., Wilborn, W. H., Granger, D. N. 1981. Effect of atropine on bile-oleic acid-induced alterations in dog jejunal hemodynamics, oxygenation, and net transmucosal water movement. *Gastroenterology* 80:31–38
23. Lee, J. S. 1979. Lymph capillary pressure of rat intestinal villi during absorption. *Am. J. Physiol.* 237:E301–7
24. MacFerran, S. N., Mailman, D. 1977. Effects of glucagon on canine intestinal sodium and water fluxes and regional blood flow. *J. Physiol.* 266:1–12
25. Mailman, D. 1981. Fluid and electrolyte absorption. In *Gastrointestinal Physiology*, ed. L. R. Johnson, pp. 107–22. St. Louis, Mo: C. V. Mosby. 173 pp. 2nd ed.
26. Mailman, D. 1978. Effects of vasoactive intestinal polypeptide on intestinal absorption and blood flow. *J. Physiol. London* 279:121–32
27. Mailman, D. 1980. Effects of morphine on canine intestinal absorption and blood flow. *Br. J. Pharmacol.* 68:617–24
28. Mailman, D. 1981. Blood flow and intestinal absorption. *Fed. Proc.* In press
29. Mailman, D. 1981. Tritiated water clearance as a measure of intestinal absorptive site and total blood flow. In *The Measurement of Splanchnic Blood Flow*, ed. D. N. Granger, G. B. Bulkley. Baltimore, MD: Williams and Wilkins. In press
30. Mailman, D., Goldschmidt, M. 1980. Effects of *E. coli* enterotoxin on canine intestinal Na and H_2O fluxes and blood flow. *Microvasc. Res.* 17:S14 (Abstr.)
31. Mailman, D., Jordan, K. 1975. The effect of saline and hyperoncotic dextran infusion on canine ileal salt and water absorption and regional blood flow. *J. Physiol. London* 252:97–113
32. Morris, A. I., Turnberg, L. A. 1980. The influence of a parasympathetic agonist and antagonist on human intestinal transport in vivo. *Gastroenterology* 79:861–66
33. Padula, R. T., Noble, P. H., Camishion, R. C. 1968. Vascularity of the mucosa of the small intestine after vagotomy and splanchnicectomy. *Surg. Gynecol. Obstet.* 122:41–48
34. Patel, G. K., Whalen, G. E., Soergel, K. H., Wu, W. C., Meade, R. C. 1979. Glucagon effects on the human small intestine. *Am. J. Dig. Dis.* 24:501–8
35. Shepherd, A. P. 1979. Intestinal capillary blood flow during metabolic hyperemia. *Am. J. Physiol.* 237:E548–54
36. Sit, S. P., Nyhof, R., Gallavan, R., Chou, C. C. 1980. Mechanisms of glucose-induced hyperemia in the jejunum. *Proc. Soc. Exp. Biol. Med.* 163:273–77
37. Valleau, J. D., Granger, D. N., Taylor, A. E. 1979. Effect of solute-coupled volume absorption on oxygen consumption in cat ileum. *Am. J. Physiol.* 236:E198–203
38. Wells, H. S., Johnson, R. G. 1934. The intestinal villi and their circulation in relation to absorption and secretion of fluid. *Am. J. Physiol.* 109:387–402
39. Winne, G. 1979. Influence of blood flow on intestinal absorption of drugs and nutrients. *J. Pharm. Therap.* 6:333–93
40. Yablonski, M. D., Lifson, N. 1976. Mechanism of production of intestinal secretion by elevated venous pressure. *J. Clin. Invest.* 57:904–15

PHYSIOLOGICAL REGULATION OF THE HEPATIC CIRCULATION

Peter D. I. Richardson and Peter G. Withrington

Department of Physiology, The Medical College of St. Bartholomew's Hospital, London EC1M 6BQ, England

INTRODUCTION

The principal determinants of hepatic perfusion are the hepatic arterial vascular resistance, which governs hepatic arterial blood flow, the vascular resistance of the preportal vascular beds governing the inflow of blood to the hepatic portal vein, and the intrahepatic portal vascular resistance. In this review, we are principally concerned with the regulation of hepatic arterial and portal venous vascular resistances and how these affect total liver blood flow and its distribution between the hepatic arterial and portal venous vascular beds.

We discuss the intrinsic regulation of the hepatic arterial and portal venous beds, and the interrelationships between these two inflow circuits to the liver. We then consider the effects of alterations in portal venous and systemic blood composition, blood gas tensions, pH and osmolarity, followed by the relationship between liver function and metabolism, and blood flow. Finally, we discuss the principal extrinsic regulators of liver blood flow, the innervation and the effects of hormones.

We have concentrated on aspects of hepatic circulatory physiology reported after the last major review of this subject (14).

INTRINSIC VASOREGULATION

Pressure-Flow Autoregulation

Controversy exists over the existence of pressure-flow autoregulation in the hepatic arterial bed. Some studies have demonstrated autoregulation, although its degree may be small or not seen in all experiments (18, 19). Other studies have indicated that autoregulation is absent in the hepatic arterial bed, the hepatic arterial resistance either remaining constant or decreasing with increasing arterial pressure (46). This controversy is not explained by technical differences in the experiments or by the state of innervation of the liver. Hepatic arterial autoregulation may thus be more apparent in the metabolically active liver than in the unstimulated preparation, a situation analogous to that in the intestine (33).

Pressure-flow autoregulation does not occur in the hepatic portal vasculature (46). These studies include a variety of species and types of experimental preparation, and the absence of portal autoregulation probably reflects the fact that the principal determinants of portal blood flow are the outflows from the intestine and spleen, and not the intrahepatic portal resistance.

Elevated Venous Pressure

Venous pressure elevation increases hepatic arterial resistance in the dog (19) and to a small extent in the cat (28). In the dog, the portal resistance decreases (19) on venous pressure elevation to a greater extent than can be accounted for by a portal response to the increased hepatic arterial resistance (see below). The responses to venous pressure elevation are different according to the metabolic and myogenic hypotheses, which predict vasodilation and vasoconstriction respectively (9). The hepatic arterial bed, therefore, exhibits myogenic and the portal vasculature weak metabolic responses to venous pressure elevation.

Hepatic Arterial–Portal Interactions

Total occlusion of one inflow circuit reduces the vascular resistance of the other circuit by about 20% (18, 19, 21, 46). Graded increases in the perfusion pressure of one inflow circuit affect the vascular resistance of the other; thus, a 10% increase in hepatic arterial perfusion pressure from the mean control value of 115 mmHg increases the portal vascular resistance by 2.4%, and a 10% increase in portal venous perfusion pressure gradient from the mean control value of 5.3 mmHg increases the hepatic arterial resistance by 3.6% (46).

Occlusion of the hepatic artery leads to a reduced portal venous pressure, and this procedure has been used in the treatment of portal hypertension; if patients or animals survive the initial period of reduced liver perfusion, the prognosis for eventual survival is good, although rearterialization of the liver from the phrenic arteries may be a factor in this improved prognosis (24).

Although portacaval bypass increases the hepatic arterial blood flow, total liver blood flow is reduced in dogs (1) and humans (2). The prognosis for survival is directly related to the absolute increase in hepatic arterial blood flow following portacaval bypass (2).

Quantitatively, arterioportal interactions are insufficient to maintain a constant liver blood flow; however, in addition to physical interrelationships between the hepatic artery and the portal vein, it is now clear that vasoactive substances present in only one of the inflow circuits can gain access to the sites that control the inflow resistance of both the hepatic arterial and portal venous vasculatures, without necessarily entering the systemic circulation. For example, intraportal isoproterenol is without direct effect on the portal vasculature but reduces hepatic arterial resistance almost as much as it would if it were present in the hepatic arterial bloodstream in the same concentration (46).

CHANGES IN PORTAL VENOUS AND SYSTEMIC ARTERIAL BLOOD COMPOSITION

Systemic Blood Gas Tensions

In the dog, severe hypoxia (mean $pO_2 = 25$ mmHg) approximately doubles hepatic arterial resistance, though more 'moderate' hypoxia (mean $pO_2 = 45$ mmHg) is without measurable effect on the hepatic arterial or portal venous vascular beds (22, 54). In the cat, systemic hypoxia resulting in mean arterial pO_2 values of 36 mmHg and portal pO_2 values of 3 mmHg is without effect on the hepatic arterial or portal venous vascular conductances (27).

Hypercapnia and the associated acidosis increase hepatic arterial and portal venous blood flows, the latter effect being due principally to decreases in mesenteric vascular resistance (9, 54).

Systemic hyperoxia with pO_2 values up to 400 mmHg is without significant effect on canine liver blood flow (22).

Hypocapnia and the associated alkalosis reduce hepatic arterial and portal venous blood flows: With systemic arterial pCO_2 of 24 mm Hg and pH of 7.82, the changes in portal venous blood flow were due to mesenteric vasoconstriction (6).

Portal Venous Blood Gas Tensions

One study has reported the effects of alterations in portal venous blood gas tensions in the absence of systemic changes on the liver circulation (7). Acidosis reduces hepatic arterial resistance and increases portal venous resistance (a fall in pH of 0.1 unit increases portal resistance by 40% and reduces arterial resistance by 9%), although the arterial changes could have been due in part to a physical response to the increased portal pressure. Alkalosis has the converse effects. Portal hyperoxia increases hepatic arterial resistance but is without significant effect on the portal inflow resitance.

Alterations in portal venous blood gas tensions and pH may be of greater significance in hepatic vasoregulation than changes in systemic blood gas tensions, since alterations in pH, pO_2 and pCO_2 of portal venous blood may occur during digestion and absorption. Reductions in portal pO_2 and pH with increasing pCO_2 are the expected changes during increased intestinal oxygen extraction and carbon dioxide liberation. These changes would increase hepatic arterial blood flow at a time when hepatic oxygen demands are elevated.

Systemic and Portal Venous Osmolarity

Hyperosmolarity of portal venous plasma induced by infusing hypertonic solutions of mannitol, glucose, or sodium chloride into the portal vein cause increases in hepatic arterial blood flow in the cat (30) and the dog (51). Relatively small increases in portal venous osmolarity of the same order of magnitude as those occurring postprandially [up to about 30–35 mOsm/l: (4)] are associated with small increases (about 5 mOsm/l) in systemic arterial osmolarity and with increases in both hepatic arterial and portal venous blood flow of about 10% (51).

Since increased portal venous osmolarity prompts an hepatic arterial vasodilation, it is possible that such a mechanism contributes to postprandial hepatic arterial hyperemia. Systemic hyperosmolarity increases total liver blood flow by increasing both hepatic arterial and portal venous blood flows, the latter by intestinal vasodilation (9). Such changes in osmolarity would be expected in several abnormal conditions such as chronic diarrhea, untreated diabetic hyperglycemia, dehydration, burns, or diabetes insipidus. In addition, one should consider the possibility that physiologic changes in hepatocyte function, particularly glycogenolysis and the consequent release of glucose, produce a local hyperosmolar environment for the vascular resistance sites, and thereby evoke vasodilation. Such effects might well contribute to the local metabolic regulation of liver blood flow.

RELATIONSHIP OF BLOOD FLOW TO LIVER FUNCTION

Postprandial hyperemia clearly occurs in the intestine (9); as a consequence, the portal venous and total liver blood flow increases after a meal. In the conscious sheep (23) postprandial hyperemia occurs in the hepatic arterial bed. The mechanism of any postprandial hepatic arterial hyperemia is at present unknown. Two likely mechanisms are a direct local vascular response to increased hepatocyte oxygen demand or a vasodilator response to material released from the gastrointestinal tract into the portal vein producing hepatic arterial vasodilation (46, 48). It is improbable that a single gastrointestinal hormone acting alone mediates postprandial hepatic arterial hyperemia, as the vasodilator concentrations exceed those occurring physiologically (38), though increased plasma osmolarity could contribute to these vasodilator effects.

Controversy surrounds the effects of metabolic stimulation on liver blood flow; in the isolated perfused sheep liver, metabolic stimulation increases total and hepatic arterial blood flows (31). In the dog, metabolic stimulation with amino acid infusions roughly doubles hepatic arterial blood flow (53), while in the cat, hepatic metabolic stimulation increases portal but slightly reduces hepatic arterial blood flow (29).

Hepatic enzyme induction increases total liver mass and liver blood flow, though the changes appear to be specific for certain types of drug-induced enzyme induction: Barbiturates, for example, increase liver mass and blood flow, whereas benzpyrine induces hepatic enzyme synthesis but does not augment liver blood flow (32).

In general, increased hepatic activity is associated with increased total liver blood flow; the contribution to such increases of the hepatic arterial circulation remains uncertain, and it is possible that species differences influence these responses. An additional point is that an increased portal venous blood flow, for example during postprandial vascular changes, would tend to diminish hepatic arterial blood flow because of the physical or hydrodynamic interaction between the two circuits, unless there were a concomitant hepatic arterial vasodilation (46).

EXTRINSIC VASOREGULATION

Nervous

Electrical stimulation of the periarterial nerves increases inflow resistance in both the hepatic arterial and portal circuits with a reduction in total liver blood flow (3, 12, 14, 42, 44). These responses are maintained throughout the period of stimulation in the dog. In the cat the hepatic arterial flow returns towards the control value during continued stimulation (autoregulatory escape) despite maintained changes in portal inflow resistance. No marked redistribution of blood flow in either inflow circuit occurs during nerve stimulation (12), and it has been proposed that a vasodilator substance is released to antagonize the arterial response. Since nerve stimulation increases hepatic metabolism an increased extracellular osmolarity may relax arterial smooth muscle and contribute to the secondary reduction in arterial resistance during nerve stimulation in the cat.

Sympathetic nerve stimulation reduces liver blood volume abruptly and extensively in both the cat and dog (3, 15). In the dog a major proportion (60%) of the hepatic blood volume can be mobilized by hepatic nerve stimulation, and 80% of this mobile store is expelled within 19 seconds (3). The liver represents a major reservoir of whole blood which can be rapidly redistributed in a precise and controlled manner in response to the activity of its autonomic innervation.

Reflex increases in hepatic arterial and portal inflow resistances and reductions in liver volume result from graded declines in carotid sinus pressure (CSP) in the dog (3). A rise in CSP to 240 mmHg, representing maximal baroreceptor inhibition of vasomotor drive, evokes reductions in hepatic arterial and portal inflow resistances concomitant with a rapid increase in hepatic blood volume, 80% of which occurs within 25 seconds. The volume of blood expelled reflexly from the liver by sinus hypotension is very much smaller (26%) than the maximum volume released by stimulation of the hepatic periarterial nerves at 15–20 Hz. Undoubtedly systemic baroreceptors and cardiopulmonary vagal afferents influence hepatic hemodynamics by modulation of sympathetic efferent activity; however, this influence may be significantly modified either by local events within the liver or by substances entering the liver in the portal flow.

The existence of a functional vagal innervation to the hepatic vasculature has been reinvestigated with transillumination techniques (25, 26). In the rat the vagus exerts vasodilator control over the caliber of sinusoids, partly through a direct innervation of the presinusoidal sphincter. Vagal stimulation opens sinusoids previously closed, and such recruitment of patent sinusoids may, unaccompanied by change in overall liver blood flow, cause extensive redistribution of hepatic blood. Confirmatory studies are required in other species, but these observations provide further experimental evidence for the view (35) that cyclic opening and closing of presinusoidal sphincters can achieve substantial hepatic microcirculatory changes undetected by measurements of overall blood flow.

Catecholamines

Hepatic arterial smooth muscle constricts when the pure alpha-adrenergic stimulant phenylephrine is administered (41); this action is antagonized by alpha-adrenergic antagonists. Hepatic arterial vasodilation is elicited by established beta-adrenergic stimulants such as isoproterenol and salbutamol (10, 18, 21, 37, 41, 46); this response is blocked by propranolol (18, 21, 41) but not by atenolol (41). $Beta_2$-adrenergic receptors, therefore, exist in the hepatic arterial bed and when stimulated by the appropriate agonists induce vasodilation; they do not appear to be activated by sympathetic innervation (11). The vascular response of the hepatic arterial bed to any circulating catecholamines depends upon at least three factors: the relative density of alpha- and $beta_2$-adrenergic receptors at the arterial smooth muscle resistance sites, the concentration of alpha- and $beta_2$-adrenergic agonists gaining access to the appropriate arterial receptors from both arterial and portal routes, and the efficacy of individual catecholamines for activating each vascular adrenergic receptor.

Epinephrine and norepinephrine are both mixed adrenergic agonists. Epinephrine is a potent agonist of both alpha- and $beta_2$-receptors, and its injection or infusion into the hepatic artery induces, particularly at lower doses, a marked arterial vasodilation (41, 50) preceded by transient vasoconstriction. Vasoconstriction predominates at higher doses. The dilation is abolished by propranolol allowing the alpha-agonist activity of epinephrine to be manifest unopposed, with significantly enhanced vasoconstriction potency. Intra-arterial injections of norepinephrine, a potent alpha-adrenergic agonist but weak $beta_2$ agonist, induces a marked initial reduction of hepatic arterial inflow succeeded by very slight vasodilation. Inhibition of these secondary vasodilator responses by propranolol does not affect the hepatic arterial vasoconstrictor potency of norepinephrine (41).

The adrenoceptor population of the portal resistance sites appears to be uniform. Intraportal administration of pure beta-adrenergic agonists does not significantly affect portal inflow resistance, which indicates the absence of beta receptors (18, 21, 43, 46). Marked portal vasoconstriction is evoked by pure alpha-adrenergic agonists (30, 43), and both epinephrine and norepinephrine (whether administered to the liver by intraportal or intra-arterial routes) induce portal vasoconstriction unaffected by propranolol (17, 39, 43) with epinephrine being the more potent agonist. Epinephrine, through its potent alpha- and $beta_2$ agonist actions, elicits a pattern of hepatic vascular responses characterized by hepatic arterial vasodilation and portal vasoconstriction. Norepinephrine and sympathetic nerve stimulation cause increases in inflow resistance at both sites. In the intact animal these differential actions may be further modified by concomitant changes in the mesenteric outflow into the portal tract.

Any action of dopamine on the liver circulation is of little physiological significance since any vascular effects it possesses alone would be overwhelmed by the concomitant presence of epinephrine, norepinephrine, and sympathetic activation. In the dog dopamine is a mild alpha-adrenoceptor stimulant (21, 45), but it evokes hepatic arterial vasodilation through activation of specific dopamine receptors that are present in the arterial bed but not in the portal vasculature (45).

Gastrointestinal Hormones

There have been no hepatic vascular studies with any of the pure gastrins. The synthetic analog pentagastrin, infused either i.v. (34) or i.a. (38), produces hepatic arterial vasodilation and a reduction in portal vascular resistance.

Preparations of secretin of different purity produce either hepatic arterial constriction in the cat (52), no appreciable effect in the dog (34) or hepatic arterial vasodilation (37, 38). Differences in the route of administration and contamination by other hormones contribute to the variation in results with secretin.

Systemic injection of cholecystokinin evokes no changes in either hepatic arterial or portal vascular resistance (34), while intra-arterial injections produce hepatic arterial vasodilation (38).

Satisfactory conclusions about the hepatic vascular roles of these hormones and any vascular interactions that may occur due to multiple and sequential secretion will be possible when adequate amounts of high purity samples are available.

More definite information is available about the hepatic vascular actions of glucagon. When infused or injected intra-arterially into the liver it causes a graded and long-lasting hepatic arterial vasodilation; on a molar basis its intrinsic vasodilator activity is relatively weak (37). A property of glucagon unique among the gastrointestinal hormones is that when injected or infused intra-arterially or intraportally it significantly antagonizes the hepatic arterial vasoconstrictor responses to a wide range of physiological stimuli (36), including sympathetic nerve stimulation (42). The concomitant portal venous vasoconstrictor responses to these stimuli are not affected (44, 49). The release of glucagon occurs during various stress procedures and is partly controlled by the sympathetic innervation to the pancreas. The proposal has been made (49) that glucagon, in addition to its metabolic function, ensures an adequate arterial supply to the hepatic parenchyma by direct hepatic arterial vasodilation and also by antagonizing hepatic arterial vasoconstriction during stress caused by increases in efferent sympathetic activity and circulating levels of vasoconstrictor substances.

Autacoids

A number of substances, potent in their effects upon vascular smooth muscle, have not been considered as candidates for significant roles in systemic hemodynamics because of rapid inactivation in the blood or during passage through the liver. The demonstration of a route of access to the hepatic arterial resistance sites directly from the portal vein necessitates a reassessment of their possible hepatic vascular role, particularly if evidence is available of their production within the gastrointestinal tract.

Bradykinin is released from the digestive organs mainly under pathological circumstances. Intra-arterial injections and intraportal injections and infusions of bradykinin elicit profound hepatic arterial vasodilation (40, 55); on a molar basis it is the most potent dilator of the hepatic arterial bed yet examined with an ED_{50} of 2.7×10^{-13} mol. Whether injected intra-arterially or intraportally it produces no changes in the portal inflow resistance.

Serotonin is released from the gastrointestinal tract into the portal vein. There is a considerable variation in the vascular responses of both inflow circuits to serotonin. Intra-arterial or portal injection induces a weak, long-lasting hepatic arterial vasoconstrictor response, although occasionally there may be an initial transient hepatic arterial vasodilation (40, 48). The hepatic portal effects of serotonin are also small and variable, consisting of vasodilation at very small doses and either no effect or very slight vasoconstriction at higher doses. Any portal vasodilator effects may be the passive result of concomitant hepatic arterial vasoconstriction. Because of its weak and variable hepatic vascular actions, serotonin seems unlikely to possess, by itself, any significant role in hepatic hemodynamics.

When injected or infused intra-arterially or intraportally histamine evokes graded hepatic arterial vasodilation; on a molar basis it is among the less potent substances investigated (13, 37, 48). Concomitant with the reduction in hepatic arterial resistance is a graded increase in portal inflow resistance (48); histamine contracts the hepatic veins in the dog but not in the cat (13). The increase in portal resistance is the same whether the histamine is administered intra-arterially or intraportally (48), a finding consistent with the view that the principal resistance increased by histamine injection is postsinusoidal after mixing of the hepatic arterial and portal blood. Actions of histamine receptor blocking agents suggest that the hepatic vascular responses are due to H_1-receptor activation, although a small population of H_2 receptors may exist (40). There is little evidence that histamine is released into the portal vein except in pathological situations and anaphylaxis.

Prostaglandins of the E type are released into the portal vein from both the intestine and the spleen; prostaglandins E_1 and E_2 both evoke hepatic arterial vasodilation (8, 37) when administered into the hepatic artery without any effects in the portal inflow tract. Prostaglandins A_1, A_2 and B_1 dilate the hepatic artery of the dog while inducing mild portal vasoconstriction (20).

Vasoconstrictor Peptides

Administered into either the hepatic artery or portal vein, angiotensin II causes marked vasoconstriction of both hepatic arterial and portal beds; this, together with a significant reduction in the mesenteric outflow into the portal vein, causes a substantial reduction in total liver blood flow (5, 36, 39, 55). Angiotensin II when injected into the artery is the most potent vasoconstrictor of the hepatic arterial bed yet examined. Both hepatic inflow circuits exhibit tachyphylaxis to angiotensin II; the portal venous vasculature is especially sensitive (39). The construction of dose-response curves of both hepatic beds to increasing doses of angiotensin gives a conventional sigmoid shape for the hepatic arterial system but a bell-shaped curve for the portal vascular bed. This reduction in vasoconstrictor response to consecutive doses of angiotensin is not exhibited by other substances that are potent portal vasoconstrictors (39); it represents a true receptor tachyphylaxis.

Vasopressin evokes hepatic arterial vasoconstriction whether administered by intra-arterial (36), intraportal (47), or intravenous (16, 47, 56) routes. The response is of short duration and no tachyphylaxis is observed. Reports of arterial vasodilation to intravenous vasopressin are almost certainly the consequence of systemic adjustments resulting from the rise in blood pressure. In contrast, the response of the portal vascular bed to vasopressin administered by any route is a reduction in portal inflow resistance (39, 47). This portal vasodilation does not simply arise as a consequence of the concomitant hepatic arterial vasoconstriction since it cannot, quantitatively, be explained on the basis of the reciprocal physical relationship established between the two hepatic inflow circuits (46).

Vasopressin, therefore, has at least three vascular actions that contribute to its efficacy in alleviating portal hypertension. These effects are marked hepatic arterial vasoconstriction, marked mesenteric vasoconstriction with a consequent reduction in outflow into the portal system, and a direct dilator action on the intrahepatic portal resistance sites causing a reduction in portal inflow resistance. It is unlikely that vasopressin exerts a significant vascular action under physiological conditions or even stress situations when it is released in large amounts. Any of its vascular actions would be masked by the concomitant actions of the efferent sympathetic innervation and elevated levels of circulating catecholamines. However, vasopressin secretion may play a significant role in promoting reabsorption of water from the small intestine since its dual action of mesenteric precapillary vasoconstriction and postcapillary hepatic portal vasodilation would alter the pressure profile across the capillaries to assist water movement from the pericapillary spaces to the intravascular compartment.

CONCLUSIONS

The liver vasculature exhibits several unique features important in the regulation of both total liver blood flow and the ratio between hepatic arterial and portal venous blood flows. Not only is there partial reciprocity between the hepatic artery and portal vein, tending to maintain a constant total liver blood flow, but also material present in one inflow can influence the vascular resistance of both inflow circuits. Substances released from or absorbed from the gastrointestinal tract and which attain vasoactive concentrations in the portal venous but not the systemic arterial blood can, therefore, modulate hepatic arterial blood flow, and may be regarded as candidates for hepatic vasoregulation. Additionally, material released from metabolically active hepatocytes may either be vasoactive or may influence liver blood flow indirectly, for example by increasing extracellular osmolarity.

Intrinsic vasoregulation of the portal blood flow is absent or exceptionally weak, and while the hepatic arterial bed exhibits intrinsic regulation this may be present only under some conditions. It is to be expected that in a 'vital' organ, such as the liver, the vasoconstrictor effects of sympathetic nervous activation do not predominate, particularly since epinephrine, at most concentrations, dilates the hepatic arterial bed. The quantitative contribution to physiological regulation of hepatic arterial blood flow by gastrointestinal hormones continues to be actively investigated.

Literature Cited

1. Baker, P. R., Shields, R. 1974. Effect of portacaval transposition on hepatic blood flow during obstruction to the venous outflow from the canine liver. *Ann. Surg.* 180:89–94
2. Burchell, A. R., Moreno, A. H., Panke, W. F., Nealon, T. F. 1976. Hepatic artery flow improvement after portacaval shunt: a single hemodynamic correlate. *Ann. Surg.* 194:289–300
3. Carneiro, J. J., Donald, D. E. 1977. Change in liver blood flow and blood content in dogs during direct and reflex alteration of hepatic sympathetic nerve activity. *Circ. Res.* 40:150–58
4. Carr, D. H., Titchen, D. A. 1978. Postprandial changes in parotid salivary secretion and plasma osmolality and the effects of intravenous infusions of saline solutions. *Q. J. Exp. Physiol.* 63:1–21
5. Cohen, M. M., Sitar, D. S., McNeill, J. R., Greenway, C. V. 1970. Vasopressin and angiotensin on resistance vessels of spleen, intestine and liver. *Am. J. Physiol.* 218:1704–6
6. Cohn, R., Kountz, S. 1963. Factors influencing control of arterial circulation in the liver of the dog. *Am. J. Physiol.* 205:1260–64
7. Gelman, S., Ernst, E. A. 1977. Role of pH, Pco_2 and O_2 content of portal blood in hepatic circulatory autoregulation. *Am. J. Physiol.* 233:E255–62
8. Geumei, A., Bashour, F. A., Swamy, B. V., Nafrawi, A. F. 1973. Prostaglandin E_1: its effects on hepatic circulation in dogs. *Pharmacology* 9:336–47
9. Granger, D. N., Richardson, P. D. I., Kvietys, P. R., Mortillaro, N. A. 1980. Intestinal blood flow. *Gastroenterology* 78:837–63
10. Greenway, C. V., Lawson, A. E. 1969. Beta adrenergic receptors in the hepatic arterial bed of the anesthetized cat. *Can. J. Physiol. Pharmacol.* 47:415–19
11. Greenway, C. V., Lawson, A. E., Mellander, S. 1967. The effects of stimulation of the hepatic nerves, infusion of noradrenaline and occlusion of the carotid arteries on liver blood flow in the anaesthetized cat. *J. Physiol. London* 192:21–41
12. Greenway, C. V., Oshiro, G. 1972. Comparison of the effects of hepatic nerve stimulation on arterial flow, distribution of arterial and portal flows and the blood content of the livers of anaesthetized cats and dogs. *J. Physiol. London* 227:487–501
13. Greenway, C. V., Oshiro, G. 1973. Effects of histamine on hepatic volume (outflow block) in anaesthetized dogs. *Br. J. Pharmacol.* 47:282–90
14. Greenway, C. V., Stark, R. D. 1971. Hepatic vascular bed. *Physiol. Rev.* 51:23–65
15. Greenway, C. V., Stark, R. D., Lautt, W. W. 1969. Capacitance responses and fluid exchange in the cat liver during stimulation of the hepatic nerves. *Circ. Res.* 25:277–84
16. Hanson, K. M. 1970. Vascular response of intestine and liver to intravenous infusion of vasopressin. *Am. J. Physiol.* 219:779–84
17. Hanson, K. M. 1972. Escape of the liver vasculature from adrenergic vasoconstriction. *Proc. Soc. Exp. Biol. Med.* 141:385–90

18. Hanson, K. M. 1973. Dilator responses of the canine hepatic vasculature. *Angiologica* 10:15–23
19. Hanson, K. M., Johnson, P. C. 1966. Local control of hepatic arterial and portal venous flow in the dog. *Am. J. Physiol.* 211:712–20
20. Hanson, K. M., Post, J. A. 1976. Splanchnic vascular responses to the infusion of prostaglandins A_1, A_2 and B_1. *Pharmacology* 14:166–81
21. Hirsch, L. J., Ayabe, T., Glick, G. 1976. Direct effects of various catecholamines on liver circulation in dogs. *Am. J. Physiol.* 230:1394–99
22. Hughes, R. L., Mathie, R. T., Campbell, D., Fitch, W. 1979. Systemic hypoxia and hyperoxia, and liver blood flow and oxygen consumption in the greyhound. *Pflügers Arch.* 381:151–57
23. Katz, M. L., Bergman, E. N. 1969. Simultaneous measurements of hepatic and portal venous blood flow in the sheep and dog. *Am. J. Physiol.* 216:946–52
24. Kim, D. K., Kinne, D. W., Fortner, J. G. 1973. Occlusion of the hepatic artery in man. *Surg. Gynecol. Obst.* 136:966–68
25. Koo, A., Liang, I. Y. S. 1979. Stimulation and blockade of cholinergic receptors in terminal liver microcirculation in rats. *Am. J. Physiol.* 236:E728–32
26. Koo, A., Liang, I. Y. S. 1979. Microvascular filling pattern in rat liver sinusoids during vagal stimulation. *J. Physiol. London* 295:191–99
27. Larsen, J. A., Krarup, N., Munck, A. 1976. Liver haemodynamics and liver function in cats during graded hypoxic hypoxemia. *Acta Physiol. Scand.* 98:257–62
28. Lautt, W. W. 1977. Effects of acute, passive hepatic congestion on blood flow and oxygen uptake in the intact liver of the cat. *Circ. Res.* 41:787–90
29. Lautt, W. W. 1980. Control of hepatic arterial blood flow: independence from liver metabolic activity. *Am. J. Physiol.* 239:H559–64
30. Lautt, W. W., MacLachlan, T. L., Brown, L. C. 1977. The effect of hypertonic infusions on hepatic blood flows and liver volume in the cat. *Can. J. Physiol. Pharmacol.* 55:1339–44
31. Linzell, J. L., Setchell, B. P., Lindsay, D. B. 1971. The isolated, perfused liver of the sheep: an assessment of its metabolic, synthetic and excretory functions. *Q. J. Exp. Physiol.* 56:53–71
32. Nies, A. S., Wilkinson, G. R., Rush, B. D., Strother, J. T., McDevitt, D. G. 1976. Effects of alteration of hepatic microsomal activity on liver blood flow in the rat. *Biochem. Pharmacol.* 25:1991–93
33. Norris, C. P., Barnes, G. E., Smith, E. E., Granger, H. J. 1979. Autoregulation of superior mesenteric blood flow in fasted and fed dogs. *Am. J. Physiol.* 237:H174–77
34. Post, J. A., Hanson, K. M. 1975. Hepatic, vascular and biliary responses to infusion of gastrointestinal hormones and bile salts. *Digestion* 12:65–77
35. Rappaport, A. M., Schneiderman, J. H. 1976. The function of the hepatic artery. *Rev. Physiol. Biochem. Pharmacol.* 76:130–86
36. Richardson, P. D. I., Withrington, P. G. 1976. The inhibition by glucagon of the vasoconstrictor actions of noradrenaline, angiotensin and vasopressin on the hepatic arterial vascular bed of the dog. *Br. J. Pharmacol.* 57:93–102
37. Richardson, P. D. I., Withrington, P. G. 1976. The vasodilator actions of isoprenaline, histamine, prostaglandin E_2, glucagon and secretin on the hepatic arterial vascular bed of the dog. *Br. J. Pharmacol.* 57:581–88
38. Richardson, P. D. I., Withrington, P. G. 1977. The effects of glucagon, secretin, pancreozymin and pentagastrin on the hepatic arterial vascular bed of the dog. *Br. J. Pharmacol.* 59:148–56
39. Richardson, P. D. I., Withrington, P. G. 1977. The effects of intraportal injections of noradrenaline, adrenaline, vasopressin and angiotensin on the hepatic portal vascular bed of the dog; marked tachyphylaxis to angiotensin. *Br. J. Pharmacol.* 59:293–302
40. Richardson, P. D. I., Withrington, P. G. 1977. A comparison of the effects of bradykinin, 5-hydroxytryptamine and histamine on the hepatic arterial and portal venous vascular beds of the dog: histamine H_1 and H_2 receptors. *Br. J. Pharmacol.* 60:123–33
41. Richardson, P. D. I., Withrington, P. G. 1977. The role of beta-adrenoceptors in the responses of the hepatic arterial vascular bed of the dog to phenylephrine, isoprenaline, noradrenaline and adrenaline. *Br. J. Pharmacol.* 60:239–49
42. Richardson, P. D. I., Withrington, P. G. 1977. Glucagon inhibition of hepatic arterial responses to hepatic nerve stimulation. *Am. J. Physiol.* 233:H647–54
43. Richardson, P. D. I., Withrington, P. G. 1978. Alpha- and beta-adrenocep-

tors in the hepatic portal venous vascular bed of the dog. *Br. J. Pharmacol.* 62:376–77P
44. Richardson, P. D. I., Withrington, P. G. 1978. The effects of intraportal infusions of glucagon on the responses of the simultaneously-perfused hepatic arterial and portal venous vascular beds of the dog to periarterial nerve stimulation. *J. Physiol. London* 284:102–3P
45. Richardson, P. D. I., Withrington, P. G. 1978. Responses of the canine hepatic arterial and portal venous vascular beds to dopamine. *Europ. J. Pharmacol.* 48:337–49
46. Richardson, P. D. I., Withrington, P. G. 1978. Pressure-flow relationships and the effects of noradrenaline and isoprenaline on the simultaneously-perfused hepatic arterial and portal venous vascular beds of the dog. *J. Physiol. London* 282:451–70
47. Richardson, P. D. I., Withrington, P. G. 1978. Effects of intra-arterial and intraportal injections of vasopressin on the hepatic arterial and portal venous vascular beds of the dog. *Circ. Res.* 43:496–503
48. Richardson, P. D. I., Withrington, P. G. 1978. Responses of the simultaneously-perfused hepatic arterial and portal venous vascular beds of the dog to histamine and 5-hydroxytryptamine. *Br. J. Pharmacol.* 64:581–88
49. Richardson, P. D. I., Withrington, P. G. 1978. The effects of intraportal infusions of glucagon on the hepatic arterial and portal venous vascular beds of the dog: inhibition of hepatic arterial vasoconstrictor responses to noradrenaline. *Pflügers Arch.* 378:135–40
50. Richardson, P. D. I., Withrington, P. G. 1979. Responses of the hepatic arterial and portal venous vascular beds of the dog to intra-arterial infusions of noradrenaline and adrenaline: inhibition of hepatic arterial vasoconstrictor responses by intraportal infusions of glucagon. *Br. J. Pharmacol.* 66:82P
51. Richardson, P. D. I., Withrington, P. G. 1980. Effects of intraportal infusions of hypertonic solutions on hepatic haemodynamics in the dog. *J. Physiol. London* 301:82–83P
52. Ross, G. 1970. Cardiovascular effects of secretin. *Am. J. Physiol.* 218:1166–70
53. Scholtholt, J. 1970. Das Verhalten der Durchblutung der Leber bei Steigerung des Sauerstoffverbrauches der Leber. *Pflügers Arch.* 318:202–16
54. Scholtholt, J., Shiraishi, T. 1970. The reaction of liver and intestinal blood flow to a generalized hypoxia, hypocapnia and hypercapnia in the anaesthetized dog. *Pflügers Arch.* 318:185–201
55. Scholtholt, J., Shiraishi, T. 1968. The action of acetylcholine, bradykinin and angiotensin on the liver blood flow of the anaesthetized dog and on the pressure in the ligated ductus choledochus. *Pflügers Arch.* 300:189–201
56. Schuurkes, J. A. J., Brouwers, H. A. A., Beijer, H. J. M., Charbon, G. A., Schapiro, H. 1976. Lysine-vasopressin: hemodynamic effects in the anesthetized dog. *Digest. Dis.* 21:1012–19

MEASUREMENT OF GASTROINTESTINAL BLOOD FLOW

Barry L. Tepperman

Department of Physiology, Faculty of Medicine, Health Sciences Centre, University of Western Ontario, London, Ontario, Canada N6A 5C1

Eugene D. Jacobson

Department of Physiology, College of Medicine, University of Cincinnati, Cincinnati, Ohio 45267

CRITERIA FOR SELECTION OF A TECHNIQUE

The choice of a method of measurement of gastrointestinal blood flow depends upon the tissue and/or vessel through which blood flow will be measured as well as the conditions under which flow is to be determined. Ideally, the method should fulfill a number of criteria including accuracy, sensitivity and reproducibility over a wide range of blood flows, lack of need for additional measurements, safety to the subject, freedom from effects upon tissue metabolism or physiological events, noninvasiveness, low expense, speed, and ease of performance (18). Each of the techniques dealt with in this chapter fulfills some but not all of these criteria, perhaps reflecting the fact that most methods for measuring blood flow have been developed to investigate a specific biomedical problem. We describe the advantages as well as specific pitfalls inherent in each technique during its usual applications.

DIRECT VISUALIZATION TECHNIQUES

Blood passes to the hollow organs of the gastrointestinal tract through an extensive plexus of smaller vessels in the muscularis, submucosa, and mucosa, eventually forming both in-parallel and in-series microvascular circuits. Consequently, methods that measure total blood flow to an organ with several tissues do not usually provide information on specific changes in flow to a particular tissue layer. Techniques that allow direct observation of the microvasculature do enable evaluation of vascular behavior in separate tissues and even in individual blood vessels.

In Vivo Fluorescence Microscopy

This technique involves direct microscopic observation of vessels in the living animal. The tissue is transilluminated with a high-intensity light source and the vasculature is visualized with a microscope. Fluorescein isothiocyanate is injected intravenously. The fluorescent agent emits light when excited by filtered light from the microscope system. The emitted light is visualized either by the microscope or on a TV monitor. Vessel diameter may be measured using an image splitting technique via a microscope recording system. This technique has been applied to investigations of the morphology and flow patterns in the gastric microcirculation, including vascular responses to stimuli and changes in microvascular permeability (17).

A major limitation of this technique is that it only provides information for a limited portion of tissue, from which one must extrapolate findings to the whole organ. Thus an additional technique is required to measure whole organ blood flow. The method has not been adapted for use in conscious animals and requires specialized equipment. It has also not been shown that the fluorescent material is inert.

Photometric Velocity Method

Photometric velocity measurements detect the nearly identical patterns of changes in light intensity caused by a red cell (or group of red cells) as it passes two measuring points along a blood vessel. The time of delay is the transit time of the blood over the known distance between two light detectors (3); the distance divided by the transit time is the velocity of the moving red blood cells.

This technique may be applied to observing neural and local control of blood flow and vessel dimensions (4) and has the advantage of allowing direct visualization of the enteric vasculature. However, this technique is invasive, can only be used to observe a limited part of the total circulation of the organ, and requires specialized equipment and skilled experimenters.

DIRECT METHODS

Most of the direct methods for the study of blood flow in an organ have been critically reviewed (13). Basically, these techniques require arterial cannulation with an extra corporal circuit with or without a pump. However, the unphysiological nature of these procedures and the striking effects of the preparation on vascular tone make them unsuitable.

Venous Outflow

This is likely the oldest method for the measurement of blood flow in an organ and requires collection of the venous effluent. The basic assumption of the technique is that venous drainage reflects arterial inflow. Its major virtues are ease of measurement, reproducibility, and accuracy. It has been employed effectively to confirm blood flow measured using other techniques (29). The obvious disadvantage of this procedure is the trauma to the experimental animal; it can only be used in acute preparations. The technique yields no information about distribution of blood flow.

Morphologic Studies

The premise underlying these approaches is that the vascular architecture reflects its function. Much morphologic information has been obtained from injection procedures using India ink, gelatin, graphite, radioopaque materials, starch granules, various colors of latex preparations, and various plastic agents.

A recent adaptation of this method involves injection of a silicone elastomer into which a pigment has been milled (Microfil). Once the injection is complete and surrounding tissues have been cleared, specimens are scanned using a dissecting microscope. This technique has been applied to studies of the small intestine (32).

There are a number of disadvantages with this technique. First, injection of the substance alters blood flow and may distort the geometry of the vessels even when injection pressure is kept within physiological limits. Second, this technique only allows one estimation and is limited by the resolution and magnification of the dissecting microscope. To overcome the latter difficulty, scanning electron microscopic examinations of casts of the microcirculation have been applied (14). Again, infusion of the casting material may influence vessel geometry.

These techniques are also costly. They do, however, provide accurate representations of the microvascular architecture.

NONCANNULATING FLOW METERS

The most widely used flowmeter is electromagnetic (23). The method depends upon the induction of voltage in a conductor (the electrolytes of blood) moving through a magnetic field at right angles to the lines of force. The magnetic field is produced by a small electromagnet that wraps around the blood vessel. The induced voltage is proportional to the flow rate. An extensive discussion of the theory, history, techniques, advantages, and disadvantages of the electromagnetic flowmeter has recently appeared (9).

The use of electromagnetic flow probes has the substantial advantage of providing continuous measurement of total blood flow, although distribution of flow within tissues cannot be ascertained by this method. This method records both mean and phasic flow, calibration is linear, and there is a high frequency response. No cannulation is required, and chronic implantation of these transducers permits measurement of blood flow in a conscious animal (38).

The reliability *in situ* of noninvasive flow meters is problematic. Although they are calibrated under undisturbed conditions, distortions in their measurements may occur with the changes in the spatial relationship of transducer to blood vessels that occasionally develop during experimental conditions. Furthermore, the flowmeter must be calibrated at zero flow. This requires transient occlusion of the vessel at least once before the experiment, a maneuver that may produce reflex, hormonal, or paracrine effects upon blood flow. The use of electrometric determination of the zero flow value obviates mechanical occlusion of the vessel, but the validity of this practice is uncertain.

INDICATOR DILUTION METHODS

If a marker is infused into the arterial blood of an organ at a constant rate and is thoroughly mixed, then an increase in blood flow will decrease the concentration of the marker (31). From the Fick principle, blood flow to the organ can be determined if the arterial and venous concentrations of the marker are determined simultaneously with the total amount of marker extracted by the organ. The Fick equation, which solves for organ blood flow, becomes the quotient of extraction rate/difference between inflow and outflow concentrations of indicator.

Dyes are the most frequently used markers for indicator dilution measurements of blood flow. The concentration of dye is determined by circulating blood into a flow-through cuvette. The absorption of the dye is recorded on an optical densitometer. Similar methodology may be applied to radioactive indicators, in which case detection of the marker is performed by continuous blood sampling or by the use of external detectors (39). Thermal markers have been studied using fine-gauge needle thermocouples to demonstrate a logarithmic relation between temperature and flow; from such determinations gastro-intestinal mucosal blood flow has been extrapolated (15). Infused cold solutions can be employed in place of a dye or isotopic marker (31).

Indicator dilution techniques have been used extensively to measure mucosal hemodynamics (27), to separate villous and mucosal blood flows (2, 34), and to determine capillary exchange capacity of the intestine (34).

These techniques are fairly accurate, harmless, and in some situations noninvasive. Rarely is laparotomy or anesthetization necessary to measure large vessel or organ blood flow. If the chemical indicator is completely cleared from the tissues on a single pass, background accumulation is not a problem and determinations can be performed repeatedly. General shortcomings of indicator dilution techniques have been addressed at length elsewhere (31). The primary difficulty has been the disappearance of marker from the circulation at rates that are not known to the investigator under all experimental conditions. In addition, any recirculation of marker substances will confound the estimation of vascular concentrations. Instrumentation for continuous determination of marker concentrations in blood is complex and costly.

WASHOUT TECHNIQUES

Flow-dependent washout of radiolabelled inert gases (predominantly xenon and krypton) to measure organ blood flow is not a new method (22). These noble gases are highly soluble in lipid membranes; hence, their transcapillary exchange after intraarterial injection or after direct injection into tissues is rapid. Removal of inert gas from a site ("washout") is assumed to be proportional to and dependent upon blood flow (26). Mathematical analysis of washout curves for total gastric or intestinal blood flows may be calculated according to a standard formula (40) after measurement of the gamma radiation from the inert gas. Concurrent determination of beta radiation with a Geiger-Müller tube placed at the antimesenteric border and plotting of a monoexponential decay curve allows calculation of blood flow in the muscularis according to other equations (21). The weights of the mucosa-submucosa and muscularis can be determined directly. With this information blood flow in the mucosa-submucosa may be calculated. Methods based on these principles have been used for the study of gastrointestinal blood flow (1, 26). Use of carbon monoxide washout to measure blood flow to the villus, villus countercurrent exchange, and effective villus flow have also been reported (6). Although invasive, these techniques allow repeated measurements and can be applied clinically. Because of their high lipid solubility, inert gases permeate tissues quickly, thereby reducing much of the error caused by variations in surface area, permeability, and capillary filtration pressure. The obvious advantage of this method is the ability to measure total blood flow concomitantly with the distribution of blood flow. Inspired nonradioactive xenon can be excited and estimated with fluorimetric methods, providing a safe clinical application of the washout technique (30).

One problem with this technique arises from the high affinity of noble gases for air, which may cause marker to be trapped in the gut lumen, thereby introducing errors that could not be factored out with current equations. In addition, the instrumentation needed for measurement is not inexpensive, and the analytic problems associated with interpretation of washout curves of inert gases are formidable (16).

CLEARANCE METHODS

Clearance methods are another special application of the Fick principle in which the marker is cleared from the blood in one passage. The most widely used clearance method is based on the pH partition hypothesis (19, 36). Its use depends on the ready clearance by the stomach of the un-ionized form of a weak base (like aminopyrine) from the plasma into the gastric lumen. When the molecule comes into contact with the low pH of gastric juice, the base dissociates, thereby losing its lipid-soluble character and being prevented from diffusing readily back into the blood. The trapping of ionized base at low pH accounts for the difference in concentrations of aminopyrine between the gastric juice and plasma phases; the extent of this difference depends upon the pK_a of the compound. The rate limiting step in the accumulation of aminopyrine in the gastric secretions is the route of delivery of the agent, namely mucosal blood flow. Given the assumption that aminopyrine is completely cleared on one passage through the circulation of that tissue which is exposed to acid gastric juice, this technique can be used to estimate gastric mucosal blood flow. This assumption has been questioned (37); it is possible that three additional factors influence clearance of aminopyrine, namely a diffusion limit between blood and the gastric lumen, facilitated excretion of marker by bulk flow, and storage within the mucosa.

Other substances have been employed to estimate gastric mucosal blood flow including aniline, iodoantipyrine, phenol red, noble gases, and neutral red. Intestinal blood flow may be estimated by a similar technique using weak acids such as barbital (12).

The primary advantage of the clearance technique is that it provides concurrent information regarding blood flow and tissue function, namely gastric acid secretion. Furthermore, the technique can be performed continuously in the anesthetized or conscious subject. The most obvious disadvantage is that it may be used only in an organ with a significant pH gradient between plasma and lumen.

FRACTIONATION OF ISOTOPES AND MICROSPHERES

This method is based on the assumption that the fractional distribution of an isotope injected into an organ is proportional to the distribution of blood flow within the organ. The technique is derived from the indicator dilution method. As initially used in the gut with ^{42}K and ^{86}Rb (11), the method is limited in the number of determinations possible. Furthermore, the assumption that all tissues have the same extraction ratio of ^{42}K or ^{86}Rb is questionable.

A theoretically more attractive method is the study of the tissue distribution of radioactively labelled microspheres (15±5 μm diameter, labelled with ^{85}Sr, ^{141}Ce, ^{51}Cr, etc). The use of this technique for the measurement of gastric and intestinal blood flows has been reviewed extensively (35). Briefly, the microspheres are injected into the artery in question. The microspheres are presumed to distribute according to the precapillary distribution of blood flow through the microcirculation and should not be recoverable in the venous effluent. The spheres become trapped in vessels of smaller diameter than the spheres. After injection the animal is killed and the stomach and intestine are separated into mucosa, submucosa, muscularis, and serosa. From the radioactivity of each tissue layer and the cardiac output, the flow to the tissue can be calculated. This method has been used to measure blood flow distribution in the splanchnic circulation (16). Nonradioactive spheres used to measure blood flow are counted microscopically, a procedure much less convenient than by radioactive determination. The nonradioactive technique has been used to measure villus capillary blood flow (5).

The use of spheres labelled with different isotopes permits more than one determination in each animal. However, serious questions have been raised (16) about the validity of the microsphere technique in assessing the distribution of blood flow when the vessels of the area studied are arranged in series with vessels of another tissue rather than in parallel (as occurs with mucosa and submucosa). Other problems inherent in the technique include non-uniform distribution due to sedimentation of the microspheres, rouleaux formation or plasma skimming, entrapment, shunting of spheres through the layers of the gut into venous blood, and movement of previously lodged spheres. These problems have been discussed extensively (35). Finally, the microsphere fractionation technique appears to be restricted to measurement in species in which the necessary blunt dissection of mucosa, submucosa, and muscularis can be easily performed.

FUNCTIONAL TECHNIQUES

These techniques are based on evidence suggesting that transmucosal movement of nutrients such as O_2, ions, and water can be a function of blood flow to the site of secretion or absorption. Flow per se is not measured.

Capillary Filtration Coefficient

The major determinant of the rate of delivery of nutrients and O_2 to the parenchyma is the tone of the precapillary sphincters and arterioles. Hence it is critical to determine both blood flow and capillary exchange capacity in order to understand effects of physiological changes or experimental interventions on O_2 and nutrient exchange. One measure of the functional exchange capacity is the capillary filtration coefficient (CFC). This index has been employed in studies of capillary fluid exchange and the control of blood flow and oxygen intake in a variety of gastrointestinal tissues (20, 24).

The methodologies used to determine the CFC include: volumetric or gravimetric techniques for measuring the rate of accumulation or loss of fluid within the tissue, measurement of the rate of lymph flow and exudation of fluid into the intestinal lumen following venous occlusion, indicator dilution techniques using labelled red blood cells or plasma proteins, measurement of the components of the Starling equilibrium (capillary filtration rate, capillary plasma oncotic pressure, interstitial oncotic pressure), and measurement of total weight gained by a tissue after venous pressure elevation (33). Measurement of CFC gives information about the functional state of exchange vessels in a tissue in terms of the density of perfused capillaries and the availability of blood flow for transvascular fluid and O_2 exchange. The primary disadvantages of these methods are the difficulty in the separation of measurements of perfused capillary density from microvessel permeability and the invasive nature of most of the techniques. Furthermore, the CFC gives no indication of where within a tissue the transvascular exchange is occurring. Finally, it has been demonstrated that values of the CFC determined by the different methodologies described are variable.

Clearance of Tritiated Water

This technique has been used to measure blood flow at the site of mucosal absorption and involves perfusion of the gut lumen with ^3H_2O (28). The absorption rate of ^3H_2O and the concentration of ^3H_2O in the intestinal effluent, mesenteric vein, and artery must be measured. The basic assumption of the method is that the absorption of ^3H_2O depends upon blood flow. The equilibrium concentration of ^3H_2O in the venous blood would be the same as the luminal ^3H_2O concentration. Hence absorptive-site blood flow can be calculated from the luminal clearance of ^3H_2O. The amount of ^3H_2O absorbed is very nearly proportional to blood flow. This conclusion has been supported by comparing ^3H_2O clearance with the clearance of other highly permeable compounds such as ^{14}C dimethylsulfoxide, ^{22}Na, and $^{14}CO_2$.

Clearance of ^3H_2O has been used to assess the relationship between absorption and blood flow in the small intestine of a number of species under a variety of experimental conditions (28). The major disadvantage of the method is that ^3H_2O clearance is probably not an exclusive function of absorptive-site blood flow and may include such factors as capillary surface area, capillary permeability and filtration pressure, surface area of villi, permeability of epithelial cells, metabolic activity, and osmotic differences.

Oxygen Tension

Determination of gastric mucosal pO_2 is another assessment of nutrient blood flow and involves the use of a gold-filled oxygen microelectrode (7). This approach assumes that the presence of oxygen in tissues is a direct reflection of nutrient microvascular blood flow and the partial pressure of oxygen in the arterial blood. When arterial oxygen tension (pO_2) is held constant, any change in intracellular pO_2 can be assumed to reflect change in either the nutrient microcirculation or in the diffusion of O_2 from capillary to cell. In either case, the available concentration of O_2 is an indication of the effectiveness of nutrient microcirculatory blood flow. With the gold-filled microelectrode one can also measure transcellular electrical potentials in the gastric epithelium.

The electrode used in this procedure appears not to interfere with blood flow nor to damage the tissue. However, questions have been raised about the functional significance of the cellular locus routinely under examination and the validity of the assumption that intracellular O_2 tension is a reflection of nutrient blood flow to the mucosa.

In Vitro Vascular Segments

Although this in vitro approach does not directly measure blood flow through a particular tissue, changes in vascular tone in response to a vasoactive substance or other experimental manipulation may be taken as presumptive evidence for dilator or constrictor properties of the drug. A number of tissue preparations have been used including 2–5 mm long ring segments, helical strips, or an intact segment of vessel several cm in length. Alterations in vascular tone are measured using force or pressure transducers.

CLINICAL TECHNIQUES

A number of methods have been developed for assessment of pathological alterations in the splanchnic circulation, including occlusive and nonocclusive ischemic disease, shock, and hemorrhage. These techniques should possess a number of the advantages enumerated earlier, but their primary attributes must be safety for the patient and ease of performance.

Angiography

Selective angiography is the most widely used clinical estimate of blood flow to splanchnic organs. Quantification of mesenteric artery blood flow has been achieved by the use of reflux angiography employing high-frequency filming equipment and a suitably forceful injector system (10). Radioopaque medium is delivered at increasing rates that eventually exceed the blood flow; at this point the injectate fills the vessel completely, having momentarily displaced the blood, and some of the medium will reflux and be detected radiographically. The rate of injection at which reflux occurs approximately equals the rate of blood flow.

The method has major disadvantages: patient risk from an invasive method and from the nephrotoxicity of the medium, the need for a highly skilled operator, and the expense of the procedure. The method is not fully quantitative, only a few determinations can be performed with one catheterization, and the medium is vasoactive.

Video Dilution Technique

This procedure involves use of intraarterial injection of contrast medium as a dye dilution indicator during fluoroscopy (25). The angiographic densities in the video image are expressed as a specific voltage and may then be calibrated by measuring the total density of the blood vessel within an arbitrary area of the television image. By positioning the densimeter over the vessel in question, the passage of injected medium is recorded as a mass versus time curve. Data accumulated so far correspond to estimations of blood flow with an electromagnetic flowmeter.

The major disadvantages of the method include its invasiveness, the toxicity and vasoactivity of the injectate, the need for a skilled professional, its expense, and the problem of uneven mixing of the contrast medium with blood.

Washout of Intraperitoneal Xenon

The washout of intraperitoneal xenon provides an estimate of total splanchnic flow by measuring the effective perfusion of the entire peritoneal surface after intraperitoneal injection of ^{133}Xe (8). The lipid soluble gas is absorbed into the circulation by passive diffusion. Since ischemic tissue should selectively retain ^{133}Xe activity, this method should highlight areas of hypoperfusion. Washout of ^{133}Xe is determined as described previously, and the washout curves may be analyzed by standard procedures.

Because it is based on the principles of inert gas washout, this method is subject to all the limitations and reservations discussed previously. Furthermore, this modified method does not have the potential for precise quantitation provided by the conventional approach. Its usefulness lies in its direct applicability to the diagnosis of occlusive and nonocclusive mesenteric ischemia.

Fluorescent Excitation Analysis

Nonradioactive xenon may be administered in inspired air. When excited by X rays or gamma rays it emits a unique radiation (30). Once a constant tracer input has been achieved, kinetic analysis allows derivation of parameters that can then be empirically compared to established blood flow values.

The primary advantage of this technique is its safety; it is noninvasive, and xenon is both inert and nonradioactive. However, the technique is at present costly and does require a detailed analysis of the time versus activity xenon curves generated.

Literature Cited

1. Bell, P. R. F., Battersby, C. 1968. Effect of vagotomy on gastric mucosal blood flow. *Gastroenterology* 54: 1032–37
2. Biber, G., Lundgren, O., Svanvik, J. 1973. Intramucosal blood flow and blood volume in the small intestine of the cat as analyzed by an indicator dilution technique. *Acta Physiol. Scand.* 87:391–403
3. Bohlen, H. G. 1981. Microcirculatory methods for studying the intestinal circulation. In *Measurement of Splanchnic Blood Flow*, ed. G. B. Bulkey, D. N. Granger. Baltimore: Williams & Wilkins
4. Bohlen, H. G., Henrich, H., Gore, R. W., Johnson, P. C. 1978. Intestinal muscle and mucosal blood flow during sympathetic stimulation. *Am. J. Physiol.* 235:H40–45
5. Bond, J. H., Levitt, M. D. 1979. Use of microspheres to measure small intestine villus blood flow in the dog. *Am. J. Physiol.* 236:E577–83
6. Bond, J. H., Levitt, D. G., Levitt, M. D. 1977. Quantitation of countercurrent exchange during passive absorption from the dog small intestine. *J. Clin. Invest.* 59:308–18
7. Bowen, J. C., Garg, D. K., Salvato, P, D., Jacobson, E. D. 1978. Differential oxygen utilization in the stomach during vasopressin and torniquet ischemia. *J. Surg. Res.* 25:15–20
8. Bulkey, G. B. 1981. Washout of intraperitoneal xenon: Effective peritoneal perfusion as an estimation of splanchnic blood flow. See Ref. 3
9. Charbon, G. A., van der Mark, F. 1981. Use of electromagnetic flowmeters for the study of splanchnic blood flow. See Ref. 3
10. Clark, R. A., Colley, D. P., Jacobson, E. D., Herman, R. Tyler, G., Stahl, D., 1980. Superior mesenteric angiography and blood flow measurement following intra-arterial injection of prostaglandin E. *Radiology* 134:327–33
11. Delaney, J. P., Grim, E. 1974. Canine gastric blood flow and its distribution. *Am. J. Physiol.* 207:1195–202
12. Fara, J. W. 1981. Use of clearance of pH trapped compounds for measurement of intestinal blood flow. See Ref. 3
13. Folkow, B. 1952. A critical study of some methods used in investigations on the blood circulation. *Acta Physiol. Scand.* 27:118–29
14. Gannon, B. J., Gore, R. W., Rogers, P. A. W. 1980. Dual blood supplies to intestinal villi of rabbit and rat. *Microvasc. Res.* 19:247 (Abstr.)
15. Grayson, J. 1949. Vascular reactions in the human intestine. *J. Physiol.* 109:439–47
16. Greenway, C. V., Murthy, V. S. 1972. Effects of vasopressin and isoprenaline infusions on the distribution of blood flow in the intestine; criteria for the validation of microsphere studies. *Br. J. Pharm.* 46:177–88
17. Guth, P. G., Moler, T. L., Wayland, H. 1980. Study of the gastric microcirculation by in vivo fluorescence microscopy In *Gastrointestinal Mucosal Blood Flow*, ed. L. P. Fielding, pp. 17–26. London/NY: Churchill Livingston, 240 pp.

18. Jacobson, E. D. 1981. Criteria for an ideal method of measuring blood flow to a splanchnic organ. See Ref. 3
19. Jacobson, E. D., Linford, R. H., Grossman, M. I. 1966. Gastric secretion in relation to mucosal blood flow studied by a clearance technique. *J. Clin. Invest.* 45:1–13
20. Jansson, G., Lundgren, O., Martinson, J. 1970. Neurohormonal control of gastric blood flow. *Gastroenterology* 58:424–29
21. Kety, S. S. 1951. The theory and applications of the exchange of inert gas at the lungs and tissues. *Pharm. Rev.* 3:1–41
22. Kety, S. S., Schmidt, C. F. 1945. The determination of cerebral blood flow in man by the use of nitrous oxide in low concentrations. *Am. J. Physiol.* 143:53–66
23. Kolin, A. 1936. An electromagnetic flowmeter. Principle of the method and its application to blood flow measurements. *Proc. Soc. Exp. Biol. NY* 35:53–56
24. Kvietys, P. R., Miller, T., Granger, D. N. 1980. Intrinsic control of colonic blood flow and oxygenation. *Am. J. Physiol.* 238:G478–84
25. Lantz, B. M. T., Link, D. P., Holcroft, S. W., Forrster, J. M. 1981. Video dilution technique: angiographic determination of splanchnic blood flow. See Ref. 3
26. Lundgren, O. 1980. Determination of blood flow distribution in the gastrointestinal tract with inert gases. See Ref. 17, pp. 59–65
27. Lundgren, O., Svanvik, J. 1973. Mucosal hemodynamics in the small intestine of the cat during reduced perfusion pressure. *Acta Physiol. Scand.* 88:551–63
28. Mailman, D. 1981. Tritiated water clearance as a measure of intestinal absorptive site and total blood flow. See Ref. 3
29. Moody, F. G. 1967. Gastric blood flow and acid secretion during direct intraarterial histamine administration. *Gastroenterology* 52:216–24
30. Nelson, J. A., Staubus, A. E., Weinrib, A. B., Westenskow, D. R., Rikkers, L. F. 1981. Flourescent excitation analysis of non-radioactive xenon: A noninvasive technique for estimating splanchnic perfusion. See Ref. 3
31. Perry, M. A., Parker, J. C. 1981. Indicator dilution measurements of splanchnic blood flow. See Ref. 3
32. Reynolds, D. G., Swan, K. G. R. 1972. Intestinal microvascular architecture in endotoxin shock. *Gastroenterology* 63:601–10
33. Richards, P. D. I., Granger, D. N. 1981. Capillary filtration coefficient as a measure of perfused capillary density. See Ref. 3
34. Shepherd, A. P. 1981. Intestinal capillary exchange capacity and oxygen uptake rate: Applications of indicator dilution principles. See Ref. 3
35. Shepherd, A. P., Maxwell, L. C., Jacobson, E. D. 1981. Limitations of the microsphere technique to fractionate intestinal blood flow. See Ref. 3
36. Shore, P. A., Brodie, B. B., Hogben, C. A. M. 1957. The gastric secretion of drugs: a pH partition hypothesis. *J. Pharmacol. Exp. Ther.* 119:361–67
37. Sonnenberg, A., Blum, A. L. 1980. Limitations to measurement of gastric mucosal blood flow by ^{14}C-aminopyrine clearance. See Ref. 17, pp. 43–58
38. Swan, K. G., Jacobson, E. D. 1967. Gastric blood flow and secretion of conscious dogs. *Am. J. Physiol.* 212:891–96
39. Wolgast, M. 1968. Studies on regional renal blood flow with P^{32}-labelled red cells and small beta-sensitive semiconductor detectors. *Acta Physiol. Scand.* 313 (Suppl.):1
40. Zierler, K. L. 1965. Equations for measuring blood flow by external monitoring of radioisotopes. *Circ. Res.* 16:309–21

COMPARATIVE AND INTEGRATIVE PHYSIOLOGY

INTRODUCTION, James Edward Heath, *Section Editor*

This Section reviews the energetic cost of locomotion in endothermic animals. For these animals, which carefully sequester heat for thermoregulatory processes, the sudden increase in heat production due to locomotion requires both increased heat dissipation and special mechanisms to protect thermally sensitive organs, notably the brain in birds and mammals. Both requirements are treated in this Section.

A substantial heat production due to locomotion is not only a problem for endotherms, but may have initiated the evolution of endothermy. There is a parsimonious relationship between the evolutionary changes in structure and function required for rapid and sustained locomotion, and those that could preadapt an evolutionary line to endothermy. The enormous energy production of flight, of upright galloping, or of high sustained swimming speed generates a heat excess. The ancestors of modern groups may have simultaneously developed new locomotor patterns and capitalized on the heat excess to gain that additional independence of environmental conditions apparently so advantageous to endotherms.

Investigators approach energetics of locomotion in animals by estimating and summing the forces generated by a moving animal or by measuring the metabolic expenditure of moving animals. Since animals do not behave ideally, the first approach requires sensitivity to the behavior and ecological needs of the subject. Refinements in behavioral analysis lead refinement in physical technique and calculation. The second and more direct approach taxes the ingenuity of the researcher. The problem is to contain an animal within a laboratory environment while inducing the animal to move at velocities of as much as 27 m sec^{-1} while assuring that the forces and motion generated by the animal approach a natural pattern. As the following chapters indicate, we have just begun to understand the energetics of locomotion in its broadest terms.

BRAIN COOLING IN ENDOTHERMS IN HEAT AND EXERCISE

M. A. Baker

Division of Biomedical Sciences and Department of Biology, University of California, Riverside, California 92521

INTRODUCTION

The upper limit of tolerable body temperatures in most land vertebrates is 40°–45°C. Although excessive temperatures have effects on virtually every organ, brain function appears to be especially vulnerable to heat (16, 17, 61). If the brain is kept cool, tolerance to elevated deep body temperature is extended (24). Many species of mammals, birds, and reptiles can keep the temperature of the brain below the temperature of the rest of the body core during heat stress. In most cases, this brain cooling is accomplished by vascular arrangements that allow venous blood cooled by evaporation to exchange heat with arterial blood supplying the brain or to cool the base of the brain conductively. The process has been studied most intensively in panting carnivores and artiodactyls in which the structure of the nasal cavity and the arrangement of the blood vessels of the nasal cavity and the brain are well-suited for brain cooling. In these animals, the extent to which the brain is cooled depends upon the rate of evaporation and the rate of blood flow through the evaporative surfaces and reaches highest levels during exercise.

CONTROL OF BRAIN TEMPERATURE IN MAMMALS

The temperature of the mammalian brain is determined by the rate of heat production of brain cells, the rate of blood flow through the brain, and the temperature of the blood supplying the brain. In addition, the temperature of surface regions can be influenced by direct heat exchange through the scalp or through the base of the skull. In the animals that have been studied most thoroughly, the monkey (37, 38), cat (2), sheep (7), and dog (4), the temperature of the cerebral arterial blood is the single factor that is most variable and most likely to produce significant changes in brain temperature. The arteries supplying the mammalian brain arise from the circle of Willis on the floor of the cranial cavity and sweep around the outside of the brainstem and hemispheres, giving off branches that run deep into the nervous tissue. The cerebral arterial blood is cooler than the heat-producing nervous tissue, and the warmest brain regions are those near the center of the brain. Changes in cerebral arterial blood temperature are followed within seconds by similar temperature changes throughout the brain. The hemispheric surfaces can be affected by ambient temperature, even when the scalp is intact (38), and the ventral surface of the brain may be cooled by conduction to venous sinuses inside and outside of the cranial cavity.

Arteries Supplying the Circle of Willis

In man and other primates, the circle of Willis is supplied by the paired internal carotid arteries and the vertebral-basilar system (1, 26). This is also the case in the monotremes, marsupials, insectivores, chiropterans, rodents, lagomorphs, and the perrisodactyls (1). In the carnivores and artiodactyls, most arterial blood destined for the circle of Willis traverses the carotid rete (1, 28). The rete is a plexus of medium-sized arteries (about 250 μm in diameter) that is large in artiodactyls and felids but rudimentary in the canids and other fissiped carnivores outside of the cat family (28, 29). The rete is usually supplied by branches of the external carotid artery, but in the domestic dog, a rete-like anastomosis between the internal carotid artery and branches of the external carotid is present (40).

The arteries supplying the circle of Willis are surrounded by venous sinuses or plexuses. In mammals with no carotid rete, the internal carotid artery runs through the cavernous sinus, an intracranial venous sinus at the base of the brain. When a carotid rete is present, the vessels of the rete lie in a lake of venous blood, either in the cavernous sinus or in the pterygoid plexus, which is outside of the cranial cavity and communicates with the cavernous sinus. Within the venous spaces, arterial blood is separated from venous blood only by the arterial wall and by a single layer of venous endothelium. In the cat, the vessels of the carotid rete have very thin walls, with only one or two layers of smooth muscle (3, 55), but in sheep the walls of the retial arteries are thicker (52).

Venous Drainage to the Cavernous Sinus

The cavernous sinus receives venous blood draining the nasal mucosa and the skin of the face. Anteriorly, the nasal mucosal veins empty into veins of the palate and into the subcutaneous dorsal and lateral nasal veins, which also drain the skin of the face. Blood in these veins can enter the cavernous sinus via the angularis oculi and ophthalmic veins (30, 53) or can flow into the facial vein and then into the external or internal jugular, bypassing the cavernous sinus. This pattern of venous drainage appears to be similar in most of the mammals studied (32), including humans. Since there are no valves in the angularis oculi vein, flow in this vessel can be in either direction. Posteriorly, veins draining the nasal mucosa join the ethmoidal and sphenopalatine veins (30, 53), which communicate with the cavernous sinus via the infraorbital vein. Magilton & Swift (49) found that the dorsal nasal, angularis aculi, and facial veins in the dog had thick muscular walls and suggested that flow could be diverted either to the cavernous sinus or to the external jugular vein by constriction of one or the other pathways. The direction and rate of blood flow in the angularis oculi vein in humans is dependent upon the thermal state of the subject (23). Blood flow was low and directed toward the face in cool subjects. In subjects with elevated deep body temperatures, blood flow was higher and directed away from the face toward the cavernous sinus. Venous blood from the tongue probably does not enter the cavernous sinus, at least in the dog, but rather drains into the facial vein and from there to the heart (3, 53).

Structure and Vascular Pattern of the Nasal Cavity in Panting Mammals

The mammalian nasal cavity contains a posterior olfactory chamber and an anterior respiratory chamber. The respiratory portion of the nasal cavity is very long in some panting mammals, and both its bony arrangements and the arrangement of blood vessels in the mucosa are well-suited for a heat-exchange surface. The cavity is occupied through much of its extent by turbinate bones protruding into it from the bony walls. The largest of these, the maxillary turbinate, is formed of a series of thin bones that arise from the medial wall of the maxilla (54). In primates it is a simple, curved plate of bone. In ungulates, it has the form of a scroll with two or more turns. In carnivores, it is an elongated branching structure that fills the nasal cavity. The mucosa covering the surfaces of the respiratory portion of the nasal cavity, including the maxillary turbinate, contains large numbers of arteriovenous anastomoses and large venous spaces immediately below the surfaces, which are exposed to the respiratory airstream (57, 58). The mucosa is kept moist by epithelial goblet cells and subepithelial glands and by a large lateral nasal gland in the panting ungulates and carnivores (54). Lateral nasal gland secretions increase as much as ten-fold in dogs at high ambient temperatures (15).

Brain-Body Temperature Relationships

In resting, conscious monkeys, the brain was always warmer than arterial blood in the aorta or the common carotid artery (37, 38). Temperature measured in the basilar artery or near the arteries of the circle of Willis was the same as carotid blood temperature. A similar pattern was found in the resting rabbit at 25°C (6). In contrast to this, measurements of brain and arterial blood temperatures in resting goats (63), cats (2), sheep (7), and dogs (4) showed that the temperatures of the cerebral arteries and of the brain could fall below central arterial blood temperature when upper respiratory heat loss was elevated. This suggested that heat exchange was occurring between warm arterial blood in the carotid rete and venous blood cooled by evaporation in the nasal cavity.

BRAIN COOLING IN A WARM ENVIRONMENT In sheep and cats exposed to heat, the temperatures of cerebral arterial blood and brain fell below central blood temperature as deep body temperature rose and the rate of panting increased. The temperature difference between central and cerebral arterial blood in panting cats and sheep ranged from 0.7 to 1.1°C (2, 7). In resting, panting dogs cerebral arterial blood cooled only 0.5°C below central blood temperature (4). The dependence of this brain cooling on upper respiratory heat loss was shown in the dog by experiments in tracheostomised animals in warm environments. Tracheostomy breathing caused an immediate rise in brain temperature (4). In sheep at 40°C ambient temperature, hypothalamic temperature was from 0.6–1.5°C below carotid blood temperature during normal panting but rose rapidly if nasal airflow was blocked or if the rate of panting was reduced by cerebroventricular injections of carbochol (66). Goats at 33°C ambient temperature had hypothalamic temperatures about 0.7°C below aortic arterial blood temperature. When the humidity of the air around the head was increased, hypothalamic temperature rose toward aortic temperature (39). In anesthetized cats, measurements of arterial temperatures on the input and output sides of the carotid rete during experimental manipulations of nasal heat loss showed a trans-rete temperature drop of over 1°C at high levels of nasal evaporation and blood flow (14).

Brain cooling during thermal stress can occur in the rabbit, a panting mammal with no carotid rete (20, 21, 22). While hypothalamic and other brain temperatures were higher than internal carotid temperature at 20–25°C ambient temperature, exposure to heat was accompanied by cooling of most brain regions below the temperature of extracranial arterial blood. The brain cooling was clearly related to cooling of the venous blood draining the nose. Basal brain regions near the venous plexuses showed the greatest cooling (0.6°C below carotid temperature in the presubiculum), suggesting conductive cooling of the base of the brain and of the cerebrospinal fluid. Cooling of the nasal venous blood and the base of the brain in the rabbit could be abolished by reducing nasal evaportion or nasal blood flow.

Experiments in tracheostomised rabbits support the suggestion that nasal venous blood can cool the base of the brain (46, 62).

BRAIN COOLING IN EXERCISE   The rate of evaporative heat loss in panting mammals during exercise can reach levels that are more than double those occurring during rapid, shallow panting at rest in the heat (31, 34, 64). In dogs, respiratory minute volume increases rapidly at the onset of heavy exercise and continues to rise during a 15 min run (31). This elevated respiratory evaporation, coupled closely to high rates of blood flow in the nasal and oral cavities (discussed below), must account for the observation that panting mammals show the greatest brain cooling during exercise.

In an antelope, brain temperature, about 0.4°C higher than carotid blood temperature at rest, fell to 2.7°C below carotid temperature after 7 min of running at high speed (65). Hypothalamic temperature in beagles running for 10 min (5.34 km/hr, 10% slope) rose only about 0.1°C while spinal cord temperature rose about 0.8°C (25). The influence of respiratory evaporative heat loss on hypothalamic temperature was evident when the spinal cord in running animals was heated. This elicited an increase in respiratory evaporation and a drop in hypothalamic temperature. In two dogs running at 7.2 km/hr up a 14% slope at 30°C ambient temperature, carotid blood temperature rose steeply at the onset of exercise and continued to rise during a 15 min run. Hypothalamic temperature dropped at the beginning of the run and then rose at about the same rate as carotid temperature, but was maintained about 1.3°C below carotid temperature during the run (5). Similar studies on a larger group of animals (3) have shown that most dogs have a drop in brain temperature at onset of exercise, but some animals show a slight rise in brain temperature and then a drop or a period of steady brain temperature a few minutes into the run. During continued exercise for up to one hour, with carotid blood temperatures above 42°C, hypothalamic temperature remains more than 1°C below carotid blood temperature. The greatest brain cooling that we have observed in any dog was in a 30 kg animal that kept the hypothalamic temperature about 1.8°C below carotid blood temperature during heavy exercise. The thalamus, a site near the middle of the brain and one of the warmest brain regions, does not cool as much as the hypothalamus in running dogs. Thalamic temperature falls, with respect to carotid blood temperature, during exercise but the fall is never as great as the fall in hypothalamic temperature. This suggests that brain cooling in the exercising dog may have two components: (*a*) conduction of heat from ventral brain regions directly to the large, flat surfaces of the cavernous sinuses, and (*b*) cooling of arterial blood in the rudimentary carotid rete by countercurrent exchange of heat with venous blood. Simultaneous measurements of temperature at many different brain sites would be needed to clarify this.

Even though the resting, heat-stressed rabbit can cool the brain below carotid blood temperature (20, 21, 22), in exercising New Zealand white rabbits there is no marked cooling of the brain as there is in the antelope and the dog; rather, brain temperature rises at about the same rate as rectal (46) or aortic arterial blood (3) temperature during exercise. This may be because brain cooling during exercise appears to depend not only on high levels of upper respiratory evaporation but also on the capacity to increase the cardiac output enough to provide a high rate of blood flow to the upper respiratory passages, the respiratory muscles, and the muscles involved in the exercise. This capacity may be developed best in animals such as the dog and the antelope, which normally have a high exercise tolerance.

Coordinated Control of Upper Respiratory Evaporation and Blood Flow During Panting

Blood flow to the tongue and the nasal mucosa in the dog is closely related to the body temperature and to the rate of breathing (45, 47, 48, 56). In anesthetized dogs, whole-body heating or heating of the hypothalamic thermosensitive region elicits high levels of flow in the arteries supplying the tongue and nasal cavity. Studies of blood flow using radioactive microspheres demonstrate increased blood flow to the tongue, the nasal mucosa, and the maxillary turbinate in unanesthetized dogs (35) and sheep (36) in the heat, and in dogs during exercise (60). Since microspheres are not trapped in arteriovenous anastomoses (AVAs), they measure only capillary flow and underestimate the total flow when AVAs are present. It is likely that the AVAs of the tongue and nasal cavity are dilated in heat and exercise and that both capillary flow and AVA flow are increased under these conditions. Blood flow through lingual AVAs in anaesthetized dogs is related inversely to tongue surface temperature, which suggested that dilation of the AVAs may be initiated by tongue cooling during panting (48).

The rate of blood flow in the common carotid artery of conscious dogs is positively correlated with the level of heat stress and with the rate of respiratory evaporation (Figure 1). This artery supplies almost the entire head of the dog. The only tissues of the head that show increased blood flow in heat-stressed and exercising dogs are the nose, tongue, and skin of the ear (33, 35, 60), and measurements of common carotid flow in heat-stressed dogs provides an index of flow to these regions. In resting dogs, common carotid flow rises and falls with the rate of breathing. When the ambient temperature is raised or when the hypothalamic thermosensitive region is heated, parallel elevations in carotid flow and in evaporation occur. The highest levels of both common carotid flow and respiratory evaporation in dogs occur during heavy exercise in a warm environment (8). Bilateral

ENDOTHERM BRAIN TEMPERATURE 91

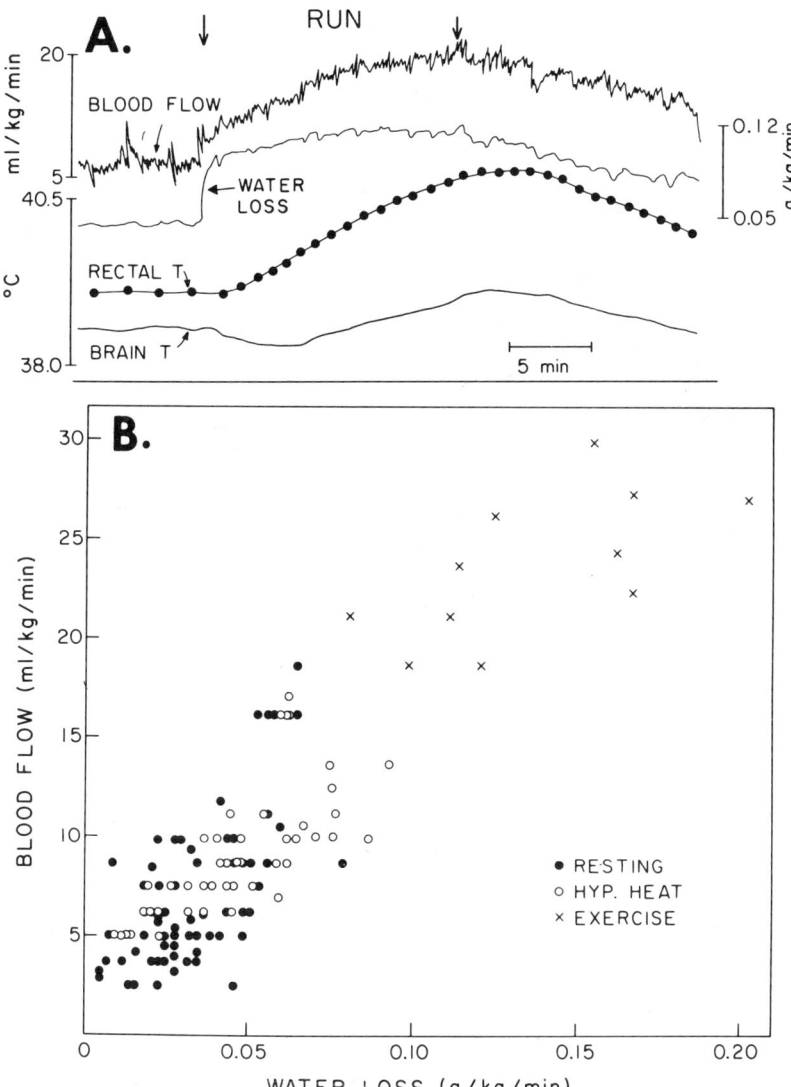

Figure 1. Relationship between cephalic blood flow and upper respiratory evaporation in the dog. Blood flow is left common carotid flow measured with an implanted ultrasonic probe. Water loss was measured with a flow-through mask. A. Carotid blood flow, evaporative water loss, rectal temperature, and brain (hypothalamic) temperature during exercise. In the period marked RUN (between the vertical arrows), the dog ran at 7.5 km/hr on a 20% slope. Ambient temperature 25°C. B. Steady-state levels of carotid blood flow and evaporative water loss in a dog at rest at ambient temperatures from 25°C to 45°C (closed circles), during heating of the hypothalamic thermosensitive zone (open circles), and during the last minute of 15 min of exercise at different workloads and different ambient temperatures. Highest rates of carotid blood flow and evaporative water loss occur during heavy exercise in a warm environment.

common carotid flow probably represents from 10–15% of the cardiac output in heat-stressed or exercising dogs. Simultaneous elevations in upper respiratory evaporation and blood flow allow high rates of heat transfer to the surfaces of the nose and mouth and also high rates of flow of venous blood into the cranial cavity for brain cooling.

Evidence for Brain Cooling in Humans

The suggestion that venous blood cooled by sweating on the face in humans could cool the cerebral arterial blood was first made (50) to explain the observation that heating or cooling of the skin of the head led to increased or decreased sweating over the rest of the body. This effect had previously been attributed to a nervous input from facial cutaneous thermoreceptors (27). McCaffrey et al postulated extracranial heat exchange between venous blood draining the face and arterial blood in the common, external, and internal carotid arteries, since they observed that tympanic membrane temperature rose or fell when localized regions of the skin of the ipsilateral face were heated or cooled (51). The tympanic membrane is supplied by the external carotid artery in the human. The relationship of tympanic temperature to hypothalamic temperature when changes in facial skin temperature occur remains unclear, although tympanic temperature and hypothalamic temperature show similar patterns of change in conscious cats and monkeys in a cool environment (9).

Cabanac & Caputa (18) used perceptual ratings of the pleasantness or unpleasantness of a peripheral thermal stimulus as an index of hypothalamic temperature in humans. Fanning the face in hyperthermic subjects caused ratings to change from those typical of hyperthermia to those typical of normothermia. Furthermore, esophageal temperature rose during face fanning, suggesting the triggering of heat-conserving mechanisms by a cool hypothalamus. Temperature over the angularis oculi vein fell during face fanning. These investigators suggested that the hypothalamus was being cooled by venous blood returning via the angularis oculi and ophthalmic veins to the cavernous sinus. They extended their studies to exercising subjects (19), in whom they observed a rise in esophageal temperature during face fanning and a decrease during head insulation. Tympanic membrane temperature fell during face fanning and rose when the head was insulated. Perceptual ratings of a cutaneous thermal stimulus were hypothermic when the face was fanned and hyperthermic when the head was insulated. These studies in the human show that changes in heat loss from the face have a powerful effect on thermoregulation that may be caused by a change in hypothalamic temperature. The possibility that the thermoregulatory responses associated with warming or cooling of the face are influenced by input from facial cutaneous thermoreceptors still cannot be ruled out.

BRAIN COOLING IN BIRDS

Birds maintain brain temperatures around 1°C below body core temperature at rest over a wide range of ambient temperatures, and this brain cooling is enhanced during flying and running (12, 13, 41–44). The rhea (30.9 kg) has a brain temperature about 1°C lower than cloacal temperature and 1.1°C lower than temperature of carotid arterial blood measured either near the heart or at the head (42). No change occurred in the temperature of carotid arterial blood coursing up the long neck of the rhea, and it was postulated that cerebral arterial blood was cooled in the *rete mirabile ophthalmicum*. This rete is an extracranial network of arteries and veins, usually supplied by branches of the internal carotid artery and by venous blood from the front of the head, including the eyes and the upper respiratory passages. The brain is supplied by the two internal carotid arteries that join each other on its ventral surface, and there is no circle of Willis (59). Branches from the ophthalmic rete join intracranial branches of the internal carotids and can provide the entire arterial supply to the brain if the internal carotid supply is occluded (44).

In resting lesser nighthawks, mallards, pigeons, white-necked ravens, roadrunners, and American kestrels (12, 41), hypothalamic temperature was lower than colonic or cloacal temperature at ambient temperatures between 23°C and 54°C with core temperatures as high as 43°C. Average core–brain temperature differences ranged from 0.70°C in the kestrel to 1.29°C in the mallard and did not change as colonic temperature rose at high ambient temperatures except in the kestrel. In this animal, the core–brain temperature difference increased to 1°C at the highest core temperatures. In running quail (43) and in flying kestrels (12), brain cooling was greater than in the same animals at rest, core–brain temperature differences reaching 1.45°C in quail and 1.2°C in kestrels during exercise. Strong evidence that brain cooling in birds depends upon the ophthalmic rete comes from studies in heat-stressed pigeons in which the core–brain temperature difference was reversed (i.e. brain became warmer than core) when blood flow through the rete was blocked by surgical occlusion of the arteries supplying it (44). Experiments in which cranial evaporative cooling was blocked in pigeons suggest that brain cooling depends upon evaporation from both the upper respiratory passages and the corneas of the eyes (13). Corneal evaporation may increase more than upper respiratory evaporation in flying birds as compared to heat-stressed birds, since total ventilation in exercising pigeons is about the same as it is in heat-stressed pigeons (13). Blood flow to the head increases in heat-stressed birds just as it does in heat-stressed dogs. Carotid blood flow in the duck increases during thermal panting and when the spinal cord is heated (10, 11). The rate of carotid flow is related closely to the respiratory frequency, as in the dog.

Acknowledgments

Work in the author's laboratory was supported by NSF Grant BNS-7901006 and PHS BRDG Grant RR 09070.

Literature Cited

1. Ask-Upmark, E. 1935. The carotid sinus and the cerebral circulation. An anatomical, experimental and clinical investigation, including some observations on rete mirabile caroticum. *Acta Psychiat. Neurol. Suppl.* 5–7: 1–374
2. Baker, M. A. 1972. Influence of the carotid rete on brain temperature in cats exposed to hot environments. *J. Physiol. London* 220: 711–28
3. Baker, M. A. 1981. Anatomical and physiological adaptations of panting mammals to heat and exercise. In *Environmental Physiology: Aging, Heat and Altitude*, ed. S. M. Horvath, M. Yousef. NY: Elsevier-North Holland
4. Baker, M. A., Chapman, L. W., Nathanson, M. 1974. Control of brain temperature in dogs: Effects of tracheostomy. *Respir. Physiol.* 22: 325–33
5. Baker, M. A., Chapman, L. W. 1977. Rapid brain cooling in exercising dogs. *Science* 195: 781–83
6. Baker, M. A., Hayward, J. N. 1967. Autonomic basis for the rise in brain temperature during paradoxical sleep. *Science* 157: 1586–88
7. Baker, M. A., Hayward, J. N. 1968. The influence of the nasal mucosa and the carotid rete upon hypothalamic temperature in sheep. *J. Physiol. London* 198: 561–79
8. Baker, M. A., Rader, R. D., Kirtland, W. H. 1979. Control of brain temperature in running dogs. *Fed. Proc.* 38: 944
9. Baker, M. A., Stocking, R. A., Meehan, J. P. 1972. Thermal relationship between tympanic membrane and hypothalamus in conscious cat and monkey. *J. Appl. Physiol.* 32: 739–42
10. Bech, C., Johansen, K. 1980. Blood flow changes in the duck during thermal panting. *Acta Physiol. Scand.* 110: 351–55
11. Bech, C., Rautenberg, W., May, B., Johansen, K. 1980. Effect of spinal cord temperature on carotid blood flow in the Pekin duck. *Pflügers Arch.* 385: 269–71
12. Bernstein, M. H., Curtis, M. B., Hudson, D. M. 1979. Independence of brain and body temperatures in flying American kestrels, *Falco sparverius*. *Am. J. Physiol.* 237(1): R58–62
13. Bernstein, M. H., Sandoval, I., Curtis, M. B., Hudson, D. M. 1979. Brain temperature in pigeons: Effects of anterior respiratory bypass. *J. Comp. Physiol.* 129: 115–18
14. Berry, D. C. 1973. *An investigation of heat exchange between blood flowing through the carotid rete and pterygoid venous plexus in the cat.* PhD thesis. Indiana University, Bloomington
15. Blatt, C. M., Taylor, C. R., Habal, M. B. 1972. Thermal panting in dogs: The lateral nasal gland, a source of water for evaporative cooling. *Science* 177: 804–5
16. Bowler, K., Tirri, R. 1974. The temperature characteristics of synaptic membrane ATPases from immature and adult rat brain. *J. Neurochem.* 23: 611–13
17. Burger, F. J., Fuhrman, F. A. 1964. Evidence of injury by heat in mammalian tissues. *Am. J. Physiol.* 206(5): 1057–61
18. Cabanac, M., Caputa, M. 1979. Natural selective cooling of the human brain: Evidence of its occurrence and magnitude. *J. Physiol. London* 286:255–64
19. Cabanac, M., Caputa, M. 1979. Open loop increase in trunk temperature produced by face cooling in working humans. *J. Physiol. London* 289: 163–74
20. Caputa, M., Kadziela, W., Narebski, J. 1976. Significance of cranial circulation for the brain homeothermia in rabbits. I. The brain-arterial blood temperature gradient. *Acta Neurobiol. Exp.* 36: 613–24
21. Caputa, M., Kadziela, W., Narebski, J. 1976. Significance of cranial circulation for the brain homeothermia in rabbits. II. The role of the cranial venous lakes in the defence against hyperthermia. *Acta Neurobiol. Exp.* 36: 625–38
22. Caputa, M., Kadziela, W., Narebski, J., Tyczyński, M. 1977. Brain-cranial venous blood heat exchanger and hyperthermia in rabbits. *Bull. Acad. Pol. Sci.* 25 (10) 695–98
23. Caputa, M., Perrin, G., Cabanac, M. 1978. Ecoulement sanguin réversible dans la veine ophtalmique: méchanisme de refroidissement sélectif du cerveau humain. *C.R. Acad. Sci.* 87D: 1011–14
24. Carithers, R. W., Seagrave, R. C. 1976. Canine hyperthermia with cerebral pro-

tection. *J. Appl. Physiol.* 40(4): 543–48
25. Clough, D. P., Jessen, C. 1974. The role of spinal thermosensitive structures in the respiratory heat loss during exercise. *Pflügers Arch.* 347: 235–48
26. Coceani, F., Gloor, P. 1966. The distribution of the internal carotid circulation in the brain of the macaque monkey (*Macaca mulatta*). *J. Comp. Neurol.* 128: 419–29
27. Crawshaw, L. I., Nadel, E. R., Stolwijk, J. A. J., Stamford, B. A. 1975. Effect of local cooling on sweating rate and cold sensation. *Pflügers Arch.* 347: 235–48
28. Daniel, P. M., Dawes, J. D. K., Prichard, M. M. L. 1953. Studies of the carotid rete and its associated arteries. *Philos. Trans. R. Soc.* B237: 173–215
29. Davis, D. D., Story, H. E. 1943. The carotid circulation in the domestic cat. *Publ. Field. Mus. (Zool. Ser.)*. 28: 1–47
30. Dawes, J. D. K., Prichard, M. M. L. 1953. Studies of the vascular arrangements of the nose. *J. Anat.* 87: 311–26
31. Flandrois, R., Lacour, J. R., Osman, H. 1971. Control of breathing in the exercising dog. *Respir. Physiol.* 13: 361–71
32. Godynicki, S. 1975. Blood vessels of the nasal cavity in the rabbit. *Folia Morphol. (Warsaw)*. 34(1): 69–76
33. Gross, P. M., Marcus, M. L., Heistad, D. D. 1980. Regional distribution of cerebral blood flow during exercise in dogs. *J. Appl. Physiol.* 48: 213–47
34. Hales, J. R. S., Bligh, J. 1969. Respiratory responses of the conscious dog to severe heat stress. *Experientia* 25: 818–19
35. Hales, J. R. S., Dampney, R. A. C. 1975. The redistribution of cardiac output in the dog during heat stress. *J. Thermal Biol.* 1: 29–34
36. Hales, J. R. S., Fawcett, A. A., Bennett, J. W., Needham, A. D. 1978. Thermal control of blood flow through capillaries and arteriovenous anastomoses in skin of sheep. *Pflügers Arch.* 378: 55–63
37. Hayward, J. N. 1967. Cerebral cooling during increased cerebral blood flow in the monkey. *Proc. Soc. Exp. Biol. Med.* 124: 555–57
38. Hayward, J. N., Baker, M. A. 1968. The role of the cerebral arterial blood in the regulation of brain temperature in the monkey. *Am. J. Physiol.* 215: 389–403
39. Jessen, C., Pongratz, H. 1979. Air humidity and carotid rete function in thermoregulation of the goat. *J. Physiol. London* 292: 469–79
40. Jewell, P. A. 1952. The anastomoses between internal and external carotid circulations in the dog. *J. Anat.* 86: 83–94
41. Kilgore, D. L. 1976. Brain temperature in birds. *J. Comp. Physiol.* 110: 209–15
42. Kilgore, D. L., Bernstein, M. H., Schmidt-Nielsen, K. 1973. Brain temperature in a large bird, the rhea. *Am. J. Physiol.* 225(3): 739–42
43. Kilgore, D. L., Birchard, G. F., Boggs, D. F. 1978. Body-to-brain temperature difference in running quail. *Physiologist* 21(4): 64
44. Kilgore, D. L., Boggs, D. F., Birchard, G. F. 1979. Role of the rete mirabile ophthalmicum in maintaining the body-to-brain temperature difference in pigeons. *J. Comp. Physiol.* 129: 119–22
45. Kindermann, W., Pleschka, K. 1973. Local blood flow and metabolism of the tongue before and during panting in the dog. *Pflügers Arch.* 340: 251–62
46. Kluger, M. J., D'Alecy, L. G. 1975. Brain temperature during reversible upper respiratory bypass. *J. Appl. Physiol.* 38(2): 268–71
47. Krönert, H., Pleschka, K. 1976. Lingual blood flow and its hypothalamic control in the dog during panting. *Pflügers Arch.* 367: 25–31
48. Krönert, H., Wurster, R. D., Pierau, Fr.-K., Pleschka, K. 1980. Vasodilatory response of arteriovenous anastomoses to local cold stimuli in the dog's tongue. *Pflügers Arch.* 388: 17–19
49. Magilton, J. H., Swift, C. S. 1969. Response of veins draining the nose to alar-fold temperature changes in the dog. *J. Appl. Physiol.* 27: 18–20
50. McCaffrey, T. V., Geis, G. S., Chung, J. M., Wurster, R. D. 1975. Effect of isolated head heating and cooling on sweating in man. *Aviat. Space Environ. Med.* 46(11): 1353–57
51. McCaffrey, T. V., McCook, R. D., Wurster, R. D. 1975. Effect of head skin temperature on tympanic and oral temperature in man. *J. Appl. Physiol.* 39(1): 114–18
52. McGrath, P. 1977. Observations on the intracranial carotid rete and the hypophysis in the mature female pig and sheep. *J. Anat.* 124: 689–99
53. Miller, M. E., Christensen, G. C., Evans, H. E. 1964. *Anatomy of the Dog.* Philadelphia: Saunders
54. Negus, V. 1958. *The Comparative Anatomy and Physiology of the Nose and Paranasal Sinuses.* Edinburgh/London: Livingstone
55. Paule, W. J., Baker, M. A. 1978. Structural basis for the heat-exchanger function of the carotid-rete in the cat. *Anat. Rec.* 190: 505

56. Pleschka, K., Kuhn, P., Nagai, M. 1979. Differential vasomotor adjustments in the evaporative tissues of the tongue and nose in the dog under heat load. *Pflügers Arch.* 382: 255–62
57. Prichard, M. M. L., Daniel, P. M. 1953. Arterio-venous anastomoses in the tongue of the dog. *J. Anat.* 87: 66–74
58. Prichard, M. M. L., Daniel, P. M. 1954. Arterio-venous anastomoses in the tongue of the sheep and the goat. *Am. J. Anat.* 95: 203–25
59. Richards, S. A. 1970. Brain temperature and the cerebral circulation in the chicken. *Brain Res.* 23: 265–68
60. Sanders, M., White, F., Bloor, C. 1977. Cardiovascular responses of dogs and pigs exposed to similar physiological stress. *Comp. Biochem. Physiol.* 58A: 365–70
61. Shibolet, S., Lancaster, M. C., Danon, Y. 1976. Heat stroke: a review. *Aviat. Space Environ. Med.* 47(3): 280–301
62. Sugano, Y., Nagasaka, T. 1980. Effects of tracheal bypass breathing on heat balance in rabbits. *Jpn. J. Physiol.* 30: 701–8
63. Taylor, C. R. 1966. The vascularity and possible thermoregulatory function of the horns in goats. *Physiol. Zoöl.* 39(2): 127–39
64. Taylor, C. R. 1977. Exercise and environmental heat loads: different mechanisms for solving different problems? In *MTP International Review of Science, Physiology Series, Environmental Physiology II,* Vol. 7, ed. D. Robertshaw. Baltimore: University Park Press
65. Taylor, C. R., Lyman, C. P. 1972. Heat storage in running antelopes: Independence of brain and body temperatures. *Am. J. Physiol.* 222: 114–17
66. Young, B. A., Bligh, J., Louw, G. 1976. Effect of thermal tachypnoea and of its mechanical or pharmacological inhibition on hypothalamic temperature in the sheep. *J. Thermal Biol.* 1: 195–98

ENERGETICS AND MECHANICS OF TERRESTRIAL LOCOMOTION

C. Richard Taylor and Norman C. Heglund

Museum of Comparative Zoology, Harvard University, Cambridge, Massachusetts 02138

INTRODUCTION

This review examines the link between the energetics and the mechanics of terrestrial locomotion. It addresses a simple question: How do muscles use the metabolic energy they consume during locomotion? It has been generally assumed that once an animal has accelerated and reached a constant speed, most of the energy it consumes is used by its muscles as they shorten and perform mechanical work. It has also been assumed that the muscles perform this work at close to their maximal efficiency—i.e. that the muscles convert ~25% of the metabolic energy liberated (as carbohydrates, fats, and/or proteins are oxidized) into mechanical work (1–3, 7, 10, 28, 31). If these two assumptions are correct, then the rate at which the muscles of animals perform mechanical work should be approximately 25% of the rate at which they consume metabolic energy. In this review we use a comparative approach to evaluate these assumptions, and it leads us to the conclusion that they are incorrect. We propose an alternative explanation for how muscles use metabolic energy to move animals along the ground.

COMPARATIVE APPROACH

Two variables, speed and body size, are used to examine the relationship between the metabolic energy consumed and the mechanical work performed by muscles. Large differences in the rate of metabolic energy consumption by muscles can be obtained with either variable. For example, the rate at which metabolic energy is consumed by muscles of an individual animal increases by more than ten-fold with increasing speed of locomotion. Also the rate of metabolic energy consumption by each gram of muscle of animals moving at the same speed differs by more than ten-fold with body size (over the size range where measurements have been made). The rate of mechanical work should change in parallel with rate of metabolic energy consumption if the assumptions linking the two are correct. Ten-fold differences should be large enough either to establish or to disprove the link between the energetics and mechanics of locomotion.

RATE OF METABOLIC ENERGY CONSUMPTION AS A FUNCTION OF SPEED AND BODY SIZE

Twelve years ago, Taylor et al (38) observed that the rate at which metabolic energy was consumed by running mammals increased nearly linearly with speed (V) and scaled as a regular function of body mass (M_b). A simple empirically based equation was formulated that allowed the rate of energy consumption of a mammal moving along the ground to be calculated from these two easily measured parameters. The initial data base for this equation was small, but during the last twelve years it has been expanded and now includes at least 63 mammalian species (representing 9 mammalian orders and 28 families) ranging in body mass from 7 g to 250 kg (20, 35, 36). Similar data have been obtained from birds (19). Recently, Fedak & Seeherman (20) reported no difference in energetic cost of terrestrial locomotion between birds and mammals of the same body mass. This finding allows the data from the birds and mammals to be combined into a single relationship.

A single equation has been formulated for predicting energetic cost of locomotion for terrestrial birds and mammals using data from 73 avian and mammalian species (36):

$$\dot{V}_{O_2}/M_b = 0.533 \ M_b^{-0.316} \cdot V + 0.300 \ M_b^{-0.303} \qquad 1.$$

where $\dot{V}_{O_2}/M_b$ has the units ml $O_2 \cdot s^{-1} \cdot kg^{-1}$; M_b is in kg; and V is in m·s^{-1}. Rates of oxygen consumption can be converted into rates of energy consumption, $\dot{E}_{metab}/M_b$ using the conversion factor of 1 ml O_2 = 20.1 J:

$$\dot{E}_{metab}/M_b = 10.7 \ M_b^{-0.316} \cdot V + 6.03 \ M_b^{-0.303} \qquad 2.$$

where $\dot{E}_{metab}/M_b$ has the units of watts·kg^{-1}. Lions and red kangaroos were not included in this allometric equation because their energy consumption did not increase linearly with speed over a wide range of speeds (11, 14). Birds that waddled were not included because there was a large additional component to their energetics that was unique (33).

Two thirds of the values calculated from this general equation fall within 25% of the observed value at the middle of the speed range where measurements were made. This agreement is impressive when one considers that the mass specific rate of oxygen consumption differed by more than 1,400% over this size range of animals. It is interesting to note that the values calculated using the new equation are changed little from those calculated using the equation reported for mammals 12 years ago, despite the large amount of additional data. It seems safe to conclude that one equation predicts reasonably well the rate at which birds and mammals consume energy as they move along the ground, and that this equation will change little as additional data are included.

RATE AT WHICH MUSCLES PERFORM MECHANICAL WORK AS A FUNCTION OF SPEED AND BODY SIZE

What "work" do the muscles do when animals move along the ground at a constant speed? The mechanical energy of an animal oscillates during each stride, and muscles must perform mechanical work to replace energy lost as heat during these oscillations. Oscillations occur both as the center of mass of the animal rises and falls and accelerates and decelerates during each stride; and also as the limbs and various segments of the animal accelerate and decelerate relative to its center of mass during the stride.

The total mechanical energy of a running animal can be described at any particular instant as the sum of: the kinetic and gravitational potential energy of the center of mass, $E_{CM_{(tot)}}$; plus the kinetic energy of elements of the body relative to the center of mass, $E_{KE_{(tot)}}$; plus the elastic strain energy (elastic potential energy) of the system, E_{ES}. This can be written as the equation:

$$E_{tot} = E_{CM_{tot}} + E_{KE_{tot}} + E_{ES}. \qquad 3.$$

Measurements have been made of $E_{CM_{tot}}$ and $E_{KE_{tot}}$ of an animal running at a constant speed (1–4, 7–10, 12, 16–18, 22, 24, 25, 30). As yet, no accurate method has been devised for measuring E_{ES}. Therefore, it is necessary to neglect E_{ES} and to assume that it does not change during a stride. However, we must keep this assumption in mind when interpreting the data. This assumption is discussed below in the section on elastic storage and recovery of energy.

If one neglects E_{ES}, then the only way animals can increase E_{tot} during a stride is by using their muscles to convert stored chemical energy into mechanical work. Limits can be set on the rate at which muscles perform this work.

The upper limit for $\dot{E}_{tot}$ is obtained by assuming that there is no transfer of energy between $E_{CM_{tot}}$ and $E_{KE_{tot}}$ during a stride. It is calculated by: adding all of the increments in $E_{CM_{tot}}$ and $E_{KE_{tot}}$ separately over a stride; dividing each sum by the stride period to obtain $\dot{E}_{CM_{tot}}$ and $\dot{E}_{KE_{tot}}$; and then adding these two terms. $\dot{E}_{tot}$ for human locomotion has been calculated in this manner (8, 9, 17, 21, 22).

It is possible to obtain a general equation relating the upper limit for $\dot{E}_{tot}/M_b$ as a function of speed and body size for birds and mammals, because both $\dot{E}_{CM_{tot}}/M_b$ and $\dot{E}_{KE_{tot}}/M_b$ have been measured as functions of speed and body size. One simply has to add the equations for $\dot{E}_{CM_{tot}}/M_b$ and $\dot{E}_{KE_{tot}}/M_b$ to obtain an equation for $\dot{E}_{tot}/M_b$.

$\dot{E}_{CM_{tot}}/M_b$ has been measured in a variety of large (7–10) and small (24) birds and mammals. In all of these animals, it increased with approximately the same linear function of speed and was independent of the size of the animals. It could be represented by the equation (24):

$$\dot{E}_{CM_{tot}}/M_b = 0.685\ V + 0.072 \qquad 4.$$

where $\dot{E}_{CM_{tot}}/M_b$ has the units of watts kg^{-1} and V is the speed in m·s^{-1}.

$\dot{E}_{KE_{tot}}$ has also been measured in a variety of birds and mammals ranging in body mass from 30 g to 100 kg (18). It increased with approximately the same curvilinear function of speed in all of the animals and was also independent of body size. It could be represented by the equation:

$$\dot{E}_{KE_{tot}}/M_b = 0.478\ V^{1.53} \qquad 5.$$

where $\dot{E}_{KE_{tot}}/M_b$ has the units of watts kg^{-1} and V is speed in m·s^{-1}.

Equations 4 and 5 can be added to obtain an equation for the upper limit of $\dot{E}_{tot}/M_b$ as a function of speed:

$$\dot{E}_{tot}/M_b = 0.478\ V^{1.53} + 0.685\ V + 0.072 \qquad 6.$$

where $\dot{E}_{tot}/M_b$ has the units of watts kg^{-1} and V is the speed in m·s^{-1}. This equation, like equations 4 and 5, applies equally well to birds and mammals over the entire size range where measurements have been made (30 g–100 kg).

The lower limit for $\dot{E}_{tot}$ is obtained by assuming complete transfers of energy between $E_{CM_{tot}}$ and $E_{KE_{tot}}$ within each stride. It is calculated by adding $E_{CM_{tot}}$ and $E_{KE_{tot}}$ at each instant during the stride to obtain E_{tot}. All of the increments in E_{tot} during the stride are then added, and the sum is divided by the stride period to obtain $\dot{E}_{tot}$. This procedure has been used for humans and animals (4, 12, 16). No transfer of energy between $E_{CM_{tot}}$

and $E_{KE_{tot}}$ can take place during the aerial phase of a stride. However, during the stance phase some exchange probably occurs. For example, when the foot lands, some of the decrease in energy as the center of mass slows (a decrease in $E_{CM_{tot}}$) is probably used to accelerate the limb forwards relative to the center of mass (an increase in $E_{KE_{tot}}$).

Heglund et al (25) found the difference between the two limits for $\dot{E}_{tot}$ was small at low to modest speeds, but reached $\cong 35\%$ of the upper limit during high-speed galloping. The magnitude of the difference was similar in galloping chipmunks and dogs and appeared to be independent of body size. The rate at which muscles must convert chemical energy into work to increase E_{tot} during a stride falls somewhere within these limts.

In addition to performing work to increase E_{tot} within a stride, muscles also perform work against friction, and when antagonistic muscles work against each other. The frictional losses have been shown to be small at all but the highest speeds in terrestrial locomotion (27, 34). For example, in humans, which present a large frontal area to the air during running, wind resistance accounts for less than 2% of the total mechanical power expended at 2.8 m·s^{-1} and less than 8% at 8.3 m·s^{-1} (8, 27). Losses against the ground are zero unless the animal is slipping (e.g. running on sand). Alexander & Vernon (4) have calculated that the work by antagonistic muscles could account for only 15% of the total positive work performed by a kangaroo hopping at 5.5 m·s^{-1}. Because these forms of work appear to be small in comparison with E_{tot}, they can reasonably be ignored for the purposes of this discussion.

EFFICIENCY OF MUSCLES DURING LOCOMOTION

Muscular efficiency, calculated as a ratio of the rate at which muscles perform work ($\dot{E}_{tot}/M_b$) to the rate at which they consume metabolic energy ($\dot{E}_{metab}/M_b$), increases both with increasing speed and with increasing body size (Figure 1, right). These increases occur because $\dot{E}_{tot}/M_b$ increases curvilinearly with ground speed and is independent of body size (Figure 1, middle), while $\dot{E}_{metab}/M_b$ increases linearly with ground speed and is proportional to $M_b^{-0.33}$ (Figure 1, left). Thus each gram of tissue of a 100 g quail or chipmunk running at 4 m·s^{-1} consumes metabolic energy at a rate about 10 times that of a 100 kg ostrich, horse, or human running at the same speed, while their muscles are performing work at about the same rate. The highest efficiency observed in the 44 g quail was about 7%, while the efficiency of the 70 kg human reached 73% (25). Therefore the changes in metabolic energy consumption observed with changing speed and body size are not the result of parallel changes in mechanical work performed by muscles. It should be noted that the results of Alexander and his colleagues are consistent with this conclusion (1, 3).

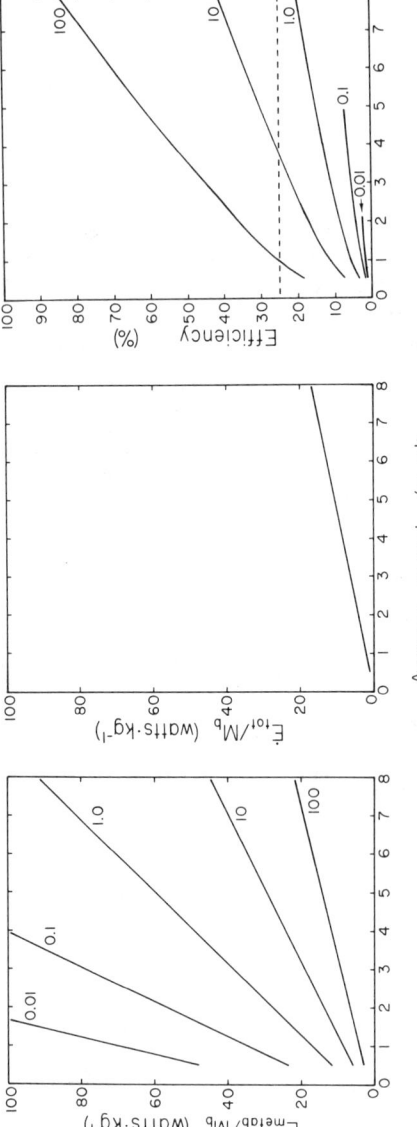

Figure 1 Mass-specific rates at which animals consume metabolic energy ($\dot{E}_{metab}/M_b$, left graph), and their muscles perform mechanical work ($\dot{E}_{tot}/M_b$, center graph) are plotted as a function of speed of locomotion for animals of different size. Percent efficiency, calculated as the ratio of $\dot{E}_{tot}/M_b$ to $\dot{E}_{metab}/M_b$ times 100, is also plotted as a function of speed (right-hand graph). The values for $\dot{E}_{metab}/M_b$ and $\dot{E}_{tot}/M_b$ were calculated using equations 2 and 6, respectively, in the text.

STORAGE AND RECOVERY OF ENERGY IN ELASTIC ELEMENTS DURING LOCOMOTION

An efficiency of greater than 25% (dotted line in Figure 1) can be interpreted as demonstrating that storage of energy in elastic elements occurs in one part of the stride as E_{tot} decreases and that this energy is recovered as useful work in another (as E_{tot} increases). The maximal efficiency that muscles can achieve while performing work without a prestretch (i.e. without the help of elastic energy storage and recovery) is about 25% for most muscles. These values have been obtained in experiments on both isolated muscles (6, 23, 28, 29) and whole animals (15, 32, 35).

Muscle physiologists and biochemists often consider efficiency of muscles as the ratio of mechanical work performed to the energy obtained from ATP instead of from carbohydrates, fats, and/or proteins. In these terms, muscles are much more efficient than 25%, because ½–⅔ of the energy contained in the carbohydrates, proteins, and fats is lost as heat during the formation of ATP via aerobic pathways. If one considers the efficiencies given in Figure 1 in terms of the energy contained in ATP that is converted into mechanical work, we would have to multiply the values by 2–3, obtaining efficiencies of more than 200% for large animals running at high speed. Obviously, it is impossible to achieve an efficiency of greater than 100%. Under these circumstances more than half of the observed energy decreases in E_{tot} must be stored as elastic energy in one part of the stride and recovered as increases in E_{tot} in another.

These values demonstrate the importance of storage and recovery of elastic energy in larger animals, but they do not tell us anything about elastic storage of energy in small animals. The same relative amount of kinetic and gravitational energy may be stored as elastic energy in one part of the stride and recovered in another in small animals as in large. Other factors may be responsible for the higher rates of metabolic energy consumption by the muscles of smaller animals. However, a recent study (5) suggests that elastic storage may increase with increasing size: The tendons of small kangaroo rats are relatively thicker than those of the larger wallabies and kangaroos; as a result, the kangaroo rats store a much smaller fraction of the decrements in E_{tot} when they land, than has been observed in the larger hopping animals (4). The size dependency of storage and recovery of elastic energy clearly needs more study.

ENERGETIC COST OF GENERATING FORCE DURING LOCOMOTION

How do muscles use the energy they consume? It seems clear that the rate at which muscles perform mechanical work does not determine the rate at which animals consume energy during locomotion. What else could do so? Muscles are active, generate force, and consume energy not only when they shorten and perform mechanical work (work = force X distance), but also when their length is unchanged as they stabilize joints (distance is zero and therefore work is zero), and when they are stretched (work is done on the active muscles, distance has a negative sign, and work is negative). It seems possible that the metabolic cost of generating muscular force, irrespective of whether mechanical work is performed by the muscle, determines the rate at which animals consume energy during locomotion.

Does the metabolic cost of generating muscular force change with speed and body size in a manner parallel to the metabolic cost of locomotion? Taylor et al (37) have recently investigated the metabolic cost of generating force in running animals (rats, dogs, humans, and horses) by measuring the metabolic cost of carrying loads. They found that at any speed, the metabolic rates increased in direct proportion to the mass supported by the muscles (Figure 2). They also found that the average accelerations of the center of mass of the animals did not change when they carried the load. Thus the muscular forces generated by the muscles increased in nearly direct proportion to the mass of the load (Force = mass X acceleration).

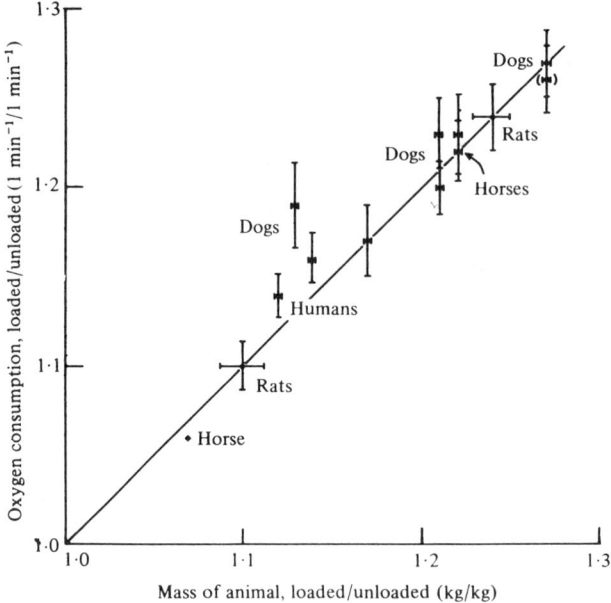

Figure 2 Oxygen consumption increases in direct proportion to the mass of a load carried by large and small animals irrespective of speeds of locomotion. Reproduced from (37) with the permission of the publisher.

It follows that the rate of metabolic energy consumption by the muscles of an animal as it runs along the ground at any speed is nearly directly proportional to the force exerted by its muscles.

Determined in this manner, the metabolic cost of generating force by each gram of muscle over an interval of time ($\int F dt$) increases linearly with speed in exactly the same way as metabolic cost of locomotion. Thus cost of generating a given force by the muscles of an individual animal increases by more than ten-fold with increasing speed in the same way as cost of locomotion.

The metabolic cost of generating muscular force over time also increases with decreasing body size, being proportional to $M_b^{-0.33}$. Thus cost of generating a given force increases by more than ten-fold with decreasing body size in the same way as cost of locomotion.

Why does energetic cost of generating force increase with increasing speed? Muscular force must be generated and decay more rapidly as an animal increases its speed. This is accomplished, at least in part, by recruiting faster muscle fibers that have more rapid rates of actin-myosin cross-bridge cycling. Since each cross-bridge cycle consumes a unit of energy, the increase in energy cost of locomotion with speed could be the result of the recruitment of fibers with faster cycling times at higher speeds.

Similar reasoning could be used to explain the higher cost of locomotion in smaller animals. Muscular force must be generated and decay more rapidly in small animals than in large ones because the small animal takes more steps per unit time to move at the same speed (26). Equivalent muscles of small animals have faster fibers with more rapid cross-bridge cycling rates than those of large animals (13). This decrease in rate of cross-bridge cycling with increasing body size could be responsible for the higher metabolic costs for locomotion in smaller animals.

The suggestion that the intrinsic velocity of shortening of the active muscle motor units involved in locomotion determines metabolic cost of locomotion is testable. Future studies will determine whether or not it is valid and should more completely identify the factors that determine the metabolic cost of locomotion in terrestrial birds and mammals.

SUMMARY

This review addresses a simple question: How do muscles use the energy they consume during terrestrial locomotion? Using a comparative approach, it was found that the mass-specific rate of metabolic energy consumption changed by more than ten-fold with body size, while the mass-specific rate at which the muscles performed mechanical work did not change at all. It was also found that the rate of metabolic energy consumption increased linearly with speed, while the rate at which muscles performed mechanical work increased curvilinearly with speed ($\propto V^{1.53}$). We conclude from these observations that the rate at which animals consume metabolic energy during terrestrial locomotion is not determined by the rate at which their muscles perform mechanical work.

Instead, the metabolic cost of generating muscular force over time ($\int F\, dt$) appears to determine the metabolic cost of terrestrial locomotion. The cost of generating force increases with increasing speed and decreases with increasing body size in exactly the same manner as cost of locomotion.

It is suggested that the metabolic cost of generating muscular force may be determined by the intrinsic velocity of shortening (i.e proportional to rates at which the cross-bridges between actin and myosin cycle) of the muscle motor units that are active during locomotion. Faster motor units are used both as animals increase speed and in equivalent muscles of smaller animals moving at the same speed. This suggestion is testable and future studies should determine whether or not it explains the higher costs of generating muscular force with increasing speed and decreasing body size.

ACKNOWLEDGMENTS

The writing of this review and the work from the authors' laboratory were supported by NSF grant #PCM7823319, NIH grant #AM18140, and NIH postdoctoral grant #AM06022.

Literature Cited

1. Alexander, R. McN. 1977. Terrestrial locomotion. In *Mechanics and Energetics of Animal Locomotion,* ed. R. McN. Alexander, G. Goldspink, pp. 168–203. London: Chapman & Hall. 346 pp.
2. Alexander, R. McN. 1980. Optimum walking techniques for quadrupeds and bipeds. *J. Zool. Lond.* 192:97–117
3. Alexander, R. McN., Jayes, A. S., Ker, R. F. 1980. Estimates of energy cost for quadrupedal running gaits. *J. Zool. Lond.* 190:155–92
4. Alexander, R. McN., Vernon, A. 1975. The mechanics of hopping by kangaroos (Macropodidae). *J. Zool. Lond.* 177:265–303
5. Biewener, A. A., Alexander, R. McN., Heglund, N. C. 1981. Elastic energy storage in the hopping kangaroo rats (*Dipodomys spectibilis*). *J. Zool. Lond.* In press
6. Cavagna, G. A., Citterio, G., Jacini, P. 1980. The two-fold elastic behavior of striated muscle. *Proc. Int. Union Physiol. Sci.* 14:351
7. Cavagna, G. A., Heglund, N. C., Taylor, C. R. 1977. Mechanical work in terrestrial locomotion: two basic mechanisms for minimizing energy expenditure. *Am. J. Physiol.* 233:R243–61
8. Cavagna, G. A., Kaneko, M. 1977. Mechanical work and efficiency in level walking and running. *J. Physiol. London* 268:467–81
9. Cavagna, G. A., Saibene, F. P., Margaria, R. 1964. Mechanical work in running. *J. Appl. Physiol.* 19:249–56
10. Cavagna, G. A., Thys, H., Zamboni, A. 1976. Sources of external work in level walking and running. *J. Physiol. London* 262:639–57
11. Chassin, P., Taylor, C. R., Heglund, N. C., Seeherman, H. J. 1976. Locomotion in lions: energetic cost and maximum aerobic capacity. *Physiol. Zool.* 49:1–10
12. Clark, J., Alexander, R. McN. 1975. Mechanics of running by quail (*Coturnix*). *J. Zool. Lond.* 176:87–113
13. Close, R. I. 1972. Dynamic properties of mammalian skeletal muscles. *Physiol. Rev.* 52:129–97

14. Dawson, T. J., Taylor, C. R. 1973. Energetic cost of locomotion in kangaroos. *Nature* 246:313–14
15. Dickinson, S. 1929. The efficiency of bicycle pedalling, as affected by speed and load. *J. Physiol. London* 67:242–55
16. Elftman, H. 1939. The force exerted by the ground in walking. *Arbeitsphysiologie* 10:485–91
17. Elftman, H. 1966. Biomechanics of muscle with particular application to studies of gait. *J. Bone Joint Surg.* 48A:363–77
18. Fedak, M. A., Heglund, N. C., Taylor, C. R. 1982. Energetics & mechanics of terrestrial locomotion. II. Kinetic energy changes of the limbs & body as a function of speed and body size in birds & mammals. *J. Exp. Biol.* In press
19. Fedak, M. A., Pinshow, B., Schmidt-Nielsen, K. 1974. Energy cost of bipedal running. *Am. J. Physiol.* 227:1038–44
20. Fedak, M. A., Seeherman, H. J. 1979. Reappraisal of energetics of locomotion shows identical cost in bipeds & quadrupeds including ostrich & horse. *Nature* 282:713–16
21. Fenn, W. O. 1930. Frictional & kinetic factors in the work of sprint running. *Am. J. Physiol.* 92:583–611
22. Fenn, W. O. 1930. Work against gravity due to velocity changes in running. *Am. J. Physiol.* 93:433–62
23. Heglund, N. C., Cavagna, G. A. 1981. Oxygen consumption, mechanical work & efficiency in isolated muscle. In press
24. Heglund, N. C., Cavagna, G. A., Taylor, C. R. 1982. Energetics & mechanics of terrestrial locomotion. III. Energy changes of the center of mass as a function of speed & body size in birds & mammals. *J. Exp. Biol.* In press
25. Heglund, N. C., Fedak, M. A., Taylor, C. R., Cavagna, G. A. 1982. Energetics & mechanics of terrestrial locomotion. IV. Total mechanical energy changes as a function of speed & body size in birds & mammals. *J. Exp. Biol.* In press
26. Heglund, N. C., Taylor, C. R., McMahon, T. A. 1974. Scaling stride frequency & gait to animal size: mice to horses. *Science* 186:1112–13
27. Hill, A. V. 1927. The air resistance to a runner. *Proc. R. Soc. London Ser. B* 102:380–85
28. Hill, A. V. 1950. The dimensions of animals and their muscular dynamics. *Sci. Prog.* 38:209
29. Kushmerick, M. J., Larson, R. E., Davies, R. E. 1969. The chemical energies of muscle contraction. 1. Activation heat, heat of shortening and ATP utilization of activation-relaxation process. *Proc. R. Soc. London Ser. B* 174:293–314
30. Manter, J. T. 1938. The dynamics of quadrupedal walking. *J. Exp. Biol.* 15:522–40
31. McMahon, T. A. 1975. Using body size to understand the structural design of animals: quadrupedal locomotion. *J. Appl. Physiol.* 39:619–27
32. Margaria, R. 1976. *Biomechanics and Energetics of Muscular Exercise.* Oxford: Clarendon. 146 pp.
33. Pinshow, B., Fedak, M. A., Schmidt-Nielsen, K. 1977. Terrestrial locomotion in penguins: it costs more to waddle. *Science* 195:592–94
34. Pugh, L. G. C. E. 1971. The influence of wind resistance in running and walking and the efficiency of work against horizontal or vertical forces. *J. Physiol. London* 213:255–76
35. Taylor, C. R. 1980. Mechanical efficiency of terrestrial locomotion: a useful concept? In *Aspects of Animal Movement*, ed. H. Y. Elder, E. R. Trueman, pp. 235–44. Cambridge: Cambridge Univ. Press. 250 pp.
36. Taylor, C. R., Heglund, N. C., Maloiy, G. M. O. 1982. Energetics and mechanics of terrestrial locomotion. I. Metabolic energy consumption as a function of speed and body size in birds and mammals. *J. Exp. Biol.* In press
37. Taylor, C. R., Heglund, N. C., McMahon, T. A., Looney, T. R. 1980. Energetic cost of generating muscular force during running: a comparison of large and small animals. *J. Exp. Biol.* 86:9–18
38. Taylor, C. R., Schmidt-Nielsen, K., Raab, J. L. 1970. Scaling of energy cost of running to body size in mammals. *Am. J. Physiol.* 219:1104–7

AVIAN FLIGHT ENERGETICS

J. M. V. Rayner

Department of Zoology, University of Bristol, Woodland Road, Bristol
BS8 1UG, England.

INTRODUCTION

Advances in the study of energy consumption in the flying bird have been limited by the difficulties involved in assessing flight performance. Only with high-speed cinematography, improved aerodynamic theory, and other improvements in technology and experimental technique has it become possible to move towards a balanced overview of the energetics of avian flight adaptations. Flight places substantial metabolic and mechanical demands on a bird. The energy rate (power) required for flight activity, the available fuel, and the rate of acquisition of new fuel reserves are significant items in the energy budget of a bird: Limitations on performance set by energetic factors also limit ecological success, and there is substantial pressure to optimize flight performance through physiological and morphological adaptations and appropriate flight strategies (32, 45). This paper reviews recent insights into flight energetics and flight performance optimization. Space does not permit general coverage of flight biology, which is widely discussed elsewhere [mechanics (27, 36, 44), physiology (5, 9), energetics (26, 58)]. Berger & Hart (5) discuss many aspects of internal energy metabolism that are crucial to flight but are not discussed here; they provide a comprehensive introduction to the literature on physiological flight energetics up to about 1972.

FLIGHT ENERGETICS

Flight energy can be measured by assessing material changes (5). Mass loss (1, 5, 25) and labelled isotope (5, 8, 19) experiments allow no control of flight behavior and generally give unusually low results. Gas exchange while the animal flies in controlled conditions can be recorded (3-6, 9, 10, 15, 23, 47, 49-53) but the problems involved are considerable, and flight physiology may be unrealistic (9, 10, 18, 55). Respiration in free-flying birds can be telemetered (5, 11) and it should soon be possible to obtain better measurements on large birds by this method (11). Since experimental determination of flight energy is difficult and provides only limited information, there has been much interest in theoretical models aiming to estimate energy consumption from readily measured characteristics. Realistic models allow flight performance to be assessed freely, but are of necessity abstractions, and their practical value is limited by the balance between the breadth of the phenomena they aim to describe, their own intrinsic accuracy, and their complexity. Present aerodynamic theories model a flying bird by appropriate features of aircraft design. Analysis is limited to simple extensions of blade-element theory (see 44), for little is known of the unsteady air flows involved in propulsion by flapping wings; however, such effects do not appear significantly to affect performance estimates (41, 42).

Forward Flight

The typical flight mode of birds is steady flapping flight, but many species have evolved adaptations enabling them to economize in their flight performance. Power consumption in flight represents the rate of transfer of energy from a bird's fuel reserves into the environment; this flow involves a variety of processes, and power can be partitioned into a number of components (36, 42). The most significant are the mechanical or aerodynamic powers: parasite power (work against body drag), profile power (work against form and friction drag on the wings), and induced power (work to generate lift and thrust). Energy transferred as profile and parasite powers appears as random turbulent air flows in the bird's wake, but induced power generates an ordered momentum-carrying structure (42); this mechanical work is produced entirely by contraction of the pectoral muscles. Parasite and profile powers should both increase with speed, V, whereas at high speeds forward momentum assists weight support, and induced power falls, approximately as V^{-1}(34, 42). The total mechanical power varies with speed as a U-shaped curve with a shallow minimum in the region around which there is little change in total power; at both extremes of the speed range a significant rise cannot be avoided. Further energy is required for the bird's internal functions [the basal metabolic rate is used, perhaps inappropriately, to estimate this (36, 54)] and for any "postural" cost of flight [estimated as 10% of mechanical work (54)]. When a stable body temperature has been reached, this energy, together with a significant proportion of the energy from fuel combustion, is dissipated as heat, partly by radiation and partly by respiration (5, 50). The muscular heat energy is usually described as the portion wasted in inefficient muscle contraction; but in part it is valuable since it maintains the high body temperatures necessary for efficient energy transfers in flight (5, 23, 50, 51), and the term "inefficient" is inappropriate.

There is as yet no accurate means of estimating the changes in internal metabolism associated with flight; simple estimates may be highly misleading, and it is hoped that future experiments will provide more information.

A number of theoretical models of flight energy have been proposed: Pennycuick (34, 36, 37), Tucker (54, 55), and Greenewalt (18) have developed simple formulae that use minimum input data to predict metabolic flight power as it changes with speed in an individual bird. Pennycuick's model, based on a detailed analysis of flapping flight in the pigeon (33), oversimplifies the problem by assuming that systematic variations in kinematics and lift coefficient indirectly result in constant profile power, and that interspecific differences in wing area do not affect power owing to compensating changes in kinematics. This simple model can readily be used, for instance, to illustrate optimal migration strategies (1, 34, 36, 37); it has found frequent use (31, 39, 40, 56, 57). Tucker's model is closely based on Pennycuick's, but contains a number of untested assumptions about size variation of aerodynamic parameters; it agrees well with some metabolic measurements, but is probably no more accurate than Pennycuick's when applied to birds as a whole. Greenewalt's model (18), based on the dynamics of fixed-wing aircraft, uses unreliable experimental results to estimate unknown coefficients and is too unrealistic for practical use.

To give a more complete description of the forces involved in flapping flight, I have proposed a model, with similar energy partitioning to those described above, that calculates the mechanical power components more accurately (42, 43, 44). Its main disadvantage is that the computations cannot be simplified, as many variables are used to describe kinematics. With unchanging wing kinematics the power-velocity $[P(V)]$ curve has a more pronounced minimum than those of the simpler models, but kinematics should (and do) vary with speed so that power consumption is always minimum (42). Allowing for this, the true $P(V)$ curve has a very flat minimum (45); in agreement with Pennycuick's prediction, profile power is near constant over much of the size range for an individual bird.

For comparison with these models there are a limited number of metabolic power curves determined from wind tunnel experiments (6, 12, 49–53). If muscular efficiency is assumed to be around 25%, experimental flight power agrees with predictions from the Pennycuick/Tucker model, but the shape of the experimental curves is usually rather flatter than expected from theory; in some experiments there is little or no change in power over a range of speeds. Torre-Bueno (51) attributes this to deficiencies in the theoretical model, but there is little reason to expect that the curves should necessarily correspond (45). Scope for error in both theory and experiment is great, and there are too few experimental results available to indicate either their accuracy or their generality. Metabolic power measurements record the total energy used by the bird, comprising both mechanical and thermal energy flows. The quantity of heat released by muscle activity depends on muscular efficiency, which has been estimated as 22–29% (6), so that excess heat is about three times mechanical work. Efficiency variations within this range could explain the discrepancy in shape between theoretical and experimental curves (e.g. 51), for birds flying at different speeds have different wing beat kinematics (33, 45) and muscle contraction rates, and therefore efficiency (16) probably varies with speed (45). Variations in efficiency also result in variations in heat production, thereby possibly altering body temperature and in turn the efficiency of internal energy flows (50). Lack of data on metabolic efficiency remains a significant barrier to understanding the energy processes occurring in flight, but it is not easy to suggest means of determining it experimentally. Clearly the processes involved in the production and regulation of internal energy flows must be studied in detail before an accurate estimate of efficiency can be included in theoretical models of flight energetics.

The configuration of the $P(V)$ curve has a number of important consequences. At low speeds induced power dominates, while at high speeds profile and parasite powers are greater; this places conflicting demands on body and wing shape: birds flying slowly or hovering (see below) must be small, ideally with long wings to minimize induced power, while at the other extreme fast flight demands a streamlined body and relatively small wings (18, 43, 45). Flight design reflects pressures to optimize energy consumption: dimensions are determined by behavior, and should therefore be related to the animal's ecology and feeding behavior. This is undoubtedly true both for birds in general [(32, 34); J. M. V. Rayner, in preparation] and at the specific level: differences in feeding behavior in related species are systematically reflected in wing morphology (13, 14, 31, 32). The $P(V)$ curve also defines characteristic speeds [(34)]: V_{mp} is the speed at which power is minimum and a bird can remain airborne for the longest time for given energy reserves; the maximum range speed, V_{mr}, allows the bird to cover the biggest air distance for available energy (least cost of transport) and is the typical speed of migrating birds. In other situations optimum flight speed varies: To maximize energy gain, birds foraging from a central point should fly faster than V_{mr} (30), while aerial feeding species should fly slightly faster than V_{mp} (32).

Dependence on Size

Many physiological variables depend on size, and it is usual to relate them to body mass, M, by a scaling equation. For instance, basal metabolism in birds is approximately proportional to $M^{0.72}$ (26). Flying birds occupy a size range from 2 g to about 12 kg, and are well suited to allometric analysis;

aerodynamic constraints on flight are strict and dictate near-isometric changes in body proportions with size (18, 27, 32, 36), with wing area $\propto M^{2/3}$ and span $\propto M^{1/3}$. Such relations describe the "average" bird as having constant proportions; they emphatically do not mean that all birds of the same size will have the same shape. As a consequence, flight speed scales as $M^{1/6}$, and mechanical power as $M^{7/6}$, assuming that all birds behave similarly (36). (Flight metabolism includes a size-related component for basal metabolism, and will have a slightly lower index than mechanical power.) Owing to the various selection pressures acting on birds of different size and different ecology there are systematic departures from isometric scales (27): Individual species can have radically different wing proportions than expected, and morphologic scaling indexes are usually higher than isometric [(18, 43); J. M. V. Rayner, in preparation]. These departures occur primarily as a response to energetic constraints to optimize performance, and have the result that $V_{mr} \propto M^{0.127}$ and $P_f = P(V_{mr}) \propto M^{1.038}$; the indexes are significantly lower than those for isometric design (J. M. V. Rayner, unpublished results). Flight performance is limited by maximum available power, P_a: Pennycuick (34) argues that this scales as $M^{2/3}$, approximately a constant multiple of resting metabolism. The difference between P_f and P_a represents the scope for increased activity above normal flight (hovering, rapid manoeuvres, etc); as mass increases, this difference drops and falls to zero at the maximum mass at which flight is possible [$\sim$12 kg; (36)]. As this mass is approached, the range of available speeds is reduced, and large birds may only be able to fly at V_{mp} or rely on soaring (35).

Measurements of metabolic power loosely support these arguments, but data are scanty, badly scattered, and include no birds larger than ducks (1 kg). With most recent reliable data I have calculated P_f as 64.7 $M^{0.882}$ (units W and kg). The index is somewhat greater than previous estimates (5, 9, 26), which suggested that P_f was a constant multiple of basal metabolism. As much of this energy is mechanical work, an index approaching 1 is to be expected: A lower value would suggest that efficiency varies with size to favor larger birds. Although the data are few, it seems clear that the power/mass ratio, P_f/M, and the cost of transport both fall as size increases, and despite assertions to the contrary (5, 26) the ratio of activity metabolism to resting (metabolic scope) is not constant, but increases with size. The best estimate is:

$$P_f/P_b = (64.7\ M^{0.88}) / (3.79 M^{0.72}) = 17.1\ M^{0.16} \qquad 1.$$

in non-passerines (10.4 $M^{0.16}$ in passerines, but size range is limited). Small birds may only be able to achieve 4 times basal metabolism in steady flight;

the largest birds might reach 25 times basal metabolism, but still be incapable of sustained powered flight. However, equations of this kind only describe the broad effect of size variation, and flight adaptations in an individual can represent substantial deviations from the average; predictions from their use can be highly misleading.

FLIGHT STRATEGIES

Flight energetics is far more complex than can be indicated in this brief statement of the major factors involved. In this final section I briefly discuss ways in which performance can be improved and in which flight energy influences foraging behavior.

Gliding and Soaring

Gliding is the predominant response of larger birds to energetic demands. In a steady glide the bird descends relative to the air, but the wings generate no thrust and little energy is used. Pennycuick (35, 36) has estimated the cost of isometric tension in the pectoral muscles as equal to basal metabolism; this is confirmed by measurements of respiration in herring gulls (3) and barnacle geese (11) and by recordings of electrical activity in the muscles of gulls (17). In large birds, flapping flight demands a greater multiple of resting energy than in small birds, and thus for them gliding can be very much cheaper than flapping flight. Flights in which glides alternate with climbs can be very economical (40); if the bird exploits rising air masses such as convecting thermals it needs little energy—only that required to hold the wings outstretched—to cover great distances (35, 36). Very large birds (e.g. vultures) can forage in this way only (38); other species use thermal soaring when feasible, but revert to flapping when thermals are insufficient (39).

Hovering Flight

Hovering—i.e. flight stationary relative to the surrounding air—is the most energy-intensive mode of flight, and is atypical of birds as a whole. The air movements that support the bird's weight in this mode must be generated entirely by the wing beats, and induced power is large. Total mechanical power is much larger in hovering than in forward flight; and unless they can use it to exploit energy-rich food resources, birds hover only for very brief periods. The airflow created by the wing beat is primarily a downward jet of air, which supports the animal's weight. It has been modelled ideally as the so-called momentum jet and is similar to the airflow generated by a propellor or a helicopter rotor (12, 13, 18, 20); but a flapping wing cannot generate a continuous jet, and a more complicated theoretical description is required (41, 43). In most birds, induced power is substantial and precludes sustained hovering, which is only possible with highly specialized morphological and physiological adaptations to minimize induced power and to sustain the high power outputs required. Hummingbirds (mass $\sim$ 2–30 g) are the best known group capable of sustained hovering. Their metabolism has been measured (4, 5, 12, 47, 58), and their feeding behavior and morphology have been shown to be determined largely by energetic factors (13, 14, 18, 20, 58). Raptors and petrels often appear stationary in the air and are frequently described as hovering, but they are invariably flying slowly into wind or in an updraft; like most other birds, they are incapable of true hovering (41, 43, 56). Apart from hummingbirds, the only known bird capable of sustained hovering is the pied kingfisher *Ceryle rudis* ($M = 90$ g), which hovers for long periods over open water where it exploits plentiful food supplies (J. M. V. Rayner et al, in preparation). Although continuous hovering is rare, many birds hover briefly while landing or feeding. This suggests that metabolic rather than aerodynamic considerations are limiting (36, 43). Flight design must accommodate the anaerobic metabolism required for short periods by such manoeuvers (36), but no information is presently available on the maximum aerobic and anaerobic energy rates available to flying birds (5).

Bounding Flight

Many small birds have relatively large, broad wings that enable them to hover and fly slowly within aerodynamic constraints (32, 43). As a result, profile drag in fast flight is high. By adopting bounding flight (40), however, such birds can save energy in fast flight. In bounding flight the wings are folded against the body for about one half of the flight time, and the excess lift and thrust needed during the flapping phase are more than compensated by the saving in profile and induced drag (27, 40, 43, 52). Such flight also has physiological benefits that relate to energy economy: muscle in small birds has limited flexibility, and bounding flight may allow control of flight speed without impairing muscular efficiency (16, 37). If muscles are designed for maximum effort (e.g. manoeuver or take-off) with very high energy demand, then bounding would permit lower energy output with maximum efficiency (40). In long-winged birds (larks, swifts), energy used for forward flight is comparable to that used for take-off (43), and these species do not bound.

Flocking and Formation Flight

Group flight has many advantages (21, 22, 58). Some authors consider the benefits entirely behavioral, but there can be sizeable energy savings for both flocks (34) and V-shaped formations (24, 28) [the estimated 70% increase in range (28) may be over-generous]. Formation flight benefits best those birds with short wings (ducks, geese) that fly at high induced power: Vortex movements in the wake are vigorous, and birds flying in upwash created by preceding animals obtain big reductions in profile and induced power. Conventional aerodynamic theory indicates that flight in line would need more, rather than less, energy for there is no upwash immediately behind each bird (28); but a vigorously flapping wing introduces updraughts directly behind the bird (43), and therefore line formations could benefit birds.

Foraging Strategies and Energetics

Many birds use flight extensively for feeding, either foraging in or from the air or flying to food sources. Wolf & Hainsworth (58) have reviewed the extensive literature on foraging strategies and energetics but have given little attention to flight strategies. Most models of foraging behavior concentrate on maximization of energy intake (32), but flight energy spent in foraging is considerable and can have an important effect on strategy. R. A. Norberg (29) has modelled this situation and shows that the best strategy is closely interrelated with the ecological character of the habitat, with the bird's physiological and ecological requirements, and with flight morphology (32). The most energy-intensive strategies should occur with the highest prey-densities (equivalently highest food energy) and should be most efficient. Foraging flight has been considered in detail for hummingbirds and other nectar-feeders (13, 14, 58), but other birds have been studied less thoroughly. Sympatric woodland insectivores employ different feeding movements according to their use of habitat, and these are reflected in morphology (31, 32). Pennycuick (35, 38) has shown how vultures use thermal soaring to follow ungulates in East Africa, and introduced the concept of foraging radius to describe the area over which an animal can range while searching for food. Evidence that optimal behavior is actively selected is provided by birds that break food objects open by dropping them: Lammergeiers [dropping bones (7)], gulls [clams and mussels (2, 48)], and crows [shells (59)] all consistently select an optimal drop height, determined no doubt by the minimum height for successful breaking and the cost of carrying the load to that height.

Flight energy can be related to other demands on the bird's resources by including a realistic measure of it in an energy budget. Since metabolic scope for flight is not constant (see above), but varies with size and probably also with other adaptations, and because also energy consumed in flight depends on the type of flight, it is incorrect to assume that flight costs can be accurately estimated as a constant multiple of resting metabolism. Better estimates can be obtained by using theoretical models of power to quantify alternative strategies (e.g. 57), or by measuring true energy costs and relating them to flight behavior. The latter has been done for hummingbirds (see 58) and for hirundines (8, 19), in which flight occupies 57% of total active time and therefore a very high proportion of daily energy, and variations in measured daily energy costs relate well to observed flight behavior. In less active birds, smaller proportions of total energy are used in flight, but this does not preclude strong pressures to optimize flight strategy, which will be reflected in both morphology and behavior.

SUMMARY

Flight energy is an important factor in the lives of birds. Many strategies and adaptations serve to minimize energy cost and to allow a range of performance consistent with a bird's ecological needs. Theoretical methods can produce good estimates of flight energy that suggest why flight adaptations occur; but remarkably little is known of the physiological adaptations required by flight, or of how these change, as I believe they must, in relation to ecology and flight behavior. More data on the metabolic power consumption of birds in natural flight would be valuable, but it is more important to determine the changes in internal metabolic processes associated with different levels of flight activity. Muscle efficiency in flight, in particular, may have substantial implications for our understanding of the energetic performance of birds. This is but one of a variety of unknown quantities, and only when the mechanisms that determine these are more deeply investigated can flight adaptations be completely understood.

ACKNOWLEDGMENTS

I thank many friends and colleagues for help and ideas, and in particular Brian Follett, Colin Pennycuick, and Keith Scholey for their comments on this manuscript. I also thank the Science Research Council for financial support.

Literature Cited

1. Alerstam, T. 1979. Wind as selective agent in bird migration. *Ornis Scand.* 10:76–93
2. Barnash, D. P., Donovan, P., Myrick, P. 1975. Clam dropping behaviour of the glaucous-winged gull (*Larus glaucescens*). *Wilson Bull.* 87:60–64
3. Baudinette, R. V., Schmidt-Nielsen, K. 1974. Energy cost of gliding flight in herring gulls. *Nature* 248:83–84
4. Berger, M. 1974. Energiewechsel von Kolibris beim Schwirrflug unter Hohenbedingungen. *J. Ornithol.* 115:273–88
5. Berger, M., Hart, J. S. 1974. Physiology and energetics of flight. In *Avian Biology*, ed. D. S. Farner, J. R. King, 4:260–415. London & NY: Academic. 528 pp.
6. Bernstein, M. H., Thomas, S. P., Schmidt-Nielsen, K. 1973. Power input during flight in the Fish Crow *Corvus ossifragus*. *J. Exp. Biol.* 58:401–10
7. Boudoint, Y. 1976. Techniques de vol et de cassage d'os chez le gypaète barbu *Gypaetes barbatus*. *Alauda* 44:1–21
8. Bryant, D. M., Westertep, K. R. 1980.

The energy budget of the house martin (*Delichon urbica*). *Ardea* 68:91–102

9. Butler, P. J. 1981. Respiration during flight. *Adv. Physiol. Sci.* 10:155–64
10. Butler, P. J., West, N. H., Jones, D. R. 1977. Respiratory and cardiovascular responses of the pigeon to sustained level flight in a wind tunnel. *J. Exp. Biol.* 71:7–26
11. Butler, P. J., Woakes, A. J. 1980. Heart rate, respiratory frequency and wing beat frequency of free-flying barnacle geese. *J. Exp. Biol.* 85:213–26
12. Epting, R. J. 1980. Functional dependence of the power for hovering on wing disc loading in hummingbirds. *Physiol. Zool.* 53:347–57
13. Feinsinger, P., Chaplin, S. B. 1975. On the relationship between wing disc loading and foraging strategy in hummingbirds. *Am. Nat.* 109:217–24
14. Feinsinger, P., Colwell, R. K., Terborgh, J., Chaplin, S. B. 1979. Elevation and the morphology, flight energetics, and foraging ecology of tropical hummingbirds. *Am. Nat.* 113:481–97
15. Gessaman, J. A. 1980. An evaluation of heart rate as an indirect measure of daily energy metabolism of the American kestrel. *Comp. Biochem. Physiol.* 65A:273–89
16. Goldspink, G. 1977. Muscle energetics and animal locomotion. In *Mechanics and Energetics of Animal Locomotion*, ed. R. McN. Alexander, G. Goldspink, pp. 57–81. London: Chapman and Hall. 346 pp.
17. Goldspink, G., Mills, C., Schmidt-Nielsen, K. 1978. Electrical activity of the pectoral muscles during gliding and flapping flight in the herring gull (*Larus argentatus*). *Experientia* 34:862–65
18. Greenewalt, C. H. 1975. The flight of birds. *Trans. Am. Philos. Soc.* 65(4):1–67
19. Hails, C. J. 1979. A comparison of flight energetics in hirundines and other birds. *Comp. Biochem. Physiol.* 63A:581–86
20. Hainsworth, F. R., Wolf, L. L. 1975. Wing disc loading: implications and importance for hummingbird energetics. *Am. Nat.* 109:229–33
21. Heppner, F. H. 1974. Avian flight formations. *Bird-Banding* 45:160–69
22. Higdon, J. J. L., Corrsin, S. 1978. Induced drag of a bird flock. *Am. Nat.* 112:727–44
23. Hudson, D. M., Bernstein, M. H. 1981. Temperature regulation and heat balance in flying white-necked ravens, *Corvus cryptoleucos*. *J. Exp. Biol.* 90:267–81
24. Hummel, D. 1978. Die Leistungsersparnis in Flugformation von Vögeln mit Unterschieden in Grösse, Form und Gewicht. *J. Ornithol.* 119:52–73
25. Hussell, D. J. T., Lambert, A. B. 1980. New estimates of weight loss in birds during nocturnal migration. *Auk* 97:547–58
26. Kendeigh, S. C., Dolnik, V. R., Gavrilov, V. M. 1977. Avian energetics. In *Granivorous Birds in Ecosystems*, ed. J. Pinowski, S. C. Kendeigh pp. 127–203. (IBP 12). Cambridge: Cambridge Univ. Press. 431 pp.
27. Lighthill, M. J. 1977. Introduction to the scaling of aerial locomotion. In *Scale Effects in Animal Locomotion*, ed. T. J. Pedley, pp. 365–404. London & NY: Academic. 545 pp.
28. Lissaman, P. B. S., Shollenberger, C. A. 1970. Formation flight of birds. *Science* 168:1003–5
29. Norberg, R. A. 1977. An ecological theory on foraging time and energetics and choice of optimal food-searching method. *J. Anim. Ecol.* 46:511–29
30. Norberg, R. A. 1981. Optimal flight speeds in birds when feeding young. *J. Anim. Ecol.* 50:473–77
31. Norberg, U. M. 1979. Morphology of the wings, legs and tail of three coniferous forest tits, the goldcrest and the treecreeper in relation to locomotor pattern and feeding station selection. *Philos. Trans. R. Soc. Lond.* B 287:131–65
32. Norberg, U. M. 1982. Flight, morphology and the ecological niche in some bats and birds. *Symp. Zool. Soc. Lond.* In press
33. Pennycuick, C. J. 1968. Power requirements for horizontal flight in the pigeon *Columba livia*. *J. Exp. Biol.* 49:527–55
34. Pennycuick, C. J. 1969. The mechanics of bird migration. *Ibis* 111:525–56
35. Pennycuick, C. J. 1972. Soaring behaviour and performance of some East African birds, observed from a motor glider. *Ibis* 114:178–218
36. Pennycuick, C. J. 1975. Mechanics of flight. See Ref. 5,5:1–73
37. Pennycuick, C. J. 1978. Fifteen testable predictions about bird flight. *Oikos* 30:16–76
38. Pennycuick, C. J. 1979. Energy costs of locomotion and the concept of foraging radius. In *Serengeti, Dynamics of an Ecosystem*, ed. A. R. E. Sinclair, M. Norton-Griffiths, pp. 164–84. Chicago: University Chicago Press. 387 pp.

39. Pennycuick, C. J., Alerstam, T., Larson, B. 1979. Soaring migration of the common crane (*Grus grus* L.) observed by radar and from an aircraft. *Ornis Scand.* 10:241–51
40. Rayner, J. M. V. 1977. The intermittent flight of birds. See Ref. 27, pp. 437–43
41. Rayner, J. M. V. 1979. A vortex theory of animal flight. Part 1. The vortex wake of a hovering animal. *J. Fluid Mech.* 91:697–730
42. Rayner, J. M. V. 1979. A vortex theory of animal flight. Part 2. The forward flight of birds. *J. Fluid Mech.* 91:731–63
43. Rayner, J. M. V. 1979. A new approach to animal flight mechanics. *J. Exp. Biol.* 80:17–54
44. Rayner, J. M. V. 1980. Vorticity and animal flight. *Semin. Ser. Soc. Exp. Biol.* 5:177–99
45. Rayner, J. M. V. 1982. Flight adaptations in vertebrates. *Symp. Zool. Soc. Lond.* In press
46. Schmidt-Nielsen, K. 1972. Locomotion: energy cost of swimming, flying, and running. *Science* 177:222–28
47. Schuchmann, K. L. 1979. Metabolism of flying hummingbirds. *Ibis* 121:85–86
48. Siegfried, W. R. 1977. Mussel-dropping behaviour of kelp gulls. *S. Afr. J. Sci.* 73:337–41
49. Thomas, S. P. 1975. Metabolism during flight in two species of bats, *Phyllostomus hastatus* and *Pteropus gouldii*. *J. Exp. Biol.* 63:273–93
50. Torre-Bueno, J. R. 1976. Temperature regulation and heat dissipation during flight in birds. *J. Exp. Biol.* 65:471–82
51. Torre-Bueno, J. R., Larochelle, J. 1978. The metabolic cost of flight in unrestrained birds. *J. Exp. Biol.* 75:223–29
52. Tucker, V. A. 1968. Respiratory exchange and evaporative water loss in the flying budgerigar. *J. Exp. Biol.* 48:67–87
53. Tucker, V. A. 1972. Metabolism during flight in the laughing gull, *Larus atricilla*. *Am. J. Physiol.* 222:237–45
54. Tucker, V. A. 1973. Bird metabolism during flight: evaluation of a theory. *J. Exp. Biol.* 58:589–709
55. Tucker, V. A. 1974. Energetics of natural avian flight. In *Avian Energetics*, ed. R. A. Paynter, pp. 298–334. Cambridge, Mass: Nuttall Ornithol. Club.
56. Withers, P. C. 1979. Aerodynamics and hydrodynamics of the "hovering" flight of Wilson's storm petrel. *J. Exp. Biol.* 80:83–91
57. Withers, P. C., Timko, P. L. 1977. The significance of ground effect to the aerodynamic cost of flight and energetics of the black skimmer (*Rhyncops nigra*). *J. Exp. Biol.* 70:13–26
58. Wolf, L. L., Hainsworth, F. R. 1978. Energy: expenditures and intakes. In *Chemical Zoology, X (Aves)*, pp. 307–58, ed. M. Florkin, B. T. Scheer. London & NY: Academic. 436 pp.
59. Zach, R. 1979. Shell-dropping: decision making and optimal foraging in northwestern crows. *Behaviour* 68:106–17

ENERGETICS OF LOCOMOTION IN WARM-BODIED FISH

E. Don Stevens

Department of Zoology, University of Guelph, Guelph, Ontario N1G 2W1, Canada

Andrew E. Dizon

National Marine Fisheries Service, NOAA, Honolulu, Hawaii 96822

INTRODUCTION

Most fishes are poikilothermic—i.e. their body temperature is within a few degrees of ambient unless ambient is changing rapidly. The exceptions are certain sharks and true tunas. Little is known about the locomotion and energetics of warm-bodied sharks. Our review focuses on a few species of tunas but draws on information from other fish (especially salmonids) to fill in the gaps in our understanding.

Of interest here is a measure of the metabolic cost of producing known amounts of work per unit time; but there are problems with both sides of this equation, especially in the aquatic environment. Webb (30) has used a scheme similar to that in Figure 1 to show how total metabolic costs are partitioned. Since this approach is appropriate for sustained swimming in steady-state (i.e. oxygen supply keeping up with demand), especially when experiments are performed for short periods (hours), it is used here.

Energy input or metabolic cost has been successfully estimated by measuring oxygen uptake and calories ingested. We focus our discussion on estimates based on these measurements.

Other techniques for measuring metabolic costs have either been tried unsuccessfully or are yet to be tried. For example, few measurements have been made of CO_2 production in fish (15, 16) and none for tuna. Knowledge of heat exchange rates and excess temperature has permitted some specula-

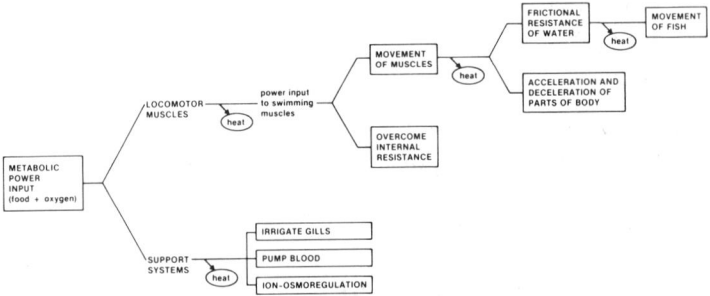

Figure 1 Schematic of energy distribution to components involved in propulsion at sustained swim speeds.

tion about energetics (19), but little is known about how heat exchange rates change during swimming. Guppy and co-workers have studied biochemical adaptations of burst swimming but not of sustained swimmming (11).

Known amounts of work per unit time have never been measured for water breathers. Metabolic costs are usually related to swimming velocity. Some workers have used arbitrary estimates—e.g. measured oxygen uptake or temperature after "feeding frenzy." Others have measured time to swim a specific distance when the tuna is controlling its own speed. Still other estimates are based on hydrodynamic theory, which is used to estimate the work to overcome drag or produce the thrust necessary to swim at a particular speed.

We review recent contributions on tuna energetics based on measurements of oxygen uptake, bioenergetic calculations from feeding and growth rates, and hydrodynamic calculations. We discuss the partitioning of this energy in tunas. Finally, we compare the cost of locomotion in tunas to that in birds and mammals and comment on the adaptive advantage of being warm-bodied.

We stress the two main differences between locomotion on land and in water. First, water as a medium offers much resistance (i.e. it has high viscosity and high density) so that almost all locomotive work goes to overcome drag. Second, water supports the fish's weight, so that almost no work goes to overcome gravity. Warm-bodied fishes swim continuously; thus inertial losses are small when they swim at constant velocity.

TOTAL COST OF LOCOMOTION

Estimates of the total cost of locomotion are based on measurements of the gross cost. There have been many attempts to calculate net cost by estimating maintenance costs (i.e. of ion-osmoregulation, of obtaining oxygen from the medium, and of delivering oxygen to cells that need it). We first discuss total and then net costs.

Oxygen Uptake Used to Estimate Total Cost of Locomotion

The only measurements of oxygen uptake in free-swimming tunas at known swim speeds are those of Gooding et al (10). In their experiments, groups of two to six skipjack tuna (*Katsuwonus pelamis*) were placed in large respirometers (4.6 m diameter, 1 m deep; or 2.4 m diameter, 0.6 m deep) where oxygen depletion and swim speed were measured. Thus over the long term the experiment provided a measure of total oxygen uptake at known swim speeds (i. e. in cases for which the fish maintained the same velocity for hours). Swim speed was controlled by the fish and not the experimenters. The data appear in Figure 2. Three aspects of their results are unique. First, there was no statistically demonstrable effect of body weight among oxygen uptake rates for skipjack tunas over a weight range of 0.60–4.0 kg. Second, the metabolic rate at any speed over the range studied (i.e. 0.9–2.2 $l \cdot \text{sec}^{-1}$) (l=body lengths) was higher than that of any other fish species studied. For example, at 1 $l \cdot \text{sec}^{-1}$, oxygen uptake in the skipjack is 469 $\text{mgO}_2 \cdot \text{kg}^{-1} \cdot \text{h}^{-1}$ whereas in eight other species it ranged from 90–300 (1). At 2 $l \cdot \text{sec}^{-1}$ oxygen uptake in skipjack is 603 $\text{mgO}_2 \cdot \text{kg}^{-1} \cdot \text{h}^{-1}$ whereas in the eight other species it ranged from 200–440 (1). These data show that tuna are less efficient than other fish species when swimming at these speeds and that the obvious streamlining of tunas does not lead to an obvious saving of total metabolic costs at swim speeds about 1–2 $l \cdot \text{sec}^{-1}$. "Presumably, the evolution of skipjack tuna (like that of fast cars) has involved sacrifice of energetic efficiency at low speeds in favour of increased efficiency at high speeds, permitting a dramatic increase in maximum attainable speed" (10).

Unfortunately, there are no oxygen uptake measurements of tunas swimming at higher speeds. However, in skipjack tuna immediately after capture at sea the median rate of 15 measurements was 1300 $\text{mgO}_2 \cdot \text{kg}^{-1} \cdot \text{h}^{-1}$ (range 900–2500) (10). Swim speed ranged from 2–5 $l \cdot \text{sec}^{-1}$. These measurements are important because they suggest the tuna's maximum aerobic scope. They are higher than any values for any other fish (or amphibian or reptile) under any exercise, temperature, or other experimental condition.

The third unique aspect of Gooding et al's data is the fact that the slope of the line relating log metabolic cost to swim speed is less than that for other fish. When the logarithm of metabolic rate is plotted versus swim speed, a straight line can usually be fitted to the data. The slope of this line for skipjack tuna is 0.21 (Figure 2) (10), less than that for the other eight species tabulated by Beamish (1). "The rate of increase in the logarithm of oxygen uptake with relative swimming speed is surprisingly similar among species despite obvious variation in methodology, size, and temperature and is reasonably well represented by a coefficient of 0.36. Thus for each increase in relative swimming speed of $l \cdot \text{sec}^{-2}$ there is a corresponding 2.3-fold eleva-

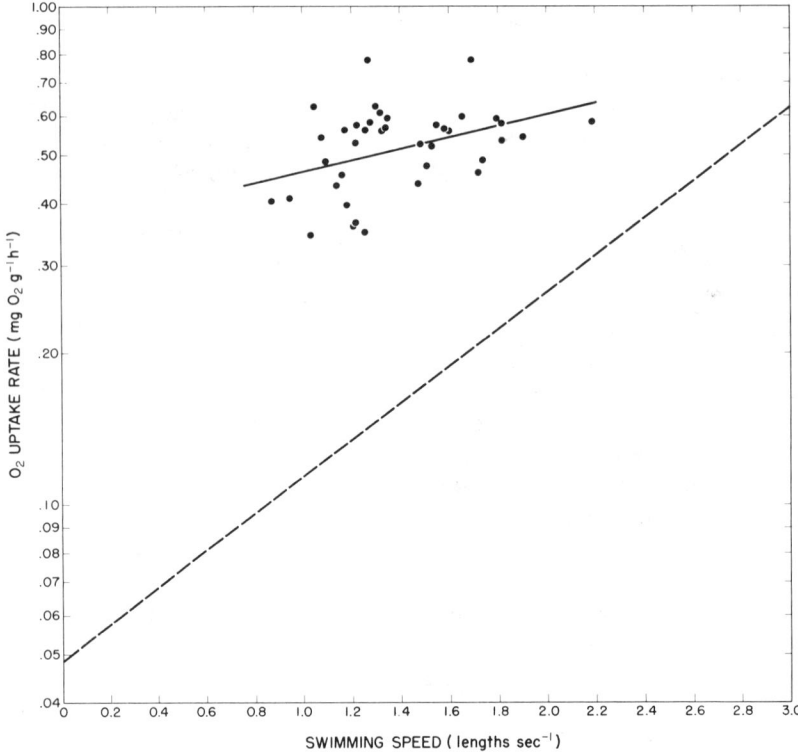

Figure 2 Relation between oxygen uptake and swim speed of skipjack tuna. The dashed line is for 1.8 kg sockeye salmon (2). Figure adapted from (10).

tion in metabolic rate" (1). But this coefficient is 0.21 for skipjack, involving only a 1.6-fold elevation in metabolic rate. The tuna becomes more efficient than the salmon at a speed of less than 5 $l \cdot sec^{-1}$, and tuna can sustain speeds greater than 5 $l \cdot sec^{-1}$, whereas salmon cannot (7).

NET COST OF LOCOMOTION

Calculation of the net cost of locomotion requires an estimate of basal metabolic rate (or at least the rate at zero activity). The major fraction of these costs is usually attributed to irrigation of the gills with water, circulation of the blood, and ion-osmoregulation.

Stevens (24) measured respiration in skipjack tuna restrained in a chamber. The average oxygen uptake was 692 $mgO_2 \cdot kg^{-1} \cdot h^{-1}$ and the mean of the lowest value recorded for each of five skipjack was 457 $mgO_2 \cdot kg^{-1} \cdot h^{-1}$. Clearly these values are not resting or basal and thus cannot be used to estimate net cost of locomotion. Brill (3) attempted to overcome the problem of obtaining resting values by measuring "stasis" metabolism (metabolism of restrained skipjack injected with a muscle relaxant, gallamine triethiodide, and then spinalectomized to stop all overt muscular movement). Smaller animals had higher mass-specific stasis metabolic rates. For a mass range of 0.32–4.7 kg, the relation was $M = 8431\ W^{-0.437}$, where M = oxygen uptake $(mgO_2 \cdot kg^{-1} \cdot h^{-1})$ and W = mass (g). Thus stasis metabolic rate for a 2 kg skipjack is 304 $mgO_2 \cdot kg^{-1} \cdot h^{-1}$ or about one half the value reported for a nonspinalectomized fish and about one half the value from the swimming fish when extrapolated to zero activity (529 $mgO_2 \cdot kg^{-1} \cdot h^{-1}$). We assume that this measurement is a reasonable estimate of the metabolic cost of the support systems in tuna at zero activity. Most evidence shows that these costs increase in proportion to swim speed.

Cost of Irrigating The Gills

The typical pattern of rhythmic movements of respiratory muscles to force water over the gills is not seen in tunas. Tunas, swimming with an open mouth, irrigate the gills by ram pressure. The literature on irrigation of the gills by open-mouth swimming has been reviewed (23).

Thus the metabolic cost of irrigating the gills is a small fraction (of the order of 1%) of total metabolic cost of swimming in tunas (i.e. it is about one order of magnitude less than that in fish irrigating their gills with buccal and opercular pumps).

Cost of Circulating The Blood

The cost of circulating the blood must be higher in tunas than in other fishes. The absolute cost must be higher because the metabolic rates are so high—i.e. because more oxygen must be delivered to the tissues. The relative cost is probably also high because of the unusual circulatory system of tunas. Tunas have elaborate counter-current heat exchangers that keep their body temperatures substantially higher than that of the water. The imposition of the heat-exchanger vascular beds in series with the typical vascular beds must result in a greater work load on the heart. The cost of circulation is estimated at 3.5% of total oxygen uptake at rest and 4.5% of total oxygen uptake at maximum activity in salmon (13). In tuna it is probably greater than 5%.

Cost of Ion-Osmoregulation

There are two elaborate studies on the metabolic cost of ion-osmoregulation in fish (8, 22). Both used euryhaline species and estimated the cost by measuring oxygen uptake at a variety of salinities. Costs increased exponentially with swimming speed, reaching 17% of total metabolism at maximum swim speed.

Thus the costs of maintaining the support systems during swimming as fractions of the total metabolic cost are probably about 1% to irrigate the gills, 5% to circulate the blood, and 15–20% for ion-osmoregulation.

METABOLIC COSTS BASED ON CALORIC INPUTS

Kitchell et al (14) constructed energy budgets for two species of tuna using the principles of bioenergetics in order to estimate the scope for growth.

Small Tropical Tuna

From (10), M (routine metabolic rate in $mgO_2 \cdot h^{-1} \cdot fish^{-1}$) is $0.288 \cdot W^{1.08}$, where mass is in g. Thus for a 2 kg skipjack routine metabolic rate is 1058 $mgO_2 \cdot h^{-1}$ or 86 kcal·day^{-1}. Given the caloric density of skipjack (1.46 kcal·g^{-1}), this is about 3.7% of total energy content per day. Measured values show that actual losses of energy content during 10 days starvation are 3.6% per day. Because this calculation is based on Gooding's estimate of routine metabolic rate, the result depends on body mass. However, because the exponent is small (0.08), changes with mass are small. The decrease in energy content per day is 2.012 $W^{0.08}$ (where mass is in g) or 3.2% for the smallest skipjack used by Gooding (0.32 kg) and 4.0% for the largest (4.7 kg).

Kitchell et al extended their calculations to estimate the energy required to account for observed growth rates of skipjack in the field. The observed growth rate is 0.7% mass per day or 10 kcal·fish^{-1}·day^{-1} for a 1 kg skipjack. In the laboratory small skipjack can consume food equivalent to 28–35% of their body mass·day^{-1}. Kitchell et al deduced that small skipjack (less than 7–10 kg) grow at rates substantially lower than maximal and thus appear limited by food availability and/or their efficiency as predators. Growth of large skipjack (greater than 7–10 kg) appears limited by the rate at which food can be consumed and physiologically processed. This upper limit appears to account for the upper size limit of skipjack observed in the field (25 kg). Observed growth rates in the field and maximum size in the field are concordant with a metabolic rate slightly more than twice the routine rate observed by Gooding (14), i.e. about 577· $W^{0.08}$. For a 0.32 kg fish this is 915 $mgO_2 \cdot kg^{-1} \cdot h^{-1}$; for a 2 kg fish it is 1060 $mgO_2 \cdot kg^{-1} \cdot h^{-1}$; and for a 4.7 kg fish it is 1134 $mgO_2 \cdot kg^{-1} \cdot h^{-1}$. All of these values are higher than rates measured at maximum sustained swimming speeds for any other species at any temperature. We can calculate the approximate speed for a particular mass when oxygen uptake = 0.577 $W^{0.08}$ (i.e. when it is twice the routine rate measured by Gooding) by substituting into the original relation (10).

$$S = (22.35 - 2.558 \log W) \cdot W^{0.2831},$$

where S is speed in km·day^{-1} and W is mass in g. To estimate fuel used we convert metabolic rate to grams of fat used per day. Fuel economy (FE) can now be calculated by dividing km travelled per day by fat used per day. $FE = (4511 - 516.4 \log W) \cdot W^{-0.7969}$, where FE is in km·g fat^{-1} when swimming at speeds equivalent to twice the routine rates.

Thus the 0.32 kg tuna travels 82 km·day^{-1} using 2.52 g fat (32 km·g fat^{-1}) whereas the 4.7 kg tuna travels 142 km using 45.8 g fat (3.10 km·g fat^{-1}). Large tuna must consume and physiologically process ten times as much fuel to swim less than twice as far. For each doubling of mass, fuel economy is approximately halved, a relation analogous to that between the fuel consumption rates of small and large cars.

Pennycuick has made similar calculations for birds (20, 21). A 3 g hummingbird travels 880 km per g fat, whereas the larger bird, a 384 g pigeon, travels only 11.7 km per g fat. The 384 g pigeon and the 320 g skipjack get similar mileage. Similar calculations can be made for a 384 g white rat (29). At maximum speed (2.25 km·h^{-1}) it could travel 54 km at a total cost of 13.8 g fat·day^{-1}. These relations are summarized in Table 1.

Giant Bluefin Tuna

The approach of Kitchell can be applied to the largest tuna, the giant bluefin. This animal's oxygen uptake has not been measured, but we do know something about its feeding and growth rates. Giant bluefin arrive in Nova Soctia in July weighing about 350 kg. They gain about 20% in body mass (all fat) in about 60 days when fed a rate of about 4% body mass per day. The caloric content of the food, mackerel, also a scombroid, is probably about the same as that of the tuna. Assuming a caloric density the same as skipjack, then mass-specific metabolic rate is:

$$\frac{0.0227 \text{ kg}}{\text{kg·day}} \times 1000 \frac{\text{g}}{\text{kg}} \times 1.46 \frac{\text{kcal}}{\text{g}} \times \frac{1 \text{ mgO}_2}{0.0034 \text{ kcal}} \times \frac{1 \text{ day}}{24 \text{ h}} = 406 \text{ mgO}_2 \cdot \text{h}^{-1} \cdot \text{kg}^{-1}.$$

This is surprisingly similar to the maximum active rate for salmon extrapolated over two orders of magnitude: Active salmon metabolism = $(1.772 \times 385000^{0.888})/385 = 420$ mgO$_2$·h^{-1}·kg^{-1}. The speed at this metabolic rate extrapolated from the salmon equation is 1.53 l·sec^{-1}. This is slightly more than the observed speed (1 l·sec^{-1} or 3 m·sec^{-1}) for which the bluefin metabolic estimate is made.

We can also compare the giant bluefin's net cost to that of a similar-sized mammal (28). Net mammal metabolism is: $M = 760 \cdot W^{-0.4} = 760 \cdot 385^{-0.4} = 70.3$ mgO$_2$·kg^{-1}·km^{-1}. The net cost for bluefin is almost exactly half the value for a mammal of equivalent mass. Thus the fuel economy of the tuna is twice that of the mammal (Table 1).

NET COSTS BASED ON THRUST AND DRAG ESTIMATES

Independent estimates of net energy costs can be made in a straight-forward manner from the thrust/drag relationships of a swimming fish (6, 7). These estimates can then be compared with the laboratory values from the respirometers. The estimates are made as follows: In steady-state swimming, total thrust force (dynes) equals total drag force (dynes). By first principles (30), drag force is $D = 0.5 \cdot S \cdot \rho \cdot V^2 \cdot C_D$, where $S =$ wetted surface area (cm^2), $\rho =$ fluid density (g·cm^{-3}), $V =$ velocity (cm·sec^{-1}), and $C_D =$ coefficient of total drag. Similarly, thrust force is $T = 0.5 \cdot S \cdot \rho \cdot V^2 \cdot C_T$ where $S =$ surface area of caudal fin (cm^2), $V^2 =$ lateral velocity of fin (cm·sec^{-1}), and $C_T =$ coefficient of total thrust.

Power is the time rate of doing work, and work is the application of force through a given distance. But since the propulsion system is not 100% efficient, more power input is required for the power output necessary to propel a given fish at a given speed. Total aerobic efficiency is generally taken (30) as 0.2, which means of course that about 80% of the input power is lost as heat and does no usable work (Figure 1). Thus $P_i = (D \cdot V \cdot 10^{-7})/n$, and $P_i = (T \cdot V \cdot 10^{-7})/n$, where $P_i =$ input power (watts) and $n =$ total aerobic efficiency $= 0.2$. To this input power must be added the power necessary to fuel the nonswimming processes described in the previous section.

Table 1 The metabolic cost of locomotion of tunas compared to other endotherms of similar mass

Animal	Mass	Speed (km · day^{-1})	Fat used (g fat · day^{-1})	Fuel economy (km · g fat^{-1})
Small endotherms				
Skipjack	320 g	82	2.52	32
Pigeon	384 g	1374	117	11.7
Rat	384 g	54	13.8	3.9
Large endotherms				
Bluefin	385 Kg	259	1237	0.21
Salmon[a]	385 Kg	397	1283	0.31
Mammal	385 Kg	259	2510	0.10

[a] Extrapolated from salmon data (2) at the maximum activity level

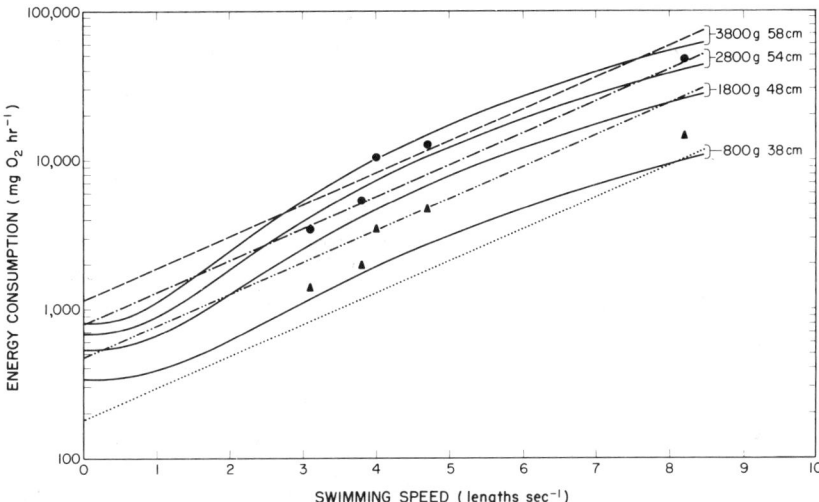

Figure 3 Comparison between measured oxygen uptake extrapolated to 8.5 $l \cdot \text{sec}^{-1}$ (broken lines) and the theoretically calculated power consumption based on hydrodynamic considerations (solid lines). Triangles (▲) are the theoretical power consumption based on an analysis of drag forces, and points (●) are based on an analysis of thrust forces for a 40 cm, 1003 g skipjack tuna (17). Figure adapted from (10).

Precise estimates of body surface area exist for many species of tuna, along with regression relationships based on length (18). Remaining are estimates of the coefficients of drag and thrust; these are the most questionable and subject to error. For the construction of the solid lines in Figure 3, a simple relationship was used (6,10). The simple drag model produces conservative values.

The measured oxygen-uptake relationship in Figure 2 extrapolated to 8.5 $l \cdot \text{sec}^{-1}$ compares well to the theoretical projections of energy consumption based on the simple drag model (Figure 3). Magnuson's (17) estimates of the theoretical power consumption of a 40 cm, 1003 g skipjack tuna, based on either model of thrust forces or based on an accurate estimate of total drag, are also presented in the figure. These compare well and are conservative, predicting even larger expenditures of energy.

THERMOCONSERVATION, LOCOMOTION, AND ENERGETICS IN TUNAS

The feature that most sets tunas apart from other fishes is the fact that they are warm-bodied. Is this fact related to locomotion and/or the energetics of locomotion? Arguments concerning the adaptive advantage of being warm relative to the water have been discussed in detail (27). Recently Stevens & Carey (25) developed a new argument.

Tuna are especially well adapted to sustain high speeds (they have many morphological features that reduce drag) rather than to achieve high burst speeds (which require low mass, and for which frictional resistance is less important). Yuen (31) reported a school of skipjack tuna that travelled 28 km in 107 min—i.e. fish 40–50 cm long can swim at speeds of about 10 $l \cdot sec^{-1}$ for at least an hour.

Stevens & Carey (25) argue that being warm-bodied confers an adaptive advantage because it increases the amount of oxygen that can be delivered to active cells. Passive diffusion of oxygen has a Q_{10} of about 1. The Q_{10} of transport of oxygen by myoglobin is higher, probably between 1.5 and 2 (E. D. Stevens, unpublished). Skipjack tuna muscle temperature changes with activity level, at least when we compare extreme activity levels. Tuna muscle has much red muscle that contains a high concentration of myoglobin. Thus the warmth may increase the rate at which myoglobin delivers oxygen to the mitochondria of active cells.

Although the above reasonably explains the advantage of warm muscle, it does not explain that of a warm stomach. Telemetric observations of free-swimming bluefin tuna (F. G. Carey, E. D. Stevens, and J. W. Kanwisher, unpublished data) show that these tunas increase stomach temperature after a cold meal and keep it warm for hours during digestion (stomach temperature 25–30°C, water temperature 12°C). In this case it seems that being warm-bodied confers an adaptive advantage because it permits an increase in the rate at which food can be physiologically processed.

All tunas, from the small tropical tunas to the giant bluefin, also have elevated brain temperatures. Brain temperature of skipjack increases from about 0.1°C to 4°C above ambient during rapid swimming (26).

Thus although being warm is what makes tuna unique among fishes, no single function can be attributed to the warmth. Tunas inhabit parts of the ocean where food is dilute and patchily distributed. They maximize energy gain by "gambling" large energy expenditures (high sustained activity) on the "expectation" of proportionately large energy returns (25, 27). Being warm is one aspect of this "gamble."

ACKNOWLEDGMENTS

We thank J. M. Renaud and J. Galbraith for comments on the first draft and Mary Anne Finkbeiner for transcribing the manuscript. One of us (E.D.S.) takes this opportunity to point out that the financial support for research on tunas provided by Fisheries and Oceans of Canada would not buy peanuts for a pet squirrel.

Literature Cited

1. Beamish, F. W. H. 1978. Swimming capacity. In *Fish Physiology,* ed. W. S. Hoar, D. J. Randall, 7:101–87. NY: Academic. 576 pp.
2. Brett, J. R., Glass, N. R. 1973. Metabolic rates and critical swimming speeds of sockeye salmon, *Oncorhynchus nerka,* in relation to size and temperature. *J. Fish. Res. Bd. Can.* 30:379–87
3. Brill, R. W. 1979. The effect of body size on the standard metabolic rate of skipjack tuna, *Katsuwonus pelamis. Fish Bull.* 77:494–98
4. Deleted in proof
5. Deleted in proof
6. Dizon, A. E., Brill, R. W. 1979. Thermoregulation in tunas. *Am. Zool.* 19:249–65
7. Dizon, A. E., Brill, R. W., Yuen, H. S. H. 1978. Correlations between environment, physiology, and activity and the effects of thermoregulation in skipjack tuna. In *The Physiological Ecology of Tunas,* ed. G. D. Sharp, A. E. Dizon, pp. 233–59. NY: Academic. 485 pp.
8. Farmer, G. J., Beamish, F. W. H. 1969. Oxygen consumption of *Tilapia nilotica* in relation to swimming speed and salinity. *J. Fish. Res. Bd. Can.* 26:2807–21
9. Deleted in proof
10. Gooding, R. M., Neill, W. H., Dizon, A. E. 1981. Respiration rates and low oxygen tolerance limits in skipjack tuna, *Katsuwonus pelamis. Fish Bull.* 79(1): In press
11. Guppy, M. 1978. Skipjack tuna white muscle: a blueprint for integration of aerobic and anaeobic carbohydrate metabolism. See Ref. 7, pp. 175–81
12. Deleted in proof
13. Jones, D. J., Randall, D. J. 1978. The respiratory and circulatory systems during exercise. See Ref. 1, pp. 425–501
14. Kitchell, J. F., Neill, W. H., Dizon, A. E., Magnuson, J. J. 1978. Bioenergetic spectra of skipjack and yellowfin tuna. See Ref. 7, pp. 357–68
15. Kutty, M. N. 1968. Respiratory quotients in goldfish and rainbow trout. *J. Fish. Res. Bd. Can.* 25:1689–728
16. Kutty, M. N. 1972. Respiratory quotient and ammonia excretion in *Tilapia mossambica. Mar. Biol. (NY)* 16:126–33
17. Magnuson, J. J. 1978. Locomotion by scombrid fishes: hydromechanics, morphology, and behavior. See Ref. 1, pp. 240–313
18. Magnuson, J. J., Weininger, D. 1978. Estimation of minimum sustained speed and associated body drag of Scombrids. See Ref. 7, pp. 293–311
19. Neill, W. H., Chang, R. K. C., Dizon, A. E. 1976. Magnitude and ecological implications of thermal inertia in skipjack tuna. *Environ. Biol. Fish.* 1:61–80
20. Pennycuick, C. J. 1968. Power requirements of horizontal flight in the pigeon. *J. Exp. Biol.* 49:527–55
21. Pennycuick, C. J. 1969. The mechanics of bird migration. *Ibis* 111:525–56
22. Rao, G. M. M. 1968. Oxygen consumption of rainbow trout (*Salmo gairdneri*) in relation to activity and salinity. *Can. J. Zool.* 46:781–86
23. Roberts, J. L. 1978. Ram gill ventilation in fish. See Ref. 7, pp. 83–88
24. Stevens, E. D. 1972. Some aspects of gas exchange in tuna. *J. Exp. Biol.* 56:809–23
25. Stevens, E. D., Carey, F. G. 1981. One why of the warmth of warm-bodied fishes. *Am. J. Physiol.* 240:R151–55
26. Stevens, E. D., Fry, F. E. J. 1971. Brain and muscle temperatures in ocean caught and captive skipjack tuna. *Comp. Biochem. Physiol.* 38A:203–11
27. Stevens, E. D., Neill, W. H. 1978. Body temperature relations of tunas, especially skipjack. See Ref. 1, pp. 315–59
28. Taylor, C. R. 1977. The energetics of terrestrial locomotion and body size in vertebrates. In *Scale Effects in Animal Locomotion,* ed. T. J. Pedley, pp. 127–41. NY: Academic. 545 pp.
29. Taylor, C. R., Schmidt-Nielsen, K., Raab, J. L. 1970. Scaling of the energetic cost of running to body size in mammals. *Am. J. Physiol.* 219:1104–7
30. Webb, P. W. 1975. Hydrodynamics and energetics of fish propulsion. *Bull. Fish. Res. Bd. Can.* 190:1–159
31. Yuen, H. S. H. 1966. Swimming speeds of yellowfin and skipjack tuna. *Trans. Am. Fish. Soc.* 95:203–9

ENERGETICS OF LOCOMOTION IN ENDOTHERMIC INSECTS

James Edward Heath and Maxine S. Heath

Department of Physiology and Biophysics, University of Illinois, Urbana, Illinois 61801

INTRODUCTION

Endothermic insects are those adult members of several insect orders that rely upon internally produced heat to regulate body temperature (31, 53). Known endotherms include some dragonflies (Odonata) (21, 53), a few katydids (Orthoptera) (30, 43), some flies (Diptera) (51), many bees and wasps (Hymenoptera) (25, 32–34, 49), some beetles (Coleoptera) (11, 36, 37, 53, 58), many moths (6, 8–10, 15, 16, 29, 31) and some butterflies (64) (Lepidoptera), and some owl flies (Neuroptera) (51). These animals are among the largest of their kind, $>$ 100 mg (6), and have insulation to decrease the rate of heat loss from the thorax (1, 6, 9). They generate and conserve heat during flight and possess special warm-up behavior patterns to elevate thoracic temperature prior to critical stages of activity (45, 46).

Endothermic insects both walk and fly. Walking is a relatively inexpensive process, but flight makes enormous demands on the muscle system. Many of these insects are capable of sustained hovering, which is probably the most energetic biological process known; it is near the upper theoretical limit for power generation by striated muscle (2, 43, 79). A few insects have developed dependence on the heat production of flight muscles for functions other than flight, such as warm-up prior to and during singing (22, 30, 69) and regulation of nest temperature (49, 68).

Comprehensive reviews of temperature regulation (53) and endothermy (31) in insects, flight energetics (46) and locomotion (41, 62) in insects, and related topics (21), and recent symposia and books (4, 14, 33, 35, 61, 63, 66, 82) deal with many aspects of our topic. We focus on the current understanding of the cost of locomotion in insects and refer to the older literature only where necessary to clarify the present position.

LOCOMOTION

Calculation and measurement of the power requirements for locomotion in insects are plagued by experimental difficulties. With notable exceptions, insects are aerobatic in flight, executing frequent large changes in velocity and direction (5, 59). An animal stroking at normal frequency and amplitude can feather its flight system and generate no aerodynamic force, or it can lift twice its body weight by changing the angle of attack of the wing and the plane of stroking (54, 75, 81). It is capable of producing large torque about the body axis (25, 81). It may well show each of these behaviors in three or four consecutive strokes. The flight of the migratory locust has been most extensively studied, but because the locust is among the steadiest and least aerobatic of flyers (73, 74, 76, 77), its generality as a model system of insect flight is limited.

No one has successfully trained an insect to perform continuous free flight in a wind tunnel, a technique spectacularly successful with birds (72). Insects tethered or attached to flight mills generally perform at levels well below the requirements for free flight (31, 39). Flight of insects attached to carefully tared instantaneous force balances generate stroke by stroke records that are subject to a series of resonance artifacts (20, 81). Attempts to measure metabolic rate in free flight are restricted by container volumes and thus become approximations of rates during hovering (8, 46). Nevertheless, the ingenuity of researchers has often reached or exceeded that of their insect subjects, and some general principles have emerged.

Terrestrial Locomotion

WALKING AND RUNNING Insect locomotory muscles have a wide range of contraction rates, and thus locomotion can occur over a wide range of speeds (41). Within the range of speeds available to a given species, velocity is dependent on body temperature. Rising body temperatures are reflected by changes in gait. Despite drastic differences in their range of speeds, both the walking stick, *Carausius morosus* (80), and the roach, *Periplaneta americana* (23), reach an alternating tripod gait at frequencies of about 3Hz ($\sim$6–8cm·sec^{-1}), suggesting a minimum speed for coordinated leg movements that may be applicable to many insect species. Maximum velocity appears to be species-specific. An Arrhenius plot of maximum velocity of *P. americana* (48) at given temperatures shows a straight line relationship between 13° and 27°–28°C. Above 28°C there is no increase in maximum velocity. While no endothermy is implied, recent reports on energetic costs of terrestrial locomotion in the cockroach, *Gromphadorhina portentosa* (38), and in 3 species of ants, Formicidae (42), show net costs of transport are comparable to those of vertebrates of the same size or are near predicted values obtained by extrapolation of vertebrate data to the size of the insects.

Endothermy during terrestrial activity has been shown to exist only in large tropical beetles (7, 11, 36, 37). In African dung beetles (Scarabaeidae) and in *Stenodontes molarium* (Cerambycidae), rates of specific energy metabolism are similar to those of active mammals of similar size at the same ambient temperature.

JUMPING While large energy expenditures are required for a single jump [9–11 mJ for locusts (13)], the intermittent nature of jumping locomotion precludes the development of endothermy from the energy expended from jumping alone. None of the jumping insects is known to be endothermic.

Flight

MECHANICS Power requirements for flight are usually estimated by application of quasi-steady-state aerodynamic principles to flapping flight. Major reviews of these applications (3, 75) are available. For analytic purposes, insect flight may be divided between hovering and fast flight, to which two different sets of aerodynamic principles may apply (76, 77). Analysis of transition or slow flight may involve both sets of principles (24).

In practice, measurements of aerodynamic forces generated by an excised wing in a wind stream are applied to the apparent position and velocity of the wing as determined by high speed photographs of the flapping wing (55, 56, 75, 77, 78). A force—lift—is generated across the wind when a differential velocity of air flow above and below the wing causes a differential pressure (Bernoulli effect). In animal flight the calculations of lift become complex because the velocity of air flow increases from attachment to tip, and velocity varies from a low value at the top and bottom of a stroke to high values at intermediate positions. Nevertheless, calculations of the aerodynamic forces tend to agree with sustained net force generated by animals suspended from a balance that resolves both lift and thrust forces (75–77). Power expended as aerodynamic force by a locust (*Shistocerca gregaria*) flying at 3.5 m·sec^{-1} if 0.84 Watts per Newton (75, 76).

The musculature of a flying insect also produces power to accelerate and decelerate the mass of the oscillating wing. The magnitude of this inertial force approaches the magnitude of aerodynamic force (17, 20, 55, 56, 76, 77) and would double the cost of flapping flight, as it may in birds, except that much of the energy expended in accelerating the wing is stored in the elasticity of the thorax and wing hinges and is recovered later in the wing stroke (3, 75). The recoil efficiency of a protein component of wing hinges (resilin) is 0.97. The mechanical efficiency of several flying insects would be beyond thermodynamic expectations, as high as 58% in a hoverfly (3, 75), unless inertial force could be captured and returned later in the wing stroke. The relatively constant stroking frequency of endothermic flying insects may match the resonance of the thorax-wing elastic system, thus reducing inertial power losses (3).

Power requirements for flight can also be calculated from measurement of drag (75, 76). Since power (P) equals drag (D) times velocity (V), three components of power in a flapping insect can be segregated. Parasite drag, the drag induced by the resistance to movement of the body and appendages, increases with V^2. Thus the power required equals DV^3. A second component, profile power, is generated by the movement of the wing irrespective of lift generated. Under ideal circumstances stroke frequency and angle of stroke remain constant. Because of the relatively high velocity of the wing compared to the forward velocity of the animal, profile drag, and thus power, has been taken as constant regardless of forward velocity (65, 76). The third component, induced power, is the energy expended in accelerating a downward draft of air to produce lift, plus the energy lost as vortices shed from the wing tips. Induced power diminishes with velocity. At 0 velocity it becomes very large (24, 65, 76). The sum of these power requirements generates a broad U-shaped curve. This theoretical curve predicts high energy expenditure in hovering, the expenditure diminishing exponentially with increasing velocity. Since the U has a rather broad base, modest increases in velocity above the minimum power point increase the power requirement slowly. The result is a range of velocities in which the cost of transport (km $W^{-1}kg^{-1}$) rises minimally—i.e. the optimal flight range (76).

The curve generated for the locust agrees in two respects with direct measures and observation. The aerodynamic power predicted by the curve should be minimal at about 3.5 $m \cdot sec^{-1}$ at a cost of about 0.8–0.9 $W \cdot N^{-1}$. This agrees closely with the flight speed of locusts in the laboratory and the measured force of 0.84 $W \cdot N^{-1}$. The curve also predicts a velocity of minimum cost of transport of about 4.5–5.5 $m \cdot sec^{-1}$, which corresponds to measured flight speeds in migrating locusts (75, 76). The general features of this curve also emerge from power generated, measured by O_2 consumption of birds trained to fly in a wind tunnel (72).

Another calculation of the power required for hovering is based on momentum jet theory (24). This approach requires few assumptions about the size and aerodynamics of a helicopter rotor or actuator disc. However, the continuous motion of a rotor differs so much from the flapping wing of an animal that power may be underestimated by up to four-fold. There is a question whether steady-state aerodynamic principles are applicable to insect flight. Calculations based upon direct measures of wing motion and aerodynamic properties generate coefficients of lift that exceed theoretical maxima (60, 65, 76, 81) for steady-state aerodynamics. Further, instantaneous direct measures of force generated in a single stroke do not agree with calculations of forces generated by moving wings. The magnitude of lift produced in downstroke is larger than calculated, while upstroke may generate no lift at all (20, 81).

It now appears that unsteady aerodynamic effects apply to hovering and flight in many insects (12, 57, 60, 65, 81). Although no general theory is yet applicable to flapping flight, several special mechanisms have been proposed (3, 65, 67, 75). In general, unsteady aerodynamics require more power at any velocity than steady-state, but return remarkable maneuverability (75).

Rayner (65) abandoned aerodynamic power estimates and proposed a general approach that estimates the momentum in the fluid flow of the wake of a flying animal. Since changes in fluid flow arise from the action of the wings, changes in momentum in these fluid flows must equal the combined lift and thrust forces generated by the animal. For many hovering insects, calculation by either momentum jet theory or wake kinetics produces similar estimates of power requirement.

GLIDING AND SOARING Considerable energy savings in locomotion can be effected by substituting periods of gliding for active flight. The forward momentum achieved in flight is rapidly dissipated during gliding by parasite and profile drag. To glide at constant velocity an animal can balance energy obtained from falling against losses from parasite and profile drag. For any given aerodynamic design, the higher the forward velocity the faster the sinking rate required. In general, animals of the size and Reynolds number of insects require large sinking rates (1–$3\text{m} \cdot \text{sec}^{-1}$) for even modest forward velocities (3–$6\text{m} \cdot \text{sec}^{-1}$) unless they possess large wings and light wingloading (3). Thus gliding is available to butterflies, large-winged moths (Saturniidae), and some dragonflies. Soaring, using rising air currents to provide lift, is only known in butterflies (26).

Small passerine birds gain advantages from bounding flight, alternating powered climbs with intermittent gliding dives (65). Similar flight behavior in insects has been reported only in some sphingid moths; its energetic consequences have not been analyzed.

HEAT TRANSFER An independent means of estimating the cost of activity is by assessing the rate of heat transfer from an insect to its surroundings. Most heat generated by insects is dissipated by natural or forced convective heat loss. Because there is no conductive loss, and radiative losses are minimal in flying insects, an approximation of natural convective heat transfer, Newton's Law of Cooling, is widely used to estimate heat loss and insulation (6, 9, 10, 18, 19, 46, 50).

During warm-up behaviors, mechanical losses are small. With few assumptions about the specific heat of insect tissue and heat distribution within the insect, reasonable estimates of the total energy expenditure can be calculated. In flight, as much as 20% of the power generated is dissipated in aerodynamic work (46, 76), and excess heat generated may be shunted to a poorly insulated abdomen or dissipated by auxiliary evaporative mechanisms (28, 31, 32, 34). Thus estimates of power requirement in flight from the heat dissipated by convection may underestimate total energy consumption by 50% or more (8, 46). Weis-Fogh found substantial agreement in the energy expended in flight as calculated from heat dissipation (74) and from rate of consumption of fat reserves (73). However, both measures were based on data from locusts that were probably not in true flight (31).

Direct calorimetry has not been applied to insects in flight or warm-up.

INDIRECT MEASURES OF POWER The total power expended in flight can be calculated from the oxygen consumption of flying animals and from the rate of utilization of flight fuels.

Oxygen consumption (or production of CO_2) is readily measured from resting insects. However, providing satisfactory confined space for free, uniform flight is apparently not feasible for animals moving $1-10 m \cdot sec^{-1}$, since no successful attempts have been reported. The metabolic rates of insects confined to a large jar and induced to hover, or at least prevented from landing, provide estimates of the total cost of hovering. The metabolic rate in flight increases 50–100 times above the resting rate. The following lists the range of power ($W \cdot N^{-1}$) in contained free flight for various insect groups: Odonata, Aeschnidae: 10–195 (46, 52); Lepidoptera, Nymphalidae: 32 (46); Megalopygidae: 32 (8); Sphingidae: 27–70 (8, 46); Saturniidae: 17–102 (8, 29, 46); Notodontidae: 37.8 (8); Noctuidae: 4–63 (8, 46); Lasiocampidae: 25–52 (8); Lymantriidae: 229 (16); Diptera, Syrphidae: 14 (46); Hymenoptera, Apidae: 33–65 (46); Vespidae: 10–14 (46). The total power requirement increases with the 0.8 power of body size, the 0.6–0.8 power of thoracic weight, and the 0.7–1.1 power of wing loading (8). Both the slope and intercept of log power on log body mass approximate those of flying birds (8). However, the scatter about such a regression is such that the power generated by animals of the same size appears to vary by a factor of 2–4-fold.

Among individuals of a given species—*Hyles lineata* for example—mass-specific power in hovering flight decreases with increasing body size as the –0.5 power of mass (15). Again, power varies 2–4-fold among individuals of the same size. This large range of measured power for hovering corresponds in magnitude to the range of aerodynamic power required for all flight speeds—i.e. from the minimum to maximum requirement for the U-shaped power-flight speed curve. If flight is considered uniform in these conditions, efficiency must vary from 2–20% among all groups or from 5–20% among animals of one size in the same species. This underscores the variability of response, and possibly efficiency, of insects in flight.

The rate of utilization of flight fuels as a function of voluntarily selected flight speeds provides another means of generating a power-speed relationship for flying insects. While the disappearance of flight fuels at intervals after flight begins can be measured readily, the factor of flight speed requires monitoring. Some insects fly consistently enough over intervals of 30 min or more when suspended from a flight mill to permit an estimation of power expended at approximately uniform velocity. Unfortunately, flight performance is affected by the inertia of the mill's suspending arm, the inflexibility of the fixed-mounted position, and the support of body weight by the round-about. Restriction to a repetitive, circular path may also interfere with the animal's guidance control.

Hocking (40) measured the utilization of glucose during prolonged flights of several insects mounted on flight mills. Uptake depended on duration of flight and exponentially on the self-selected flight speed. Calculations based on rates of glucose utilization showed power requirements of 36–58 $W \cdot N^{-1}$ for the bee, *Apis mellifera*, and 13–37 $W \cdot N^{-1}$ for the fly, *Tabanis affinus*, at flight speeds of 2–6 $m \cdot sec^{-1}$.

Different insect groups use carbohydrate (40, 46), lipids (27, 44, 73), and amino acids, notably proline (47), as circulating flight fuels (46). The dynamics of mobilization of reserves and of switching from one fuel type to another are generally measured in insects attached to a flight mill (46). Carefully conceived biochemical studies may be weakened because the insects on a flight mill are not in true flight (39). A recent study, for example, reports the effect on flight speed in the locust of partitioning flight fuel between carbohydrate and lipid sources (27). However, the maximum flight speed attained was only 1.75 $m \cdot sec^{-1}$ which is below the stall speed of the locust ($\sim$2.5 $m \cdot sec^{-1}$) (76).

Weis-Fogh (73) measured the rate of utilization of fat by locusts during extended flights on a large flight mill; 80–85% of the power was supplied from lipid reserves. His locusts flew at velocities of 2.3–3.6 $m \cdot sec^{-1}$ and expended 6–18 $W \cdot N^{-1}$; power requirement varied as the square of velocity. However, over that flight-speed range, aerodynamic calculations would suggest a difference in power requirement of about 10% rather than three fold (76, 77). In fact, aerodynamic considerations suggest a required range in sustainable power of only 150% for the entire range of flight velocity in the locust (77). The two estimates cannot be reconciled.

WARM-UP Some endothermic insects are able to shift the alternating neural control pattern of flight to synchronous activation of antagonistic flight muscles during warm-up (45, 46). Those with fibrillar or asynchronous flight muscles uncouple the flight muscles mechanically during warm-up (45). In either case, all the energy of muscle contraction is dissipated as heat. The rate of warm-up increases with ambient temperature in all groups and varies from about 1–7°C min^{-1} (35). In the final 30 sec of a warm-up bout, metabolic rate approaches 50–100% of that measured in the flying insect (35, 46). Although warm-up lasts only a few minutes, it is a significant additional cost of flight activity. In animals showing intermittent flight, the body temperature may be maintained between flights by the warm-up mechanism, which is an additional cost of activity (45, 46).

In summary, the data on power requirements of flight in insects large enough to be endothermic are few and fragmentary, owing primarily to the inconsistent aerobatic flight behavior of such insects. Fine-grain studies of the significance of special aerodynamic features in particular insect groups will depend on the application of appropriate aerodynamic principles. Further, instantaneous force measurement (20, 81), radioisotope labeling of metabolic processes (70, 71), and improvement in wind tunnel or flight mill (25) design may ultimately result in less variation in measured power at all flight speeds.

ACKNOWLEDGMENT

M. S. Heath is supported by DHHS Training Grant 5T32CA 09067.

Literature Cited

1. Adams, P. A. 1969. How moths keep warm. *Discovery* 4:83–88
2. Alexander, R. McN. 1973. Muscle performance in locomotion and other strenuous activities. See Ref. 14, pp. 1–21
3. Alexander, R. McN. 1977. Flight. See Ref. 4, pp. 249–78
4. Alexander, R. McN., Goldspink, G., eds. 1977. *Mechanics and Energetics of Animal Locomotion*, London: Chapman and Hall. 346 pp.
5. Balciunas, J., Knopf, K. 1977. Orientation, flight speeds, and tracks of three species of migrating butterflies. *Fla. Entomol.* 60:37–39
6. Bartholomew, G. A. 1981. A matter of size: An examination of endothermy in insects and terrestrial vertebrates. See Ref. 35, pp. 45–78
7. Bartholomew, G. A., Casey, T. M. 1977. Endothermy during terrestrial activity in large beetles. *Science* 195:882–83
8. Bartholomew, G. A., Casey, T. M. 1978. Oxygen consumption of moths during rest, pre-flight warm-up, and flight in relation to body size and wing morphology. *J. Exp. Biol.* 76:11–25
9. Bartholomew, G. A., Epting, R. J. 1975. Allometry of post-flight cooling rates in moths: A comparison with vertebrate homeotherms. *J. Exp. Biol.* 63:603–13
10. Bartholomew, G. A., Epting, R. J. 1975. Rates of post-flight cooling in sphinx moths. In *Perspectives of Biophysical Ecology*, ed. D. M. Gates, R. B. Schmerl, 12:405–15. NY: Springer. 609 pp.
11. Bartholomew, G. A., Heinrich, B. 1978. Endothermy in African dung beetles during flight, ball making and ball rolling. *J. Exp. Biol.* 73:65–83
12. Bennett, L. 1975. Insect aerodynamics near hovering. See Ref. 82, 2:815–28
13. Bennet-Clark, H. C. 1975. The energetics of the jump of the locust, *Schistocerca gregaria*. *J. Exp. Biol.* 63:53–83
14. Bolis, L., Schmidt-Nielsen, K., Maddrell, S. H. P., eds. 1973. *Comparative Physiology: Locomotion, Respiration, Transport and Blood*, Amsterdam: North Holland. 634 pp.

15. Casey, T. M. 1976. Flight energetics of sphinx moths: power input during hovering flight. *J. Exp. Biol.* 64:529–43
16. Casey, T. M. 1980. Flight energetics and heat exchange of gypsy moths in relation to air temperature. *J. Exp. Biol.* 88:133–45
17. Casey, T. M. 1981. A comparison of mechanical and energetic estimates of flight cost for hovering sphinx moths. *J. Exp. Biol.* 91:117–129
18. Church, N. S. 1960. Heat loss and the body temperature of flying insects. I. Heat loss by evaporation of water from the body. *J. Exp. Biol.* 37:171–85
19. Church, N. S. 1960. Heat loss and the body temperature of flying insects. II. Heat conduction within the body and its loss by radiation and convection. *J. Exp. Biol.* 37:186–213
20. Cloupeau, M., Devillers, J. F., Devezeaux, D. 1979. Direct measurements of instantaneous lift in desert locust; comparison with Jensen's experiments on detached wings. *J. Exp. Biol.* 80:1–15
21. Corbet, P. S. 1980. Biology of Odonata. *Ann. Rev. Entomol.* 25:189–217
22. Counter, S. A. Jr. 1977. Bioacoustics and neurobiology of communication in the tettigoniid *Neoconocephalus robustus. J. Insect Physiol.* 23:993–1008
23. Delcomyn, F. 1971. The locomotion of the cockroach *Periplaneta americana. J. Exp. Biol.* 54:443–52
24. Ellington, C. P. 1978. The aerodynamics of normal hovering flight: three approaches. In *Comparative Physiology: Water, Ions and Fluid Mechanics*, ed. K. Schmidt-Nielsen, L. Bolis, S. H. P. Maddrell, pp. 327–45. Cambridge: Cambridge Univ. Press. 360 pp.
25. Esch, H. 1976. Body temperature and flight performance of honey bees in a servo-mechanically controlled wind tunnel. *J. Comp. Physiol.* 109:265–77
26. Gibo, D. L., Pallett, M. J. 1979. Soaring flight of monarch butterflies, *Danaus plexippus* (Lepidoptera, Danaidae), during the late summer migration in southern Ontario, Canada. *Can. J. Zool.* 57:1391–401
27. Goldsworthy, G. J., Jutsum, A. R., Robinson, N. L. 1979. Substrate utilization and flight speed during tethered flight in the locust. *J. Insect Physiol.* 25:183–86
28. Hanegan, J. L. 1973. Control of heart rate in cecropia moths: response to thermal stimulation. *J. Exp. Biol.* 59:67–76
29. Hanegan, J. L., Heath, J. E. 1970. Activity patterns and energetics of the moth, *Hyalophora cecropia. J. Exp. Biol.* 53:611–27
30. Heath, J. E., Josephson, R. K. 1970. Body temperature and singing in the katydid, *Neoconocephalus robustus* (Orthoptera, Tettigoniidae). *Biol. Bull.* 138:272–85
31. Heinrich, B. 1974. Thermoregulation in endothermic insects. *Science* 185:747–56
32. Heinrich, B. 1976. Heat exchange in relation to blood flow between thorax and abdomen in the bumblebee, *Bombus vosnesenskii. J. Exp. Biol.* 64:561–85
33. Heinrich, B. 1979. *Bumblebee Economics.* Cambridge, MA: Harvard Univ. Press. 245 pp.
34. Heinrich, B. 1979. Keeping a cool head: Honeybee (*Apis mellifera*) thermoregulation. *Science* 205:1269–71
35. Heinrich, B., ed. 1981. *Insect Thermoregulation,* NY: Wiley. 328 pp.
36. Heinrich, B., Bartholomew, G. A. 1979. Roles of endothermy and size in interspecific and intraspecific competition for elephant dung in an African dung beetle, *Scarabaeus laevistriatus. Physiol. Zool.* 52:484–96
37. Heinrich, B., Bartholomew, G. A. 1979. The ecology of the African dung beetle. *Sci. Am.* 241:146–56
38. Herreid, C. F. II, Prawel, D. A., Full, R. J. 1981. Energetics of running cockroaches. *Science* 212:331–33
39. Hersch, M. I., Hepburn, H. R., Skews, B. W. 1980. Some tethered flight characteristics of a rose chafer beetle, *Pachynoda sinuata* (Coleoptera: Scarabaeidae). *Comp. Biochem. Physiol.* 65A:505–8
40. Hocking, B. 1953. The intrinsic range and speed of flight of insects. *Trans. R. Entomol. Soc. Lond.* 104:223–345
41. Hughes, G. M., Mill, P. J. 1974. Locomotion: Terrestrial. See Ref. 66, 3:335–79
42. Jensen, T. F., Holm-Jensen, I. 1980. Energetic cost of running in workers of 3 ant species, *Formica fusca, Formica rufa* and *Camponotus herculeanus* (Hymenoptera, Formicidae). *J. Comp. Physiol.* 137B:151–56
43. Josephson, R. K. 1981. Temperature and the mechanical performance of insect muscle. See Ref. 35, pp. 19–44
44. Jutsum, A. R., Goldsworthy, G. J. 1976. Fuels for flight in *Locusta. J. Insect Physiol.* 22:243–49
45. Kammer, A. E. 1981. Physiological mechanisms of thermoregulation. See Ref. 35, pp. 115–58

46. Kammer, A. E., Heinrich, B. 1978. Insect flight metabolism. *Adv. Insect Physiol.* 13:133–229
47. Khan, M. A., De Kort, C. A. D. 1978. Further evidence for the significance of proline as a substrate for flight in the Colorado potato beetle. *Comp. Biochem. Physiol.* 60B:407–11
48. McConnell, E., Richards, A. G. 1955. How fast can a cockroach run? *Bull. Brooklyn Entomol. Soc.* 50:36–43
49. Makino, S., Yamane, S. 1980. Heat production by the foundress of *Vespa simillima*, with description of its embryo nest (Hymenoptera: Vespidae). *Insecta Matsumurana* 0(19):89–101
50. May, M. L. 1976. Thermoregulation and adaptation to temperature in dragonflies (Odonata: Anisoptera). *Ecol. Monogr.* 46:1–32
51. May, M. L. 1976. Warming rates as a function of body size in periodic endotherms. *J. Comp. Physiol.* 111B:55–70
52. May, M. L. 1979. Energy metabolism of dragonflies (Odonata: Anisoptera) at rest and during endothermic warm-up. *J. Exp. Biol.* 83:79–94
53. May, M. L. 1979. Insect thermoregulation. *Ann. Rev. Entomol.* 24:313–49
54. May, M. L., Wilkin, P. J., Heath, J. E., Williams, B. A. 1980. Flight performance of the moth, *Manduca sexta*, at variable gravity. *J. Insect Physiol.* 26:257–66
55. Nachtigall, W. 1973. Investigation of wing movements and the generation of aerodynamic forces in flying Diptera. See Ref. 14, pp. 77–97
56. Nachtigall, W. 1976. Wing movements and the generation of aerodynamic forces by some medium-sized insects. See Ref. 63, pp. 31–47
57. Newman, B. G., Savage, S. B., Schouella, D. 1977. Model tests on a wing section of an *Aeschna* dragonfly. See Ref. 61, pp. 445–77
58. Nicolson, S. W., Louw, G. N. 1980. Preflight thermogenesis, conductance and thermoregulation in the protea beetle, *Trichostetha fascicularis* (Scarabaeidae, Cetoniinae). *S. Afr. J. Sci.* 76:124–26
59. Nikolaev, N. A. 1974. Speeds and types of flight of some Pieridae and Nymphalidae (Lepidoptera: Pieridae, Nymphalidae). *Vestn. Mosk. Univ. Biol.* 29:(4):59–62 (In Russian)
60. Norberg, R. A. 1975. Hovering flight of the dragonfly *Aeschna juncea L.*, kinematics and aerodynamics. See Ref. 82, 2:763–81
61. Pedley, T. J., ed. 1977. *Scale Effects in Animal Locomotion.* London: Academic. 545 pp.
62. Pringle, J. W. S. 1974. Locomotion: Flight. See Ref. 66, 3:433–76
63. Rainey, R. C., ed. 1976. *Insect Flight.* Oxford: Blackwell Scientific. 287 pp.
64. Rawlins, J. E. 1980. Thermoregulation by the black swallowtail butterfly, *Papilio polyxenes* (Lepidoptera, Papilionidae). *Ecology* 61:345–57
65. Rayner, J. M. V. 1979. A new approach to animal flight mechanics. *J. Exp. Biol.* 80:17–54
66. Rockstein, M., ed. 1974. *The Physiology of Insecta,* Vols. 3, 4. NY: Academic. 517 pp.; 448 pp. 2nd ed.
67. Savage, S. B., Newman, B. G., Wong, D. T.-M. 1979. The role of vortices and unsteady effects during the hovering flight of dragonflies. *J. Exp. Biol.* 83:59–77
68. Seeley, T., Heinrich, B. 1981. Regulation of temperature in the nests of social insects. See Ref. 35, pp. 159–234
69. Stevens, E. D., Josephson, R. K. 1977. Metabolic rate and body temperature in singing katydids. *Physiol. Zool.* 50:31–42
70. Taylor, P. 1978. Radioisotopes as metabolic labels for *Glossina* (Diptera: Glossinidae). I. Laboratory determination of the relationship between radioisotope metabolism, respiration and temperature. *Bull. Entomol. Res.* 68:1–10
71. Taylor, P. 1978. Radioisotopes as metabolic labels for *Glossina* (Diptera: Glossinidae). II. The excretion of ^{137}Cs under field conditions as a means of estimating energy utilisation, activity and temperature regulation. *Bull. Entomol. Res.* 68:331–40
72. Tucker, V. A. 1973. Aerial and terrestrial locomotion: a comparison of energetics. See Ref. 14, pp. 63–76
73. Weis-Fogh, T. 1952. Fat combustion and metabolic rate of flying locusts (*Schistocerca gregaria* Forskal). *Philos. Trans. R. Soc. Lond. Ser. B* 237:1–36
74. Weis-Fogh, T. 1964. Biology and physics of locust flight. VIII. Lift and metabolic rate of flying locusts. *J. Exp. Biol.* 41:257–71
75. Weis-Fogh, T. 1973. Quick estimates of flight fitness in hovering animals, including novel mechanisms for lift production. *J. Exp. Biol.* 59:169–230
76. Weis-Fogh, T. 1975. Flapping flight and power in birds and insects, conventional and novel mechanisms. See Ref. 82, 2:729–62

77. Weis-Fogh, T. 1976. Energetics and aerodynamics of flapping flight: a synthesis. See Ref. 63, pp. 48–72
78. Weis-Fogh, T. 1977. Dimensional analysis of hovering fligbt. See Ref. 61, pp. 405–20
79. Weis-Fogh, T., Alexander, R. McN. 1977. The sustained power output obtainable from striated muscle. See Ref. 61, pp. 511–25
80. Wendler, G. 1966. The co-ordination of walking movements in arthropods. In *Nervous and Hormonal Mechanisms of Integration. Symp. Soc. Exp. Biol.* 20:229–49
81. Wilkin, P. J. 1971. *A preliminary study of the instantaneous forces on the thorax of a sphingid moth and the desert locust flying in a wind tunnel.* PhD thesis. University of Illinois, Urbana. 90 pp.
82. Wu, T. Y.-T., Brokaw, C. J., Brennan, C., eds. 1975. *Swimming and Flying in Nature,* 2:412–1005. NY: Plenum.

RENAL AND ELECTROLYTE PHYSIOLOGY

Introduction, Thomas E. Andreoli, *Section Editor*

Two of the more fascinating aspects of renal physiology—and ones few of us might have envisioned two decades ago—are the striking degrees of heterogeneity of nephron structure and function that occur either (*a*) axially within a given nephron segment (e.g. for S_1, S_2, and S_3 regions of the proximal nephron) or (*b*) depending on nephron topography (i.e. whether a given nephron is juxtamedullary or superficial). Indeed, the proliferation of experimental data on axial and topographical nephron heterogeneity—both in structural and functional terms—has occurred at a rate sufficiently rapid to befuddle those of us accustomed to thinking of the nephron in simpler terms: a proximal tubule, loop of Henle, distal convolution, and collection duct, all conveniently invariant, either structurally or functionally, with respect to nephron topography. Consequently, it seems desirable to summarize briefly some of the facets and biologic implications of nephron heterogeneity.

The three articles in this Section are focussed on these issues. In the first, R. E. Bulger and D. C. Dobyan have provided a detailed description of various types of nephrons; of the architectural detail of the various regions of the nephron, as well as of renal cells in tissue culture; and of some of the variations of nephron types among different species. In the second paper, C. A. Berry has summarized the transport properties of various segments of the nephron, as they are currently proposed, and how these transport characteristics vary, for a given nephron segment, with in vivo nephron topography. Finally, L. A. Walker and H. Valtin reflect—concretely and with prudent speculation—on the significance of nephron heterogeneity to the integrated function of the kidney. These authors illustrate their argument by considering in detail the implications of nephron heterogeneity for three cardinal aspects of renal function: phosphate homeostasis, the renal response to extracellular fluid volume expansion, and the renal water repletion reaction.

RECENT ADVANCES IN RENAL MORPHOLOGY

Ruth Ellen Bulger and Dennis C. Dobyan

Department of Pathology and Laboratory Medicine, University of Texas Medical School, Houston, Texas 77030

> To all students for whom the ever-widening horizons and increasing details of medicine make the art of healing more difficult and yet more certain.
>
> Homer W. Smith, *Principles of Renal Physiology* New York 1956

During the last two decades powerful new instruments (e.g. the transmission and scanning electron microscopes) have provided exciting new insights into the architecture of the mammalian kidney. Precise morphological details have recently been described for a variety of species, and numerous structural differences among the species are becoming evident. We must correlate these new observations with known physiological differences in order to better understand the structural-functional relationships of the nephron. The precise details of human renal anatomy are still illusive owing to the unavailability of well-fixed normal tissue. This review reflects on several important new observations regarding kidney structure.

Three basic principles underlie modern morphological studies. First, the physiological state of the animal at the time of study must be accurately defined. Second, the combined use of light microscopy, transmission and scanning electron microscopy, freeze-fracture, and histochemistry allows one to eliminate artifacts that may inhere in any one technique. Finally, application of stereological techniques avoids experimental bias and allows the presentation of quantitative results.

Detailed studies have led to a more precise definition of terms. For example, three families of renal corpuscles are being commonly described in dog (13, 14, 15), human (14), rabbit (49), and *Psammomys* (8) on the basis of their position in the renal cortex and the patterns formed by their efferent vessels (Figure 1). Superficial renal corpuscles frequently have long efferent vessels that course to the kidney surface and divide to form stellate vessels. Efferent vessels from midcortical renal corpuscles either divide abruptly to form peritubular capillary plexuses near the glomerulus or they form a long-meshed capillary network in the medullary ray. Juxtamedullary efferent arterioles most often course downward to form vascular bundles, though a few divide near the glomerulus. Renal corpuscles are therefore best defined as superficial, midcortical, or juxtamedullary.

Nephrons are best classified as short-looped or long-looped on the basis of whether their loops of Henle turn in the outer or inner medulla. Most

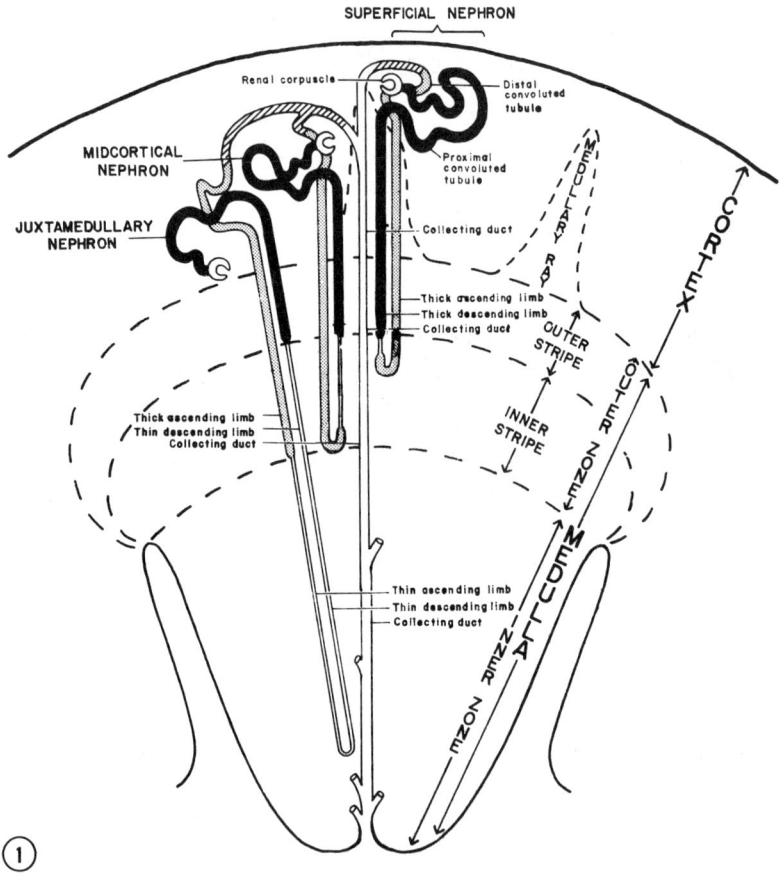

Figure 1 Schematic diagram showing the 3 types of renal corpuscles and the positions of short- and long-looped nephrons.

species have both types; superficial renal corpuscles give rise to short-looped nephrons, juxtamedullary renal corpuscles give rise to long-looped nephrons, and midcortical renal corpuscles give rise to either type. Certain species have only short-looped or long-looped nephrons (91). In these, the nephrons with the shortest loops share certain morphologic features with the short-looped nephrons of other species. For example, in the cat, two separate types of descending thin limbs can be distinguished, and one is structurally similar to the short-looped nephrons of other species. Figure 2 is a schematic representation of the segments from the urinary tubule.

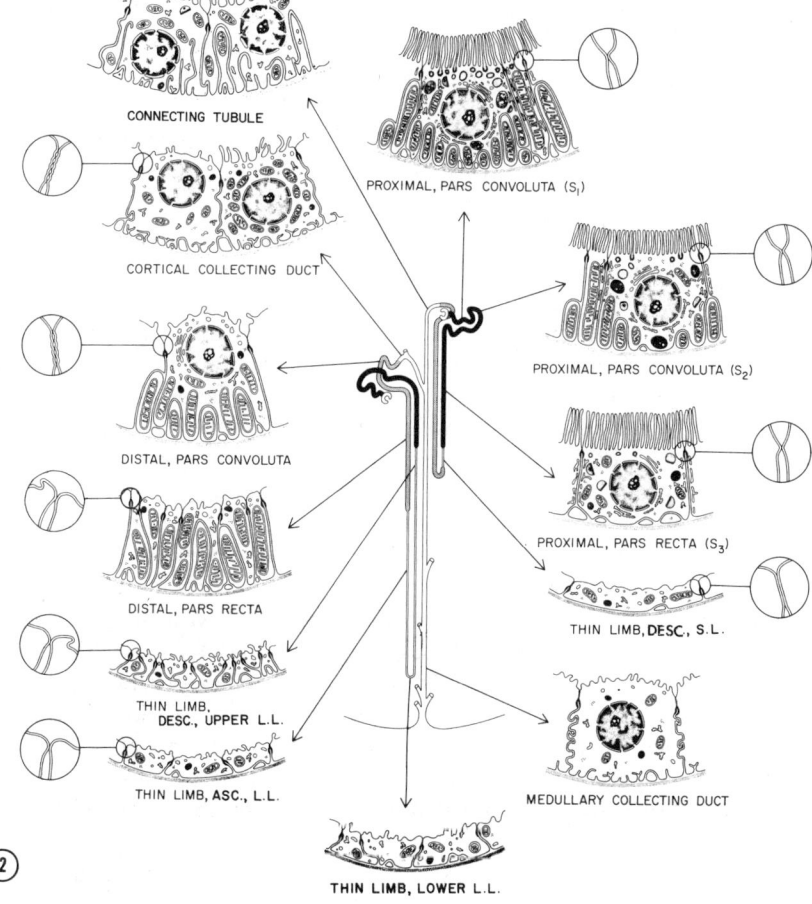

Figure 2 Schematic representation of short- and long-looped nephrons and their various cellular components.

GLOMERULAR BASEMENT MEMBRANE OF THE RENAL CORPUSCLE (Figures 3–5).

The glomerular basement membrane (GBM) is a complex structure containing collagenous and noncollagenous glycoproteins and proteoglycans (Figure 3). The collagenous part of the GBM and mesangial matrix consists of several Type IV collagens and a Type V (AB_2) collagen (83). Using affinity-purified rabbit antibodies, Madri et al (65) have recently identified the glycoproteins, laminin and fibronectin, in the glomerulus. Fibronectin, a molecule thought to mediate cell adhesion of mesenchymal cells, was localized in the mesangial matrix around mesangial cell processes by light and electron microscopy. Laminin was localized in the mesangial matrix, in the lamina rara interna, and in adjacent regions of the lamina densa of the GBM, as well as throughout the entire thickness of the tubular basement membranes (65).

Kanwar & Farquhar (52) found sulfated glycosaminoglycans (GAG) concentrated in particles distributed in a regular lattice-like network in the GBM. To assess the role of this GAG in the permeability of the GBM, Kanwar et al (53) subjected glomerular basement membranes to digestion in situ by vascular perfusion of glycosaminoglycan-degrading enzymes and then subsequently perfused native ferritin (diameter 11 nm). The enzyme heparinase removes most of the glycosaminoglycans, including heparin sulfate. The perfusion of heparinase resulted in a dramatic increase in GBM permeability to native ferritin. Ferritin was seen in large amounts in the urinary spaces and could be identified within the lamina densa and lamina rara externa of the GBM. The permeability increase seemed to be related to the removal of heparin sulfate, since no permeability increase to ferritin was noted after removal of hyaluronic acid (by *Streptomyces* hyaluronidase treatment) or after removal of most other glycosaminoglycans (chondroitins 4- and 6-sulfates and dermatan sulfate) by chondroitinase-ABC.

A number of recent studies have been concerned with the glomerular visceral epithelial cell (podocyte) (Figures 3,4) and the maintenance of its elaborate cell shape. Seiler et al (90) had demonstrated structural changes in podocyte shape after the in vivo administration of polycations. Using an in vitro model, Andrews & Coffey (3) examined the response of glomerular epithelium, exposed on the surface of kidney slices, to various agents that might effect their shape. Agents that disrupted microtubules (vinblastine or colchicine) induced a loss in the morphological integrity of the podocyte cell body and major cell processes. Cytochalasin B, on the other hand, appeared to inhibit cation-induced loss of podocyte foot processes.

Cox et al (26) using a model of norepinephrine-induced acute renal failure, and Stein et al (93) using a model of uranyl nitrate-induced acute

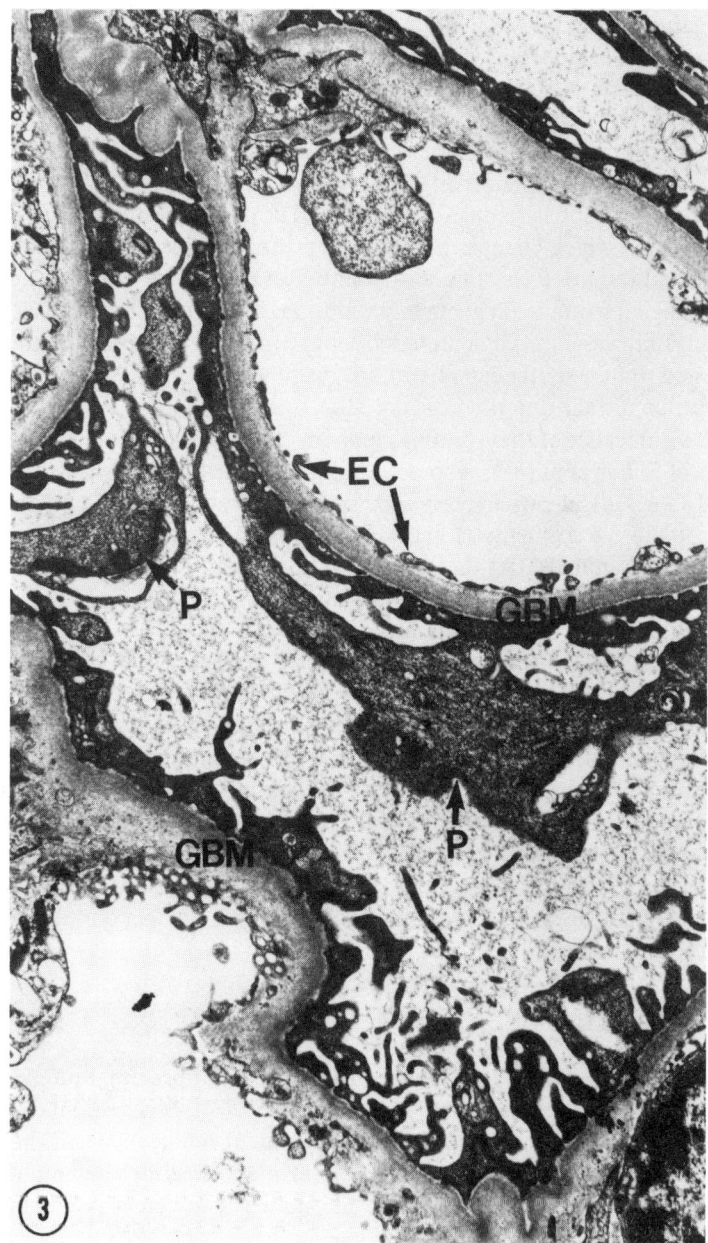

Figure 3 Transmission electron micrograph of human renal corpuscle showing endothelial cells (EC), podocytes (P), mesangium (M), and glomerular basement membrane (GBM). × 7,080.

renal failure proposed that dramatic changes in the shape of glomerular podocytes were important in the loss of normal glomerular function in these models. Similar changes were recently reported in ischemic acute renal failure (9). Other investigators have seen either no change or only minor alterations in similar models of acute renal failure (7,22,28). These divergent results are unexplained.

Similar interest has recently been directed at the glomerular endothelium (Figure 5), in which changes have been reported in experimental models of acute renal failure. Reporting a dramatic decrease in the number and size of endothelial fenestrae in gentamicin-induced acute renal failure (5,32) and in uranyl nitrate-induced acute renal failure (6) in rats, these investigators suggested that endothelial alterations were related to a change in the glomerular ultrafiltration coefficient (K_f).

The significance of these findings remains questionable in light of a recent report of Schor et al (87), who detected no significant differences in endothelial fenestrae in rats treated with gentamicin or tobramycin. They did occasionally see segments of endothelial cells with sparse fenestrae in the drug-treated animals, but these were infrequent and difficult to quantitate. Similar areas can be seen in control animals (R. E. Bulger, G. Eknoyan, D. C. Dobyan, unpublished observations).

The mesangial cells may effect glomerular permeability by contraction, thereby causing a change in effective capillary surface area. In this respect, three types of cells were cloned from glomerular epithelial cells and maintained in culture (56): glomerular epithelial cells, contractile mesangial cells, and a cell type that produced renin. Arginine vasopressin and angiotensin II caused purified cultures of these rat mesangial cells to contract in vitro (4). Barnes et al (10) concluded that the binding sites present in cultured mesangial cells represent physiologically important receptors and may regulate parameters of filtration by binding to and inducing contraction in these cells.

THE PROXIMAL TUBULE (Figures 6–8)

Detailed descriptions of the three segments of the rat proximal tubule have been available for several years [Figure 6, (19,67,98)]. More recent descriptions of the three segments of the rabbit proximal tubule (49) and the four segments of the dog proximal tubule (21) make it clear that numerous species variations occur, and hence physiological differences among the species should be anticipated.

The proximal convoluted tubule transports large quantities of salt and water isosmotically while maintaining concentration gradients for certain specific solutes such as glucose. Such findings have focused interest on the role of the paracellular pathway, which is comprised of the tight junction

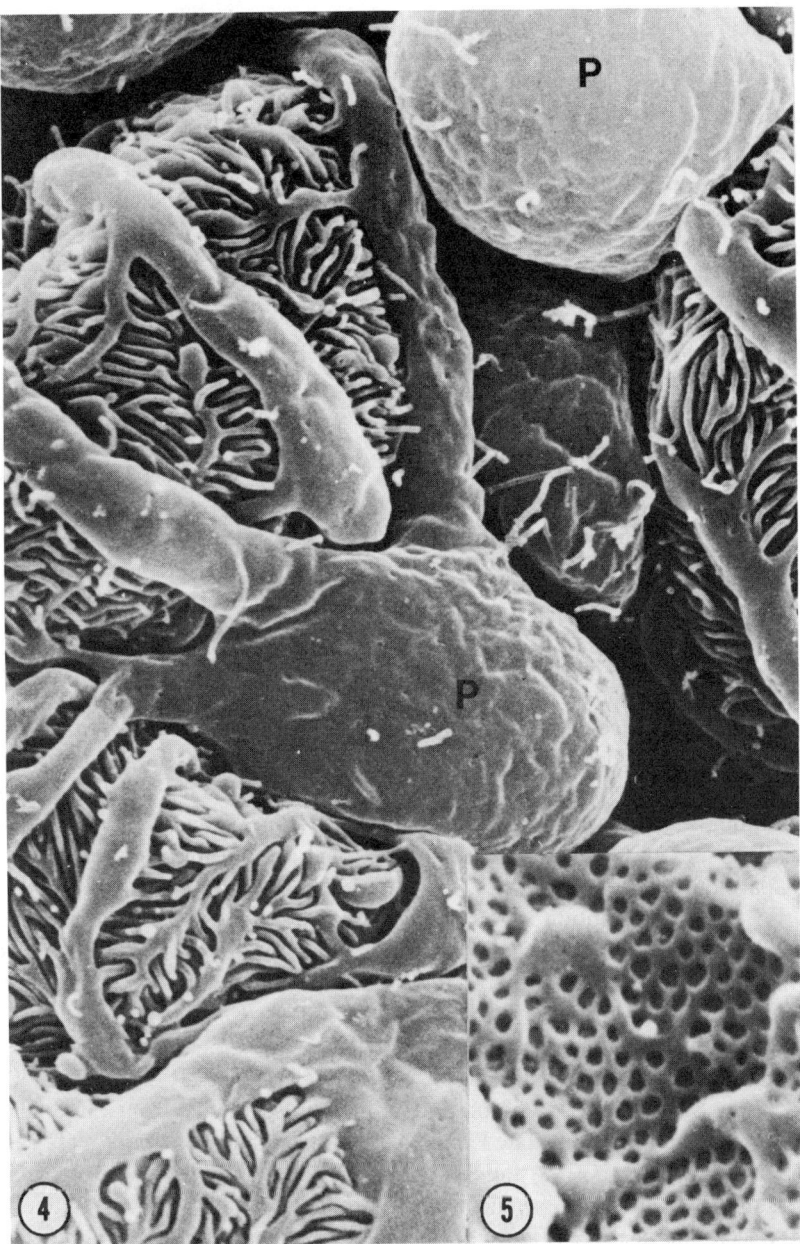

Figure 4 Scanning electron micrograph of a rabbit renal corpuscle showing podocytes (P) and interdigitating pedicels. X 9,000.

Figure 5 (Inset) Scanning electron micrograph showing the fenestrated endothelium of the glomerular capillary. X 32,000.

(zonula occludens) and the lateral intercellular space. It is well known that the proximal convoluted tubule, with its many lateral interdigitating processes, establishes a complicated system of lateral intercellular spaces (Figure 7). The tight junctions of the proximal convolutions of rat kidneys have been described as comprising a single (or double) tenuous fibril(s) with some discontinuities (98). A careful comparative study of the tight junctions of pars convoluta and pars recta in a variety of species (82) indicated important differences in junctional morphology. Tight junctions from pars recta cells were characterized by a larger number of junctional fibrils than pars convoluta cells of the same species. The tight junctions of pars convoluta and pars recta in dog, tree shrew (*tupaia*), and cat have a greater depth, which suggest functional differences in these species.

Maunsbach & Boulpaep (68) studied the effect of hydrostatic pressure changes on the ultrastructure of the paracellular shunt pathway in the proximal tubule of *Necturus*. In control and volume-expanded stopped-flow tubules at high luminal pressure, the average width of the lateral intercellular space significantly decreased. In contrast, in control stopped-flow tubules at low luminal pressure, those lateral intercellular spaces widened. These ultrastructural changes correlated with applied transepithelial pressure gradients but not with transepithelial volume fluxes. Under these conditions, the electrical resistance increases (17). Maunsbach & Boulpaep described no change in the width or length of tight junctions in these experiments. Since an increase in the permeability for sodium chloride and raffinose was induced by volume expansion with no widening in intercellular spaces, they postulated a change in the junctional complex in this situation. Earlier freeze-fracture preparations from rat and *Necturus* proximal tubules showed an increase in junctional discontinuities after volume expansion (42,45).

A careful description of the rabbit proximal tubule from superficial and juxtamedullary nephrons (103) demonstrated that the cortical part of the rabbit pars recta was largely composed of S_2 segments while the outer medullary portion was comprised of S_3 segments. Using isolated perfused tubules, these authors demonstrated that the S_2 segment had the highest secretory rate for para-aminohippuric acid (PAH) ($S_1= 281 \pm 21$; $S_2= 1,508 \pm 104$; $S_3= 318 \pm 46$ fmol $mm^{-1}min^{-1}$ in segments from superficial glomeruli). It was speculated that earlier studies had used pars recta segments containing both S_2 and S_3 regions. This may have led to the incorrect assumption that the S_3 segment of the proximal tubule was most active in the secretion of PAH (Figure 8). Thus it is important to use morphological and physiological techniques in concert.

Maack et al (64) reviewed the role of the proximal tubule in the plasma turnover and in the maintenance of adequate levels of low molecular weight

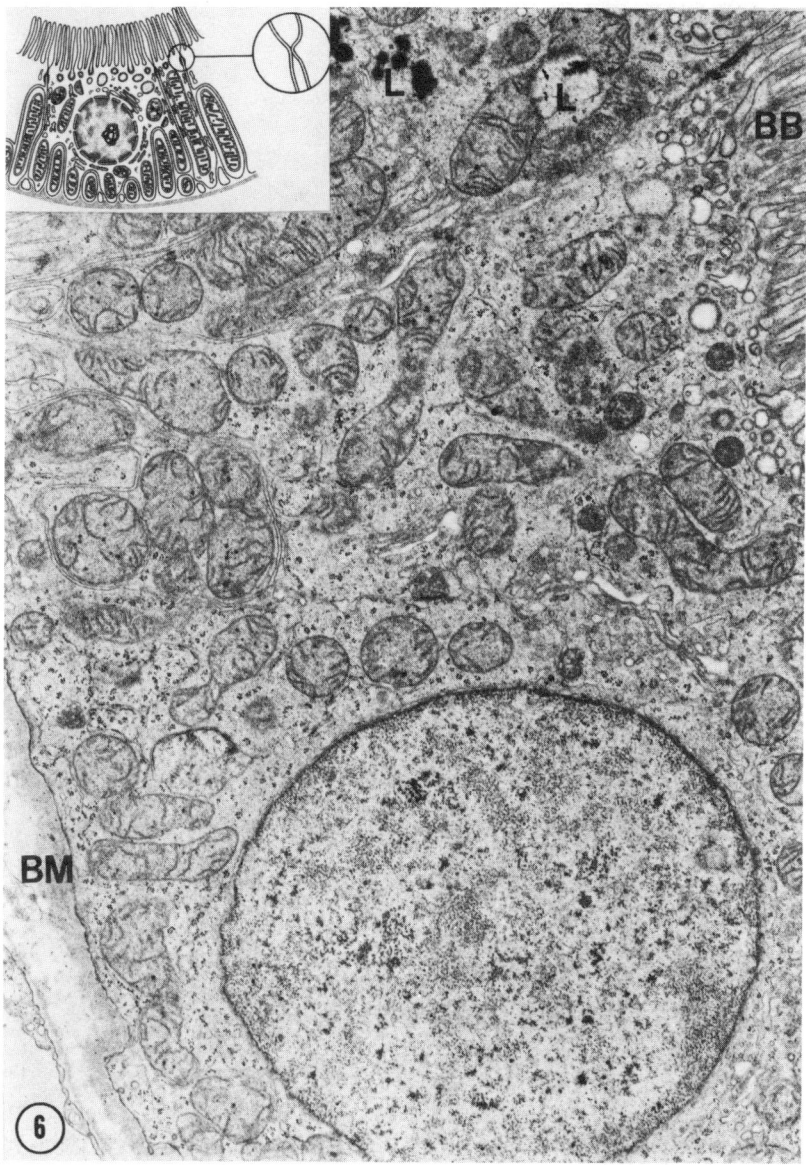

Figure 6 Transmission electron micrograph of a human proximal convoluted tubule. Note microvillus brush border (BB), large lysosomes (L) and tubular basement membrane (BM). × 8,800.

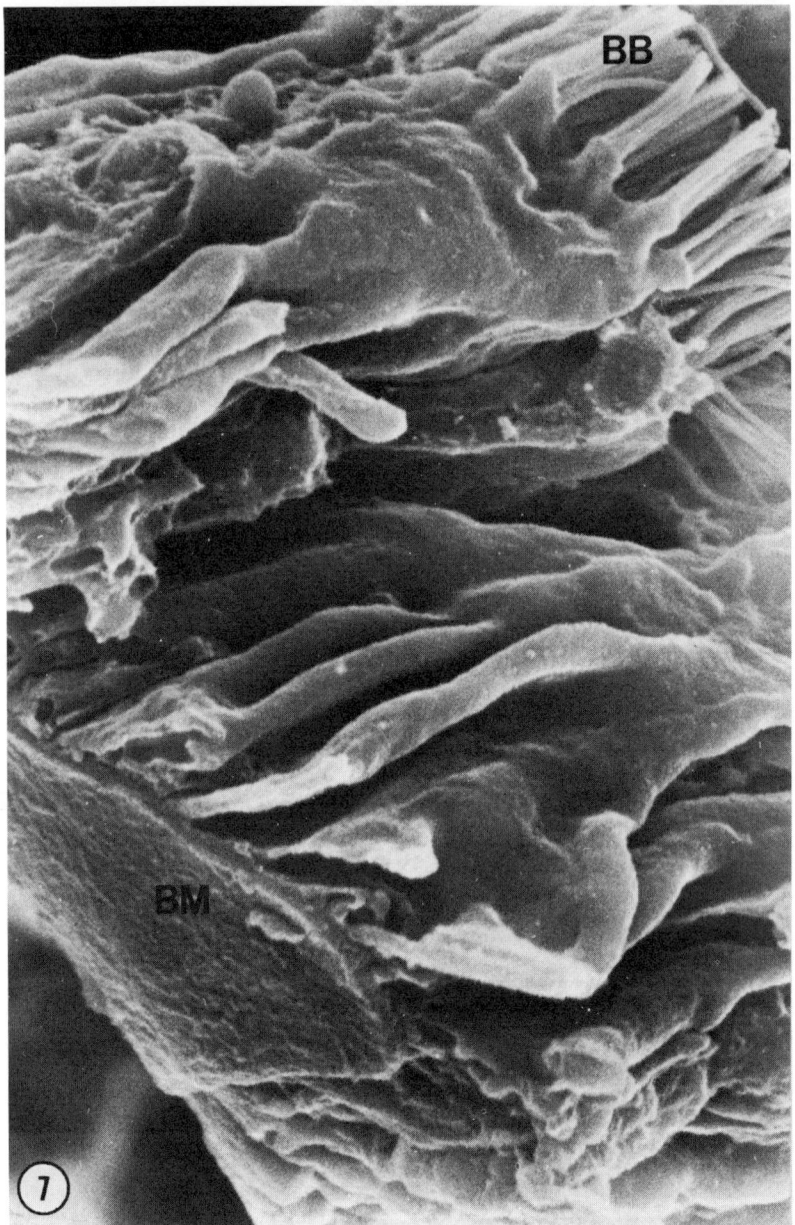

Figure 7 Scanning electron micrograph of a rat proximal convoluted tubule. Note the microvillus brush border (BB), extensive lateral cell processes and tubular basement membrane (BM). X 15,000. *J. Am. Med. Assoc.* 1980. 244:704–5. Copyright American Medical Association.

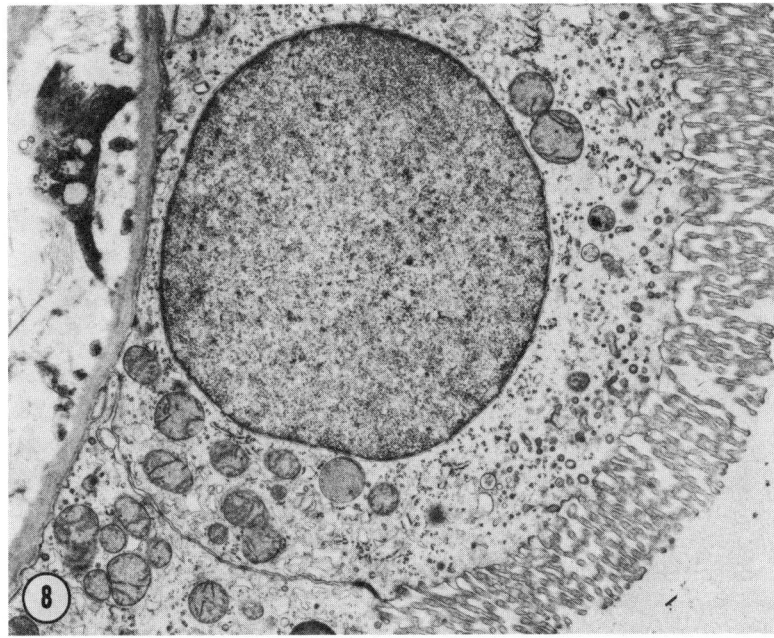

Figure 8 Transmission electron micrograph of a human pars recta segment of the proximal tubule. × 6,800.

proteins such as lysozyme, immunoproteins, and peptide hormones. The degree of renal uptake was inversely related to molecular size as would be expected if uptake was related to glomerular filtration. In general, these small proteins were taken up by the endocytotic apparatus of the proximal tubule at the apical border and then were catabolized in the lysosomes. Although this appeared to be the major pathway, some proteins may interact with receptors and be modified on peritubular sites.

THE THIN LIMB SEGMENTS (Figures 9–11)

Four regions of the thin limbs have been described on the basis of their positions in the kidney and their differing morphology. They are the descending thin limb of short-looped nephrons, the upper (outer medullary) descending thin limb of long-looped nephrons (Figure 10), the lower (inner medullary) descending thin limb of long-looped nephrons, and the ascending thin limb of long-looped nephrons (Figure 10). The morphology of the thin limbs has been described in a variety of species but as yet little is known about the various thin limb segments in humans (Figure 9).

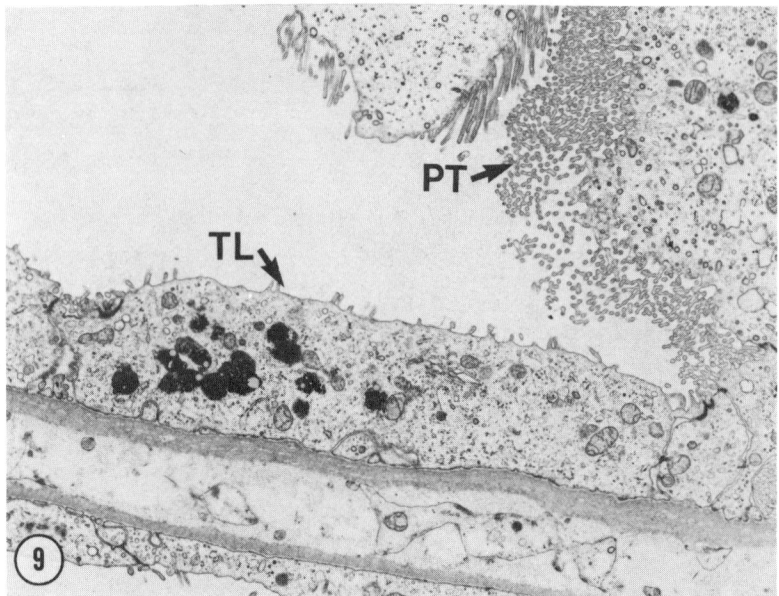

Figure 9 Transmission electron micrograph of a transition from a human pars recta (P_3 segment) of the proximal tubule (PT arrow) to descending thin limb of Henle (TL arrow). X 4,900.

The descending thin limbs of short-looped nephrons have a small luminal diameter and are lined by a flat noninterdigitated epithelium that contains sparse cell organelles. Their morphology is similar in all species studied (49,88). In fact, in cat, a species characterized only by long-looped nephrons, many thin limbs resemble the descending thin limbs of short-looped nephrons seen in other species. In certain species characterized by their high ability to concentrate urine—e.g. rat (59), mouse (60), and *Psammomys* (51)—these thin limbs travel in vasa recta bundles separated from the descending thin limbs of long-looped nephrons, whereas in other species—e.g. Syrian hamster (62), rabbit (49), and cat (61)—they travel in the interbundle region. They have a deep tight junction with the highest mean number of junctional fibrils of any region of the thin limbs [4.8 ± 1.1 fibrils in rabbit (86), 3.75 ± 0.19 fibrils in rat (89)]. This may reflect a lower permeability to sodium chloride in this segment. Few intramembranous particles were seen in luminal and basal-lateral membranes in the rat (89). Studies of toad bladder suggested that intramembranous particles may be related to transepithelial water flux under antidiuretic hormone (ADH) influence (47,48). This epithelium appears physiologically to have a relatively unimportant paracellular pathway (with several junctional fibrils and a simple lateral extracellular cell boundary), a relatively low water permea-

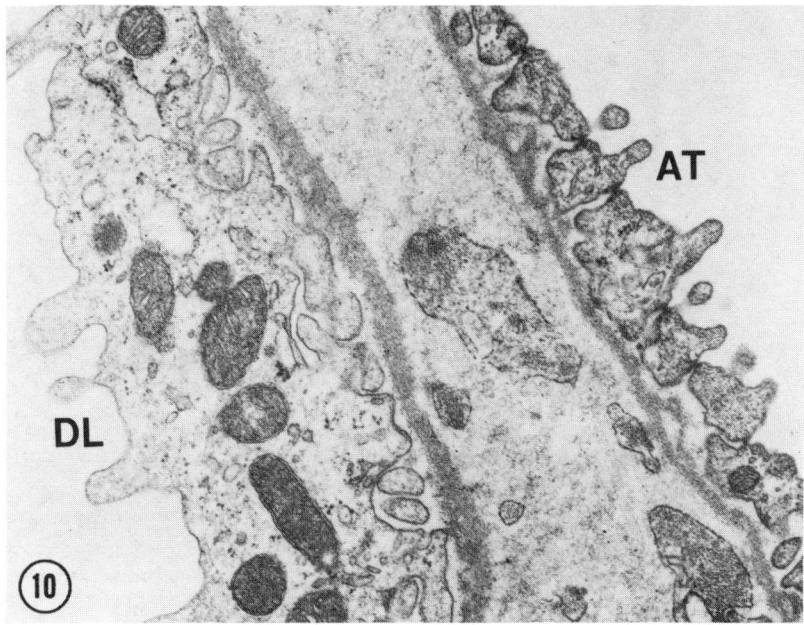

Figure 10 Transmission electron micrograph showing the descending (DL) and ascending (AL) thin limbs of Henle of long-looped nephrons in the dog kidney. X 5,900.

bility (low number of intramembranous particles), and little cell machinery for active transport (few cell organelles such as mitochondria). A clue to the function of short-looped thin limbs may lie in the observation that, in certain species, they travel through the inner stripe in the vascular bundles adjacent to the venous vasa recta of the inner medulla. Such a position would maximize the exchange of substances, such as urea, between the blood returning from the inner medulla and the short-looped nephrons. Net addition of urea into thin limbs of cortical nephrons would greatly enhance the medullary accumulation of urea, whereas net addition of urea into loops of Henle within the inner medulla would tend to defeat such accumulation of urea. This would oppose the urinary concentrating process(95). Parenthetically, Kaissling et al (51) also stressed the role of the vascular bundles in urea recycling from the elaborate renal pelvis seen in *Psammomys* back into the lumen of cortical nephrons.

Considerable morphological differences have been noted among species in the upper region of the descending thin limb of long-looped nephrons. In the rabbit (49) this region had a fairly simple epithelium with tight junctions of intermediate length. In most other species—e.g. rat (59, 88) and *Psammomys* (11)—the epithelium was highly interdigitated with shallow tight junctions. In these species, which are good concentrators, the epithelium contained many mitochondria and numerous microvilli and basal

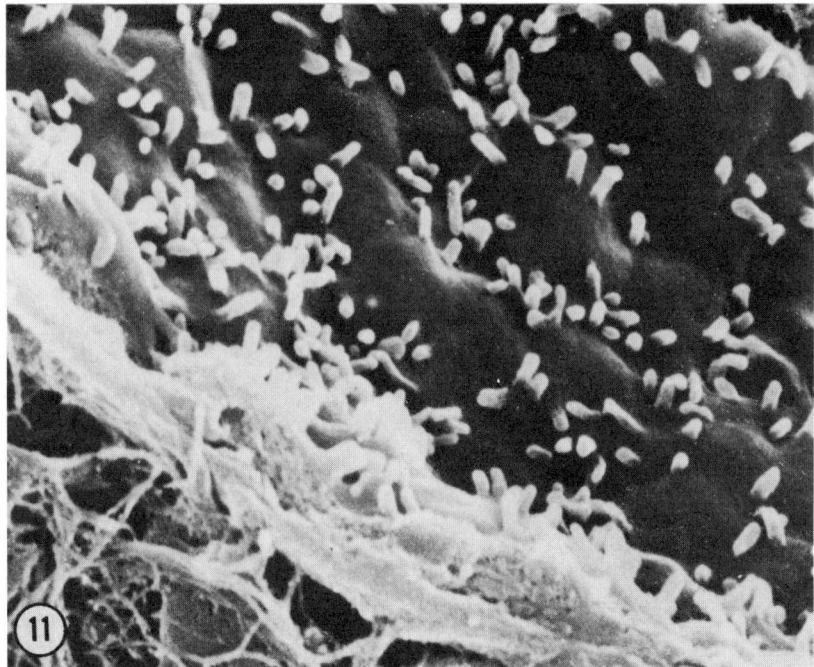

Figure 11 Scanning electron micrograph of a rabbit thin limb of Henle. Note the thin epithelium and occasional microvilli on the luminal surface. X 14,100.

lateral cell projections. The elaborate morphology of this region in certain species and the extremely positive staining with carbonic anhydrase (R. E. Bulger, unpublished observations) are inconsistent with a purely passive roll for this region. Future physiological studies will likely delineate a more active role in cell transport in this zone than has been presently postulated.

In rat (89) this zone of the thin limb had a tight junction composed of a single fibril of tenuous composition sometimes appearing as only a strand of particles. The complex shape of the cell provides an elaborate paracellular pathway. The cell membranes of the upper descending thin limb of long-looped nephrons had an unusually high density of intramembraneous particles when compared with any other thin limb region.

For the long-looped nephrons to function as postulated in the passive models of urine concentration, the inner medullary descending thin limbs should be relatively permeable to water but should not permit the free passage of sodium chloride. This, in fact, is reflected by the simplified morphology of the thin limb as it descends into the inner medulla. Cell shape was less elaborate, with only scattered microvilli and fewer basal-lateral cell processes. Schwartz et al (89) found a mean of 3.13 ± 0.14 fibrils that branched frequently in the tight junction of this thin limb region in the rat. Here intramembranous particles were fewer than in the outer region of the descending thin limb but still outnumbered the particles seen in the short-looped nephrons. Abramow & Orci (1) perfused rabbit thin descending limb from inner medulla in vitro and reported a high transepithelial electrical resistance (with a mean exceeding 700 Ωcm^2), a value typical of tight epithelia. The sodium permeability from lumen to bath was very low (1.7×10^{-6} cm·sec^{-1}). Although 1–4 fibrils were seen in freeze-fracture replicas from these tubules, the junctions usually had 3–4 uninterrupted fibrils with a moderately high number of particles in the membranes. (Some profiles contained very many particles and fewer junctional fibrils; these may have come from another region of the thin limb.)

The ascending thin limb began prior to the bend and was remarkably similar in all species studied. This segment had an extensively interdigitated epithelium and shallow tight junctions. Schwartz et al (89) reported 1.31 ± 0.09 fibrils in the tight junction. The junctional fibril was continuous and uncommonly thick, differing from the interrupted fibril in the upper descending portion but similar to that seen in certain fish chloride cells (54). The elaborate cell shape created a large lateral intercellular pathway, but the cells had few microvilli or secondary basal-lateral processes and contained few mitochondria.

These cells thus appeared to have a large paracellular pathway consistent with their proposed function in the passive efflux of sodium chloride. They did not appear to be heavily involved in active transport. The basal-lateral membranes of this zone had many intramembraneous particles but still fewer than in the upper descending part of long-looped nephrons.

THE DISTAL TUBULE (Figures 12–15)

The distal tubule of the mammalian nephron had long been separated into three distinct regions: the ascending thick limb (pars recta) of the loop of Henle, the macula densa region, and the distal convoluted tubule. In view of recent morphological and physiological studies we concluded that this is no longer a suitable division. It now appears that a more acceptable separation of the distal tubule would be into a medullary ascending thick limb of the loop of Henle (MAT), a cortical ascending thick limb of the loop of Henle (CAT), and the distal convoluted tubule (DCT). The macula densa region would now be incorporated into the CAT.

The MAT begins at the junction of the inner and outer medulla in most species (Figure 13). Although Allen & Tisher (2) reported two cell types lining the ascending thick segment, all other investigators have described the epithelium lining this segment as of one cell type. These cells are connected by relatively shallow tight junctions. The cells possess elaborate lateral interdigitations extending from the basal region of the cell towards the cell apex. The extent to which they approach the luminal membrane depends on the species. Lying within these processes are long rod-shaped mitochondria characteristic of an epithelium functioning in active transport. As the MAT passes through the outer medulla, and the epithelium becomes gradually shorter in the rabbit (49), the rat (50), the dog (21). The MAT converts to the CAT as it enters the medullary ray of the cortex (Figure 14). The CAT constitutes the thinnest portion of the the distal tubule. The cells comprising the CAT were considerably shorter than those of the medullary portion [e.g. in rat, CAT = 3–6 μm and MAT = 12–15 μm (50)]. The orderly arrangement of mitochondria within interdigitating processes is lost and the cells contain a haphazard arrangement of basal and lateral processes and possess a few stubby mitochondria [rat (50); rabbit (49)]. Eventually the CAT passes by the parent glomerulus, where it forms the specialized plaque of cells known as the macula densa. Continuing on for a short distance it abruptly gives rise to the distal convoluted tubule.

The short post-macula densa segment at the end of the CAT has been described in rabbit (49,71), dog (21), mouse (49), rat (50), and human (77). The demonstration of this segment now allows the macula densa to be redefined as a specialized area within the CAT.

The morphological separation of the pars recta of the distal tubule into MAT and CAT has been supported by a number of physiological studies. Several investigators (23,24,70) measured the parathyroid hormone (PTH)-sensitive adenyl cyclase (AC) activity in both the MAT and CAT of several different species. Significant PTH-sensitive AC activity was demonstrated in the CAT but not in the MAT. Consistent with these findings was the observation that PTH stimulated calcium transport in the CAT but not in the MAT in isolated perfused rabbit tubules (18,96). Heterogeneity was further supported by studies examining the effects of ADH on isolated segments of MAT and CAT from mouse kidneys. ADH increased the lumen positive transepithelial voltage as well as increased the net chloride absorption in the MAT (38,43) but not in the CAT (43). In short, the pars recta of the distal tubule appeared to possess two morphologically and physiologically distinct segments.

The DCT arises abruptly from the post-macula densa segment of the CAT (Figure 15). This segment of the nephron has been difficult to describe because of marked species differences. In the rabbit (49) the DCT is short and the end is sharply defined. The DCT is composed of a homogeneous population of cells considerably taller than those lining the CAT. The cell processes are extensively interdigitated and contain long mitochondria. The DCT then gives rise to the next segment, known as the connecting tubule.

RENAL MORPHOLOGY 163

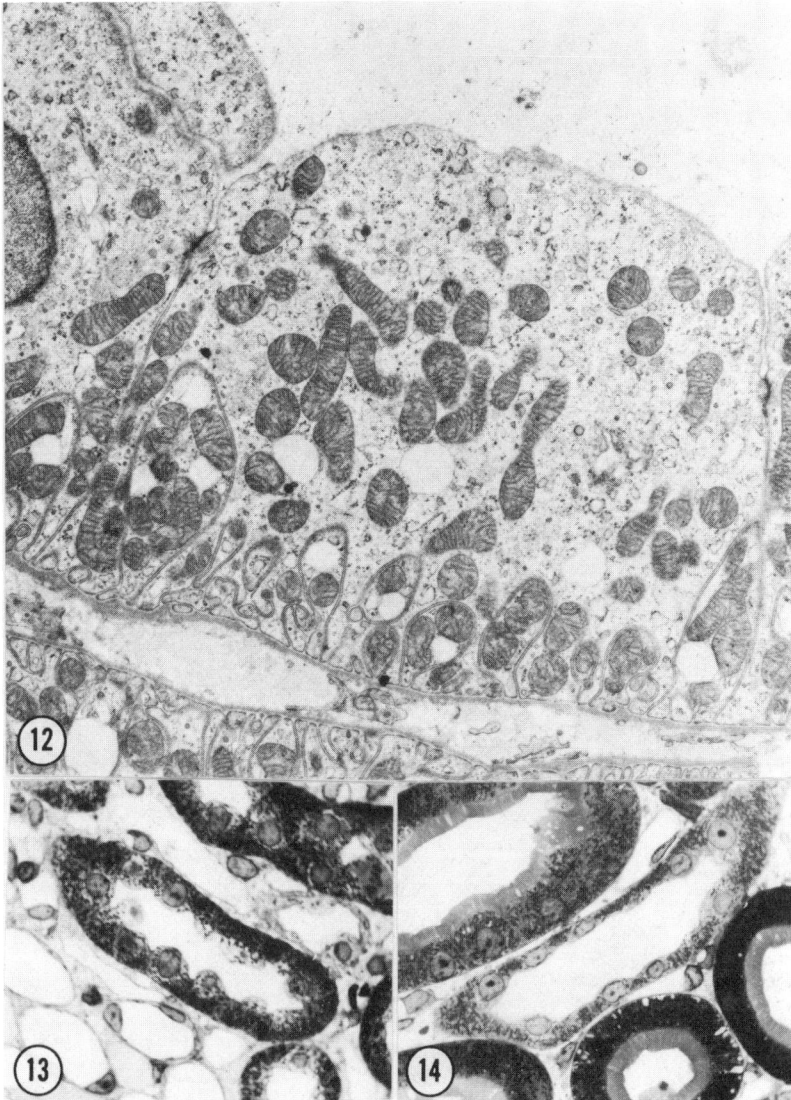

Figure 12 Transmission electron micrograph of a human ascending thick segment of the distal tubule. X 6,800.

Figure 13 Light micrograph of an Epon section showing the medullary ascending thick portion of the distal tubule of the rat kidney. Toluidine blue. X 200.

Figure 14 Light micrograph of an Epon section showing the cortical ascending thick portion of the distal tubule of the rat kidney. (Figures 13 & 14 are similar magnifications to show differences in cell height) Toluidine blue. X 200.

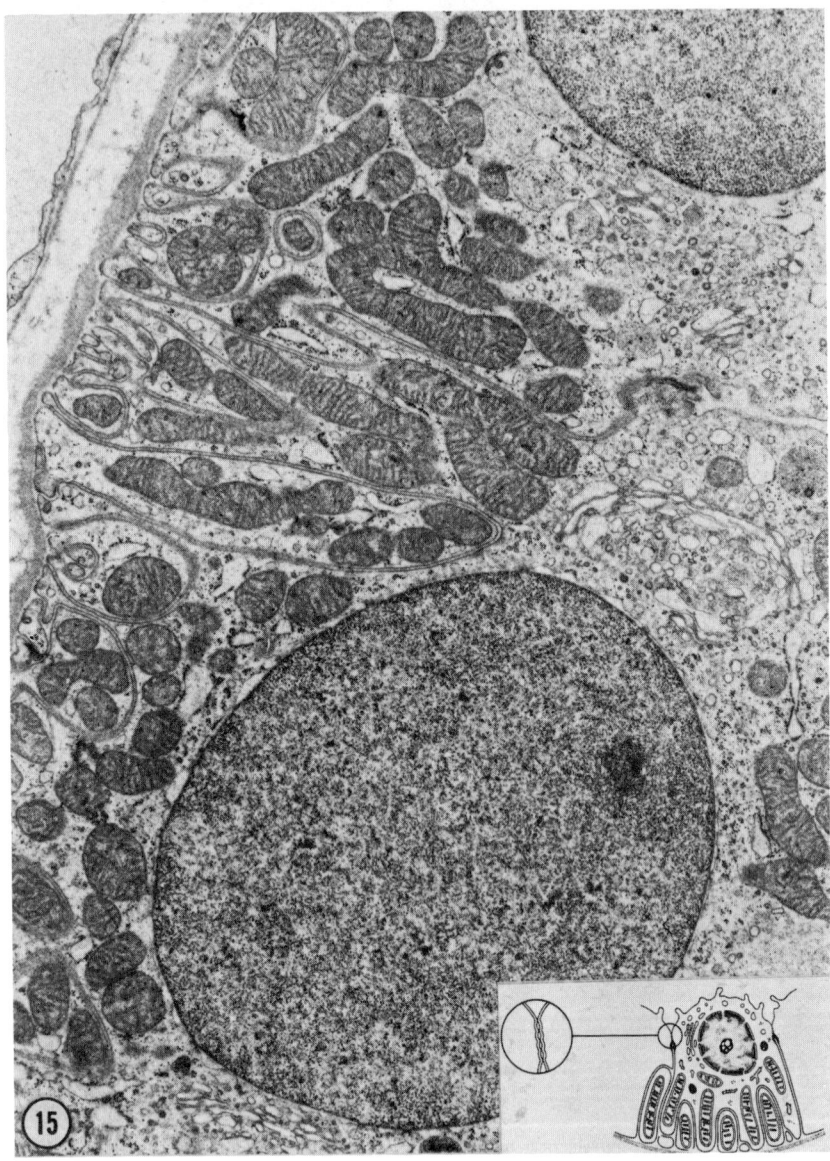

Figure 15 Transmission electron micrograph of a human distal convoluted tubule. X 8,700.

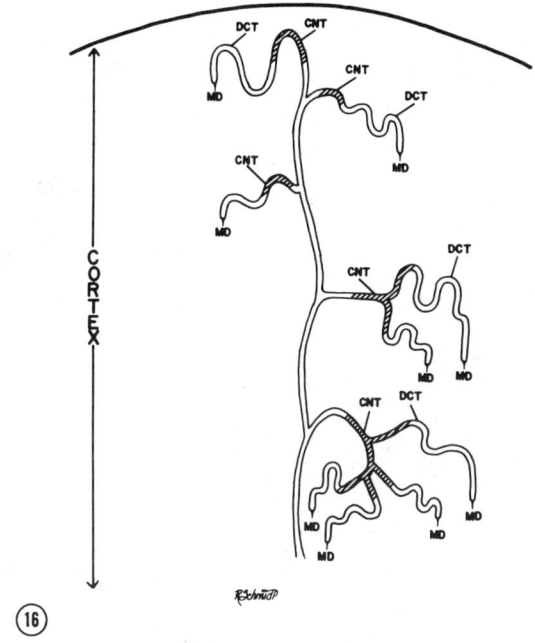

Figure 16 Schematic diagram of the distal nephron showing relationship between distal convoluted tubule (DCT), connecting tubule (CNT), and cortical collecting duct (MD = macula densa).

THE CONNECTING TUBULE (Figures 16,17)

In the rabbit nephron, the connecting tubule has been described as a well-defined tubular portion interposed between the DCT and the collecting duct (CD) (Figure 16) (49). Superficial nephrons of the rabbit drain individually via the connecting tubule (CnT) into the cortical collecting duct (CCD), while in juxtamedullary nephrons and most mid-cortical nephrons the connecting tubule begins 15–20 cells before joining the arcades. The arcades are therefore composed of connecting tubules.

Kaissling & Kriz (49) described two types of cells in the connecting tubules. The intercalated cells were similar to those discussed below in the section on the collecting ducts. The connecting tubule cell, the second cell type, had a smooth luminal surface and true infoldings of the basal cell membrane that penetrated all regions of the cell even to the cell apex. The cells did not extensively interdigitate but had short lateral folds between adjacent connecting tubule cells. The tight junctions were deep (152 ± 47.6 nm).

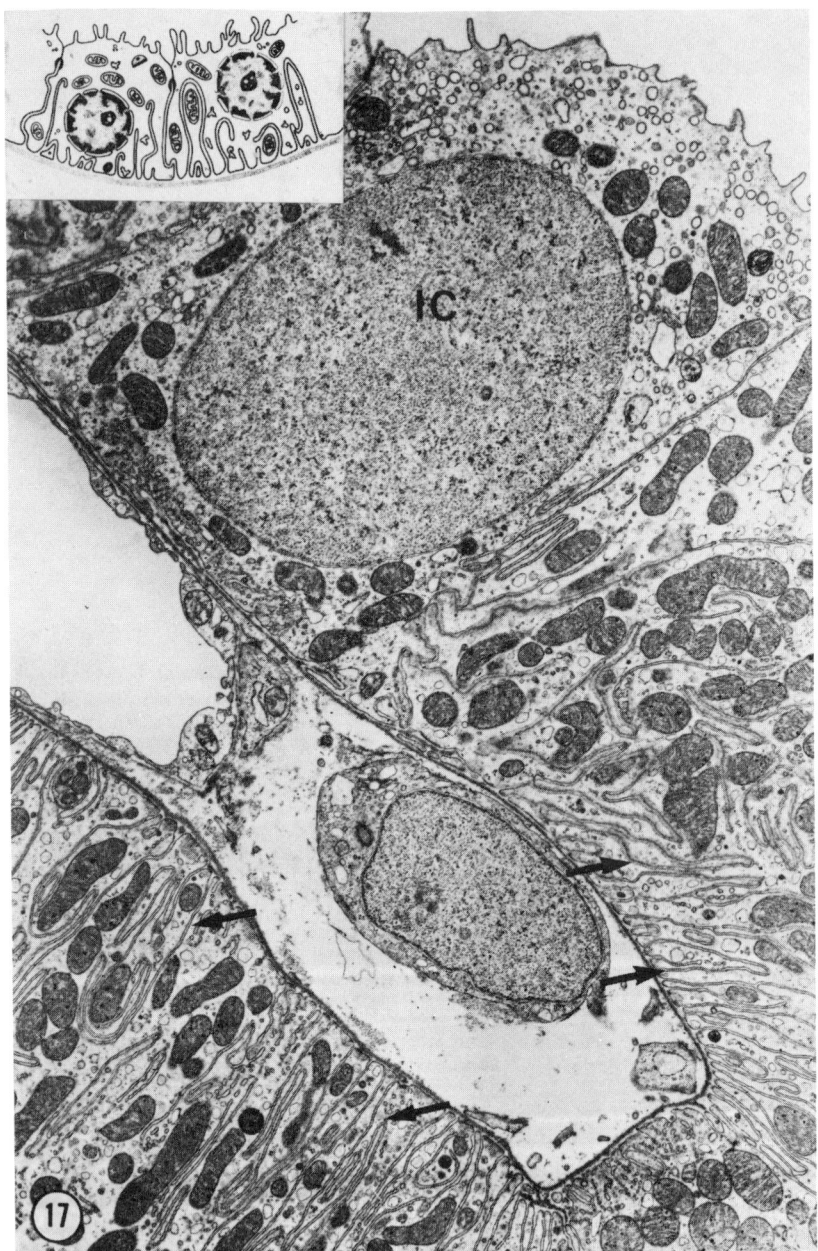

Figure 17 Transmission electron micrograph of human connecting tubule. Note the presence of an intercalated cell (IC) and the basal infoldings of the connecting tubule cells (arrows). × 6,500.

The connecting tubule has been classified as part of the distal convoluted tubule because it appeared to arise from the metanephrogenic blastema in the human (75,78) and because in several species, including the rat (27, 36,61, 104), the mouse (49) and the human (74,99), it appeared to begin by a gradual transformation of the epithelium of the distal tubule. Morel (70) did not see clear-cut segmentation with sharp transitions in these species when the DCT was assayed for hormone sensitivities. All portions responded to five of the six hormones tested. Others have defined this segment as part of the collecting duct system because of the presence of intercalated cells within this zone and because of the noninterdigitating nature of the connecting tubule cell (49). Peter (77) maintained that the arcades in humans and rabbits originate from the ureteric bud.

From micrographs of rat and human (Figure 17) connecting tubules, it is clear that connecting tubule cells do exist in these species in association with intercalated cells prior to the conversion to the CCD. These connecting tubule cells of both species have deep infoldings of the basal cell membrane much more complex than those seen in rabbit. This may explain why they have not been clearly described in these species before. Owing to the morphological similarity of connecting tubular cells and distal tubular cells, the issue of whether the former begin abruptly or change gradually from the latter will have to be resolved by careful serial reconstructions of this region. Embryological studies to determine whether the connecting tubule is derived from the metanephric blastema or the ureteric bud would also help in classifying this region.

THE COLLECTING DUCTS (Figures 18, 19)

The collecting duct system of the mammalian kidney has been the focus of intense morphological and physiological investigation in recent years. It has become increasingly apparent that this final portion of the urinary tubule functions as more than just a passive conduit for the elimination of products generated by the more proximal portions of the nephron. The collecting duct system now appears to be the final mediator of normal fluid and electrolyte balance. It may play a vital role in the control of acid-base balance and potassium homeostasis in addition to being the primary site of action of ADH. We thus review recent advances in the description of collecting duct morphology and match these new observations with their possible functional correlates (Figures 18,19).

The mammalian collecting duct system can be divided into cortical (Figure 16), medullary, and papillary portions, each possessing distinctive morphological and physiological characteristics. The CD begins in the cortex where it forms from the CnT of predominantly superficial nephrons or is continuous with the arcades of the midcortical and juxtamedullary nephrons. It enters the medullary ray and travels unbranched through the outer medulla. Once entering the inner medulla the CDs begin to coalesce and ultimately give rise to the large papillary collecting ducts (ducts of Bellini). The epithelial cells that line the collecting duct are of two main types. The more common principal cell is found along the entire collecting duct tree, while the intercalated cell appears to be largely localized in the cortex and outer medulla in most species. Since the morphology of these cell types has been extensively reviewed (19,98), we discuss only recent developments.

A significant new hypothesis centers on the ability of the intercalated cell to exist in a number of morphological variations. These cells, which are often referred to as dark cells, can be distinguished from the principal cells by their darker cytoplasmic matrix and abundance of ribosomes and mitochondria, as seen with both light and transmission electron microscopy. Their elaborate pattern of surface projections or microplicae (seen with scanning electron microscopy) is an additional distinction (Figure 19). Intercalated cells have been identified in the CnT, CCD, and outer medullary CD for most species and are only rarely found in the inner medulla [with the notable exception of the desert rodent *Psammomys obesus* (49)]. Recent quantitative studies have indicated that the intercalated cell comprised about 33% of the cell population in the cortex and 50% in the outer medulla of the rabbit collecting system (49) and from 24–40% of the cell population in the cortex and 18–22% in the outer medulla of the rat kidney (39,92,94).

Kaissling and Kriz (49) and Kriz et al (61) have recently reported two types of intercalated cells in the rabbit kidney. Beginning with the CnT they described a "black" and "grey" manifestation of the intercalated cell on the basis of differences in cytoplasmic density and cytologic components. The two varieties were never found adjacent; intermediate cell types were identified. The grey cell type had a lighter appearance and contained fewer ribosomes. A more definitive separation between the cell types could be made by examining the numerous cytoplasmic vesicles, which were almost exclusively confined to the grey manifestation. These vesicles were of two types: spherical vesicles with coated invaginated outer membranes, or flat vesicles that were often piled up in the apical region of the cell. The intercalated cells in the CCD had a dark appearance and became progressively lighter as the CD entered the outer medulla. These observations suggested that the black and grey manifestations may represent different functional states of the intercalated cell.

Additional evidence for the existence of two types of intercalated cells has come from experiments utilizing freeze-fracture techniques. An early study (46) reported numerous rod-shaped particles (290 X 160 Å) in the luminal

RENAL MORPHOLOGY 169

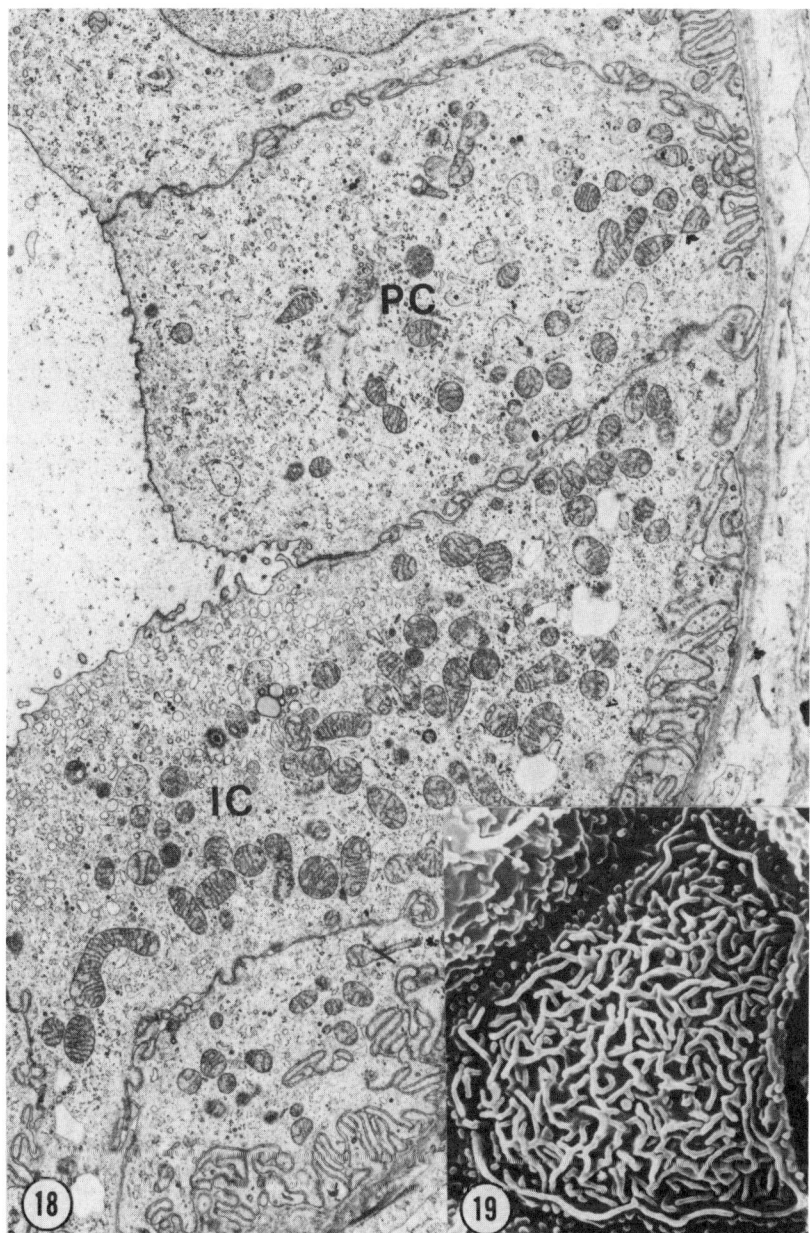

Figure 18 Transmission electron micrograph of a human cortical collecting duct. Note the intercalated (IC) and principal (PC) cells. × 7,000.

Figure 19 (Inset) Scanning electron micrograph showing an intercalated cell from the rabbit kidney. × 8,000.

membrane of the intercalated cell in the rat CD. A recent extension of this study (94) considered not only the luminal membrane features but also those of the cytoplasmic vesicles known to be present within the intercalated cell. The authors demonstrated two varieties of intercalated cells: a Type I cell, which contained a high density of rod-shaped particles associated with the luminal membrane; and a Type II cell, in which such particles were associated not with the luminal membrane but within the membrane of the cytoplasmic vesicles. These observations in the rat CD correlated well with findings (49) in the rabbit CnT and strongly suggested that the intercalated cell can exhibit widely divergent variations in ultrastructure.

The suggestion that several types of intercalated cells exist was of particular interest in light of the persistent controversy concerning the exact nature of the cell types that line the CD system. It has been suggested that principal cells and intercalated cells represent different functional states of a single cell line. Early studies (37) reported a 100% increase in the incidence of intercalated cells in the distal nephron of rats subjected to bicarbonate loading and a 200% increase in cell number after induction of respiratory acidosis. Increased numbers of intercalated cells were also reported by Ordonez & Spargo (76) and more recently by Evan et al (31). These studies, which relied heavily on scanning electron microscopy of the luminal surfaces of CDs, reported numerous intermediate cell types and concluded that the intercalated cell was a functional modification of the more common principal cell.

A number of recent studies have reached opposite conclusions and argue strongly in favor of the intercalated cell as a distinct and constant cell population. Hansen et al (39) subjected rats to conditions of acute respiratory acidosis, acute metabolic alkalosis, chronic metabolic acidosis, and chronic hypo- or hyperkalemia and then quantitated the incidence of intercalated cells in the CCD and outer MCD. They detected no change in the incidence of intercalated cells by light and scanning electron microscopy in any functional state examined. Stanton et al (92) quantitated the number of intercalated cells in the CnT and initial portion of the CCD in rats subjected to prolonged potassium depletion and also failed to observe any changes in the number of dark cells. Finally, Stetson et al (94) examined the MCD in potassium depleted rats by thin section and freeze-fracture electron microscopy and reached similar conclusions. The authors showed that while potassium depletion resulted in marked changes in the structural makeup of the intercalated cell (including a 3.3-fold increase in luminal cell membrane and an 80% decrease in the volume of cytoplasmic vesicles), no change in the total number of intercalated cells could be noted. Furthermore, the number of high density particle cells (as determined by freeze-fracture) increased from 22% in controls to 36% in potassium-depleted animals. This value was nearly identical to that obtained by thin section electron microscopy. The authors concluded that potassium deprivation caused a shift in the intercalated cell population such that all cells were now Type I, with a high density of rod-shaped particles in the luminal membrane. Since Type II cells (containing vesicles with rod-shaped particles) were absent in the potassium-deprived state, the suggestion was made that the studded vesicles could fuse with the luminal membrane and hence provide not only luminal membrane volume but also increases in rod-shaped particles.

These results suggested that the intercalated cell represents a distinct cell population. Since changes in intercalated cell structure were seen in potassium depletion, a condition known to stimulate potassium reabsorption along the distal nephron (29,66), these cells may function in the *reabsorption* of potassium.

Recent morphological evidence suggests possible roles for the principal cells of the CD. Physiological studies indicate that the CD may participate in the control of the urinary excretion of potassium. These studies have shown that potassium secretion occurs along the collecting duct (16,35,81). For a detailed discussion see (34).

Several morphologic studies have attempted to discern how the principal cell functions in potassium transport. Wade et al (101) isolated renal CCD from rabbits chronically treated with deoxycorticosterone acetate (DOCA) or dexamethasone (DMS), agents known to increase potassium secretion in the CCD. The authors quantitated the basolateral membrane surface areas of intercalated and principal cells. They observed an increase of 140% (DOCA) and 90% (DMS) in only the basolateral membrane area of principal cells. No changes could be observed in the intercalated cells. Similar observations were made in rats after potassium adaptation. Stanton et al (92) examined the CnT and initial portion of the CCD and Rastegar et al (80) evaluated the outer medullary collecting duct in animals chronically given potassium. Both groups of investigators reported a dramatic increase in the basolateral membrane surface density of the principal cells.

The basolateral uptake of potassium has been suggested to play an important role in transepithelial potassium transport (69). Katz et al (55) have recently demonstrated significant Na,K-ATPase activity in the CCD and MCD of mice, rats, and rabbits. This activity was markedly enhanced by potassium loading in rats (30). Thus a correlation might exist between the increase in basolateral membrane surface and Na,K-ATPase activity. In this respect, Rastegar et al (80) have postulated that the amplification of basolateral membrane related to an increase in the number of potassium pump sites.

The CD plays an important role in acid-base balance and contributes significantly to urinary acidification. Using histochemistry, Lonnerholm & Riddlestrahle (63) demonstrated significant carbonic anhydrase activity in cortical and outer medullary CDs. Carbonic anhydrase activity was particularly high in the intercalated cell, suggesting a possible role for this cell type in bicarbonate reabsorption or hydrogen ion secretion.

Finally, since the collecting duct is the primary site of action of ADH, it is important to mention a number of recent studies that have examined structural changes induced by the administration of this hormone. Harmanci and coworkers (40) infused ADH into homozygous Brattleboro rats and then used freeze-fracture techniques to quantitate the number of intramembrane particle clusters in the luminal membrane of the papillary CD. Significantly higher numbers of these clusters were present in rats given ADH when compared to control rats not treated with the hormone. As the dose of ADH was increased so did the incidence of intramembrane particle clusters. In addition, when ADH was removed both urinary osmolality and the number of intramembrane clusters decreased. Since similar particle clusters have been described in amphibian urinary bladder epithelium after ADH and have been related to increases in water permeability, a similar relationship might also exist in the mammalian collecting duct.

INSIGHTS USING NEW TECHNIQUES

Tissue culture techniques are rapidly gaining popularity as aids in the study of renal function. A number of approaches are being used to isolate individual cell types either from a particular segment of the nephron or from cell lines derived from the kidney. Precise morphological identification enables extrapolation of functional parameters obtained in culture to their in vivo counterparts.

One approach has been to use established renal epithelial cell lines. Mullin et al (73) have shown that the LLC-PK$_1$ cell line derived from the pig kidney could actively transport glucose and that this process was inhibited by phlorizin. Transmission electron micrographs of confluent cultures contained cells with microvilli on the luminal surfaces, and with apical tight junctions. These findings suggested that this cell line was similar to the mammalian proximal tubule. Valentich (100) presented convincing evidence that the MDCK cell line derived from the cocker spaniel kidney was quite similar to the mammalian CD. Using light microscopy and scanning and transmission electron microscopy (Figures 20, 21), he demonstrated two cell types quite similar to the principal and intercalated cells of the medullary CCT.

A second method of examining different cell types has been the isolation, separation, and cloning of homogeneous cell populations derived from kidney. Using free-flow electrophoresis and density gradient centrifugation, several investigators (57,58,79) isolated proximal tubule cells from rat and hamster kidneys. These cells had the histochemically demonstrable alkaline

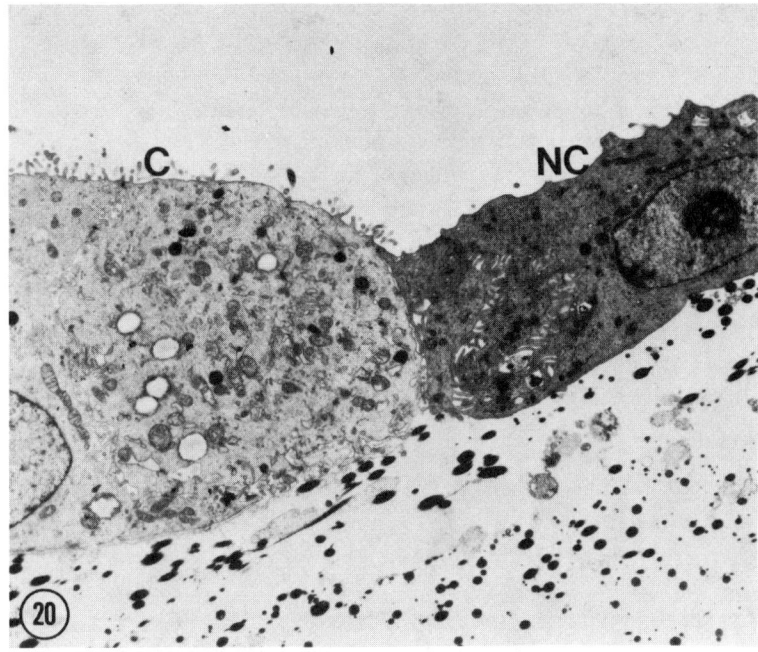

Figure 20 Transmission electron micrograph of a ciliated (C) and a nonciliated (NC) cell from the MDCK cell line. The nonciliated cells are similar to the intercalated cells and the ciliated cells are similar to the principal cells of the mammalian cortical collecting duct. × 5,200. (Courtesy of Dr. John D. Valentich).

phosphatase activity and brush border characteristic of the proximal tubule. Using enzymatic digestion and gradient centrifugation, Eveloff et al (33) isolated single cells from the thick ascending limb of Henle's loop. Morphologically, these cells showed polarity and had a few stubby microvilli on their luminal surfaces; they also exhibited deep infoldings of the basal-lateral cell membranes. These cells were rich in Na,K-ATPase activity and their consumption of oxygen was inhibited by furosemide, suggesting functions characteristic of the ascending thick limb of Henle. Kreisberg et al (56) and Ausiello et al (4) have successfully cloned and maintained in vitro three distinct homogeneous cell types from isolated, dissociated glomeruli and from explants of whole isolated glomeruli.

Finally, Horster (44) has described a technique for the in vitro culture of segmental cell populations from the mammalian nephron. In this study, individual nephron segments including the proximal straight tubule, the medullary and cortical ascending thick limbs of Henle, and the cortical and medullary collecting ducts were dissected from rabbit kidney and placed in primary culture. Nephron segments were successfully cultured to form epithelial monolayers or tubules. Electron microscopy of these cultures

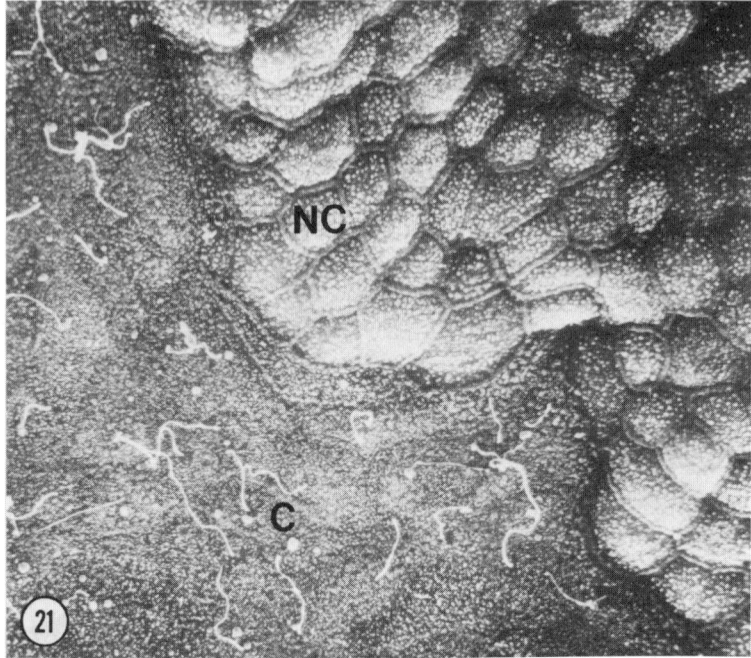

Figure 21 Scanning electron micrograph showing the luminal surface of a culture of MDCK derived cells. Note that the ciliated (C) and nonciliated (NC) cells are similar to the mammalian cortical collecting duct. X 1,950. (Courtesy of Dr. John D. Valentich).

showed that the cells retained their polarity, their basal membrane, cytomembranes, cellular junctions, and microvilli.

X-RAY MICROANALYSIS

Use of X-ray microanalysis in conjunction with electron microscopy allows the morphological localization at the molecular level of both soluble and insoluble substances. X-ray microanalysis can be performed in a scanning or transmission microscope or in a microprobe. Qualitative or quantitative data can be obtained in a nondestructive manner for elements with atomic numbers greater than sodium. Since minute quantities can be measured (10^{-18} grams) if they are present in a minimum concentration, renal physiologists are using this machine to analyze microdroplet samples obtained by micropuncture or microperfusion (72).

Quantitation of electrolytes in frozen-dried kidney cryosections have been recently described (12,97) in superficial proximal and distal tubules of the rat kidney. Some results from these studies (12,97) are presented in Table 1.

Table 1 Nuclear concentrations mM kg^{-1} wet weight (mean ± SEM)

	Na	Cl	P	K
Proximal tubule nucleus	19.6 ± 0.4	22.8 ± 0.7	149.5 ± 1.6	144.4 ± 2.0
Distal tubule nucleus	11.4 ± 0.8	12.5 ± 1.1	175.2 ± 3.4	143.3 ± 2.8

The use of frozen-hydrated sections with subsequent drying in the microscope allowed for the determination of water content in cells as well as their elemental concentrations (84,85). These techniques have been tested on renal medulla of rats (20). Intracellular sodium values were high in cells present in the concentrating rat renal medulla (collecting duct cells, 344± 122 mM kg^{-1} wet weight; papillary epithelial cells, 287±105; and interstitial cells, 898±194). Mean interstitial sodium was 590±119 mM kg^{-1} wet weight. Churchill et al (25) have extended these studies using rats in well-defined states of antidiuresis and diuresis. In diuresis, the sodium and chlorine levels fell dramatically while the water content rose from about 53% to approximately 81%.

ACKNOWLEDGMENTS

The authors would like to express their thanks to Ms. Teresa Lewis and Ms. Cherie Gorman for their excellent technical assistance and to Ms. Flo Evans and Ms. Toni Green for their efforts in the typing of this manuscript.

Literature Cited

1. Abramow, M., Orci, L. 1980. On the "tightness" of the rabbit descending limb of the loop of Henle—physiological and morphological evidence. *Int. J. Biochem.* 12:23–27
2. Allen, F., Tisher, C. C. 1976. Morphology of the ascending thick limb of Henle. *Kidney Int.* 9:8–22
3. Andrews, P. M., Coffey, A. K. 1980. In vitro studies of kidney glomerular epithelial cells. In *SEM II*, Chicago: SEM, Inc., pp. 179–91
4. Ausiello, D., Kreisberg, J. I., Roy, C., Karnovsky, M. J. 1980. Contraction of cultured rat glomerular cells of apparent mesangial origin after stimulation with angiotensin II and arginine vasopressin. *J. Clin. Invest.* 65:754–60
5. Avasthi, P. S., Huser, J., Evan, A. P. 1979. Glomerular endothelial cells in gentamicin-induced acute renal failure (ARF) in rats. *Kidney Int.* 16:771A (Abstr.)
6. Avasthi, P. S., Evan, A. P., Hay, D. 1980. Glomerular endothelial cells in uranyl nitrate-induced acute renal failure in rats. *J. Clin. Invest.* 65:121–27

7. Baehler, R. W., Kotchen, T. A., Burke, J. A., Galla, J. H., Bhathena, D. 1977. Considerations of the pathophysiology of mercuric chloride–induced acute renal failure. *J. Lab. Clin. Med.* 90:330–40
8. Bankir, L., Kaissling, B., de Rouffignac, C., Kriz, W. 1979. The vascular organization of the kidney of *Psammomys obesus*. *Anat. Embryol.* 155:149–90
9. Barnes, J. L., Osgood, R. W., Reineck, H. J., Stein, J. H. 1979. Glomerular alterations in the ischemic model of acute renal failure. *Kidney Int.* 16:771A (Abstr.)
10. Barnes, L. D., Guy, M. N., Lifschitz, M. D., Kreisberg, J. I. 1981. Angiotensin II receptors in mesangial cells cultured from rat renal glomeruli. *Kidney Int.* 19:163A (Abstr.)
11. Barrett, J. M., Kriz, W., Kaissling, B., Rouffignac, C. de. 1978. The ultrastructure of the nephrons of the desert rodent (*Psammomys obesus*) kidney: II. Thin limbs of Henle of long-looped nephrons. *Am. J. Anat.* 151:499–514

12. Beck, F., Bauer, R., Bauer, U., Mason, J., Dorge, A., Rick, R., Thurau, K. 1980. Electron microprobe analysis of intracellular elements in the rat kidney. *Kidney Int.* 17:756–63
13. Beeuwkes, R. III. 1971. Efferent vascular patterns and early vascular-tubular relations in the dog kidney. *Am. J. Physiol.* 221:1361–74
14. Beeuwkes, R. III. 1980. Vascular-tubular relationships in the human kidney. *International Symposium on Renal Pathophysiology,* ed. A. Leaf, G. Giebisch, pp. 155–63. NY: Raven
15. Beeuwkes, R. III. 1980. The vascular organization of the kidney. *Ann. Rev. Physiol.* 42:521–42
16. Bengele, H. H., McNamara, E. R., Alexander, E. A. 1979. Potassium secretion along the inner medullary collecting duct. *Am. J. Physiol.* 236:F278–82
17. Boulpaep, E. L. 1972. Permeability changes of the proximal tubule of *Necturus* during saline loading. *Am. J. Physiol.* 222:517–31
18. Bourdeau, J. E. L., Burg, M. B. 1980. Effect of PTH on calcium transport across the cortical thick ascending limb of Henle's loop. *Am. J. Physiol.* 239:F121–26
19. Bulger, R. E. 1979. Kidney morphology. In *Strauss and Welt's Diseases of the Kidney,* ed. L. E. Early, C. W. Gottschalk. Boston: Little, Brown and Company
20. Bulger, R. E., Beeuwkes, R. III, Saubermann, A. J. 1981. Application of scanning electron microscopy to X-ray analysis of frozen-hydrated sections III. Elemental content of cells in the rat renal papillary tip. *J. Cell Biol.* 88:274–80
21. Bulger, R. E., Cronin, R. E., Dobyan, D. C. 1979. Survey of the morphology of the dog kidney. *Anat. Rec.* 194:41–66
22. Bulger, R. E., Cronin, R. E., Dobyan, D. C. 1981. Glomerular architectural changes in norepinephrine induced acute renal failure. *Am. J. Anat.* 159:379–84
23. Chabardes, D., Imbert, M., Clique, A., Montegut, M., Morel, F. 1975. PTH-sensitive adenyl cyclase activity in different segments of the rabbit nephron. *Pfluegers Arch.* 354:229–39
24. Chabardes, D., Imbert-Teboul, M., Gagon-Brunette, M., Morel, F. 1978. Different hormonal target sites along the mouse and rabbit nephrons. In *Biochemical Nephrology,* ed. W. G. Guder, U. Schmidt, pp. 447–54. Bern: Hans Huber

25. Churchill, S., Beeuwkes, R., Kinter, L., Riley, W., Bulger, R. E., Saubermann, A. 1981. X-ray microanalysis of papillary cells from antidiuretic (AD) and diuretic (D) rats. *Int. Congr. Nephrol. Athens, Greece* (Abstr.)
26. Cox, J. W., Baehler, R. W., Sharma, H., O'Dorisio, T., Osgood, R. W., Stein, J. H., Ferris, T. G. 1974. Studies on the mechanisms of oliguria in a model of unilateral acute renal failure. *J. Clin. Invest.* 53:1546–58
27. Crayen, M., Thoenes, W. 1975. Architektur und cytologische Charakterisierung des distalen Tubulus der Rattenniere. *Fortschr. Zool.* 23:279–88
28. Cronin, R. E., DeTorrente, A., Miller, P. D., Bulger, R. E., Burke, T. J., Schrier, R. W. 1978. Pathogenic mechanisms in early norepinephrine induced acute renal failure: Functional and histological correlates of protection. *Kidney Int.* 14:115–25
29. Diezi, J., Michoud, P., Aceves, J., Giebisch, G. 1973. Micropuncture study of electrolyte transport across papillary collecting duct of the rat. *Am. J. Physiol.* 224:623–34
29a. Dobyan, D. C., Bulger, R. E. 1980. The mammalian kidney. *J. Am. Med. Assoc.* 244:704–5
30. Doucet, A., Katz, A. I. 1979. Site of potassium adaptation in the nephron *Kidney Int.* 16:811 (Abstr.)
31. Evan, A., Huser, J., Bengele, H. H., Alexander, E. A. 1980. The effect of alterations in dietary potassium on collecting system morphology in the rat. *Lab Invest.* 42:668–75
32. Evan, A. P., Rankin, L. I., Luft, F. C. 1980. Scanning and transmission electron microscopic glomerular and tubular changes following gentamicin administration in the rat. *Clin. Res.* 28:444 (Abstr.)
33. Eveloff, J., Haase, W., Kinne, R. 1980. Separation of renal medullary cells: Isolation of cells from the thick ascending limb of Henle's loop. *J. Cell Biol.* 87:672–81
34. Giebisch, G., Stanton, B. 1979. Potassium transport in the nephron. *Ann. Rev. Physiol.* 41:241–56
35. Grantham, J. J., Burg, M. B., Orloff, J. 1970. The nature of transtubular Na and K transport in isolated rabbit renal collecting tubules. *J. Clin. Invest.* 49:1815–26
36. Griffith, L. D., Bulger, R. E., Trump, B. F. 1968. Fine structure and staining of mucosubstances on "intercalated cells" from the rat distal convoluted tubule

and collecting duct. *Anat. Rec.* 160:643–62
37. Hagege, J., Gabe, M., Richet, G. 1974. Scanning of the apical pole of distal tubular cells under differing acid-base conditions. *Kidney Int.* 5:137–46
38. Hall, D. A., Varney, D. 1979. Effect of vasopressin on chloride transport by mouse medullary thick ascending limbs of Henle's loop perfused in vitro. *Kidney Int.* 16:818 (Abstr.)
39. Hansen, G. P., Tisher, C. C., Robinson, R. R. 1980. Response of the collecting duct to disturbances of acid-base and potassium balance. *Kidney Int.* 17:326–37
40. Harmanci, M. C., Kachadorian, W. A., Valtin, H., DiScala, V. A. 1978. Antidiuretic hormone-induced intramembranous alterations in mammalian collecting ducts. *Am. J. Physiol.* 235:F440–43
41. Harmanci, M. C., Stern, P., Kachadorian, W. A., Valtin, H. 1980. Vasopressin and collecting duct intramembranous particle clusters: a dose-response relationship. *Am. J. Physiol.* 239:F560–64
42. Harmanci, M. C., Wade, J. B., Di Scala, V. A. 1975. Altered structure of tight junctions in rat proximal tubules with volume expansion. *8th Ann. Meet. Am. Soc. Nephrol.*, p. 82 (Abstr.)
43. Hebert, S. C., Culpepper, R. M., Andreoli, T. E. 1981. ADH-stimulated NaCl transport in mouse medullary thick ascending limbs. 1. Evidence for ADH-stimulated neutral NaCl cotransport and functional nephron heterogeneity. *Am. J. Physiol.* 241:F412–31
44. Horster, M. 1980. Hormonal stimulation and differential growth response of renal epithelial cells cultivated in vitro from individual nephron segments. *Int. J. Biochem.* 12:29–35
45. Humbert, F., Grandchamp, A., Pricam, C., Perrelet, A., Orci, L. 1976. Morphological changes in tight junction of *Necturus maculosus* proximal tubules undergoing saline diuresis. *J. Cell Biol.* 69:90–96
46. Humbert, F., Pricam, C., Perrelet, A., Orci, L. 1975. Specific plasma membrane differentiations in the cells of the kidney collecting tubule. *J. Ultrastruct. Res.* 52:13–20
47. Kachadorian, W. A., Levine, S. D., Wade, J. B., DiScala, V. A., Hays, R. M. 1977. Relationship of aggregated intramembranous particles to water permeability in vasopressin-treated toad urinary bladder. *J. Clin. Invest.* 59:576–81
48. Kachadorian, W. A., Wade, J. B., DiScala, V. A. 1975. Vasopressin: induced structural change in toad bladder luminal membrane. *Science* 190:67–69
49. Kaissling, B., Kriz, W. 1979. Structural analysis of the rabbit kidney. *Advances in anatomy. Embryol. Cell Biol.* 56:1–123
50. Kaissling, B., Peter, S., Kriz, W. 1977. The transition of the thick ascending limb of Henle's loop into the distal convoluted tubule in the nephron of the rat kidney. *Cell Tiss. Res.* 182:111–18
51. Kaissling, B., Rouffignac, C. de, Barrett, J. M., Kriz, W. 1975. The structural organization of the kidney of the desert rodent, *Psammomys obesus. Anat. Embryol.* 148:121–43
52. Kanwar, Y. S., Farquhar, M. G. 1979. Anionic sites in the glomerular basement membranes. In vivo and in vitro localization to the laminae rarae by cationic probe. *J. Cell Biol.* 81:137–53
53. Kanwar, Y. S., Linker, A., Farquhar, M. G. 1980. Increased permeability of the glomerular basement membrane to ferritin after removal of glycosaminoglycans (heparan sulfate) by enzyme digestion. *J. Cell. Biol.* 86:688–93
54. Karnaky, K. J. Jr. 1980. Ion-secreting epithelia: Chloride cells in the head region of *Fundulus heteroclitus. Am. J. Physiol.* 7:R185–98
55. Katz, A. I., Doucet, A., Morel, F. 1979. Na-K-ATPase activity along the rabbit, rat and mouse nephron. *Am. J. Physiol.* 6:F114–20
56. Kreisberg, J. I., Hoover, R. L., Karnovsky, M. J. 1978. Isolation and characterization of rat glomerular epithelial cells in vitro. *Kidney Int.* 14:21–30
57. Kreisberg, J. I., Pitts, A. M., Pretlow, T. G. II. 1977. Separation of proximal tubule cells from suspensions of rat kidney cells in density gradients of ficoll in tissue culture medium. *Am. J. Pathol.* 86:591–602
58. Kreisberg, J. I., Sachs, G., Pretlow, T. G. II, McGuire, R. A. 1977. Separation of proximal tubule cells from suspensions of rat kidney cells by free-flow electrophoresis. *J. Cell Physiol.* 93:169–72
59. Kriz, W., Schnermann, J., Koepsell, H. 1972. The position of short and long loops of Henle in the rat kidney. *Z. Anat. Entwickl. Gesch.* 138:301–19
60. Kriz, W., Koepsell, H. 1974. The structural organization of the mouse kidney. *Z. Anat. Entwickl. Gesch.* 144:137–63

61. Kriz, W., Kaissling, B., Pszolla, M. 1978. Morphological characterization of the cells in Henle's loop and the distal tubule. In *New Aspects of Renal Function,* ed. H. G. Vogel, K. J. Ullrich, pp. 67–79. Amsterdam/Oxford: Excerpta Medica
62. Kriz, W., Barrett, J. M., Peter, S. 1976. The renal vasculature: Anatomical functional aspects. Kidney and urinary tract physiology II. *Int. Rev. Physiol.* 11:1–21
63. Lonnerholm, G., Ridderstrale, Y. 1980. Intracellular distribution of carbonic anhydrase in the rat kidney. *Kidney Int.* 17:162–74
64. Maack, T., Johnson, V., Kau, S. T., Figueredo, J., Sigulem, D. 1979. Renal filtration, transport, and metabolism of low-molecular-weight proteins: A review. *Kidney Int.* 16:251–70
65. Madri, J. A., Roll, F. J., Furthmayr, H., Foidart, J. M. 1980. Ultrastructural localization of fibronectin and laminin in the basement membranes of the murine kidney. *J. Cell Biol.* 86:682–87
66. Malnic, G., Klose, R. M., Giebisch, G. 1964. Micropuncture study of renal potassium excretion in the rat. *Am. J. Physiol.* 206:674–86
67. Maunsbach, A. B. 1966. Observations on the segmentation of the proximal tubule in the rat kidney. Comparison of results from phase contrast fluorescence and electron microscopy. *J. Ultrast. Res.* 16:239–58
68. Maunsbach, A. B., Boulpaep, E. L. 1980. Hydrostatic pressure changes related to paracellular shunt ultrastructure in proximal tubule. *Kidney Int.* 17:732–48
69. Mello-Aires, M., Giebisch, G., Malnic, G. 1973. Kinetics of potassium transport across single distal tubules of rat kidney. *J. Physiol. London* 232:47–70
70. Morel, F. 1981. Sites of hormone action in the mammalian nephron. *Am. J. Physiol.* 240:F159–64
71. Morel, F., Chabardes, D., Imbert, M. 1976. Functional segmentation of the rabbit distal tubule by microdetermination of hormone-dependent adenylate cyclase activity. *Kidney Int.* 9:264–77
72. Morel, F., Roinel, N., LeGrumellec, C. 1969. Electron probe analysis of tubular fluid composition. *Nephron* 6:350–64
73. Mullin, J. M., Weibel, J., Diamond, L., Kleinzeller, A. 1980. Sugar transport in the LLC-PK$_1$ renal epithelial cell line: similarity to mammalian kidney and the influence of cell density. *J. Cell Physiol:* 104:375–89
74. Myers, C. H., Bulger, R. E., Tisher, C. C., Trump, B. F. 1966. Human renal ultrastructure. IV. Collecting duct of healthy individuals. *Lab. Invest.* 15:1921–50
75. Oliver, J. 1968. *Nephrons and Kidneys.* NY/Evanston/London: Harper and Row, Hoeber Med. Div.
76. Ordonez, N. G., Spargo, B. H. 1976. The morphologic relationship of light and dark cells of the collecting tubules in potassium-depleted rats. *Am. J. Pathol.* 84:317–26
77. Peter, K. 1909, 1927. *Untersuchungen über Bau und Entwicklung der Niere.* Jena: Gustav Fischer. Vols. 1, 2
78. Potter, E. L. 1972. *Normal and Abnormal Development of the Kidney.* Chicago: Year Book Med. Publ.
79. Pretlow, T. G. II, Jones, J., Dow, S. 1974. Separation of cells having histochemically demonstratable glucose-6-phosphatase from suspensions of hamster kidney cells in an isokinetic density gradient of Ficoll in tissue culture medium. *Am. J. Pathol.* 74:275–86
80. Rastegar, A., Biemesderfer, D., Kashgarian, M., Hayslett, J. P. 1980. Changes in membrane surfaces of collecting duct cells in potassium adaptation. *Kidney Int.* 18:293–301
81. Reineck, H. J., Osgood, R. W., Stein, J. H. 1978. Net potassium addition beyond the superficial distal tubule of the rat. *Am. J. Physiol.* 235:F104–10
82. Roesinger, B., Schiller, A., Taugner, R. 1978. A freeze-fracture study of tight junctions in the pars convoluta and pars recta of the renal proximal tubule. *Cell Tiss. Res.* 186:121–33
83. Roll, F. J., Madri, J. A., Albert, J., Furthmayr, H. 1980. Codistribution of collagen type IV and AB2 in the basement membranes and mesangium of the kidney. *J. Cell Biol.* 85:597–616
84. Saubermann, A. J., Echlin, P., Peters, P. D., Beeuwkes, R. III. 1981. Application of scanning electron microscopy to X-ray analysis of frozen-hydrated sections. I. Specimen handling techniques. *J. Cell Biol.* 88:257–267
85. Saubermann, A. J., Beeuwkes, R. III, Peters, P. D. 1981. Application of scanning electron microscopy to X-ray analysis of frozen-hydrated sections. II. Analysis of standard solutions and artificial electrolyte gradients. *J. Cell Biol.* 88:268–73
86. Schiller, A., Taugner, R., Kriz, W. 1980. The thin limbs of Henle's Loop in the rabbit. A freeze fracture study. *Cell Tiss. Res.* 207:249–65

87. Schor, N., Ichikawa, I., Rennke, H. G., Troy, J. L., Brenner, B. M. 1981. Pathophysiology of altered glomerular function in aminoglycoside-treated rats. *Kidney Int.* 19:288–96
88. Schwartz, M. M., Venkatachalam, M. A. 1974. Structural differences in thin limbs of Henle: Physiological implications. *Kidney Int.* 6:193–208
89. Schwartz, M. M., Karnovsky, M. J., Venkatachalam, M. A. 1979. Regional membrane specialization in the thin limbs of Henle's loops as seen by freeze-fracture electron microscopy. *Kidney Int.* 16:577–89
90. Seiler, M. W., Venkatachalam, M. A., Cotran, R. S. 1975. Glomerular epithelium: Structural alterations induced by polycations. *Science* 189:390–93
91. Sperber, I. 1944. Studies on the mammalian kidney. *Zool. Bidrag. Uppsala* 22:249–431
92. Stanton, B. A., Biemesderfer, D., Wade, J. B., Giebisch, G. 1981. Structural and functional study of the rat distal nephron: Effects of potassium adaptation and depletion. *Kidney Int.* 19:36–48
93. Stein, J. H., Gottschall, J., Osgood, R. W., Ferris, T. F. 1975. Pathophysiology of a nephrotoxic model of acute renal failure. *Kidney Int.* 8:27–41
94. Stetson, D. L., Wade, J. B., Giebisch, G. 1980. Morphologic alterations in the rat medullary collecting duct following potassium depletion. *Kidney Int.* 17:45–56
95. Stewart, J. 1975. Urea handling by the renal countercurrent system: Insights from computer simulation. *Pfluegers Arch.* 356:133–51
96. Suki, W. N., Rouse, D., Ng, Roland C.K., Kokko, J. P. 1980. Calcium transport in the thick ascending limb of Henle. Heterogeneity of functions in the medullary and cortical segments. *J. Clin. Invest.* 66:1004–9
97. Thurau, K., Dorge, A., Mason, J., Beck, F., Rick, R. 1979. Intracellular elemental concentrations in renal tubular cells: An electron microprobe analysis. *Klin. Wochenschr.* 57:993–99
98. Tisher, C. C. 1981. Anatomy of the kidney. In *The Kidney*, ed. B. M. Brenner, F. C. Rector, Jr. Philadelphia/London/Toronto: W. B. Saunders. 2nd ed.
99. Tisher, C. C., Bulger, R. E., Trump, B. F. 1968. Human renal ultrastructure III. The distal tubule in healthy individuals. *Lab. Invest.* 18:655–68
100. Valentich, J. D. 1982. Morphological similarities between the dog kidney cell line MDCK and the mammalian cortical collecting tubule. *Ann. N.Y. Acad. Sci.* In press
101. Wade, J. B., O'Neil, R. G., Pryor, J. L., Boulpaep, E. L. 1979. Modulation of cell membrane area in renal collecting tubules by corticosteroid hormones. *J. Cell Biol.* 81:439–45
102. Woodhall, P. B., Tisher, C. C. 1973. Response of the distal tubule and cortical collecting duct to vasopressin in the rat. *J. Clin. Invest.* 52:3095–108
103. Woodhall, P. B., Tisher, C. C., Simonton, C. A., Robinson, R. R. 1978. Relationship between para-aminohippurate secretion and cellular morphology in rabbit proximal tubules. *J. Clin. Invest.* 61:1320–29
104. Young, D., Wissig, S. L. 1965. A histologic description of certain epithelial and vascular structures in the kidney of the normal rat. *Am. J. Anat.* 115:43–70

HETEROGENEITY OF TUBULAR TRANSPORT PROCESSES IN THE NEPHRON

Christine A. Berry

Department of Physiology and Cardiovascular Research Institute, University of California, San Francisco, California 94143

INTRODUCTION

This review describes the current state of knowledge regarding intranephron and internephron heterogeneity of tubular transport function. Anatomists, micropuncturists, and microperfusionists have, in general, agreed upon the definitions of intranephron and internephron heterogeneity. Intranephron heterogeneity is defined as the comparative properties of different segments of an individual nephron. The nephron has been subdivided into twelve distinct segments: early and late proximal convoluted tubule, proximal straight tubule, descending and ascending thin limbs of Henle's loop, medullary and cortical thick ascending limbs of Henle's loop, distal tubule, connecting tubule, cortical collecting tubule, and medullary and papillary collecting ducts.

Internephron heterogeneity is defined as the comparative properties of analogous structures in superficial (SF) and juxtmedullary (JM) nephrons. Anatomists offer the most precise definition of the SF and JM nephron populations: SF are nephrons whose efferent arterioles run straight to the renal surface before splitting off into capillaries; JM are nephrons whose efferent arterioles give rise to the arterial vasa recta. By this definition, in the rabbit 28% of the nephrons are SF and 9% are JM. The remaining 63%, belonging to neither group, are called midcortical. Micropuncturists approximate this anatomical definition by defining SF nephrons as those accessible from the surface and JM as those possessing loops that extend into the tip of the papilla. Microperfusionists often do not discriminate between SF and JM nephron segments. Those who have identified SF and JM nephrons have studied only the proximal segments, leaving the thin limbs and distal segments undefined. For microperfusionists, SF proximal segments can be defined as nephrons where at least one loop reaches the surface; JM proximal segments can be defined as nephrons whose glomeruli are immediately adjacent to the corticomedullary junction. Many microperfusionists (11, 63), however, have extended the definition of SF nephrons to include those with glomeruli in the midcortex.

PROXIMAL SEGMENTS (TABLE 1)

In the past, the proximal tubule was divided into only two segments. The initial and convoluted portion of the proximal tubule was referred to as the pars convoluta or proximal convoluted tubule (PCT). The more distal and straighter segment was known as the pars recta or proximal straight tubule (PST). Recent observations, however, have divided the proximal tubule into three distinct morphological segments: S_1, S_2, and S_3. S_1 is the early PCT in both SF and JM proximal tubules. S_2 is the late SFPCT, early SFPST, and late JMPCT. S_3 is located principally in the outer medulla and is the terminal SFPST and the entire JMPST. Investigators working in the isolated perfused tubule have identified S_1 by its attachment to its glomeruli (49) or by its low rate of PAH secretion (63, 100). They have defined S_2 by the failure to observe glomeruli (9, 11, 96) or by its high rate of PAH secretion (63, 100). S_3 is identified by its medullary location.

The transport processes that occur in the proximal tubule are determined to a large extent by the composition of the luminal fluid. When a proximal tubule, regardless of origin, is perfused with a glomerular ultrafiltrate, sodium-coupled transport processes predominate (10). An organic solute–sodium cotransport system reabsorbs sugars and amino acids electrogenically. An organic anion–sodium cotransport system reabsorbs phosphate, acetate, citrate, and lactate in an electrically neutral manner. A hydrogen ion–sodium countertransport system acidifies the luminal fluid and effects the reabsorption of buffer anions such as bicarbonate by an electrically neutral process. Preferential reabsorption of solutes in the early proximal tubule results in a late proximal tubular fluid that is low in organic solutes and buffer anions and high in chloride. Unique and different transport processes occur when a proximal tubule, regardless of origin, is perfused with a sodium chloride solution. In the presence of this perfusate approximately two thirds of the sodium chloride is reabsorbed by active processes (10). This transport might be by "simple" electrogenic sodium transport or by electroneutral sodium chloride transport. Approximately one third of

Table 1 Transport properties of the proximal tubule

Property		Early PCT	Mid-late PCT	PST
PD (mV)	SF	−4.3 (50)	−5.3 (49), −4.3 (50), −2.1 (11)	−2.1 (53)
	JM	−5.3 (50)	−7.5 (49), −5.3 (50), −2.9 (11)	−1.8 (53)
Fluid Absorption	SF	0.74 (63), 0.82 (50)	0.96 (63), 1.25 (40), 1.4 (9)	0.47 (53, 95)
(nl/mm min)	JM	0.94 (63), 1.13 (50)	1.37 (63), 1.13 (40)	0.56 (53, 95)
Glucose Flux, 1−b	SF	88.8 (63)	57.8 (63), 73 (9), 84 (91)	6 (91)
(pmol/mm min)	JM	98.4 (63)	58.2 (62), 73 (9)	
Glycine Tm	SF		28.5 (4)	2.5 (4)
(pmol/mm min)	JM			
Net Bicarbonate Flux	SF	44.4 (49)	89 (40), 81 (9), 84 (17)	19.2 (95)
(pmol/mm min)	JM	86.7 (49)	111 (40), 81 (9)	29.4 (95)
Phosphate Flux, 1−b	SF	11.4 (63)	3.0 (63), 6.6 (20)	2.2 (20)
(pmol/mm min)	JM	10.7 (63)	4.6 (63)	
Net PAH Flux	SF	−0.3 (0, 100)	−1.5 (90, 100)	−0.3 (100)
(pmol/mm min)	JM	−0.35 (100)	−1.4 (100)	−0.2 (100)
P_{Na}/P_{Cl}	SF	1.56 (50)	0.6−0.3 (50), 1.2 (11, 86)	0.47 (53), 0.43 (95), 0.3 (76)
	JM	2.0 (50)	2.0 (50), 2.9 (11)	1.25 (53), 2.0 (95)
P_{HCO_3}/P_{Cl}	SF		0.40 (11), 0.44 (27)	0.35 (95)
	JM		0.47 (11)	0.53 (95)
Sodium Permeability	SF		9.3 (58), 10.7 (27)	2.6 (53), 2.3 (76)
(10^{-5} cm/s)	JM			5.8 (53)
Chloride Permeability	SF		3.6 (40, 96), 8.8 (27)	5.6 (53), 7.5 (76)
(10^{-5} cm/s)	JM		3.3 (96), 2.6 (40)	2.1 (53)
Bicarbonate Permeability	SF		1.3 (96), 2.1 (40), 2.8 (2)	2.0 (95), 0.04 (76)
(10^{-5} cm/s)	JM		1.3 (96), 2.3 (40)	1.1 (95)
Water Permeability	SF		4,400−8,800 (58), 35,000 (3)	30−40,000 (77)
(10^{-5} cm/s)	JM			
Specific Resistance	SF	12 (79)	6 (79), 7 (60)	8.6 (60)
(Ω cm^2)	JM			

the sodium chloride is reabsorbed by the passive processes of diffusion and convection (10). The extent to which these transport processes differ along the length of the proximal tubule, and between SF and JM proximal nephrons, is discussed below.

Reabsorption of a Glomerular Ultrafiltrate

COMPARISON OF PCT AND PST Fluid absorption in PCT perfused with an ultrafiltrate of rabbit serum is two to three times greater than in PST (9, 48, 49, 53, 58, 63, 90, 95). These differences in fluid absorption are associated with similar differences in other manifestations of transport. The lumen-negative potential difference (PD) (11, 28, 49, 50, 53), the basolateral cell membrane PD (12, 88) and its glucose-induced hyperpolarization (12), the cell surface to volume ratio (97), and the Na,K-ATPase activity (30, 52) are approximately two times greater in PCT than in PST.

The two- to three-fold difference in fluid absorption between PCT and PST perfused with glomerular ultrafiltrate qualitatively, but not quantitatively, correlates with observed differences in solute transport. Glucose (91), glycine (4), and bicarbonate (17, 95) reabsorption are all five to ten times less in PST than in PCT. Fluid absorption is not correspondingly five to ten times lower in PST because PST possess a "simple" electrogenic sodium reabsorptive mechanism (78) that can account for approximately 50% of the observed rate of fluid transport.

Studies designed to examine the axial heterogeneity of the organic solute transport have found important differences along the length of the proximal tubule. Turner & Moran (92) have isolated brush-border membrane vesicles from rabbit cortex and medulla, and have shown that the K_m of the sodium-glucose symporter is greater in cortical PCT (S_1 and S_2) than in medullary PST (S_3). In the isolated rabbit proximal tubule Barfuss & Schafer (4) have found that the apparent permeability coefficient for glycine in SFPCT is five times greater than in SFPST, and that the apparent affinity of the sodium-glycine symporter is greater in SFPST than in SFPCT.

In summary, a comparison of PCT and PST shows a consistent correlation among various transport functions and parameters, with each segment uniquely suited to carry out its own transport function. In the PCT the bulk of substances are reabsorbed by solute reabsorptive mechanisms that proceed at high rates, but low affinities. In the PST the last remaining substances are removed by solute reabsorptive mechanisms that proceed at low rates but high affinities. The differences in rate might be attributed to a difference in the amount of surface membrane per unit cell volume (97) and thus might not be due to intrinsic differences in the luminal membrane carriers or in the Na,K-ATPase pump systems; however, the differences in solute affinities can only be attributed to intrinsic differences in the organic solute symporters or their accessibility in the membrane.

COMPARISON OF EARLY AND LATE PCT Recent studies by McKeown et al (63), have shown that early SFPCT and early JMPCT have lower fluid absorptive rates than their respective late PCT. Jacobson (49) has also found that early SFPCT have considerably lower rates of fluid absorption than have been reported for late SFPCT (9, 17, 40, 58, 63). In contrast, Corman et al (18) find that in the rat the rate of fluid absorption is essentially constant along the length of the accessible SFPCT.

Observations in early and late PCT regarding the intranephron heterogeneity of solute transport rates are inconsistent. In some studies, solute transport rates are greater in early PCT than in late PCT. Glucose (9, 63, 91) and phosphate (63) transport follow this pattern. Acidification, however, does not. In the rabbit, bicarbonate reabsorption is approximately 50% lower in early SFPCT than in late SFPCT (49), whereas in rabbit JMPCT there are no differences in acidification rates between early and late segments (49). In the rat, bicarbonate concentration decreases exponentially with length along the accessible SFPCT, suggesting that there is one rate constant for bicarbonate reabsorption (18).

The mechanism of acidification differs along the length of the rat SFPCT. In early SFPCT there appears to be an electrogenic acidification mechanism along with the more universal electroneutral sodium-for-hydrogen antiporter. Frömter & Gessner (29) have observed a lumen-positive, acetazolamide-sensitive hydrogen ion secretory PD of 1 mV in early SFPCT and of only 0.2 mV in late SFPCT perfused with bicarbonate Ringers. This five-fold difference in the magnitude of the acidification PD cannot be accounted for by differences in transepithelial specific resistance. In early rat SFPCT resistance is only twice that of late SFPCT (79). These data, therefore, suggest that the rate of electrogenic acidification is greater in early than in late SFPCT.

COMPARISON OF SF AND JM PROXIMAL TUBULES Measurements of fluid absorption, solute transport, and transepithelial PD in isolated rabbit SFPCT and JMPCT have indicated that there is internephron heterogeneity of fluid absorption and PD. The rate of fluid absorption (49, 63) and the transepithelial PD (11, 63, 90) in SFPCT are lower than in JMPCT. Studies on phosphate and organic solute transport in the rabbit generally show that there are not significant differences in rates between the two populations (9, 40, 63). On the other hand, the absorption of sodium, bicarbonate (49), and chloride is less in early SFPCT than in JMPCT.

There is some evidence that JMPCT, but not SFPCT, might possess a "simple" electrogenic sodium reabsorptive mechanism similar to that found in SFPST (78). In the absence of organic solutes there is a lumen-negative PD in the JMPCT, but not in SFPCT (8, 48). In addition, in JMPCT but not SFPCT, fluid absorption continues in the absence of luminal organic solutes (replaced by raffinose), buffer anions (replaced by cyclamate), and a chloride concentration gradient (48). However, there has not been direct evidence of "simple" electrogenic sodium transport in JMPCT. Direct evidence would be the demonstration of fluid absorption from symmetrical sodium chloride solutions.

Reabsorption of a Sodium Chloride Solution

It is now generally agreed that approximately one third of the fluid and solute transport from a high chloride–low bicarbonate solution simulating late proximal tubular fluid is attributable to passive processes (10). Such passive sodium chloride and water reabsorption has been unequivocally demonstrated in rat SFPCT (10), rabbit SFPCT (48), and rabbit SFPST (10). However, few studies have examined the heterogeneity of salt and water transport under these conditions. Jacobson has reported that JMPCT behave differently from SFPCT (48). When JMPCT are perfused with high chloride–low bicarbonate solutions simulating late proximal fluid in the presence of ouabain, fluid absorption essentially ceases. In contrast, fluid absorption in SFPCT under the same conditions continues at approximately two thirds the control rate. This study suggests that passive sodium chloride and volume absorption occur in SFPCT but not in JMPCT. This apparent difference might be attributable to observed differences in the passive permeability properties of the paracellular pathway (11, 40, 96) (see below).

COMPARISON OF RELATIVE ION PERMEABILITIES Considerable variation exists in the electrophysiologically determined relative permeabilities to sodium, chloride, and bicarbonate between SFPCT and JMPCT, and along the length of SFPCT. The effect of a membrane on ion permeability ratios is best considered in comparison to the ion mobility ratio. In solution, the sodium-to-chloride mobility ratio is 0.63. If a membrane has a sodium-to-chloride permeability ratio (P_{Na}/P_{Cl}) of less than 0.63, then it is chloride-selective; if it has a ratio greater than 0.63, then it is sodium-selective. In the rabbit kidney, early SFPCT (50), all late SFPCT without convolutions that reach the surface (11), and the entire JM proximal tubule (11, 50, 53, 95) are sodium-selective. In general, JM proximal segments have greater P_{Na}/P_{Cl} than SF proximal segments. In contrast, only late SFPCT with convolutions that reach the surface (50) and SFPST (53, 76, 95) are chloride selective. In the rat SFPCT, sodium permeability is also slightly greater than chloride permeability (27). Thus, the great majority of proximal tubules in the kidney are sodium-selective.

In solution, the bicarbonate-to-chloride mobility ratio is 0.5. The average bicarbonate-to-chloride permeability ratio (P_{bicarb}/P_{Cl}) is similar to this value in all nephron segments regardless of origin. However, there is a distinct and direct relationship between P_{Na}/P_{Cl} and P_{bicarb}/P_{Cl} (11, 96). Tubules with low P_{Na}/P_{Cl} have low P_{bicarb}/P_{Cl}; tubules with high P_{Na}/P_{Cl} have higher P_{bicarb}/P_{Cl}. The variation in ion selectivity is due to changes in both P_{Na} and P_{Cl} (40, 53, 96), but not to changes in P_{bicarb} (40, 96).

The variation in ion selectivity in the proximal tubule has two important consequences for proximal tubular reabsorption. First, the P_{Na}/P_{Cl} will determine the fate of the sodium ions that are coupled to the reabsorption of organic solutes (10). Sodium-coupled sugar or amino acid transport generates a lumen-negative transepithelial PD. As a result of this PD, either sodium ion can diffuse back into the lumen or chloride ion can diffuse out of the lumen through the paracellular pathway. Specifically, if P_{Na}/P_{Cl} is 0.5, then two thirds of the paracellular current flow is in the form of chloride ions moving from lumen to blood and one third is in the form of sodium ions moving from blood to lumen. On the other hand, if the P_{Na}/P_{Cl} is 2.0, then two thirds of the paracellular current flow is in the form of sodium ions moving from blood to lumen and one third is in the form of chloride moving from lumen to blood. Thus in chloride-selective tubules (such as the late SFPCT with loops reaching the surface and the SFPST), organic solute reabsorption will effect substantial additional sodium chloride reabsorption. In contrast, in sodium selective tubules (such as mid-cortical PCT, JMPCT, and JMPST), organic solute reabsorption will effect little additional sodium chloride reabsorption, but instead will result in the recycling of sodium ions.

Second, P_{Na}/P_{Cl} and P_{bicarb}/P_{Cl} determine the transepithelial diffusion PD in proximal tubules perfused with late proximal tubular fluid high in chloride and low in bicarbonate ion concentration. This PD provides the electrochemical driving force for diffusive sodium reabsorption. According to the Goldman equation, there will be a lumen-positive PD whenever P_{bicarb}/P_{Cl} is less than one. The magnitude of the lumen-positive PD will be modified by P_{Na}/P_{Cl}. The higher the value of P_{Na}/P_{Cl}, the lower the lumen-positive PD will be. A small lumen-positive PD, however, does not necessarily indicate low diffusive sodium absorption. Diffusive sodium absorption is determined by both PD and absolute sodium permeability. If P_{Na}/P_{Cl} varies because of changes in P_{Cl}, then total paracellular conductance (approximately equal to the sum of the absolute sodium and chloride permeabilities) and the diffusive component of sodium chloride absorption would vary directly with P_{Cl}, being greater in chloride-selective and lower in sodium-selective tubules. Conversely, if P_{Na}/P_{Cl} varies because of changes in P_{Na}, then total paracellular conductance and the diffusive component of sodium chloride absorption would vary directly with P_{Na}, being greater in sodium-selective and lower in chloride-selective tubules. In SFPST and JMPST (53) and in undefined S_2 (96), the sum of the absolute sodium and chloride permeabilities is the same in chloride- and sodium-selective proximal tubules. This observation suggests that the total paracellular conductance and the diffusive component of sodium chloride absorption of these segments could also be similar. This analysis may not hold in JMPCT. Jacobson (48) has reported no fluid absorption from a sodium chloride solution in JMPCT in the absence of active transport, suggesting that total paracellular conductance may be lower in JMPCT than in other proximal segments.

COMPARISON OF REFLECTION COEFFICIENTS   The other important determinants of passive sodium chloride absorption that could be heterogeneous in SFPCT and JMPCT are the reflection coefficients for sodium chloride, sodium bicarbonate, and organic solutes. An effective osmotic gradient will be generated in tubules perfused with a high chloride–low bicarbonate solution by a lower reflection coefficient for sodium chloride than for sodium bicarbonate and organic solutes. Recent measurements of these parameters (42), however, do not suggest that SFPCT and JMPCT differ.

Secretion into the Proximal Tubule

PAH   PAH, although not endogenous, is the prototype for organic acid secretion. In their classical study on intratubular heterogeneity of solute transport, Tune, Burg & Patlak (90) showed that PAH was actively secreted by the entire proximal tubule, but that SFPST secreted PAH at least four times more rapidly than SFPCT. Grantham et al (33) found that PAH secretion could induce fluid secretion in PST, but not in PCT. Woodhall et al (100) have further defined the axial heterogeneity of PAH transport. S_2 avidly secrete PAH; whereas, S_1 and S_3 transport PAH at a rate five times slower.

UREA   Urea is not actively transported in the PCT (57). However, Kawamura & Kokko (54) have offered evidence that urea is actively secreted in SFPST and JMPST.

THIN LIMBS OF HENLE'S LOOP

Virtually all that is known regarding internephron heterogeneity of the thin descending limb of Henle's loop (DLH) and the thin ascending limb of

Henle's loop (ALH) is anatomical. SF nephrons have short loops devoid of an ALH. JM nephrons have long loops that possess both DLH and ALH. Thus all ALH are JMALH; DLH may be either SF or JM. In microperfusion studies the DLH and ALH are identified by their attachment to other segments. The DLH is attached to the terminal PST; the JMALH is attached to the thick ascending limb of Henle's loop.

The thin limbs of Henle's loop participate in the alteration of urine osmolality. In the DLH the tubular fluid is concentrated; in the ALH it is diluted. Whether these processes are active or passive is the subject of considerable debate. Micropuncture studies in the rat (68) and in the hamster (61) support the possibility of active transport. On the other hand, microperfusion studies provide no evidence for active transport processes in either segment in the rat (44), hamster (44), or rabbit (47, 56). In addition, Garg et al (30) find that the Na,K-ATPase activity of both DLH and ALH is not significantly different from zero. The above microperfusion studies (44, 47, 56) have also provided evidence in all three species for intrinsic differences in the permeability characteristics of the DLH and ALH that would enable these segments to concentrate and dilute the tubular fluid utilizing the energy derived from the sodium chloride and urea concentration gradients set up by more distal portions of the nephron. Specifically, the DLH is highly permeable to water, but impermeable to sodium chloride and urea. Passive concentration of the fluid in the DLH is achieved by water abstraction as the medullary interstitium becomes more concentrated from base to tip. The ALH is highly permeable to sodium chloride, moderately permeable to urea, and impermeable to water. Passive fluid dilution of the fluid in the ALH is achieved by more rapid sodium chloride exit than urea entry down their respective concentration gradients in a water-impermeable epithelium.

Some aspects of the permeability characteristics of the DLH are disputed. Stoner & Roch-Ramel (86) find that if rabbit DLH are perfused at low pressures, they are impermeable to water; if they are perfused at high pressures, they are permeable to water. This controversy awaits resolution.

DISTAL SEGMENTS

The exact definition of the distal convoluted tubule has undergone considerable change in recent years. Micropuncturists originally defined the distal convoluted tubule on the basis of its accessibility to puncture. It included anything from the macula densa to the first branch point with collecting tubules. More recently, however, it has become apparent that the distal convoluted tubule is heterogeneous, and, furthermore, that the details of heterogeneity differ among species. According to Morel and coworkers' (66) examination of adenylate cyclase activity, the rabbit distal convoluted tubule comprises four divisions: cortical thick ascending limb of Henle's loop, distal tubule, connecting tubule, and initial collecting tubule. In the rat distal convoluted tubule, the distal tubule and the connecting tubule cell types are intermixed. For this review the distal segments (DS) include those located beyond the junction of the thin and thick ascending limb of Henle's loop and before the collecting segments. Accordingly, there are three subdivisions of the DS: the thick ascending limbs of Henle's loop, distal tubule, and connecting tubule. The main transport function of the DS is to dilute the tubular fluid by reabsorbing sodium in excess of water.

THICK ASCENDING LIMB OF HENLE'S LOOP

The thick ascending limb of Henle's loop (TALH) can be divided into morphologically distinct segments. The cells of the medullary TALH (MTALH) are columnar and thicker than the cells of the flatter cortical TALH (CTALH) owing to the presence of microvilli. The transition between the MTALH and the CTALH occurs in the outer medulla. The CTALH extends beyond the macula densa. Although SF and JM nephrons have both MTALH and CTALH, no attempt has been made to differentiate between the TALH in the two populations. In the rabbit, however, the CTALH is approximately twice as long in SF as in JM nephrons, while the MTALH is approximately twice as long in JM as in SF nephrons (16). Length considerations, therefore, suggest that microperfusion studies would tend to favor SF CTALH and JM MTALH. Thus differences in TALH transport function might be due either to intranephron or to internephron heterogeneity. For this review nephron heterogeneity in the TALH is considered axial in nature.

The principle transport function of the TALH is dilution of the tubular fluid. Roughly 30–40% of the filtered sodium chloride is reabsorbed by this segment. By increasing the salt concentration of the medullary interstitium, the MTALH provides part of the energy for the formation of a concentrated urine. Both similarities and differences in basal transport function between MTALH and CTALH have been observed in the rabbit. Both segments reabsorb sodium chloride, generate a lumen-positive PD, have exceedingly low osmotic water permeabilities, and are more permeable to sodium than to chloride (16, 73). There are differences in the rate of sodium chloride transport and Na,K-ATPase activities between MTALH and CTALH. The MTALH has a greater capacity to transport sodium chloride at high flow rates (73) and has three to four times the Na,K-ATPase activity of the CTALH (30, 52). The CTALH can generate steeper sodium chloride concentration gradients at low flow rates than the MTALH (16). These differences in transport may be related to their individual functions. The main function of the MTALH is to provide for maximal medullary hypertonicity against minimal concentration gradients. A high capacity transport system best effects this result. The main function of the CTALH is to maximally dilute the tubular fluid. The ability to generate steep concentration gradients best effects this result. In the mouse some differences also exist. The MTALH is sodium-selective (38), and salt transport is bicarbonate-independent (26). The CTALH is unselective (38), and salt transport is bicarbonate-dependent (26).

Differences may exist between MTALH and CTALH in the mechanism of calcium transport in the rabbit. Suki et al (87) and Imai (45) find evidence for active calcium absorption in the CTALH and none in the MTALH (87). However, Bourdeau & Burg (13) can account for all calcium transport in CTALH by passive processes.

Salt absorption in the TALH is an active process. In contrast to other tubular segments that generate a lumen-negative PD while reabsorbing salt, the MTALH and the CTALH generate a lumen-positive PD (16, 73). The mechanism responsible for active salt absorption was considered to be active chloride transport because the positive PD was eliminated by removal of chloride, but maintained after removal of sodium, from the perfusion solutions. However, two findings have always cast doubt on the primacy of active chloride transport and suggested that chloride absorption might be secondary and attributable to a sodium chloride cotransport system. First, the TALH has an exceptionally high concentration of Na,K-ATPase (30, 52). Second, ouabain eliminates the positive PD and salt transport (16, 73). Until quite recently efforts to demonstrate primary active sodium transport had been unsuccessful (15). Now, however, several groups of investigators have reported evidence for a sodium chloride cotransport system in the TALH (24, 25, 26, 34, 38). Eveloff et al (24, 25) have shown that the oxygen consumption of cells isolated from rabbit MTALH is inhibited by removal of sodium or chloride and by addition of furosemide or ouabain to the incubation medium. In addition, plasma membrane vesicles isolated from these same cells have a chloride-dependent sodium uptake mechanism. More importantly, Greger & Frömter (34) have presented data that explain the previous findings of a positive PD after sodium removal. They found that normally prepared sodium-free solutions contained 0.5 mM sodium, presumably leached from glass. When this solution was used to perfuse rabbit CTALH, the positive PD was maintained. When sodium was totally eliminated, the PD was completely abolished. The I_{50} was 0.16 mM sodium.

The precise mechanism of the sodium chloride cotransport process in the TALH has not been determined. Two possibilities exist. First, as advanced by Andreoli and coworkers, the transcellular sodium chloride cotransport process could be neutral, effected by either a sodium chloride symporter (38) or by the combination of a sodium-hydrogen and chloride-hydroxyl antiporter (26). In this case the lumen-positive PD would be generated by sodium chloride diffusion from the lateral intercellular space to the lumen across sodium selective junctional complexes (38). A lateral intercellular space to lumen sodium concentration gradient of approximately 150 mM sodium chloride would be required to generate the positive PD observed in the mouse MTALH by this mechanism (38). Second, as advanced by Greger & Frömter, the transcellular sodium chloride cotransport process could be either directly or indirectly electrogenic. A direct electrogenic process would be a sodium chloride symporter that utilized two chlorides and one sodium. An indirect electrogenic process would be a symporter that utilized two chlorides, one sodium, and one potassium, with parallel back diffusion of potassium into the lumen. Electrophysiologic data in the rabbit CTALH from Greger & Frömter (34) show that luminal furosemide abolishes the lumen-positive PD immediately, while bath ouabain requires time to exert its effect. If the PD were due to intercellular space hypertonicity, the washout time should be the same for furosemide and ouabain. Thus these data support an electrogenic, rather than neutral, sodium chloride symporter. Studies on the electrogenicity and the potassium-dependence of the sodium chloride cotransport system in TALH-membrane vesicles are needed.

Distal Tubule

What is now referred to as the distal tubule (DT) bears no resemblance to the micropuncture concept of the distal convoluted tubule. In the rabbit this segment is less than one millimeter in length and possesses few convolutions (35, 36). It can best be identified by its bright appearance, single cell type, and attachment somewhat distal to the macula densa. Little is known about transport in the DT. However, transepithelial PD and antidiuretic hormone and mineralocorticoid responsiveness have been examined. No distinctions have been made between SFDT and JMDT. In the rat, the DT and connecting tubule segments are merged. The DT/connecting tubule is discussed below.

To date, only two groups of investigators have examined transport in the DT in vitro. Gross & Kokko (35, 36) and Imai (46) identified DT by their attached glomeruli. These authors found that the PD was 20–40 mV lumen-negative, suggesting active sodium transport. The hormone responsiveness of the DT was almost nonexistent. Of all the agents tested, only calcitonin has been found to induce an adenylate cyclase response (66), but its physiologic significance has not been investigated directly. Clearance studies have shown that the predominant effect of calcitonin in the rat was to increase phosphate and decrease magnesium excretion (69); in the rabbit, it was to increase sodium chloride excretion (14). In the rabbit DT (35, 36, 48), PD is unaffected by mineralocorticoids, in accordance with the lack of mineralocorticoid-stimulated Na,K-ATPase activity (30). Furthermore, the rabbit DT (35) is water-impermeable, and in accordance with its lack of antidiuretic hormone stimulated adenylate cyclase activity (66), it does not respond to antidiuretic hormone with an increase in water permeability.

Connecting Tubule

The connecting tubule (CNT) consists of dark or intercalated cells and granular cells. The CNT must be described differently for the rabbit and the rat nephron. In the rabbit the SFCNT does not branch and lies between the DT and the initial cortical collecting tubule. In the rabbit mid-cortical and JM nephrons, the CNT branches several times and comprises the arcades. In the rat, the DT and the CNT cell types are intermixed. This DT/CNT segment comprises some 75–80% of the micropuncturist's distal convoluted tubule in Sprague-Dawley and some 50% in Munich-Wistar rats (99).

Much remains to be known about CNT transport function and its control. CNT adenylate cyclase is stimulated by isoproterenol and parathyroid hormone (66); its Na,K-ATPase activity is doubled by mineralocorticoids (30). The only data on pure CNT transport function have been obtained by Imai in the rabbit (46). No distinction was made between SFCNT and JMCNT. The PD was approximately 30 mV lumen-negative, suggesting active sodium reabsorption. Antidiuretic hormone had no effect on water permeability. PD was reversibly decreased with isoproterenol, but not changed by physiologic concentrations of antidiuretic hormone or when the rabbits were pretreated with mineralocorticoid. The physiologic responses of isoproterenol and antidiuretic hormone are in accordance with those predicted by its enzyme responsiveness (66). The failure to observe an influence of mineralocorticoid on PD, however, is not. An effect of mineralocorticoid on transport processes other than sodium should be investigated.

Other studies have been done on the tubule located between the macula densa and the first branch point of the collecting tubule (the distal convoluted tubule). These studies are predominately on DT/CNT, but include initial cortical collecting tubule. Such studies have shown active sodium reabsorption (19, 80) and potassium secretion (19, 80), but no basal bicarbonate reabsorption (59). Active sodium transport probably increases along the length of the distal convoluted tubule in the rat since resistance decreases (7) and PD becomes more lumen-negative (98). The sign of the transepithelial PD in the rat DT/CNT, however, is controversial, perhaps owing to differences in electrodes, liquid junction correction, or tip localization. Wright (98) and Hayslett et al (37) find a low lumen-negative value in the early distal convoluted tubule. Allen & Barratt (1) find a lumen-positive value. Allen & Barratt (1) have also observed that aldosterone decreases the lumen-positive PD in the early distal convoluted tubule. No data are available on the mineralocorticoid responsiveness of DT/CNT Na,K-ATPase activity in the rat.

Examination of calcium transport and the effect of parathyroid hormone in the distal convoluted tubule can be assumed to represent the CNT, since this is the segment where parathyroid hormone receptors have been described (66). In the rabbit distal convoluted tubule, calcium reabsorption was strongly stimulated by parathyroid hormone without changing PD and thus sodium transport (80). Comparable observations have been made in the rat (19).

COLLECTING SEGMENTS

The Collecting Segments (CS) begin in the outer cortex before the confluence of two adjacent distal convoluted tubules and extend into the calyx of the renal pelvis. Since the CNT of SF and JM nephrons empty into the CS in the outer cortex, there is not a distinction between SFCS and JMCS. The CS consist of two cell types: the light or principal cells and the dark or intercalated cells. Since there is a gradual rather than an abrupt decrease in the number of dark cells from cortex to papilla, the CS cannot be subdivided on the basis of cell type. There are, however, distinguishable subdivisions based on their location in the kidney: cortical collecting tubule (CCT), medullary collecting duct (MCD), and papillary collecting duct (PCD).

The CS are important sites for the final regulation of urinary sodium, potassium, hydrogen ion, and water excretion. Direct comparisons between CCT, MCD, and PCD are difficult to make because few investigators have examined like properties in more than one segment. However, whether the comparisons are direct or indirect, it appears that the CS are heterogeneous with regard to some of their transport functions and homogenous with regard to others. Important qualitative differences are found in sodium and potassium transport, in the control of sodium and potassium transport by mineralocorticoids and diet, and in urea permeability. Minor quantitative differences are found in water transport and its control by antidiuretic hormone. No differences are found in hydrogen or bicarbonate ion transport and their control by mineralocorticoids and diet.

Sodium and Potassium Transport and Their Control

A direct comparison of transport function and its control can be made between CCT and MCD (82–85). The magnitude of the transepithelial PD varied from cortex to papilla in the isolated rabbit CS, in accordance with the variation in the dark to light cell ratio (85). In the CCT the PD was highly lumen-negative, in the outer MCD (OMCD) it was slightly lumen-negative, and in the inner MCD (IMCD) it was lumen-positive. The lumen-negative PD suggests active sodium absorption. The lumen-positive PD could be due to active chloride reabsorption or to active potassium or hydrogen secretion. More recently, Stokes has found that in CCT the negative PD was ouabain-sensitive (83) and associated with spontaneous sodium and chloride reabsorption and potassium secretion (82). In addition, the PD and ion transport rates were enhanced by pretreatment with mineralocorticoids (82) and inhibited by prostaglandins (84). In the OMCD the small negative PD was not ouabain-sensitive (83); sodium and potassium transport rates were low (82) and insensitive to mineralocorticoids (82), but were inhibited by prostaglandins (84). In the IMCD there was no spontaneous or mineralocorticoid-induced sodium, chloride or potassium transport (82). The ion selectivity also differs between CCT and OMCD (83). The CCT are ten times more permeable to potassium than to sodium, while the OMCD are unselective with respect to sodium and potassium. In summary, Stoke's data suggest that the CCT possess mineralocorticoid and prostaglandin-sensitive sodium and potassium transport processes. In contrast, the IMCD appear to transport only hydrogen.

Other studies have examined sodium and potassium transport in CS. In general, results in the CCT agree with, and those in the MCD and PCD differ from, those of Stokes (82–85). The CCT can only be studied using microperfusion. These studies universally report active sodium reabsorption and potassium secretion (67, 75, 82, 84) which is mineralocorticoid (1, 35, 36, 67, 75, 82) and prostaglandin (43, 84) sensitive. The MCD and PCD have been studied using micropuncture and microcatheterization in addition to microperfusion. In contrast to data from microperfusion, data from micropuncture and microcatheterization indicate that both the IMCD (6) and PCD (21, 93) actively reabsorb sodium and that the IMCD and PCD secrete (5, 41) and reabsorb (21, 72) potassium. The cause of these discrepancies in data with regard to sodium and potassium transport in the MCD may be methodological (microperfusion versus micropuncture and microcatheterization) or species differences (rabbit versus rat and hamster).

Sodium and potassium transport in the CS are controlled in part by mineralocorticoids. However, the segmental localization of mineralocorticoid target sites is uncertain. Functional studies in the rat from Ullrich & Papavassiliou (93) indicate that sodium reabsorption from the PCD decreased following adrenalectomy, suggesting that it is controlled by mineralocorticoids. On the other hand, Stokes's data in the rabbit show a mineralocorticoid effect on sodium, potassium, and chloride transport in CCT, but not in IMCD or OMCD (82). Biochemical studies find that mineralocorticoids dramatically increase Na,K-ATPase activity in CCT, but not in MCD (30); that adrenalectomy significantly reduces citrate synthase in CCT and IMCD, but not in OMCD (62); and that there are mineralocorticoid receptors in both CCT and MCD (51). In summary, although data exist supporting each CS as a mineralocorticoid target site, there is definite heterogeneity whenever two or more segments have been examined in the same study. The biochemical and functional studies that examined the CCT and the one functional study that investigated the PCD identify these segments as sites of action for mineralocorticoid. However, the role for mineralocorticoids in the MCD is not clear.

Potassium transport in CS is also controlled by diet. The mechanism appears to be independent of mineralocorticoids (22). Chronic potassium loading increases the Na,K-ATPase activity in the CCT and MCD (22). It also decreases the number of dark cells and increases the basalateral membrane surface area of light cells in the MCD, but not in the PCD (70). Chronic potassium depletion increases the number of dark cells as assayed by freeze fracture techniques (81). In summary, dietary control of potassium excretion occurs in CCT and MCD, but not in PCD.

Hydrogen and Bicarbonate Transport and Their Control

Acidification has been observed in CCT using microperfusion (55, 64, 65), in the IMCD using microcatheterization (31), and in PCD using micropuncture (41, 94). In theory, the ability of each segment to establish a hydrogen ion concentration gradient would be expected to increase from cortex to papilla. Measurements of transepithelial resistance support such a view. The resistance of the PCD in the rat (71) is twice that of the CCT in the rabbit (39). However, no study to date has examined pH gradients in more than one CS. Thus there are no direct observations to support this contention. Where comparative data are available, all CS appear to behave similarly. Acidification is active (55, 94), is independent of sodium (55, 94), generates a lumen-positive PD (55, 64), reabsorbs bicarbonate (64, 65), and is inhibited by acetazolamide (55, 94) and SITS (55, 94).

McKinney & Burg (64, 65) have shown that the CCT is capable of both bicarbonate secretion and reabsorption, depending on the preexisting acid-base status of the rabbit. Ullrich & Papavassiliou (94) have measured both bicarbonate and glycodiazine transport in the rat PCD. They found that although bicarbonate transport varied with the acid-base state, glycodiazine transport, which represents hydrogen ion transport, was unaffected by it. The variation in net bicarbonate transport might be explained by variations in the rate of bicarbonate secretion, without alteration of the absorptive component.

Mineralocorticoid did not affect acidification in the rat PCD (94) but has been reported to increase the magnitude of the lumen-positive PD in the rabbit CCT perfused without sodium (55). This observation could represent an effect of mineralocorticoid on acidification that is independent of its effect on sodium transport. However, since mineralocorticoid has been shown to decrease dramatically the conductance of the rabbit CCT (67), it is more likely that the increase in the lumen-positive PD is due to an increase in transepithelial resistance.

Water Transport and its Control

The CS all participate in the formation of a concentrated urine and in the maintenance of a dilute urine. In the absence of antidiuretic hormone, they are all poorly permeable to water; in the presence of antidiuretic hormone, they all become highly permeable to water (74). However, there appears to be a quantitative difference in the magnitude of the antidiuretic hormone–induced adenylate cyclase and in the antidiuretic hormone–induced percentage increase in water permeability from cortex to papilla, both being greater in the CCT than in the MCD or the PCD (66, 74).

Endogenous prostaglandins modulate the action of antidiuretic hormone in the CS. Grantham & Orloff (32) showed that prostaglandin E_1 inhibited the antidiuretic hormone–induced change in water permeability in the CCT but was inactive alone. Similarly, in PCD prostaglandin E_2 inhibits the antidiuretic hormone–induced adenylate cyclase response (23).

Urea Transport

One important difference between CS is in their permeability to urea. Urea permeability of the CS is unaltered by antidiuretic hormone. Both CCT and the MCD are essentially impermeable to urea (74), while the PCD are moderately permeable to urea (74). The low urea permeability of the CCT and the OMCD is important in preventing the loss of recycled urea in the cortex and insuring its delivery to the papilla. The high urea permeability of the PCD enables the recycled urea to enter the medullary interstitium where it facilitates the concentrating function of the thin limbs of Henle's loop. Interestingly, the lanthanum permeability of junctional complexes in the rat follows a permeability pattern similar to that of urea: It is low in the CCT and IMCD but high in the IMCD and PCD (89).

Acknowledgments

This work was supported by Public Health Service Research Grants HL-06285 and RO1-AM 07219 from the National Institute of Arthritis, Metabolism and Digestive Diseases. The author would like to thank Drs. Floyd Rector, Jr., Robert Alpern, and Martin Cogan for critical reading of the manuscript.

Literature Cited

1. Allen, G. G., Barratt, L. J. 1981. Effect of aldosterone on the transepithelial potential difference of the rat distal tubule. *Kidney Int.* 19:678–86
2. Alpern, R. J., Cogan, M. G., Rector, F. C. 1981. Proximal tubule bicarbonate permeability in the rat: Effect of volume expansion and metabolic alkalosis. *Kidney Int.* 19:229A
3. Andreoli, T. E., Schafer, J. A., Troutman, S. 1978. Perfusion rate-dependence of transepithelial osmosis in isolated proximal convoluted tubules. Estimation of hydraulic conductance. *Kidney Int.* 14:263–69
4. Barfuss, D. W., Schafer, J. A. 1979. Active amino acid absorption by proximal convoluted and proximal straight tubules. *Am. J. Physiol.* 236:F149–62
5. Bengele, H. H., McNamara, E. R., Alexander, E. A. 1979. Potassium secretion along the inner medullary collecting duct. *Am. J. Physiol.* 236:F278–82
6. Bengele, H. H., Lechene, C., Alexander, E. A. 1980. Sodium and chloride transport along the inner medullary collecting duct: effect of saline expansion. *Am. J. Physiol.* 238:F504–8
7. de Bermudez, L., Windhager, E. E. 1975. Osmotically induced changes in electrical resitance of distal tubules of rat kidneys. *Am. J. Physiol.* 229:1536–46
8. Berry, C. A. 1981. Electrical effects of acidification in rabbit proximal convoluted tubule. *Am. J. Physiol.* 240:F459–70
9. Berry, C. A., Cogan, M. G. 1981. Influence of peritubular protein on solute absorption in the rabbit proximal tubule: A specific effect on NaCl transport. *J. Clin. Invest.* 68:506–16
10. Berry, C. A., Rector, F. C. Jr. 1980. Active and passive sodium transport in the proximal tubule. *Miner. Electrol. Metab.* 4:149–60
11. Berry, C. A., Warnock, D. G., Rector, F. C. Jr. 1978. Ion selectivity and proximal salt reabsorption. *Am. J. Physiol.* 235:F234–45
12. Biagi, B., Kubota, T., Sohtell, M., Giebisch, G. 1981. Intracellular potentials in rabbit proximal tubules perfused in vitro. *Am. J. Physiol.* 240:F200–10
13. Bourdeau, J. E., Burg, M. B. 1979. Voltage dependence of calcium transport in the thick ascending limb of Henle's loop. *Am. J. Physiol.* 5:F357–64
14. Brendt, T. J., Knox, F. G. 1980. Effects of parathyroid hormone and calcitonin on electrolyte excretion in the rabbit. *Kidney Int.* 17:473–78
15. Burg, M. G., Bourdeau, J. E. 1978. Function of the thick ascending limb of Henle's loop. In *New Aspects of Renal Function.*, ed. H. G. Vogel., K. J. Ullrich, pp. 91–102. Amsterdam-Oxford: Excerpta Medica
16. Burg, M., Green, N. 1973. Function of the thick ascending limb of Henle's loop. *Am. J. Physiol.* 224:659–68
17. Burg, M. G., Green, N. 1977. Bicarbonate transport by isolated perfused rabbit proximal convoluted tubules. *Am. J. Physiol.* 233:F307–14
18. Corman, B., Thomas, R., McLeod, R., de Rouffignac, C. 1980. Water and total CO_2 reabsorption along the rat proximal convoluted tubule. *Pflügers Arch.* 389:45–53
19. Costanzo, L. S., Windhager, E. E. 1980. Effects of PAH, ADH and cyclic AMP on distal tubular Ca and Na reabsorption. *Am. J. Physiol.* 239:F478–85
20. Dennis, V. W., Woodhall, P. B., Robinson, R. R. 1976. Characteristics of phosphate transport in isolated proximal tubules. *Am. J. Physiol.* 231:979–85
21. Diezi, J., Michoud, P., Aceves, J., Giebisch, G. G. 1973. Micropuncture study of electrolyte transport across papillary collecting duct of the rat. *Am. J. Physiol.* 224:623–34
22. Doucet, A., Katz, A. 1980. Renal potassium adaption: Na-K-ATPase activity along the nephron after chronic potas-

sium loading. *Am. J. Physiol.* 238:F380–86
23. Edwards, R. M., Jackson, B. A., Dousa, T. P. 1981. Vasopressin (VP) sensitive cAMP system in the isolated papillary collecting duct (PCD). *Kidney Int.* 19:239A
24. Eveloff, J., Bayerdörffer, E., Haase, W., Kinne, R. 1980. Biochemical and physiological studies on cells isolated from the medullary thick ascending limb of Henle's loop. *Int. J. Biochem.* 12:55–59
25. Eveloff, J., Kinne, R. K. H. 1981. Sodium-chloride cotransport in plasma membranes isolated from thick ascending limb of Henle's loop (TALH). *Fed. Proc.* 49:356A
26. Friedman, P. A., Andreoli, T. E. 1981. Bicarbonate-stimulated transepithelial voltage and NaCl transport in the mouse renal cortical thick ascending limb. *Clin. Res.* 29:462A
27. Frömter, E. 1974. Electrophysiology and isotonic fluid absorption of proximal tubules of mammalian kidney. In *Kidney and Urinary Tract Physiology*, ed. K. Thurau, pp. 1–38. London: Butterworth & Co. Ltd
28. Frömter, E., Gessner, K. 1974. Active transport potentials, membrane diffusion potentials and streaming potentials across rat kidney proximal tubule. *Pflügers Arch.* 351:85–98
29. Frömter, E., Gessner, K. 1975. Effect of inhibitors and diuretics on electrical potential differences in rat kidney proximal tubule. *Pflügers Arch.* 357:209–24
30. Garg, L., Knepper, M., Burg, M. 1981. Mineralocorticoid stimulation of Na-K-ATPase in nephron segments. *Kidney Int.* 19:241A
31. Graber, M. L., Bengele, H., Caflesch, C. R., Mroz, E., Lechene, C., Schwartz, J. M., Alexander, E. A. 1980. Acidification by inner medullary collecting duct of the rat. *Clin. Res.* 28:533A
32. Grantham, J. J., Orloff, J. 1968. Effect of prostaglandin E_1 on the permeability response of the isolated collecting tubule to vasopressin, adenosine 3'5'-monophosphate and theophylline. *J. Clin. Invest.* 47:1154–61
33. Grantham, J. J., Qualizza, P. B., Irwin, R. I. 1974. Net fluid secretion in the proximal straight renal tubules in vitro: Role of PAH. *Am. J. Physiol.* 226:191–97
34. Greger, R., Frömter, E. 1980. Time course of ouabain and furosemide effects on transepithelial potential differences in cortical thick ascending limbs of rabbit nephrons. *Proc. Int. Union Physiol. Sci.* 14:445A
35. Gross, J. B., Imai, M., Kokko, J. P. 1975. A functional comparison of the cortical collecting tubule and the distal convoluted tubule. *J. Clin. Invest.* 55:1284–94
36. Gross, J. B., Kokko, J. P. 1977. Effects of aldosterone and potassium-sparing diuretics on electrical potential differences across the distal nephron. *J. Clin. Invest.* 59:82–89
37. Hayslett, J. P., Boulpaep, E. L., Giebisch, G. H. 1978. Factors influencing potential differences in mammalian distal tubule. *Am. J. Physiol.* 234:F182–91
38. Hebert, S. C., Culpepper, R. M., Andreoli, T. E. 1981. Functional heterogeneity and ADH-stimulated NaCl co-transport. *Am. J. Physiol.* In press
39. Helman, S. I., Grantham, J. I., Burg, M. B. 1971. Effect of vasopressin on electrical resistance of renal cortical collecting tubules. *Am. J. Physiol.* 220:1825–32
40. Holmberg, C., Kokko, J. P., Jacobson, H. R. 1981. Determination of chloride and bicarbonate permeability in proximal convoluted tubules. *Am. J. Physiol.* In press
41. Hierholzer, K. 1961. Secretion of potassium and acidification in collecting ducts of mammalian kidney. *Am. J. Physiol.* 201:318–24
42. Hierholzer, K., Kawamura, S., Seldin, D. W., Kokko, J. P., Jacobson, H. R. 1980. Reflection coefficients of various substrates across superficial and juxtamedullary proximal convoluted segments of rabbit nephrons. *Miner. Electrol. Metab.* 3:172–80
43. Iino, Y., Imai, M. 1978. Effects of prostaglandins on Na transport in isolated collecting tubules. *Pflügers Arch.* 373:125–32
44. Imai, M. 1977. Function of the thin ascending limb of Henle of rats and hamsters perfused in vitro. *Am. J. Physiol.* 232:F201–9
45. Imai, M. 1978. Calcium transport across rabbit thick ascending limb of Henle's loop perfused in vitro. *Pflügers Arch.* 374:255–63
46. Imai, M. 1979. The connecting tubule: A functional subdivision of the rabbit distal nephron. *Kidney Int.* 15:346–56
47. Imai, M., Kokko, J. 1974. Sodium, urea, and water transport in the thin ascending limb of Henle. Generation of osmotic gradients by passive diffusion of solutes. *J. Clin. Invest.* 53:393–402

48. Jacobson, H. R. 1979. Characteristics of volume reabsorption in rabbit superficial and juxtamedullary proximal convoluted tubules. *J. Clin. Invest.* 63:410–18
49. Jacobson, H. R. 1981. Effects of CO_2 and acetazolamide on bicarbonate and fluid transport in rabbit proximal tubules. *Am. J. Physiol.* 240:F54–62
50. Jacobson, H. R., Kokko, J. P. 1976. Intrinsic differences in various segments of the proximal convoluted tubule. *J. Clin. Invest.* 57:818–25
51. Katz, A. I., Doucet, A. 1981. Aldosterone binding along the rabbit nephron. *Kidney Int.* 19:246A
52. Katz, A. I., Doucet, A., Morel, F. 1979. Na-K-ATPase activity along the rabbit, rat and mouse nephron. *Am. J. Physiol.* 237:F114–20
53. Kawamura, S., Imai, M., Seldin, D. W., Kokko, J. P. 1975. Characteristics of salt and water transport in superficial and juxtamedullary straight segments of proximal tubules. *J. Clin. Invest.* 55:1269–77
54. Kawamura, S., Kokko, J. P. 1976. Urea secretion by the straight segment of the proximal tubule. *J. Clin. Invest.* 58:604–12
55. Koeppen, B. M., Helman, S. I. 1981. Acidification of the luminal fluid by the rabbit cortical collecting tubule perfused *in vitro*. *Am. J. Physiol.* Submitted
56. Kokko, J. 1970. Sodium chloride and water transport in the descending limb of Henle. *J. Clin. Invest.* 49:183–46
57. Kokko, J. P. 1972. Urea transport in the proximal tubule and descending limb of Henle. *J. Clin. Invest.* 51:1999–2008
58. Kokko, J. P., Burg, M. B., Orloff, J. 1971. Characteristics of NaCl and water transport in the renal proximal tubule. *J. Clin. Invest.* 50:69–76
59. Lucci, M. S., Pucacco, L. R., Carter, N. W., DuBose, T. D. 1981. Effect of altered acid-base status on bicarbonate reabsorption by the rat distal convoluted tubule. *Clin. Res.* 29:469A
60. Lutz, M., Cardinal, J., Burg, M. B. 1973. Electrical resistance of renal proximal tubule perfused in vitro. *Am. J. Physiol.* 225:729–34
61. Marsh, D., Azen, S. 1975. Mechanism of NaCl reabsorption by hamster thin ascending limbs of Henle's loop. *Am. J. Physiol.* 228:71–79
62. Marver, D., Schwartz, M. J. 1980. Identification of mineralocorticoid target sites in the isolated rabbit cortical nephron. *Proc. Natl. Acad. Sci. USA* 77:3672–76
63. McKeown, J., Brazy, P. C., Dennis, V. W. 1979. Intrarenal heterogeneity for fluid phosphate and glucose absorption in the rabbit. *Am. J. Physiol.* 234:F312–18
64. McKinney, T. D., Burg, M. B. 1978. Bicarbonate absorption by rabbit cortical collecting tubules in vitro. *Am. J. Physiol.* 234:F141–45
65. McKinney, T. D., Burg, M. B. 1978. Bicarbonate secretion by rabbit cortical collecting tubules in vitro. *J. Clin. Invest.* 61:1421–27
66. Morel, F., Imbert-Teboul, M., Chabardès, G. 1981. Distribution of hormone-dependent adenylate cyclase in the nephron and its physiological significance. *Ann. Rev. Physiol.* 43:569–83
67. O'Neil, R. G., Helman, S. I. 1977. Transport characteristics of renal collecting tubules. Influence of DOCA and diet. *Am. J. Physiol.* 233:F544–58
68. Pennell, J., Lacy, F., Jamison, R. 1974. An in vivo study of the concentrating process in the decending limb of Henle's loop. *Kidney Int.* 5:337–41
69. Poujeol, P., Touvay, C., Roinel, N., de Rouffignac, C. 1980. Stimulation of renal magnesium reabsorption by calcitonin in the rat. *Am. J. Physiol.* 239:F524–32
70. Rastegar, A., Biemesderfer, D., Kashgarian, M., Hayslett, J. P. 1980. Changes in membrane surfaces of collecting duct cells in potassium adaptation. *Kidney Int.* 18:293–301
71. Rau, W., Frömter, E. 1974. Electrical properties of the medullary collecting ducts of the golden hamster kidney. II. The transepithelial resistance. *Pflügers Arch.* 351:113-31
72. Reineck, H. J., Osgood, R. W., Stein, J. H. 1978. Net potassium addition beyond the superficial distal tubule of the rat. *Am. J. Physiol.* 235:F104–10
73. Rocha, A. S., Kokko, J. P. 1973. Sodium chloride and water transport in the medullary thick ascending limb of Henle. Evidence for active chloride transport. *J. Clin. Invest.* 52:612–23
74. Rocha, A. S., Kokko, J. P. 1974. Permeability of medullary nephron segments to urea and water: Effect of vasopressin. *Kidney Int.* 6:379-87
75. Schwartz, G. J., Burg, M. B. 1978. Mineralocorticoid effects on cation transport by cortical collecting tubules in vitro. *Am. J. Physiol.* 235:F576–85
76. Schafer, J. A., Patlak, C. S., Andreoli, T. E. 1977. A mechanism for isotonic fluid absorption linked to active and passive ion flows in the mammalian su-

perficial pars recta. *Am. J. Physiol.* 233:F154–67
77. Schafer, J. A., Patlak, C. S., Troutman, S. L., Andreoli, T. E. 1978. Volume absorption in the pars recta. II. Hydraulic conductivity coefficient. *Am. J. Physiol.* 234:F340–48
78. Schafer, J. A., Troutman, S. L., Watkins, M. L., Andreoli, T. E. 1978. Volume absorption in the pars recta. I. "Simple" active Na^+ transport. *Am. J. Physiol.* 234:F332–39
79. Seeley, J. 1973. Variation in electrical resistance along the length of rat proximal convoluted tubule. *Am. J. Physiol.* 255:48–57
80. Shareghi, G. R., Stoner, L. C. 1978. Calcium transport across segments of the rabbit distal nephron in vitro. *Am. J. Physiol.* 235:F367–75
81. Stetson, D. L., Wade, J. B., Giebisch, G. 1980. Morphologic alterations in the rat medullary collecting duct following potassium depletion. *Kidney Int.* 17:45–56
82. Stokes, J. B. 1980. Localization of mineralocorticoid response along the cortical and outer medullary collecting tubule of the rabbit. *Clin. Res.* 463A
83. Stokes, J. B. 1981. Na and K transport across the rabbit cortical and outer medullary collecting tubule. *Clin. Res.* 29:477A
84. Stokes, J. B., Kokko, J. P. 1977. Inhibition of sodium transport by prostaglandin E_2 across the isolated, perfused rabbit collecting tubule. *J. Clin. Invest.* 59:1099–104
85. Stokes, J. B., Tisher, C. G., Kokko, J. P. 1978. Structural functional heterogeneity along the rabbit collecting tubules. *Kidney Int.* 14:585–93
86. Stoner, L. C., Roch-Ramel, F. 1979. The effects of pressure on the water permeability of the descending limb of Henle's loops of rabbit. *Pflügers Arch.* 382:7–15
87. Suki, W. N., Rouse, D., Ng, R. C. K., Kokko, J. P. 1980. Calcium transport in the thick ascending limb of Henle. Heterogeneity of function in the medullary and cortical segments. *J. Clin. Invest.* 66:1004–9
88. Terreros, D. A., Grantham, J. A., Tarr, M., Grantham, J. J. 1981. Axial heterogeneity of transmembrane electrical potential in isolated renal tubules. *Kidney Int.* 19:259A
89. Tisher, C. C., Yarger, W. E. 1975. Lanthanum permeability of tight junctions along the collecting duct of the rat. *Kidney Int.* 7:35–43
90. Tune, B. M., Burg, M. B., Patlak, C. S. 1969. Characteristics of p-aminohippurate transport in proximal renal tubules. *Am. J. Physiol.* 217:1057–63
91. Tune, B., Burg, M. B. 1971. Glucose transport by proximal renal tubules. *Am. J. Physiol.* 221:580–85
92. Turner, R. J., Morau, A. 1981. Heterogeneity of glucose transport along the proximal tubule—evidence from vesicle studies. *Fed. Proc.* 40:371A
93. Ullrich, K. J., Papavassiliou, F. 1979. Sodium reabsorption in the papillary collecting duct of rats. *Pflügers Arch.* 379:49–52
94. Ullrich, K. J., Papavassiliou, F. 1981. Bicarbonate reabsorption in papillary collecting duct of rats. *Pflügers Arch.* 389:271–75
95. Warnock, D. G., Burg, M. G. 1977. Urinary acidification: CO_2 transport by the rabbit proximal straight tubule. *Am. J. Physiol.* 232:F20–25
96. Warnock, D. G., Yee, V. J. 1981. Anion permeabilities of the isolated, perfused rabbit proximal tubule. *Am. J. Physiol.* Submitted
97. Welling, L. W., Welling, D. J. 1975. Surface areas of brush border and lateral cell walls in the rabbit proximal nephron. *Kidney Int.* 8:343–48
98. Wright, F. S. 1971. Increasing magnitude of electrical potential along the renal distal tubule. *Am. J. Physiol.* 220:624–38
99. Woodhall, P. B., Tisher, C. C. 1973. Response of the distal tubule and cortical collecting duct to vasopressin in the rat. *J. Clin. Invest.* 52:3095-108
100. Woodhall, C., Tisher, C., Simonton, C. A., Robinson, R. R. 1978. Relationship between para-aminohippurate secretion and cellular morphology in rabbit proximal tubules. *J. Clin. Invest.* 61:1320–29

BIOLOGICAL IMPORTANCE OF NEPHRON HETEROGENEITY

Larry A. Walker and Heinz Valtin

Department of Physiology, Dartmouth Medical School, Hanover, New Hampshire 03755

INTRODUCTION

Nephron heterogeneity probably reflects adaptations that subserve important physiological functions; such functions are likely to involve the maintenance of balance for water and solutes. Although even the early anatomists inferred function from structure (72), recognition of the biological importance of certain findings often lags behind the descriptive phase. Thus we are just beginning to integrate the many facts about nephron heterogeneity to posit systems that fulfill critical functions (15, 16, 21, 33, 38, 39, 41, 42, 67, 72).

Here we shall try to attach biological meaning to some of the differences along and between nephrons that have been described in the preceding two chapters. Although heterogeneity is probably important for many functions, we limit our effort to three examples, showing how differences appear to subserve: (*a*) the maintenance of phosphate balance; (*b*) the adjustment to expansion of the extracellular fluid volume; and (*c*) the conservation of water. Heterogeneity along a given part of a nephron, such as the proximal tubule (sometimes called axial heterogeneity), is becoming increasingly recognized. But because the functional meaning of this type of heterogeneity is not yet clear in most instances, we discuss examples that involve primarily heterogeneity between different populations of nephrons.

MAINTENANCE OF PHOSPHATE BALANCE

Intact rats, eating a normal diet, reabsorb approximately 80% of the phosphate that is filtered (15). The fractional reabsorption is increased to nearly

100% when rats are placed on a restricted intake of phosphate, and the fraction is reduced to approximately 50% when the dietary intake of phosphate is increased (69). The bulk of reabsorption under all three circumstances occurs in the proximal convolutions (38, 42), and the reabsorptive rate in this part of the nephron is altered by changes in dietary phosphate (8, 49, 70). Questions remain, however, regarding what relative contribution various other tubular segments may make to the changes in phosphate excretion. Most data on localization of phosphate transport have been obtained through cortical micropuncture, which directly evaluates only surface nephrons. In the case of phosphate loading, such studies have left uncertainty as to the site of tubular modulation. Several reports agree that in rats given a surfeit of phosphate, with (7, 37) or without (49) infusion of parathyroid hormone (PTH), the fraction of filtered phosphate excreted exceeds the fraction flowing in late distal tubules at the surface of the kidney. Two possible explanations for these findings exist and they are not necessarily mutually exclusive: (*a*) There might be tubular secretion of phosphate into the distal nephron, most likely into collecting ducts (7)—reminiscent of the regulation of potassium excretion; or (*b*) there might be decreased reabsorption of phosphate preferentially in juxtamedullary nephrons—which, if correct, would invoke heterogeneity. Knox et al (37) sampled fluid from the bend of long loops of Henle in the papilla. They found that, at the same time that 72% of the filtered phosphate was excreted and only 51% was delivered to the collecting system out of superficial tubules (point A, Figure 1), 78% was delivered into ascending thin limbs at the papilla (point g); note that the results in Figure 1 do not refer to phosphate.

The evidence is not yet conclusive concerning the role of juxtamedullary nephrons. On the one hand, studies have indicated reabsorption in segments beyond the bend of Henle's loop, albeit under conditions other than phosphate loading (24, 54). On the other hand, ascending limbs of Henle appear to transport virtually no phosphate (43, 50, 57), and there is no evidence for secretion of phosphate in the collecting system, certainly not of the required magnitude ($> 20\%$ of the filtered load). Therefore, although the combination of reabsorption from juxtamedullary distal nephrons and secretion by collecting ducts cannot be excluded, the available data strongly suggest that at least part of the high fractional excretion during phosphate-loading is attributable to greater delivery out of juxtamedullary nephrons.

Several other studies have shown quantitative differences in the handling of phosphate by superficial as opposed to deep nephrons. In contrast to phosphate-loaded animals, those on a normal or low phosphate intake (or after thyroparathyroidectomy) exhibit a smaller mean fractional excretion than the fraction flowing at late distal micropuncture sites [references cited in (15, 24)]. Although this finding may reflect reabsorption of phosphate from collecting ducts, it has been difficult to demonstrate such reabsorption under many conditions, and controversy over the magnitude of the reabsorptive capacity of this part of the nephron remains. Several studies utilizing tracer microinjections into superficial distal tubules have failed to demonstrate any reabsorption in the terminal nephron (9, 23, 61). Also, Dennis et al (14) did not observe net phosphate transport in isolated cortical collecting tubules from the rabbit. Moreover, Bengele and coworkers (5) found, by microcatheterization of collecting ducts, that only after acute thyroparathyroidectomy could reabsorption of phosphate be detected, and this amounted to only a small fraction of the filtered load (approximately 1%).

In contrast, Poujeol et al (53) reported that, in Wistar rats, 18% of the tracer injected into superficial distal tubules was reabsorbed (equivalent to 8% of the filtered load). They did not observe this reabsorption in rats of the Munich-Wistar strain, and argued that strain differences and technical difficulties might account for the controversy over reabsorption of phosphate in the terminal nephron (53, 54). The same group (54) also showed that, in juxtamedullary nephrons of Munich-Wistar rats, considerable reabsorption occurs between the bend of Henle's loop and the collecting duct. This portion includes the juxtamedullary distal tubules as well as the arcades or connecting tubules. It is interesting that the arcades contain an adenylate cyclase very sensitive to PTH, at least in rabbits (46), though this fact may relate more to calcium transport than to phosphate.

Thus the reabsorption of phosphate in collecting ducts, if any, is very limited, and cannot account completely for the difference between the fraction excreted and that delivered out of superficial distal tubules.

These considerations suggest that, in animals on a low or normal phosphate intake, or where reabsorption of phosphate is otherwise enhanced, as by thyroparathyroidectomy, heterogeneity of nephron function may be important in the determination of phosphate excretion. A body of evidence favors this view. Goldfarb (22) took advantage of the separate venous drainage systems for superficial and deep portions of the renal cortex in cats. Using both micropuncture and arteriovenous differences, he found that, in cats on a normal diet, juxtamedullary nephrons reabsorb phosphate more avidly than do superficial nephrons. Poujeol et al (53) compared reabsorption of tracer phosphate after injections into the whole kidney with that after injection into single nephrons. In rats on a normal diet, the reabsorption by the whole kidney (as determined by injections into the renal artery) was much greater than that for superficial nephrons (as determined by injections into Bowman's space). This question was also examined (24) by cortical and papillary micropuncture in thyroparathyroidectomized rats, in which both absolute and fractional proximal reabsorption of phosphate appeared to be greater in juxtamedullary than in superficial nephrons.

The factors that account for the enhanced reabsorption have not been identified. McKeown and co-workers (45) found no differences in the rate of phosphate transport, per unit length, between isolated perfused superficial and juxtamedullary proximal tubules. However, juxtamedullary proximal convoluted tubules are longer than superficial ones, at least in the rat (1) and rabbit (73). Therefore, part of the difference in reabsorptive capacity between superficial and deep nephrons may be ascribable to tubular length. The same argument might apply to the arcades (46).

Summary. It is hard to compare some data because they were obtained not only in different species and strains but also under varying circumstances, such as diverse dietary intakes of phosphate, exogenous PTH, and/or thyroparathyroidectomy. Nevertheless, the evidence probably justifies the following speculative view, advanced at least in part by Haas et al (24). Balance for phosphate appears to be maintained by alterations in phosphate transport in two major portions of the nephron, and these changes may differ quantitatively in superficial and juxtamedullary nephrons. First, in the proximal tubules, bulk reabsorption can be elevated or depressed, and the range of responsiveness is greater for deep nephrons. Second, finer adjustments, especially during avid phosphate reabsorption, are observable in those segments that lie between the bend of loops of Henle and the point of entry into a common collecting duct. This component is also more pronounced in deep nephrons. The major modulation of reabsorption may occur preferentially in the deep nephrons because they are longer than the superficial nephrons, so that either enhancement or inhibition of transport can be expressed over a greater distance.

MAINTENANCE OF FLUID BALANCE WITH EXPANSION OF EXTRACELLULAR SPACE

There are two general ways (not absolutely independent) in which heterogeneity may play a role in the renal response to enlargement of the extracellular fluid volume: (*a*) through alteration of the glomerular filtration rate in, and the distribution of blood flow between superficial and juxtamedullary nephrons; and (*b*) through differences in the tubular handling of salt and water between these two populations of nephrons.

Barger proposed (2) that, since nephrons arising in the deep cortex have longer loops of Henle, and thus a larger reabsorptive area, than do those originating in the outer cortex, a change in the distribution of blood flow and/or filtration toward the inner cortex would favor sodium retention; or conversely that a shift in the opposite direction would promote greater sodium excretion. Several investigators have suggested that when the dietary intake of salt is increased, or when saline is infused, perfusion of outer cortical nephrons is selectively increased, thereby contributing to the natriuresis that is observed. It is to be emphasized that such increases do not necessarily arise from a redistribution or shift of blood from one part of the cortex to another. Considerable controversy remains as to the role of changes in regional flow in maintenance of sodium balance; the many studies that have addressed this question have been reviewed (16, 41). In general, it appears that alterations in single nephron filtration rates must be distinguished from those of blood flow. A selective increase in the filtration rate of superficial nephrons likely plays a minor role in the natriuresis of volume expansion, if indeed it occurs at all. At the same time, blood flow may be preferentially increased in the inner cortex. Lameire et al (41) have proposed that such changes may reduce sodium reabsorption during volume expansion by decreasing filtration fraction in juxtamedullary nephrons, and/or by washout of medullary solutes (see below). Thus if a change in hemodynamics is involved in the response to expansion of the extracellular space, it is not simply a matter of increasing delivery of filtrate to the superficial nephrons with their lower reabsorptive area.

Evidence concerning the second mode of involvement of heterogeneity during volume expansion—i.e. a difference between superficial and deep nephrons in the tubular handling of sodium and chloride—is less controversial. The issue here is analogous to that discussed for the maintenance of phosphate balance in the preceding section, and it is shown diagrammatically in Figure 1. Under hydropenic conditions, the fraction of filtered sodium flowing in the late distal tubule of superficial nephrons (point A in Figure 1) exceeds that in the urine (point C) (51). During volume expansion to 10% of body weight with saline or Ringer's solution, this difference between points A and C is diminished or abolished; and with more severe expansion to 15% of body weight, the fraction at point A may even be lower than that at point C (4). Such results could be interpreted to mean that net reabsorption in the last portion of the nephron is diminished, abolished, or even converted to net secretion during volume expansion. However, more recent experiments employing micropuncture at the "base" of the papillary collecting duct (point B) strongly suggest that the changes are due to enhanced delivery of sodium selectively out of juxtamedullary nephrons. (The term "base" refers to the most proximal point of the collecting duct that is accessible to micropuncture in the extrarenal portion of the papilla; in Munich-Wistar rats, this point is situated approximately 2 mm from the papillary tip.) During volume expansion, the fraction of sodium flowing at point A (Figure 1) was significantly lower than that at point B; in other words, there is addition of sodium to the collecting system prior to the papillary collecting duct (51, 62). Essentially identical results have been obtained for chloride (29). Several considerations suggest that this addition is unlikely to reflect secretion into the cortical or outer medullary collecting ducts.

In these studies, sodium and chloride were reabsorbed rather than secreted from the papillary collecting duct—i.e. between points B and C (Figure 1)—even during volume expansion; evidence for net secretion of sodium into collecting ducts has been obtained only by the technique of microcatheterization and only under conditions of massive volume expansion amounting to nearly 20% of body weight [(60); for conflicting views on the handling of salt and water by collecting ducts during volume expansion, see (34)]. Although it is not safe to extrapolate results obtained on the

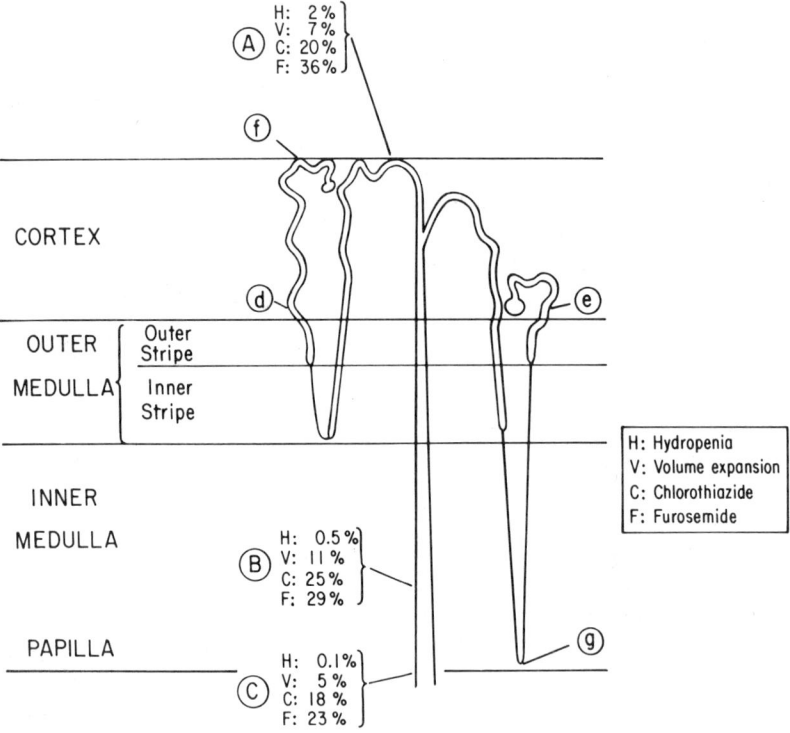

Figure 1 Fractions of filtered sodium flowing at three points of micropuncture in Munich-Wistar rats: A, late distal tubule at the surface of the kidney; B, base of a papillary collecting duct; and C, tip of that same duct. Results are shown for four conditions: hydropenia (H); volume expansion to 10% of body weight with Ringer's solution (V); volume expansion plus chlorothiazide (C); and volume expansion plus furosemide (F). Values were taken from (51) and rounded off; similar results have been reported for chloride (29).

last portion of collecting ducts coursing through the inner medulla to those lying in the cortex and outer medulla, there is additional evidence that secretion by the latter two portions is unlikely. In cortical collecting ducts, electrical resistance is high (28), and the permeability of tight junctions to lanthanum is lower than in papillary ducts (66). Consistent with these observations is the finding that cortical collecting tubules perfused in vitro show little back diffusion of sodium (19). Furthermore, no secretion of chloride could be demonstrated in these segments when they were studied in vitro (26).

In addition to this somewhat inferential, negative evidence in regard to secretion, several results argue that the addition of sodium between points A and B results from increased delivery selectively out of juxtamedullary nephrons. Furosemide abolished—in fact, reversed—the addition of sodium between points A and B (Figure 1) during volume expansion (51). Since the site of action of this diuretic is in the thick ascending limb and not the collecting tubule, the authors interpreted these results to suggest that the pattern of addition is not due to collecting duct secretion. The same group (55) came to the same conclusion using a different approach. They induced papillary (but not outer medullary or cortical) necrosis in rats with bromoethylamine hydrobromide, thereby disrupting function only of juxtamedullary nephrons. In contrast to sham-treated rats, the group with papillary necrosis failed to show net addition of sodium between points A and B (Figure 1) during volume expansion with Ringer's solution, even though delivery of sodium out of superficial late distal tubules was similar in the two groups. The fact that the addition of potassium between points A and B was not simultaneously abolished renders the interpretation of these experiments particularly credible. Thus, taken together, the results suggest that during volume expansion, juxtamedullary nephrons contribute a larger fraction of the filtered loads of sodium chloride to the collecting ducts than do superficial nephrons.

The sites for reduced reabsorption within juxtamedullary nephrons have not been identified conclusively. Volume expansion leads to decreased fractional reabsorption in the proximal tubule. There is some evidence in rats (3) and dogs (10) that during expansion with saline, filtration fraction falls to a greater extent in the inner than in the outer cortex. Since proximal reabsorption is believed to be sensitive to alterations in filtration fraction and the consequent changes in peritubular oncotic pressure, proximal reabsorption may be inhibited more in juxtamedullary than in superficial nephrons. The question cannot be answered conclusively because the terminal portions of superficial and juxtamedullary proximal tubules (points d and e of Figure 1, respectively) are not accessible to micropuncture. Indirect approaches to the problem—based on comparison of fractional delivery at points f and g (51) as well as on expansion with hyperoncotic albumin (52)—have not resolved the issue. In the latter study (52), it was demonstrated that during plasma volume expansion with hyperoncotic albumin—a situation in which the addition of sodium between points A and B (Figure 1) is absent—proximal reabsorption of sodium in juxtamedullary nephrons is not suppressed. This is in contrast to the findings during equivalent expansion of plasma volume with Ringer's solution, when proximal reabsorption in all nephrons is decreased and the pattern of addition is observed. On balance, the findings suggest that disproportionate reduction of proximal reabsorption may partly account for the greater fractional delivery of filtrate out of juxtamedullary nephrons during volume expansion.

In addition, evidence implicates juxtamedullary loops of Henle in this response. The addition of sodium chloride between points A and B (Figure 1) that is seen with volume expansion is reversed by furosemide (29, 51) but not by chlorothiazide (51). Since a major site of action for furosemide is the thick ascending limb of Henle, that structure may be the locus of disproportionate inhibition within juxtamedullary nephrons.

Regarding mechanisms that might mediate the greater inhibition of sodium chloride transport in juxtamedullary nephrons, an interesting and plausible proposal has been advanced by Osgood and his associates (51), which involves the passive model of countercurrent multiplication in the inner medulla. They postulate that when the corticopapillary interstitial osmotic gradient has been reduced or eliminated, as it is when a large load of Ringer's solution is infused, the consequent reduced reabsorption of water from descending long limbs of Henle abolishes the difference in concentration for sodium and chloride between ascending thin limbs and the interstitium (20, 35, 51), and hence the driving force for passive reabsorption of sodium and chloride from these limbs. The mechanism would apply preferentially to juxtamedullary nephrons because only these have *thin* ascending limbs of Henle (Figure 1). The reversal by furosemide of the addition between points A and B might be explained by a greater effect of the diuretic in superficial than in juxtamedullary ascending limbs, a reasonable suggestion since the thick ascending limbs of outer cortical nephrons are longer than those of deep nephrons. In support of their proposal, Osgood et al (51) have shown that the concentration gradient for sodium between thin ascending limb and ascending vasa recta is, in fact, abolished when Ringer's solution is infused or when furosemide is given on top of the volume expansion.

The mechanisms that selectively affect the juxtamedullary nephrons during volume expansion may involve the prostaglandins. The net addition of chloride between points A and B (Figure 1) is reduced after inhibition of prostaglandin synthesis (30). This effect might be mediated through possible influence of the prostaglandins on medullary blood flow (12, 59), resulting in diminished "washout" of the corticopapillary osmotic gradient during inhibition, and/or by affecting reabsorption of chloride selectively from medullary thick ascending limbs of Henle (65). The higher capacity of the inner medulla for synthesizing prostaglandins, as compared to cortex (44), may be relevant here since the deeper nephrons traverse areas rich in the hormone while superficial nephrons do not. It should be emphasized, however, that this role for prostaglandins in volume expansion remains controversial, and other factors are likely also involved.

Summary. The renal response to isotonic expansion of extracellular fluid involves a decrease in the reabsorption of sodium and chloride from all nephrons, with a much more marked inhibition in juxtamedullary than in superficial nephrons. The site of the greater suppression may be mainly the medullary (i.e., long) ascending limbs of Henle, although a contribution from deep proximal tubules has not been excluded. As a result of the selective inhibition, a larger fraction of the filtered sodium and chloride is delivered into collecting ducts from deep than from superficial nephrons. The mechanisms that effect the preferential decrease may include: a proportionately greater decrease in the filtration fraction of inner cortical as opposed to outer cortical glomeruli; and a decrease in the corticopapillary interstitial osmotic gradient leading, ultimately, to decreased passive reabsorption of sodium and chloride from thin (long) ascending limbs of Henle. The decreased osmotic gradient could result from an augmented medullary blood flow and/or from diminished reabsorption of chloride from thick ascending limbs of Henle—and, in turn, both of these effects might be mediated by renal prostaglandins, which are synthesized mainly in the medulla.

CONSERVATION OF WATER

It has been appreciated for many years that the concentration of urine by the mammalian kidney is a complex process, which undoubtedly depends in large measure on heterogeneity among nephrons. The presence of long loops of Henle and the ability to concentrate urine has been discussed since at least 1909, when Karl Peter published his classic monograph (72). More recently, other aspects of the countercurrent system have come to light, and in this section, we discuss three of them, in which heterogeneity appears to play a critical role. These are: (*a*) a mechanism for medullary recycling of urea preferentially via short loops of Henle; (*b*) the stimulation by vasopressin of sodium chloride transport in medullary, but not cortical, thick ascending limbs; and (*c*) the induction, by vasopressin, of differences in the glomerular filtration rate between superficial and deep nephrons.

Medullary Recycling of Urea

In an earlier review (72), one of us discussed the importance of nephron heterogeneity in this process, which returns to the inner medulla urea that has been dissipated from it. As judged from computer simulations (63, 64), a high concentration of urea is best maintained in the inner medulla if the urea leaving that area through diffusion and via ascending vasa recta is reclaimed by secretion preferentially into partes rectae and/or descending *short* limbs of Henle. Anatomical findings consistent with this concept have been reported (17, 39, 40). In the rat, as well as in several other species that concentrate their urine well, the vascular bundles of the outer medulla are of the complex type, in which thin descending limbs of Henle intermingle with vasa recta. Most important with regard to heterogeneity, these descending limbs belong exclusively to short-looped nephrons. The arrangement is ideal for recycling urea, since urea leaving the inner medulla in the ascending vasa recta can immediately be trapped in descending short limbs by countercurrent exchange. Beyond the bend of Henle's loop, tubular structures of superficial nephrons are impermeable to urea. In the collecting system, the permeability to urea is heterogeneous; the cortical (11) and outer medullary (56) collecting ducts are impermeable to urea, but the papillary collecting duct is highly permeable (47, 56). Consequently, the bulk of urea is trapped within the tubular lumen until it reaches the latter segment and can again be deposited in the interstitium of the inner medulla.

This scheme stresses the importance of short-looped, rather than long-looped, nephrons in the concentrating process, whereas historically the emphasis has been the other way around. The well-known postnatal increase in the ability to concentrate urine may also involve mainly short loops (18). In newborn rats, a rapid rise in concentrating capacity is closely correlated with the maturation of superficial nephrons and hence with the penetration of their loops of Henle into the outer medulla. The contribution of urea to the papillary interstitial osmolality increases during the period of penetration but remains relatively constant thereafter. Whether these events are causally related cannot be established at this time, and certainly there are other aspects of renal development to be considered. But the results are consistent with an important role for superficial nephrons in the maintenance of the corticopapillary gradient through medullary recycling of urea.

Vasopressin and Transport of Salt in the Thick Ascending Limb of Henle

Imbert-Teboul and associates (31, 32) have demonstrated that, in several species, the thick ascending limb contains a vasopressin-sensitive adenylate cyclase. They suggested that this finding might indicate a vasopressin-stimulated increase in transport of sodium chloride, a phenomenon that has now been shown by several workers (25, 27, 58). The stimulatory effect of vasopressin is limited to the medullary portion of thick ascending limbs, and cannot be demonstrated in the cortical part (27). This demonstration of heterogeneity is consistent with the finding (31) that, in the rabbit, the adenylate cyclase in cortical thick ascending limbs is much less sensitive to vasopressin than is that in medullary thick limbs.

The heterogeneity of responsiveness in the thick ascending limb may promote enhanced urinary concentration in several interrelated ways. First, for any given amount of salt reabsorption from the entire thick limb, the fraction reabsorbed in the medullary portion is more likely than that in the cortical portion to contribute to countercurrent multiplication. This conclusion may follow largely from the higher peritubular capillary flow in the cortex as compared to the medulla.

Second, there may be an effect that involves tubular-vascular relationships for short-looped and long-looped nephrons. Kriz and associates (40) demonstrated that although thin descending limbs of short-looped nephrons are incorporated into vascular bundles in the outer medulla of the rat kidney (See Medullary Recycling of Urea, above), those of long-looped nephrons are surrounded by thick ascending limbs of both long and short loops. This arrangement allows for direct interaction of thick ascending limbs from all nephrons with descending limbs of long-looped nephrons. Sodium chloride transported out of thick ascending limbs thus might enter descending long limbs and be transferred to the inner medulla, ultimately to be utilized as the single effect by passive means in the thin ascending limbs. Such preferential entry of reabsorbate from thick ascending limbs into descending long limbs has been incorporated into a modification of the passive model (6).

Third, abstraction of water from the collecting duct of the outer medulla, which will be directly enhanced by salt reabsorption from the medullary thick ascending limb, is critical to maximal concentration of urine. This conclusion follows because: during antidiuresis, the volume of water that must be reabsorbed in the cortex and outer medulla is large in comparison to that which must be reabsorbed in the inner medulla; and this reabsorption of water will promote the deposition of urea in the inner medulla. Thus any enhancement by vasopressin of sodium chloride transport in the medullary thick ascending limb would by these three means stimulate the operation of the countercurrent system.

Vasopressin and Heterogeneity of Filtration Rates in Single Nephrons

The work of Davis & Schnermann (13), expanded by Trinh et al (67), has added a fascinating new twist to the subject of heterogeneity: Not only does vasopressin utilize heterogeneity in carrying out its functions, it also induces the phenomenon. The first authors showed that when vasopressin is given to homozygotes of the Brattleboro strain, which lack the hormone (71), there is a selective increase in the single nephron filtration rate (sGFR) only of juxtamedullary nephrons. The results of Trinh et al (67), shown in Table 1, confirm and extend these findings. Compared to Brattleboro heterozygotes (which have vasopressin and served as controls for the study), Brattleboro homozygotes lacked heterogeneity between superficial and juxtamedullary nephrons, in respect not only to sGFR but also to glomerular volume and proximal tubular length. The lack was due entirely to a reduction of the values in juxtamedullary nephrons, not to an increase in those of superficial nephrons. In turn, the induction of heterogeneity when Brattleboro homozygotes were treated with the synthetic analog of vasopressin, DDAVP, was due entirely to increased values in juxtamedullary nephrons (Table 1). The fact that such treatment induced heterogeneity even when it was instituted in adult rats (M.-M. Trinh, personal communication) shows that this effect of vasopressin does not depend on concurrent maturation of the kidney.

Davis & Schnermann (13) postulated that the selective increase in juxtamedullary sGFR when vasopressin is given might be accounted for by changes in postglomerular vascular resistance that result indirectly as a consequence of the antidiuretic state rather than from a direct vasoconstric-

Table 1 Absence of nephron heterogeneity in Brattleboro homozygous rats and induction of heterogeneity through treatment with 1-desamino-8-D-arginine-vasopressin (DDAVP)[a]

	sGFR[b] (nl/min/gkW)		Ratio: S/JM		
	S[b]	JM[b]	sGFR	Glomerular volume	Proximal tubular length
Brattleboro heterozygote	47	67	0.71	0.71	0.78
Brattleboro homozygote (DI)	43	41*	1.04*	1.08*	0.96*
DI plus DDAVP from 2 weeks[c]	45	56	0.78	0.84	0.83
DI plus DDAVP from 12 weeks[c]	38	47	0.67	0.83	0.86

[a] From (67).
[b] Abbreviations: sGFR = glomerular filtration rates in single nephrons; S = superficial cortial nephron; JM = juxtamedullary nephron; DI = Brattleboro homozygotes.
[c] Young homozygotes were treated for 6–8 wks beginning at 2 wks of age; older homozygotes for 6–8 wks beginning at 12 wks of age.
*Significantly different from values for Brattleboro heterozygotes.

tor action. Their hypothesis invokes the rise in medullary interstitial osmolality that is known to occur during antidiuresis. This rise, by withdrawing water from descending vasa recta, might sufficiently increase the viscosity of blood in these vessels to lead to greater resistance to flow in postglomerular vessels and hence to an increase in net ultrafiltration pressure. The effect would be exerted selectively on juxtamedullary nephrons because only this type gives rise to vasa recta that course through the hypertonic inner medulla. The view that an indirect influence, rather than a vasoconstrictor effect, might account for the induction of heterogeneity is strengthened by preliminary data (68) showing that heterogeneity is markedly reduced in a strain of mice with nephrogenic diabetes insipidus, which have high plasma concentrations of vasopressin (F. T. LaRochelle, Jr., P. Stern, personal communication) but low medullary tonicity (36). Of course, the hypothesis does not account for the anatomical change induced by vasopressin, especially the increase in proximal tubular length (Table 1). Trinh and associates (67) have suggested that this "growth" may be somehow analogous to compensatory renal hypertrophy, in which filtering and reabsorptive areas are adapted in response to increased loads.

Whatever the mechanism for the vasopressin-induced changes in juxtamedullary glomeruli, the alterations will augment the antidiuretic response. First, the reduction in medullary blood flow will decrease the amount of solute that is carried away from the medullary interstitium. And second, the increase of sGFR in deep nephrons will result in greater amounts of solute delivered to the ascending limb of Henle for countercurrent multiplication. The net result will be a more efficient build-up and maintenance of the corticopapillary osmotic gradient, and thus of the driving force for the abstraction of water.

Summary. Of the 3 topics we have considered in this article, the last (conservation of water) may have the clearest relationship to nephron heterogeneity. There are several ways in which heterogeneity subserves this function:

1. It used to be said that superficial nephrons are parasitic on the deep ones, in the sense that the latter lay down the corticopapillary osmotic gradient, which reabsorbs water entering the inner medulla largely from superficial nephrons. This view is no longer tenable, for at least two reasons: (*a*) Far more water needs to be withdrawn in the cortex, to go from hypotonicity at the beginning of the distal tubule to isotonicity at its end, than needs to be reabsorbed to go from isotonicity to hypertonicity in the collecting ducts; and (*b*) active transport of sodium chloride in the cortex and outer medulla ultimately supplies the energy that makes possible the presumed passive countercurrent multiplication in the inner medulla.

2. Medullary recycling of urea operates most efficiently if urea reenters

the tubular system in the pars recta and descending limb of Henle belonging to superficial rather than juxtamedullary nephrons.

3. Vasopressin can enhance sodium chloride transport selectively in medullary thick ascending limbs, thereby stimulating a critical step in the operation of the countercurrent system.

4. Vasopressin is at least partly responsible for some aspects of heterogeneity by inducing functional and anatomical enlargement selectively in juxtamedullary nephrons. Once this heterogeneity is established, it aids the urinary concentrating process by promoting the buildup and maintenance of the corticopapillary gradient.

ACKNOWLEDGMENTS

Dr. Walker was supported by National Research Service Award 5T 32 AM-07301 and is a Research Fellow of the National Kidney Foundation. Work in the authors' laboratory was supported mainly by U.S.P.H.S. Research Grant AM-08469 from the National Institute of Arthritis, Metabolism, and Digestive Diseases. The authors would like to thank Ms. Ethel Garrity for help with preparation of the manuscript, and Drs. Franklyn G. Knox and Vincent W. Dennis for reviewing the paper and offering valuable suggestions.

Literature Cited

1. Baines, A. D., de Rouffignac, C. 1969. Functional heterogeneity of nephrons. II. Filtration rates, intraluminal flow velocities and fractional water reabsorption. *Pfluegers Arch.* 308:260–76
2. Barger, A. C. 1966. Renal hemodynamic factors in congestive heart failure. *Ann. NY Acad. Sci.* 139:276–84
3. Barratt, L. J., Wallin, J. D., Rector, F. C. Jr., Seldin, D. W. 1973. Influence of volume expansion on single-nephron filtration rate and plasma flow in the rat. *Am. J. Physiol.* 224:643–50
4. Bengele, H. H., McNamara, E. R., Alexander, E. A. 1977. Volume expansion natriuresis: nephron function beyond the superficial late distal tubule. *Am. J. Physiol.* 232:F566–70
5. Bengele, H. H., Lechene, C. P., Alexander, E. A. 1979. Phosphate transport along the inner medullary collecting duct of the rat. *Am. J. Physiol.* 237: F48–54
6. Bonventre, J. V., Lechene, C. 1980. Renal medullary concentrating process: an integrative hypothesis. *Am. J. Physiol.* 239:F578–88
7. Boudry, J.-F., Troehler, U., Touabi, M., Fleisch, H., Bonjour, J.-P. 1975. Secretion of inorganic phosphate in the rat nephron. *Clin. Sci. Mol. Med.* 48: 475–89
8. Brazy, P. C., McKeown, J. W., Harris, R. H., Dennis, V. W. 1980. Comparative effects of dietary phosphate, unilateral nephrectomy, and parathyroid hormone on phosphate transport by the rabbit proximal tubule. *Kidney Int.* 17:788–800
9. Brunette, M. G., Taleb, L., Carriere, S. 1973. Effect of parathyroid hormone on phosphate reabsorption along the nephron of the rat. *Am. J. Physiol.* 225: 1076–81
10. Bruns, F. J., Alexander, E. A., Riley, A. L., Levinsky, N. G. 1974. Superficial and juxtamedullary nephron function during saline loading in the dog. *J. Clin. Invest.* 53:971–79
11. Burg, M., Helman, S., Grantham, J., Orloff, J. 1970. Effect of vasopressin on the permeability of isolated rabbit cortical collecting tubules to urea, acetamide and thiourea. In *Urea and the Kidney*, ed. B. Schmidt-Nielsen, pp. 193–99. Amsterdam: Excerpta Medica
12. Chuang, E. L., Reineck, H. J., Osgood, R. W., Kunau, R. T. Jr., Stein, J. H. 1978.

Studies of the mechanism of reduced urinary osmolality after exposure of the renal papilla. *J. Clin. Invest.* 61:633–39
13. Davis, J. M., Schnermann, J. 1971. The effect of antidiuretic hormone on the distribution of nephron filtration rates in rats with hereditary diabetes insipidus. *Pfluegers Arch.* 330:323–34
14. Dennis, V. W., Bello-Reuss, E., Robinson, R. R. 1977. Response of phosphate transport to parathyroid hormone in segments of rabbit nephron. *Am. J. Physiol.* 233:F29–38
15. Dennis, V. W., Stead, W. W., Myers, J. L. 1979. Renal handling of phosphate and calcium. *Ann. Rev. Physiol.* 41:257–71
16. de Rouffignac, C., Bonvalet, J. P. 1974. Heterogeneity of nephron population. In *Kidney and Urinary Tract Physiology,* ed. K. Thurau, pp. 391–409. Baltimore: University Park Press
17. de Rouffignac, C., Imbert, M. 1977. Role of sodium and urea in renal concentrating mechanism. *Proc. Eur. Colloq. Renal Physiol., 2nd., Balatonfured, Hungary*
18. Edwards, B. R., Mendel, D. B., LaRochelle, F. T. Jr., Stern, P., Valtin, H. 1982. Postnatal development of urinary concentrating ability in rats: changes in renal anatomy and neurohypophysial hormones. *Proc. First Int. Workshop Devel. Renal Physiol.* In press
19. Frindt, G., Burg, M. B. 1972. Effect of vasopressin on sodium transport in renal cortical collecting tubules. *Kidney Int.* 1:224–31
20. Gelbart, D. R., Battilana, C. A., Bhattacharya, J., Lacy, F. B., Jamison, R. L. 1978. Transepithelial gradient and fractional delivery of chloride in thin loop of Henle. *Am. J. Physiol.* 235:F192–98
21. Giebisch, G. 1980. Methods of localizing transport processes using micropuncture techniques—evidence for nephron heterogeneity. *Int. J. Biochem.* 12:3–8
22. Goldfarb, S. 1980. Juxtamedullary and superficial nephron phosphate reabsorption in the cat. *Am. J. Physiol* 239:F336–42
23. Greger, R. F., Lang, F., Marchand, G., Knox, F. G. 1977. Site of renal phosphate reabsorption—micropuncture and microinfusion study. *Pfluegers Arch.* 369:111-18
24. Haas, J. A., Berndt, T., Knox, F. G. 1978. Nephron heterogeneity of phosphate reabsorption. *Am. J. Physiol.* 234:F287–90
25. Hall, D. A., Varney, D. M. 1980. Effect of vasopressin on electrical potential difference and chloride transport in mouse medullary thick ascending limb of Henle's loop. *J. Clin. Invest.* 66:792–802
26. Hanley, M. J., Kokko, J. P. 1978. Study of chloride transport across the rabbit cortical collecting tubule. *J. Clin. Invest.* 62:39–44
27. Hebert, S. C., Culpepper, R. M., Misanko, B. S., Andreoli, T. E. 1980. ADH induces anti-saluresis in medullary but not cortical thick ascending limbs (tALH). *Clin. Res.* 28:533A (Abstr.)
28. Helman, S. I., Grantham, J. J., Burg, M. B. 1971. Effect of vasopressin on electrical resistance of renal cortical collecting tubules. *Am. J. Physiol.* 220:1825–32
29. Higashihara, E., DuBose, T. D. Jr., Kokko, J. P. 1978. Direct examination of chloride transport across papillary collecting duct of the rat. *Am. J. Physiol.* 235:F219–26
30. Higashihara, E., Stokes, J. B., Kokko, J. P., Campbell, W. B., DuBose, T. D. Jr. 1979. Cortical and papillary micropuncture examination of chloride transport in segments of the rat kidney during inhibition of prostaglandin production: A possible role for prostaglandins in the chloruresis of acute volume expansion. *J. Clin. Invest.* 64:1277–87
31. Imbert, M., Charbardes, D., Montegut, M., Clique, A., Morel, F. 1975. Vasopressin dependent adenylate cyclase in single segments of rabbit kidney tubule. *Pfluegers Arch.* 357:173–86
32. Imbert-Teboul, M., Chabardes, D., Montegut, M., Clique, A., Morel, F. 1978. Vasopressin-dependent adenylate cyclase activities in the rat kidney medulla: evidence for two separate sites of action. *Endocrinology* 102:1254–61
33. Jamison, R. L. 1973. Intrarenal heterogeneity. The case for two functionally dissimilar populations of nephrons in the mammalian kidney. *Am. J. Med.* 54:281–89
34. Jamison, R. L., Sonnenberg, H., Stein, J. H. 1979. Questions and replies: role of the collecting tubule in fluid, sodium, and potassium balance. *Am. J. Physiol* 237:F247–61
35. Johnston, P. A., Battilana, C. A., Lacy, F., Jamison, R. 1977. Evidence for a concentration gradient favoring outward movement of sodium from the thin loop of Henle. *J. Clin. Invest.* 59:234–40

36. Kettyle, W. M., Valtin, H. 1972. Chemical and dimensional characterization of the renal countercurrent system in mice. *Kidney Int.* 1:135–44
37. Knox, F. G., Haas, J. A., Berndt, T., Marchand, G. R., Youngberg, S. P. 1977. Phosphate transport in superficial and deep nephrons in phosphate loaded rats. *Am. J. Physiol.* 233:F150–53
38. Knox, F. G., Osswald, H., Marchand, G. R., Spielman, W. S., Haas, J. A., Berndt, T., Youngberg, S. P. 1977. Phosphate transport along the nephron. *Am. J. Physiol.* 233:F261–68
39. Kriz, W., Barrett, J. M., Peter, S. 1976. The renal vasculature: anatomical-functional aspects. *Int. Rev. Physiol.* 11:1–21
40. Kriz, W., Schnermann, J., Koepsell, H. 1972. The position of short and long loops of Henle in the rat kidney. *Z. Anat. Entwicklungsges.* 138:301–19
41. Lameire, N. H., Lifschitz, M. D., Stein, J. H. 1977. Heterogeneity of nephron function. *Ann. Rev. Physiol.* 39:159–84
42. Lang, F. 1980. Renal handling of calcium and phosphate. *Klin. Wochenschr.* 58:985–1003
43. Lang, F., Greger, R., Marchand, G. R., Knox, F. G. 1977. Stationary microperfusion study of phosphate reabsorption in proximal and distal nephron segments. *Pfluegers Arch.* 368:45–48
44. Larsson, C., Anggard, E. 1973. Regional differences in the formation and metabolism of prostaglandins in the rabbit kidney. *Eur. J. Pharmacol.* 21:30–36
45. McKeown, J. W., Brazy, P. C., Dennis, V. W. 1979. Intrarenal heterogeneity for fluid, phosphate, and glucose absorption in the rabbit. *Am. J. Physiol.* 237:F312–18
46. Morel, F., Chabardes, D., Imbert, M. 1976. Functional segmentation of the rabbit distal tubule by microdetermination of hormone-dependent adenylate cyclase activity. *Kidney Int.* 9:264–77
47. Morgan, T., Berliner, R. W. 1968. Permeability of the loop of Henle, vasa recta, and collecting duct to water, urea and sodium. *Am. J. Physiol.* 215:108–15
48. Deleted in proof
49. Muehlbauer, R. C., Bonjour, J. P., Fleisch, H. 1977. Tubular localization of adaptation to dietary phosphate in rats. *Am. J. Physiol.* 233:F342–48
50. Murayama, Y., Morel, F., Le Grimellec, C. 1972. Phosphate, calcium, and magnesium transfers in proximal tubules and loops of Henle, as measured by single nephron microperfusion experiments in the rat. *Pfluegers Arch.* 333:1–16
51. Osgood, R. W., Reineck, H. J., Stein, J. H. 1978. Further studies on segmental sodium transport in the rat kidney during expansion of the extracellular fluid volume. *J. Clin. Invest.* 62:311–20
52. Osgood, R. W., Reineck, H. J., Stein, J. H. 1979. Effect of hyperoncotic albumin on superficial and juxtamedullary nephron sodium transport. *Am. J. Physiol.* 237:F34–37
53. Poujeol, P., Corman, B., Touvay, C., de Rouffignac, C. 1977. Phosphate reabsorption in rat nephron terminal segments. Intrarenal heterogeneity and strain differences. *Pfluegers Arch.* 371:39–49
54. Poujeol, P., Jamison, R. L., de Rouffignac, C. 1980. Phosphate reabsorption in juxtamedullary nephron terminal segments. *Pfluegers Arch.* 387:27–31
55. Reineck, H. J., Parma, R., Barnes, J. L., Osgood, R. W. 1980. Nephron heterogeneity in the renal excretion of sodium and potassium in the rat. *Am. J. Physiol.* 239:F187–93
56. Rocha, A. S., Kokko, J. P. 1974. Permeability of medullary nephron segments to urea and water: Effect of vasopressin. *Kidney Int.* 6:379–87
57. Rocha, A. S., Magaldi, J. B., Kokko, J. P. 1977. Calcium and phosphate transport in isolated segments of rabbit Henle's loop. *J. Clin. Invest.* 59:975–83
58. Sasaki, S., Imai, M. 1980. Effects of vasopressin on water and NaCl transport across the in vitro perfused medullary thick ascending limb of Henle's loop in mouse, rat, and rabbit kidneys. *Pfluegers Arch.* 383:215–21
59. Solez, K., Fox, J. A., Miller, M., Heptinstall, R. H. 1974. Effects of indomethacin on renal inner medullary plasma flow. *Prostaglandins* 7:91–98
60. Sonnenberg, H. 1975. Secretion of salt and water into the medullary collecting duct of Ringer-infused rats. *Am. J. Physiol.* 228:565–68
61. Staum, B. B., Hamburger, R. J., Goldberg, M. 1972. Tracer microinjection study of renal tubular phsophate reabsorption in the rat. *J. Clin. Invest.* 51:2271–76
62. Stein, J. H., Osgood, R. W., Kunau, R. T. Jr. 1976. Direct measurement of papillary collecting duct sodium transport in the rat. *J. Clin. Invest.* 58:767–73
63. Stewart, J. 1975. Urea handling by the renal countercurrent system: insights

from computer simulation. *Pfluegers Arch.* 356:133–51
64. Stewart, J., Valtin, H. 1972. Computer simulation of osmotic gradient without active transport in renal inner medulla. *Kidney Int.* 2:264–70
65. Stokes, J. B. 1979. Effect of prostaglandin E_2 on chloride transport across the rabbit thick ascending limb of Henle: Selective inhibition of the medullary portion. *J. Clin. Invest.* 64:495–502
66. Tisher, C. C., Yarger, W. E. 1975. Lanthanum permeability of tight junctions along the collecting duct of the rat. *Kidney Int.* 7:35–43
67. Trinh-Trang-Tan, M.-M., Diaz, M., Gruenfeld, J.-P., Bankir, L. 1981. ADH-dependent nephron heterogeneity in rats with hereditary diabetes insipidus. *Am. J. Physiol.* 40 (*Renal Fluid Electrol. Physiol.* 9): F372–80
68. Trinh-Trang-Tan, M.-M., Sokol, H. W., Bankir, L., Valtin, H. 1982. Homozygous Brattleboro rats lack normal nephron heterogeneity as a consequence of their urine concentrating defect. *Ann. NY Acad. Sci.* In press

69. Troehler, U., Bonjour, J.-P., Fleisch, H. 1976. Inorganic phosphate homeostasis: renal adaptation to the dietary intake in intact and thyroparathyroidectomized rats. *J. Clin. Invest.* 57:264–73
70. Ullrich, K. J., Rumrich, G., Kloss, S. 1977. Phosphate transport in the proximal convolution of the rat kidney. I. Tubular heterogeneity, effect of parathyroid hormone in acute and chronic parathyroidectomized animals and effect of phosphate diet. *Pfluegers Arch.* 372:269–74
71. Valtin, H. 1976. Animal model of human disease: hereditary diabetes insipidus. *Am. J. Pathol.* 83:633–36
72. Valtin, H. 1977. Structural and functional heterogeneity of mammalian nephrons. *Am. J. Physiol.* 233:F491–501
73. Woodhall, P. B., Tisher, C. C., Simonton, C. A., Robinson, R. R. 1978. Relationship between para-aminohippurate secretion and cellular morphology in rabbit proximal tubules. *J. Clin. Invest.* 61:1320–29

RESPIRATORY PHYSIOLOGY

Introduction, Alfred P. Fishman, *Section Editor*

Until two decades ago, the lung was explored almost exclusively in terms of its respiratory functions. Since then, the scope of research has enlarged to include its nonrespiratory activities. Study has focused on the alveolar-capillary barrier—both on the alveolar aspect of this barrier, particularly with respect to the generation, functional roles, and disposition of surfactant, and on the structure and functions of pulmonary vascular endothelium.

The pulmonary vascular endothelium is only one segment of a vascular lining that extends from one end of the body to the other. This lining is remarkable on several accounts. It is noteworthy for its remarkable biological indifference to the blood that runs continuously, for a lifetime, over its luminal aspect. Because of the innate properties of this surface, blood does not clot and proteins are not denatured. So far, despite intensive search, no material has been synthesized that is as tolerant as endothelium of the plasma constituents and of the formed elements in the blood. Yet the endothelium is not only an extraordinary passive lining. It has long been known that under appropriate stimulation, endothelium can change its character. Indeed, its potential for becoming phagocytic led to its inclusion in the designation "reticulo-endothelial system."

In addition, the endothelium constitutes an important metabolic organ, operative under normal conditions. It is not customary to regard it as such because it is distributed throughout the body as a part of blood vessels rather than concentrated in a mass. Nor is it impressive by conventional histologic criteria since organelles are sparse and there is little anatomical evidence of an extensive metabolic machinery.

However, the disposition and appearance of endothelium have proved deceptive. The magnitude of the deception can be judged from the chapters in this section on the nonrespiratory functions of the lungs. Teleologically, these functions are not surprising. As the lining of the vascular tree of the lungs, pulmonary vascular endothelium is strategically situated between the venous return to the heart and the systemic circulation. In this situation, its metabolic activities can act to protect the heart and the brain from a variety of humoral, as well as particulate, insults. It can operate efficiently not only because of its location but also because of its extent: It is the largest endothelial surface in the body. That the metabolic functions of the endothelium do live up to expectations is illustrated by its important role in the regulation of systemic arterial blood pressure by way of the angiotensin converting enzyme; this enzyme, located at the endothelial surface of the pulmonary vessels, helps to sustain systemic arterial blood pressure by converting angiotensin I to angiotensin II while preventing bradykinin, a powerful systemic vasodilator, from gaining access to the systemic circulation. The sampler of research provided by the following chapters illustrates the types of metabolic functions performed by the lungs and their contributions to homeostasis in the body as a whole.

In the first chapter, Una Ryan considers both the anatomical dispositions of enzymes that handle vasoactive substances in the blood and the conditions, such as hypoxia, that may influence the activity of these enzymes. James Ryan then focuses on the endogenous polypeptides, particularly the angiotensin converting enzyme, and considers in biochemical terms how they operate and how their activities can be influenced. Said covers current thinking about the genesis, uptake, and inactivation of prostaglandins and touches on the physiological implications of these important compounds. Gillis & Pitt describe the handling of circulating biogenic amines by the lungs and provide a succinct overview of the complicated interplay by which some (such as norepinephrine and 5-hydroxytryptamine) are removed whereas others (such as dopamine and epinephrine, which are structurally related) are ignored. Finally, Fantone, Kunkel & Ward deal with vasoactive mediators and chemotactic factors that result from invasion of the lung by inflammatory cells and by mast cells that are normal residents of the lungs.

Clearly these few chapters do not provide an exhaustive view of current thinking about the nonrespiratory functions of the lung. Nonetheless, enough material is reviewed to provide an inkling of the many unanswered questions. Two in particular stand out: 1. Is endothelium in the lungs and other organs the same functionally, and 2. how important is the pulmonary processing of circulating materials? These questions have stimulated considerable discussion and research, and more of both is needed to settle them.

The material in these chapters and the questions they pose should illustrate how the nonrespiratory functions of the lungs are currently being investigated and foretell the direction of inquiry in the coming decade.

STRUCTURAL BASES FOR METABOLIC ACTIVITY

Una S. Ryan

Department of Medicine, University of Miami School of Medicine, Miami, Florida 33101

INTRODUCTION

By virtue of its strategic position, interposed between the mixed venous blood and the systemic arterial blood, the pulmonary circulation provides a mechanical sieve for the venous drainage from virtually the entire body. In addition, the pulmonary vasculature acts as a biochemical sieve in that it modifies in a specific or highly selective manner the properties, and therefore the activities, of circulating vasoactive substances.

Structural characteristics are as important in the selective metabolic activities of the lungs as they are in gas exchange. In fact the very architecture that suits the lungs so well for the exchange of gases is also ideal for the bulk processing of blood-borne substrates. Thus, as shown in Figures 1-4, the extensiveness of the pulmonary capillary bed and the small caliber of the vessels favor the interaction of circulating substances with lung cells. The pulmonary endothelium is the first line of contact between blood-borne substrates and the cellular machinery of the lungs. Here I review pulmonary endothelial cells and the structural specializations that befit their role in the metabolic functions of the lungs.

Pulmonary endothelial cells possess enzymes, inhibitors, receptors, and transport mechanisms important in (a) the regulation of the hormonal composition and fluidity of blood, (b) the provision of volatile and nonvolatile solutes to parenchymal cells, and (c) the specific responses of the endothelial cells themselves, and perhaps also of other lung cell types. This review treats chiefly the first category.

It was, in fact, in studies of the role of the lungs in blood pressure homeostasis that we began to probe the structure-function relationships

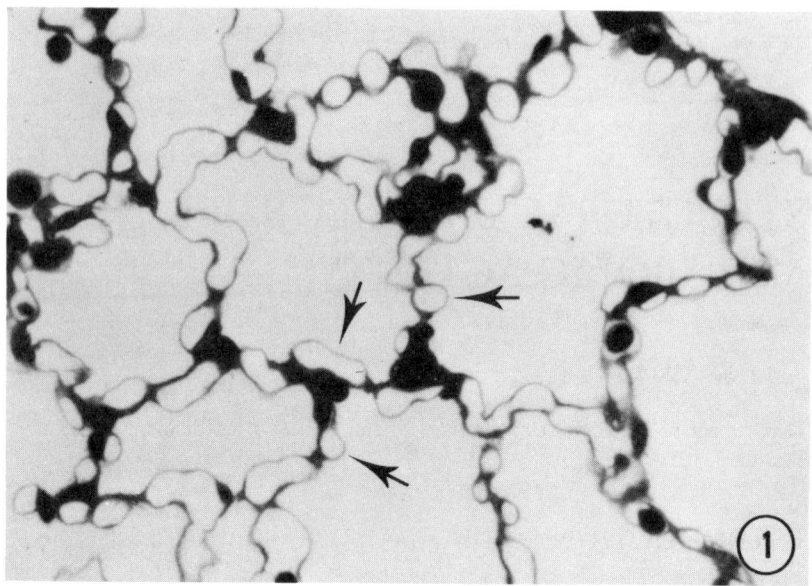

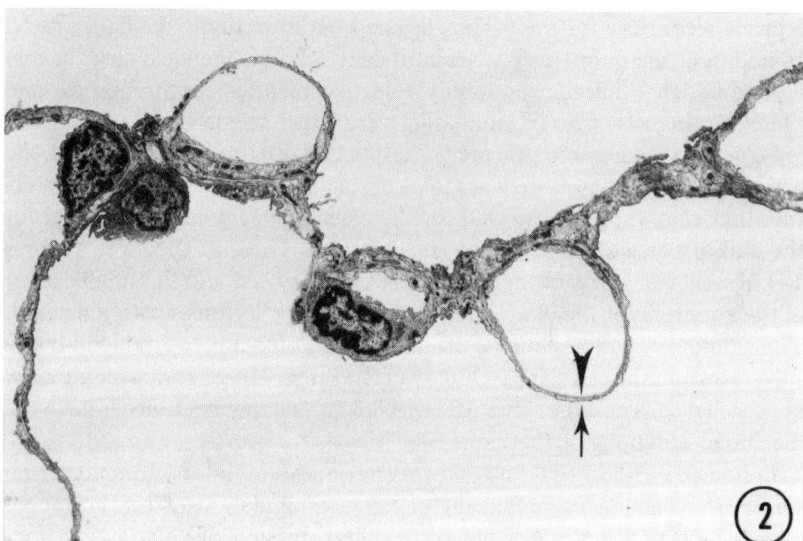

Figures 1 & 2 1: Light micrograph (× 325) of a portion of the capillary bed of a blood-free rat lung. The vast extent and extreme thinness of the alveolar capillary unit (arrows) can be appreciated, but not its cellular nature (compare Figure 2).

2: Low-power electron micrograph (× 2500) from a blood-free rat lung similar to that shown in Figure 1. The alveolar-capillary unit is composed of type 1 alveolar epithelial cells (arrows) and endothelial cells (arrowheads) with the respective basal laminae intervening.

STRUCTURE & METABOLIC ACTIVITY 225

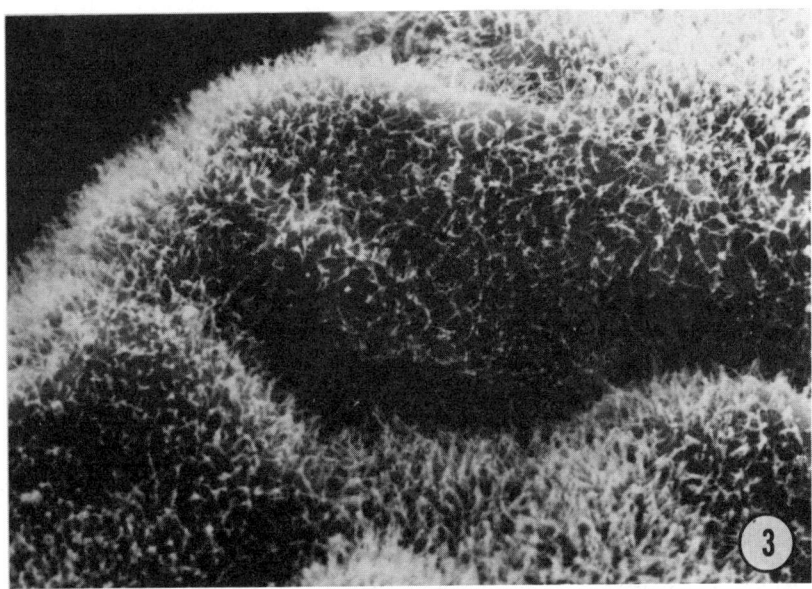

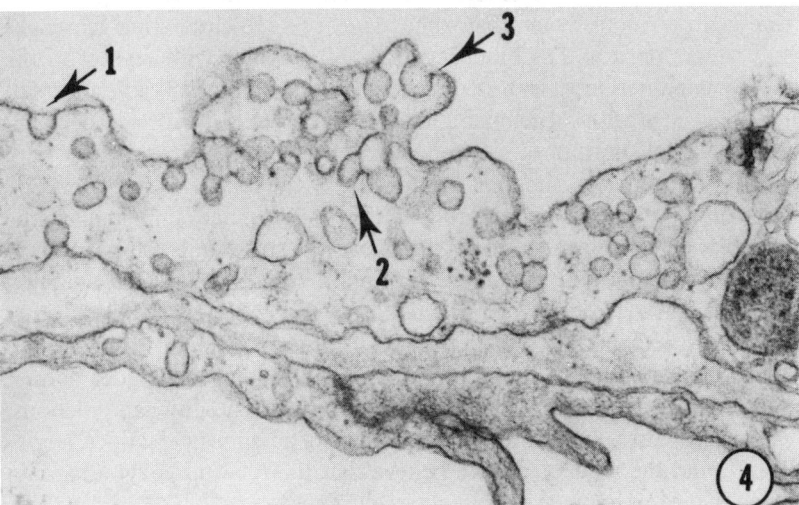

Figures 3 & 4 3: Scanning electron micrograph (X 3000) of the luminal surface of the pulmonary artery of the rat. Endothelial projections protrude towards the lumen with the effect of increasing the surface area of the endothelial cell that is exposed to circulating substrates. [From (26)]

4: Transmission electron micrograph (X 47,000) showing a capillary endothelial cell with large numbers of caveolae. Caveolae may occur at the luminal surface and may be spanned by a delicate diaphragm (e.g. arrow 1), may occur in the cytoplasm singly or fused in groups (arrow 2), or may show evidence of a supportive stomal ring (arrow 3).

underlying the specific metabolic activities of the lungs. Such studies set the scene for our fine focus on endothelial cells.

I consider, in turn, the underlying structural bases for metabolic activities under three interrelated, but arbitrarily designated, headings: anatomical, cellular, and subcellular.

Anatomical Bases of Metabolic Activities

In processing hormones, hormone precursors, and other excitatory substances, the lungs handle a broad range of chemical compounds: polypeptides, biogenic amines, fatty acid derivatives, and steroids. An equally broad range of biological and chemical mechanisms are involved. At first glance this broad coverage belies the specificity with which the lungs can distinguish within chemical groups of hormones and drugs removing, for example, 5-HT but not histamine, hydrolyzing bradykinin but not oxytocin, inactivating PGE but not prostacyclin. How is this selectivity achieved? The answer must lie at least in part with the structural organization of the lungs: If their structural integrity is destroyed (e.g. by homogenization) the lungs possess an abundance of enzymes capable of inactivating the substances mentioned above without discrimination. Thus within intact lungs the enzymes are partitioned such that some have access to circulating substrates while others do not. The exact locations of enzymes with respect to the blood or air interfaces, with respect to different lung cell types, and with respect to the precise subcellular compartment are therefore major determinants of which hormones or excitatory substances are metabolized and which are not, and of the final destination of the metabolic products.

The lung is composed of two cooperating systems, one supplying air, the other blood. Therefore the most fundamental structural determinant of which cells will come into contact with circulating substrates is the anatomy of the pulmonary vascular bed. Pulmonary vessels of all sizes, both arterial and venous, are lined with endothelial cells of the continuous type (Figure 2), the cells being linked by tight junctions that do not leak under normal circumstances, and, except at the level of the capillaries, by gap junctions.

Several features peculiar to the metabolism of adenine nucleotides, angiotensin I, and the kinins led us to believe that the relevant enzymes are on the surface of pulmonary endothelial cells (24).

In view of the great efficiency of metabolism of vasoactive substances (e.g. bradykinin is completely inactivated in the less than 2 sec transit time through the pulmonary circulation), processing seemed likely to occur most prominently at the level of the smallest vessels, where circulating substrates would have their greatest exposure to endothelium. Considering the vast surface area represented by the pulmonary capillary bed, the layer of blood coming into contact with the endothelial cell surface must be very thin. This situation would favor interactions between substrate-laden plasma and endothelial enzymes, binding sites and transport processes, just as it favors gas exchange across the alveolar capillary unit.

We therefore began to study the structural characteristics of the endothelial cells of the pulmonary microvasculature. At the level of the alveolar-capillary unit the air is separated from the blood by a barrier, until recently thought to be relatively inert; its cellular nature cannot be discerned by light microscopy (Figure 1). Even with the magnification and resolving power of the electron microscope (Figure 2) capillary endothelial cells are more remarkable for their extreme thinness than for being well-endowed with the subcellular organelles normally associated with active metabolism.

Nevertheless, the processing of the biogenic amines, adenine nucleotides, bradykinin, and angiotensin has been localized to endothelial cells using a variety of approaches.

Using lungs perfused with [^{3}H]5-HT, autoradiographic techniques indicate that the silver grains are associated with endothelial cells of arterioles and capillaries (2, 32). Noradrenaline is taken up by endothelium but the distribution of silver grains is not uniform, some endothelial cells being more heavily labelled than others (10). Although inactivation of 5-HT and noradrenaline is ultimately dependent on enzymic deamination by monoamine oxidase, uptake of the substrate may be the crucial step (33) that imparts the selectivity that differentiates pulmonary monoamine inactivation from hepatic inactivation in spite of identical enzymic content. The importance of uptake under normal conditions and during modulations (e.g. by the estrus cycle) emphasizes the likelihood that metabolism of amines by lung homogenates is a misleading guide to the situation in vivo (33).

Techniques exist for the cytochemical localization of adenine nucleotides (11). Using electron microscope cytochemistry we showed that caveolae of pulmonary endothelial cells contain an enzyme that inactivates ATP by hydrolysis of one or more phosphate ester bonds (24). Isolated perfused rat lungs degrade adenosine-5'-monophosphate (AMP) to form adenosine and small amounts of inosine (19, 24). When isolated lungs are perfused with both AMP and Pb (NO$_3$)$_2$, insoluble, electron-dense lead phosphate is deposited on those endothelial caveolae directly facing the vascular lumen (Figure 5). Studies using blocks of fixed tissue showed that caveolae throughout endothelial cells possessed activity, but in intact lungs only those caveolae open to the vascular lumen have access to substrate (24). Recently, Crutchley et al (3) have shown that the potent platelet-aggregating agent, adenosine-5'-diphosphate (ADP), is inactivated during passage through the lungs. As described below, we have shown that the metabolism of ADP is accomplished by pulmonary endothelial cells in culture (4). Thus the site of its inactivation is probably similar to that of ATP and AMP.

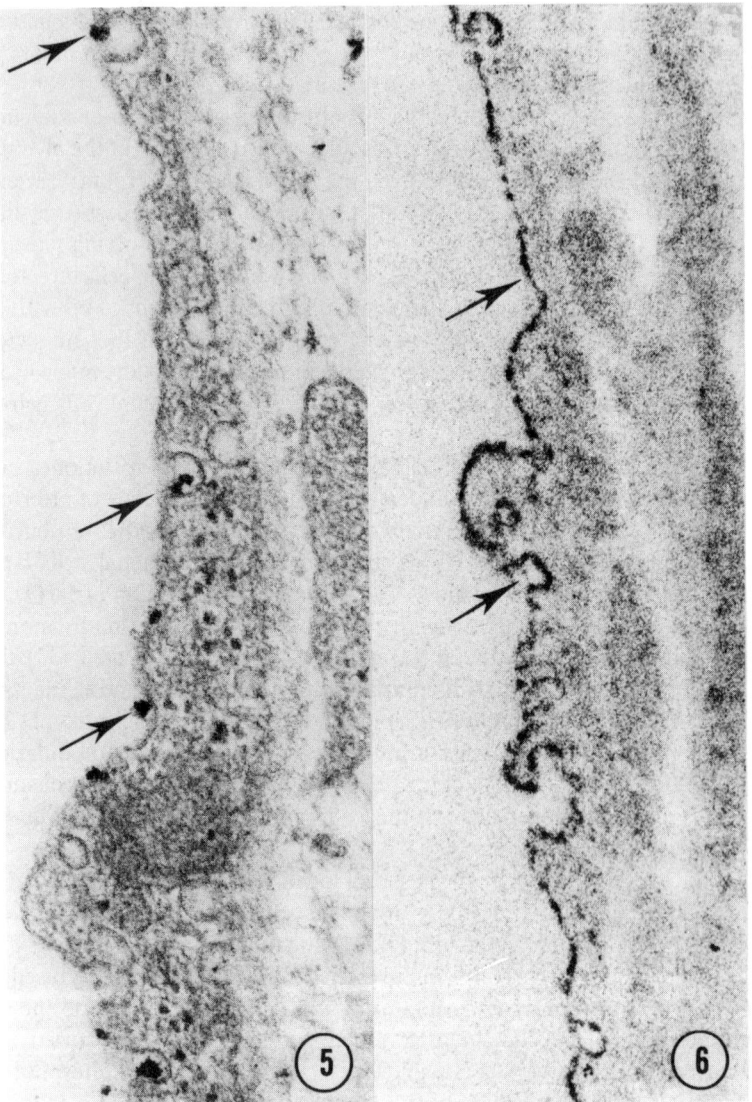

Figures 5 & 6 5: Sites of 5'-nucleotidase activity shown by electron microscope cytochemistry (X 85,000). Reaction product is localized on the endothelial plasma membrane directly exposed to the vascular lumen, preferentially on caveolae (arrows). [See (24)]

6: Sites of angiotensin converting enzyme (kininase II) activity demonstrated by immunocytochemistry (X 74,000). Deposition of reaction product occurs along the full extent of the luminal plasma membrane, including caveolae (arrows). [From (28)]

Intact lungs eliminate prostaglandins of the E and F series but not those of the A series nor PGI_2 (7, 13, 14). Initially, it was assumed that elimination of prostaglandins could be accounted for by the action of enzymes on or near the luminal surface of endothelial cells (14). However, cellular uptake is required (18, 19). We showed that lungs perfused with $[^3H]PGF_{1a}$ take up the prostaglandin and slowly release the relatively inactive 15-keto-dihydro derivative of PGF_{1a}. Clearly, 15-OH-dehydrogenase and C_{13}-reductase exist somewhere within the lungs.

Prostaglandins of the E and F series, but not those of the A series, are taken up via an active transport mechanism (1, 6). The transport system may exist on endothelial cells, and degradation of the prostaglandins may occur within these cells. However, pulmonary artery endothelial cells in culture do not appear to possess enzymes capable of forming the dihydro or 15-keto derivatives of $PGF_{1\alpha}$ (16).

Pulmonary endothelial cells present an apparent paradox. On the one hand these cells participate in the elimination of prostaglandins of central venous blood; on the other hand, the same cells synthesize and release PGE_2, PGI_2, and perhaps PGB_2 (e.g. 18) (also see below). The synthesis and release of the inhibitor of platelet aggregation, PGI_2, may well be an important function in maintaining the fluidity of pulmonary and arterial blood.

Endothelial cells, especially those of the pulmonary vascular bed, play a central role in regulating the overall functioning of the renin-angiotensin-aldosterone system and the kallikrein-kinin system (16, 17). Endothelial cells possess peptidase enzymes capable of metabolizing angiotensin I and bradykinin (16, 17), of which the best characterized is a dipeptidyl carboxypeptidase known as angiotensin converting enzyme (ACE), or kininase II (5, 31). Other relevant peptidase enzymes on endothelial cells include one capable of converting angiotensin II into angiotensin III (des-Asp^1-angiotensin II) (16).

Because of its importance to both the renin-angiotensin and kallikrein-kinin systems, there is much to be gained by understanding how ACE is disposed on endothelial cells, how it reacts with circulating peptide hormones, and how it can be inhibited. When Dorer et al (5) purified ACE to homogeneity from pig lung, it became possible to examine for sites of the enzyme by immunocytochemistry at the ultrastructural level. We prepared antibodies to the Dorer ACE and conjugated them to microperoxidase (11-MP or 8-MP) (15, 27). The antibody-microperoxidase conjugates were reacted with blocks of rat lung and incubated with 3,3' diaminobenzidine in the presence of H_2O_2. The tissue was examined by electron microscopy. The resulting electron-dense reaction product indicated that ACE is situated along the entire luminal plasma membrane of pulmonary endothelial cells including the caveolae (Figure 5) and endothelial projections (24, 25,

28). Use of immunocytochemistry and immunofluorescence showed the enzyme to be similarly situated on pulmonary endothelial cells in culture (22, 27). ACE is not unique to pulmonary endothelium. Nonetheless, bulk conversion of angiotensin I into angiotensin II (and presumably bulk degradation of bradykinin) is likely to occur within the pulmonary vascular bed (17).

The picture emerges of a variety of enzymes situated on the surface of pulmonary endothelial cells strategically poised for interaction with the appropriate substrates delivered by the blood and equally strategically placed to determine the quantities of active substances allowed to pass downstream or into the extravascular space. However, still to be resolved are the questions whether endothelial cells alone can account for the processing achieved by intact lungs and whether the endothelial cells synthesize the enzymes or merely act as a depository for enzymes washed up or bound to receptors on the walls of the pulmonary vessels but synthesized elsewhere.

One direct way to study the specific metabolic activities of pulmonary endothelial cells is to use pure isolates or lines of pulmonary endothelial cells in culture.

Cellular Bases for Metabolic Activities

In earlier studies, we had shown [(19) and review in this volume by J. W. Ryan] that the caveolae-plasma membrane fraction of lung homogenate converts angiotensin I to angiotensin II and degrades bradykinin to yield the characteristic products formed by intact lungs. Furthermore, the preparation could degrade ATP and 5'-AMP to adenosine and free phosphate (19). We then showed that pulmonary artery endothelial cells removed as pure intact monolayers on cellulose acetate paper could also degrade bradykinin and convert angiotensin I to angiotensin II (20).

In the 1970s pulmonary endothelial cells in culture became a laboratory resource reviewed in 21. Endothelial cells can now be obtained from pulmonary artery and vein of large animals such as cow and pig (21, 22, 27, 28) and from the pulmonary microvasculature of small animals such as rat, guinea pig, and rabbit—animals commonly used in studies of nonventilatory functions of the lungs (8, 21). Our original aim was to determine if pure lines of endothelial cells could metabolize bradykinin and angiotensin I to yield the same products as those produced by intact lungs. Using a sensitive radioassay, we have shown that pulmonary endothelial cells in primary culture and after many passages do indeed contain abundant ACE (22, 27, 28) and that the enzyme can be demonstrated by immunofluorescence microscopy and electron microscope immunocytochemistry (27, 28). Furthermore, recent studies indicate that ACE is synthesized by these cells

(22). The cells contain the full range of intracellular organelles presumed necessary for the synthesis of a complex glycoprotein such as ACE; these are disposed within the flattened dimensions of the cell.

Endothelial cells in culture grow as a single sheet with no tendency to form multiple layers. The characteristic "cobblestone" appearance (Figure 7) has become the hallmark of a healthy endothelial monolayer. Endothelial cells in culture possess endothelial projections and caveolae, which demonstrates that these structures are independent of contractions of the vessel wall. Moreover, both caveolae and projections stain prominently with anti-ACE conjugates (28).

A spectrum of morphological, biochemical, and immunochemical tests now exist to identify pulmonary endothelial cells (21) and to estimate the maintenance of differentiated characteristics after long periods in culture. Having highly characterized lines of pulmonary endothelium available on a routine basis has enabled us to examine for other properties of endothelial cells.

All substances processed by lung cells and all substances released into the circulation at the level of the lungs must cross the endothelium in one direction or the other. Since lungs release prostaglandins (13), we examined for the ability of pulmonary endothelial cells to synthesize prostaglandins and prostaglandin-related substances. Pulmonary endothelial cells metabolize 1-^{14}C-arachidonic acid to yield a variety of ^{14}C-labelled products, including PGI_2 and PGE_2 (16, 18). Synthesis of these metabolites is inhibited by aspirin at 50 μM. ^3H-Aspirin, tritiated in the acetyl moiety, forms a bond with a protein of endothelial cells, and the acetylated protein co-purifies with prostaglandin synthetase. Using endothelial cells incubated with ^3H-acetyl salicylate, we prepared electron microscope autoradiographs to examine for the subcellular sites of prostaglandin synthetase (18). The predominant labelling (77%) occurred over cisternae of the endoplasmic reticulum, a location entirely consistent with the biochemical finding that prostaglandin synthetase is associated with the microsomal fraction. Thus endothelial cells can contribute to the efflux of prostaglandins by lung (18).

The metabolism of ADP by pulmonary endothelial cells is complex (4). It is degraded to AMP and then to adenosine by enzymes on the exterior of the endothelial cells. Much of the adenosine is then taken intracellularly where it is converted in part to inosine. Some of the inosine is returned to the extracellular space. The remainder of the adenosine is incorporated into intracellular ADP and ATP. The uptake of adenosine by endothelial cells is strongly inhibited by dipyridamole (4, 12). Dipyridamole may exert some of its antithrombotic effects by inhibiting cellular uptake of adenosine. Thus endothelial cells can inactivate the platelet aggregating agent, ADP, and can release into the vascular space an agent, PGI_2, capable of preventing

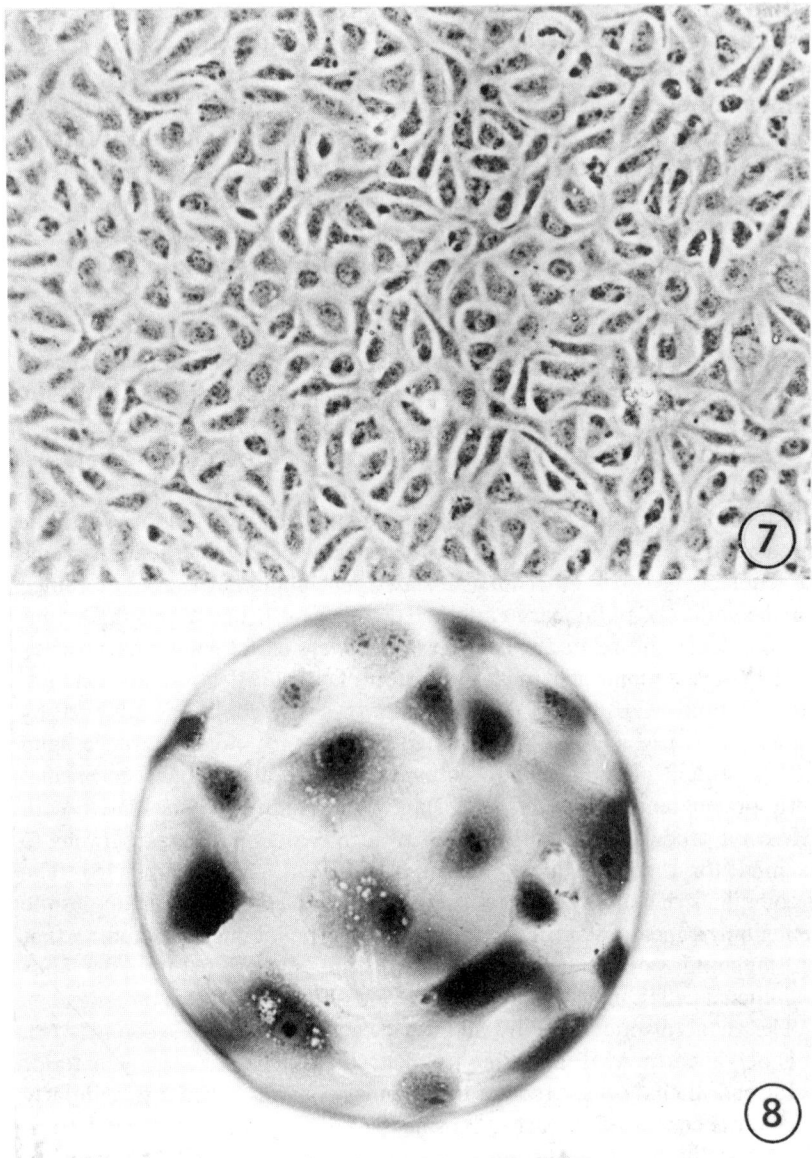

Figures 7 & 8 7: Bovine pulmonary endothelial cells of the 13th passage in monolayer culture (X 170). The regular 'cobblestone' appearance has become the hallmark of a healthy endothelial culture.

8: Endothelial cells cultured on microcarrier beads (X 285). This system allows for large-scale, long-term culture of endothelial cells that have never been exposed to proteolytic enzymes. [From (21)] platelet aggregation and of disaggregating platelet clumps. Further, adenosine, a product of the degradation of ADP, is anti-aggregatory. However, ADPase and PGI_2 are not the only endothelial components active in hemostasis, and it would be a mistake to overlook the hemostatic potential of endothelium.

Current success in the culture of pulmonary endothelial cells has contributed greatly to our understanding of specific metabolic activities of endothelial cells. However, few laboratories have succeeded in establishing long-term cultures of endothelial cells that maintain the cobblestone monolayer morphology, division rates, and differentiated characteristics of endothelium. Further there remain a number of endothelial properties not yet demonstrated in culture. For example, much is known of the "paraendocrine" functions of endothelial cells—i.e. their ability to act on passing hormones—but little is known of the "transducing" functions of endothelium—i.e. the effects of hormones on endothelial cells. Furthermore, collaborative activities along the vessel wall, particularly during changes in tone, presume endothelial cell-cell junctions of the communicating type. Elaborate junctional complexes occur between endothelial cells in situ and in fresh isolates (23), but only vestigial gap junctions persist in culture. Study of certain properties awaits the raising of large numbers of cells; such scale-up implies long-term culture. The factors involved in the aging process of cells in culture are not fully understood, but isolation and passaging by the use of protease enzymes may yield cells in culture that do not have all of the surface enzymes and other proteins believed to exist on endothelial cells in situ (29). We have therefore developed means to obtain endothelial cells in long-term culture, avoiding exposure to enzymes at both the isolation step and during subculture (29).

By gentle scraping of the luminal surface of bovine pulmonary artery with a scalpel, endothelial cells can be harvested as pure intact sheets composed of up to several hundred cells (23, 29). The endothelial cells of the initial isolates retain both the polarity and junctional contacts that they maintained in the vessel (23).

Cells can be seeded directly onto microcarriers from fresh isolates or from monolayer cultures. Microcarriers in roller bottles provide a substratum suitable for obtaining high yields of endothelial cells in culture that can be passaged without using enzymes, simply by dividing cell-covered beads among new bottles and adding new beads and fresh medium. Endothelial cells attach to the beads rapidly, reach confluence quickly, and exhibit the cobblestone morphology characteristic of these cells in monolayer culture (Figure 8).

The theoretical surface area for cell attachment using microcarriers is enormous: 2.2×10^9 cells per liter of culture medium (23). Cell lines grown on microcarriers retain ACE activity, Factor VIII antigen, and junctional contacts at high passage numbers (23). It is too early to assess the impact of these new culture techniques on studies of metabolic activities of pulmonary endothelium. However, the ability to culture large numbers of endothelial cells that have never been exposed to exogenous proteolytic enzymes may be critical in studies designed to map the topography of endothelial surface enzymes (Figure 9).

Subcellular Bases of Metabolic Activities

Despite the recent highlighting of endothelial cells, especially the luminal surface, it is still hard to imagine from sections such as Figure 2 that endothelial cells have the requisite subcellular apparatus for sustained and active metabolic activities and production of substances for export.

Because of their extremely flattened shape, endothelial cells are perhaps victims of the cross-section more than any other cell type. These cells appear to lack the organelles ordinarily associated with high metabolic or secretory activities. However, closer examination indicates that many subcellular specializations of the endothelial cells make them remarkably well-suited not only for the exchange of gases but also for the processing of circulating vasoactive substances.

Scanning electron micrographs of the luminal surface of endothelial cells of the pulmonary artery indicate that the surface is covered with finger-like projections (Figure 3) (see also 30). The projections vastly increase the surface area available for processing of blood solutes by enzymes situated on the plasma membrane and produce an eddy flow of cell-free plasma. Endothelial projections can be recognized in sectioned material and are seen in smaller vessels of the lungs, including capillaries. They are also evident on pulmonary endothelial cells in culture (9, 16) and are therefore independent of contractions of the vessel wall.

Thin sections of pulmonary vessels reveal large numbers of caveolae intracellulares, many of which open to the vascular lumen (24). Indeed, at the level of capillaries, caveolae are frequently the only prominent cellular organelles in the profiles of endothelium (Figure 4). Like endothelial projections, caveolae vastly increase surface area. In addition, their position, directly facing the vascular lumen, would facilitate the rapid return of the products of specific metabolic reactions to the circulation. The luminal stoma of the caveola is spanned by a diaphragm apparently composed of a single protein lamina (Figure 4); this structure may provide a specialized microenvironment for processing vasoactive substances (24). Caveolae are of interest because of their large numbers and their position in direct communication with the vascular space (Figure 4). However, they become even more interesting by virtue of their involvement in the processing of adenine nucleotides (Figure 5), bradykinin, and angiotensin I (Figure 6).

Replicas of cultured endothelial cells freeze-fractured as monolayers indicate that surface specializations of endothelial cells are not distributed uniformly over the cell surface. Caveolae appear to be organized in tracts, reaching densities of 600 micron^{-2} in some areas. While the relationship of caveolae to the pulmonary processing of adenine nucleotides, bradykinin, and angiotensin I has been shown (reviewed in 16, 24), the significance of unequal distribution of caveolae is not known.

Our studies on the localization of 5'-nucleotidase by cytochemical means and on the localization of ACE by immunocytochemical methods indicated that the disposition of reaction product was not uniform but showed preferential deposition on globular structures of the plasma membrane; this finding prompted us to search for the precise sites of the enzyme within the plasma membrane (24). Replicas of freeze-fractured pulmonary endothelial cells indicate intramembranous particles. However, the intramembranous particles are grouped in rosettes or plaques at the sites of the caveolae and may correspond, both in size and position, to enzyme sites (24). While we know ACE exists on the luminal surface of pulmonary endothelial cells, at present we do not know the precise topographical distribution of ACE or, indeed, of any of the other endothelial enzymes capable of processing circulating vasoactive substances.

Intramembranous particles can move in the plane of the plasma membrane, they do not represent a homogeneous population, and some of them project to the exterior of the membrane. Clearly, some are related to the formation of endothelial cell-cell junctions. It will be a challenge for the future to examine for the precise distribution of particles on the exterior surface of the endothelial cells and to determine which, if any, of the particles can be identified as enzymes or enzyme-binding sites.

To this end, we have begun to develop the technology for visualizing the true outer surface of endothelial cells with the full resolving power of the transmission electron microscope. We have developed techniques for obtaining true surface replicas of cells grown in monolayer culture (9). Such surface replica techniques present the view of the endothelial cells that is exposed to circulating vasoactive substances. As Figure 9 shows, the endothelial surface contains a bewildering array of surface particles. We plan to procced to immunocytochemical localization of ACE and other surface enzymes, to begin to map the topography of the endothelial plasma membrane. In order to understand modulations of enzyme activity (e.g. by hypoxia or by specific inhibitors) we must know the precise location and environment of the enzyme in the cell membrane.

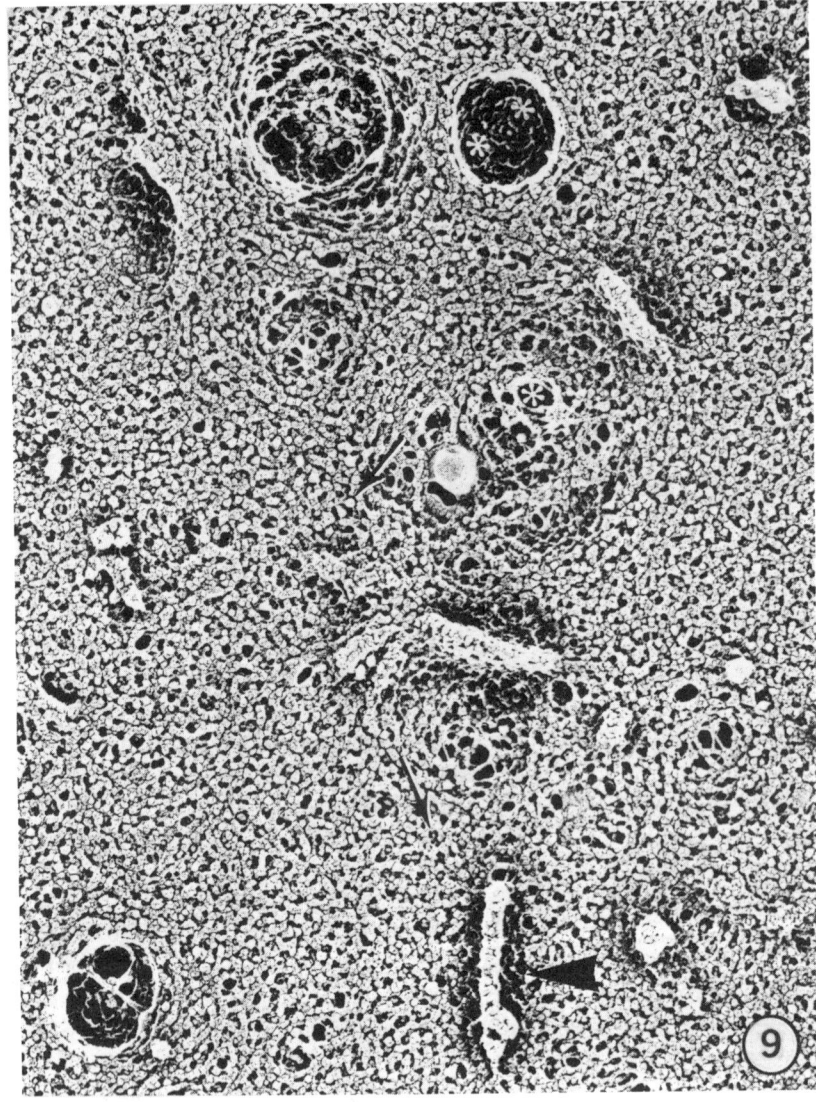

Figure 9 Surface replica of pulmonary endothelial cell, rotary shadowed with platinum and carbon and printed in reverse contrast (X 34,000). The true (unfractured) outer surface of the endothelial plasma membrane is studded with particles of approximately 350 Å (arrows) that could represent enzymes or binding sites. Endothelial projections (arrowhead), caveolae (*), and larger depressions (**) are also evident.

CONCLUSIONS AND FUTURE DIRECTIONS

It appears that the vascular bed of the lungs is studded with enzymes that can continuously process selected vasoactive substances as they pass in circulating blood. For other vasoactive substances, access to intracellular enzymes must involve carrier systems, placed such that they distinguish among members of closely related classes of compounds. As we have suggested before (16, 17, 24), small amounts of cell-bound enzyme could metabolize relatively large amounts of substrate.

The unique features of pulmonary enzymes may relate to the structure of the lungs and to the position of the lung in the circulatory system. Thus the total amount of angiotensin converting enzyme (ACE) within the lungs may greatly exceed the amount needed to process the concentrations of angiotensin I and bradykinin usually found in pulmonary artery blood. However, it is not clear that the total amount of enzyme is available to circulating substrates. While we know that ACE occurs on the luminal surface of pulmonary endothelial cells, it is becoming evident that the membrane bound enzyme is sensitive to factors such as oxygen concentration, factors that do not affect the enzyme in soluble form. The effects of components of the plasma membrane on the molecular configuration and hence on the activity of the enzyme are not known.

Due to technical advances in cell culture, immunocytochemistry, and electron microscope surface replication techniques, it has now become timely to undertake studies of the precise sites and environment of enzymes, binding sites, and transport proteins within the endothelial plasma membrane. In this way it should be possible to improve understanding of factors that may modulate the activity of the enzymes in vivo and to explore more directly and in yet greater detail the structural bases of metabolic activity.

ACKNOWLEDGMENTS

Many colleagues have collaborated in the work described here. Particular thanks are due to J. W. Ryan, D. S. Smith, D. R. Schultz, A. Chung, C. Whitaker, M. A. Hart, G. Maxwell, L. White, and D. Bielefeldt. The work is supported by grants HL21568 and HL22896 from the National Institutes of Health, and by The Council for Tobacco Research, Inc., U.S.A.

Literature Cited

1. Bito, L. Z., Baroody, R. A., Reitz, M. E. 1977. Dependence of pulmonary prostaglandin metabolism on carrier-mediated transport processes. *Am. J. Physiol.* 232(4): E382–87
2. Cross, S. A. M., Alabaster, V. A., Bakhle, Y. S., Vane, J. R. 1974. Sites of uptake of ^{3}H-5-hydroxytryptamine in rat isolated lung. *Histochemie* 39:83–91
3. Crutchley, D. J., Eling, T. E., Anderson, M. W. 1978. ADPase activity of isolated perfused rat lungs. *Life Sci.* 22:1413–20
4. Crutchley, D. J., Ryan, U. S., Ryan, J. W. 1980. Effects of aspirin and dipyridamole on the degradation of adenosine diphosphate by cultured cells derived from bovine pulmonary artery. *J. Clin. Invest.* 66:29–35
5. Dorer, F. E., Kahn, J. R., Lentz, K. E., Levine, M., Skeggs, L. T. 1972. Purification and properties of angiotensin converting enzyme from hog lungs. *Circ. Res.* 31:356–66
6. Eling, T. E., Anderson, M. W. 1976. Studies on the biosynthesis, metabolism and transport of prostaglandins by the lung. *Agents & Actions* 6(4):543–47
7. Ferreira, S. H., Vane, J. R. 1967. Prostaglandins: their disappearance from and release into the circulation. *Nature* 216:868–73
8. Habliston, D. L., Whitaker, C., Hart, M. A., Ryan, U. S., Ryan, J. W. 1979. Isolation and culture of endothelial cells from the lungs of small animals. *Am. Rev. Resp. Dis.* 119:853–68
9. Hart, M. A., Ryan, U. S. 1978. Surface replicas of pulmonary endothelial cells in culture. *Tiss. Cell* 10:441–49
10. Hughes, J., Gillis, C. N., Bloom, F. E. 1969. The uptake and disposition of DL-norepinephrine in perfused rat lung. *J. Pharmacol. Exp. Ther.* 169:237–48
11. Marchesi, V. T., Barrnett, R. J. 1963. The demonstration of enzymatic activity in pinocytic vesicles of blood capillaries with the electron microscope. *J. Cell Biol.* 17:547–56
12. Pearson, J. D., Carleton, J. S., Hutchings, A., Gordon, J. L. 1978. Uptake and metabolism of adenosine by pig aortic endothelial and smooth muscle cells in culture. *Biochem. J.* 170:265–71
13. Piper, P. J., Vane, J. R. 1971. The release of prostaglandins from lung and other tissues. *Ann. N.Y. Acad. Sci.* 180:363–85
14. Piper, P. J., Vane, J. R., Wyllie, H. J. 1970. Inactivation of prostaglandins by the lungs. *Nature* 225:600–4
15. Ryan, J. W., Day, A. R., Ryan, U. S., Chung, A., Marlborough, D. I., Dorer, F. E. 1976. Localization of angiotensin converting enzyme (kininase II). I. Preparation of antibody-heme-octapeptide conjugates. *Tiss. Cell* 8:111–24
16. Ryan, J. W., Ryan, U. S. 1977. Pulmonary endothelial cells. *Fed. Proc.* 36:2683–91
17. Ryan, J. W., Ryan, U. S. 1980. Biochemical and morphological aspects of the actions and inactivation of kinins and angiotensins. In *Enzymatic Release of Vasoactive Peptides,* ed. F. Gross, H. G. Vogel, pp. 259–74. NY: Raven
18. Ryan, J. W., Ryan, U. S., Habliston, D., Martin, L. 1978. Synthesis of prostaglandins by pulmonary endothelial cells. *Trans. Assoc. Am. Physicians* 91:343–50
19. Ryan, J. W., Smith, U. 1971. Metabolism of adenosine-5'-monophosphate during circulation through the lungs. *Trans. Assoc. Am. Physicians* 84:297–306
20. Ryan, J. W., Smith, U. 1973. The metabolism of angiotensin I by endothelial cells. In *Protides of the Biological Fluids,* ed. H. Peeters, pp. 379–84. Oxford: Pergamon
21. Ryan, U. S. 1981. Isolation and culture of pulmonary endothelial cells. In *Pulmonary Toxicology,* ed. G. E. R. Hook. In press
22. Ryan, U. S., Clements, E., Habliston, D., Ryan, J. W. 1978. Isolation and culture of pulmonary artery endothelial cells. *Tiss. Cell* 10:535–54
23. Ryan, U. S., Mortara, M., Whitaker, C. 1980. Methods for microcarrier culture of bovine pulmonary artery endothelial cells avoiding the use of enzymes. *Tiss. Cell* 12(4):619–35
24. Ryan, U. S., Ryan, J. W. 1977. Correlations between the fine structure of the alveolar-capillary unit and its metabolic activities. In *Metabolic Functions of the Lung,* ed. Y. S. Bakhle, J. R. Vane, pp. 197–232. NY/Basel: Marcel Dekker
25. Ryan, U. S., Ryan, J. W. 1977. Specific metabolic activities of pulmonary endothelial cells. In *Pulmonary Macrophage and Epithelial Cells,* ed. C. L. Sanders, R. P. Schneider, G. E. Dagle, H. A. Ragan, pp. 115–140. Springfield, VA: Energy Res. Dev. Admin.
26. Ryan, U. S., Ryan, J. W. 1979. Vasoactive substances and the lungs: Cellular mechanisms. In *The Microembolism*

Syndrome, ed. T. Saldeen, pp. 223-32. Stockholm: Amqvist & Wiksell
27. Ryan, U. S., Ryan, J. W., Chiu, A. T. 1976. Kininase II (angiotensin converting enzyme) and endothelial cells in culture. *Adv. Exp. Med. Biol.* 70:217-27
28. Ryan, U. S., Ryan, J. W., Whitaker, C., Chiu, A. 1976. Localization of angiotensin converting enzyme (kininase II). II. Immunocytochemistry and immunofluorescence. *Tiss. Cell* 8:125-46
29. Ryan, U. S., Schultz, D. R., Del Vecchio, P., Ryan, J. W. 1980. Endothelial cells of bovine pulmonary artery lack receptors for C3b and for the Fc portion of IgG. *Science* 208:748-79
30. Smith, U., Ryan, J. W., Michie, D. D., Smith, D. S. 1971. Endothelial projections: As revealed by scanning microscopy. *Science* 173:925-27
31. Soffer, R. L. 1976. Angiotensin converting enzyme and the regulation of vasoactive peptides. *Ann. Rev. Biochem.* 45:73-94
32. Strum, J. M., Junod, A. F. 1972. Radioautographic demonstration of 5-hydroxytryptamine-^{3}H uptake by pulmonary endothelial cells. *J. Cell. Biol.* 54:456-67
33. Youdim, M. B. H., Bakhle, Y. S., Ben-Harari, R. R. 1980. Inactivation of monoamines by the lung. In *Metabolic Activities of the Lung,* pp. 105-222. *Ciba Found. Symp. 78.*

PROCESSING OF ENDOGENOUS POLYPEPTIDES BY THE LUNGS

James W. Ryan

Department of Medicine, University of Miami School of Medicine, Miami, Florida 33101

INTRODUCTION

The lungs in effect clear a number of hormones and prohormones such that their concentrations in systemic arterial blood are significantly altered in comparison with those of the same substances in central venous blood (28, 70–72, 94). The clearance of a given hormone or prohormone may greatly exceed the quantity likely to be needed to support lung function. In some instances, neither the hormone nor prohormone is retained by the lungs but is metabolized such that the reaction products are released into the pulmonary venous effluent and therefore into the systemic arterial circulation (60, 66, 67, 81). Through such clearance or processing of hormones, especially polypeptide hormones, the lungs can be said to regulate the hormonal composition of arterial blood (70, 71, 74, 75).

Little is known about which polypeptide hormones and prohormones are altered during passage through the pulmonary vascular bed and which are not (28, 72). Our knowledge is limited by techniques of assay and by the availability of polypeptides labelled intrinsically at high specific radioactivity. Polypeptides capable of affecting blood pressure or of stimulating smooth muscle are relatively easy to study, and much is known of the clearances, and even the immediate metabolic fates, of bradykinin, the angiotensins, substance P, and vasopressin (cf 28–30, 71, 94). However, precise functional assays are not easily developed for other hormones of interest such as parathormone, ACTH, gastrin, proinsulin, C-peptide, and insulin. Similarly, little is known of the fates of other excitatory polypeptides (e.g. the anaphylatoxins) during passage through the lungs (cf 17). Radioimmunoassays are available, but these do not help to define function or modulation of function that may arise through limited hydrolysis. A relatively large polypeptide such as insulin could undergo limited hydrolysis without yielding a product whose reaction with antibody is perceptibly changed. As a case in point, relatively few antibody preparations can distinguish angiotensin I from its lower homolog, des-Asp^1-angiotensin I (26, 49).

The larger polypeptide hormones can be labelled extrinsically, and, in principle, their fates can be examined. However, there are problems with the use of such labels, not the least of which are (a) possible interference by, for example, iodinase enzymes, and (b) the potential for a large extrinsic label to modify the interaction of a given polypeptide substrate with one or more of its metabolic enzymes. Morgat and colleagues (48) have described simple means (catalytic dehalogenation in $^3H_2$ gas) for converting extrinsically labelled polypeptides into their intrinsically labelled analogs. However, the procedure has not been widely used and may in fact be relatively unsuitable for use with hormones (e.g. insulin) where two peptide chains are bonded by disulfide bridges.

This review concerns primarily those polypeptide hormones and related compounds known to be, in effect, 'cleared' by metabolic enzymes located on or near the luminal surface of pulmonary endothelial cells. The scope of such pulmonary processing of polypeptides is not known. Further, the mechanisms by which angiotensin I is converted into angiotensin II and by which bradykinin and its higher homologs are inactivated may not be the only mechanisms operative.

IMMEDIATE METABOLIC FATES OF BRADYKININ AND ANGIOTENSIN I

Bradykinin and angiotensin I are degraded almost quantitatively during a single passage through the pulmonary vascular bed. Five of the eight peptide bonds of bradykinin are hydrolyzed and apparently five or more of the nine peptide bonds of angiotensin I are cleaved (66–69, 80, 81). Hydrolysis of any of the peptide bonds of bradykinin eliminates biological activity. However, hydrolysis of one of two bonds in angiotensin I suffices to form either angiotensin II or another potential prohormone, des-Asp^1-angiotensin I, a presumed precursor of angiontensin III [des-Asp^1-angiotensin II (1, 6–9, 33, 49, 57, 60, 81, 95)]. Hydrolysis of both bonds yields angiotensin III. In addition, a number of biologically inactive lower homologs of angiotensin I are formed.

Circulating angiotensin I and bradykinin are hydrolyzed without leaving the pulmonary vascular space (25, 55, 64, 66–74, 89). The hydrolytic reactions proceed briskly when the lungs are perfused with blood or artificial salt solutions (67–69, 80). Blood enzymes appear to make little if any contribution. Thus it has been postulated that the relevant enzymes are on or near the luminal surface of pulmonary endothelial cells (67, 70–75). Angiotensin converting enzyme (E.C. 3.4.15.1; also known as kininase II) is known to be thus situated (64, 76, 77, 79, 83). There is growing reason to believe that other of the relevant kininase and angiotensinase enzymes are also on or near the luminal surface of pulmonary endothelial cells. Bovine endothelial cells in monolayer culture degrade [^{3}H]Phe^8-bradykinin and its [^{3}H]Phe^5-analog to form radioactive products like those formed by isolated blood-free rat lungs perfused with Krebs-Henseleit solution [(67–69) and J. W. Ryan, unpublished observations]. The sites of hydrolysis of bradykinin are shown in Figure 1, and those for angiotensin I are shown in Figure 2.

The enzymes implied by the hydrolytic reactions shown in Figures 1 & 2 have in some instances been assayed by using substrates simpler and less subject to side-reactions than bradykinin and angiotensin I. A number of such substrates exist for angiotensin converting enzyme (62,63), and pulmonary endothelial cells in culture have been shown to react with [^{3}H]hippuryl-Gly-Gly, [^{3}]hippuryl-His-Leu, and [^{3}H]benzoyl-Phe-Ala-Pro to produce, respectively, [^{3}H]hippuric acid and [^{3}H]benzoyl-Phe (61, 62, 82). Similarly, the cells react with Arg-Pro-[^{3}H]benzylamide to produce

```
       1   2   3   4   5   6   7   8   9
      Arg-Pro-Pro-Gly-Phe-Ser-Pro-Phe-Arg
            △   △       △       △   △
```

Figure 1 Hydrolysis of bradykinin during a single passage through isolated rat lungs. The numbers indicate residue numbers of amino acids. Hydrolytic sites are indicated by arrows. It is not yet known whether the hydrolysis of the Ser^6-Pro^7 bond occurs when angiotensin converting enzyme is not inhibited (69). The latter enzyme, a dipeptidyl carboxypeptidase, can hydrolyze bradykinin in two steps in vitro. First the Phe^8-Arg^9 dipeptide is released and then the Ser^6-Pro^7 dipeptide is released [e.g. see (18)]. Under conditions of low pulmonary blood flow in vivo the mean transit time may be sufficiently slow to permit the second reaction to occur.

```
       1   2   3   4   5   6   7   8   9  10
      Asp-Arg-Val-Tyr-Ile-His-Pro-Phe-His-Leu
            △                   △   △
```

Figure 2 Hydrolysis of angiotensin I during a single passage through isolated rat lungs. At least two hydrolytic sites, in addition to those shown, have not been identified with certainty. Two as yet unidentified lower homologs of angiotensin I are formed. Of the reactions indicated, none goes to completion (cf 81). More His-Leu is formed than angiotensin II; possibly because lower homologs of angiotensin I (e.g. des-Asp^1-angiotensin I and the 3–10 and 4–10 lower homologs) are fair to good substrates for angiotensin converting enzyme (8,9). Similarly, des-Asp^1-angiotensin I is a major product in some lung perfusion studies and a minor product in others. Interspecific effects have not yet been evaluated.

Pro-[³H]benzylamide (J. W. Ryan, unpublished observations). However, results obtained by using the simpler substrates can be highly misleading. Bovine pulmonary endothelial cells react with [³H]benzoyl-Phe-Arg, the acylated C-terminal dipeptide of bradykinin, to yield [³H] benzoyl-Phe. However, this carboxypeptidase N-like enzyme does not react with bradykinin, even when perfused through isolated rat lungs or incubated with bovine pulmonary endothelial cells in culture, the lungs and cells having previously been treated with an inhibitor of angiotensin converting enzyme (J. W. Ryan, unpublished observations). The affinity of the enzyme may be much higher for the smaller substrate than for the larger.

ENZYME KINETICS IN VITRO

Much has been written about the interactions of angiotensin converting enzyme and oligopeptides or acylated oligopeptides in vitro (cf 4, 5, 12–15, 18, 19, 38). Similarly, there are some data on the kinetics of interactions of the enzyme with bradykinin and its higher homologs and angiotensin I and its lower homologs in vitro (12, 15, 18). To a certain extent, some of the in vitro studies have helped to elucidate the processing of kinins and angiotensins by intact lungs. For example, each of the N-terminal higher homologs of bradykinin is degraded by rat lungs in vivo at a rate slower than that for bradykinin itself (58, 69). Lys-bradykinin is inactivated at approximately one half the rate and Met-Lys bradykinin is degraded at approximately 1/10th to 1/20th the rate for bradykinin. Curiously, angiotensin converting enzyme, an enzyme that acts at a site far from the N-terminus of bradykinin, may account for the differences between the hydrolytic rates for bradykinin and its higher homologs. Highly purified pig lung angiotensin converting enzyme is more reactive with bradykinin than with Lys-bradykinin and is even less reactive with Met-Lys-bradykinin (20, 21). The N-terminal extensions of the higher homologs may interfere with binding to the enzyme.

Less is known about the ability of angiotensin converting enzyme of intact lungs to discriminate among angiotensin I and its C-terminal lower homologs. In vitro, des-Asp¹-angiotensin I has a higher affinity and higher specificity constant (K_{cat}/K_m) for the enzyme than does angiotensin I itself (7–9, 91, 92). If these data are indicative of conditions in vivo, it is conceivable that the conversion of des-Asp¹-angiotensin I into angiotensin III would be favored over the conversion of angiotensin I into angiotensin II (cf 8). Presumably, such discrimination would have implications for aldosterone secretion because angiotensin III has been reported to be a somewhat more potent aldosterone secretogogue than is angiotensin II (6, 57). However, there appears to be marked variability among species in terms of the enzyme, an aminopeptidase A-like enzyme, necessary to convert angiotensin I into des-Asp^1-angiotensin I or angiotensin II into angiotensin III (cf 1, 26, 33, 40, 41, 49, 74, 85). An immunoreactive angiotensin III–like substance occurs in relative abundance in the plasma (arterial and venous) of rats but is far less abundant in human plasma. Recently, Garcia del Rio and colleagues (33) have reported that the concentrations of an angiotensin III–like substance in arterial plasma rise in rats infused intravenously with des-Asp^1-angiotensin I. Captopril (2 mg/kg), an inhibitor of angiotensin converting enzyme, markedly decreases the plasma concentrations of angiotensins II and III of animals not infused. Garcia del Rio et al (33) noted, as have previous investigators (cf 95), that the biological half-life of angiotensin III is probably shorter than that of angiotensin II. Very likely, the half-life of des-Asp^1-angiotensin I is shorter than that of angiotensin I, a phenomenon that may explain why des-Asp^1-angiotensin I infused intravenously into humans has effects on aldosterone secretion no stronger (and often weaker) than those of intravenously administered angiotensin I. Given the brief transit time of blood passing from the lungs to the adrenal gland, the conversion within the lungs of angiotensins I and II into their respective des-Asp^1 analogs should favor the possibility that some angiotensin III will reach its receptors in the cortex. It may be pertinent to add here that capillaries of rat adrenal cortex have been shown to possess both aminopeptidase A and angiotensin converting enzyme (16, 74, 75).

ENZYME KINETICS IN VIVO

Given a pure enzyme and a suitable, well-characterized substrate, one can readily define kinetic constants. However, efforts to define the kinetics of interaction between a circulating endogenous polypeptide and peptidase enzymes disposed at fixed points along a capillary wall are more difficult (53, 54, 70–75). Kinetic constants bearing on association and disassociation of substrate and enzyme are likely to be heavily influenced by the rate of blood flow and the diameter of blood vessels. Surface area of the endothelial lining is also likely to be important. Endogenous inhibitors (including alternative substrates) require consideration (65, 74). In the case of the microvascular bed of the lungs, the quality of inhalants and the efficiency of exchange of gases may also be important (45, 70–75, 87–89).

One probably cannot deduce much about the interactions of a given polypeptide with a lung peptidase in vivo by examining the interactions of the polypeptide and pure enzyme in vitro. Interactions of angiotensin converting enzyme and its natural and synthetic substrates may be taken as cases in point. The hydrolysis in vitro of angiotensin I, but to a far smaller degree that of des-Asp[1]-angiotensin I, is strongly influenced by the concentration of chloride in the assay buffer (cf 7). The hydrolysis of bradykinin does not require chloride but proceeds optimally in the presence of relatively low concentrations, 0.01M NaCl (7–9, 20, 21).

The K_m of the reaction of hippuryl-Gly-Gly with angiotensin converting enzyme is lowered and the V_{max} is increased by adding Na_2SO_4 to the reaction mixture (19). Phosphate, Tris, and borate buffers inhibit the reaction in comparison to reaction rates measured using a Hepes buffer. Inhibition by phosphate is dependent on pH (19). The effects of Na_2SO_4 are even more dramatic on the reaction of hippuryl-His-Leu with enzyme. The K_m is reduced 20-fold and the ratio of V_{max}/K_m is raised almost 2-fold (62, 63). One synthetic substrate, benzoyl-Phe-Ala-Pro, has a K_m lower than those of angiotensin I and des-Asp[1]-angiotensin I. Its reaction with enzyme is supported by low salt concentrations but it is not influenced by Na_2SO_4 (62).

Enzymes are usually assayed in vitro under conditions in which the reaction obeys (or almost obeys) zero-order enzyme kinetics; i.e. conditions in which enzyme is saturated with substrate and reaction velocity approaches V_{max} (12, 19). Such conditions are seldom if ever obtained in vivo. Again, the interaction of angiotensin I with angiotensin converting enzyme can be taken as a case in point. Assuming that the K_m in vitro is approximately the same (or same order of magnitude) as that in vivo, the blood concentration of angiotensin I is far below K_m. The K_m reported by Chiu et al (8) is 33 μM, and the concentration of angiotensin I is seldom greater than 0.0001 μM (36, 47, 50, 85). A further complicating factor is that the concentration of angiotensin converting enzyme in blood [0.01 μM; computed from data in (14, 43)] is greater than that of the substrate, angiotensin I. Concentrations of bradykinin in blood are believed to be lower than those of angiotensin I. The content of the enzyme of lungs is far higher than that of an equivalent weight of blood (cf 12); thus the molecular ratio of enzyme/substrate is higher still.

Relatively little is known of the kinetics of enzyme: substrate reactions when the concentration of enzyme exceeds that of substrate. Nonetheless, there may be some interesting implications. First, at any given instant, angiotensin converting enzyme is unlikely to be saturated with substrate. Assuming that the interaction of the two obeys first-order enzyme kinetics, the percent utilization of substrate (or formation of angiotensin II as a percent of angiotensin I in central venous blood) should not change when the concentration of substrate is quartered, halved, doubled, or quadrupled so long as the substrate concentration remains well-below K_m. Excluding considerations of alternative substrates, endogenous inhibitors and other potential modulating factors, such as side-reactions (e.g. formation of des-Asp^1-angiotensin I), blood pH and chloride concentration, the quantity of angiotensin II formed should be a direct function of the quantity of angiotensin I entering the pulmonary artery.

Second, given the fact that the blood concentration of angiotensin I is approximately 1/300,000th of K_m, it is inconceivable that one would encounter concentrations of angiotensin I in vivo (under physiologic or pathologic conditions) sufficient to saturate the enzyme or even move the reaction into the range of mixed first- and zero-order enzyme kinetics. Thus if modulation of the rate of conversion of angiotensin I into angiotensin II is observed using a whole organ or an in vivo preparation, one must look to factors other than enzyme kinetics to explain the result. Dilation or contraction of blood vessels, changes in perfusion pressures, and shunting of blood are far more likely to be determinants of the efficiency of hydrolysis. Similarly, changes in the shape and surface area of the endothelial cells may be important. Further, an enzyme such as angiotensin converting enzyme has alternative substrates, and endogenous inhibitors are known to occur in variable concentrations (cf 65, 86).

Given the fact that enzyme per se is not limiting (although net enzyme activity is probably variably limiting), it is tempting to suggest that the lungs may have an enormous capacity for varying the amount of, for example, angiotensin II formed per unit time. The concept is made more interesting by the facts that the lungs receive the entire cardiac output, can tolerate extremely large changes in blood flow (e.g. from 7–40 liters/min in a well-trained athlete), and empty their venous effluent directly into the systemic arterial circulation. In this way, the lungs might contribute to the 'physiologic hypertension' of exercise. Further, in conditions such as congestive heart failure, the slow, low-volume pulmonary blood flow might contribute to secondary hyperaldosteronism, a frequent complication. On the latter point, it may be pertinent that Captopril has been used to advantage in congestive heart failure (22, 93). In addition to improving pulmonary blood flow, at lower pressures, Captopril reduces aldosterone secretion.

PEPTIDASE INHIBITORS AND ALTERNATIVE FATES OF ENDOGENOUS PEPTIDES

Through efforts to develop more effective antihypertensive drugs with fewer side-effects, a number of new orally effective inhibitors of angiotensin converting enzymes have been developed. The first of these was Captopril [SQ 14,225; 2-D-methyl-3-mercaptopropanoyl-L-proline; cf (13, 52)]. More recently, Merck has announced the discovery of an even more potent inhibitor, the precursor of which is called MK 421 (56). A number of other drugs of the same class are in earlier stages of development. Although inhibitors such as these exert their effects on angiotensin converting enzyme of all tissues including that of the lungs, it is still the enzyme (or inhibition of the enzyme) of the lungs that, to a large extent, determines the quantities of angiotensin II and bradykinin likely to enter the systemic arterial circulation.

The efficacy of inhibition of angiotensin converting enzyme as a means of reducing blood pressure is well-established (13, 34, 35, 42, 47, 49, 52). However, little or nothing is known about the long-term effects on intrinsic lung function or on the immediate metabolic fates of kinins and angiotensins. On the former point, it would not be surprising to find that some patients treated with Captopril develop pulmonary fibrosis or Goodpasture's syndrome. These side-effects, if encountered, may not be caused by inhibition of angiotensin converting enzyme but by the free mercapto-group of Captopril. Indeed the use of Captopril is already known to be associated with the development of disorders like those caused by D-penicillamine, another mercapto-bearing compound (39). Side-effects of MK 421, a compound that does not contain sulfur, have not yet been reported. To my knowledge, no one has yet investigated the effects of an inhibitor of angiotensin converting enzyme on normal lung function. However (and as noted above), inhibition of the enzyme in patients with congestive heart failure can promptly and profoundly improve lung, cardiac, and renal performance (22, 93). Similarly, the postoperative pulmonary hypertension of patients who have undergone cardiopulmonary bypass procedures can be thus relieved (50).

The anticipated long-term treatment of patients with a drug such as Captopril means that a large segment of the population (perhaps 20-60% of all hypertensive patients) may live many years under circumstances in which they have little or no functioning angiotensin converting enzyme. Available data indicate that angiotensin I, to a small extent, accumulates in blood (cf 22, 33, 35, 49, 93). It is less clear whether kinins accumulate. Further, it appears that the body mass of enzyme increases with treatment (32). The fact that angiotensin I does not accumulate enormously may be caused in part to the fact that angiotensin I is degraded through alternative pathways (68, 74, 75, 81). Since one cannot readily measure accumulation of kinins in blood, one can be more certain that other peptidase enzyme suffice to effect inactivation. Indeed, inhibition of angiotensin converting enzyme of isolated rat lungs is known to change the means of inactivation of bradykinin but has little directly discernable (as yet) effect on the rate of inactivation (67-69). However, indirect evidence indicates that the rate of inactivation is slightly reduced; otherwise, it would not be possible to potentiate the blood pressure effects of intravenously administered bradykinin (27, 58).

Inhibitors of angiotensin converting enzyme are only the first class of peptidase inhibitors to be introduced. Other classes are in development. Our group has begun to develop an inhibitor of the enzyme that converts angiotensin II into angiotensin III (10, 74). Another group has described a drug that inhibits an enkephalinase. The latter enzyme is a dipeptidyl carboxypeptidase distinguishable from angiotensin converting enzyme (2, 37; cf 23). In addition, a number of protease inhibitors, designated for clinical use, are in development (e.g. elastase inhibitors). Their effects on lung peptidase enzymes have yet to be examined.

Over the next 5–10 years, a variety of highly selective, potent enzyme inhibitors may well be in common clinical use. Precisely what this new pharmacology will mean to the processing of endogenous polypeptides by the lungs remains to be seen. Conversely, as it becomes clearer which of the lung peptidase (and protease) enzymes participate in the processing of endogenous polypeptides, opportunities will arise to design new inhibitors. Inhibitors of angiotensin converting enzyme modulate both the renin-angiotensin and the kallikrein-kinin systems. If the enzyme has other naturally occurring substrates, the scope of effects may be even greater than that yet recognized. It is not inconceivable that an inhibitor that modulates the renin-angiotensin system or the kallikrein-kinin system (but not both) would be useful clinically. Certainly, such an inhibitor would be useful for research purposes.

Inhibitors have already begun to help define which peptidase enzymes are functionally important in the degradation of bradykinin and angiotensin II and which are not. The α-chloromethylketone of L-aspartic acid clearly potentiates the effects of angiotensin II on rat arterial blood pressure (10, 74). Thus an aminopeptidase A–like enzyme is likely to be important in inactivating angiotensin II (cf 40, 41). Recently, Ondetti and colleagues (51) have synthesized two highly potent inhibitors, one selective for carboxypeptidase B and related enzymes and the other for carboxypeptidase A–like enzymes. An enzyme of the latter class, called 'angiotensinase C', has been reported to be important in the inactivation of angiotensin II (96). An enzyme of the former class, called 'kininase I', has been reported to be important in the degradation of bradykinin (24). However, the inhibitor of carboxypeptidase A–like enzymes does not potentiate angiotensin II, and the other inhibitor does not potentiate bradykinin (M. A. Ondetti, B. Rubin, personal communication). The latter finding is consistent with results of studies performed in this laboratory: Isolated rat lungs and bovine pulmonary endothelial cells in culture do not degrade bradykinin by removal of its C-terminal arginine [(67, 68) and unpublished observations].

CONCLUDING REMARKS AND OUTLOOK FOR FUTURE RESEARCH

The range and variety of endogenous polypeptides processed during passage through the pulmonary vascular bed are not known. Further, only one relevant lung peptidase enzyme has been characterized in any detail—namely, angiotensin converting enzyme (86). Clearly, there are enormous opportunities for further definition of the scope of the processing of peptides, not only in terms of the range and variety but also in terms of the mechanisms employed. Hydrolysis of peptide bonds need not be the only means. Relatively little attention has been paid to specific or selective transport systems, yet one knows a priori that such systems are likely to exist. Conceivably, the fractional clearances of some polypeptides by the lungs are too small for precise measurement, yet are critical for normal lung function. As pointed out by Tierney (90), the fractional clearance of glucose by the lungs is often too small for reliable detection. It deserves emphasis that the accuracy and precision of assays for glucose greatly exceed those for the measurement of endogenous polypeptides.

Hydrolysis of, for example, angiotensin I or bradykinin and then release of products back into the vascular lumen does not exclude the possibility that some of either peptide (or fragment) is taken up for use within the lungs. Indeed, both angiotensin II and bradykinin have pharmacologic effects on the lung vasculature (3, 50, 59). Further, it has been my experience that one need not long perfuse isolated lungs with radioactive bradykinin before the outflow valve of the respirator becomes contaminated with the same isotope. Nonetheless, it is the inactivation of bradykinin and the formation of angiotensin II at sites adjacent to the vascular lumen that make a critical contribution to the concentrations of these polypeptides in systemic arterial blood.

Although the ability of the lungs selectively to process endogenous polypeptides has been known since the late 1960s, relatively few clinical or clinically related studies have been performed to examine the pathophysiologic consequences of the processing. Because bradykinin tends to lower blood pressure and angiotensins II and III act to raise blood pressure, it is believed that the "clearances" of bradykinin and angiotensin I by the lungs have implications for blood pressure homeostasis (73). Whether there are pulmonary disorders that contribute to systemic hypertension is not clear. One can more readily envision lung diseases (e.g. those characterized by a reduced microvascular bed) in which the inactivation of bradykinin and the formation of angiotensin II are relatively inefficient (44). Presumably, hypotension would be favored, especially among patients with reduced effective blood volumes.

Hypoxia affects the ability of the lungs (and other tissues) to degrade pharmacologic quantities of bradykinin and angiotensin I (45, 87, 88). Severe hypoxia significantly reduces the ability of the lungs to metabolize low concentrations of an extrinsically labeled analog of angiotensin I (89).

The apparent effects of hypoxia on the net activity of angiotensin converting enzyme are unlikely to be specific. As shown previously, inhibition of angiotensin converting enzyme with a Bothrops peptide or with Captopril does not preserve bradykinin to any readily measurable extent (67, 68). Thus it is likely that hypoxia reduces the net activity of a number of endothelial associated peptidase enzymes. The net activities of other enzymes, such as ADPase and 5'-nucleotidase, may also be affected, a possibility that remains untested. Given the apparent roles of the latter enzymes in defending the systemic arterial circulation from platelet aggregation and intravascular coagulation (11, 78), it would appear timely to test empirically the scope of effects of hypoxia on enzymes of the surface of pulmonary endothelium.

The effects of hypoxia on net activity of angiotensin converting enzyme are likely to arise from many different factors (cf 89). Mean transit time of blood across the pulmonary vascular bed is known to be reduced in acute hypoxia (31), and one can postulate decreased exposure to enzyme. However, endothelial cells in culture respond to hypoxia by a reduction in ability to degrade bradykinin (45). Thus changes in the conformation of the enzyme or adjacent cell structures may play a part.

Future work might well focus on the response of endothelial cells to, for example, enzymes released by other cell-types, to intercurrent infections, and to excitatory substances such as the anaphylatoxins, prostaglandin-related substances (including the leukotrienes), and platelet activating factor. Thromboplastin occurs in a latent state in association with pulmonary endothelial cells (46) and may be activated in some forms of lung injury. Recently, we have shown that pulmonary endothelial cells in culture do not possess Fc receptors (84). However, in work underway at present, we have found that endothelial cells exposed to lysates of white blood cells express Fc receptors. Thus the composition of active enzymes and other factors of the endothelial cell surface may be subject to modulation.

Perhaps the most meaningful clinical findings obtained so far are those implied by results of the extensive clinical trials on Captopril as an antihypertensive agent: Complete inhibition of angiotensin converting enzyme is compatible with life. If the same is true for other lung peptidase enzymes, we may safely conclude that the pulmonary processing of endogenous polypeptides is functionally, but not vitally, important. However, we may do well to keep in mind the experience of those who study coagulation disorders: Often the absence of one functionally significant clotting factor is compatible with normal coagulation. Deletion (or inhibition) of two such factors may not be compatible with life.

Literature Cited

1. Ackerly, J. A., Peach, M. J., Vaughan, E. D. Jr., Glenn, A. W. 1980. Formation of [des-Asp^1]angiotensin I by the perfused rat lung. *Endocrinology* 107(6)80:1699–704
2. Benuck, M., Marks, N. 1980. Characterization of a distinct membrane-bound dipeptidyl carboxypeptidase inactivating enkephalin in brain. *Biochem. Biophys. Res. Comm.* 95(2):822–28
3. Berkov, S. 1974. Hypoxic pulmonary vasoconstriction in the rat. The necessary role of angiotensin II. *Circ. Res.* 35:256–61
4. Cheung, H. S., Cushman, D. W. 1973. Inhibition of homogeneous angiotensin-converting enzyme of rabbit lung by synthetic venom peptides of *Bothrops jararaca*. *Biochim. Biophys. Acta,* 293:451–63
5. Cheung, H. S., Wang, F. L., Ondetti, M. A., Sabo, E. F., Cushman, D. W. 1980. Binding of peptide substrates and inhibitors of angiotensin converting enzyme. Importance of the COOH- terminal dipeptide sequence. *J. Biol. Chem.* 255(2):401–7
6. Chiu, A. T., Peach, M. J. 1973. Stimulation of aldosterone biosynthesis by angiotensin heptapeptide. *Fed. Proc.* 32:765 (Abstr.)
7. Chiu, A. T., Ryan, J. W., Stewart, J. M., Dorer, F. E. 1976. Conversion of des-Asp^1-angiotensin I to angiotensin III. *Fed. Proc.* 34:704 (Abstr.)
8. Chiu, A. T., Ryan, J. W., Stewart, J. M., Dorer, F. E. 1976. Formation of angiotensin III by angiotensin converting enzyme. *Biochem. J.* 155:189
9. Chiu, A. T., Ryan, J. W., Stewart, J. M., Dorer, F. E. 1976. Inhibition of angiotensin converting enzyme (ACE) by angiotensin- II (AII) antagonists. *Circulation* 53 & 54:II–167 (Abstr.)
10. Chung, A., Ryan, J. W. 1980. Drugs that potentiate angiotensin II. *Circulation* 62(III):89 (Abstr.)
11. Crutchley, D., Ryan, U. S., Ryan, J. W. 1980. Effects of aspirin and dipyridamole on the degradation of adenosine diphosphate by cultured cells derived from bovine pulmonary artery. *J. Clin. Invest.* 66:29–35
12. Cushman, D. W., Cheung, H. S. 1971. Spectrophotometric assay and properties of the angiotensin-converting enzyme of rabbit lung. *Biochem. Pharmac.* 20:1637–48
13. Cushman, D. W., Cheung, H. S., Sabo, E. F., Ondetti, M. A. 1977. Design of potent competitive inhibitors of angiotensin converting enzyme. Carboxylalkanoyl and mercaptoalkanoyl amino acids. *Biochemistry* 16(25):5484–91
14. Das, M., Hartley, J. L., Soffer, R. L. 1977. Serum angiotensin converting enzyme. Isolation and relationship to the pulmonary enzyme. *J. Biol. Chem.* 252(4):1316–19
15. Das, M., Soffer, R. L. 1975. Pulmonary angiotensin converting enzyme. Structural and catalytic properties. *J. Biol. Chem.* 250:6762–68
16. Del Vecchio, P. J., Ryan, J. W., Chung, A., Ryan, U. S. 1980. Capillaries of the adrenal cortex possess aminopeptidase A and angiotensin converting enzyme activities. *Biochem. J.* 186:605–8
17. Denny, J. B., Johnson, A. R. 1979. Uptake of ^{125}I-labelled C3a by cultured human endothelial cells. *Immunology* 36(2):169–77
18. Dorer, F. E., Kahn, J. R., Lentz, K. E., Levine, M., Skeggs, L. T. 1974. Hydrolysis of bradykinin by angiotensin converting enzyme. *Circ. Res.* 34:823–27
19. Dorer, F. E., Kahn, J. R., Lentz, K. E., Levine, M., Skeggs, L. T. 1976. Kinetic properties of pulmonary angiotensin converting enzyme. Hydrolysis of hippurylglycylglycine. *Biochim. Biophys. Acta* 429:220–28
20. Dorer, F. E., Ryan, J. W., Stewart, J. M. 1974. Hydrolysis of bradykinin and its higher homologues by angiotensin converting enzyme. *Biochem. J.* 141:915–17
21. Dorer, F. E., Stewart, J. M., Ryan, J. W. 1978. Studies on the hydrolysis of

bradykinin by angiotensin converting enzyme (kininase II). *Experientia* 34: 1436
22. Dzau, V. J., Colucci, W. S., Williams, G. H., Curfman, G., Meggs, L., Hollenberg, N. K. 1980. Sustained effectiveness of converting enzyme inhibition in patients with severe congestive heart failure. *New Engl. J. Med.* 302(25): 1373–79
23. Erdos, E. G., Johnson, A. R., Boyden, N. T. 1978. Hydrolysis of enkephalin by cultured human endothelial cells and by purified peptidyl dipeptidase. *Biochem. Pharmac.* 27:843–48
24. Erdos, E. G., Yang, H. Y. T. 1970. Kininases. In *Handbuch der Experimentalische Pharmakologie, Bradykinin, Kallidin und Kallikrein*, ed. E. G. Erdos, pp. 289–323. NY: Springer
25. Fanburg, B. L., Glazier, J. B. 1973. Conversion of angiotensin 1 to angiotensin 2 in the isolated perfused dog lung. 1973. *J. Appl. Physiol.* 35:325–31
26. Fei, D. W., Graham, W. F., McDougal, J. G., Scoggins, B. A., Coghlan, J. P. 1980. [des-Asp^1]angiotensin II in the sheep: Blood levels and its effects on plasma renin concentration. *Life Sci.* 27(16):1495–502
27. Ferreira, S. H. 1965. A bradykinin-potentiating factor (BPF) present in the venom of *Bothrops jararaca*. *Br. J. Pharmacol. Chemother.* 24:163–69
28. Ferreira, S. H., Bakhle, Y. S. 1977. Inactivation of bradykinin and related peptides in the lung. In *Metabolic Functions of the Lung*, ed. Y. S. Bakhle, J. R. Vane, 4:33–53. NY Basel: Marcel Dekker
29. Ferreira, S. H., Vane, J. R. 1967. Half-lives of peptides and amines in the circulation. *Nature* 215:1237–40
30. Ferreira, S. H., Vane, J. R. 1967. Prostaglandins: Their disappearance and release into the circulation. *Nature* 216:868–73
31. Fishman, A. P. 1976. Hypoxia on the pulmonary circulation: How and where it acts. *Circ. Res.* 38:221–31
32. Fyhrquist, F., Forslund, T., Tikkanen, I., Gronhagen-Riska, C. 1980. Induction of angiotensin I converting enzyme in rat lung with captopril (SQ 14,225). *Eur. J. Pharmacol.* 67(4)80:473–75
33. Garcia del Rio, C., Smellie, W. S. A., Morton, J. J. 1981. des-Asp-angiotensin I: Its identification in rat blood and confirmation as a substrate for converting enzyme. *Endocrinology* 108(2):406–12
34. Gavras, H., Brunner, H. R., Laragh, J. H., Sealey, J. E., Gavras, I., Vukovich, R. A. 1974. An angiotensin converting enzyme inhibitor to identify and treat vasoconstrictor and volume factors in hypertensive patients. *New Engl. J. Med.* 291:817–21
35. Gavras, H., Brunner, H. R., turini, G. A., Kershaw, G. R., Tifft, C. P., Cuttelod, S., Gavras, I., Vukovich, R. A., McKinstry, D. N. 1978. Antihypertensive effect of the oral angiotensin converting enzyme inhibitor SQ 14,225 in man. *New Engl. J. Med.* 298:991–95
36. Gocke, D. J., Gerten, J., Sherwood, L. M., Laragh, J. H. 1969. Physiological and pathological variations of plasma angiotensin II in man. *Circ. Res.* 24 (Suppl. 1): I–131, I–146
37. Gorenstein, C., Snyder, S. H. 1979. Two distinct enkephalinases: Solubilization, partial purification and separation from angiotensin converting enzyme. *Life Sci.* 25(24–25):2065–70
38. Harris, R. B., Ohlsson, J. T., Wilson, I. B. 1981. Inhibition and affinity chromatography of human serum angiotensin converting enzyme with cysteinyl-proline derivatives. *Arch. Biochem. Biophys.* 206(1):105–12
39. Hoorntje, S. J., Kallenberg, C. G., Weening, J. J., Donker, A. J., The, T. H., Hoedemaeker, P. J. 1980. Immune complex glomerulopathy in patients treated with captopril. *Lancet* 1:1212–14
40. Khairallah, P. A., Page, I. H. 1967. Plasma angiotensinases. *Biochem. Med.* 1:1–8
41. Khairallah, P. A., Page, I. H., Bumpus, F. M., Smeby, R. R. 1962. Angiotensin II. Its metabolic fate. *Science* 138: 523–25
42. Krieger, E. M., Salgado, H. C., Assan, C. J., Greene, L. L. J., Ferreira, S. H. 1971. Potential screening test for detection of overactivity of renin-angiotensin system. *Lancet* 1:269–71
43. Lanzillo, J. J., Polsky-Cynkin, R., Fanburg, B. L. 1980. Large scale purification of angiotensin I converting enzyme from human plasma utilizing an immunoadsorbent affinity gel. *Anal. Biochem.* 103(2):400–7
44. Leuenberger, P. J., Lipset, J. S., Osman, M. M., Cerreta, J. M., Mellins, R. B., Turino, G. M. 1981. Impaired angiotensin conversion and bradykinin clearance in experimental canine pulmonary emphysema. *J. Clin. Invest.* 67(1): 201–9
45. Leuenberger, P. J., Stalcup, S. A., Mellins, R. B., Greenbaum, L. M., Turino, G. M. 1978. Influence of acute hypoxia

on in vivo conversion of angiotensin I to II in dogs. *Proc. Soc. Exp. Biol. Med.* 156:586–89
46. Maynard, J. R., Dreyer, B. E., Pitlick, F. A. 1977. Tissue factor activity of cultured human endothelial and smooth muscle cells and fibroblasts. *Blood* 50:387–96
47. Millar, J. A., Hammat, M. T., Johnston, C. I. 1980. Effect of inhibition of converting enzyme on inactive renin in the circulation of salt-replete and salt-deplete normal subjects. *J. Endocrinol.* 86(2):329–35
48. Morgat, J. L., Hung, L. T., Fromageot, P. 1970. Preparation of highly labelled [^{3}H]-angiotensin II. *Biochim. Biophys. Acta* 207:374–76
49. Morton, J. J., Tree, M., Casals-Stenzel, J. 1980. The effect of captopril on blood pressure and angiotensin I, II, and III in sodium-depleted dogs: Problems associated with the measurement of angiotensin II after inhibition of converting enzyme. *Clin. Sci.* 58(6):445–50
50. Niarchos, A. P., Roberts, A. J., Laragh, J. H. 1979. Effects of the converting enzyme inhibitors (SQ 20,881) on the pulmonary circulation in man. *Am. J. Med.*, 67(5):785–91
51. Ondetti, M. A., Condon, M. E., Reid, J., Sabo, E. F., Cheung, H. S., Cushman, D. W. 1979. Design of potent and specific inhibitors of carboxypeptidase A and B. *Biochemistry* 18:1427–30
52. Ondetti, M. A., Rubin, B., Cushman, D. W. 1977. Design of specific inhibitors of angiotensin converting enzyme: A new class of orally active antihypertensive agents. *Science* 196:441–44
53. Oparil, S., Koerner, T., O'Donoghue, J. K. 1974. Structural requirements for substrates and inhibitors of angiotensin I converting enzyme in vivo and in vitro. *Circ. Res.* 34:19–26
54. Oparil, S., Koerner, T., Tregear, G. W., Barnes, B. A., Haber, E. 1973. Substrate requirements for angiotensin I conversion in vivo and in vitro. *Circ. Res.* 32:415–23
55. Oparil, S., Tregear, G. W., Koerner, T., Barnes, B. A., Haber, E. 1971. Mechanism of pulmonary conversion of angiotensin I to angiotensin II in the dog. *Circ. Res.* 29:628–90
56. Patchett, A. A. et al. 1980. A new class of angiotensin converting enzyme inhibitors. *Nature* 288:280–83
57. Peach, M. J., Chiu, A. T. 1974. Stimulation and inhibition of aldosterone. Biosynthesis in vitro by angiotensin II and analogs. *Circ. Res.* 34 (Suppl. I): I7–I13
58. Roblero, J., Ryan, J. W., Stewart, J. M. 1973. Assay of kinins by their effects on blood pressure. *Res. Commun. Chem. Pathol. Pharmacol.* 6(1):207–12
59. Rowe, G. G., Afonson, S., Castillo, C. A., Lioy, F., Lugo, J. E., Crumpton, C. W. 1963. The systemic and coronary hemodynamic effects of synthetic bradykinin. *Am. Heart J.* 65:656
60. Ryan, J. W. 1974. The fate of angiotensin II. In *Handbook of Experimental Pharmacology*, ed. I. H. Page, F. M. Bumpus, 37:81–110. NY: Springer
61. Ryan, J. W., Chung, A., Ammons, C., Carlton, M. L. 1977. A simple radioassay for angiotensin converting enzyme. *Biochem. J.* 167:501–4
62. Ryan, J. W., Chung, A., Martin, L. C., Ryan, U. S. 1978. New substrates for the radioassay of angiotensin converting enzyme of endothelial cells in culture. *Tissue & Cell* 10:555–62
63. Ryan, J. W., Chung, A., Ryan, U.S. 1980. Angiotensin converting enzyme. I. New strategies for assay. *Environ. Health Persp.* 35:165–70
64. Ryan, J. W., Day, A. R., Ryan, U. S., Chung, A., Marlborough, D. I., Dorer, F. E. 1976. Localization of angiotensin converting enzyme (kininase II). I. Preparation of antibody-heme octapeptide conjugates. *Tissue Cell* 8:111–24
65. Ryan, J. W., Martin, L. C., Chung, A., Pena, G. 1979. Mammalian inhibitors of angiotensin converting enzyme (kininase II). *Adv. Exp. Med. Biol.* 120B: 599–606
66. Ryan, J. W., Niemeyer, R. S., Goodwin, D. W., Smith, U., Stewart, J. M. 1971. Metabolism of (8-L-[^{14}C]-phenylalanine)- angiotensin I in the pulmonary circulation. *Biochem. J.* 125: 921–23
67. Ryan, J. W., Roblero, J., Stewart, J. M. 1969. Inactivation of bradykinin in rat lung. *Pharmacol. Res. Commun.* 1:192 (Abstr.)
68. Ryan, J. W., Roblero, J., Stewart, J. M. 1968. Inactivation of bradykinin in the pulmonary circulation. *Biochem. J.* 110:795–97
69. Ryan, J. W., Roblero, J., Stewart, J. M. 1970. Inactivation of bradykinin in rat lung. *Adv. Exp. Med. Biol.* 8:263–72
70. Ryan, J. W., Ryan, U. S. 1975. Metabolic activities of plasma membrane and caveolae of pulmonary endothelial cells, with a note on pulmonary prostaglandin synthetase. In *Lung Metabolism*, ed. A. F. Junod, R. de Haller, pp. 399–424. NY: Academic

71. Ryan, J. W., Ryan, U. S. 1977. Is the lung a para-endocrine organ? *Am. J. Med.* 63:595–603
72. Ryan, J. W., Ryan, U. S. 1977. Pulmonary endothelial cells. *Fed. Proc.* 36:2683–91
73. Ryan, J. W., Ryan, U. S. 1978. Humoral control of arterial blood pressure: A role for the lung? *Cardiovasc. Med.* 3:531–52
74. Ryan, J. W., Ryan, U. S. 1980. Biochemical and morphological aspects of the actions and inactivation of kinins and angiotensins. In *Enzymatic Release of Vasoactive Peptides*, ed. F. Gross, H. G. Vogel, pp. 259–74. NY: Raven
75. Ryan, J. W., Ryan, U. S. 1981. Processing of angiotensins, kinins, adenine nucleotides and prostaglandins by the lungs. *Z. Atemw. Lungenerkr.* In press
76. Ryan, J. W., Ryan, U. S., Schultz, D. R., Whitaker, C., Chung, A., Dorer, F. E. 1975. Subcellular localization of pulmonary angiotensin converting enzyme (kininase II). *Biochem. J.* 146:497–99
77. Ryan, J. W., Smith, U. 1971. A rapid, simple method for isolating pinocytotic vesicles and plasma membrane of lung. *Biochim. Biophys. Acta* 249:177–80
78. Ryan, J. W., Smith, U. 1971. Metabolism of adenosine-5'-monophosphate during circulation through the lungs. *Trans. Assoc. Am. Physicians* 84:297–306
79. Ryan, J. W., Smith, U. 1973. The metabolism of angiotensin I by endothelial cells. In *Protides of the Biological Fluids*, ed. H. Peeters, 20:379–84. Oxford: Pergamon
80. Ryan, J. W., Smith, U., Niemeyer, R. S. 1972. Angiotensin I: Metabolism by plasma membrane of lung. *Science* 176:64–66
81. Ryan, J. W., Stewart, J. M., Leary, W. P., Ledingham, J. G. 1970. Metabolism of angiotensin I in the pulmonary circulation. *Biochem. J.* 120:221–23
82. Ryan, U. S., Ryan, J. W. 1979. Angiotensin converting enzyme. II. Pulmonary endothelial cells in culture. *Env. Health Persp.* 35:171–80
83. Ryan, U. S., Ryan, J. W., Whitaker, C., Chiu, A. 1976. Localization of angiotensin converting enzyme (kininase II). II. Immunocytochemistry and immunofluorescence. *Tissue & Cell* 8:125–46
84. Ryan, U. S., Schultz, D. R., Del Vecchio, P. J., Ryan, J. W. 1980. Endothelial cells of bovine pulmonary artery lack receptors for C3b and for the Fc portion of IgG. *Science* 208:748–49
85. Semple, P. F., Morton, J. J. 1975. Angiotensin II and its heptapeptide and hexapeptide fragments in arterial and venous blood of man. *Clin. Sci. Mol. Med.* 48:2 (Abstr.)
86. Soffer, R. L. 1976. Angiotensin converting enzyme and the regulation of vasoactive peptides. *Ann. Rev. Biochem.* 45:73–94
87. Stalcup, S. A., Lipset, J. S., Legant, P. M., Leuenberger, P. J., Mellins, R. B. 1979. Inhibition of converting enzyme activity by acute hypoxia in dogs. *J. Appl. Physiol.* 46(2):227–34
88. Stalcup, S. A., Lipset, J. S., Woan, J. M., Leuenberger, P., Mellins, R. B. 1979. Inhibition of angiotensin converting enzyme activity in cultured endothelial cells by hypoxia. *J. Clin. Invest.* 63:966–76
89. Szidon, P., Bairey, N., Oparil, S. 1980. Effect of acute hypoxia on the pulmonary conversion of angiotensin I to angiotensin II in dogs. *Circ. Res.* 46:221–26
90. Tierney, D. F. 1974. Lung metabolism and biochemistry. *Ann. Rev. Physiol.* 36:209–31
91. Tsai, B. S., Peach, M. J. 1977. Angiotensin homologs and analogs as inhibitors of rabbit pulmonary angiotensin converting enzyme. 1977. *J. Biol. Chem.* 252(1):4674–81
92. Tsai, B. S., Peach, M. J., Khosla, M. C., Bumpus, F. M. 1975. Synthesis and evaluation of [des-Asp^1]-angiotensin I as a precursor for [des-Asp^1]-angiotensin II ("Angiotensin III"). *J. Med. Chem.* 18:1180–83
93. Turini, G. A., Brunner, H. R., Gribic, M., Waeber, B., Gavras, H. 1979. Improvement of chronic congestive heart failure by oral Captopril. *Lancet* 1:1213–15
94. Vane, J. R. 1969. The release and fate of vaso-active hormones in the circulation. *Br. J. Pharmacacol.* 35:209–42
95. Vaughan, E. D. Jr., Ackerly, J. A., Lantz, C. H., Glenn, A. W., Peach, M. J. 1977. Formation and circulating level of (Des-Asp^1)-angiotensin (ANG) I in the rat. *Circulation* 56:215 (Abstr.)
96. Yang, H. Y. T., Erdos, E. G., Chiang, T. S. 1968. New enzymatic route for the inactivation of angiotensin. *Nature* 218:1224–26

PULMONARY METABOLISM OF PROSTAGLANDINS AND VASOACTIVE PEPTIDES

Sami I. Said[*]

V.A. Medical Center, and University of Texas Health Science Center, Dallas, Texas 75235

PROSTAGLANDINS AND RELATED LIPIDS

It was early recognized that the lung plays a special role in the formation and degradation of the prostaglandins PGE_2 and $PGF_{2\alpha}$ (1,2). Today many more biologically active compounds are known to be derived from arachidonic acid, the common precursor of PGs and related lipids, and the lung retains a dominant place in the metabolic transformations of these compounds.

Biosynthesis

The generation of arachidonic acid, through the action of phospholipase A_2 and other enzymes on membrane phospholipids, is the first and rate-limiting step in the biosynthesis of PGs and related compounds. In one major biosynthetic pathway, which leads to the formation of prostaglandins, thromboxanes, and prostacyclin, the initial transformation of arachidonic acid (to PG endoperoxides PGG_2 and PGH_2) is catalyzed by a cyclooxygenase (Figure 1a). [For more details on the biosynthesis of these products see (3,4).]

Experiments on lung homogenates, isolated perfused lungs, and lung microsomes demonstrate the ability of mammalian lung to produce prosta-

---

[*]Present address: Department of Medicine, University of Oklahoma Health Sciences Center and V.A. Medical Center, P.O. Box 26307, Oklahoma City, OK 73126

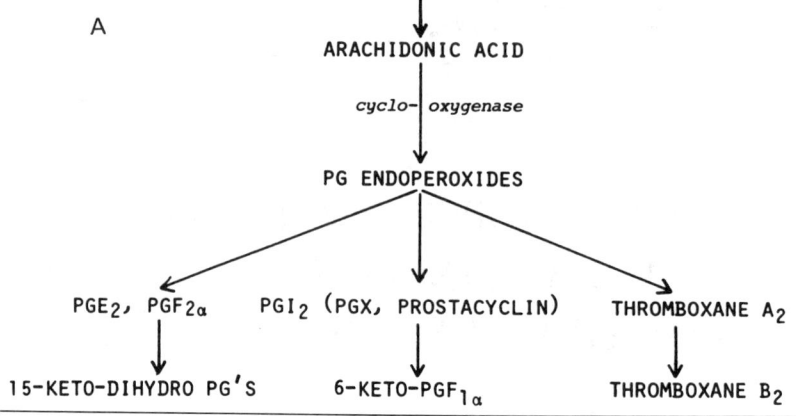

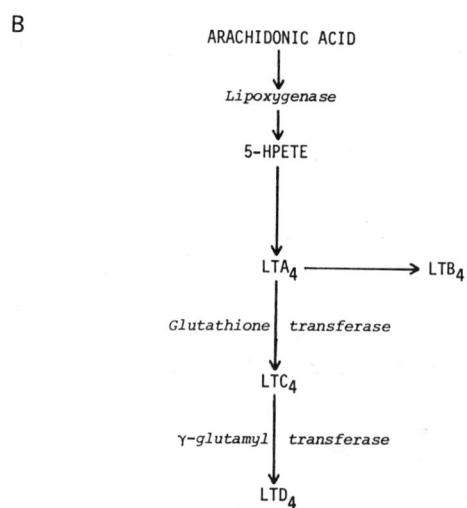

Figure 1 A: Simplified outline of biosynthetic pathways of arachidonic acid initiated by cyclooxygenase-catalyzed reaction. For abbreviations see text. B: Simplified outline of metabolic pathways initiated by lipoxygenase-catalyzed oxygenation. For abbreviations see text.

glandins (PGE_2, $PGF_{2\alpha}$, PGD_2), prostacyclin (PGI_2), and thromboxanes. The relative amounts of these products vary in different species (5–7). Thus, TXB_2, the metabolite of TXA_2, is the dominant cyclooxygenase product in guinea-pig lung, while human and rat lungs form equivalent amounts of TXB_2 and of 6-keto $PGF_{1\alpha}$, the spontaneous hydrolysis product of PGI_2. The biological activities of prostaglandins, thromboxanes, and prostacyclin have recently been reviewed (8,9).

The second major biosynthetic pathway from arachidonic acid is initiated with a lipoxygenase-catalyzed oxygenation at C-5 (Figure 1b). This pathway, also active in the lung, leads to the formation of the unstable 5-HPETE (5-hydroperoxy-6,8,11,14-eicosatetraenoic acid), and subsequently to other compounds, including the newly discovered group of substances known as the leukotrienes (derived from leukocytes and containing a conjugated triene structure) (10). Leukotriene A (LTA_4), an unstable epoxide, is converted by glutathione transferase to leukotriene C (LTC_4 or 5-hydroxy-6-S-glutamylcysteinylglycyl (glutathionyl)-7,9,11,14-icosatetraenoic acid). Through the action of γ-glutamyl transferase (transpeptidase), a widely occurring membrane-bound enzyme, LTC_4 is converted to leukotriene D (LTD_4 or 5-hydroxy-6-S-cysteinylglycyl-7,9,11,14-icosatetraenoic acid). LTD_4 has been identified as the slow-reacting substance of anaphylaxis (SRS-A), and these chemical transformations have been documented in rat basophilic leukemia cells (11), rabbit polymorphonuclear leukocytes (12), cat paws perfused with compound 48/80 (13), sensitized guinea-pig lungs (14), and human lungs (15).

SRS-A is inactivated by arylsulfatase (16). This inactivation has been found to result from the cleavage of the cysteinylglycine peptide bond in LTD_4 (13). SRS-A (17, 18) is believed to be an important mediator of anaphylaxis and bronchial asthma. Its activities include contractions of guinea-pig trachea and ileum that are slow, sustained, and resistant to atropine and mepyramine. The leukotrienes elicit these effects, as well as increased microvascular permeability in guinea-pig skin (19).

Leukotrienes LTC_4 and LTD_4 are at least 1,000 times more potent than histamine in causing contraction of isolated human bronchi (19), and LTD_4 is more potent than LTC_4 on isolated guinea-pig lung strips (20).

Inhibition of Synthesis

The liberation of arachidonic acid can be inhibited by corticosteroids (21), which apparently induce the synthesis of a peptide that inhibits the activation of phospholipase A_2 (22). Aspirin, indomethacin, and other nonsteroidal anti-inflammatory agents (e.g. meclofenamate and ibuprofen) inhibit the cyclooxygenase enzyme, and consequently inhibit the biosynthesis of PGs, prostacyclin, and thromboxanes. Thromboxane synthesis is selectively inhibited by imidazole and its N-substituted derivatives and by pyridine and its derivatives (7). Thromboxane synthesis inhibition is effective in platelet as well as lung enzyme systems, and is associated with increased synthesis of PGE_2 and prostacyclin (7). Inhibition of lung prostacyclin synthetase has not been described, but tranylcypromine and lipid hydroperoxides inhibit this enzyme in aortic endothelium (23). Phenidone (1-phenyl-3-pyrizolidone) and other experimental agents, as well as methylester derivatives of

HETEs, appear to have some lipoxygenase inhibitory activity, but specific and nontoxic inhibitors of this enzyme system are still lacking (24).

Release

Since PGs and related compounds are not normally stored (except in seminal fluid), their release must result from increased synthesis. Stimulated PG synthesis can be triggered by a variety of stimuli that activate phospholipase A_2 or other membrane-bound enzymes responsible for liberating arachidonic acid. These stimuli may be mechanical, chemical, or hormonal, and may be related to pathologic conditions. Mechanical stimuli, such as stretching or stroking of the lung surface, hyperinflation and hyperventilation provoke the release of systemic and pulmonary vasodilator prostaglandins (25) and prostacyclin (26). This release accounts for approximately 40% of the systemic hypotension that complicates hyperventilation in dogs (27). Thrombin and collagen stimulate arachidonic acid release. Bradykinin and angiotensin II also release PGs and prostacyclin from the lung (28). The bronchoconstrictor action of bradykinin in vitro is attributable to a released product of cyclooxygenase activity, as this action is abolished with indomethacin (29).

Among the pathologic states shown to be associated with release of PGs and other products of the cyclooxygenase pathway are anaphylaxis, alloxan-induced pulmonary edema, pulmonary embolism, endotoxin shock (due to *E. coli, Pseudomonas,* or other gram-negative bacteria) (30,31), hemorrhagic shock (32), aspiration of HCl (33), and hyperoxia (34).

In some experimental models of lung injury, including acid aspiration and alloxan pulmonary edema, the release of arachidonic acid metabolites (and of other mediators) is more readily demonstrated in pulmonary lymph than in arterial blood. In these conditions, both the concentrations of these metabolites and the flow of lymph are greatly increased (35).

The sites of PG synthesis in the lung include endothelial cells, which are particularly active in generating prostacyclin (36), alveolar macrophages, fibroblasts (37), and type II pneumonocytes. Platelets are major producers of thromboxane A_2 (38); as mentioned above, polymorphonuclear leukocytes, basophils, and mast cells are active sites of lipoxygenase-catalyzed pathways.

Stimulated Release of Arachidonic Acid as a Factor in Lung Injury

As the above examples illustrate, the stimulated release of arachidonic acid and its biosynthetic products is a common denominator in many forms of lung injury. The possibility that this release contributes to pulmonary injury is suggested by the following observations: (*a*) Pretreatment with imidazole reduced the mortality from endotoxin shock (due to *Salmonella enteritidis*) in rats from 100% to 40% (31); (b) imidazole also improved metabolic and hemodynamic functions in anesthetized cats injected with E. coli endotoxin (32); and (c) methylprednisolone in large doses prevented the increase in pulmonary vascular permeability due to E. coli endotoxin (39).

The mechanisms by which the stimulation of PG synthesis may contribute to lung injury include: (a) the direct action of final biosynthetic products of cyclooxygenase (e. g. thromboxane A_2 or $PGF_{2\alpha}$); (b) the direct action of intermediate, unstable products (e.g. the endoperoxides); (c) the toxic effects of free oxygen radicals generated during the conversion of PGG_2 to PGH_2 (40); and (d) the chemotactic effect of lipoxygenase products, attracting leukocytes that may release their lysosomal enzymes as well as additional free radicals.

Despite the likelihood that many arachidonic acid metabolites may contribute to lung injury, there is some evidence that prostacyclin may serve to modulate (i.e. prevent or reduce) lung injury. Infusions of prostacyclin (100 ng kg^{-1} min^{-1}) begun within 30 min of the injection of E. coli endotoxin and continued for 3 hr markedly improved the morbidity and mortality of lethal endotoxemia in anesthetized dogs (41).

Uptake and Inactivation

Early in the investigation of pulmonary inactivation of prostaglandins, Ferreira & Vane (2) reported that the lung removed 90% of PGE_1 or $PGF_{2\alpha}$ during one circulation. Shortly thereafter, McGiff et al (42) showed that the pulmonary removal of PGs was selective, and that PGA compounds passed through the lung with little loss in activity. This selectivity was first ascribed to a selectivity of the PG-metabolizing enzymes in the lung, 15-hydroxy-prostaglandin dehydrogenase and 13,14-reductase, both intracellular enzymes. More recently, however, the dependence of pulmonary removal of PGs on a carrier-mediated, energy-requiring transport process has been demonstrated (43,44). Further, the key inactivating enzyme, 15-OH-PG dehydrogenase, exhibits no significant substrate specificity (45). The pulmonary handling of PGs, therefore, like that of biogenic amines, requires an uptake process before enzymic inactivation (Figure 2). The substrate specificity for the inactivation of PGs appears to reside with the transport carrier; the structural requirements for this transport have been defined (44,46). Many pharmacological agents that suppress the removal of PGs by the lung probably do so by inhibiting PG transport. These inhibitors include highly ionized organic molecules (e.g. bromcresol green and diphloretin phosphate) and less polar compounds capable of reacting with sulfhydryl groups (e.g. furosemide and ethacrynic acid) (47).

The cellular sites of PG uptake and inactivation remain unknown. Porcine endothelial cell preparations derived from pulmonary artery or aorta

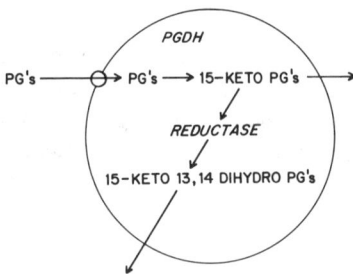

Figure 2 Schema of mechanisms involved in pulmonary removal of PGs from the circulation. Arrow penetrating large circle (cell) at upper left represents carrier-mediated transport. This is followed by chemical inactivation by the enzymes 15-OH PG dehydrogenase (PGDH) and 13,14-reductase [modified from (6)].

failed to degrade either $PGF_{2\alpha}$ or PGA_1 (48), though they formed the previously described PGA-glutathione conjugate (49).

Leukotrienes C_4 and D_4 appear to be inactivated (as assayed on guinea-pig ileum) during a single passage through the pulmonary circulation (50), but the mechanism of this inactivation remains unknown. Because of its instability and its unavailability as a pure compound (8), the pulmonary metabolism of thromboxane A_2 has not been adequately investigated.

The biological activity of prostacyclin is not significantly reduced during passage through the lung (51–53). This conclusion is consistent with biochemical evidence of only slight metabolic degradation of this compound in the lung (54). In contrast to PGE_2, $PGF_{2\alpha}$, and the leukotrienes, therefore, prostacyclin qualifies as a circulating hormone (26,55).

The efficient and selective pulmonary removal of prostaglandins and related lipids has important implications for pulmonary and systemic function. Because they are effectively removed from the circulation, compounds like PGE_2, $PGF_{2\alpha}$, and the leukotrienes are rendered essentially "local hormones," acting mainly at the sites of their synthesis and release. On the other hand, PGA_2 and prostacyclin, which escape pulmonary inactivation, are circulating hormones capable of influencing organs beyond those where they are released. Circulating prostacyclin levels may serve an important physiologic function by maintaining a general vasodilator influence, inhibiting platelet aggregation, and protecting the integrity of pulmonary and systemic endothelium.

Physiologic and Pathologic Influences

The pulmonary handling of prostaglandins is subject to physiologic and pathologic influences. The inactivation of PGE_2 by rabbit lung is enhanced 3-fold during pregnancy and following progesterone treatment (56,57). Exposure to 100% O_2 for 48 hr inhibits the metabolism of $PGF_{2\alpha}$ by guinea-pig lungs in vitro (58), and of PGE_2 by perfused rat lung (59). The inhibition of prostaglandin metabolism by oxygen is possibly related to inactivation of enzymic SH groups. Indirect evidence suggests that 15-hydroxy prostaglandin dehydrogenase contains SH groups essential for its activity (60). The effect of high oxygen pressure on the transport process for prostaglandins has not been evaluated. Other conditions that can suppress the pulmonary inactivation of PGs are exposure to cigarette smoke, prolonged (>3 hr) cardiopulmonary bypass, and treatment with bleomycin, monocrotaline, or paraquat (61).

VASOACTIVE PEPTIDES IN THE LUNG

Normal lung contains or may generate several active peptides (29). Some of these are briefly reviewed here.

Vasoactive intestinal peptide (VIP), a 28-residue neuropeptide occurring in the central and peripheral nervous systems, relaxes airway, pulmonary vascular, gastrointestinal, and uterine smooth muscle, and induces systemic vasodilation, hypotension, and other effects (62). Pulmonary VIP (or a closely related peptide) is principally located in nerves supplying the airways and pulmonary vessels (63) as well as in mast cells (64).

A "spasmogenic lung peptide" distinct from other peptides with similar actions is extractable from normal lung. It dilates systemic vessels but contracts airway, pulmonary vascular, and other smooth muscle (65). Its chemical composition has yet to be determined.

Substance P, another neuropeptide with wide distribution in the body, is localized in the lung in nerves supplying airways and pulmonary vessels (66). It is a systemic vasodilator but contracts tracheal and pulmonary vascular smooth muscle.

A bombesin-like peptide has been demonstrated in endocrine cells of fetal lungs (67) but not yet in lungs of adult animals. Bombesin, a 14-residue peptide from the skin of the frog, *Bombina bombina,* has spasmogenic activity on airway, pulmonary vascular, and other smooth muscle. The mammalian counterpart of the amphibian peptide is probably the 27-residue gastrin-releasing peptide isolated by McDonald et al (68).

Bradykinin may be formed in the lung through the action of activated kallikrein on tissue kininogens. A potent systemic vasodilator and vasodepressor, bradykinin is capable of increasing systemic (including bronchial) vascular permeability. However, its ability to increase pulmonary microvascular permeability has not been established. In combination with hypoxia (69), however, bradykinin can induce pulmonary edema.

Angiotensin II, generated in pulmonary endothelium, constricts pulmonary as well as systemic vessels. Inhibition of the conversion of angiotensin I to angiotensin II has been reported to reduce pulmonary arterial pressure and vascular resistance (70). A role for angiotensin II in the pulmonary vasoconstriction induced by hypoxia was proposed (71) but has not been confirmed.

Almost certainly, other active peptides will be identified in the lung in the months ahead. Because the lung develops from the embryonic foregut, all peptides found in the upper gastrointestinal tract are likely to occur in the lung, too.

Metabolism: Bradykinin and Angiotensin

Bradykinin is effectively inactivated, and angiotensin I is converted to angiotensin II, during a single passage through the lung. Both reactions are catalyzed by the same enzyme, variously termed kininase II, angiotensin I-converting enzyme, or dipeptidyl carboxy-peptidase (E.C. 3.4.15.1). This membrane-bound enzyme, located in pulmonary endothelium (72,73) and elsewhere, breaks peptidyl-dipeptide bonds from the carboxyl terminal end. In addition to bradykinin and angiotensin I, other endogenous substrates of this enzyme are the enkephalins and insulin, at least in vitro (74). Peptides that are not hydrolyzed by this enzyme include those with a penultimate proline residue (e.g. angiotensin II), those lacking a free carboxy-terminal carboxyl group, and those smaller than a tripeptide (75).

Angiotensin-converting enzyme also occurs in the endothelium of peripheral vascular beds (76), and in some epithelial cells, such as the brush border of the renal proximal tubule (77) and intestinal mucosa (78). Although the concentration of the enzyme in pulmonary endothelium is lower than in some other tissues, the large area of the pulmonary vascular bed, to which the entire cardiac output is accessible, gives special physiological significance to the angiotensin conversion in the pulmonary circulation. Acute hypoxia decreases pulmonary kininase II activity in dogs (79).

The past few years have witnessed a lively interest in developing inhibitors of converting enzyme for therapeutic use. Inhibitors are now available that have greater specificity and potency (75,80), and are effective when given orally. These inhibitors promise to be valuable in the treatment of systemic hypertension, and can also reduce pulmonary hypertension (70).

Metabolism of Other Peptides

Substance P is inactivated by cultured endothelial cells from human umbilical cord and lung (73), but inactivation of substance P by whole lung appears to be inefficient. Other peptides that are not appreciably inactivated in the pulmonary circulation include oxytocin, vasopressin, angiotensin II (81), and VIP (82). Some pulmonary degradation of angiotensin II may occur during pulmonary edema, possibly because of the release of intracellular metabolizing angiotensinases from pulmonary endothelial and other cells (83).

Physiological Role

As outlined above, biologically active peptides present in the lung have potent effects on tracheobronchial, pulmonary, and systemic vascular smooth muscle. Although the cellular distribution of these peptides has not been fully defined, at least two (VIP, substance P) are neuropeptides localized mainly in nerve fibers and nerve terminals that innervate the airways and pulmonary vessels. Serving as neurotransmitters or neuromodulators (84), these two peptides, therefore, can influence smooth muscle in airways and blood vessels, as well as bronchial glands and other structures.

Early data suggest a role for VIP as a transmitter of nonadrenergic relaxation of airways (85). Since the nonadrenergic system is the principal inhibitory influence in human airways (86), the neurotransmitter of this system has obvious physiological significance; conversely, its deficiency could have important implications for bronchial asthma and other disease states.

Acknowledgment

Supported in part by NIH Lung Center Award HL-14187 and Research Grant CA-21205. My thanks to Vicky Usry for preparing the manuscript.

Literature Cited

1. Änggard, E., Greén, K., Samuelsson, B. 1965. Synthesis of tritium-labelled prostaglandin E_2 and studies on its metabolism in guinea-pig lung: prostaglandins and related factors. *J. Biol. Chem.* 240:1932–40
2. Ferreira, S. H., Vane, J. R. 1967. Prostaglandins. Their disappearance from and release into the circulation. *Nature* 216:868–73
3. Samuelsson, B., Goldyne, M., Granstrom, E., Hamberg, M., Hammarström, S., Malsten, C. 1978. Prostaglandins and thromboxanes. *Ann. Rev. Biochem.* 47:997–1029
4. Lands, W. E. 1979. The biosynthesis and metabolism of prostaglandins. *Ann. Rev. Physiol.* 41:633–52
5. Al-Ubaidi, F., Bakhle, Y. S. 1980. Differences in biological activation of arachidonic acid in perfused lungs from guinea pig, rat and man. *Eur. J. Pharmacol.* 62:89–96
6. Eling, T. E., Ally, A. I. 1981. Pulmonary biosynthesis and metabolism of prostaglandins and related substances. *Bull. Eur. Physiopathol. Resp.* 17:1–15
7. Tai, H.-H. 1981. Biosynthesis and metabolism of pulmonary prostaglandins, thromboxanes and prostacyclin. *Bull. Eur. Physiopathol. Resp.* 17:1–20
8. Moncada, S., Vane, J. R. 1979. Pharmacology and endogenous roles of prostaglandin endoperoxides, thromboxane A_2, and prostacyclin. *Pharmacol. Rev.* 30:293–331
9. McGiff, J. C. 1981. Prostaglandins, prostacyclin, and thromboxanes. *Ann. Rev. Pharmacol. Toxicol.* 21:479–509
10. Samuelsson, B., Borgeat, P., Hammarström, S., Murphy, R. C. 1979. Introduction of a nomenclature: Leukotrienes. *Prostaglandins* 17:785–87
11. Orning, L., Hammarström, S., Samuelsson, B. 1980. Leukotriene D: A slow reacting substance from rat basophilic leukemia cells. *Proc. Natl. Acad. Sci. USA* 77:2014–17
12. Borgeat, P., Samuelsson, B. 1979. Arachidonic acid metabolism in polymorphonuclear leukocytes: Unstable intermediate in formation of dihydroxy acids. *Proc. Natl. Acad. Sci. USA* 76:3213–17

13. Houglum, J., Pai, J.-K., Atrache, V., Sok, D.-E., Sih, C. J. 1980. Identification of the slow reacting substances from cat paws. *Proc. Natl. Acad. Sci. USA* 77:5688-92
14. Morris, H. R., Taylor, G. W., Piper, P. J., Tippins, J. R. 1980. Structure of slow-reacting substance of anaphylaxis fom guinea-pig lung. *Nature* 285:104-6
15. Bach, M. K., Brashler, J. R., Hammarström, S., Samuelsson, B. 1980. *Biochem. Biophys. Res. Commun.* 93:1121-26
16. Orange, R. P., Murphy, R. C., Austen, K. F. 1974. Inactivation of slow reacting substance of anaphylaxis (SRS-A) by arylsulfatases. *J. Immunol.* 113:316-22
17. Feldberg, W., Kellaway, C. H. 1938. Liberation of histamine and formation of lysocithin-like substances by cobra venom. *J. Physiol.* 94:187-226
18. Kellaway, C. H., Trethewie, E. R. 1940. The liberation of a slow-reacting smooth muscle–stimulating substance in anaphylaxis. *J. Physiol.* 30:121-45
19. Dahlén, S.-E., Hedqvist, P., Hammarström, S., Samuelsson, B. 1980. Leukotrienes are potent constrictors of human bronchi. *Nature* 288:484-86
20. Lewis, R. A., Austen, K. F., Drazen, J. M., Clark, D. A., Marfat, A., Corey, E. J. 1980. Slow reacting substances of anaphylaxis: Identification of leukotrienes C-1 and D from human and rat sources. *Proc. Natl. Acad. Sci. USA* 77:3710-14
21. Gryglewski, R. J., Panczenko, B., Korbut, R., Grodzinska, L., Ocetkiewics, A. 1975. Corticosteroids inhibit prostaglandin release from perfused mesenteric blood vessels of rabbit and from perfused lungs of sensitized guinea pig. *Prostaglandins* 10:343-55
22. Flower, R. J., Blackwell, G. J. 1979. Anti-inflammatory steroids induce biosynthesis of a phospholipase A_2 inhibitor which prevents prostaglandin generation. *Nature* 278:456-59
23. Salmon, J. A., Smith, D. R., Flower, R. J., Moncada, S., Vane, J. R. 1978. Further studies on the enzymatic conversion of prostaglandin endoperoxide into prostacyclin by porcine aorta microsomes. *Biochim. Biophys. Acta* 523:250-62
24. Goetzl, E. J. 1980. Mediators of immediate hypersensitivity derived from arachidonic acid. *N. Engl. J. Med.* 303:822-25
25. Said, S. I. 1977. Release of biologically active materials from the lung. Release induced by physical and chemical stimuli. In *Metabolic Functions of the Lung*, ed. Y. S. Bakhle, J. R. Vane, pp. 297-320. NY: Marcel Dekker
26. Gryglewski, R. J. 1979. Prostacyclin as a circulatory hormone. *Biochem. Pharmacol.* 28:3161-66
27. Said, S. I. 1973. The lung in relation to vasoactive hormones. *Fed. Proc.* 32:1972-76
28. Gryglewski, R. J. 1979. Prostacyclin as a circulating hormone. *Biochem. Pharmacol.* 28:3161-66
29. Said, S. I., Mutt, V., Erdös, E. G. 1980. The lung in relation to vasoactive polypeptides. In *Ciba Found. Symp. 78: Metabolic Activities of the Lung.* Amsterdam; Elsevier, pp. 217-37
30. Anderson, F. L., Jubiz, W., Tsagaris, T. J., Kuida, M. 1975. Endotoxin-induced prostaglandin E and F release in dogs. *Am. J. Physiol.* 223:410-14
31. Cook, J. A., Wise, W. C., Halushka, P. V. 1980. Elevated thromboxane levels in the rat during endotoxic shock. *J. Clin. Invest.* 65:227-30
32. Matsuzaki, Y., Tai, H.-H., Sakio, H., Said, S. I. 1979. Release of thromboxane A_2 in hemorrhagic shock. Presented at the International Prostaglandin Conference, May 30-31, 1979, Washington DC.
33. Aoyagi, M., Hamasaki, Y., Mojarad, M., Michael, J., Said, S. I. 1980. Instillation of HCl in airways induces pulmonary release of thromboxane, prostacyclin and prostaglandin $F_{2\alpha}$. *Fed. Proc.* 39:279
34. Crutchley, D. J., Boyd, J. A., Eling, T. E. 1980. Enhanced thromboxane B_2 release from challenged guinea pig lung after oxygen exposure. *Am. Rev. Resp. Dis.* 121:695-99
35. Matsuzaki, Y., Oyamada, M., Hara, N., Tai, H.-H., Said, S. I. 1980. Increased thromboxane generation by the lung during alloxan-induced pulmonary edema. *Clin. Res.* 28:530A
36. Weksler, B. B., Macus, A. J., Jaffe, E. A. 1977. Synthesis of prostaglandin I_2 (prostacyclin) by cultured human and bovine endothelial cells. *Proc. Natl. Acad. Sci. USA* 74:3922-26
37. Hammarström, S. 1977. Prostaglandin production by normal and transformed 3T3 fibroblasts in cell culture. *Eur. J. Biochem.* 74:7-12
38. Needleman, P., Moncada, S., Bunting, S., Vane, J. R., Hamberg, M., Samuelsson, B. 1976. Identification of an enzyme in platelet microsomes which generates thromboxane A_2 from prosta-

glandin endoperoxides. *Nature* 261: 558–60
39. Brigham, K. K., Bowers, R. E., McKeen, C. R. 1981. Methylprednisolone prevention of increased lung vascular permeability following endotoxemia in sheep. *J. Clin. Invest.* 67:1103–10
40. Kuehl, F. A., Egan, R. W. 1980. Prostaglandins, arachidonic acid, and inflammation. *Science* 210:978–84
41. Krausz, M. M., Utsunomiya, T., Feuerstein, G., Wolfe, J. H. N., Shepro, D., Hechtman, H. B. 1981. Prostacyclin reversal of lethal endotoxemia in dogs. *J. Clin. Invest.* 67:1118–25
42. McGiff, J. C., Terragno, N. A., Strand, J. C., Lee, J. B., Lonigro, A. J., Ng, K. K. F. 1969. Selective passage of prostaglandins across the lung. *Nature* 216: 762–66
43. Bito, L. Z., Baroody, R. A., Reitz, M. E. 1977. Dependence of pulmonary prostaglandin metabolism on carrier-mediated transport processes. *Am. J. Physiol.* 232:E382–87
44. Anderson, M., Eling, T. 1976. Prostglandin removal and metabolism by insulated perfused rat lung. *Prostaglandins* 11:645–77
45. Nakano, J., Anggard, E., Samuelsson, B. 1969. 15-Hydroxyprostanoate dehydrogenase. Prostaglandins as substrates and inhibitors. *Eur. J. Biochem.* 11:386–89
46. Hawkins, H., Smith, J., Nicolaou, K., Eling, T. 1978. Studies on the mechanism involved in the fate of prostacyclin (PGI_2) and 6-keto-$PGF_{1\alpha}$ in the pulmonary circulation. *Prostaglandins* 16: 871–84
47. Bakhle, Y. S. 1979. Action of prostaglandin dehydrogenase inhibitors on prostaglandin uptake in rat isolated lung. *Br. J. Pharmacol.* 65:635–39
48. Ody, C., Dieterle, Y., Wand, I., Stalder, H., Junod, A. F. 1979. PGA_1 and PGF_2 metabolism by pig pulmonary endothelial, smooth muscle, and fibroblasts. *J. Appl. Physiol.* 46:211–16
49. Gross, K. B., Gillis, C. N. 1976. The formation of prostaglandin A_1-glutathione adduct in the lungs. *Biochim. Biophys. Acta* 450:266–68
50. Piper, P. J., Tippins, J. R., Samhoun, M. N., Morris, H. R., Taylor, G. W., Jones, C. M. 1981. SRS-A and its formation by the lung. *Bull. Euro. Physiopathol. Resp. (Clin. Resp. Physiol.)* 17:571–83
51. Armstrong, J. M., Lattimer, N., Moncada, S., Vane, J. R. 1978. Comparison of the vasodepressor effects of prostacyclin and 6-oxo-PGF_1 with those of PGE_2 in rats and rabbits. *Br. J. Pharmacol.* 62:125–30
52. Waldman, H. M., Alter, I., Kot, P. A., Rose, J. C., Ramwell, P. W. 1978. Effect of lung transit on systemic depressor responses to arachidonic acid and prostacyclin in dogs. *J. Pharmacol. Exp. Ther.* 204:289–93
53. Dusting, G. J., Moncada, S., Vane, J. R. 1978. Recirculation of prostacyclin (PGI_2) in the dog. *Br. J. Pharmacol.* 64:315–20
54. Wong, P. Y.-K., McGiff, J. C., Sun, F. F., Malik, K. U. 1978. Pulmonary metabolism of prostacyclin (PGI_2) in the rabbit. *Biochem. Biophys. Res. Commun.* 83:731–38
55. Moncada, S., Korbut, R., Bunting, S., Vane, J. R. 1978. Prostacyclin is a circulating hormone. *Nature* 273:767–68
56. Bedwani, J. R., Marley, P. B. 1975. Enhanced inactivation of prostaglandin E_2 by the rabbit lung during pregnancy or progesterone treatment. *Br. J. Pharmacol.* 53:547–54
57. Boura, A. L. A., Murphy, R. D. 1978. Some factors affecting inactivation of prostaglandin E_2 (PGE_2) by the rat isolated perfused lung. *Br. J. Pharmacol.* 62:411P
58. Chaudhari, A., Sivarajah, K., Warnock, R., Eling, T. E., Anderson, M. W. 1979. Inhibition of pulmonary prostaglandin metabolism of exposure of animals to oxygen or nitrogen dioxide. *Biochem. J.* 184:51–57
59. Klein, L. S., Fisher, A. B., Soltoff, S., Coburn, R. F. 1978. Effect of O_2 exposure on pulmonary metabolism of prostaglandin E_2. *Am. Rev. Resp. Dis.* 118:622–25
60. Hamsen, H. S. 1976. 15-hydroxyprostaglandin dehydrogenase—a review. *Prostaglandins* 12:647–79
61. Gillis, C. N., Catravas, J. D. 1981. Clearance measurements as a means of detecting lung microvascular injury. *Ann. NY Acad. Sci.* In press
62. Said, S. I. 1980. Vasoactive intestinal peptide (VIP): isolation, distribution, biological actions, structure-function relations, and possible functions. In *Gastrointestinal Hormones*, ed. G. B. Jerzy Glass, pp. 245–73. NY: Raven
63. Dey, R. D., Shannon, W. A. Jr., Said, S. I. 1982. Localization of VIP-immunoreactive nerves in airways and pulmonary vessels of dogs, cats, and human subjects. *Cell & Tiss. Res.* In press
64. Cutz, E., Chan, W., Track, N. S., Goth, A., Said, S. I. 1978. Release of vasoac-

tive intestinal polypeptide in mast cells by histamine liberators. *Nature* 275: 661–62
65. Said, S. I., Mutt, V. 1977. Relationship of spasmogenic and smooth muscle relaxant peptide from normal lung to other vasoactive compounds. *Nature* 265:84–86
66. Dey, R. D., Said, S. I. 1981. Localization of substance P-like immunofluorescence in nerves within dog airways and pulmonary vessels. *Fed. Proc.* 40:595
67. Wharton, J., Polak, J. M., Bloom, S. R., Ghatei, M. A., Solcia, E., Brown, M. R., Pearse, A. G. E. 1978. Bombesin-like immunoreactivity in the lung. *Nature* 273:769–70
68. McDonald, T. J., Jornvall, H., Nilsson, G., Vagne, M., Ghatei, M., Bloom, S. R., Mutt, V. 1979. Characterization of a gastrin releasing peptide from porcine non-antral gastric tissue. *Biochem. Biophys. Res. Commun.* 90:227–33
69. O'Brodovich, H. M., Stalcup, S. A., Pang, L. M., Lipset, J. S., Mellins, R. B. 1981. Bradykinin production and increased pulmonary endothelial permeability during acute respiratory failure in unanesthetized sheep. *J. Clin. Invest.* 67:514–22
70. Niarchos, A. P., Roberts, A. J., Laragh, J. H. 1979. Effects of the converting enzyme inhibitor (SQ 20881) on the pulmonary circulation in man. *Am. J. Med.* 67:785–91
71. Berkov, S. 1974. Hypoxic pulmonary vasoconstriction in the rat; the necessary role of angiotensin. *Circ. Res.* 35:256
72. Ryan, U. S., Ryan, J. W., Whitaker, C., Chiu, A. 1976. Localization of angiotensin converting enzyme (kininase II). II. Immunocytochemistry and immunofluorescence. *Tiss. Cell* 8:125–45
73. Johnson, A. R., Erdös, E. G. 1977. Metabolism of vasoactive peptides by human endothelial cells in culture. Angiotensin I converting enzyme (kininase II) and angiotensinase. *J. Clin. Invest.* 59:684–95
74. Erdös, E. G., Johnson, A. R., Boyden, N. T. 1978. Hydrolysis of enkephalin by cultured human endothelial cells and by purified peptidyl dipeptidase. *Biochem. Pharmacol.* 27:843–48
75. Cushman, D. W., Cheung, H. S., Sabo, E. F., Rubin, B., Ondetti, M. A. 1979. Development of specific inhibitors of angiotensin I converting enzyme (kininase II). *Fed. Proc.* 38:2778–82
76. Caldwell, P. R. B., Seegal, B. C., Hsu, K. C., Das, M., Soffer, R. L. 1976. Angiotensin-converting enzyme: vascular endothelial localization. *Science* 191: 1050–51
77. Ward, P. E., Schultz, W., Reynolds, R. C., Erdös, E. G. 1977. Metabolism of kinins and angiotensins in the isolated glomerulus and brush border of rat kidney. *Lab Invest.* 36:559–606
78. Ward, P. E., Sheridan, M. A., Hammon, K. J., Erdös, E. G. 1980. Angiotensin I converting enzyme (kininase II) of the brush border of human and swine intestine. *Biochem. Pharmacol.* 29: 1525–29
79. Stalcup, S. A., Lipset, J. S., Woan, J.-M., Leuenberger, P., Mellins, R. B. 1979. Inhibition of angiotensin converting enzyme activity in cultured endothelial cells by hypoxia. *J. Clin. Invest.* 63:966–76
80. Patchett, A. A., Harris, E., Tristram, E. W., Wyvratt, M. J., Wu, M. T., Taub, D., Peterson, E. R., Ikeler, T. J., ten Broeke, J., Payne, L. G., Ondeyka, D. L., Thorsett, E. D., Greenlee, W. J., Lohr, N. S., Hoffsommer, R. D., Joshua, H., Ruyle, W. V., Rothrock, J. W., Aster, S. D., Maycock, A. L., Robinson, F. M., Hirschmann, R., Sweet, C. S., Ulm, E. H., Gross, D. M., Vassil, T. C., Stone, C. A. 1980. A new class of angiotensin-converting enzyme inhibitors. *Nature* 288:280–83
81. Bakhle, Y. S., Vane, J. R. 1974. Pharmacokinetic function of the pulmonary circulation. *Physiol. Rev.* 54:1007–45
82. Kitamura, S., Yoshida, T., Said, S. I. 1975. Vasoactive intestinal polypeptide: inactivation in liver and potentiation in lung of anesthetized dogs. *Proc. Soc. Exp. Biol. Med.* 148:25–29
83. Kumamoto, K., Stewart, T. A., Johnson, A. R., Erdös, E. G. 1981. Prolylcarboxypeptidase (angiotensinase C) in human lung and cultured cells. *J. Clin. Invest.* 67:210–15
84. Said, S. I. 1980. Peptides common to the nervous system and the gastrointestinal tract. In *Frontiers in Neuroendocrinology*, ed. L. Martini, W. F. Ganong, 6:293–311. NY: Raven
85. Matsuzaki, Y., Hamasaki, Y., Said, S. I. 1980. Vasoactive intestinal peptide: a possible transmitter of non-adrenergic relaxation of guinea pig airways. *Science* 210:1252–53
86. Richardson, J., Beland, J. 1976. Nonadrenergic inhibitory nervous system in human airways. *J. Appl. Physiol.* 41: 764–71

THE FATE OF CIRCULATING AMINES WITHIN THE PULMONARY CIRCULATION

C. N. Gillis and B. R. Pitt

Departments of Pharmacology and Anesthesiology, Yale University School of Medicine, New Haven, Connecticut 06510

In 1925 it was reported (70) that vasoconstriction prevented adequate perfusion of an isolated kidney unless the lung was included in the extracorporeal circuit. Almost 30 years later, Gaddum et al (25) identified 5-hydroxytryptamine as the likely vasoconstrictor substance inactivated during its passage through the pulmonary circulation. Since then, there has been extensive investigation of both the mechanisms and possible physiological significance of this and other nonrespiratory metabolic functions of lung. Interest has focused especially on the disposition of prostaglandins, peptides, amines, steroids, and xenobiotic substances within the pulmonary circulation (4, 23, 35, 46). Discussion of many of these functions appears in other chapters of this volume; accordingly, we deal mainly with the fate of circulating biogenic amines within the pulmonary circulation.

DISPOSITION OF AMINES WITHIN THE PULMONARY CIRCULATION

A number of recent reviews have summarized much of the pertinent information regarding the handling of biogenic amines by the lung (4, 29, 47). It is convenient to categorize knowledge of pulmonary amine disposition according to whether the latter is limited, predominantly, by transport from the vascular space or by intrapulmonary metabolism.

Limited by Transport

The neurotransmitters, norepinephrine (NE) and 5-hydroxytryptamine (5-HT), are the only endogenous amines known to be affected, in vivo, by the lung of several animal species (35, 46) and humans (1, 32). During a single pass through the pulmonary circulation, both compounds are rapidly extracted and extensively degraded by monoamine oxidase (MAO) and, in the case of NE, also by catechol-O-methyl transferase, to their corresponding deaminated and/or O-methylated products (29, 35). Inhibition of intrapulmonary metabolism by pargyline or tropolone preserves amine transported from the vascular space, but does not change the rate of transport (29, 35). In these cases, therefore, amine transport, rather than metabolism within lung, is the rate-limiting step in overall removal. The primary morphologic site for removal of both NE and 5-HT has been shown by fluorescence histochemistry (44) and autoradiography (41, 73) to be endothelial cells of the microvasculature. Pulmonary hemodynamic evidence (outflow occlusion technique) concerning the longitudinal distribution of 5-HT vasoconstrictor activity also indicates that much of the 5-HT removal probably takes place in the capillary bed (61). Removal of 5-HT and NE is saturable, temperature sensitive, sodium-dependent, ouabain-sensitive, and altered by inhibitors of oxidative metabolism, suggesting a carrier-mediated process (43–46, 55, 71). Pharmacologically, NE (and 5-HT) disposition possesses characteristics of adrenergic neuronal uptake and thus shows high affinity and sensitivity to inhibition by cocaine, imipramine, or hypothermia (29, 43, 44). However, prominent features of an extraneuronal uptake process are also evident in the *absence* of both stereospecificity (44) and intrapulmonary storage of unchanged amine (35, 47), the absence of effect of 6-hydroxydopamine pretreatment (43), and the fact that NE uptake is inhibited by normetanephrine (44). A similar pharmacologic profile has been shown for 5-HT and NE uptake by isolated endothelium of pig pulmonary artery (48), further strengthening the association of uptake and metabolism of 5-HT and NE with endothelial cells.

Limited by Intrapulmonary Metabolism

The capacity of lung to remove and inactivate other compounds structurally related to 5-HT and NE has received relatively little attention. However, it is known that phenylethylamine (5, 63), mescaline (64), tyramine, and octopamine (36) are removed on passage through the pulmonary vasculature of isolated perfused lungs. In these cases, inactivation is dependent solely on the activity of functionally distinct forms of intrapulmonary MAO (5, 63). Kinetic transport parameters of mescaline and phenylethylamine removal in perfused lungs resemble Michaelis-Menten kinetics of enzymatic degradation of these amines in homogenized lung tissue (36). Thus it is appropriate to study effects of intrapulmonary MAO on phenylethylamine and mescaline in vivo. This may be contrasted with the difficulty in modeling pulmonary 5-HT transport parameters from kinetic data of MAO activity in vitro (76) and reinforces the observation that 5-HT removal is transport-limited and therefore dependent on structural integrity of the lung.

Amines Unaffected by Intrapulmonary Passage

Several amines similar in structure to 5-HT and NE (and similar in substrate specifity for intrapulmonary forms of MAO) are neither taken up nor metabolized within the pulmonary circulation. Thus both epinephrine (14) and dopamine (15) escape degradation during their intrapulmonary transit in vivo. In the perfused lung however, dopamine is deaminated (36), suggesting the need for caution in extrapolation of such data to the lung in vivo. Both amines are substrates for MAO and both have been shown to be profoundly affected by their passage through extrapulmonary vascular beds (66). Although histamine is removed from liver and renal circulations, there is no significant pulmonary removal (13) despite the presence of considerable imidazole-N-methyl transferase and diamine oxidase activity in the lung. However, histamine is extensively metabolized by minced preparations of animal (6) and (49) lung, suggesting that lack of histamine metabolism in the intact lung is due to a selective barrier for this amine within the pulmonary circulation. Other amines that are unaffected in their pulmonary transit include isoproterenol (13) and imidazole (52). Based on these observations, it is apparent that (*a*) pulmonary amine uptake is a highly selective process; and (*b*) estimates of amine metabolism made with homogenized lung tissue preparations may be misleading in predicting pulmonary amine removal in vivo.

Pulmonary Amine Capacitance Function

The lung accumulates a number of exogenous lipophilic basic amines with a variety of pharmacological effects. These drugs include propranolol (26), lidocaine (7), and imipramine, amphetamine, methadone, and chlorcyclizine (2, 52). Many of these substances leave the pulmonary vascular space by simple diffusion and then are slowly released in unchanged form. Thus the lung may function as a "capacitor" for these drugs. Geddes et al (26) recently reported that 75% of injected propranolol was taken up by lungs of conscious patients. However, in patients receiving the drug orally prior to study, uptake was only 33%, suggesting a saturable intrapulmonary compartment for propranolol storage.

PHYSIOLOGIC SIGNIFICANCE OF PULMONARY AMINE CLEARANCE

The magnitude of biogenic amine clearance by lung ranges from 80–95% for 5-HT to 25–50% for NE. However, in considering the possible physiological significance of these processes, it is important to consider the amount *surviving* transpulmonary passage, rather than solely the magnitude of pulmonary clearance per se because it is amine surviving and thus reaching the left atrium that can exert effects on the systemic (including the bronchial and coronary) circulation. Furthermore, it is evident that, especially for amines extensively cleared (e.g. 5-HT), small changes in pulmonary inactivation can effect dramatic quantitative changes in left atrial and hence systemic arterial blood concentrations. This is illustrated in Figure 1. It can be seen that inhibition of normal 5-HT clearance (assumed to be 95%) by only 5% results in doubling of amine surviving transpulmonary transit; in contrast, to achieve a similar doubling of NE survival (assuming a normal clearance of 50%) would require over 98% inhibition of clearance.

Extrapulmonary Significance

Although much is known about the mechanisms of amine inactivation in the lung, the physiologic role of these functions remains undefined. It was originally suggested (4) that pulmonary degradation of a vasoactive substance was a common feature of hormones with essentially "local" action, whereas circulating hormones were unaffected by the lung. In this frame-

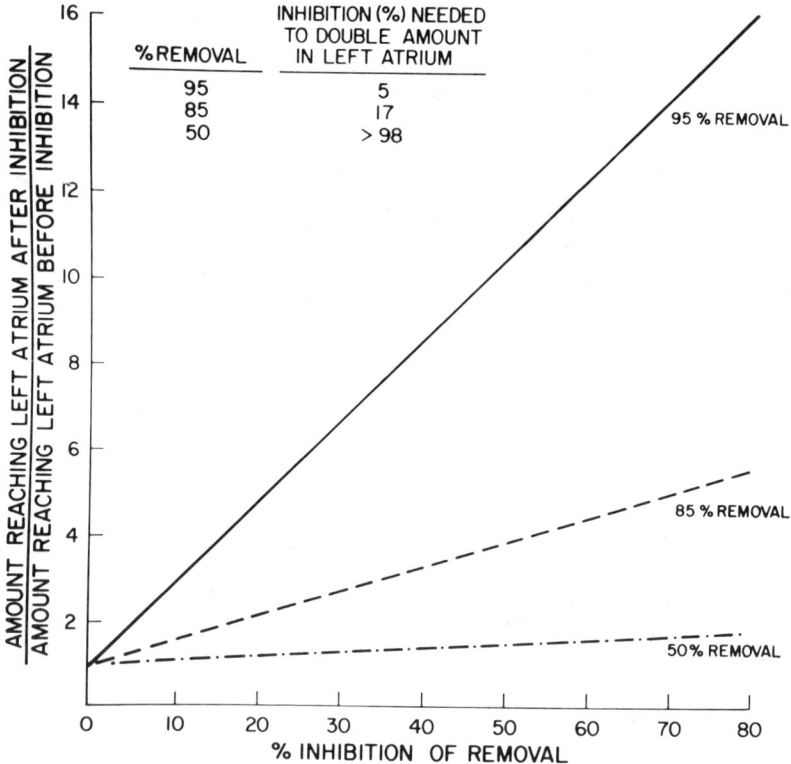

Figure 1 Predicted effect of altered pulmonary removal on systemic arterial concentration of circulating amine.

work, the lung can be viewed as a "metabolic filter" that protects the arterial circulation from effects of locally acting substances such as 5-HT and NE, while leaving unaffected circulating hormones such as epinephrine, dopamine, and histamine. The notion of the lung as a biochemical filtration system is in part a function of its anatomical location (interposed between the right and left heart, thereby receiving the entire cardiac output) and its capacious capillary endothelial cell surface area (approximately 70mm^2 in humans). Although the prototype of metabolizing organs, the liver, has considerably greater specific MAO activity than the lung, it has been shown by mathematical modeling and experimental observations that 5-HT clearance by lung may be quantitatively more important than that of the liver with respect to the disposition of circulating 5-HT (76). Aside from these considerations, little evidence presently links pulmonary amine clearance with extrapulmonary physiology and cardiovascular homeostasis. However, it has been reported that 5-HT clearance (both uptake and metabolism) in rat lungs was increased by changes in the levels of endogenous or exogenous ovarian steroids (3). Thus 5-HT metabolism by lung may be under physiologic regulation. We recently reported that the development of specific pulmonary monoamine oxidase subtypes is influenced by age in rabbits from 28 days of gestation to 28 days postnatal (27). Thus pulmonary amine metabolism has been shown to be under some extrapulmonary regulation.

A potential extrapulmonary role of pulmonary 5-HT uptake may involve prevention of arterial thrombogenesis and subsequent hemostasis secondary to platelet aggregation. Vascular injury often results in disruption of endothelium and exposure of subendothelial cell tissue. Platelet adherence then ensues and subsequent platelet aggregation occurs. This aggregation is mediated by a number of substances, including 5-HT, some of which are derived from platelet exocytosis stimulated by adherence. Since the concentration of 5-HT directly influences platelet aggregation (21), pulmonary uptake of the amine may modulate thrombogenesis and arterial hemostasis. It has been reported that 5-HT uptake is an ubiquitous process in endothelial cells (67). A saturable uptake process for 5-HT has been described for endothelial cells of the aorta (48, 68), epididymal fat pad (62), and brain (40). Therefore, several authors (30, 35, 47) have questioned whether the striking pulmonary metabolic functions of the lung circulation reflect not unique properties of pulmonary endothelium but rather the large number of such cells in the lung, and especially the enormous surface area presented to circulating blood.

Intrapulmonary Significance

Since 5-HT and NE are well known pulmonary vasoconstrictors, intrapulmonary removal of these amines may be involved in the neurohumoral control of pulmonary blood flow. However, the site of 5-HT and NE induced changes in pulmonary vascular resistance is the arterial side of the lung (37), and the site of amine uptake is most likely capillary endothelium (41, 51, 61, 73). Consequently, during moderate and brief 5-HT infusions, it appears unlikely that 5-HT metabolism is involved with regulation of pulmonary hemodynamics. However, with high concentrations of 5-HT that lead to pulmonary venoconstriction in intact dogs (42) and sheep (16), removal may be important since pulmonary venous resistance has more effect on capillary pressure than arterial resistance (61). Therefore, 5-HT removal may be an important physiologic process regulating pulmonary venomotor activity during periods of elevated 5-HT delivery to lung or increased intrapulmonary release of 5-HT.

Pulmonary amine clearance may affect lung fluid balance since 5-HT and NE can influence transcapillary water and solute exchange. Serotonin has been reported to increase lung lymph flow in sheep (16) and electron-microscopic evidence (50) indicates that 5-HT increases inter-endothelial cell gaps. Sympathetic nerves appear to innervate vessels as small as 30 μm in the lung (22), and activation of the adrenergic nervous system by experimental intracranial hypertension (74) or stellate ganglion stimulation (38) leads to increased lymph and protein flow in lungs of sheep. The latter effects were inhibited by the alpha-adrenergic antagonist, phentolamine, suggesting direct neuronal (possibly noradrenergic) influence on exchange vessels. Therefore, pulmonary endothelial cell uptake and metabolism of amines may be related to neurohumoral regulation of transcapillary fluid and solute movement. A similar possibility may exist in brain, in which the microcirculation is innervated by noradrenergic (53, 59) and serotonergic (58) neurons; neuronal control of capillary permeability has been suggested (53, 57). Acceptance of these proposals requires demonstration that inhibition of 5-HT or NE removal alters lung fluid balance in the expected manner. Such experiments remain to be done.

In addition to the proposal that pulmonary amine clearance is involved in regulating intrapulmonary downstream vascular resistance and fluid exchange, it is conceivable that interactions between pulmonary endothelial cells and circulating amines are part of an acceptor (or perhaps receptor) coupling to a pulmonary physiologic function. Some support for this notion arises from data indicating the presence of catecholamine receptors in cultured intimal endothelial cells (17). Radioligand binding studies in isolated cerebral microvessels suggest the presence of receptors as defined by kinetic analysis (54). Perhaps endothelial cell receptor binding of circulating amines represents the initial step in a series of reactions (perhaps mediated by cyclic nucleotides) that results in promotion of as-yet-unknown endothelial cell functions or intrapulmonary extra-endothelial cell coupling (i.e. to vascular smooth muscle, mast cells, neuroepithelial cell bodies, etc).

PATHOPHYSIOLOGY OF PULMONARY AMINE CLEARANCE

In recent years, attention has turned to efforts to describe pulmonary amine clearance during varied pathologic states and environmental conditions. In one sense, these efforts underscore an attempt to further our understanding of the physiologic significance of nonrespiratory lung function, since, by studying these systems in abnormal circumstances, we may uncover clues concerning their normal function. In addition, it is hoped that further study of pulmonary metabolism will provide new insights into the etiology of pulmonary dysfunction and clinical correlates such as adult respiratory distress syndrome (30, 32).

A number of drugs and conditions associated with pulmonary endothelial cell damage and altered amine removal are summarized in Table 1. Pretreatment of rats with monocrotaline, a pyrrolizidine alkaloid that causes endothelial cell swelling and vacuolization, is associated with depressed removal of both 5-HT and NE in isolated perfused rat lungs (34). Decreased removal of 5-HT by isolated perfused lungs of rats pretreated with alpha napthylthiourea preceded the development of pulmonary edema and was reversible (11), demonstrating that 5-HT removal may be an early and a sensitive biochemical marker of endothelial cell damage, as previously suggested (30, 32). Pretreatment of rabbits with subcutaneous injections of bleomycin, an antineoplastic agent whose clinical use is associated with pulmonary fibrosis, led to a decrease in pulmonary 5-HT and NE removal (19). Depression of amine clearance *preceded* biochemical (hydroxyproline content) evidence of lung damage, again supporting the contention that endothelial cell metabolic function may be a useful early index of imminent lung pathology. Noteworthy is that these studies were performed in vivo. Previous work has shown that indicator dilution measurement of pulmonary amine removal is applicable to intact animals (18) and man (32). Removal, estimated by indicator dilution, is drug-sensitive (18) and saturable in vivo (J. D. Catravas, N. Gillis, unpublished) and in perfused dog lung (60). Removal of *trace* concentrations of amine appears to be independent of pulmonary blood flow (56, 60) and surface area (10, 24) over a wide range of changes in these parameters and therefore may reflect only uptake kinetics. Agents that produce endothelial cell damage via oxygen-free radical production also affect pulmonary metabolic functions. Exposure to high partial pressures of oxygen depresses pulmonary removal of 5-HT (9) and NE (8) in isolated perfused rat lungs. Imipramine uptake was unaffected, suggesting selective damage to pulmonary endothelial cells (8). Perfusion of isolated lungs with a mixture of xanthine oxidase and hypoxanthine (72), or xanthine oxidase alone (R. Cook, R. E. Howell, C. N. Gillis, unpublished observations) depresses 5-HT clearance but does not affect the disposition of phenylethylamine, suggesting selective damage to amine transport mechanisms. Two recent reports have increased our understanding of pulmonary metabolism and endothelial cell damage associated with hyperoxia. First, exposure of isolated pulmonary endothelial cells to high partial pressures of oxygen decreased the ability of these cells to remove 5-HT from their medium (12). Second, preliminary evidence from our laboratory indicates that after as little as 24 hr exposure to partial pressures of oxygen in excess of 600 mm Hg, conscious rabbits with indwelling cannulae show depressed 5-HT clearance as assessed by multiple indicator dilution techniques (20).

Other entities associated with altered pulmonary amine clearance are summarized in section C of Table 1. In primary or secondary pulmonary hypertension in adults (69) or children (28), there is no transpulmonary gradient of endogenous norepinephrine (measured radioenzymatically). We previously reported that in adult patients undergoing cardiac surgery, clearance of radiolabelled NE and 5-HT was greater in the group with pulmo-

Table 1 Pulmonary amine removal by the injured lung

Mechanism of pulmonary injury	Experimental conditions	5-HT removal	NE removal
A. Drug Induced			
Monocrotaline	Isolated perfused lung	↓(34)[a]	↓(34)
Paraquat	Isolated perfused lung	↓(65)	
Alpha napthylthiourea	Isolated perfused lung	↓(11)	
Bleomycin	Intact anesthetized animal	↓(19)	↓(19)
Elastase	Isolated perfused lung	→(10)	
B. Oxygen-Free Radical			
Hyperbaric hyperoxia	Isolated perfused lung	↓(9)	↓(8)
Normobaric hyperoxia	Cultured endothelial cell	↓(12)	
Normobaric hyperoxia	Conscious animal	↓(20)	
Xanthine oxidase/ hypoxanthine	Isolated perfused lung	↓(72)	
C. Pulmonary Vascular Disease			
Pulmonary hypertension	Adult man	↑(31)	↑(31) ↓(69)
	Children		↓(28)
Pulmonary microembolization	Intact anesthetized animal	↓(24)	→(24)
Post-cardiopulmonary bypass	Adult man	↑(33) →(32) ↓(31)	↑(33) ↓(31)[b]

[a] ↑ = increased; ↓ = decreased; → = unchanged. Appropriate references are given in parentheses.
[b] In patients with pulmonary hypertension.

nary hypertension than in those without (31). However, it is difficult to compare these studies owing to differences in underlying disease, anesthesia used, and techniques employed to estimate pulmonary amine extraction. Experimental pulmonary hypertension due to injection of glass beads into lungs of intact anesthetized dogs led to a selective depression in 5-HT removal but did not alter NE removal, as estimated by indicator dilution studies (24). White et al (75) reported that isolated perfused dog lobes with elevated pulmonary vascular resistance and elevated shunt fractions do not take up radiolabelled 5-HT released by platelets, although a higher incidence of platelet entrapment occurs in these lobes compared to lobes with normal pulmonary physiology. Pulmonary extraction of 5-HT and NE was depressed by 12% after restoration of normal pulmonary blood flow in dogs that had undergone 1–4 hr of total cardiopulmonary bypass (B. R. Pitt, C. N. Gillis, G. L. Hammond, unpublished observations), a procedure associated with endothelial cell damage. Recently, we found no significant change in pulmonary 5-HT extraction following cardiac surgery in 5 patients (32). Significantly, none of these individuals had serious postoperative pulmonary complications. It remains to be seen what changes occur in pulmonary metabolism of amines or other substances in patients who develop acute respiratory distress, a complex syndrome that seems to be associated in its earliest stages with lung endothelial malfunction (39).

SUMMARY

Pulmonary biogenic amine clearance is a carrier-mediated drug sensitive process associated with uptake into endothelial cells and subsequent metabolism by monoamine oxidase and other enzymes. Its function seems to be maintenance of arterial circulatory homeostasis by biochemically regulating circulating vasoactive hormones. Numerous pulmonary pathologic conditions are associated with alterations in these functions. However, a more precise physiologic role remains to be defined. Areas for further research include:

1. Comparative studies on the ability of endothelium from extrapulmonary and pulmonary vasculatures to extract and metabolize vasoactive amines and other humoral substances. This work must be done both in intact animals and in isolated cultured endothelium derived from organ microvasculatures.

2. Study of factors (e.g. steroids, development, etc) that may regulate pulmonary metabolic functions.

3. Measurements in humans under various clincal conditions. To this end, indicator dilution estimates and transpulmonary gradients have pro-

vided intriguing promise of means to assess endothelial metabolic functions in both normal and injured lung.

ACKNOWLEDGMENTS

Work from the authors' laboratory discussed in this review was supported in part by grants HL 13315, HL 22244, HL 23245, and HL 07410 from the National Institutes of Health. We are grateful to Ms. Suzanne Darrach for her help in preparing the manuscript.

Literature Cited

1. Al-Ubaidi, F., Bakhle, Y. S. 1980. Metabolism of vasoactive hormones in human isolated lung. *Clin. Sci.* 58:45–51
2. Anderson, M. W., Orton, T. C., Pickett, R. D., Eling, T. E. 1974. Accumulation of amines in the isolated perfused rabbit lung. *J. Pharmacol. Exp. Ther.* 189:456–66
3. Bakhle, Y. S., Ben-Harari, R. R. 1979. Effects of oestrus cycle and exogenous ovarian steroids on 5-hydroxytryptamine metabolism in rat lung. *J. Physiol. London* 291:11–18
4. Bakhle, Y. S., Vane, J. R. 1974. Pharmacokinetic function of the pulmonary circulation. *Physiol. Rev.* 54:1007–45
5. Bakhle, Y. S., Youdim, M. B. H. 1976. Metabolism of phenylethylamine in rat isolated perfused lung: evidence for monoamine oxidase 'type B' in lung. *Br. J. Pharmacol.* 56:125–27
6. Bennett, A. 1965. The metabolism of histamine by guinea pig and rat lung *in vitro. Br. J. Pharmacol.* 24:147–55
7. Bertler, A., Lewis, D. H., Lofstrom, J. B., Post, C. 1978. *In vivo* uptake of lidocaine in pigs. *Acta Anesthesiol. Scand.* 22:530–36
8. Block, E. R., Cannon, J. 1978. Effect of oxygen exposure on lung clearance of amines. *Lung* 155:287–95
9. Block, E. R., Fisher, A. 1977. Depression of serotonin clearance with rat lungs during oxygen exposure. *J. Appl. Physiol.* 42:33–38
10. Block, E. R., Ryerson, G., Harris, J. O., Cannon, J. K. 1979. Effect of pulmonary emphysema on 5-hydroxytryptamine uptake by hamster lungs. *Am. Rev. Resp. Dis.* 119:290 (Abstr.)
11. Block, E. R., Schoen, F. J. 1981. Effect of alpha napthylthiourea on uptake of 5-hydroxytryptamine. *Am. Rev. Resp. Dis.* 123:69–73
12. Block, E. R., Stalcup, S. A. 1981. Depression of serotonin uptake by cultured endothelial cells exposed to high O_2 tensions. *J. Appl. Physiol.* 50:1212–19
13. Boileau, J. C., Campeau, L., Biron, P. 1970. Pulmonary fate of histamine, isoproterenol, physoalaemine and substance P. *Can. J. Physiol. Pharmacol.* 48:681–84
14. Boileau, J. C., Campeau, L., Biron, P. 1971. Comparative fate of intravenous epinephrine. *Revue Can. Biol.* 30:281–86
15. Boileau, J. C., Crexells, C., Biron, P. 1972. Free pulmonary passage of dopamine. *Revue Can. Biol.* 31:65–72
16. Brigham, K. L., Owen, P. J. 1975. Increased sheep lung vascular permeability caused by histamine. *Circ. Res.* 37:647–57
17. Buonassisi, V., Venter, J. C. 1976. Hormone and neurotransmitter receptors in an established vascular endothelial cell line. *Proc. Natl. Acad. Sci. USA* 73:1612–16
18. Catravas, J. D., Gillis, C. N. 1980. Pulmonary clearance of ^{14}C-5-hydroxytryptamine and ^{3}H-norepinephrine *in vivo:* effects of pretreatment with imipramine or cocaine. *J. Pharmacol. Exp. Ther.* 213:120–27
19. Catravas, J. D., Lazo, J. S., Gillis, C. N. 1981. Biochemical markers of bleomycin toxicity: Removal of ^{14}C-5-hydroxytryptamine and ^{3}H-norepinephrine by rabbit lung *in vivo. J. Pharmacol. Exp. Ther.* 217:524–29
20. Dobuler, K. J., Catravas, J. D., Gillis, C. N. 1981. Hyperoxic lung injury and altered metabolic functions in conscious rabbits. *Am. Rev. Resp. Dis.* 123:229 (Abstr.)
21. Fernandez-Madrid, F. 1977. Serotonin, other biologically active principals, and the lung. *Int. Anesthesiol. Clin.* 15:168–188
22. Fillenz, M. 1970. Innervation of pulmonary and bronchial blood vessels. *J. Anat.* 106:449–61

23. Fishman, A. P., Pietra, G. G. 1974. Handling of bioactive materials by the lung. *N. Engl. J. Med.* 290:884–90, 953–59
24. Flink, J. R., Pitt, B. R., Hammond, G. L., Gillis, C. N. 1981. Selective effects of microembolization on pulmonary clearance of biogenic amines. *Am. Rev. Respir. Dis.* 123:235
25. Gaddum, J. H., Hebb, C. O., Silver, A., Swan, A. A. B. 1953. 5-Hydroxytryptamine: Pharmacological action and destruction in perfused lungs. *Q. J. Exp. Physiol.* 38:255–62
26. Geddes, D. M., Nesbitt, K., Traill, T., Blackburn, J. P. 1979. First pass uptake of ^{14}C-propranolol by the lung. *Thorax* 34:810–13
27. Gewitz, M. H., Gillis, C. N. 1981. Uptake and metabolism of biogenic amines in the developing rabbit lung. *J. Appl. Physiol.* 50:118–22
28. Gewitz, M., Pitt, B., Laks, H., Hammond, G. L., Talner, N., Gillis, C. N. 1980. Reversible effect of pulmonary hypertension on extraction of endogenous catecholamines by the pulmonary vasculature. In *Pediatric Cardiology,* M. Godman. London: Churchill Livingston. (Abstr.)
29. Gillis, C. N. 1976. Extraneuronal transport of noradrenaline in the lung. In *The Mechanism of Neuronal and Extraneuronal Transport of Catecholamines,* ed. D. M. Paton, pp. 281–297. NY: Raven Press
30. Gillis, C. N. 1980. Metabolism of vasoactive hormones by pulmonary vascular endothelium: possible functional significance. In *Vascular Neuroeffector Mechanisms,* ed. J. A., Bevan, T. Godfriend, R. A. Maxwell, P. M. Vanhoutte, pp. 304–15. NY: Raven Press.
31. Gillis, C. N., Cronau, L. H., Greene, N. M., Hammond, G. L. 1974. Removal of 5-hydroxytryptamine and norepinephrine from the pulmonary vascular space of man: influence of cardiopulmonary bypass and pulmonary arterial pressure on these processes. *Surgery* 76:608–16
32. Gillis, C. N., Cronau, L. H., Mandel, S., Hammond, G. L. 1979. Indicator dilution measurement of 5-hydroxytryptamine clearance by human lung. *J. App. Physiol.* 46:1178–83
33. Gillis, C. N., Greene, N. M., Cronau, L. H., Hammond, G. L. 1972. Pulmonary extraction of 5-hydroxytryptamine and norepinephrine before and after cardiopulmonary bypass in man. *Circ. Res.* 30:666–74

34. Gillis, C. N., Huxtable, R. J., Roth, R. A. 1978. Effects of monocrotaline pretreatment of rats on removal of 5-hydroxytryptamine and noradrenaline by perfused lung. *Br. J. Pharmacol.* 63:435–43
35. Gillis, C. N., Roth, J. A. 1976. Pulmonary disposition of circulating vasoactive hormones. *Biochem. Pharmacol.* 25:2547–53
36. Gillis, C. N., Roth, J. A. 1977. The fate of biogenic monoamines in perfused rabbit lung. *Br. J. Pharmacol.* 59:585–90
37. Grimm, D. J., Dawson, C. A., Hakim, T. S., Linehan, J. H. 1978. Pulmonary vasomotion and the distribution of vascular resistance in a dog lung lobe. *J. Appl. Physiol.* 45:545–50
38. Hakim, T. S., vander Zee, H., Malik, A. B. 1979. Effects of sympathetic nerve stimulation on lung fluid and protein exchange. *J. Appl. Physiol.* 47:1025–30
39. Hammond, G. L. 1980. Acute respiratory failure. *Surg. Clin. North America* 60:1133–49
40. Hardebo, J. E., Owman, C. 1980. Characterization of the *in vitro* uptake of monoamines into brain microvessels. *Acta Physiol. Scand.* 108:223–29
41. Hughes, J., Gillis, C. N., Bloom, F. E. 1969. The uptake and disposition of dl-norepinephrine in perfused rat lung. *J. Pharmacol. Exp. Ther.* 169:237–48
42. Hyman, A. L. 1969. The direct effects of vasoactive agents on pulmonary veins. Studies of response to acetylcholine, serotonin, histamine and isoproterenol in intact dogs. *J. Pharmacol. Exp. Ther.* 168:96–105
43. Iwasawa, Y., Gillis, C. N. 1974. Pharmacological analysis of norepinephrine and 5-hydroxytryptamine removal from the pulmonary circulation: differentiation of uptake sites for each amine. *J. Pharmacol. Exp. Ther.* 188:386–93
44. Iwasawa, Y., Gillis, C. N., Aghajanian, G. 1973. Hypothermic inhibition of 5-hydroxytryptamine and norepinephrine uptake by lung: Cellular location of amines after uptake. *J. Pharmacol. Exp. Ther.* 186:498–507
45. Junod, A. F. 1972. Uptake, metabolism and efflux of ^{14}C-5-hydroxytryptamine in isolated perfused rat lungs. *J. Pharmacol. Exp. Ther.* 183:341–55
46. Junod, A. F. 1975. Mechanism of uptake of biogenic amines in the pulmonary circulation. In *Lung Metabolism,* ed. A. F. Junod, R. deHaller, pp. 387–56. London: Academic

47. Junod, A. F. 1977. Metabolism of vasoactive agents in lung. *Am. Rev. Respir. Dis.* 115:51–57
48. Junod, A. F., Ody, C. 1977. Amine uptake and metabolism by endothelium of pig pulmonary artery and aorta. *Am. J. Physiol.* 232:C88–94
49. Lilja, B., Lindell, S. E., Saldeen, T. 1961. Formation and destruction of C^{14}-histamine in human lung tissue *in vitro*. *J. Allergy* 31:492–96
50. Majno, G., Palade, G. E. 1961. Studies on inflammation I. The Effect of histamine and serotonin on vascular permeability: An electron microscopic study. *J. Biophys. Biochem. Cytol.* 11:571–605
51. Nicholas, T. E., Strum, J. M., Angelo, L. S., Junod, A. F. 1974. Site and mechanism of uptake of H^3-1-norepinephrine by isolated perfused rat lungs. *Circ. Res.* 35:670–80
52. Orton, T. C., Anderson, M. W., Pickett, R. D., Eling, T. E., Fouts, J. R. 1973. Xenobiotic accumulation and metabolism by isolated perfused rabbit lungs. *J. Pharmacol. Exp. Ther.* 186:482–97
53. Owman, C., Edvinsson, L., Hardebo, J. E. 1980. Amine mechanisms and contractile properties of the cerebral microvascular endothelium. In *Vascular Neuroeffector Mechanisms*, ed. J. A. Bevan, T. Godfriend, R. A. Maxwell, P. M. Vanhoutte, pp. 277–90. NY: Raven Press
54. Peroutka, S. J., Moskowitz, M. A., Reinhard, J. F., Snyder, S. H. 1980. Neurotransmitter receptor binding in bovine cerebral microvessels. *Science* 208:610–12
55. Pickett, R. D., Anderson, M. W., Orton, T. C., Eling, T. E. 1975. The pharmacodynamics of 5-hydroxytryptamine uptake and metabolism by the isolated perfused rabbit lung. *J. Pharmacol. Exp. Ther.* 194:545–53
56. Pitt, B. R., Gillis, C. N., Hammond, G. L. 1980. Effect of altered cardiac output on pulmonary amine extraction. *Fed. Proc.* 39:964 (Abstr.)
57. Raichle, M. E., Hartman, B. K., Eichling, J. O., Sharpe, L. G. 1975. Central noradrenergic regulation of cerebral blood flow and vascular permeability. *Proc. Natl. Acad. Sci. USA* 72:3726–30
58. Reinhard, J. F., Liebman, J. E., Schlosberg, A. J., Moskowitz, M. A. 1979. Serotonin neurons project to small blood vessels in the brain. *Science* 206:85–87
59. Rennels, M. L., Nelson, E. 1975. Capillary innervation in the mammalian cerebral nervous system: An electron microscopic demonstration. *Am. J. Anat.* 144:234–41
60. Rickaby, D. A., Dawson, C. A., Linehan, J. H. 1981. Effect of blood flow rate and cell free perfusate on the kinetics of serotonin uptake in the dog lung. *Fed. Proc.* 40:621 (Abstr.).
61. Rickaby, D. A., Dawson, C. A., Maron, M. B. 1980. Pulmonary inactivation of serotonin and site of serotonin pulmonary vasoconstriction. *J. Appl. Physiol.* 48:606–12
62. Robinson-White, A., Peterson, S., Hechtman, H. B., Shapro, D. 1981. Serotonin uptake by isolated adipose capillary endothelium. *J. Pharmacol. Exp. Ther.* 216:125–28
63. Roth, J. A., Gillis, C. N. 1975. Multiple forms of amine oxidase in perfused rabbit lung. *J. Pharmacol. Exp. Ther.* 194:537–44
64. Roth, R. A., Gillis, C. N. 1978. Effect of ventilation and pH on removal of mescaline and biogenic amines by rabbit lung. *J. Appl. Physiol.* 44:553–58
65. Roth, R. A., Wallace, K. B., Alper, R. H., Bailie, M. D. 1979. Effect of paraquat treatment of rats on disposition of 5-hydroxytryptamine and angiotensin I by perfused lung. *Biochem. Pharmacol.* 28:2349–55
66. Sharman, D. F. 1975. The metabolism of circulating catecholamines. In *Handbook of Physiology. Endocrinology VI*, ed. H. Blaschko, A. D. Smith, G. Sayers, pp. 699–712. Washington DC: Am. Physiol. Soc.
67. Shepro, D. 1980. Nonrespiratory, metabolic functions of pulmonary endothelial cells. In *Acute Respiratory Failure: Etiology and Treatment*, ed. H. B. Hechtman, pp. 20–53. Boca Raton, Fla: CRC Press
68. Shepro, D., Batbouta, J. C., Robblee, L. S., Carson, M. P., Belamarich, F. A. 1975. Serotonin transport by cultural bovine aortic endothelium. *Circ. Res.* 36:799–806
69. Sole, M. J., Drobac, M., Schwartz, L., Hussain, M. N., Vaughan-Neil, E. F. 1979. The extraction of circulating catecholamines by the lungs in normal man and in patients with pulmonary hypertension. *Circulation* 60:160–63
70. Starling, E. H., Verney, E. D. 1925. The secretion of urine as studied on the isolated kidney. *Proc. R. Soc. London Ser. B* 97:321–23
71. Steinberg, H., Bassett, D. J. P., Fisher, A. B. 1975. Depression of pulmonary 5-hydroxytryptamine uptake by meta-

bolic inhibitors. *Am. J. Physiol.* 228: 1298–1303
72. Steinberg, H., Das, D. K., May, W., Eglow, R., Greenwald, R. 1979. Injury to pulmonary endothelium induced by superoxide radical. *Am. Rev. Resp. Dis.* 119:364 (Abstr.)
73. Strum, J. M., Junod, A. F. 1972. Radioautographic demonstration of 5-hydroxytryptamine-^{3}H uptake by pulmonary endothelial cells. *J. Cell Biol.* 54:456–67
74. vander Zee, H., Malik, A. B., Lee, B. C., Hakim, T. S. 1980. Lung fluid and protein exchange during intracranial hypertension and role of sympathetic mechanisms. *J. Appl. Physiol.* 48:273–80
75. White, M. K., Hechtman, H. B., Shepro, D. 1975. Canine lung uptake of plasma and platelet serotonin. *Microvasc. Res.* 9:230–41
76. Wiersma, D. A., Roth, R. A. 1980. Clearance of 5-hydroxytryptamine by rat lung and liver: The importance of relative perfusion and intrinsic clearance. *J. Pharmacol. Exp. Ther.* 212:97–102

CHEMOTACTIC MEDIATORS IN NEUTROPHIL-DEPENDENT LUNG INJURY

Joseph C. Fantone, Steven L. Kunkel, and Peter A. Ward

Department of Pathology, University of Michigan, Ann Arbor, Michigan 48109

Acute inflammatory reactions in the lung can be divided into two general classes. Reactions mediated by antigen activation of IgE-sensitized mast cells resulting in degranulation with release of both vasoactive mediators and eosinophil chemotactic factors are commonly referred to as anaphylaxis. A second category of acute inflammation results from the local generation of phlogistic mediators that are chemotactic for polymorphonuclear leukocytes, causing their local recruitment and the subsequent injury of lung tissue. Here we focus on chemotactic mechanisms involved in neutrophil-mediated lung injury. The reader is referred to two recent reviews of the mechanisms involved in anaphylaxis (35,58).

THE ORIGIN AND REGULATION OF CHEMOTACTIC FACTORS IN THE LUNG

Complement-Derived Chemotactic Factors

Direct injury to the lung by either immunologic or nonimmunologic mechanisms is frequently followed by an influx of acute inflammatory cells, most notably the polymorphonuclear leukocyte (neutrophil). The process by which this cell migrates from the intravascular space to the site of injury is assumed to involve the mechanism of chemotaxis, or cell-directed migration. When neutrophils are exposed in vitro to a chemical gradient of a chemotactic substance, the cell orients itself and migrates in the direction of increasing concentration of chemical stimulus. An abundant literature discusses in detail the biochemical responses of neutrophils to chemotactic stimuli (reviewed in 63). In the following discussions we emphasize the various sources of chemotactic factors in the lung and their regulation.

The most extensively studied source of chemotactic factor is the human complement system. Through the local activation of either the classical or alternative complement pathways, two potent anaphylatoxins are generated: C3a ($M_r = 9,000$) and C5a ($M_r = 12,000$) (24). These biologically active agents are defined by their ability to cause in vitro the release of vasoactive amines from mast cells and basophils as well as their effects on smooth muscle. In addition to its anaphylatoxin activity, C5a is the most potent serum-derived chemo-attractant for neutrophils. This peptide is active in nanomolar concentrations and can cause unidirectional movement not only of neutrophils but also monocytes, eosinophils, and basophils (16). Earlier studies suggested that C3a may also be chemotactically active for neutrophils. However, recent studies using highly purified C3a have failed to show significant chemotactic activity at physiologic concentrations (25).

Recently a specific serum protein, carboxypeptidase (SCPN) has been shown to be enzymatically active, cleaving the C-terminal arginyl or lysyl residues from several phylogistic mediators including C3a, C5a, and bradykinin. Present in human plasma at a concentration of 30–40 μg/ml, SCPN cleaves the C-terminal arginine from C5a producing C5a-des-arg. (7). C5a-des-arg is approximately 1/40 as active a chemotactic agent as C5a. However, an acidic protein has been identified in serum that can interact with C5a-des-arg and restore its full chemotactic activity (3, 46). Since relatively large quantities of SCPN are present in serum, this recent evidence suggests that the C5a-des-arg "helper factor" complex, and not C5a per se, may be responsible for most chemotactic activity generated in vivo by complement activation.

Spontaneous activation of the complement system and persistent generation of complement-derived chemotactic factors are modulated by two different mechanisms. The first involves the inhibition of both the classical and alternative pathways, thus preventing the generation of C5a. C1-esterase inhibitor (C1-INA) is an α_2-macroglobulin (M_r circa 100,000) present in serum; it forms an irreversible complex with activated C1 inhibiting the esterolytic sites of the C1s molecule (18, 44). In addition, C1-INA inhibits other proteases involved in coagulation and fibrinolysis, including plasmin, kallikrein, activated Hageman Factor, and Factor XIa (17). Three mechanisms can overcome inhibition by C1-INA and allow the successful activation of the classical complement pathway. These include an intense activating stimulus in which large quantities of C1 are generated to react with C4 before inactivation by C1-INA; consumption of C1-INA by activation of the kinin, fibrinolytic, and/or coagulation systems; and, finally, a genetic deficiency of C1-INA that manifests itself as the clinical syndrome of angioedema (12).

The generation of C5a by the alternative pathway is inhibited by two serum proteins, B1H and C3b inactivator (C3b-INA). B1H competitively binds to C3b, blocking the binding of Factor B to C3b. C3b-INA functions to cleave C3b into two inactive fragments, C3c and C3d, inhibiting the activation of C5 by the C5 convertase (C3bBb) (29). Both proteins are required to inhibit the spontaneous activation of the alternative pathway.

Recently a second group of activities defined as chemotactic factor inactivators have been identified in serum; these irreversibly inactivate chemotactic factor activity derived from bacteria, lymphocytes, and the complement system (60). The chemotactic factor inactivator activity (CFI) inactivates C5a and C5a-des-arg and is distinct from SCPN, having a molecular weight of approximately 150,000 (32). Although its precise physical properties have not been characterized, preliminary evidence suggests that CFI functions through enzymatic cleavage of the C5a molecule. Experimental studies have shown that intra-tracheal instillation of CFI purified from human serum into animal lungs will inhibit immune complex induced lung injury (see below) (27).

Cell-Derived Chemotactic Factors

Leukocyte-derived products have the capacity to generate chemotactically active metabolites from both serum and structural components of the lung. When C3 and C5 are incubated in vitro with cathepsin G and elastase, which have been purified from human neutrophils, anaphylatoxin activity for C3 and chemotactic activity from C5 are generated (43, 62). Further studies have suggested that greater than 90% of the C5-derived chemotactic activity generated in the dermis in immune complex vasculitis is derived from lysosomal enzyme cleavage of C5 and not from activation of the complement system (65).

Additional studies have shown that a chemotactically active substance can be generated in serum after incubation with a plasmin-generating system (plasminogen and streptokinase) (59). Therefore, local activation of either the coagulation or the fibrinolytic stystm can generate chemotactic activity in serum. Since neutrophils (21) and macrophages (61) can secrete plasminogen activator in response to specific stimuli, they can also amplify the phlogistic response by the local activation of plasminogen to plasmin and further production of chemotactic activity.

Chemotactically active products can be generated by the action of proteolytic enzymes on structural substrates in the lung. Collagen derived peptides have chemotactic activity for monocytes and fibroblasts in vitro (48, 49). Also, biologically active fragments that are chemotactic for monocytes, neutrophils, and alveolar macrophages have been generated from human elastin after incubation with human neutrophil elastase (55). These chemotactically active fragments from elastin are high in desmosine cross-links and have a molecular weight between 14,000 and 20,000.

Since alveolar macrophages of smokers and alveolar macrophages from nonsmokers exposed to smoke in vitro secrete increased amounts of elastase (51), the local generation of chemotactic factors with an influx of inflammatory cells may play an important role in the structural damage present in the lungs of smokers. Neutrophils and macrophages, when exposed to phagocytic (e.g. immune complexes, bacteria, inert particulate matter) or chemotactic stimuli also secrete lysosomal enzymes extracellularly, which could generate additional chemotactic factors by the proteolysis of tissue substrates (62, 65). This provides an additional mechanism by which an inflamatory reaction may be initiated or amplified in vivo.

The generation of oxygen metabolites by neutrophils and macrophages after phagocytic or chemotactic stimulus has been well established. The primary products secreted extracellularly after stimulation of these cells appear to be superoxide anion (O_2^-) and hydrogen peroxide (H_2O_2) (reviewed in 2, 31). Additional studies suggest that other reactive metabolites may be generated by the metal catalyzed nonenzymatic conversion of O_2^-:

$$M^{n+} + O_2^- = O_2 + M^{(n-1)+}$$

$$\underline{M^{(n-1)+} + H_2O_2 = OH^- + OH\cdot + M^{n+}}$$

$$O_2^- + H_2O_2 = O_2 + OH^- + OH\cdot$$

and the myeloperoxidase-H_2O_2 oxidation of halide ions to hypohalite and subsequent singlet oxygen (O_2^1) production. Several of the products of oxygen metabolism have been directly implicated in neutrophil-mediated cell and tissue injury both in vitro (8, 30, 39, 67) and in vivo (28, 37). These reactive oxygen metabolites have also been implicated in tissue injury secondary to toxic levels of oxygen (50, 53), anti-neoplastic drugs (13), and certain toxins such as paraquat (5).

The possible role of oxygen metabolites in the generation of chemotactically active substances has recently been investigated. Human plasma incubated with xanthine–xanthine oxidase (an O_2^- and H_2O_2 generating system) produces a biologically active lipid that is chemotactic for neutrophils (47). Although the precise physical and structural characteristics of this lipid are not known, the biologic activity is heat labile and the lipid appears to be bound to albumin in its active form. Additional studies have demonstrated the generation of a chemotactic lipid after incubation of arachadonic acid with an O_2^- generating system (45). This material is

MEDIATORS IN LUNG 287

chromatographically distinct from arachadonic acid and is functionally active at a concentration of 3.0 ng/ml. Thus potent oxidants derived from inflammatory cells can generate additional chemotactic factors locally. The evidence also suggests that oxidants derived from other sources such as cigarette smoke and other inhaled noxious gases may initiate or amplify a local inflammatory response in the lung.

Oxidants may also augment the local generation of chemotactic factors in the lung via the inactivation of the normal serum inhibitors (a_1 antiprotease) of leukocyte and monocyte proteases. Oxygen metabolites (52) as well as cigarette smoke (6) inhibit the serum elastase inhibitor (a_1-antiprotease) in vitro. Since the phenolic anti-oxidants hydroquinone and thymol block the effects of the smoke-derived inactivation of a_1-antiprotease, it is thought that the oxidants in cigarette smoke play a pivotal role in this phenomenon. Thus the local delivery or generation of potent oxidants in the lung may add to the inactivation of the naturally occurring inhibitors of neutrophil- and monocyte-derived proteases, causing increased destruction of the structural components of the lung (e.g. collagen and elastin) and increasing the local concentration of the chemotactic factors derived from these substrates.

A recent report has further complicated our interpretation of the role of oxygen metabolites in tissue injury. When either C5a or the synthetic chemotactic peptide N-Formyl-Methionyl-Leucyl-Phenylalanine (F-met-leu-phe) are incubated with a myeloperoxidase halide system in vitro, the methione residues appear to be oxidized and the chemotactic activity of each abolished (9). Therefore, clarification and definition of the precise phlogistic role of oxidants and oxygen metabolites in the lung awaits further study.

Another source of chemotactic factors in the lung are specific cellular products that appear to be derived primarily from inflammatory cells. When macrophages, neutrophils, mast cells, basophils, or eosinophils are activated by phagocytic, chemotactic, or other immunologic stimuli, arachadonic acid is generated from phospholipids in the cell membrane and metabolized by either specific cyclooxygenases (producing prostaglandins and thromboxanes) or lipoxygenases, generating a class of biologically active hydroxy-eicostetranoic acids (HETES) and intermediate compounds. The primary intermediate metabolite in the generation of lipoxygenase products from arachadonic acid that has been implicated in inflammatory reactions is 5-hydroperoxy-eicosa-6,8,11,14-tetraenoic acid (5-HPETE) (20). The specific products derived from 5-HPETE are distinct for individual cell types and include a class of biologically active complex HETES known as leukotrienes. Several of the mono-HETES and complex-HETES including certain leukotrienes have been shown to be chemotactic for neutrophils and eosinophils in vitro. The most potent of these arachadonic acid metabolites are 5-HETE, and 5,12-di-HETE (leukotriene B_4), the latter having chemotactic activity in a dose range similar to that for C5a. Therefore, local activation of inflammatory cells can activate a membrane-associated phospholipase generating arachadonic acid from membrane phospholipids, with further metabolism by activation of a lipoxygenase enzyme system, resulting in generation of potent biologically active products. These products can amplify the inflammatory response and, under certain conditions, may be the principal mediators of the inflammatory response.

Additional products of the alveolar macrophage have also been suggested as potential sources of chemotactic activity in vivo. Recent studies have demonstrated the presence of hemolytically active C5 in bronchial lavage fluids, and in vitro studies suggest that mononuclear phagocytes may secrete several complement components including C5 into the intra-alveolar space (68). Therefore, without prior vascular injury and exudation of plasma, C5-derived chemotactic peptides could theoretically be generated in the alveolus following alveolar insult from the environment. Alveolar macrophages have also been shown to secrete a low molecular weight chemotactic factor for neutrophils following stimulation by immune complexes, microorganisms, and particulate matter (23, 68). The chemotactic factor derived from human alveolar macrophages attracts neutrophils preferentially and has a molecular weight of 400–600 daltons. It also has been reported to contain a lipid component distinct from the lipoxygenase products. Additional studies have shown that the stimulated alveolar macrophage also secretes unsaturated fatty acids, primarily linoleic acid, which stimulates macrophage mobility (36). In vitro studies of macrophage chemotaxis have shown that both linoleic acid, which is found in elevated concentration in certain inflammatory sites, and phosphatidylglycerol, present in bronchial lavage fluids, enhance macrophage migration. Although the relative in vivo phlogistic potential of each of these mediators is difficult to assess, these studies suggest that the alveolar macrophage may play a pivotal role in the initiation and modulation of lung inflammation.

Several other sources of chemotactic factors that may play a significant role in lung inflammation have been described, but a detailed description is beyond the scope of this review. "Lymphokines" represent a broad class of lymphocyte-derived inflammatory mediators, several of which have been reported to be chemotactic for neutrophils, monocytes, lymphocytes, basophils, eosinophils, and fibroblasts (reviewed in 1). Another group of cell-derived chemotactic factors is released during IgE-mediated activation of mast cells. These have been termed "eosinophil chemotactic factors of anaphylaxis" (19). In addition, both bacteria and viruses have been reported to secrete products chemotactic for inflammatory cells (64, 66). Thus there exist numerous sources of chemotactic agents capable of initiating or amplifying the inflammatory response. The precise in vivo role of each is not known and awaits further study.

EXPERIMENTAL STUDIES IMPLICATING CHEMOTACTIC FACTORS IN LUNG INJURY

Intravascular and Intra-alveolar Effects of Chemotactic Factors

The role of chemotactic factors in initiating lung injury has received detailed study in recent years. Intravascular activation of the complement system with the generation of chemotactic peptides and sequestration of neutrophils in the lung has been suggested as a mechanism of respiratory distress resulting from extracorporal hemodialysis, severe burns, and extensive trauma. Neutrophils in vitro, when exposed to chemotactic factors (i.e. C5a, F-met-leu-phe), respond with an increase in cell volume and formation of multicellular aggregates (26, 42). When either chemotactic factors or cobra venom factor (an activator of the alternative complement pathway) are instilled intravenously into rabbits, a rapid, marked, and transient neutropenia develops, reaching a nadir within 1–2 min and returning to normal within 15–30 min (40, 41). The neutrophils are apparently sequestered in the lung as a result of neutrophil aggregation and embolization in capillaries as well as increased adherence to vascular endothelial cells (margination). Although most in vivo studies have failed to produce significant lung injury as a result of intravascular generation of chemotactic factors, a recent report describes the development of pulmonary leukostasis and edema after intravascular infusion of zymosan-activated homologous serum into sheep (10). Additional study is necessary to define the precise relationship between these experimental studies and lung injury in humans.

Since the recruitment of neutrophils and other inflammatory cells into tissues is believed to result from the local generation of chemotactic factors, several groups have investigated the effects of intratracheal instillation of preformed chemotactic factors in lung. Intrapulmonary instillation of highly purified preparation of either C5 (11), C5a (22, 33, 38, 56), or C5a-des-arg (57), produces an acute inflammatory reaction. This reaction is dose-dependent, reaches a peak between 4 and 6 hours, and is characterized by edema, fibrin deposition, and a neutrophil infiltrate. Each reactant is active in microgram quantities, and comparative studies suggest that the phlogistic properties of C5a-des-arg are greater than those of C5a (57). Although in vitro studies show C5a to be a more potent chemotactic than C5a-des-arg, the possible formation of a C5a-des-arg "helper" factor complex could explain these findings. These observations substantiate the in

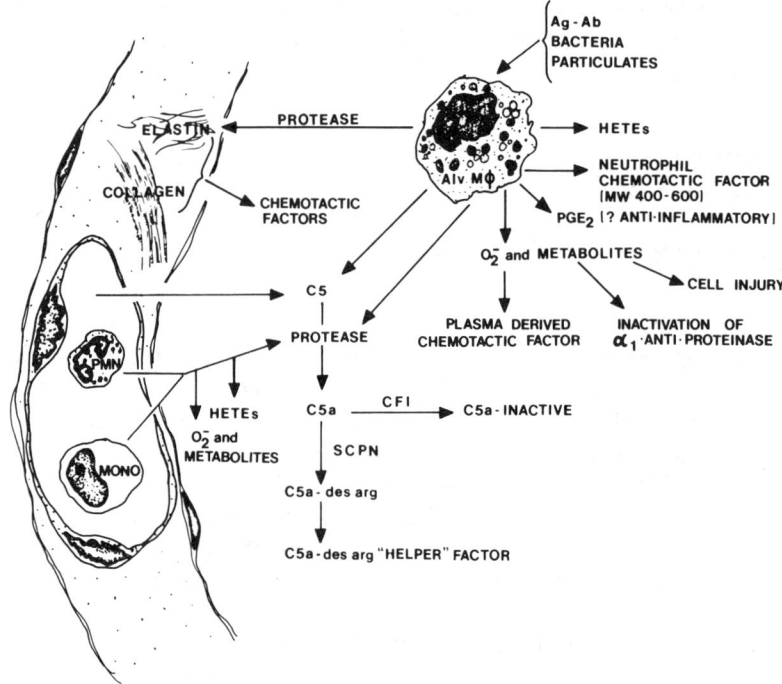

Figure 1 Mechanisms of neutrophil-dependent lung injury.

vivo phlogistic role of C5 and C5-derived chemotactic peptides in neutrophil-dependant lung injury.

Immune Complex and Neutrophil-Dependent Lung Injury

Immune complexes have been shown to cause acute lung injury in several animal models (reviewed in 15). In each model the ability of the immune complexes to cause lung injury depends on the presence of neutrophils and complement; complexes having the greatest complement-fixing capacity are the most phlogistic (54). Immunofluorescent studies of vessel walls and alveolar intersitium reveal the presence of immunoglobulin, antigen, and complement components. The mechanism of lung injury mediated by immune complexes is believed to be representative of most neutrophil-dependent lung injury and is thought to result from the local fixation of complement and generation of C5a and other neutrophil chemotactic factors (Figure 1). Neutrophils and alveolar macrophages phagocytize the complexes, releasing additional chemotactic factors (see above) as well as proteases and oxygen-free radicals and metabolites. These products generate additional chemotactic factors from both plasma and the structural components of lung. Oxygen metabolites may injure the resident lung cells and may in certain cases stimulate pulmonary fibrosis (28). Current evidence suggests that the resolution of many forms of neutrophil-dependent injury results from elimination of the inciting insult with a decrease in the generation of chemotactic factors. In addition, potent inhibitors of chemotactic factors, proteases, and oxygen radicals (ceruloplasmin) aid in the modulation of neutrophil-dependent tissue injury. Local secretion of prostaglandins of the E series (primarily E_2) may aid in the suppression and/or resolution of tissue injury (4, 14, 34, 69).

Literature Cited

1. Altman, L. C. 1978. Chemotactic lymphokines: A review. In *Leukocyte Chemotaxis, Methods, Physiology, and Clinical Implications,* ed. J. I. Gallin, P. G. Quie. NY: Raven
2. Babior, B. M. 1978. Oxygen-dependent microbial killing by phagocytes. Part I. *New Engl. J. Med.* 298:659–68
3. Beebe, D. P., Ward, P. A., Spitznagel, S. K. 1980. Isolation and characterization of an acidic chemotactic factor from complement activated human serum. *Clin. Immunol. Immunopathol.* 15:88–105
4. Bonta, I. L., Parnham, M. J. 1979. Time dependent stimulatory and inhibitory effects of prostaglandin E_1 on exudative and tissue components of granulomatous inflammation in rats. *Br. J. Pharmacol.* 65:465–72
5. Burls, R. F., Lawrence, R. A., Love, M. M. 1980. Liver necrosis and lipid peroxidation in the rat as a result of paraquat and diquat administration. Effect of selenium deficiency. *J. Clin. Invest.* 65:1024–31
6. Carp, H., Janoff, A. 1978. Possible mechanisms of emphysema in smokers. In vitro suppression of serum elastase-inhibitory capacity by fresh cigarette smoke and its prevention by antioxidants. *Am. Rev. Resp. Dis.* 118:617–21
7. Chenoweth, D. E., Hugli, T. F. 1980. Human C5a and C5a analogs as probes of the neutrophil C5a receptor. *Mol. Immunol.* 17:151–57
8. Clark, R. A., Klebanoff, S. J. 1975. Neutrophil-mediated tumor cell cytotoxicity: Role of the peroxidase system. *J. Exp. Med.* 141:1442–47
9. Clark, R. A., Szot, S., Venkatasubramanian, K., Schiffmann, E. 1980. Chemotactic factor inactivation by myeloperoxidase mediated oxidation of methionine. *J. Immunol.* 124: 2020–26

10. Craddock, P. R., Fehr, J., Brigham, K. L., Kronenberg, R. S., Jacob, H. S. 1977. Complement and leukocyte-mediated pulmonary dysfunction in hemodialysis. *New Engl. J. Med.,* 296:769–74
11. Desai, U., Kreutzer, D. L., Showell, H. J., Arroyave, C. F., Ward, P. A. 1979. Acute inflammatory pulmonary reactions induced by chemotactic factors. *Am. J. Pathol.* 96:71–83
12. Donaldson, V. H., Evans, R. R. 1963. A biochemical abnormality in hereditary angioneurotic edema. *Am. J. Med.* 35:37
13. Doroshow, J. H., Locker, G. Y., Myers, C. E. 1980. Enzymatic defenses of the mouse heart against reactive oxygen metabolites. Alterations produced by doxorubicin. *J. Clin. Invest.* 65:128–35
14. Fantone, J. C., Kunkel, S. L., Ward, P. A., Zurier, R. B. 1980. Suppression by Prostaglandin E_1 of vascular permeability induced by vasoactive inflammatory mediators. *J. Immunol.* 125:2591–96
15. Fantone, J. C., Ward, P. A. 1981. Experimental studies of immune complex injury in the lung. *Am. Rev. Resp. Dis.* In press
16. Fernandez, H. N., Henson, P. M., Otani, A., Hugli, T. E. 1978. Chemotactic response of human C3a and C5a anaplylatoxins. 1. Evaluation of C3a and C5a leukotaxis in vitro under simulated in vivo conditions. *J. Immunol.* 120: 109–15
17. Forbes, C. D., Pensky, J., Ratnoff, O. D. 1970. Inhibition of activated Hageman Factor and activated plasma thromboplastin antecedent by purified serum C1 inactivator. *J. Lab. Clin. Med.* 76:809–15
18. Gigli, I., Ruddy, S., Austen, K. F. 1968. The stoichiometric measurement of the serum inhibitor of the first component of complement by the inhibition of im-

mune hemolysis. *J. Immunol.* 100: 1154–64
19. Goetzl, E. J., Austen, K. F. 1975. Purification and sythesis of eosinophilotactic tetrapeptides of human lung tissue: Identification as eosinophil chemotactic factor of anaphylaxis. *Proc. Natl. Acad. Sci. USA* 72:4123–27
20. Goetzl, E. J. 1980. Mediators of immediate hypersensitivity derived from arachadonic acid. *New Engl. J. Med.* 303: 822–25
21. Granelli-Piperno, A., Vassalli, J. D., Reich, E. 1977. Secretion of plasminogen activator by human polymorphonuclear leukocytes modulation by glucocortiuoids and other effectors. *J. Exp. Med.* 146:1693–706
22. Henson, P. M., McCarthy, K., Larsen, G. L., Webster, R. O., Giclas, P. C., Dreisin, R. B., King, T. E., Shaw, J. O. 1979. Complement fragments, alveolar macrophages, and alveolitis. *Am. J. Pathol.* 97:93–110
23. Henson, P. M. 1980. Mechanisms of exocytosis in phagocytic inflammatory cells. *Am. J. Pathol.* 101:494–511
24. Hugli, T. E., Muller-Eberhard, H. J. 1978. Anaphylatoxins: C3a and C5a. *Adv. Immunol.* 26:1
25. Hugli, T. E. 1978. Chemical aspects of the serum anaphylatoxins. *Contemp. Top. Mol. Immunol.* 7:181–214
26. Jacob, H. S., Craddock, P. R., Hammerschmidt, D. E., Moldow, C. F. 1980. Complement-induced granulocyte aggregation. An unsuspected mechanism of disease. *New Engl. J. Med.* 302:784–94
27. Johnson, K. J., Anderson, T. P., Ward, P. A. 1977. Suppression of immune complex-induced inflammation by the chemotactic factor inactivator. *J. Clin. Invest.* 59:951–58
28. Johnson, K. J., Fantone, J. C., Kaplan, J., Ward, P. A. 1981. In vivo damage of rat lungs by oxygen metabolites. *J. Clin. Invest.* 67:983–92
29. Kazlatchkine, M. D., Fearon, D. T., Silbert, J. E., Austen, K. F. 1979. Surface-associated heparin inhibits zymosan-induced activation of the human alternative complement pathway by augmenting the regulating action of the control proteins on particle bound C3b. *J. Exp. Med.* 150:1202–15
30. Kellogg, E. W., Fridovich, I. 1977. Liposome oxidation and erythrocyte lysis by enzymatically generated superoxide and hydrogen peroxide. *J. Biol. Chem.* 252:6721–28
31. Klebanoff, S. J. 1980. Oxygen metabolism and the toxic properties of phagocytes. *Ann. Int. Med.* 93:480–89
32. Kreutzer, D. L., Claypool, W. D., Jones, M. L., Ward, P. A. 1979. Isolation by hydrophobic chromatography of the chemotactic factor inactivators from human serum. *Clin. Immunol. and Immunopathol.* 12:162–76
33. Kreutzer, D. L., Desai, U., Orr, W., Showell, H., Ward, P. A. 1979. Induction of acute inflammatory reactions in lung following intrapulmonary instillation of preformed chemotactic peptides and purified complement components. *Chest* 755:2595–625
34. Kunkel, S. L., Thrall, R. S., Kunkel, R. G., McCormick, J. N., Ward, P. A., Zurier, R. B. 1979. Suppression of immune complex vasculitis in rats by prostaglandin. *J. Clin. Invest.* 64:1525–31
35. Lichtenstein, L. M. 1979. Anaphylactic reactions. In *Mechanisms of Immunopathology*, ed. S. Cohen, P. A. Ward, R. T. McCluskey, pp. 13–28. NY: Wiley
36. Lynn, W. S., Mukherjee, C. 1979. Motility of rabbit alveolar cells. Role of unsaturated fatty acids. *Am. J. Pathol.* 96:663–72
37. McCormick, J. R., Harkin, M. M., Johnson, K. J., Ward, P. A. 1981. The effect of superoxide dismutase on pulmonary and dermal inflammation. *Am. J. Pathol.* 102:55–61
38. Merrill, W. W., Naegel, G. P., Matthey, R. A., Reynolds, H. 1980. Alveolar macrophage-derived chemotactic factor. Kinetics of in vitro production and partial characterization. *J. Clin. Invest.* 65:268–76
39. Nathan, C. F., Brukner, L., Silverstein, S. C., Cohn, Z. A. 1979. Extracellular cytolysis by activated macrophages and granulocytes. *J. Exp. Med.* 149:100–13
40. O'Flaherty, J. T., Craddock, P. R., Jacob, H. S. 1978. Effect of intravascular complement activation on granulocyte adhesiveness and distribution. *Blood* 51:731–39
41. O'Flaherty, J. T., Showell, H. J., Ward, P. A. 1977. Neutropenia induced by systemic infusion of chemotactic factors. *J. Immunol.* 118:1586–89
42. O'Flaherty, J. T., Ward, P. A. 1978. Leukocyte aggregation induced by chemotactic factors. A Review. *Inflammation* 3:177–94
43. Orr, F. W., Varani, J., Kreutzer, D. L., Senior, R. M., Ward, P. A. 1979. Digestion of the fifth component of complement by leukocytic enzymes sequential

generation of chemotactic activities for leukocytes and for tumor cells. *Am. J. Pathol.* 94:75–84
44. Pensky, J., Levy, L. R., Lepow, I. H. 1961. Partial purification of a serum inhibitor of C'l-esterase. *J. Biol. Chem.* 236:1674
45. Perez, H. D., Goldstein, J. M. 1980. Generation of a chemotactic lipid from arachadonic acid by exposure to a superoxide generating system. *Fed. Proc.* 39:1170 (Abstr.)
46. Perez, H. D., Goldstein, I. M., Chernoff, D., Webster, R. O., Henson, P. M. 1980. Chemotactic activity of C5a des arg: Evidence of a requirement for an anionic peptide 'helper factor' and inhibition by a cationic protein in serum from patients with systemic lupus erythematosus. *Mol. Immunol.* 17:163–69
47. Petrone, W. F., English, D. K., Wong, K., McCord, J. M. 1980. Free radicals and inflammation: The superoxide dependent activation of a neutrophil chemotactic factor in plasma. *Proc. Natl. Acad. Sci. USA* 77:1159–63
48. Postlethwaite, A. E., Kang, A. H. 1976. Collagen and collagen peptide-induced chemotaxis of human blood monocytes. *J. Exp. Med.* 143:1299–307
49. Postlewaite, A. E., Snyderman, R. 1975. Characterization of chemotactic activity produced in vivo by a cell-mediated immune reaction in the guinea pig. *J. Immunol.* 114:274–78
50. Rister, M., Baehner, R. L. 1976. The alteration of superoxide dismutase, catalase, glutathione peroxidase, NAD(P)H, cytochrome C reductase in guinea pig polymorphonuclear leukocytes and alveolar macrophages during hyperoxia. *J. Clin. Invest.* 58:1174–84
51. Rodriguez, R. J., White, R. R., Senior, R. M., Levine, E. A. 1977. Elastase release from human alveolar macrophages: comparison between smokers and nonsmokers. *Science* 198:313–14
52. Roos, B., Keller, H. U., Hess, M. W., Cottier, H. 1977. *Alveolar Macrophages: Phagocytosis-Induced Release of Neutrophil Chemotactic Activity, Pulmonary Macrophages and Epithelial Cells*, ed. C. L. Sanders, et al, pp. 326–39. Washington DC: ERDA Tech. Inf. Cent.
53. Saltzman, H. A., Fridovich, I. 1975. Oxygen toxicity. Introduction to a protective enzyme: Superoxide dismutase. *Circulation* 48:921–23
54. Scherzer, H., Ward, P. A. 1978. Lung injury produced by immune complexes of varying composition. *J. Immunol.* 121:947–52

55. Senior, R. M., Griffin, G. L., Mecham, R. P. 1980. Chemotactic activity of elastin-derived peptides. *J. Clin. Invest.* 66:859–62
56. Shaw, J., Henson, J., Phillips, D., Henson, P. M. 1978. Lung injury induced by intratracheal administration of a chemotactic fragment from the fifth component of complement. *Am. Rev. Resp. Dis.* 117:80
57. Stimler, N. P., Hugli, T. E., Bloor, C. M. 1980. Pulmonary injury induced by C3a and C5a anaphylatoxins. *Am. J. Pathol.* 100:327–48
58. Sullivan, T. J., Kulczycki, A. 1980. Immediate hypersensitivity responses in clinical immunology. See Ref. 63, pp. 115–42
59. Taylor, F. B., Jr., Ward, P. A. 1967. Generation of chemotactic activity in rabbit serum by plasminogen streptokinase mixtures. *J. Exp. Med.* 126:149–60
60. Till, G., Ward, P. A. 1975. Two distinct chemotactic factor inactivators in human serum. *J. Immunol.* 114:843–47
61. Unkeless, J., Gordon, S., Reich, E. 1974. Secretion of plasminogen activator by stimulated macrophages. *J. Exp. Med.* 139:834–41
62. Verge, P., Olsson, I. 1975. Cationic proteins of human granulocytes. VI. Effects on the complement system and mediation of chemotactic activity. *J. Immunol.* 115:1505–8
63. Ward, P. A. 1980. Chemotaxis. In *Textbook of Immunology*, ed. C. W. Parker, 1:272–297. Philadelphia: W. B. Saunders
64. Ward, P. A., Cohen, S., Fluragon, T. D. 1972. Leukotactic factors elaborated by virus infected tissues. *J. Exp. Med.* 135:1095–103
65. Ward, P. A., Hill, J. H. 1970. C5 chemotactic fragments produced by an enzyme in lysosomal granules of neutrophils. *J. Immunol.* 104:535–43
66. Ward, P. A., Lepow, I. H., Newman, L. J. 1960. Bacterial factors chemotactic for polymorphonuclear leukocytes. *Am. J. Pathol.* 52:725–36
67. Weiss, S. J., LoBuglio, A. F., Kessler, H. B. 1980. Oxidative mechanisms of monocyte mediated cytotoxicity. *Proc. Natl. Acad. Sci. USA* 77:584–87
68. Weissman, G., Goldstein, I., Hoffstein, S. 1976. Prostaglandins and the modulation by cyclic nucleotides of lysosomal enzyme release. In *Advances in Prostaglandin and Thromboxane Research*, ed. B. Samuelsson and R. Paoletti, 2:803–18. NY: Raven
69. Zurier, R. B., Quagliata, F. 1971. Effects of prostaglandin E_1 on adjuvant arthritis. *Nature* 234:304–6

CELL AND MEMBRANE PHYSIOLOGY

Introduction, John Gergely, *Section Editor*

The four papers in this section deal with certain aspects of two important membrane systems of muscle: the plasma membrane and the sarcoplasmic reticulum. Two are concerned with ways in which membranes are put together in the course of development or the ways in which components of membranes are broken down and replaced by new ones. The other two deal with the molecular basis of a key function for each membrane. Martonosi reviews recent studies on the development of the sarcoplasmic reticulum. Pumplin & Fambrough trace the turnover of a key component of the plasma membrane—the acetylcholine receptor. Both chapters emphasize the role of activity—mediated via the nerve impulse—in regulating developmental and metabolic processes in membranes. Ikemoto and Stefani & Chiarandini review ion transport and ion movement in sarcoplasmic reticulum and plasma membrane, respectively. Active and passive ion movements are essential in transmitting the signal from the nerve to the interior of the muscle and in linking it to the contractile system. These reviews cover important advances in their selected specific fields. They also highlight insights applicable to membranes wherever they occur and are of interest to students of a variety of problems involving membranes.

STRUCTURE AND FUNCTION OF THE CALCIUM PUMP PROTEIN OF SARCOPLASMIC RETICULUM

Noriaki Ikemoto

Department of Muscle Research, Boston Biomedical Research Institute, and Department of Neurology, Harvard Medical School, Boston, Massachusetts 02114

INTRODUCTION

According to the generally accepted view concerning the regulation of contraction and relaxation of muscle, excitation initiated at the cell surface by the nerve impulse is propagated into the cell through the transverse membrane system (T-system). It spreads further to another membrane system, the sarcoplasmic reticulum (SR), which in turn releases the Ca^{2+} accumulated in the compartment delimited by the SR membrane. Combination of Ca^{2+} with the Ca^{2+} binding subunit of troponin located in the thin filaments, or in lower organisms with myosin, releases the inhibition of the interaction between actin and myosin [for recent review see (101)]. The reaccumulation of cytoplasmic Ca^{2+} into the SR reverses the process, the result being the detachment of myosin from actin, which is manifested in relaxation. This Ca^{2+} uptake into the SR is an ATP-dependent process representing the operation of a Ca^{2+} pump located in the membrane.

The fact that preparations of fragmented SR consist of resealed vesicles of the SR membrane has made it possible to study Ca^{2+} transport in vitro. The $(Ca^{2+}Mg^{2+})$-activated ATPase (Ca^{2+} ATPase) moiety of the SR membrane plays a key role in the Ca^{2+} transport mechanism, as evidenced by the facts that vesicles capable of accumulating Ca^{2+} can be reconstituted from purified ATPase and phospholipids (e.g. 72, 112), and that during embryonic development both Ca^{2+} ATPase and Ca^{2+} transport activities increase in parallel with an increase of the Ca^{2+} ATPase moiety of the SR membrane (71). In view of the central role the ATPase plays in Ca^{2+}

transport, extensive studies on its reaction mechanism have been carried out in many laboratories [for recent review see (19, 39, 98, 118)]. Through these kinetic studies, various reaction intermediates involved in the Ca^{2+} pump mechanism have been resolved. Information has also been gained about the structure of the Ca^{2+} ATPase enzyme, such as the primary structure of the ATPase polypeptide and identification of various functionally important amino acid residues. This article reviews recent progress in studies of the kinetics and structure of the Ca^{2+} pump protein of SR and suggests directions for the eventual establishment of the correlation between the structural and functional aspects of the Ca^{2+} transport mechanism.

STRUCTURE OF THE CALCIUM ATPase MOLECULE

Primary Structure

Five polar segments of polypeptide (Segments I–V), which represent about 60% of the protein, have been sequenced (1, 2, 56, 105). (*a*) The lengths of the segments, (*b*) their location within the parent polypeptide, (*c*) the proposed disposition of those segments within the membrane (2, 56), and (*d*) the location of amino acid residues, whose relation to various enzymic activities are discussed below, are diagramatically presented in Figure 1. Limited digestion of the Ca^{2+} ATPase with trypsin (50, 64, 74, 94, 103)

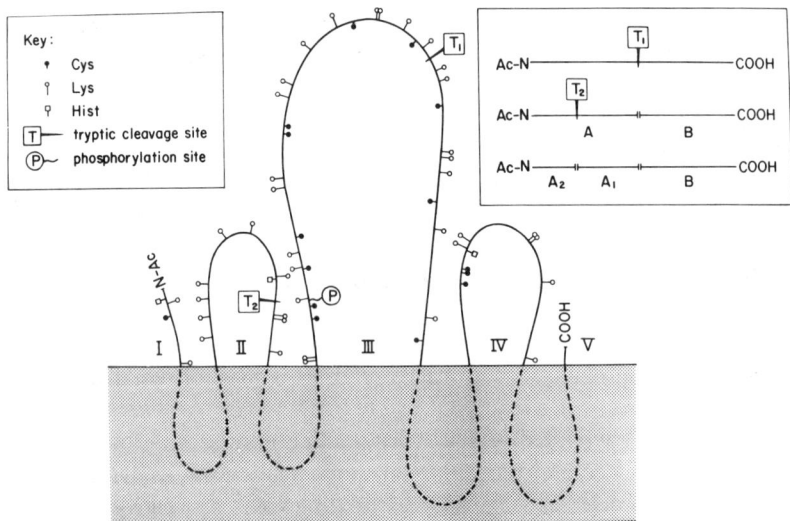

Figure 1 Tentative disposition of five polar segments (Segments I–V, solid line) and apolar segments (broken line) of the Ca^{2+}-ATPase polypeptide. Shaded area represents lipid bilayer of the sarcoplasmic reticulum membrane. For the suggested mode of protein folding, see (65).

produces in the first step two major fragments: $M_r=50{,}000-60{,}000$ (A) and $M_r=45{,}000-55{,}000$ (B); in the second step, cleavage of the A fragment produces subfragments A_1 (30,000–33,000 daltons) and A_2 (20,000–24,000 daltons). Sequence studies (2, 56, 105) have aligned the tryptic fragments as protein Ac.NH_2-A_2-A_1-B.COOH (Ac.NH_2: acetylated amino terminus).

Disposition of Various Segments within the Membrane

The first and second tryptic cleavage sites are located in Segments III and II, respectively (see Figure 1). This suggests that Segments II and III are exposed to the external side of the SR membrane (2, 56) since the membrane is impermeable to trypsin. On the basis of several criteria (56) such as accessibility of Cys^{12} to N-ethylmaleimide (MalNEt), Segment I also seems to be exposed to the external side. The two apolar segments (one between Segments I and II, and the other between Segments II and III) of the protein are unsequenced. Their length is about 100 amino acid residues, which is sufficient for the peptide to travel back and forth across the lipid bilayer (2, 56). The exact location of Segment IV in the parent molecule, and disposition of Segments IV and V, have not yet been settled. However, since more than half of the protein mass is exposed to the outside of the SR membrane and there appears to be little protein mass at the interior side of the membrane [cf (21, 29, 40) vs. (113)], it is tempting to localize both Segments IV and V at the exterior of the membrane (2, 56). However, the possibility that either Segment IV or Segment V, or both, might be located at the interior side of the membrane cannot be excluded, since the Ca^{2+} ATPase is controlled by the internal as well as the external ionic milieu (e.g. 19).

Organization of Polypeptides in the Ca^{2+} Pump

The concept that the operating unit of the Ca^{2+} pump consists of more than two '100,000-dalton' polypeptides is supported by several pieces of independent evidence. (a) There are three times as many particles on the outer surface as within the membrane. The latter, intramembranous particles represent the apolar domain of the molecules (e.g. 90). (b) The apparent molecular weight of the Ca^{2+} ATPase solubilized in non-ionic detergent is $n \times 100{,}000$ ($n > 2$) as determined by column chromatography (3, 60) and sedimentation equilibrium (75); n decreases as the detergent concentration increases, presumably owing to gradual dissociation of the oligomer to monomers (75). (c) If two batches of solubilized Ca^{2+} ATPase, presumably in the form of monomers, are labeled by two different fluorescent probes and then detergent is removed, fluorescence energy transfer is established between these probes, indicating that monomers are reassembled in close proximity (107). (d) In the presence of cross-linking reagents (6, 61, 62), or under conditions in which formation of an intermolecular disulfide bond is catalyzed (61, 76), complexes of $M_r = n \times 100,000$ ($2 < n < 6$) are produced. (*e*) Kinetic and EPR spectroscopic evidence described below also supports the above view.

Localization of Functionally Important Sites

PHOSPHORYLATION SITE During the Ca^{2+} ATPase reaction the ATPase polypeptide is covalently phosphorylated by the γ-phosphate of ATP, forming an acid stable phosphoenzyme (EP) (e.g. 66). On SDS gels, the incorporated ^{32}P is associated with the '100,000-dalton' polypeptide (68) in the case of intact SR; and ^{32}P migrates with the A-fragment, or A_1 subfragment (35, 74, 95, 103) after tryptic digestion. The amino acid residue at which phosphorylation takes place has been identified as aspartic acid (1, 7); this particular residue (Asp^{26}) is located in Segment III (2, 56) as illustrated in Figure 1.

Ca^{2+} Transport Site and Channel

The sites to which high-affinity Ca^{2+} binding takes place (11, 24, 45, 51, 73, 119) are directly involved in the transport mechanism (46), and are thus defined as Ca^{2+} transport sites. Pick & Racker (83) have found that incorporation of the covalently reacting carboxyl reagent dicyclohexylcarbodiimide (DCCD) into the Ca^{2+} ATPase moiety of SR inhibits high-affinity Ca^{2+} binding, Ca^{2+} ATPase activity, and Ca^{2+} uptake; the inhibition is prevented by Ca^{2+}, suggesting competitive binding of DCCD to the Ca^{2+} sites. The bound [^{14}C]-DCCD was localized in the A_2 subfragment of the Ca^{2+} ATPase (83). Shamoo et al (91) found that the A_2-subfragment confers Ca^{2+} conductivity on black lipid bilayers. Thus it appears that the site to which Ca^{2+} binds and the channel through which Ca^{2+} is transported across the membrane are both located in the A_2 subfragment.

Functionally Important Amino Acid Residues

CYSTEINYL RESIDUES Of 24–25 cysteinyl residues, 17–19 SH groups are titrated with 5,5'-dithiobis-(2-nitrobenzoate) (Nbs_2) in the denatured enzyme, whereas 13–15 SH groups are reactive in the native conformation (34, 47, 77, 103, 104, 114). (The number of residues is compared in terms of a mass of 110,000 daltons.) This indicates that 4 SH groups are buried in apolar regions of the membrane. As shown in Figure 1, the total number of cysteinyl residues as determined from sequence data of the polar segments is 15 (cf 2). Thus practically all of the cysteinyl residues in this region are free to react with SH reagents, although two of these might be in the form of an S-S bridge.

The free thiols described above can be grouped in three kinetically distinguishable classes in terms of the rate of reaction with MalNEt (38, 114),

Nbs$_2$ (34, 78, 104), and S-mercuric N-dansyl cysteine [Dns.S.Hg (47, 48)]. Blocking of the second most reactive class with the above reagents exerts crucial inhibition of ATPase activity. In view of earlier reports that ATP protects one thiol from blockage and prevents the inhibition of the enzyme (38, 82), it was suggested that there is only one 'crucial' SH (e.g. 77). More recently it was found that the number of 'crucial' thiols is two or more, and blocking of different thiols inhibits different elementary steps of the enzyme reaction (55, 114). For example, blocking of one thiol (designated as SH_D) results in the inhibition of the reaction step in which EP is decomposed (55, 114), whereas blocking of another (designated as SH_F) decreases the steady-state level of EP (55, 114). Reactivity of SH_D with MalNEt varies with [Ca^{2+}] in the concentration range in which high-affinity Ca^{2+} binding takes place, whereas the reactivity of SH_F is virtually independent of [Ca^{2+}] (55, 114). Such distinctions between SH_D and SH_F suggest that these thiols are located in different regions of the Ca^{2+} ATPase polypeptide, to which different roles in the reaction mechanism may be assigned. It is interesting to note in this context that thiols are clustered in a few regions of the primary structure. As first pointed out by Green et al (34), thiols densely populate the A_1 subfragment, particularly in the vicinity of the phosphorylation site (Asp^{26} of Segment III). There are three thiols in Segment IV, which are again clustered. In view of characteristic properties of some 'functionally important' thiols (e.g. SH_D and SH_F), localization of those thiols in the polypeptide structure should permit one to assess the roles various domains may play in the reaction mechanism.

LYSINE RESIDUES   Of 51 lysine residues per 110,000 daltons of the Ca^{2+} ATPase (cf. 2), 41 residues are located in the sequenced polar segments of the polypeptide (see Figure 1). According to the results of Murphy (79), one of the lysine residues located in the A_1 subfragment is highly reactive with pyridoxal-5'-phosphate which results in rapid inactivation of the ATPase activity ('essential' lysine); inactivation is prevented if Ca^{2+} and ATP are present during the modification procedure. The lysine residue located next to the phosphorylation site was tentatively identified as the 'essential' lysine [(79); cf (2)]. On the other hand, the reaction of fluorescein 5' isothiocyanate (FITC) with one of the lysines located in the B fragment results in the inhibition of ATPase and Ca^{2+} uptake activities (84, 85), which is prevented by relatively higher concentration of ATP (K_m=20 μM, 85), suggesting that this lysine is a part of a low-affinity ATP-binding, or 'regulatory,' site (85).

According to recent studies, chemical modification of lysine residues of the Ca^{2+} ATPase with fluorescamine (43) results in the inhibition of Ca^{2+} ATPase activity either by blocking EP formation or EP decomposition.

Thus there appear to be several functionally distinct lysine residues as in the case of cysteinyl residues.

HISTIDINE RESIDUES As suggested by earlier reports (120) one or more histidine residues may be involved in the function of the Ca^{2+} ATPase, since photoinactivation of the SR membrane in the presence of Rose Bengal or Methylene Blue results in a selective modification of histidine residues accompanied by inactivation of both Ca^{2+} uptake and ATPase activities. According to the studies of Tenu et al (102), ethoxyformylation of SR modifies first one lysine and next one histidine; in parallel with the modification of histidine there is a decrease of both Ca^{2+} uptake and ATPase activity. ATPase activity, but not Ca^{2+} uptake, is restored by removing the ethoxyformyl group from histidine. In contrast to the effect of ATP on the 'crucial' lysine residue previously described (79), the above histidine appears to be unaffected by ATP (102). Several interesting features emerge from the inspection of sequence data (cf Figure 1). In contrast to relatively dense distributions of cysteine and lysine, only three histidine residues are found in the polar segments.

ATPase REACTION COUPLED WITH THE CALCIUM PUMP

Various schemes of the mechanism by which the ATPase reaction is coupled with the Ca^{2+} pump have been discussed [for recent reviews see (19, 98)]. The essence of those schemes is illustrated in Figure 2. In order to facilitate the demonstration of various features, such as multiple conformational states and location of various sites at various reaction steps, the scheme is illustrated in a diagrammatic fashion. The whole reaction cycle can be described in terms of the interconversion of three classes of conformational states as indicated by *three different configurations* in the diagram. One of these, which prevails in the absence of Ca^{2+} (i.e. the one between step 9 and step 1), can be regarded as the 'ground' state. Phosphorylation of the enzyme does not take place unless the enzyme changes from the ground state to either the Ca state or the Mg state described below. The Ca state and the Mg state, which are often referred to as E_1 and E_2 states, respectively, are distinguished by the fact that the former has the ability to form EP from an M.ATP (M, metal) but not from Pi; the latter (E_2) can form EP from Pi, but not from M.ATP (8, 9, 16, 19, 37, 39, 57, 58, 87, 118). Under conventional conditions for the enzyme assays (e.g. $[Ca^{2+}] = 10^{-6}$ M; $[Mg^{2+}] = 1-5$ mM, pH = 7.0), the enzyme would be in the Ca state (19, 39). During the M.ATP-induced reaction the Ca state is eventually converted to the Mg state via several forms of EP intermediates. The Mg state can also be created directly from the ground state by adding higher concentrations of Mg^{2+} (e.g. 20 mM) in the absence of Ca^{2+} (e.g. 8). Lowering the pH and increasing the reaction temperature shift the equilibrium from the Ca state to the Mg state (e.g. 17, 20). The enzyme states can also be distinguished depending on whether the Ca^{2+} transport site is at the external or internal side of the membrane, which are aligned in the left-hand and right-hand sides of the model, respectively. In view of the facts that lower pH stabilizes the Mg state (e.g. 17) and that proton ejection takes place during the Ca^{2+} pump cycle (e.g. 12), the transport site may be occupied

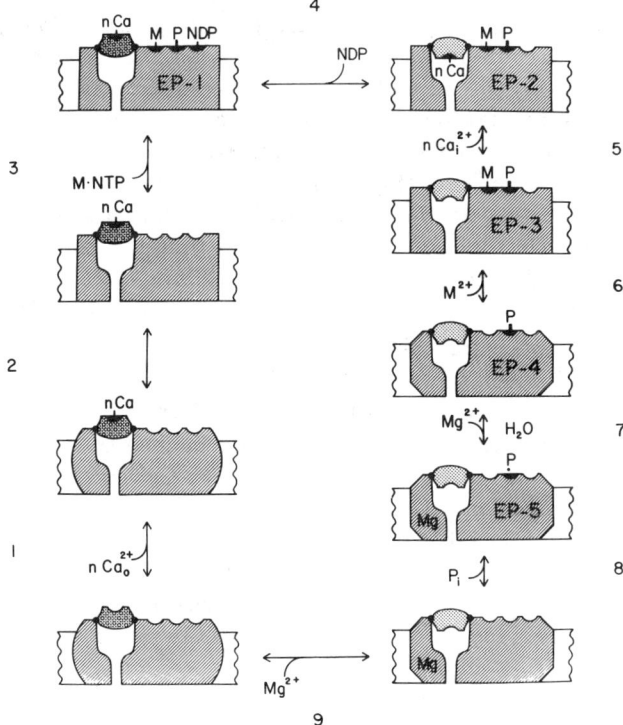

Figure 2 Diagrammatic representation of sequential formation of various intermediates in the Ca^{2+}-ATPase reaction of sarcoplasmic reticulum. Although a variety of models are proposed, the 'mobile pore' hypothesis (e.g. 98) was adopted to depict the mechanism by which externally bound Ca is translocated across the sarcoplasmic reticulum membrane. This model assumes that the 'pore,' or Ca^{2+}-channel, is located in each of the Ca^{2+} ATPase subunits. However, it may well be that the pore is formed between the subunits (69). On the basis of the literature (see text) five types of EP-intermediates (EP-1 through EP-5) have been postulated. Note the three different configurations of the enzyme protein (shaded area). Ca_i, intravesicular calcium; Ca_o, extravesicular calcium; M, metal moiety of the substrate complex; NDP, nucleotide diphosphate; NTP, nucleotide triphosphate.

by protons in the Mg state. Similarly, K^+ appears to be another species that occupies the transport site in the Mg state (13, 53). Although not indicated in the scheme, we can tentatively assume that the transport site devoid of Ca is in fact filled with protons or K, or both. In the following section, sequential formation of various reaction intermediates from the ground state are examined.

Sequential Formation of Various Reaction Intermediates

STEP 1 Two high-affinity Ca^{2+} binding, or transport, sites ($K = 3 \times 10^6$ M^{-1}) (e.g. 45) are observed per each Ca^{2+} ATPase polypeptide chain (11, 45, 73). According to equilibrium dialysis studies of Ca^{2+} binding to the purified ATPase (45), n is 2 at 22° whereas it is 1 at 0°, suggesting that the two transport sites are heterogeneous in terms of temperature dependence. According to recent observations (24), however, the amount of Ca bound to the high-affinity site (6–7 nmol per mg SR protein) as determined by a filtration technique, is roughly constant in the temperature range between –30° and 22°, suggesting that n would be 1 regardless of the reaction temperature. Thus the actual number of the transport sites operating per each polypeptide chain remains to be further investigated.

STEP 2 As demonstrated in several reports (e.g. 88, 97), the formation of EP (Step 3) is much slower when the ATPase reaction is started from an enzyme that has no bound Ca than when the reaction is started from an enzyme preincubated with Ca^{2+}. Since Ca^{2+} binding per se seems to be a rapid process, the rate-limiting step in the $E + Ca^{2+} + ATP \rightarrow EP.Ca + ADP$ reaction may involve changes in enzyme conformation (88, 97). The fact that the Ca-bound enzyme is more stable than the Ca-free form (63) also supports this view. Changes in the enzyme conformation induced by Ca^{2+} binding have been determined by various methods (see Conformational Changes of the Ca^{2+} ATPase Molecule Coupled with Enzyme Reaction, below).

ISOMERIZATION OF EP INTERMEDIATES (STEP 4–STEP 7) Upon addition of M.NTP, the terminal phosphate of NTP is covalently incorporated into the Ca^{2+} ATPase, forming an acid-stable EP. The fact that the Ca^{2+} ATPase activity (67, 111) and the initial rate of EP formation (115) are dependent primarily on [Mg.ATP], but not on [ATP], has demonstrated that the Mg.ATP complex is the true substrate for the Ca^{2+} ATPase reaction (67, 115). So far as EP formation is concerned, Mg as the metal moiety of the substrate complex can be replaced by either Ca, Mn, or Co (115). Various nucleotide species other than ATP, and also other phosphate compounds, can serve as effective substrates for the EP formation reaction in the presence of appropriate bivalent cations (e.g. 89).

That several distinct forms of EP intermediates are involved in the ATPase reaction can be deduced from various pieces of evidence (listed in Table I): (a) At least two sequential forms of EP are distinguishable in terms of affinity of the transport site for Ca^{2+} [(45); cf (108)]; i.e. the one formed first has high affinity for Ca^{2+} (EP-1, Table 1), while others formed later have a reduced affinity (EP-2 + EP-3 + EP-4). (b) Shigekawa et al (92, 93) found that one of the EP intermediates reacts with ADP to form ATP (ADP-sensitive; EP-2), whereas the other (EP-3 + EP-4) is ADP-insensitive. Recent rapid multi-mixing studies (32) show that in an early phase of the EP formation reaction the [ADP]-dependent rapidly decaying portion of the EP formed is small (EP-1), suggesting that the first EP formed (EP-1) has bound ADP. It is rapidly converted to a second EP, which is ADP-sensitive (EP-2), subsequently becoming ADP-insensitive (EP-3 + EP-4). (c) The last ADP-insensitive form of EP can be resolved further into two states (EP-3 and EP-4), as indicated by the fact that the portion of the ADP-insensitive EP that is rapidly decomposed by the addition of Mg^{2+}(Mg^{2+}-sensitive, EP-4) is formed at a much slower rate than the formation of the total ADP-insensitive EP (i.e. EP-3 + EP-4) (115). Because EP-4 is Mg^{2+}-sensitive whereas EP-3 is Mg^{2+}-insensitive, it is tentatively proposed that conformational changes from the Ca state to the Mg state take place in Step 6.

Several important hypotheses concerning various events related to the isomerization steps involving EP are included in Figure 2. It is assumed that translocation of externally bound Ca to the inside of the membrane, which according to the 'mobile gate' model [(117; cf 69)] is equivalent to a turn of the 'gate,' occurs in Step 4. The recent finding that upon EP formation the externally bound Ca is occluded, as evidenced by the fact that it becomes inaccessible to EGTA (24, 99, 100) even in the solubilized enzyme (99, 100), can be explained at least partly by the reorientation of the transport site toward the molecular pocket represented by the Ca^{2+} channel. Concurrently, the affinity of the transport site for Ca^{2+} decreases, and NDP is released from the enzyme. Release of NDP confers NDP-sensitivity on EP-2 owing to the fact that the rate of the back reaction of Step 4 is determined by [NDP], since the rate is expressed as k_{-4} [EP 2] [NDP]. The metal moiety of the M.NTP complex must remain bound after NDP has been released, since the effective substrate for the decay of EP-2 through the back reaction is free ADP, although the substrate for the formation of EP-1 is an M.NTP complex (67,115). In view of the fact that higher concentrations of Ca^{2+} prevent the decomposition of EP-2 (45, 53), release of the translocated Ca^{2+} to the intramolecular pocket, and

Table 1 Characteristics that distinguish various forms of EP-intermediates

Characteristic	EP–1	EP–2	EP–3	EP–4	EP–5
Acid-stability	stable			→	labile
Ca^{2+}-affinity	high	low			→
ADP-sensitivity	*	sensitive	insensitive		→
Mg^{2+}-sensitivity			insensitive	sensitive	
Transport site	exterior	interior			→

*EP. ADP. For details see text.

eventually to the intravesicular lumen, seems to take place in Step 5. The rate of the EP-3 → EP-4 reaction (Step 6) is inhibited by low concentrations of bivalent metal (on the order of micromolar), suggesting that the metal bound to the acceptor site for the metal moiety of M.NTP is released in this step (115). In view of the facts that decomposition of EP-4 is accelerated by Mg^{2+} and K^+ (e.g. 115), and that the EP + H_2O^{18} ↔ EP + Pi exchange is rapid (4, 18, 54), the fifth form of EP may be a noncovalent (or acid-labile) complex. However, the proposed involvement of an acid-labile EP in the Ca^{2+} ATPase reaction (30, 31) is a matter of dispute (96); thus establishment of a method that permits EP measurements in both denaturing and nondenaturing conditions is required to settle this point.

STEP 8 AND STEP 9   It has been proposed that the liberation of Pi and Mg^{2+} from the enzyme occurs in sequential fashion (19), and the enzyme returns to the 'ground state'.

BACK REACTION   A considerable body of evidence indicates that the reaction cycle described above is completely reversible (reviewed in 39). Some of the important aspects documented in the reaction scheme (Figure 2), particularly isomerization of several forms of the Mg state (Steps 7, 8 and 9), are based upon findings in studies of the back reaction (cf 19). Calorimetric studies showed that the enthalpy change of Mg^{2+} binding is much larger than that of Pi binding (26). This suggests that extensive conformational changes are produced by Mg^{2+} binding as a prerequisite for EP formation from Pi, although formally the process of E.Mg.P formation may be written in random sequence [as in (26, 58, 87)]. The nature and location of this bound Mg have not yet been settled (52, 53). However, because Mg-induced acceleration of EP decomposition in the forward reaction can be observed even after prolonged incubation in Mg-chelating agents (33), it appears that the bound Mg is occluded from the external membrane surface.

Subunit-Subunit Interaction

As described above (see Structure of the Ca^{2+} ATPase Molecule), the Ca $^{2+}$ pump appears to consist of more than two ATPase polypeptides. Recent evidence described below suggests that the kinetic behavior of the Ca^{2+} ATPase is regulated by subunit-subunit interaction. Thus it is necessary to incorporate the mode of oligomeric interaction into the reaction scheme discussed above.

A recent report (27) that the transport ratio (i.e. Ca transported / EP formed) can be as high as 6 in the GTP-driven reaction suggests that phosphorylation in one of the subunits triggers the Ca^{2+} translocation reaction of several subunits. Alternatively, incorporation of approximately 1 mol of DCCD per 4 mol of ATPase leads to almost complete inhibition of Ca^{2+} ATPase activity (83). Similarly, incorporation of 1 mol of FITC into about 2 mol of the Ca^{2+} ATPase produces almost complete inhibition of Ca^{2+} uptake (84). The above facts are consistent with the view that several interacting subunits are essential for the functional integrity of the Ca^{2+} pump.

The hypothesis that the Ca^{2+} ATPase reaction may be carried out by two or more subunits operating as a flip-flop pair was suggested by Froehlich & Taylor (30, 31), based chiefly on the finding that both the formation and decomposition of EP are accelerated by relatively low [ATP] in a cooperative fashion, although in that [ATP] range there appears to be only one ATP binding site per polypeptide chain. More recently the kinetic process of Ca^{2+} binding has been studied with the use of a variety of direct and indirect methods (J. P. Froehlich, Y. Dupont, personal communications). From these studies, it appears that Ca^{2+} binding to the enzyme occurs in two kinetically distinguishable phases (cf 51); in a first phase one Ca is rapidly bound to the enzyme with a lower affinity, and in a second phase another Ca is bound with a higher affinity but at a much slower rate. Since the Ca^{2+} binding per se must be rapid, the apparent slowness of the second Ca^{2+} binding may be attributable to the slowness of the enzyme conformational change (E→E*) that is a prerequisite for the second Ca^{2+} binding [model A, Figure 3; cf (51)]. A similar 'two-step' Ca^{2+} binding was observed in conditions in which the number of Ca^{2+} sites operating per polypeptide chain is one (Y. Dupont, in preparation). Thus model A should be modified as shown in Model B (Figure 3), in which it is assumed that the Ca^{2+}-induced conformational change in one subunit (E→E#) affects that in the other (E→E*) by oligomeric interaction. From Model B it is expected that, in the back reaction, bound Ca would be dissociated in two phases. In fact it was observed that half of the bound Ca dissociates very rapidly, whereas the remaining half dissociates slowly [(49), cf (97)]. Although in Model B

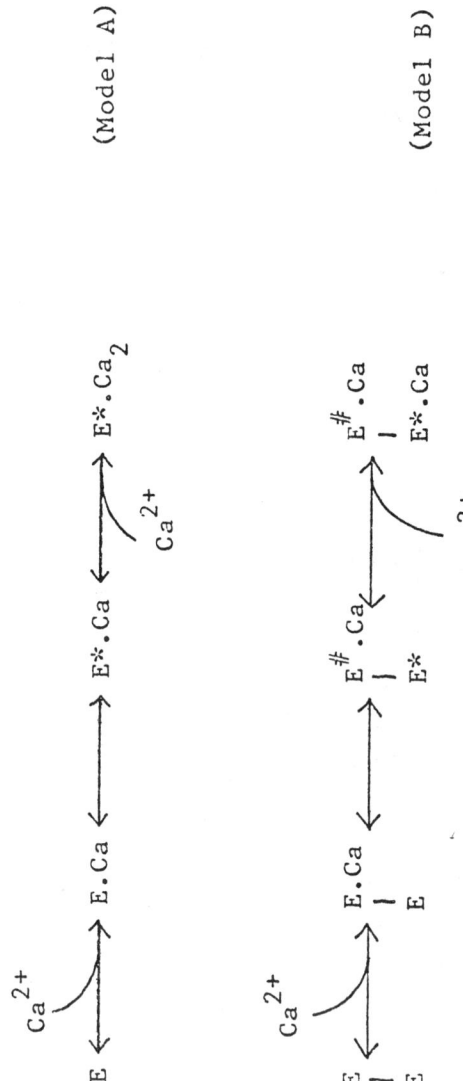

Figure 3 Two models for 'two-step' Ca^{2+} binding. Model B is most appropriate to account for recent data (see text).

the enzyme is tentatively represented as a dimer, the actual number of subunits should be regarded as 2 X n ($1 \leq n \leq 3$).

THE RELATION OF CONFORMATIONAL CHANGES IN THE CALCIUM ATPase MOLECULE TO STEPS COUPLED WITH THE ENZYME REACTION

Progress has been made towards the elucidation of this problem by studies of (a) intrinsic fluorescence intensity of tryptophan residues of the Ca^{2+} ATPase, (b) reactivity of amino-acid residues with specific reagents, (c) local or global protein motion with the use of covalently attached spin-label probes, (d) lamellar X-ray diffraction pattern of layered SR membranes, and (e) fluorescence intensity of covalently attached fluorescent probes.

  a. Dupont (22, 23) found that the intensity of intrinsic tryptophan fluorescence varies with the three enzyme states—i.e. in order from lower to higher fluorescence intensity, 1. Mg state produced in the presence of Mg^{2+} and EGTA, 2. EP formed from Pi, and 3. Ca state produced by high-affinity Ca^{2+} binding. In the light of stopped-flow fluorometric data (e.g. 23, 36), the fluorescence increase appears to be associated with the second phase of the two-step Ca^{2+} binding described above. Since most of the tryptophanyl residues (18 out of 19 tryptophan residues per 1.1×10^5 g) appear to be buried in apolar regions of the Ca^{2+} ATPase (2), the Ca^{2+}-controlled conformational changes may involve at least the apolar regions of the enzyme molecule. However, because of the large number of tryptophanyl residues it is rather difficult to pinpoint the regions where the conformational changes take place, although the distinction of Ca^{2+}-sensitive and -insensitive tryptophans is experimentally possible with the use of an appropriate fluorescence quencher (109). Ca^{2+}-induced changes in tryptophan fluorescence are observed with detergent-solubilized, monomeric, Ca^{2+} ATPase as well as with membrane-associated enzyme that is presumably oligomeric (15, 25, 110). As judged from the [Ca^{2+}] dependence of the fluorescence change, Ca^{2+} binding is cooperative in the membrane-associated enzyme, whereas it is noncooperative in the monomeric enzyme (110), further supporting the notion that cooperative Ca^{2+} binding is a result of oligomeric interaction (cf Model B, Figure 3).

  b. Reactivity of various amino acid residues with covalently or noncovalently reacting reagents varies as a function of [Ca^{2+}], [Mg^{2}], and [ATP], and a combination of these. For example, the [Ca^{2+}] affects reactivity of different thiol classes in different ways as described below. The rate of reaction of the 'fast' class with MalNEt or Dns.Cys.Hg is virtually independent of [Ca^{2+}]. However, the reaction rate of the 'second fast' class with these reagents increases with [Ca^{2+}] in the [Ca^{2+}] range in which high affinity Ca^{2+} binding takes place (47, 48, 114). In contrast, the rate of reaction of the 'slow' class with Dns.Cys.Hg (but not with MalNEt) decreases with $[Ca^{2+}]$ in the same $[Ca^{2+}]$ range (47, 48). Thus it appears that the $E \rightarrow E^\#$ or $E \rightarrow E^*$ reactions, or both (see Model B, Figure 3) involve extensive changes in the enzyme conformation (see Step 2, Figure 2). It also appears that associated with these reactions different types of structural changes (e.g. exposure versus occlusion of cysteinyl residues) occur in different regions of the molecule (48), indicating that Ca^{2+} binding produces a 'fine tuning' of the enzyme conformation. In contrast, changes produced by M.NTP binding and EP formation are relatively small, and the changes involved are rather global [(48); cf (70)]. These facts suggest that the enzyme conformation created in Step 2 would be essentially maintained throughout the subsequent reaction steps. In accord with this view, a number of thiols become less reactive with Nbs_2 in the presence of ATP, Mg^{2+}, and Ca^{2+} (34, 77, 78), whereas Ca^{2+} appears to exert a selective effect on a limited number of thiols (78). The reaction of lysine residues with TNBS has also been used for studies of various conformational states. The accessibility of five TNBS-reactive lysines shows characteristic changes corresponding to Ca^{2+} binding and EP formation, respectively (116, 117). The rate of dinitrophenylation varies with Ca^{2+} in the $[Ca^{2+}]$ range in which high-affinity binding takes place (5), as in the case of SH_D described above.

    c. Molecular motion of the Ca^{2+} ATPase protein, as monitored by EPR spectroscopy with the aid of thiol-directed spin label reagents attached to the Ca^{2+} ATPase moiety of the SR membrane, is altered as Ca^{2+} binding (51), ATP binding (59), and the Mg.ATP-induced enzyme reaction take place (14, 80, 81). Furthermore, there is a qualitative correlation between the rate of Pi liberation and the extent of protein motion (cf 42). Thus it appears that the protein motion is related to changes in the enzyme conformation associated with various reaction steps. However, it should be noted that the time scale of the protein motion detected by conventional EPR spectroscopy ($\leq 10^{-7}$sec) and by saturation transfer EPR spectroscopy (10^{-3}–10^{-5}sec) (cf 42) is much faster than the time scale for any reaction steps in which the enzyme undergoes conformational change ($\geq 10^{-3}$sec). Molecular motion of the protein-attached spin label probe appears to reflect protein-protein interactions as evidenced by the fact that protein motion increases upon addition of detergent under nonsolubilizing conditions (28). A similar conclusion has also been derived from studies of the temperature-dependence of protein rotational motion (44). Accessibility of the protein-bound spin-label derivative of thiol reagents to various paramagnetic quenchers (e.g. ascorbate, nickel, and ferricynanide) has been used for the assessment of the external-internal transition of the protein molecule (10, 106), yielding contradictory conclusions: Reorientation of the enzyme molecule is rather extensive (106) or very little (10). This contradiction may be at least partly due to the difference in the location of the investigated thiol groups to which spin probes are attached.

  d. X-ray diffraction techniques applied to studies of the density profile of layered SR membranes permit one to resolve the disposition of the Ca^{2+} ATPase molecule in the SR membrane (21, 113). Recent progress in the X-ray diffraction analysis has been achieved in studies with the use of 'caged' ATP, which permits one to compare the same preparation before and after induction of the ATPase reaction (40, 41). It appears that the Ca^{2+} ATPase molecule, which has an ovoidal shape before interacting with ATP, becomes more globular as the ATPase-coupled Ca^{2+} pump is activated (41).

  e. Extrinsic fluorescence probes attached to well-defined sites of the ATPase polypeptide should provide crucial information concerning 1. the type of conformational changes that occurs and 2. the region of the enzyme polypeptide where the changes occur. Several promising probes and methods for this purpose have begun to emerge. The fluorescence intensity of the FITC molecule attached to the Ca^{2+} ATPase, presumably to one of the lysine residues located in the B subfragment region, decreases upon Ca^{2+} binding, and increases upon chelation of bound Ca in the presence of Mg^{2+} (84, 86), suggesting that the higher and lower fluorescence levels correspond to the Mg (E_2) and Ca states (E_1), respectively. Although ATPase activity is severely inhibited by labeling, acetyl phosphatase activity remains intact; during the acetyl phosphatase reaction FITC fluorescence increases, probably reflecting the $E_1 \rightarrow E_2$ conversion (86). According to our preliminary data (K. Miki et al), the fluorescence intensity of the thiol-directed fluorescent reagent N-(1-anilinonaphtyl-4) maleimide (ANM), which is covalently attached to the second most reactive thiol (but not the one attached to the most reactive thiol), increases as EP is formed from ATP; and the fluorescence change is reversed as EP decays to form ATP upon addition of ADP. The second ANM is localized in the A_2 subfragment region. Thus concomitant with the EP reaction there is a conformational change in the area of the polypeptide chain where the transport site and channel appear to be located (see above).

SUMMARY

Recent developments concerning the structure and function of the Ca^{2+} pump protein of the sarcoplasmic reticulum have been briefly reviewed. Various new methods have become available that make it possible to monitor dynamic changes in the structure of the enzyme molecule associated with elementary steps of the enzyme reaction. In the light of information about chemical reactivity of various amino acid residues and their location in the primary structure of the ATPase polypeptide, it will be fruitful to use extrinsic conformational probes placed at specific locations to monitor the kinetics of the enzyme. Furthermore, a growing body of evidence suggests that subunit-subunit interactions of an oligomeric Ca^{2+} ATPase are involved in the regulation of the kinetics of the enzyme. Thus the kinetic mechanism has to be reinterpreted at all levels—i.e. primary, secondary, tertiary, and quarternary—of structure.

ACKNOWLEDGMENTS

I wish to thank Dr. J. Gergely for his encouragement in writing this article, and Drs. C. Hidalgo, M. S. Rosemblatt, and T. L. Scott for their comments on the manuscript. My thanks are also due to Drs. J. P. Froehlich, Y. Dupont, and D. Haynes for their valuable information. This work was supported by grants from NIH (AM 16922) and MDA.

Literature Cited

1. Allen, G., Green, N. M. 1976. A 31-residue tryptic peptide from the active site of the Ca^{2+}-transporting adenosine triphosphatase of rabbit skeletal sarcoplasmic reticulum. *FEBS Lett.* 63:188–92
2. Allen, G., Trinnaman, B. J., Green, N. M. 1980. The primary structure of the calcium ion-transporting adenosine triphosphatase protein of rabbit skeletal sarcoplasmic reticulum. *Biochem. J.* 187:591–616
3. Andersen, J. P., LeMaire, M., Moller, J. V. 1980. Properties of detergent-solubilized and membraneous $(Ca^{2+}+Mg^{2+})$-activated ATPase from sarcoplasmic reticulum as studied by sulfhydryl reactivity and ESR spectroscopy. Effect of protein-protein interactions. *Biochim. Biophys. Acta* 603:84–101
4. Ariki, M., Boyer, P. D. 1980. Characterization of medium inorganic phosphate-water exchange catalyzed by sarcoplasmic reticulum vesicles. *Biochemistry* 19:2001–4
5. Bailin, G. 1980. Dinitrophenylation of rabbit skeletal sarcoplasmic reticulum ATPase protein. *Biochim. Biophys. Acta* 623:213–24
6. Bailin, G. 1980. Crosslinking of sarcoplasmic reticulum ATPase protein with 1,5-difluoro 2,4-dinitrobenzene. *Biochim Biophys. Acta* 624:511–21
7. Bastide, F., Meissner, G., Fleischer, S., Post, R. L. 1973. Similarity of the active site of phosphorylation of the adenosine triphosphatase for transport of sodium and potassium ions in kidney to that for transport of calcium ions in the sarcoplasmic reticulum of muscle. *J. Biol. Chem.* 248:8385–91
8. Chaloub, R. M., Guimaraes-Mutta, H., Verjovski-Almeida, S., deMeis, L., Inesi, G. 1979. Sequential reaction in Pi utilization for ATP synthesis by sarcoplasmic reticulum. *J. Biol. Chem.* 254:9464–68
9. Chaloub, R. M., deMeis, L. 1980. Effect of K^+ on phosphorylation of the sarcoplasmic reticulum ATPase by either Pi or ATP. *J. Biol. Chem.* 255:6168–72
10. Champeil, P., Rigand, J. L., Garybobo, C. M. 1980. Calcium translocation mechanism in sarcoplasmic reticulum vesicles, deduced from location studies of protein-bound spin labels. *Prod. Natl. Acad. Sci. USA* 77:2405–9
11. Chevallier, J., Butow, R. A. 1971. Calcium binding to the sarcoplasmic reticulum of rabbit skeletal muscle. *Biochemistry* 10:2733–37
12. Chiesi, M., Inesi, G. 1980. Adenosine 5'-triphosphate dependent fluxes of manganese and hydrogen ions in sarcoplasmic reticulum vesicles. *Biochemistry* 19:2912–18
13. Chiu, V. C. K., Haynes, D. H. 1980. Rapid kinetic studies of active Ca^{2+} transport in sarcoplasmic reticulum. *J. Membrane Biol.* 56:219–39
14. Coan, C. R., Inesi, G. 1977. Ca^{2+} dependent effect of ATP on spin-labeled sar-

coplasmic reticulum. *J. Biol. Chem.* 252:3044–49
15. Dean, W. L., Gray, R. D. 1980. Transient kinetics of a Ca^{2+}-induced fluorescence change from membrane-associated and solubilized sarcoplasmic reticulum Ca^{2+}-ATPase. *J. Biol. Chem.* 255:7514–16
16. deMeis, L., Carvalho, M. G. C. 1976. On the sidedness of membrane phosphorylation by Pi and ATP synthesis during reversal of the Ca^{2+} pump of sarcoplasmic reticulum vesicles. *J. Biol. Chem.* 251:1413–17
17. deMeis, L., Tume, R. K. 1977. A new mechanism by which an H^+ concentration gradient drives the synthesis of adenosine triphosphate, pH jump, and adenosine triphosphate synthesis by the Ca^{2+}-dependent adenosine triphosphate of sarcoplasmic reticulum. *Biochemistry* 16:4455–63
18. deMeis, L., Boyer, P. 1978. Induction by nucleotide triphosphate hydrolysis of a form of sarcoplasmic reticulum ATPase capable of medium-phosphate-oxygen exchange in presence of calcium. *J. Biol. Chem.* 253:1556–59
19. deMeis, L., Vianna, A. L. 1979. Energy interconversion by the Ca^{2+}-dependent ATPase of the sarcoplasmic reticulum. *Ann. Rev. Biochem.* 48:275–92
20. deMeis, L., Martin, O. B., Aves, E. W. 1980. Role of water, hydrogen ion and temperature on the synthesis of adenosine triphosphate by the sarcoplasmic reticulum adenosine triphosphatase in the absence of a calcium ion gradient. *Biochemistry* 19:4252–61
21. Dupont, Y., Harrison, S. C. Hasselbach, W. 1973. Molecular organization in the sarcoplasmic reticulum membrane studied by x-ray diffraction. *Nature* 244:555–58
22. Dupont, Y. 1976. Fluorescence studies of the sarcoplasmic reticulum calcium pump. *Biochem. Biophys. Res. Commun.* 71:544–50
23. Dupont, Y., Leigh, J. B. 1978. Transient kinetics of sarcoplasmic reticulum $Ca^{2+}+Mg^{2+}$ ATPase studied by fluorescence. *Nature* 273:396–98
24. Dupont, Y. 1980. Occlusion of divalent cations in the phosphorylated calcium pump of sarcoplasmic reticulum. *Eur. J. Biochem.* 109:231–38
25. Dupont, Y., LeMaire, M. 1980. Fluorometric study of the solubilized Ca^{2+}-ATPase of sarcoplasmic reticulum. *FEBS Lett.* 115:247–52
26. Epstein, M., Kuriki, Y., Bittsonen, R., Racker, E. 1980. Calorimetric studies of ligand-induced modulation of calcium adenosine 5'-triphosphatase from sarcoplasmic reticulum. *Biochemistry* 19:5564–68
27. Fassold, E., Chak, D. V., Hasselbach, W. 1981. Variable Ca^{2+}-transport: phosphoprotein ratios in the early part of the GTP-driven calcium-transport reaction of the sarcoplasmic reticulum. *Eur. J. Biochem.* 113:611–16
28. Fellmann, P., Andersen, J., Devaux, P. F., LeMaire, M., Bienvenue, A. 1980. Photoaffinity spin-labeling of the Ca^{2+} ATPase in sarcoplasmic reticulum: evidence for oligomeric structure. *Biochem. Biophys. Res. Commun.* 95:289–95
29. Fleischer, S., Wang, C.-T., Saito, A., Pilarska, M., McIntyre, J. O. 1979. Structural studies of sarcoplasmic reticulum in vitro and in situ. In *Cation Flux Across Biomembranes* ed. Y. Mukohata, L. Packer, pp. 193–205. NY: Academic
30. Froehlich, J. P., Taylor, E. W. 1975. Transient state kinetic studies of sarcoplasmic reticulum adenosine triphosphatase. *J. Biol. Chem.* 250:2013–21
31. Froehlich, J. P., Taylor, E. 1976. Transient state kinetic effects of calcium ion on sarcoplasmic reticulum adenosine triphosphatase. *J. Biol. Chem.* 251:2307–15
32. Froehlich, J. P., Heller, P. F., Passonneau, J. V. 1980. Dephosphorylation by ADP of sarcoplasmic reticulum ATPase in the presence of KCl. *Fed Proc.* 39:2151 (Abstr.)
33. Garrahan, P. J., Rega, A. F., Alonso, G. L. 1976. The interaction of magnesium ions with the calcium pump of sarcoplasmic reticulum. *Biochim. Biophys. Acta* 448:121–32
34. Green, N. M., Allen, G., Hebdon, G. M., Thorley-Lawson, D. A. 1977. In *Calcium Binding Proteins and Calcium Function*, ed. R. H. Wasserman, R. A. Corradino, E. Carafori, R. H., Kretsinger, D. H. MacLennan, F. L. Siegel, pp. 164–72. NY: North-Holland
35. Green, N. M., Allen, G., Hebdon, G. M. 1980. Structural relationship between the calcium and magnesium-transporting ATPase of sarcoplasmic reticulum and the membrane. *Ann. NY Acad. Sci.* 358:149–58
36. Guillain, F., Gingold, M. P., Buschlen, S., Champeil, P. 1980. A direct fluorescence study of the transient steps induced by calcium binding to sarcoplasmic reticulum ATPase. *J. Biol. Chem.* 255:2072–76

37. Guimaraes-Motta, H., deMeis, L. 1980. Pathway for ATP synthesis by sarcoplasmic reticulum ATPase. *Arch. Biochem. Biophys.* 203:395–403
38. Hasselbach, W., Seraydarian, K. 1966. The role of sulfhydryl groups in calcium transport through the sarcoplasmic membranes of skeletal muscle. *Biochem. Z.* 345:159–72
39. Hasselbach, W. 1978. The reversibility of the sarcoplasmic calcium pump. *Biochim. Biophys. Acta* 515:23–53
40. Herbette, L., Marquardt, J., Scarpa, A., Blasie, J. K. 1977. A direct analysis of lamellar X-ray diffraction from hydrated oriented multilayers of fully functional sarcoplasmic reticulum. *Biophys. J.* 20:245–72
41. Herbette, L., Blasie, J. K. 1980. Static and time resolved structural studies on isolated sarcoplasmic reticulum membrane. In *Calcium-Binding Proteins: Structure and Function,* ed. F. L. Siegel, E. Carafoli, R. H. Kretsinger, D. H. MacLennan, R. H. Wasserman, pp. 115–20. NY: Elsevier North Holland
42. Hidalgo, C., Thomas, D. D., Ikemoto, N. 1978. Effect of the lipid environment on protein motion and enzymatic activity of the sarcoplasmic reticulum calcium ATPase. *J. Biol. Chem.* 253:6879–87
43. Hidalgo, C., Petrucci, D. A., Vergara, C. 1981. Uncoupling of Ca^{2+} transport in sarcoplasmic reticulum as a result of labeling lipid amino groups and inhibition of Ca^{2+} ATPase activity by modification of lysine residues of the Ca^{2+} ATPase polypeptide. *J. Biol. Chem.* In press
44. Hoffmann, W., Sarzala, M. G., Chapman, D. 1979. Rotational motion and evidence for oligomeric structure of sarcoplasmic reticulum Ca^{2+} activated ATPase. *Proc. Natl. Acad. Sci. USA* 76:3860–64
45. Ikemoto, N. 1975. Transport and inhibitory Ca^{2+} binding sites on the ATPase enzyme isolated from the sarcoplasmic reticulum. *J. Biol. Chem.* 250:7219–24
46. Ikemoto, N. 1976. Behavior of the Ca^{2+} transport sites linked with the phosphorylation reaction of ATPase purified from the sarcoplasmic reticulum. *J. Biol. Chem.* 251:7275–77
47. Ikemoto, N., Morgan, J. F., Yamada, S. 1978. Ca^{2+}-controlled conformational states of the Ca^{2+} transport enzyme of sarcoplamic reticulum. *J. Biol. Chem.* 253:8027–33
48. Ikemoto, N. 1979. Conformation of various reaction intermediates of sarcoplasmic reticulum Ca^{2+} ATPase. See Ref. 29, pp. 77–87
49. Ikemoto, N., Garcia, A. M., Kurobe, Y., Scott, T. 1981. Nonequivalent subunits in the calcium pump of sarcoplasmic reticulum. *J. Biol. Chem.* In press
50. Inesi, G., Scales, D. 1974. Tryptic cleavage of sarcoplasmic reticulum protein. *Biochemistry* 13:3298–306
51. Inesi, G., Kurzmack, M., Coan, C., Lewis, D. E. 1980. Cooperative calcium binding and ATPase activation in sarcoplasmic reticulum vesicles. *J. Biol. Chem.* 255:3025–31
52. Kalbitzer, H. R., Stehlik, D., Hasselbach, W. 1978. The binding of calcium and magnesium to sarcoplasmic reticulum vesicles as studied by manganese electron paramagnetic resonance. *Eur. J. Biochem.* 82:245–55
53. Kanazawa, T., Yamada, S., Yamamoto, T., Tonomura, Y. 1971. Reaction mechanism of the Ca^{2+}-dependent ATPase of sarcoplasmic reticulum from skeletal muscle. *J. Biochem Tokyo* 70:95–123
54. Kanazawa, T., Boyer, P. D. 1973. Occurrence and characteristics of a rapid exchange of phosphate oxygens catalyzed by sarcoplasmic reticulum vesicles. *J. Biol. Chem.* 248:3163–72
55. Kawakita, M., Yasuoka, K., Kaziro, Y. 1980. Selective modification of functionally distinct sulfhydryl groups of sarcoplasmic reticulum Ca^{2+}, Mg^{2+}-adenosine triphsphatase with N-ethylmaleimide. *J. Biochem. Tokyo* 87:609–17
56. Klip, A., Reithmeier, R. A. F., MacLennan, D. H. 1980. Alignment of the major tryptic fragments of the adenosine triphosphatase from sarcoplasmic reticulum. *J. Biol. Chem.* 255:6562–68
57. Knowles, A. F., Racker, E. 1975. Formation of adenosine triphosphate from Pi and adenosine diphosphate by purified Ca^{2+}-adenosine triphosphatase. *J. Biol. Chem.* 250:1949–51
58. Kolassa, N., Punzengruber, C., Suko, J., Makinose, M. 1979. Mechanism of calcium-independent phosphorylation of sarcoplasmic reticulum ATPase by orthophosphate. Evidence of magnesium-phosphoprotein formation. *FEBS Lett.* 108:495–500
59. Landgraf, W. C., Inesi, G. 1969. ATP dependent conformational change in 'spin labelled' sarcoplasmic reticulum. *Arch. Biochem. Biophys.* 130:111–18
60. LeMaire, M., Moller, J. P., Tanford, C. 1976. Retension of enzyme activity by

detergent solubilized sarcoplasmic Ca^{2+} ATPase. *Biochemistry* 15:2336–42
61. Louis, C. F., Saunders, M. J., Holroyd, J. A. 1977. The cross-linking of rabbit skeletal muscle sarcoplasmic reticulum protein. *Biochim. Biophys. Acta* 493:78–92
62. Louis, C. F., Holroyd, J. A. 1978. The effects of deoxycholate and trypsin on the cross-linking of rabbit skeletal muscle sarcoplasmic reticulum proteins. *Biochim. Biophys. Acta* 535:222–32
63. McIntosh, D. B., Berman, M. C. 1978. Calcium ion stabilization of the calcium transport system of sarcoplasmic reticulum. *J. Biol. Chem.* 253:5140–46
64. MacLennan, D. H., Campbell, K. P. 1979. Structure, function and biosynthesis of sarcoplasmic reticulum proteins. *Trends Biochem. Sci.* 4:148–151
65. MacLennan, D. H., Reithmeier, R. A. F., Shoshan, V., Campbell, K. P., LeBel, D., Herrmann, T. R., Shamoo, A. E. 1980. *Ann. NY Acad. Sci.* 358:138–48
66. Makinose, M. 1969. The phosphorylation of the membrane protein of the sarcoplasmic vesicles during active calcium transport. *Eur. J. Biochem.* 10:74–82
67. Makinose, M., Boll, W. 1979. The role of magnesium in the sarcoplasmic calcium pump. See Ref. 29, pp. 89–100
68. Martonosi, A. N. 1969. The protein composition of sarcoplasmic reticulum membranes. *Biochem. Biophys. Res. Commun.* 36:1039–44
69. Martonosi, A. N. 1975. The mechanism of calcium transport in sarcoplasmic reticulum. In *Calcium Transport in Contraction and Secretion*, ed. E. Carafoli et al, pp. 313–27. Amsterdam: North Holland
70. Martonosi, A. N. 1976. The effect of ATP upon the reactivity of SH groups in sarcoplasmic reticulum membranes. *FEBS Lett.* 67:153–55
71. Martonosi, A. N., Roufa, D., Boland, R., Reyes, E., Tillack, T. W. 1977. Development of sarcoplasmic reticulum in cultured chicken muscle. *J. Biol. Chem.* 252:318–32
72. Meissner, G., Fleischer, S. 1974. Dissociation and reconstitution of functional sarcoplasmic reticulum vesicles. *J. Biol. Chem.* 249:302–9
73. Meissner, G. 1973. ATP and Ca^{2+} binding by the Ca^{2+} pump protein of sarcoplasmic reticulum. *Biochim. Biophys. Acta* 298:907–26
74. Migala, A., Agostini, B., Hasselbach, W. 1973. Tryptic fragmentation of the calcium transport system in the sarcoplasmic reticulum. *Z. Naturforsch.* 28:178–82
75. Møller, J. V., Lind, K. E., Anderson, J. P. 1980. Enzyme kinetics and substrate stabilization of detergent-solubilized and membraneous $(Ca^{2+} + Mg^{2+})$-activated ATPase from sarcoplasmic reticulum. *J. Biol. Chem.* 255:1912–20
76. Murphy, A. J. 1976. Cross-linking of the sarcoplasmic reticulum ATPase protein. *Biochem. Biophys. Res. Commun.* 70:160–66
77. Murphy, A. 1976. Sulfhydryl group modification of sarcoplasmic reticulum membranes. *Biochemistry* 15:4492–96
78. Murphy, A. J. 1978. Effects of divalent cations and nucleotides on the reactivity of the sulfhydryl groups of sarcoplasmic reticulum membrane. *J. Biol. Chem.* 253:385–89
79. Murphy, A. J. 1977. Sarcoplasmic reticulum adenosine triphosphatase: labeling of an essential lysyl residue with pyridoxal-5'-phosphate. *Arch. Biochem. Biophys.* 180:114–20
80. Nakamura, H., Hori, H., Mitsui, T. 1972. Conformational change in sarcoplasmic reticulum induced by ATP in the presence of magnesium ion and calcium ion. *J. Biochem. Tokyo* 72:635–46
81. Pang, D. C., Briggs, F. N., Rogowski, R. S. 1974. Analysis of the ATP-induced conformational changes in sarcoplasmic reticulum. *Arch. Biochem. Biophys.* 164:332–40
82. Panet, R., Selinger, Z. 1970. Specific alkylation of the sarcoplasmic reticulum ATPase by N-ethyl-[1-^{14}C] maleimide and identification of the labeled protein in acrylamide gel electrophoresis. *Eur. J. Biochem.* 14:440–44
83. Pick, U., Racker, E. 1979. Inhibition of the Ca^{2+} ATPase from sarcoplasmic reticulum by dicyclohexylcabodiimide: Evidence for location of the Ca^{2+} binding site in a hydrophobic region. *Biochemistry* 18:108–13
84. Pick, U., Karlish, S. J. D. 1980. Indications for an oligomeric structure and for conformational changes in sarcoplasmic reticulum Ca^{2+} ATPase labeled selectively with fluorescein. *Biochim. Biophys. Acta* 626:255–61
85. Pick, U., Bassilian, S. 1981. Modification of the ATP binding site of the Ca^{2+} ATPase from sarcoplasmic reticulum by fluorescein isothiocyanate. *FEBS Lett.* 123:127–31
86. Pick, U. 1981. Dynamic interconversions of phosphorylated and non-phosphorylated intermediates of the Ca^{2+}

ATPase from sarcoplasmic reticulum followed in a fluorescein-labeled enzyme. *FEBS Lett.* 123:131–36
87. Punzengruber, C., Prager, R., Kolassa, N., Winker, F., Suko, J. 1978. Calcium gradient-dependent and calcium gradient-independent phosphorylation of sarcoplasmic reticulum by orthophosphate. *Eur. J. Biochem.* 92:349–59
88. Rauch, B., Chak, D. v., Hasselbach, W. 1978. An estimate of the kinetics of calcium binding and dissociation of the sarcoplasmic reticulum transport ATPase. *FEBS Lett.* 93:65–68
89. Rossi, B., deAssis Leone, F., Gache, C., Lazdunski, M. 1979. Pseudosubstrates of the sarcoplasmic Ca^{2+} ATPase as tools to study the coupling between substrate hydrolysis and Ca^{2+} transport. *J. Biol. Chem.* 254:2302–7
90. Scales, D., Inesi, G. 1976. Assembly of ATPase protein in sarcoplasmic reticulum membranes. *Biophys. J.* 16:735–51
91. Shamoo, A. E., Ryan, T. E., Stewart, P. S., MacLennan, D. H. 1976. Localization of ionophore activity in a 20,000-dalton fragment of the adenosine triphosphatase of sarcoplasmic reticulum. *J. Biol. Chem.* 251:4147–54
92. Shigekawa, M., Dougherty, J. P. 1978. Reaction mechanism of Ca^{2+} dependent ATP hydrolysis by skeletal muscle sarcoplasmic reticulum in the absence of added alkali metal salts. III. Sequential occurrence of ADP-sensitive and ADP-insensitive phosphoenzyme. *J. Biol. Chem.* 253:1458–64
93. Shigekawa, M., Akowitz, A. A. 1979. On the mechanism of Ca^{2+} dependent adenosine triphosphatase of sarcoplasmic reticulum. Occurrence of two types of phosphoenzyme intermediates in the presence of KCl. *J. Biol. Chem.* 254: 4726–30
94. Stewart, P. S., MacLennan, D. H. 1974. Surface particles of sarcoplasmic reticulum membranes. Structural features of the adenosine triphosphatase. *J. Biol. Chem.* 249:985–93
95. Stewart, P. S., MacLennan, D. H., Shamoo, A. E. 1976. Isolation and characterization of tryptic fragments of the adenosine triphosphatase of sarcoplasmic reticulum. *J. Biol. Chem.* 251:712–19
96. Sumida, M., Kanazawa, T., Tonomura, Y. 1976. Reaction mechanism of the Ca^{2+} dependent ATPase of sarcoplasmic reticulum from skeletal muscle. XI. Reevaluation of the transition of ATPase activity during the initial phase. *J. Biochem.* 79:259–64

97. Sumida, M., Wang, T., Mandel, F., Froehlich, J. P., Schwartz, A. 1978. Transient kinetics of Ca^{2+} transport of sarcoplasmic reticulum. A comparison of cardiac and skeletal muscle. *J. Biol. Chem.* 253:8772–77
98. Tada, M., Yamamoto, T., Tonomura, Y. 1978. Molecular mechanism of active calcium transport by sarcoplasmic reticulum. *Physiol. Rev.* 58:1–79
99. Takakuwa, Y., Kanazawa, T. 1981. Reaction mechanism of (Ca^{2+},Mg^{2+})-ATPase of sarcoplasmic reticulum. I. Phosphoenzyme with bound Ca which is exposed to the external medium. *J. Biol. Chem.* 256:2691–95
100. Takisawa, H., Makinose, M. 1981. Occluded bound calcium on the phosphorylated sarcoplasmic transport ATPase. *Nature* 290:271–73
101. Taylor, E. W. 1979. Mechanism of actomyosin ATPase and the problem of muscle contraction. *CRC Crit. Rev. Biochem.* 103–64
102. Tenu, J. P., Chelis, C., Saint Legar, D., Carrette, J., Chevallier, J. 1976. Mechanism of an active transport of calcium. Ethoxyformylation of sarcoplasmic reticulum vesicles. *J. Biol. Chem.* 251: 4322–29
103. Thorley-Lawson, D. A., Green, N. M. 1973. Studies on the location and orientation of proteins in the sarcoplasmic reticulum. *Eur. J. Biochem.* 40:403–13
104. Thorley-Lawson, D. A., Green, N. M. 1977. The reactivity of the thiol groups of the adenosine triphosphatase of sarcoplasmic reticulum and their location on tryptic fragments of the molecule. *Biochem. J.* 167:739–48
105. Tong, S. W. 1980. Studies on the structure of the calcium-dependent ATPase from rabbit skeletal muscle sarcoplasmic reticulum. *Arch. Biochem. Biophys.* 203:780–91
106. Tonomura, Y., Morales, M. F. 1974. Change in state of spin labels bound to sarcoplasmic reticulum with change in enzymic state, as deduced from ascorbate-quenching studies. *Proc. Natl. Acad. Sci. USA* 71:3687–91
107. Vanderkooi, J. M., Ierokomas, A., Nakamura, H., Martonosi, A. 1977. Fluorescence energy transfer between Ca^{2+} transport ATPase molecules in artificial membranes. *Biochemistry* 16: 1262–67
108. Verjovski-Almeida, S., Kurzmack, M., Inesi, G. 1978. Partial reactions in the catalytic and transport cycle of sarcoplasmic reticulum. *Biochemistry* 17: 5006–13

109. Verjovski-Almeida, S. 1981. Heterogeneity of tryptophanyl residues in the sarcoplasmic reticulum ATPase probed by fluorescence energy transfer between the protein and fluorescent ionophore X537A. *J. Biol. Chem.* 256:2662–68
110. Verjovski-Almeida, S., Silva, J. L. 1981. Different degrees of cooperativity of the Ca^{2+} induced changes in fluorescence intensity of solubilized sarcoplasmic reticlulum ATPase. *J. Biol. Chem.* 25:2940–44
111. Vianna, A. L. 1975. Interaction of calcium and magnesium in activating and inhibiting the nucleoside triphosphatase of sarcoplasmic reticulum vesicles. *Biochim. Biophys. Acta* 410:389–406
112. Warren, G. B., Toon, P. A., Birdsall, N. J. M., Lee, A. G., Metcalfe, J. C. 1974. Reversible lipid titrations of the activity of pure adenosine triphosphatase-lipid complex. *Biochemistry* 13:5501–7
113. Worthington, C. R., Liu, S. C. 1973. Structure of sarcoplasmic reticulum membranes at low resolution (17A). *Arch. Biochem. Biophys.* 157:573–79
114. Yamada, S., Ikemoto, N. 1978. Distinction of thiols involved in the specific reaction steps of the Ca^{2+} ATPase of the sarcoplasmic reticulum. *J. Biol. Chem.* 253:6801–7
115. Yamada, S., Ikemoto, N. 1980. Reaction mechanism of calcium ATPase of sarcoplasmic reticulum. Substrates for phosphorylation reaction and back reaction, and further resolution of phosphorylated intermediates. *J. Biol. Chem.* 255:3108–19
116. Yamamoto, T., Tonomura, Y. 1976. Chemical modification of the Ca^{2+} dependent ATPase of sarcoplasmic reticulum from skeletal muscle. II. Use of 2,4,6-trinitrobenzene sulfonate to show functional movements of the ATPase molecule. *J. Biochem. Tokyo* 79:693–707
117. Yamamoto, T., Tonomura, Y. 1977. Chemical modification of the Ca^{2+} dependent ATPase of sarcoplasmic reticulum from skeletal muscle. III. Changes in the distribution of exposed lysine residues among subfragments with change in enzymic state. *J. Biochem. Tokyo* 82:653–60
118. Yamamoto, T., Takisawa, H., Tonomura, Y. 1979. Reaction mechanisms for ATP hydrolysis and synthesis in the sarcoplasmic reticulum. *Curr. Top. Bioenerg.* 9:179–236
119. Yate, D. W., Duance, V. C. 1976. *Biochem. J.* 159:719–28
120. Yu, B. P., Masoro, E. J., Bertrand, H. A. 1974. The functioning of histidine residues of sarcoplasmic reticulum in Ca^{2+} transport and related activities. *Biochemistry* 13:5083–87

TURNOVER OF ACETYLCHOLINE RECEPTORS IN SKELETAL MUSCLE

D. W. Pumplin

Department of Anatomy, University of Maryland School of Medicine, Baltimore, Maryland 21202

D. M. Fambrough

Department of Embryology, Carnegie Institution of Washington, Baltimore, Maryland 21210

INTRODUCTION

The acetylcholine receptor (AChR) is an integral membrane protein that opens a cation-selective ion channel in response to the binding of acetylcholine. The AChR is a major protein of the post-synaptic membranes of neuromuscular junctions and synapses in the electroplax. Here we review the mechanisms of synthesis and degradation of AChR and the mechanisms by which these processes are regulated to determine the number and location of AChR. We emphasize data appearing since previous reviews (41, 47, 92).

AChR STRUCTURE

The structure and functions of AChR have been reviewed recently (1, 10, 60, 66, 74.) The AChR from *Torpedo* electroplax consists of five polypeptide chains: two α, and one each of β, γ and δ subunits. The amino terminal sequences of these subunits show a high order of homology, suggesting an evolutionary relationship (as also found, for example, in the α and β subunits of hemoglobin) (99). The α subunits are implicated in the binding of ACh. Both protease digestion (127) and studies with anti-AChR antibodies

(69, 118, 119) have shown that all four subunits span the lipid bilayer. A hydrophobic photoaffinity reagent labels regions presumed to lie within the bilayer (120). Deep-etch views of the outer surface of rapidly frozen electroplax membrane (62) reveal 8.5 nm-diameter projections having a ring-like structure similar to that seen by negative staining of electroplax receptors (28, 86). The packing density of the projections was about 10,000 μm^{-2}, equivalent to the number of receptors expected from α-bungaro-toxin binding data. In the same areas of membrane, conventional freeze-fracture revealed intramembranous particles, but the number of particles was about half the number of projections. The exact relation between intramembranous particles seen in freeze-fracture replicas and AChR remains undetermined.

Evidence continues to mount for molecular heterogeneity of AChRs. New evidence includes demonstration of an excitatory action of d-tubocurarine on embryonic AChR of skeletal muscle (130) and characterization of the antiserum from a myasthenic patient, which recognizes extrajunctional but not junctional AChR of rat skeletal muscle (101, 125). Electrophoretic analysis of large peptides of iodinated junctional and extrajunctional AChR of rat muscle failed to reveal any differences between the receptors (85). Although multiple post-translational modifications of AChR polypeptides are documented (122), it remains unclear whether the differences between extrajunctional and junctional AChRs are basically due to post-translational modifications or to differing primary structures of AChR polypeptides. The many precedents in which separate genes code for embryonic and adult forms of proteins (as for example, the hemoglobins) provide some basis for favoring a direct genetic origin of the difference. The possibility of molecular heterogeneity of AChRs, no matter of what origin, complicates analysis of such phenomena as AChR clustering.

BIOSYNTHESIS

Translation of mRNA

The electric organs of *Torpedo* and also the mouse BC3H-1 cell line are significant sources of AChR messenger RNAs. Using in vitro translation systems and partially purified mRNA, synthesis of AChR polypeptides has been demonstrated (2, 3, 80). The in vitro synthesized polypeptide chains were identified with the aid of antisera. Separate messages exist for each of the four kinds of polypeptide chains of the AChR, and ACh receptor polypeptides are synthesized on membrane-bound polysomes (82). In vitro translation in the presence of dog pancreas microsomes (3) resulted in processing of each of the polypeptide chains to slightly larger molecular weight, indicating glycosylation. Partial protease resistance of each of the chains was interpreted to signify a transmembrane orientation of each. Even when pancreas microsomes were used, full assembly of ACh receptor molecules has not yet been achieved. This is probably due to the very small number of receptor polypeptide chains synthesized in vitro compared to the number of microsomal vesicles involved. Thus the average vesicle probably accumulated less than the minimal number of polypeptides of receptor needed for complete assembly of receptor units of M_r 250,000.

The identification of AChR mRNA, together with amino acid sequence data on the N-terminal ends of each of the four *Torpedo* polypeptide chains (99), will facilitate isolation of AChR genes. Analysis of AChR genes should go far towards solving the question of the relation between junctional and extrajunctional AChR in skeletal muscle and also the relation of skeletal muscle AChR to AChRs of ganglionic synapses and at some CNS sites (91).

Assembly, Processing, and Intracellular Transport of AChR

Direct amino acid labeling experiments have shown that full-sized AChR are synthesized and assembled in about 15 min by cultured myotubes (35, 83). The site of biosynthesis of the AChR polypeptide chains is apparently the rough endoplasmic reticulum (ER) (82). The addition of oligosaccharide chains in the ER involves the lipid-linked oligosaccharide pathway, which is sensitive to inhibition by tunicamycin (124). Tunicamycin treatment of cultured chick myotubes has no measurable effect on the transport and processing of pre-formed intracellular AChR and their incorporation into plasma membrane. However, tunicamycin greatly inhibits further production of AChR, resulting in depletion of the intracellular pool of newly synthesized molecules (40, 53, 104). Indirect evidence suggests that there may be biosynthesis of underglycosylated AChR in the presence of tunicamycin, followed by rapid degradation of the incomplete receptors (97). The resistance of already synthesized intracellular AChR to tunicamycin effects probably indicates that the transfer of oligosaccharides from dolichol donor lipid takes place coincident with or immediately after polypeptide biosynthesis.

The intracellular residence site of most of the newly synthesized AChR is the Golgi apparatus (43). In this compartment the ACh receptors appear to be membrane-associated with ACh binding sites facing the lumen of Golgi vesicles, for the ACh binding sites remain sequestered when muscle is homogenized and are revealed only by addition of detergent at concentrations that solubilize membranes (53). AChR remain in the Golgi apparatus for approximately two hours before transport to the plasma membrane. The long intracellular residence time is characteristic of plasma membrane and secretory proteins of myotubes (42, 104), suggesting that the membrane and secretory proteins may be following a common pathway of biosynthesis, processing, and transport, and may even be packaged together for transport to the cell surface. Efforts to test this idea have led to conflicting results (104, 114). Comparing the kinetics of biosynthesis and transport to the cell surface for the AChR and the secretory acetylcholinesterase (AChE), Smilowitz reported that the ionophores monensin and nigericin selectively inhibited AChE secretion at low doses, suggesting divergent pathways of AChE and AChR from Golgi to the cell surface. However, a tight correlation between transport of AChR and AChE at all ionophore concentrations has been reported, and an alternative explanation for Smilowitz's observations has been proposed (104). Neither report rigorously excludes divergent pathways or a common vehicle. The ionophore effects upon membrane biogenesis and secretion may be related to ionophore effects upon the structure and luminal contents of the Golgi apparatus (55).

Incorporation into Plasma Membrane

ACh receptors are distributed rather evenly in the plasma membrane of young myotubes developing in tissue culture. At this stage no selective sites of insertion were found by autoradiographic analysis of the distribution of new 125I-α-bungarotoxin binding sites (59). However, in more mature myotubes in vitro AChR become nonrandomly arranged, with prominent clusters forming especially at sites of attachment of myotube to substrate. Such clusters show a high degree of positional stability in frog (4), rat (8, 9), and chick muscle (50). Axelrod and colleagues (8, 9) have shown that this positional stability is maintained despite relatively rapid turnover of the AChR within the clusters and that newly arriving AChR are inserted directly into the cluster area rather than converging to the cluster from surrounding plasma membrane. Fischbach et al (45) have demonstrated metabolic maintenance of AChR clusters on chick myotubes, and Schuetze et al (110) have estimated by autoradiography that the turnover rate of AChRs in clusters and in nonclustered regions is the same, judged by disappearance of 125I-labeled bungarotoxin-AChR complexes from these regions. The ACh receptors in the clusters are immobilized (shown in fluorescence photobleaching experiments), and the clusters are associated with cytoskeletal specializations (19, 20; see 75) and with build-up of specialized areas of extracellular matrix (26, 42, 64). The co-localization of these cellular elements might indicate regions of new membrane insertion and of secretion of extracellular matrix components. However, AChR clusters in the rat and in *Xenopus* (19, 88) are reversibly disrupted by lowering extracellular calcium ion concentration and by uncoupling oxidative phosphorylation. These observations imply a mechanism for assembly of dispersed elements, from which one might argue that nonclustered AChR in the plasma membrane might be expected to join clusters. Indeed, several clustering factors have been discovered, at least one of which causes rapid accumulation of receptors into clusters (30).

During myogenesis in vivo, AChR clusters are prominent only at developing neuromuscular junctions, where accumulation of AChR results in an enlargement of the post-synaptic specialization disc. New AChR evidently add on to the periphery of the area as it enlarges [at least in the rat (126) where turnover of the junctional AChR is very slow; see below]. McMahan and colleagues (25, 108) have shown that specialization of the junctional basal lamina endows it with a directive role in the organization of AChR during regeneration and maintenance of the post-synaptic surface of denervated and damaged muscle (and see 22, 77). This basal lamina contains molecular specializations defined by antisera (107) and by monoclonal antibodies (42). At least one of these antigens may also have a role in clustering of AChR on myotubes in tissue culture (42). Steinbach (116) has studied the high degree of stability of the post-synaptic AChR clusters in permanently denervated cat muscles. The half-time for loss of clustering was about 140 days, at least 10 times the estimated half-life for the population of ACh receptor molecules in the junction (see below). Atsumi (5) recently described the distribution of intracellular AChR in chick muscles during the period of rapid development of the junctional AChR clusters and loss of extrajunctional AChR. She reasoned that the localized insertion of AChR into the junctional area might result from very localized biosynthesis of AChR and intracellular accumulation beneath the site of insertion. However, Atsumi reported that the intracellular AChRs were widely distributed, with no obvious accumulation in the junctional area. She suggested that some unknown mechanism restricts entry of newly synthesized AChR into extrajunctional plasma membrane of innervated muscle fibers, allowing only insertion into synaptic areas.

Factors Affecting Biosynthesis and Organizations of AChR

Myotubes cocultured with spinal cord develop increased numbers of AChR even when they remain uninnervated (17, 46), an effect seemingly due to a diffusible substance released from the nerves. One can readily envision such substances being responsible, at least in part, for controlling the clustering of AChR under a portion of a neurite during formation of a neuromuscular junction, as well as helping to maintain the aggregation and low turnover rate of AChR at a mature neuromuscular junction. Several "factors" have been described. The specificity and relevance of these factors are not yet established, whereas nerve-induced clustering of AChR has been shown to be specific for certain neurons (31, 67). Explants of embryonic day 18 rat brain caused a three-fold increase in the total AChR and increased incidence of clusters of AChR in the rat muscle cell line L6 (94). The causative agent was destroyed by heating and trypsin treatment and was excluded from a G-100 column; it increased overall protein synthesis and was not shown to have a specific effect on the synthesis or degradation of AChR. A similar three-fold increase in total AChR in culture chick myotubes was induced by sciatin. This 80,000-dalton protein, purified from chicken sciatic nerves, increases protein synthesis generally (79). A small polypeptide ($M_r\sim1700$) extracted from embryonic chick brain increased both AChR and AChE 3–5-fold, with a 10–40-fold increase in the incidence of AChR clusters (65). Degradation of AChR was not inhibited, and general protein synthesis was not increased.

Medium conditioned by growth of cells of the neuroblastoma-glioma hybrid NG108-15 increased the number of AChR clusters, but not the total number of AChR in cultured rat myotubes (30). The aggregation factor has been partially purified; it is protease-sensitive, and has a molecular weight between 150,000 and 250,000, and an isoelectric point near 4.6. It occurs in the cytoplasm rather than the membrane of NG108-15 cells and is also found in embryonic, but, not adult, rat brain (11). This factor also decreased the mobility of AChR in nonclustered regions, as shown by fluorescence photobleaching recovery measurements (7). Consistent with these findings, the factor decreased the rate at which detergent extracted AChR from treated myotubes, an indication of increased association of receptors with the cytoskeleton (95).

The mechanism of action of these various factors remains to be worked out, and probably awaits further purification. The effect of the aggregation factor on AChR mobility is a useful clue in that it is independent of effects on insertion or degradation. Raising the concentration of AChR in the myotube might be expected to increase the number of visible clusters and decrease mobility by shifting the equilibrium between single and clustered receptors (33, 57). Perhaps relevant to this point is the finding that trypsinization of AChR-rich vesicles from electroplax resulted in clustering of AChR and even budding of clusters from the vesicle surface (68). However, Orida & Poo (89, 90) have reported that trypsinization of *Xenopus* somite muscle in vitro caused dispersal of AChR clusters. Increased aggregation of the same number of AChR, as caused by one of the factors (30), could be related to modification of AChR structure, increasing its affinity for self;

alternatively, such a factor could cross-link receptors in some way that does not accelerate turnover.

Muscle activity affects the numbers and distribution of AChR (76, 78). The major effect of muscle activity is to suppress biosynthesis without altering the rather rapid degradation of extrajunctional AChR, resulting in a disappearance of all but the synaptic receptors (73, 100). Agents that affect muscle activity have been tested in cultured myotubes. Tetrodotoxin (TTX) inhibits spontaneous contractions of cultured myotubes and increases both total AChR and the number of AChR clusters (32, 113). In cultured muscle, this increase in AChR was blocked by low external levels of calcium. In contrast, raising the external calcium to 10 mM doubled the number of total AChR, and no further increase occurred with the addition of TTX (18). The effect of elevated calcium concentration is apparently on synthesis of AChR since it was blocked by the inhibitors cycloheximide and actinomycin D and degradation rates did not change. Direct electrical stimulation of myotubes opposed increases in AChR levels evoked by raised external calcium, Dantrolene sodium, and TTX; this was interpreted as due to opposite effects on internal stores of calcium. Apparently in contradiction to these results, the ionophore A23187 elevated intracellular calcium and suppressed extrajunctional AChR synthesis in denervated muscle (49).

Cyclic nucleotides have also been proposed as regulators of AChR levels. The phosphodiesterase inhibitor Ro-20-1724 raised levels of cAMP and AChR in cultured chick myotubes (21). Increases in AChR up to four-fold were also found with 8-bromo cAMP and cholera toxin, both of which raise cAMP levels. Synthesis of AChR was specifically affected, but there was no change in the degradation rate. The phosphodiesterase inhibitors, caffeine and theophylline, also raised AChR levels, an effect directly opposite to that seen (18) in similar cultures. Two reports (16, 21) showed a rise in AChR in response to increased cAMP. One (16) reported that a rise in cGMP depressed AChR, but this was not confirmed (21). A rise in cAMP levels during myoblast fusion in chick cultures (129) might participate in the induction of AChR synthesis (112).

DEGRADATION

Mechanism of Turnover

Degradation of AChR takes place in secondary lysosomes (41). New evidence from electron microscope autoradiography indicates that blockade of lysosomal hydrolases with trypan blue leads to continued accumulation of labeled material after surface AChR are labeled briefly with iodinated α-bungarotoxin, whereas in controls a plateau level of about 2% of label in the lysosomal compartment was reached in about an hour (44). Also, Libby et al (72) found that blockade of cathepsins with leupeptin and antipain resulted in intracellular accumulation of undegraded AChR, thus demonstrating that internalization of AChR for degradation is not coupled directly to degradation per se. Libby et al suggested that the vehicles for transport of AChR from surface to lysosome were coated vesicles, for these were seen in increased abundance in leupeptin-treated myotubes.

Some cells that engage in high-volume uptake of exogenous material by pinocytosis and phagocytosis seem to recycle their plasma membrane proteins (that is, interiorize and re-exteriorize them repeatedly on a much faster time-scale than their turnover rates) (58). Receptors involved in rapid uptake mechanisms (such as the asialoglycoprotein receptor, the low density lipoprotein receptor, and the transferrin receptor) are reutilized and perhaps are selectively interiorized and recycled in the process of receptor-mediated endocytosis. In a series of experiments designed to detect recycling of AChR in myotubes, no evidence for recycling could be obtained (54). Thus apparently the AChR remains in the plasma membrane from the time of its incorporation to the moment when internalization initiates its degradation process.

Turnover Rates

The turnover rate of embryonic AChR of cultured myotubes has been measured by direct labeling experiments, with isotopically labeled amino acids (54, 81). The kinetics were found to be first order exponential with a half-time of about 18 hr for turnover of the AChR population. In all other cases the turnover rate of AChR has been estimated by measuring the degradation of ^{125}I-labeled α-bungarotoxin bound to AChR [summarized in (41)]. These estimates are probably reasonable approximations. While many more such measurements have been made in the past two years, only a few new findings have been reported.

A study of the effects of agonist (carbachol) and antagonist (d-tubocurarine) on AChR turnover was carried out (54) using the heavy isotope labeling technique to measure turnover of both unlabeled and labeled AChRs. No change in turnover rate was found under conditions of virtually total occupancy of AChR with agonist or antagonist. Betz & Changeux (16) examined the effects of cAMP and cGMP on AChR metabolism and found no effects upon turnover rate. Pestronk & Drachman (93) examined the effect of lithium ions and found a pronounced selective acceleration of AChR turnover.

Treatment of myotubes in culture with tunicamycin may result in formation of AChR containing less than the normal complement of sugar residues. These underglycosylated receptors seemed to be degraded at a faster than normal rate, suggesting the possibility that glycosylation plays a role in stabilizing AChR (97). Recent measurements (110) confirm less-compelling earlier evidence that the turnover rate of AChR in AChR clusters on cultured myotubes is as fast as the rate of turnover of nonclustered AChR.

Betz et al (15) have confirmed the work of Burden (24), showing that the turnover rate of junctional AChR in the chick is the same as for extrajunctional AChR up to about three weeks post-hatch, after which a slow rate begins. However, several studies have shown that the turnover rate of junctional AChR in the rat is very slow shortly after the formation of junctions (t½ ≅ 7 days) (84, 101, 117) as indicated in the pioneering work of Berg & Hall (12). The AChR of adult neuromuscular junctions are characterized by relatively rapid ion channel kinetics compared with embryonic AChR and extrajunctional AChR of denervated adult muscle. Sakmann and colleagues (84, 105) have investigated the possible relation between the development of adult endplate type channel kinetics and the clustering of AChR and slowing of AChR turnover rate that occur during synaptogenesis in the rat. They found that the change in kinetics occurred mainly during the second postnatal week (see also 48), corresponding to the time of elimination of multiple innervation and the formation of junctional folds. During this period the turnover rate of AChR was already slow and the accumulation of AChR at the junction was relatively modest. The conversion of channel kinetics involved at least 95% of all the AChR, and thus it was concluded that receptors with slow-channel kinetics were altered to have fast-channel kinetics (rather than being replaced by AChR with fast-channel kinetics.) This finding implies that some modification of the AChR or of their immediate environment occurs during that period (84). Homolgous studies of the AChR of chick neuromuscular junctions revealed that, even in the adult, there was no change in channel kinetics (111). From the studies of AChR turnover rates, pharmacological properties, organization, and channel kinetics, a complicated picture is emerging of the process of formation of neuromuscular junctions. There occurs a series of events that seem not to be directly related, including clustering of AChR, slowing of turnover, and (in mammals and frogs) shortening of channel open time.

Levitt et al (71) found that the turnover rate of junctional AChR changes fairly rapidly after denervation, approaching the faster turnover rate characteristic of extrajunctional AChR. This change occurred while post-synaptic membrane remained an area packed with AChR, meaning that rapid replacement of AChR was occurring at former junctional areas. According to Steinbach (116), the half-life of the denervated junctional area receptor population is on the order of 140 days in the cat. Given even conservative estimates of the speed of AChR turnover in these areas, one concludes that continued selective accumulation of AChR must occur at these sites. This is consonant with the finding (25) that some element of the synaptic basal lamina must have a directive role in the maintenance of high AChR packing density in post-synaptic sites. Steinbach suggests that the slow decay of post-synaptic membrane AChR accumulation after denervation may be a measure of the half-time of decay of the molecular specializations of the basal lamina that have such a directive influence. This may also explain the earlier observations (51) on numbers of AChR at denervated junctional sites in the rat.

Acceleration of Degradation by Cross-Linking

Accelerated degradation of AChR induced by binding of specific antibodies may be a major contributor to the loss of junctional AChR occuring in myasthenia gravis and its animal model, experimental autoimmune myasthenia gravis (EAMG). [In addition, antibody binding to the receptor may interfere with ACh binding (52) and/or lead to complement-mediated destruction of endplates (39).] At present the relative importance of these processes in myasthenia gravis is not well established, and, indeed, the relative importance is presumed to differ from one patient to the next. In EAMG mice, the occurrence of myasthenic symptoms was found to be strain-related (13, 29) and not correlated with titer of anti-AChR antibody which could accelerate turnover of AChR in tissue cultured cells (13, 14). Literature on myasthenia gravis has been reviewed recently (36, 123).

Antibody-induced acceleration of turnover of AChR has been demonstrated both in cultured myotubes (see 36) and at neuromuscular junctions of adult muscle (61, 102, 115). The effect is selective for those AChR carrying bound antibody (37) and requires divalent antibody, monovalent Fab fragments being ineffective (38). Divalent antibodies are capable of cross-linking adjacent receptors which apparently is a key step in inducing accelerated degradation (see below). Receptors may also be cross-linked by linking bound α-BuTX either by anti-α-BuTX antibodies (38) or by using avidin with biotinylated toxin (6); both accelerate degradation. A monoclonal antibody to AChR has been used to induce EAMG (70, 103). Since AChR have a number of antigenic determinants, it is probable that in myotubes treated with myasthenic sera, each receptor becomes linked to several other receptors via antibody molecules binding to different determinants, resulting in a receptor cluster [(34) and see below]. Would a monoclonal antibody binding to only a single site on each AChR, cross-linking each receptor to only one other receptor, still be able to accelerate degradation?

Clusters of AChR induced by cross-linkage have been seen in cultured myotubes by fluorescence (6), autoradiography (96), and freeze-fracture (98). The freeze-fracture appearance of membrane regions where clusters of AChR had been induced by cross-linkage differs from that of the natural AChR cluster sites occurring at positions of close contact between myotube and substrate. The freeze-fracture image of AChR-rich synaptic membrane is different yet. No one has compared the structure of receptor aggregations induced by antibodies that accelerate degradation with that induced by neuronal factors that do not accelerate degradation. The paradigm for AChR degradation due to cross-linking is that suggested for surface IgG receptors on lymphocytes (63, 109, 121), involving patching (clustering) of receptors due to cross-linkage followed by energy-dependent collection of these patches into larger clusters and subsequent internalization. As previously noted, clustering of AChR has been seen, but successive steps are not yet established.

The mechanisms of internalization of AChR and their delivery to lysosomes are unknown either for normal degradation or for that induced by cross-linking antibodies. Mechanisms for internalization of other intrinsic membrane proteins are being worked out (56) and will likely shed light on AChR degradation. Several surface membrane receptors, whose internalization is induced by binding appropriate ligands, either exist in the membrane within coated pits (58, 87) or are rapidly translocated to particular regions of the cell membrane where coated pits with associated actin (106) are induced to form (106, 128) by a process possibly involving a local release of calcium (23). Association of a membrane receptor with a coated pit may involve an additional recognition factor (58). Internalization rates of specific membrane proteins may be altered selectively by factors that affect association with coated pits. The possibility that AChR are internalized via coated pits (27) is suggested by the finding of AChR particles in association with pits of the correct diameter (97), as well as by the increase in coated vesicles noted in myotubes after leupeptin inhibition of lysosomal proteases (72).

ACKNOWLEDGMENTS

Research in the authors' laboratories has been supported in part by grants from the Muscular Dystrophy Association, and by grant NS15513 from the N.I.H. to D.N.P.

Literature Cited

1. Adams, P. R. 1981. Acetylcholine receptor kinetics. *J. Membr. Biol.* 58:161–74
2. Anderson, D. J., Blobel, G. 1980. In vitro synthesis and membrane integration of the subunits of Torpedo acetylcholine receptor. *Soc. Neurosci. Abstr.* 6:209
3. Anderson, D., Blobel, G. 1981. In vitro synthesis, glycosylation and membrane insertion of the four subunits of Torpedo acetylcholine receptor. *Proc. Natl. Acad. Sci. USA* 78:5598–602
4. Anderson, M. J., Cohen, M. W. 1977. Nerve-induced and spontaneous redistribution of acetylcholine receptors on cultured muscle cells. *J. Physiol.* 268:757–73
5. Atsumi, S. 1981. Localization of surface and internal acetylcholine receptors in developing fast and slow muscles of the chick embryo. *Dev. Biol.* 86:122–35
6. Axelrod, D. 1980. Crosslinkage and visualization of acetylcholine receptors on myotubes with biotinylated α-bungarotoxin and fluorescent avidin. *Proc. Natl. Acad. Sci. USA* 77:4823–27
7. Axelrod, D., Bauer, H. C., Stya, M., Christian, C. N. 1981. A factor from neurons induces partial immobilization of nonclustered acetylcholine receptors on cultured muscle cells. *J. Cell Biol.* 88:459–62
8. Axelrod, D., Ravdin, P., Koppel, D. E., Schlessinger, J., Webb, W. W., Elson, E. L., Podleski, T. R. 1976. Lateral motion of fluorescently labeled acetylcholine receptors in membranes of developing muscle fibers. *Proc. Natl. Acad. Sci. USA* 73:4594–98
9. Axelrod, D. P., Ravdin, P. M., Podleski, T. R. 1978. Control of acetylcholine receptor mobility and distribution in cultured muscle membrane. A fluorescence study. *Biochim. Biophys. Acta* 511:23–38
10. Barrantes, F. J. 1979. Endogenous chemical receptors: some physical aspects. *Ann. Rev. Biophys. Bioeng.* 8:287–321
11. Bauer, H. C., Daniels, M. P., Pudimat, P. A., Jacques, L., Sugiyama, H., Christian, C. N. 1981. Characterization and partial purification of a neuronal factor which increases acetylcholine receptor aggregation on cultured muscle cells. *Brain Res.* 209:395–404
12. Berg, D. K., Hall, Z. W. 1975. Loss of α-bungarotoxin from junctional and extrajunctional acetylcholine receptors in rat diaphragm in vivo and in organ culture. *J. Physiol.* 252:771–89
13. Berman, P. W., Patrick, J. 1980. Experimental myasthenia gravis: a murine system. *J. Exp. Med.* 151:204–23
14. Berman, P. W., Patrick, J. 1980. Linkage between the frequency of muscular weakness and loci that regulate immune responsiveness in murine experimental myasthenia gravis. *J. Exp. Med.* 152:507–20
15. Betz, H., Bourgeois, J.-P., Changeux, J.-P. 1980. Evolution of cholinergic proteins in developing slow and fast skeletal muscles in chick embryo. *J. Physiol* 302:197–218
16. Betz, H., Changeux, J. P. 1979. Regulation of muscle acetylcholine receptor synthesis in vitro by cyclic nucleotide derivatives. *Nature* 278:749–53
17. Betz, W., Osborne, M. 1977. Effects of innervation on acetylcholine sensitivity of developing muscle *in vitro*. *J. Physiol.* 270:75–88
18. Birnbaum, M., Reis, M. A., Shainberg, A. 1980. Role of calcium in the regulation of acetylcholine receptor synthesis in cultured muscle cells. *Pflügers Arch.* 395:37–43
19. Bloch, R. J. 1979. Dispersal and reformation of acetylcholine receptor clusters of cultured rat myotubes treated with inhibitors of energy metabolism. *J. Cell Biol.* 82:626–43
20. Bloch, R. J., Geiger, B. 1980. The localization of acetylcholine receptor clusters in areas of cell-substrate contact in cultures of rat myotubes. *Cell* 21:25–35
21. Blosser, J. C., Appel, S. H. 1980. Regulation of acetylcholine receptor by cyclic AMP. *J. Biol. Chem.* 255:1235–38
22. Braithwaite, A. W., Harris, A. J. 1979. Neural influences on acetylcholine receptor clusters in embryonic development of skeletal muscles. *Nature* 279:549–51
23. Braun, J., Shaafi, R. I., Unanue, E. R. 1979. Cross-linking of ligands to surface immunoglobulin triggers mobilization of intracellular 45 Ca^{2+} in B-lymphocytes. *J. Cell Biol.* 82:755–66
24. Burden, S. 1977. Acetylcholine receptors at the neuromuscular junction: developmental change in receptor turnover. *Devel. Biol.* 61:79–85
25. Burden, S. J., Sargent, P. B., McMahan, U. J. 1979. Acetylcholine receptors in regenerating muscle accumulate at original sites in the absence of the nerve. *J. Cell Biol.* 82:412–25

26. Burrage, T. G., Lentz, T. L. 1979. Surface specializations associated with high density accumulations of acetylcholine receptors in embryonic chick muscle. *Soc. Neurosci. Abstr.* 5:477
27. Bursztajn, S., Fischbach, G. D. 1979. Coated vesicles in cultured myotubes contain acetylcholine receptors. *Soc. Neurosci. Abstr.* 5:1631
28. Cartaud, J., Benedetti, E. L., Sobel, A., Changeux, J-P. 1978. A morphological study of the cholinergic receptor protein from *Torpedo marmorata* in its membrane environment and in its detergent-extracted form. *J. Cell Sci.* 29:313–37
29. Christadoss, P., Lennon, V. A., David, C. S. 1979. Genetic control of experimental autoimmune myasthenia gravis in mice. 1. Lymphocyte proliferative response to acetylcholine receptor is under H-2 linked Ir gene control. *J. Immunol.* 123:2540–43
30. Christian, C. N., Daniels, M. P., Sugiyama, H., Vogel, Z., Jacques, L., Nelson, P. G. 1978. A factor from neurons increases the number of acetylcholine receptor aggregates on cultured muscle cells. *Proc. Natl. Acad. Sci. USA* 75:4011-15
31. Cohen, M. W., Weldon, P. R. 1980. Localization of acetylcholine receptors and synaptic ultrastructure at nerve-muscle contacts in culture: dependence on nerve type. *J. Cell Biol.* 86:388–401
32. Cohen, S. A., Fischbach, G. D. 1973. Regulation of muscle acetylcholine sensitivity by muscle activity in cell culture. *Science* 181:76–78
33. Cohen, S. A., Pumplin, D. W. 1979. Clusters of intramembrane particles association with binding sites for α-bungarotoxin in cultured chick myotubes. *J. Cell Biol.* 82:494–516
34. Conti-Tronconi, B., Tsartos, S., Lindstrom, J. 1981. Monoclonal antibodies as probes of acetylcholine receptor structure. 2. Binding to native receptor. *Biochemistry* 20:2181–91
35. Devreotes, P. N., Gardner, J. M., Fambrough, D. M. 1977. Kinetics of biosynthesis of acetylcholine receptor and subsequent incorporation into plasma membrane of cultured chick skeletal muscle. *Cell* 10:365–73
36. Drachman, D. B. 1981. The biology of myasthenia gravis. *Ann. Rev. Neurosci.* 4:195–225
37. Drachman, D. B., Angus, C. W., Adams, R. N., Kao, I. 1978. Effect of myasthenic patients' immunoglobulin on acetylcholine receptor turnover selectivity of degradation process. *Proc. Natl. Acad. Sci. USA* 75:3422–26
38. Drachman, D. B., Angus, C. W., Adams, R. N., Michelson, J. D., Hoffmann, G. J. 1978. Myasthenic antibodies cross-link acetylcholine receptors to accelerate degradation. *N. Engl. J. Med.* 298:1116–22
39. Engel, A. G., Lambert, E. H., Howard, F. M. 1977. Immune complexes (IgG and C3) at the motor end-plate in myasthenia gravis: ultrastructural and light microscopic localization and electrophysiologic correlations. *Mayo Clin. Proc.* 52:267–80
40. Fambrough, D. M. 1977. Synthesis and glycosylation of ACh receptors. *Carnegie Inst. Wash. Yearb.* 76:12–13
41. Fambrough, D. M. 1978. Control of acetylcholine receptors in skeletal muscle. *Physiol. Rev.* 59:165–227
42. Fambrough, D. M., Bayne, E. K., Gardner, J. M., Anderson, M. J., Wakshull, E., Rotundo, R. L. 1981. Monoclonal antibodies to skeletal muscle cell surface. In *Neuroimmunology*, ed. J. Brockes. NY: Plenum
43. Fambrough, D. M., Devreotes, P. N. 1978. Newly synthesized acetylcholine receptors are located in the Golgi apparatus. *J. Cell Biol.* 76:237–44
44. Fambrough, D. M., Devreotes, P. N., Card, D. J., Gardner, J., Tepperman, K. 1978. Metabolism of acetylcholine receptors in skeletal muscle. *Natl. Cancer Inst. Monogr.* 48:277–93
45. Fischbach, G. D., Berg, D. K., Cohen, S. A., Frank, E. 1976. Enrichment of nerve-muscle synapses in spinal cord-muscle cultures and identification of relative peaks of ACh sensitivity at sites of transmitter release. *Cold Spring Harbor Symp. Quant. Biol.* 40:347–57
46. Fischbach, G. D., Cohen, S. A. 1973. The distribution of acetylcholine sensitivity over uninnervated and innervated muscle fibers grown in cell culture. *Dev. Biol.* 31:147–62
47. Fischbach, G. D., Frank, E., Jessell, T. M., Rubin, L. L., Schuetze, S. M. 1979. Accumulation of acetylcholine receptors and acetylcholinesterase at newly formed nerve-muscle synapses. *Pharmacol. Rev.* 30:411–28
48. Fischbach, G. D., Schuetze, S. M. 1980. A postnatal decrease in acetylcholine channel open time at rat endplates. *J. Physiol.* 303:125–37
49. Forrest, J. W., Mills, R. G., Bray, J. J., Hubbard, J. I. 1981. Calcium-dependent regulation of the membrane potential and extrajunctional acetylcholine

receptors of rat skeletal muscle. *Neuroscience* 6:741–49
50. Frank, E., Fischbach, G. D. 1979. Early events in neuromuscular junction formation in vitro: induction of acetylcholine receptors clusters in the postsynaptic membrane and morphology of newly formed synapses. *J. Cell Biol.* 83:143–58
51. Frank, E., Gautvik, K., Sommerschild, H. 1975. Cholinergic receptors at denervated mammalian motor endplates. *Acta Physiol. Scand.* 95:66–76
52. Fulpius, B. W., Miskin, R., Reich, E. 1980. Antibodies from myasthenic patients that compete with cholinergic ligands for binding to nicotinic receptors. *Proc. Natl. Acad. Sci. USA* 77:4326–30
53. Gardner, J. M. 1978. Studies on the biosynthesis of acetylcholine receptors. *Carnegie Inst. Wash. Yearb.* 77:12–16
54. Gardner, J. M., Fambrough, D. M. 1979. Acetylcholine receptor degradation measured by density labeling: effects of cholinergic ligands and evidence against recycling. *Cell* 16:661–74
55. Garfield, R. E., Somlyo, A. P. 1977. Golgi apparatus and lectin-binding sites. *Exp. Cell Res.* 109:163–79
56. Geisow, M. 1980. Pathways of endocytosis. *Nature* 288:434–36
57. Gershon, N. 1978. Model for capping of membrane receptors based on boundary surface effects. *Proc. Natl. Acad. Sci. USA* 75:1357–60
58. Goldstein, J. L., Anderson, R. G. W., Brown, M. S. 1979. Coated pits, coated vesicles and receptor mediated endocytosis. *Nature* 279:679–85
59. Hartzell, H. C., Fambrough, D. M. 1973. Acetylcholine receptor production and incorporation into plasma membranes of developing muscle fibers. *Dev. Biol.* 30:153–65
60. Harvey, A. L. 1980. Actions of drugs on developing skeletal muscle. *Pharmacol. Ther.* 11:1–41
61. Heinemann, S., Merlie, J., Lindstrom, J. 1978. Modulation of acetylcholine receptor in rat diaphragm by antireceptor sera. *Nature* 274:65–68
62. Heuser, J. E., Salpeter, S. R. 1979. Organization of acetylcholine receptors in quick-frozen, deep-etched, and rotary-replicated Torpedo postsynaptic membrane. *J. Cell Biol.* 82:150–73
63. Huet, C., Ash, J. F., Singer, S. J. 1980. The antibody-induced clustering and endocytosis of HLA antigens on cultured human fibroblasts. *Cell* 21:429–38

64. Jacob, M., Lentz, T. L. 1979. Localization of acetylcholine receptors by means of horseradish peroxidase-α-bungarotoxin during formation and development of the neuromuscular junction in the chick embryo. *J. Cell Biol.* 82:195–211
65. Jessell, T. M., Siegel, R. E., Fischbach, G. D. 1979. Induction of acetylcholine receptors on cultured skeletal muscle by a factor extracted from brain and spinal cord. *Proc. Natl. Acad. Sci. USA* 76:5397–401
66. Karlin, A. 1980. Molecular properties of nicotinic acetylcholine receptors. In *The Cell Surface and Neuronal Function*, ed. C. W. Cotman, G. Poste, G. L. Nicholson, pp. 191–260. NY: Elsevier/North-Holland
67. Kidokoro, Y., Anderson, M. J., Gruener, R. 1980. Changes in synaptic potential properties during acetylcholine receptor accumulation and neurospecific interactions in Xenopus nerve-muscle cell culture. *Dev. Biol.* 78:464–83
68. Klymkowsky, M. W., Heuser, J. E., Stroud, R. M. 1980. Protease effects on the structure of acetylcholine receptor membranes from Torpedo californica. *J. Cell Biol.* 85:823–38
69. Klymkowsky, M. W., Stroud, R. M. 1979. Immunospecific identification and three dimensional structure of a membrane-bound acetylcholine receptor from Torpedo californica. *J. Mol. Biol.* 128:319–34
70. Lennon, V. A., Lambert, E. H. 1980. Myasthenia gravis induced by monoclonal antibodies to acetylcholine receptors. *Nature* 285:238–40
71. Levitt, T. A., Loring, R. H., Salpeter, M. M. 1980. Neuronal control of acetylcholine receptor turnover rate at a vertebrate neuromuscular junction. *Science* 210:550–51
72. Libby, P., Bursztajn, S., Goldberg, A. L. 1980. Degradation of the acetylcholine receptor in cultured muscle cells: selective inhibitors and the fate of undegraded receptors. *Cell* 19:481-91
73. Linden, D. C., Fambrough, D. M. 1979. Biosynthesis and degradation of acetylcholine receptors in rat skeletal muscles: effects of electrical stimulation. *Neuroscience* 4:527–38
74. Lindstrom, J., Anholt, R., Einarson, B., Engel, A., Osame, M., Montal, M. 1980. Purification of acetylcholine receptors, reconstitution into lipid vesicles, and study of agonist-induced ca-

tion channel regulation. *J. Biol. Chem.* 255:8340–50
75. Lo, M. M. S., Garland, P. B., Lamprecht, J., Barnard, E. A. 1980. Rotational mobility of the membrane-bound acetylcholine receptor of Torpedo electric organ measured by phosphorescence depolarization. *FEBS Lett.* 111:407–12
76. Lomo, T., Rosenthal, J. 1972. Control of acetylcholine sensitivity by muscle activity in the rat. *J. Physiol.* 221:493–513
77. Lomo, T., Slater, C. R. 1980. Control of junctional acetylcholinesterase by neuronal and muscular influences in the rat. *J. Physiol. London* 303:191–202
78. Lomo, T., Westgaard, R. H. 1975. Further studies on the control of ACh sensitivity by muscle activity in the rat. *J. Physiol.* 252:603–26
79. Markelonis, G. J., Oh, T. H. 1981. Purification of sciatin using affinity chromatography on concanavalin A-agarose. *J. Neurochem.* 37:95–97
80. Mendez, B., Valenzuela, P., Martial, J. A., Baxter, J. D. 1980. Cell-free synthesis of acetylcholine receptor polypeptides. *Science* 209:695–97
81. Merlie, J. P., Changeux, J.-P., Gros, F. 1976. Acetylcholine receptor degradation measured by pulse chase labeling. *Nature* 264:74–76
82. Merlie, J. P., Hofler, J. G., Sebbane, R. 1981. Acetylcholine receptor synthesis from membrane polysomes. *J. Biol. Chem.* 256:6995–99
83. Merlie, J. P., Sebbane, R. 1981. Acetylcholine receptor subunits transit a precursor pool before acquiring α-bungarotoxin binding activity. *J. Biol. Chem.* 256:3605–8
84. Michler, A., Sakmann, B. 1980. Receptor stability and channel conversion in the subsynaptic membrane of the developing mammalian neuromuscular junction. *Dev. Biol.* 80:1–17
85. Nathanson, N. M., Hall, Z. W. 1979. Subunit structure and peptide mapping of junctional and extrajunctional acetylcholine receptors from rat muscle. *Biochemistry* 18:3392–401
86. Nickel, E., Potter, L. T. 1973 Ultrastructure of isolated membranes of Torpedo electric tissue. *Brain Res.* 57:508–17
87. Orci, L., Carpenter, J-L., Perrelet, A., Anderson, R. G., Goldstein, J. L., Brown, M. S. 1978. Occurrence of low density lipoprotein receptors within large pits on the surface of human fibroblasts as demonstrated by freeze-etching. *Exp. Cell Res.* 113:1–13
88. Orida, N. K., Poo, M.-M. 1980. Developmental changes of acetylcholine receptor mobility in embryonic muscle membrane. *Exp. Cell Res.* 130:281–90
89. Orida, N. K., Poo, M.-M. 1980. Trypsin disperses acetylcholine receptor clusters in the embryonic muscle membrane. *Nature* 275:31–35
90. Orida, N., Poo, M.-M. 1981. Maintenance and dissolution of acetylcholine receptor clusters in the embryonic muscle cell membrane. *Brain Res.* 227:293–98
91. Oswald, R. E., Freeman, J. A. 1981. Alpha-bungarotoxin binding and central nervous system nicotinic acetylcholine receptors. *Neuroscience* 6:1–14
92. Patrick, J., Berman, P. W. 1980. Metabolism of nicotinic acetylcholine receptor. See Ref. 66, pp. 157–90.
93. Prestronk, D., Drachman, D. B. 1981. Lithium reduces the number of acetylcholine receptors in skeletal muscle. *Science* 210:342–44
94. Podleski, T. R., Axelrod, D., Ravdin, P., Greenberg, I., Johnson, M. M., Salpeter, M. M. 1978. Nerve extract induces increase and redistribution of acetylcholine receptors on cloned muscle cells. *Proc. Natl. Acad. Sci. USA* 75:2035–39
95. Prives, J., Christian, C., Penman, S., Olden, K. 1980. Neuronal regulation of muscle acetylcholine receptors; role of the muscle cytoskeleton and receptor carbohydrate. In *Nerve Cells in Culture*, ed. E. Giacobini, pp. 35–52. NY: Raven
96. Prives, J. M., Hoffman, L., Tarrab-Hazdai, R., Fuchs, S., Amsterdam, A. 1979. Ligand induced changes in stability and distribution of acetylcholine receptors on surface membranes of muscle cells. *Life Sci.* 24:1713–18
97. Prives, J. M., Olden, K. 1980. Carbohydrate requirement for expression and stability of acetylcholine receptor on the surface of embryonic muscle cells in culture. *Proc. Natl. Acad. Sci. USA* 77:5263–67
98. Pumplin, D. W., Drachman, D. B. 1981. Myasthenic antibodies alter the arrangement of acetylcholine receptors in cultured rat myotubes: freeze-fracture studies. In *Diseases of the Motor Unit*, ed. D. Schotland. NY: Houghton Mifflin. In press
99. Raftery, M. A., Hunkapiller, M. W., Strader, C. D., Hood, L. E. 1980. Acetylcholine receptor: complex of homologous subunits. *Science* 208:1454–56

100. Reiness, C. G., Hall, Z. W. 1977. Electrical stimulation of denervated muscles reduces incorporation of methionine into ACh receptor. *Nature* 268:655–57
101. Reiness, C. G., Hall, Z. W. 1981. The developmental change in immunological properties of the acetylcholine receptor in rat muscle. *Dev. Biol.* 81:324–31
102. Reiness, C. G., Weinberg, C. B., Hall, Z. W. 1978. Antibody to acetylcholine receptor increases degradation of junctional and extrajunctional receptors in adult muscle. *Nature* 274:68–70
103. Richman, D. P., Gomez, C. M., Berman, P. W., Burres, S. A., Fitch, F. W., Arnason, B. G. W. 1980. Monoclonal anti-acetylcholine receptor antibodies can cause experimental myasthenia. *Nature* 286:738–39
104. Rotundo, R. L., Fambrough, D. M. 1980. Secretion of acetylcholinesterase: relation to acetylcholine receptor metabolism. *Cell* 22:595–602
105. Sakmann, B., Brenner, H. R. 1978. Change in synaptic channel gating during neuromuscular development. *Nature* 276:401–2
106. Salisbury, J. C., Condeelis, J. S., Satir, P. 1980. Role of coated vesicle microfilaments, and calmodulin in receptor-mediated endocytosis by cultured B-lymphoblastoid cells. *J. Cell Biol.* 87:132–41
107. Sanes, J. R., Hall, Z. W. 1979. Antibodies that bind specifically to synaptic sites on muscle fiber basal lamina. *J. Cell Biol.* 83:357–70
108. Sanes, J. R., Marshall, L. M., McMahan, U. J. 1978. Reinnervation of muscle fiber basal lamina after removal of myofibers: differentiation of regenerating axons at original synaptic sites. *J. Cell Biol.* 78:176–94
109. Schreiner, G. F., Unanue, E. R. 1976. Membrane and cytoplasmic changes in B lymphocytes induced by ligand-surface immunoglobulin interaction. *Adv. Immunol.* 24:38–165
110. Schuetze, S. M., Frank, E. F., Fischbach, G. D. 1978. Channel open time and metabolic stability of synaptic and extrasynaptic acetylcholine receptors on cultured chick myotubes. *Proc. Natl. Acad. Sci. USA* 75:520–23
111. Schuetze, S. M. 1980. The acetylcholine channel open time in chick muscle is not decreased following innervation. *J. Physiol.* 303:111–24
112. Shainberg, A., Brik, H. 1978. The appearance of acetylcholine receptors triggered by fusion of myoblasts in vitro. *FEBS Lett.* 88:327–31
113. Shainberg, A., Cohen, S. A., Nelson, P. G. 1976. Induction of acetylcholine receptors in muscle cultures. *Pflügers Arch.* 36:255–61
114. Smilowitz, H. 1980. Routes of intracellular transport of acetylcholine receptor and esterase are distinct. *Cell* 19:237–44
115. Stanley, E. F., Drachman, D. B. 1978. Effect of myasthenic immunoglobulin on acetylcholine receptors of intact mammalian neuromuscular junctions. *Science* 200:1285–87
116. Steinbach, J. H. 1981. Neuromuscular junctions and α-bungarotoxin binding sites in denervated and contralateral cat skeletal muscles. *J. Physiol.* 313:513–28
117. Steinbach, J. H., Merlie, J., Heinemann, S., Bloch, R. 1979. Degradation of junctional and extrajunctional acetylcholine receptors by developing rat skeletal muscle. *Proc. Natl. Acad. Sci. USA* 76:3547–51
118. Strader, C. D., Raftery, M. A. 1980. Topographic studies of Torpedo acetylcholine receptors subunits as a transmembrane complex. *Proc. Natl. Acad. Sci. USA* 77:5807–11
119. Strader, C. B. D., Revel, J.-P., Raftery, M. A. 1979. Demonstration of the transmembrane nature of the acetylcholine receptor by labeling with anti-receptor antibodies. *J. Cell Biol.* 83:499–510
120. Tarrab-Hazdai, R., Bercovici, T., Goldfarb, V., Gitler, C. 1980. Identification of the acetylcholine receptor subunit in the lipid bilayer of Torpedo electric organ excitable membranes. *J. Biol. Chem.* 255:1204–9
121. Taylor, R. B., Duffus, W. P. H., Raff, M. C., dePetris, S. 1978. Redistribution and pinocytosis of lymphocyte surface immunoglobulin molecules induced by anti-immunoglobulin antibody. *Nature New Biol.* 233:225–29
122. Vandlen, R. L., Wu, W. C.-S., Eisenach, J. C., Raftery, M. A. 1979. Studies of the composition of purified *Torpedo california* acetylcholine receptor and of its subunits. *Biochemistry* 10:1845–54
123. Vincent, A. 1980. Immunology of acetylcholine receptors in relation to myasthenia gravis. *Physiol. Rev.* 60:756–824
124. Waechter, C. J., Lennarz, W. J. 1976. The role of polyprenol linked sugars in glycoprotein synthesis. *Ann. Rev. Biochem.* 45:95–112

125. Weinberg, C. B., Hall, Z. W. 1979. Antibodies from patients with myasthenia gravis recognize determinants unique to extrajunctional acetylcholine receptors. *Proc. Natl. Acad. Sci. USA* 76:504–8
126. Weinberg, C. B., Reiness, C. G., Hall, Z. W. 1981. Topographical segregation of old and new acetylcholine receptors at developing ectopic endplates in adult rat muscles. *J. Cell Biol.* 88:215–18
127. Wennogle, L. P., Changeux, J.-P. 1980. Transmembrane orientation of proteins present in acetylcholine receptor-rich membranes from *Torpedo marmorata* studied by selective proteolysis. *Eur. J. Biochem.* 106:381–93
128. Willingham, M. C., Maxfield, F. R., Pastan, I. H. 1979. α-2 Macroglobulin binding to the plasma membrane of cultured fibroblasts: diffuse binding followed by clustering in coated regions. *J. Cell Biol.* 82:614–25
129. Zalin, R. J., Montague, W. 1974. Changes in adenylate cyclase, cyclic AMP, and protein kinase levels in chick myoblasts, and their relationship to differentiation. *Cell* 2:103–8
130. Ziskind, L., Dennis, M. J. 1978. Depolarizing effect of curare on embryonic rat muscles. *Nature* 276:622–23

THE DEVELOPMENT OF SARCOPLASMIC RETICULUM MEMBRANES

Anthony Martonosi

Department of Biochemistry, SUNY Upstate Medical Center, Syracuse, New York 13210

INTRODUCTION AND BACKGROUND

> Growth-control mechanisms are probably very complicated. Nevertheless, there is a strong drive to devise simple models. The dilemma is that the models may be misleading. Probably each of us is willing to tolerate being misled a little because we hope that our preferred model will be of more help than hindrance. Unfortunately, there is no way to know in advance which model will be a help and which will be a hindrance.
>
> R. W. Holley

Sarcoplasmic reticulum (SR) is a highly specialized network of membrane tubules and cisternae with unique protein and phospholipid composition (30, 31, 39, 76). Its primary function is the regulation of cytoplasmic Ca^{2+} concentration (13, 14, 50). A key element in this regulation is the Ca^{2+}-transport ATPase, which represents 60–80% of the total protein content in SR of adult animals (68, 69, 76, 131). The Ca^{2+}-ATPase is an intrinsic membrane protein of close to 100,000-dalton mass (69, 76) that penetrates through the lipid phase of the membrane and requires phospholipids for activity (4, 45, 79, 88, 126).

Two functional states of the SR alternate during the contraction-relaxation cycles: (*a*) The transport of calcium into the SR during relaxation is energized by the hydrolysis of ATP through the formation of an aspartyl-phosphate enzyme intermediate (20, 21, 39, 50, 93, 121, 131). (*b*) The release of Ca^{2+} from the SR during contraction is triggered by the depolarization of the transverse tubules; the transmission of the stimulus from T tubules to SR occurs at specialized junctions between the two membranes (13, 25, 30, 31, 110).

In addition to the Ca^{2+}-ATPase, SR membranes contain varying amounts of calsequestrin (67, 70, 71), a high-affinity Ca^{2+}-binding protein (69, 85, 86), an intrinsic glycoprotein (85), several proteolipids (60, 69, 89), and smaller amounts of other proteins of unknown function. The high-affinity Ca^{2+}-binding protein and calsequestrin are extrinsic proteins which are readily released from the membrane upon washing with salt solution containing EGTA (23).

Sarcoplasmic reticulum evolves from rough endoplasmic reticulum by co- or post-translational insertion of Ca^{2+}-ATPase molecules synthesized on membrane-bound polysomes (15, 36, 87a). The transverse tubules are invaginations of the surface membrane. The development of SR is coordinated with the appearance of transverse tubules and junctions form between them already at early stages of development (27, 51, 58, 108).

Even in fully developed muscle SR retains the multifunctional character of endoplasmic reticulum and contains the enzymes and functions usually associated with endoplasmic reticulum membranes of other tissues.

This chapter deals with the biosynthesis of SR and its regulation during embryonic and postnatal development. The process has been studied extensively in vivo, in tissue culture, and in cell-free translation systems. Although much of the current information relates to embryonic development, it is likely that the same principles of membrane assembly and regulation govern the adaptation of SR structure and function to physiological requirements in the adult animals.

THE Ca^{2+} TRANSPORT ATPase CONTENT OF MUSCLE DURING DEVELOPMENT IN VIVO

During early embryonic development skeletal muscles perform little work; they contain only few scattered myofilaments and no organized sarcoplasmic reticulum. The Ca^{2+}-sensitive ATPase and Ca^{2+} transport activities are barely measurable (8, 77, 81, 102, 104) and roughly correspond to the activities found in nonmuscle cells. In skeletal muscle microsomes of 10-day-old chicken embryos the Ca^{2+}-transport ATPase determined by SDS-polyacrylamide gel electrophoresis is only about 2% of the total protein content (8, 82), and the steady-state concentration of Ca^{2+}-sensitive phosphoprotein, which is a measure of the concentration of Ca^{2+}-transport sites, is about 0.1 nmol mg^{-1} protein. Within a few weeks of development the

Ca^{2+}-ATPase content of the whole muscle and of the isolated SR increase more than 20-fold with parallel changes in the Ca^{2+}-transport, Ca^{2+}-sensitive ATPase activity, and steady-state concentration of phosphoenzyme intermediate (8, 29, 37, 81). Essentially similar observations were made on developing rabbit muscle (102, 104, 134). In heart muscle microsomes of chicken embryos the Ca^{2+}-ATPase content is already high at 10–14 days of development, in accord with the early appearance of cardiac activity, and there is only a slight increase as development proceeds (8, 77).

The change in the Ca^{2+}-ATPase content arises largely from an increase in the concentration of Ca^{2+}-ATPase in the membrane (3, 81, 122), but there is also some increase in the total surface area of SR during development (17, 66). The density of 8 nm freeze-etch particles characteristic of the Ca^{2+}-ATPase (2, 18, 30, 31, 53) increases during development from about 200 μm^{-2} in 10-day-old chicken embryo muscle to 4330 μm^{-2} after 46 days of development (3, 73, 81, 82, 122). Assuming that the 8 nm freeze-etch particles represent a cluster of 4 ATPase molecules (53, 74, 76, 80, 105), the approximate Ca^{2+}-ATPase polypeptide chain density in SR of 10-day-old embryos is about 800 μm^{-2}, and increases to about 16,000 μm^{-2} in fully developed muscles. This massive increase in ATPase concentration is also reflected in an increase in the protein/lipid ratio of isolated membranes (8, 9, 102). Surprisingly, the change in ATPase concentration determined by gel electrophoresis, electron microscopy, and analysis of the active site concentration of ATPase after specific labeling with [^{32}P] ATP roughly parallels the change in Ca^{2+}-transport activity throughout development (73, 81, 82). If this correlation is valid, it would imply that the specific Ca^{2+}-transport activity of the pump molecules remains constant throughout development in spite of the marked changes in the ATPase/lipid ratio and in the fatty acid composition of membrane lipids. There is no evidence for immunological differences between Ca^{2+}-ATPases isolated from immature and adult fast rabbit muscles (122a).

THE DEVELOPMENT OF SARCOPLASMIC RETICULUM IN TISSUE CULTURE

The major conclusions derived from studies of SR development in vivo were confirmed in tissue culture with some further insight into the mechanism of regulation of the synthesis of Ca^{2+}-transport ATPase.

In chicken pectoralis or rat muscle cultures the differentiation of muscle cells proceeds through a precisely timed sequence of morphological and biochemical changes similar to those taking place in vivo; these include proliferation of presumptive myoblasts, their fusion into multinucleated myotubes, followed by the accumulation of contractile proteins and SR.

The accumulation of Ca^{2+}-ATPase measured either by active-site labeling with [^{32}P] ATP (37, 81) or by immunoprecipitation (37, 48, 133) begins during the fusion of myoblasts, and continues during the next 10 days of in vitro development accompanied by a large release in the myosin and ATP: creatine phosphotransferase content (37). The concentration of Ca^{2+}-ATPase determined by active-site labeling with [^{32}P] ATP increases from 0.01 nmole mg^{-1} total protein on the second day to about 0.04 nmol mg^{-1} protein by the 5th day of culture (81). The apparent leveling off is due to the continued synthesis of contractile and cytoplasmic proteins as indicated by the increase in cell mass. The maximum steady-state concentration of Ca^{2+}-sensitive phosphoprotein in 5-day-old cultures (0.04 nmol mg^{-1} protein) is 7–10 times smaller than in pectoralis muscles of 2–3 week-old chicks, indicating a relatively poor development of SR in tissue culture (65, 81). This may be due in part to the high rate of degradation of Ca^{2+}-ATPase in tissue culture [half life is about 20 hr (48, 133)] compared with adult rat skeletal muscle in vivo (half life about 9–11 days).

The relationship between fusion of myoblasts into myotubes and the accumulation of Ca^{2+}-ATPase is not clear. In chicken muscle cultures grown in low Ca^{2+} medium ($\simeq 100\ \mu M\ Ca^{2+}$) fusion is inhibited and little or no accumulation of the Ca^{2+}-ATPase was observed (37). On the other hand, transfer of rat muscle cultures grown for 44 hr in normal medium into low Ca^{2+} medium delayed but did not inhibit the accumulation of the Ca^{2+}-ATPase (86, 133). Differences in experimental conditions may contribute to these observations, since Cantini et al (12, 107) found poor development of the sarcotubular system in rat muscle cultures grown in Ca^{2+}-deficient media. As inhibition of the fusion of rat muscle cells with cytochalasin B did not interfere with the development of sarcotubular elements (12) the inhibition observed in Ca^{2+}-deficient medium may be a direct effect of Ca^{2+} deficiency, and cannot be attributed to the inhibition of fusion.

In embryonic chicken heart cells cultured in a medium that permitted cell proliferation, the rate of Ca^{2+}-ATPase synthesis increased steadily during the entire culture period without the initial lag phase seen in skeletal muscle (47). Therefore cessation of cell proliferation may not be a required prelude to differentiation and Ca^{2+}-ATPase synthesis in cardiac muscle. Dibutyryl cyclic AMP (1 mM) had no effect upon the synthesis of Ca^{2+}-ATPase, though it produced modest but significant stimulation of the synthesis of myosin heavy chain (47).

THE LIPID COMPOSITION OF SARCOPLASMIC RETICULUM DURING DEVELOPMENT

The Ca^{2+}-transport and ATPase activities of SR are sharply influenced by the microviscosity of the lipid environment (4, 45, 72, 88, 126). This underscores the potential importance of developmental changes in the composition of SR membrane lipids.

The phospholipid composition of whole chicken muscle or isolated chicken SR remains relatively constant between 12 and 50 days of development (8, 9). There are, however, marked changes in the fatty acid composition. During development in vivo the palmitate content decreases and the linoleate content increases (8, 9); these changes are more pronounced in isolated microsomes than in whole muscle homogenates. Other fatty acids varied to a lesser extent. The accumulation of unsaturated fatty acids occurred with an increase in average chainlength that would maintain membrane fluidity relatively constant. In accordance there is only a slight decrease of the principal transition temperature of membrane lipids detected by differential scanning calorimetry (83), and the specific Ca^{2+}-transport activity and Ca^{2+} permeability of the SR vesicles isolated at various stages of development change only slightly (73).

The simultaneous changes in fatty acid composition and in the concentration of Ca^{2+}-transport ATPase in the membrane may imply that the insertion of the ATPase requires unsaturated fatty acids, but so far there is no direct experimental evidence to support this possibility. The synthesis of lipid and protein components of the membrane probably occurs independently but it is so coordinated that the fatty acid composition becomes optimal when the transport ATPase content of the membrane reaches adult levels.

A different developmental trend in fatty acid composition is observed in cultured chicken muscle, where there is a steady increase in stearate and a decrease in linoleate and arachidonate content between 2 and 13 days of culture (7). This may contribute to the relatively poor development of SR in tissue culture. The fatty acid composition of cultured muscle cells reflects the fatty acid composition of the medium (7, 49); therefore this system could be useful for analysis of the influence of the physical properties of membrane lipids upon the synthesis and insertion of the Ca^{2+}-ATPase into the bilayer.

In developing rabbit muscle the phosphatidylcholine content of SR increases during the first week after birth with a decrease in phosphatidylethanolamine and other phospholipids (102, 104, 134). The cholesterol content is at least in part related to contamination by surface membranes and therefore it is difficult to evaluate. Contamination by other membranes may explain the relatively high Mg^{2+}-activated ATPase and low Ca^{2+}-transport activity of the light subfractions of SR isolated from developing muscle by sucrose gradient centrifugation (134).

Sarcoplasmic reticulum is able to synthesize diacyl and dialkyl phospholipids by the same pathways as endoplasmic reticulum of other tissues, and there are interesting changes in the activity of these enzymes in SR membranes isolated at various stages of embryonic development (91, 103, 123).

THE SYNTHESIS AND INSERTION OF Ca^{2+} BINDING PROTEINS INTO SR DURING DEVELOPMENT IN VIVO AND IN TISSUE CULTURE

The high-affinity Ca^{2+}-binding protein and calsequestrin are present in high concentration in microsomes isolated from 10–14-day-old chicken embryo muscle (8, 73) or muscles of fetal and newborn rabbits (102, 104), which contain only traces of Ca^{2+}-ATPase. The concentration of the two Ca^{2+}-binding proteins changes only slightly as the Ca^{2+}-ATPase accumulates during subsequent development.

These observations were extended in rat muscle cultures using antibodies directed against calsequestrin (133) and the high-affinity Ca^{2+}-binding protein (86) for the isolation and quantitation of the radioactively labeled proteins. Several-fold increase in the rate of synthesis of calsequestrin (133) and the high-affinity Ca^{2+}-binding protein (86) was observed before the fusion of myoblasts and the increased synthesis of Ca^{2+}-ATPase began. The temporal patterns of the synthesis of calsequestrin and the high-affinity Ca^{2+}-binding protein were identical. These observations support the conclusion that the Ca^{2+}-binding proteins and the Ca^{2+}-transport ATPase are synthesized under separate control.

The turnover of the high-affinity Ca^{2+}-binding protein [half life about 10 hr (86)] is greater than that of calsequestrin [half life $\simeq$ 23 hr, (133)], although the two proteins are synthesized at nearly identical rates. This may explain the relatively low concentration of the high-affinity Ca^{2+}-binding protein in mature membrane preparations. Transverse tubular vesicles isolated from heavy SR preparations after disruption in a French pressure cell are enriched in the high-affinity Ca^{2+}-binding protein (85), while calsequestrin is assumed to be localized in the cisternal portion of the SR (84, 101). Thus the two Ca^{2+}-binding proteins may be distributed in different regions of the sarcotubular system. Some of the calsequestrin found in the heavy microsomal fraction may be associated with Golgi vesicles.

The temporal relationship between the synthesis of Ca^{2+}-ATPase and calsequestrin was confirmed by immunofluorescence studies on cultured rat skeletal muscle. Calsequestrin was detected in mononucleated myoblasts well before fusion ($\simeq$ 45 hr) localized at discrete perinuclear regions, which may represent the Golgi apparatus (55). As development proceeded, the staining spread to progressively larger regions of the myotubes adjacent to the nucleus and assumed a fibrous appearance. The Ca^{2+}-ATPase appeared much later ($\simeq$ 60 hr) in granular patches throughout the cytoplasm of all fused and some unfused cells (55). Calsequestrin is a glycoprotein, and its accumulation in the Golgi region may be required for processing prior to its delivery into the SR. The distribution of calsequestrin at the interface between the A and I bands in adult rat skeletal muscle is consistent with its presumed primary localization in the cisternae of SR (56).

THE SYNTHESIS OF Ca^{2+}-TRANSPORT ATPase AND CALSEQUESTRIN IN CELL-FREE SYSTEMS

For cell-free translation of Ca^{2+}-ATPase, membrane-bound polysomes isolated from 14–16-day-old chicken embryo muscles (15) or from leg muscles of neonatal rats (36, 95) were used. Free polysomes, although capable of the synthesis of a large number of proteins with molecular weights up to 200,000, were essentially inactive in the synthesis of Ca^{2+}-ATPase (15, 36).

The electrophoretic mobility, isoelectric point, and tryptic peptide map of the in vitro translated product was similar to authentic Ca^{2+}-ATPase isolated from adult chicken muscle or from cultured muscle cells (15, 95). The NH_2-terminal methionine group of the ATPase derived from initiator methionyl $tRNA_f^{Met}$, was acetylated during translation (95). These observations suggest that the Ca^{2+}-ATPase is synthesized without an NH_2-terminal signal sequence. The ATPase polypeptide synthesized on rough microsomes was incorporated into the membrane as an intrinsic membrane protein (15). Washing of the microsomes with 0.5 M KCl and 10 mM EGTA, or with 0.05–0.1 mg deoxycholate per mg protein, was ineffective in releasing the Ca^{2+}-ATPase from the membrane; only conditions leading to solubilization of the microsomes liberated most of the ATPase into the supernatant (15). Recent observations suggest co-translational insertion of the Ca^{2+}-ATPase into the membrane (87a) It is likely that the Ca^{2+}-ATPase polypeptide crosses the membrane several times (59). Since the 30 NH_2-terminal amino acids of the Ca^{2+}-ATPase are relatively hydrophilic, the NH_2-terminal segment of the molecule may remain on the cytoplasmic side of the membrane (95a). The extra electron density (43) and the presence of 4 nm particles on the cytoplasmic surface (69, 124), together with the preferential localization of 8 nm freeze-etch particles in the cytoplasmic leaflet of the bilayer (2, 3, 18, 53, 122), suggest that a major portion of the ATPase is exposed on the cytoplasmic surface. No information is available about the location of the carboxyl-terminus or whether the acetylation of the NH_2-terminal methionine occurs before or after insertion of the polypeptide chain into the membrane. The Ca^{2+}-ATPase of SR is not glycosylated.

The symmetrical arrangement of SR in the two halves of the sarcomere may imply that the synthesis of Ca^{2+}-transport ATPase is confined to central growth regions of the membrane and the newly synthesized ATPase molecules are distributed over the surface of SR by lateral diffusion. The precise alignment of membrane elements and contractile filaments suggests some interaction between them.

The surface membrane of muscle cells contains a Ca^{2+}-ATPase with properties similar to the SR enzyme. The transfer of Ca^{2+}-ATPase molecules to the cell surface may involve interaction with a "carrier" molecule [perhaps analogous to the light subunit of the Na,K-ATPase (46, 111)] that may function in a manner similar to that of the β_2-microglobulin in the HLA system (92). Alternatively, glycosylation of some ATPase molecules may be the "zip code" that directs them to the surface membrane. Glycosylation of the surface membrane ATPase or its association with a "carrier" protein may be difficult to detect, since the total mass of the surface membrane is only 1–10% of the mass of SR.

Calsequestrin is synthesized on membrane bound polysomes in a precursor form (66,000 daltons), which is processed co-translationally to the mature calsequestrin (63,000 daltons) in the presence of pancreatic microsomes (95). Since calsequestrin is a glycoprotein it is presumably transferred from the rough endoplasmic reticulum into the Golgi apparatus for further processing before it reaches the SR (68).

REGULATION OF THE SYNTHESIS OF SR PROTEINS

The cytoplasmic free calcium concentration in muscle cells is regulated by a complex system of Ca^{2+} channels, Ca^{2+}-transport pumps, and Ca^{2+}-binding components located in the surface membranes, mitochondria, SR, contractile filaments, and cytoplasm (14, 75, 106). It is assumed that the concentration and activity of these systems are adjusted to the physiological requirements by coordinate regulation of their rate of synthesis and degradation. Some features of this regulation are intrinsic properties of muscle cells that are expressed in tissue culture; others are under neural or hormonal control, and may be studied more effectively in the intact animal.

Myogenic Mechanisms of Regulation. The Hypothetical Role of Ca^{2+} in the Regulation of Gene Expression

Membrane-bound polysomes isolated from 11–12-day-old chicken embryo skeletal muscles are less active in the translation of Ca^{2+}-ATPase than corresponding preparations from 14–16-day-old embryos. This suggests that the increase in the rate of synthesis of Ca^{2+}-transport ATPase in the late prenatal and early postnatal periods may be due to an increase in the concentration of translatable Ca^{2+}-ATPase mRNA in the cell. Therefore the synthesis of Ca^{2+}-ATPase is probably regulated at the level of transcription and/or processing of Ca^{2+}-ATPase mRNA.

The increase in the concentration of Ca^{2+}-ATPase in chicken pectoralis or rat muscle cells is initiated about the time of the fusion of myoblasts into multinucleated myotubes and slightly precedes the accumulation of contractile proteins and ATP: creatine phosphotransferase (37, 48, 81, 134). Inhibition of fusion by lowering the Ca^{2+} concentration of the growth medium below 150 μM (112) prevents or delays the accumulation of Ca^{2+}-ATPase and contractile proteins (37). It is reasonable to assume that medium Ca^{2+} concentration regulates the synthesis of intracellular enzymes via changes in cytoplasmic free Ca^{2+} concentration.

A hypothesis was proposed in which changes in intracellular free Ca^{2+} concentrations are assumed to influence the rate of synthesis of Ca^{2+}-ATPase and other Ca^{2+}-modulated proteins by Ca^{2+}-dependent regulation of the transcription or processing of the relevant classes of mRNA (78, 82). If this hypothesis is correct changes in cytoplasmic free Ca^{2+} concentration induced by various means should influence the expression of specific muscle proteins. There are several indications that this may be the case:

1. Continuous exposure of muscle cells to low concentration (10^{-8} M) of the Ca^{2+} ionophore A23187 increases the steady-state level of a Ca-sensitive, hydroxylamine-labile phosphoprotein, which is assumed to reflect the concentration of Ca^{2+}-ATPase (81). A23187 also increased the rate of synthesis and degradation of proteins in explanted soleus muscle (57).

2. Brief exposure of cultured chicken pectoralis muscle cells to the Ca^{2+} ionophores ionomycin and A23187 ($\simeq 4 \times 10^{-6}$ M for 1–3 hr) increases several-fold the rate of [^{35}S]methionine incorporation into two membrane proteins—one of 100,000 and one of 80,000 daltons—which sediment in the mitochondrial and microsomal fractions (97, 99, 129). Increased synthesis of the 80,000-dalton protein was also observed upon cell-free translation

of poly(A)-enriched RNA isolated from ionomycin-treated as compared with control cultures (129). Therefore ionomycin selectively increases the cellular concentration of mRNA that codes for the 80,000-dalton protein. The effect is presumably mediated through an increase in cytoplasmic $[Ca^{2+}]$, as evidenced by the contraction of myotubes upon the addition of the ionophore. The 100,000-dalton protein is tentatively identified as the Ca^{2+}-ATPase; the identity of the 80,000-dalton protein is unknown. An increased rate of cell-free translation of the 80,000-dalton protein was also observed using poly(A)RNA isolated 3 hr after transfer of muscle cultures grown in low-Ca^{2+} medium into a medium of normal Ca^{2+} concentration (129). Since at low medium $[Ca^{2+}]$ the Ca^{2+} permeability of the cells increases (128), such transfer is likely to produce transient rise in cytoplasmic $[Ca^{2+}]$. Further work is required to ascertain whether Ca^{2+} plays a role in the synthesis of myofibrillar proteins.

3. The total Ca^{2+} content of embryonic muscle cells is high before the accumulation of Ca^{2+}-ATPase begins and decreases with age both in vivo and in tissue culture as development proceeds (81, 132). In view of the small amount of SR (8, 81) and parvalbumin (63) and the rapid fluxes of calcium across the surface membranes in embryonic muscle (99), if a major part of the cell-associated Ca^{2+} is intracellular, the free $[Ca^{2+}]$ of the cytoplasm is likely to be elevated.

4. In chicken embryo muscles between 13 and 19 days of development the action potential is largely due to a Ca^{2+} current, indicating the presence of Ca^{2+} channels in the surface membrane (117). Fluctuation of cytoplasmic $[Ca^{2+}]$ causes spontaneous contractions.

While these observations support the proposed role of cytoplasmic Ca^{2+} in the regulation of the biosynthesis of SR a direct analysis of cytoplasmic Ca^{2+} concentration during fusion of myoblasts and their differentiation is required to strengthen the hypothesis.

The nuclear envelope of most cells is permeable to small molecules and ions (16, 38). The Ca^{2+} content of nuclei analyzed in situ by electron microprobe is similar to that of the cytoplasm (113, 114, 115) and much smaller than that of SR (116). Therefore the free Ca^{2+} concentration of the nucleoplasm is expected to follow changes in the cytoplasmic $[Ca^{2+}]$.

The effect of $[Ca^{2+}]$ upon the transcription or processing of mRNA may be mediated by nuclear Ca^{2+}-binding proteins serving as Ca^{2+} receptors. Ca^{2+}-binding proteins with a broad range of affinities for Ca^{2+} were recently observed among the nonhistone chromosomal proteins and in the insoluble protein fraction of nuclei isolated from skeletal muscles of embryonic and adult chicken or rabbits and from rat liver (109). On the basis of electrophoretic mobility the nuclear Ca^{2+}-binding proteins are distinct from parvalbumin, troponin C, or prothrombin, but one of the fractions may contain calmodulin.

Among possible modes of involvement of Ca^{2+}-binding proteins in gene regulation is the Ca^{2+}-dependent phosphorylation of histones or other chromatin components. Marked activation of the phosphorylation of histone H3 by 0.1 μM Ca^{2+} was observed in isolated nuclei of HeLa cells (127).

The postulated mechanism must account for the observation that the induction of the synthesis of the 80,000- and 100,000-dalton proteins by ionomycin was accompanied by inhibition of myosin synthesis (97). This implies that the expression of different structural genes may be under the influence of distinct Ca^{2+}-dependent regulatory mechanisms with presumably different Ca^{2+} affinities.

Ca^{2+} may play a similar role in the control of gene expression during development of fertilized eggs (26, 34, 52 96), in the transformation of lymphocytes (32, 35, 44), in the replication of hemopoietic stem cells (33), and in embryonic induction of *Rana pipiens* (1).

The idea of the ionic control of gene expression is not new (61, 64). Much information is available on the possible role of monovalent cations (61, 62, 64, 130), and Mg^{2+} (99a), Zn^{2+}, and Cd^{2+} (16a) in growth control and in the expression of specific gene loci. However, the regulatory potential of Na^+, K^+, and Mg^{2+} is limited by their relatively constant concentration in the living cell, while the concentration of Ca^{2+} varies over a wide range (10^{-7}—10^{-5} M) depending on the activity and metabolic state of the cells. Some of the effects attributed to changes in cellular Na^+, K^+, or Mg^{2+} concentrations may be actually due to secondary changes is [Ca^{2+}] via the operation of cation:Ca^{2+}-exchange pumps.

The Influence of Muscle Activity and Innervation on the SR

There is a striking correlation between the speed of muscle contraction and relaxation and the extent of development of SR in various muscles (54, 72). Some of these differences are quantitative. There is more SR in fast-twitch than in slow-twitch, cardiac, or smooth muscles; and the concentration of Ca^{2+}-ATPase in the membrane, as indicated by the density of 8 nm intramembranous freeze etch particles, is also greater in fast-twitch than in slow-twitch or smooth muscle fibers (5, 6, 10, 11, 22, 94). Large differences were noted in the Ca^{2+} transport and Ca^{2+}-ATPase activities of microsomes isolated from skeletal, cardiac, and smooth muscles (40–42, 54, 72, 118).

In addition, however, there are indications of immunochemical differences between Ca^{2+}-ATPases isolated from skeletal and cardiac muscles

(17a, 19); and differences in the rates of elementary reaction steps of ATP hydrolysis suggest (125) that kinetically and structurally distinct isoenzymes of the Ca^{2+}-ATPase may be present in phenotypically different muscle fibers of the same animal. This implies that neural influence may selectively promote the synthesis of certain muscle specific isoenzymes of the Ca^{2+}-ATPase. A relationship between nerve-muscle activity and the extent of development of SR is supported by the following observations: (a) Much of the increase in the Ca^{2+}-ATPase content of chicken muscle occurs around and after the time of hatching and coincides with vigorous muscle activity (81). (b) The concentration of Ca^{2+}-ATPase in cultured chicken muscle is only 1/3–1/10 of their innervated in vivo developing counterparts (81). (c) Repeated injections of curare into the chorioallantoic sac of chicken embryos between the 8th and 14th days of development markedly reduced the accumulation of Ca^{2+}-ATPase and other muscle proteins (98). (d) Chronic denervation of muscle decreases the Ca^{2+}-transport activity of SR and produces muscle atrophy (54, 119). (e) Cross-innervation of muscle changes the amount as well as the composition of SR membranes (54, 87, 100, 120). The activity pattern imposed by the nerve appears to be responsible for these changes (40–42, 100). Chronic electric stimulation of fast-twitch rabbit muscle with the frequency pattern of the slow nerve induces progressive transformation of the SR to the type found in slow-twitch muscle (41, 42). This is reflected in a decrease in Ca^{2+}-transport activity together with a decrease in the amount of 105,000-dalton Ca^{2+}-ATPase polypeptide and a loss of 7–9 nm diameter intramembranous freeze-etch particles in the concave fracture faces of the SR. It is unfortunate that only fast-to-slow transformation was investigated in detail, since the loss of Ca^{2+}-transport activity may be caused by necrosis, which is a possible consequence of chronic stimulation. Slow-to-fast conversion induced by stimulation would be more convincing as an increase in Ca^{2+}-transport activity is less readily attributable to secondary changes. There is as yet no evidence indicating changes in the immunospecificity of the Ca^{2+}–transport ATPase as a result of cross-innervation or chronic stimulation.

These observations represent the most important clues that muscle activity or some of its metabolic consequences influence the gene expression leading to the synthesis of the Ca^{2+}-transport ATPase and other membrane components. If the stimulus frequency defines muscle phenotype, these observations imply that the cells translate specific stimulus frequencies into selective metabolic signals for the expression of the "slow" or "fast" phenotypes. Within the framework of the Ca^{2+} hypothesis, discrimination is possible if the Ca^{2+} receptors regulating the expression of the slow and fast isoenzymes have different Ca^{2+} affinities and sense the different steady-state Ca^{2+} levels established by the two stimulus frequencies. Alternatively discrimination may be based on the relative response time of the Ca^{2+}-induced change (rate of phosphorylation or dephosphorylation, etc) in comparison with the duration and frequency of the Ca^{2+} pulses. Essentially no information is available on the possible involvement of Ca^{2+} in signal processing of this type.

While innervation and muscle activity promote the synthesis and accumulation of several muscle proteins, the density of extrajunctional acetylcholine receptors is low in the surface membranes of innervated muscle and increases 20–30-fold after functional and surgical denervation or in tissue culture. Electrical stimulation following denervation reduces or eliminates the increase in extrajunctional acetylcholine sensitivity (24, 28). Extensive proliferation of transverse tubules was observed in denervated muscles and in tissue culture, with the formation of labyrinthine networks of T tubules (51, 90). Little is known about the molecular mechanisms that regulate the assembly of surface membranes and the expansion of surface membranes into T tubules. These observations nevertheless indicate that the responses of different membrane components to physiological stimuli vary, and the assembly of surface membranes and SR probably involves distinct regulatory mechanisms.

ACKNOWLEDGMENT

Supported by research grants from the NIH (AM 26545), NSF (PCM 7919502), and the Muscular Dystrophy Association. My thanks are due to Ricardo Boland, Ted Chyn, David Cunningham, Doo Bong Ha, Takashi Morimoto, Shiro Ohnoki, Young Chul Park, Eva Reyes, Dikla Roufa, David Sabatini, Angelo Schibeci, Judy Tsai, and Fang Sheng Wu for collaboration in various phases of the work reported from our laboratory.

Literature Cited

1. Barth, L. G., Barth, L. J. 1974. Ionic regulation of embryonic induction and cell differentiation in Rana pipiens. *Devel. Biol.* 39:1–22
2. Baskin, R. J. 1971. Ultrastructure and calcium transport in crustacean muscle microsomes *J. Cell. Biol.* 48:49–60
3. Baskin, R. J. 1974. Ultrastructure and calcium transport in microsomes from developing muscle. *J. Ultrastruct. Res.* 49:348–71
4. Bennett, J. P., McGill, K. A., Warren, G. B. 1980. The role of lipids in the functioning of a membrane protein: the sarcoplasmic reticulum calcium pump. *Curr. Top. Membr. Transp.* 14:127–64
5. Beringer, T. 1976. A freeze-fracture study of sarcoplasmic reticulum from fast and slow muscle of the mouse. *Anat. Rec.* 184:647–64
6. Bertaud, W. S., Rayns, D. G., Simpson, F. O. 1970. Freeze-etch studies on fish skeletal muscle. *J. Cell. Sci.* 6:537–57
7. Boland, R., Chyn, T., Roufa, D., Reyes, E., Martonosi, A. 1977. The lipid composition of muscle cells during development. *Biochim. Biophys. Acta* 489:349–59

8. Boland, R., Martonosi, A., Tillack, T. W. 1974. Developmental changes in the composition and function of sarcoplasmic reticulum. *J. Biol. Chem.* 249:612–23
9. Boland, R., Martonosi, A. 1976. The lipid composition and Ca transport function of sarcoplasmic reticulum (SR) membranes during development in vivo and in vitro. In *Function and Biosynthesis of Lipids,* ed. N. G. Bazan, R. R. Brenner, N. M. Guisto, pp. 233–39. NY/London: Plenum
10. Bray, D. F., Rayns, D. G. 1976. A comparative freeze-etch study of the sarcoplasmic reticulum of avian fast and slow muscle fibers. *J. Ultrastruct. Res.* 57:251–59
11. Bray, D. F., Rayns, D. G., Wagenaar, E. B. 1978. Intramembrane particle densities in freeze fractured sarcoplasmic reticulum. *Can. J. Zool.* 56:140–45
12. Cantini, M., Sartore, S., Vitadello, M., Schiaffino, S. 1979. Development of the sarcotubular system in fusion-arrested myoblasts. *Cell Biol. Int. Rep.* 3:151–56
13. Caputo, C. 1978. Excitation and contraction processes in muscle. *Ann. Rev. Biophys. Bioeng.* 7:63–83
14. Carafoli, E., Crompton, M. 1978. The regulation of intracellular calcium. *Curr. Top. Membr. Transp.* 10:151–216
15. Chyn, T. L., Martonosi, A. N., Morimoto, T., Sabatini, D. D. 1979. In vitro synthesis of the Ca^{2+} transport ATPase by ribosomes bound to sarcoplasmic reticulum membranes. *Proc. Natl. Acad. Sci. USA* 76:1241–45
16. Civan, M. M. 1978. Intracellular activities of sodium and potassium. *Am. J. Physiol.* 234:F261–69
16a. Compere, S. J., Palmiter, R. D. 1981. DNA methylation controls the inducibility of the mouse metallothionein-I gene in lymphoid cells. *Cell* 25:233–40
17. Crowe, L. M., Baskin, R. J. 1977. Stereological analysis of developing sarcotubular membranes. *J. Ultrastruct. Res.* 58:10–21
17a. Damiani, E., Betto, R., Salvatori, S., Volpe, P., Salviati, G., Margreth, A. 1981. Polymorphism of sarcoplasmic reticulum adenosine triphosphatase of rabbit skeletal muscle. *Biochem. J.* 197:245–48
18. Deamer, D. W., Baskin, R. J. 1969. Ultrastructure of sarcoplasmic reticulum preparations. *J. Cell. Biol.* 42:296–307
19. De Foor, P. H., Levitsky, D., Biryukova, T., Fleischer, S. 1980. Immunological dissimilarity of the calcium pump protein of skeletal and cardiac muscle sarcoplasmic reticulum. *Arch. Biochem. Biophys.* 200:196–205
20. Degani, C., Boyer, P. D. 1973. A borohydride reduction method for characterization of the acyl phosphate linkage in proteins and its application to sarcoplasmic reticulum adenosine triphosphatase. *J. Biol. Chem.* 248:8222–26
21. De Meis, L., Vianna, A. L. 1979. Energy interconversion by the Ca^{2+}-dependent ATPase of the sarcoplasmic reticulum. *Ann. Rev. Biochem.* 48:275–92
22. Devine, C. E., Rayns, D. G. 1975. Freeze fracture studies of membrane systems in vertebrate muscle. II. Smooth muscle. *J. Ultrastruct. Res.* 51:293–306
23. Duggan, P. F., Martonosi, A. 1970. Sarcoplasmic reticulum. IX. The permeability of sarcoplasmic reticulum membranes. *J. Gen. Physiol.* 56:147–67
24. Edwards, C. 1979. The effects of innervation on the properties of acetylcholine receptors in muscle. *Neuroscience* 4:565–84
25. Endo, M. 1977. Calcium release from the sarcoplasmic reticulum. *Physiol. Rev.* 57:71–108
26. Epel, D. 1980. Ionic triggers in the fertilization of sea urchin eggs. *Ann. NY Acad. Sci.* 339:74–85
27. Ezerman, E. B., Ishikawa, H. 1967. Differentiation of the sarcoplasmic reticulum and T system in developing chick skeletal muscle in vitro. *J. Cell. Biol.* 35:405–20
28. Fambrough, D. M. 1979. Control of acetylcholine receptors in skeletal muscle. *Physiol. Rev.* 59:165–227
29. Fanburg, B. L., Drachman, D. B., Moll, D., Roth, S. I. 1968. Calcium transport in isolated sarcoplasmic reticulum during muscle maturation. *Nature* 218:962–64
30. Franzini-Armstrong, C. 1975. Membrane particles and transmission at the triad. *Fed. Proc.* 34:1382–89
31. Franzini-Armstrong, C. 1980. Structure of sarcoplasmic reticulum. *Fed. Proc.* 39:2403–9
32. Freedman, M. H., Raff, M. C., Gomperts, B. 1975. Induction of increased calcium uptake in mouse T lymphocytes by concanavalin A and its modulation by cyclic nucleotides. *Nature* 255:378–82
33. Gallien-Lartigue, O. 1976. Calcium and ionophore A-23187 as initiators of DNA replication in the pluripotent

haemopoietic stem cell. *Cell Tissue Kinet.* 9:533–40
34. Gilkey, J. C., Jaffe, L. F., Ridgway, E. B., Reynolds, G. T. 1978. A free calcium wave traverses the activating egg of the medaka, *Oryzias latipes*. *J. Cell. Biol.* 76:448–66
35. Greene, W. C., Parker, C. M., Parker, C. W. 1976. Calcium and lymphocyte activation. *Cell. Immunol.* 25:74–89
36. Greenway, D. C., MacLennan, D. H. 1978. Assembly of the sarcoplasmic reticulum. Synthesis of calsequestrin and the $Ca^{2+}+Mg^{2+}$–adenosine triphosphatase on membrane-bound polyribosomes. *Can. J. Biochem.* 56:452–56
37. Ha, D. B., Boland, R., Martonosi, A. 1979. Synthesis of the calcium transport ATPase of sarcoplasmic reticulum and other muscle proteins during development of muscle cells in vivo and in vitro. *Biochim. Biophys. Acta* 585:165–87
38. Harris, J. R. 1978. The biochemistry and ultrastructure of the nuclear envelope. *Biochim. Biophys. Acta* 515:55–104
39. Hasselbach, W. 1979. The sarcoplasmic calcium pump. A model of energy transduction in biological membranes. *Top. Curr. Chem.* 78:1–56
40. Heilmann, C., Brdiczka, D., Nickel, E., Pette, D. 1977. ATPase activities, Ca^{2+} transport and phosphoprotein formation in sarcoplasmic reticulum subfractions of fast and slow rabbit muscles. *Eur. J. Biochem.* 81:211–22
41. Heilmann, C., Muller, W., Pette, D. 1981. Correlation between ultrastructural and functional changes in sarcoplasmic reticulum during chronic stimulation of fast muscle. *J. Membr. Biol.* 59:143–49
42. Heilmann, C., Pette, D. 1979. Molecular transformations in sarcoplasmic reticulum of fast-twitch muscle by electro-stimulation. *Eur. J. Biochem.* 93:437–46
43. Herbette, L., Marquardt, J., Scarpa, A., Blasie, J. K. 1977. A direct analysis of lamellar x-ray diffraction from hydrated oriented multilayers of fully functional sarcoplasmic reticulum. *Biophys. J.* 20:245–72
44. Hesketh, I. R., Smith, G. A., Houslay, M. D., Warren, G. B, Metcalfe, J. C. 1977. Is an early calcium flux necessary to stimulate lymphocytes? *Nature* 267:490–94
45. Hidalgo, C., Thomas, D. D., Ikemoto, N. 1978. Effect of the lipid environment on protein motion and enzymatic activity of the sarcoplasmic reticulum calcium ATPase. *J. Biol. Chem.* 253:6879–87
46. Hobbs, A. S., Albers, R. W. 1980. The structure of proteins involved in active membrane transport. *Ann. Rev. Biophys. Bioeng.* 9:259–91
47. Holland, P. C. 1979. Biosynthesis of the Ca^{2+}– and Mg^{2+}–dependent adenosine triphosphatase of sarcoplasmic reticulum in cell cultures of embryonic chick heart. *J. Biol. Chem.* 254:7604–10
48. Holland, P. C., MacLennan, D. H. 1976. Assembly of sarcoplasmic reticulum. Biosynthesis of the adenosine triphosphatase in rat skeletal muscle cell culture. *J. Biol. Chem.* 251:2030–36
49. Horwitz, A. F., Wight, A., Ludwig, P., Cornell, R. 1978. Interrelated lipid alterations and their influence on the proliferation and fusion of cultured myogenic cells. *J. Cell. Biol.* 77:334–57
50. Inesi, G. 1979. Transport across sarcoplasmic reticulum in skeletal and cardiac muscle. In *Membrane Transport in Biology*, ed. G. Giebisch, D. C. Tosteson, H. H. Ussing, pp. 357–93. Berlin: Springer
51. Ishikawa, H. 1968. Formation of elaborate networks of T-system tubules in cultured skeletal muscle with special reference to the T-system formation. *J. Cell. Biol.* 38:51–66
52. Jaffe, L. F. 1980. Calcium explosions as triggers of development. *Ann. NY Acad. Sci.* 339:86–101
53. Jilka, R. L., Martonosi, A. N., Tillack, T. W. 1975. Effect of the purified (Mg^{2+} + Ca^{2+})-activated ATPase of sarcoplasmic reticulum upon the passive Ca^{2+} permeability and ultrastructure of phospholipid vesicles. *J. Biol. Chem.* 250:7511–24
54. Jolesz, F., Sreter, F. A. 1981. Development, innervation, and activity-pattern induced changes in skeletal muscle. *Ann. Rev. Physiol.* 43:531–52
55. Jorgensen, A. O., Kalnins, V. I., Zubrzycka, E., MacLennan, D. H. 1977. Assembly of the sarcoplasmic reticulum. Localization by immunofluorescence of sarcoplasmic reticulum proteins in differentiating rat skeletal muscle cell cultures. *J. Cell. Biol.* 74:287–98
56. Jorgensen, A. O., Kalnins, V, MacLennan, D. H. 1979. Localization of sarcoplasmic reticulum proteins in rat skeletal muscle by immunofluorescence. *J. Cell. Biol.* 80:372–84
57. Kameyama, T., Etlinger, J. D. 1979. Calcium-dependent regulation of pro-

tein synthesis and degradation in muscle. *Nature* 279:344–46
58. Kelly, A. M. 1971. Sarcoplasmic reticulum and T tubules in differentiating rat skeletal muscle. *J. Cell. Biol.* 49:335–44
59. Klip, A., Reithmeier, R. A. F., MacLennan, D. H. 1980. Alignment of the major tryptic fragments of the adenosine triphosphatase from sarcoplasmic reticulum. *J. Biol. Chem.* 255:6562–68
60. Knowles, A., Zimniak, P., Alfonzo, M., Zimniak, A., Racker, E. 1980. Isolation and characterization of proteolipids from sarcoplasmic reticulum. *J. Membr. Biol.* 55:233–39
61. Kroeger, H., Lezzi, M. 1966. Regulation of gene action in insect development. *Ann. Rev. Entomol.* 11:1–22
62. Leffert, H. L. 1980. Growth regulation by ion fluxes. *Ann. NY Acad. Sci.* 339:1–335
63. LePeuch, C. J., Ferraz, C., Walsh, M. P., Demaille, J. G., Fischer, E. H. 1979. Calcium and cyclic nucleotide dependent regulatory mechanisms during development of chick embryo skeletal muscle. *Biochemistry* 18:5267–73
64. Lezzi, M. 1970. Differential gene activation in isolated chromosomes. *Int. Rev. Cytol.* 29:127–68
65. Lough, J. W., Entman, M. L., Bossen, E. H., Hansen, J. L. 1972. Calcium accumulation by isolated sarcoplasmic reticulum of skeletal muscle during development in tissue culture. *J. Cell. Physiol.* 80:431–36
66. Luff, A. R., Atwood, H. L. 1971. Changes in the sarcoplasmic reticulum and transverse tubular system of fast and slow skeletal muscles of the mouse during postnatal development. *J. Cell. Biol.* 51:369–83
67. MacLennan, D. H. 1974. Isolation of a second form of calsequestrin. *J. Biol. Chem.* 249:980–84
68. MacLennan, D. H., Campbell, K. P. 1979. Structure, function and biosynthesis of sarcoplasmic reticulum proteins. *Trends Biochem. Sci.* 4:148–51
69. MacLennan, D. H., Holland, P. C. 1976. The calcium transport ATPase of sarcoplasmic reticulum. In *The Enzymes of Biological Membranes*, ed. A. Martonosi, 3:221–59. NY: Plenum
70. MacLennan, D. H., Wong, P. T. S. 1971. Isolation of a calcium-sequestering protein from sarcoplasmic reticulum. *Proc. Natl. Acad. Sci. USA* 68:1231–35
71. MacLennan, D. H., Zubrzycka, E., Jorgensen, A. O., Kalnins, V. I. 1978. Assembly of the sarcoplasmic reticulum. In *The Molecular Biology of Membranes*, ed. S. Fleischer, Y. Hatefi, D. H. MacLennan, A. Tzagoloff, pp. 309–20. NY: Plenum
72. Martonosi, A. 1972. Biochemical and clinical aspects of sarcoplasmic reticulum function. *Curr. Top. Membr. Transp.* 3:83–197
73. Martonosi, A. 1975. Membrane transport during development in animals. *Biochim. Biophys. Acta.* 415:311–33
74. Martonosi, A. 1977. Protein-protein interactions in sarcoplasmic reticulum: functional significance. In *FEBS Symp. Vol. 45, A4, Membrane Proteins*, ed. P. Nicholls et al, pp. 135–40. Oxford: Pergamon
75. Martonosi, A. 1980. Calcium pumps. *Fed. Proc.* 39:2401–2
76. Martonosi, A., Beleer, T. J. 1982. The mechanism of Ca^{2+} transport by sarcoplasmic reticulum. In *Handbook of Physiology*, ed. L. D. Peachey, R. Adrian. Bethesda, Md: Am. Physiol. Soc. In press
77. Martonosi, A., Boland, R., Halpin, R. A. 1972. The biosynthesis of sarcoplasmic reticulum membranes and the mechanism of calcium transport. *Cold Spring Harbor Symp. Quant. Biol.* 37:455–68
78. Martonosi, A., Chyn, T. L., Schibeci, A. 1978. The calcium transport of sarcoplasmic reticulum. *Ann. NY Acad. Sci.* 307:148–59
79. Martonosi, A., Donley, J. R., Pucell, A. G., Halpin, R. A. 1971. Sarcoplasmic reticulum. XI. The mode of involvement of phospholipids in the hydrolysis of ATP by sarcoplasmic reticulum membranes. *Arch. Biochem. Biophys.* 144:529–40
80. Martonosi, A., Nakamura, H., Jilka, R. L., Vanderkooi, J. M. 1977. Protein-protein interactions and the functional states of sarcoplasmic reticulum membranes. In *Biochemistry of Membrane Transport, FEBS Symp. No. 42*, ed. G. Semenza, E. Carafoli, pp. 401–15. Berlin: Springer
81. Martonosi, A., Roufa, D., Boland, R., Reyes, E., Tillack, T. W. 1977. Development of sarcoplasmic reticulum in cultured chicken muscles. *J. Biol. Chem.* 252:318–32
82. Martonosi, A., Roufa, D., Ha, D. B., Boland, R. 1980. The biosynthesis of sarcoplasmic reticulum. *Fed. Proc.* 39:2415–21
83. Martonosi, M. A. 1974. Thermal analy-

sis of sarcoplasmic reticulum membranes. *FEBS Lett.* 47:327–29
84. Meissner, G. 1975. Isolation and characterization of two types of sarcoplasmic reticulum vesicles. *Biochim. Biophys. Acta* 389:51–68
85. Michalak, M., Campbell, K. P., MacLennan, D. H. 1980. Localization of the high affinity calcium binding protein and an intrinsic glycoprotein in sarcoplasmic reticulum membranes. *J. Biol. Chem.* 255:1317–26
86. Michalak, M., MacLennan, D. H. 1980. Assembly of the sarcoplasmic reticulum. Biosynthesis of the high affinity calcium binding protein in rat skeletal muscle cell cultures. *J. Biol. Chem.* 255:1327–34
87. Mommaerts, W. F. H. M., Buller, A. J., Seraydarian, K. 1969. The modification of some biochemical properties of muscle by cross-innervation. *Proc. Natl. Acad. Sci. USA* 64:128–33
87a. Mostov, K. E., De Foor, P., Fleischer, S., Blobel, G. 1981. Co-translational membrane integration of calcium pump protein without signal sequence cleavage. *Nature* 292:87–88
88. Nakamura, H., Jilka, R. L., Boland, R., Martonosi, A. 1976. Mechanism of ATP hydrolysis by sarcoplasmic reticulum and the role of phospholipids. *J. Biol. Chem.* 251:5414–23
89. Ohnoki, S., Martonosi, A. 1980. Purification and characterization of the proteolipid of rabbit sarcoplasmic reticulum. *Biochim. Biophys. Acta* 626:170–78
90. Pellegrino, C., Franzini, C. 1963. An electron microscope study of denervation atrophy in red and white skeletal muscle fibers. *J. Cell. Biol.* 17:327–49
91. Pilarska, M., Zimniak, P., Pikula, S., Sarzala, M. G. 1980. The terminal step of the de novo synthesis of diacyl and alkylacyl phospholipids in subfractions of rabbit sarcoplasmic reticulum during ontogenesis. *FEBS Lett.* 114:21–24
92. Ploegh, H. L., Cannon, L. E., Strominger, J. L. 1979. Cell-free translation of the mRNAs for the heavy and light chains of HLA-A and HLA-B antigens. *Proc. Natl. Acad. Sci. USA* 76:2273–77
93. Racker, E. 1979. Transport of ions *Accounts Chem. Res.* 12:338–44
94. Rayns, D. G., Devine, C. E., Sutherland, C. L. 1975. Freeze fracture studies of membrane systems in vertebrate muscle. I. Striated muscle. *J. Ultrastruct. Res.* 50:306–21
95. Reithmeier, R. A. F., de Leon, S., MacLennan, D. H. 1980. Assembly of the sarcoplasmic reticulum. Cell-free synthesis of the Ca^{2+} + Mg^{2+} adenosine triphosphatase and calsequestrin. *J. Biol. Chem.* 255:11839–46
95a. Reithmeier, R. A. F., MacLennan, D. H. 1981. The NH_2 terminus of the [Ca^{2+} + Mg^{2+}] adenosine triphosphatase is located on the cytoplasmic surface of the sarcoplasmic reticulum membrane. *J. Biol. Chem.* 256:5957–60
96. Ridgway, E. B., Gilkey, J. C., Jaffe, L. F. 1977. Free calcium increases explosively in activating medaka eggs. *Proc. Natl. Acad. Sci. USA* 74:623–27
97. Roufa, D., Martonosi, A. N. 1980. The effect of Ca^{2+} ionophores upon the synthesis of muscle proteins in normal and fusion blocked cultured skeletal muscle. *Fed. Proc.* 39:954
98. Roufa, D., Martonosi, A. 1981. Effect of curare on the development of chicken embryo skeletal muscle in ovo. *Biochem. Pharmacol.* 30:1501–5
99. Roufa, D., Wu, F. S., Martonosi, A. 1981. The effect of Ca^{2+} ionophores upon the synthesis of proteins in cultured skeletal muscles. *Biochim. Biophys. Acta* 674:225–37
99a. Rubin, A. H., Terasaki, M., Sanui, H. 1979. Major intracellular cations and growth control: correspondence among magnesium content, protein synthesis, and the onset of DNA synthesis in BALB/c3T3 cells. *Proc. Natl. Acad. Sci. USA* 76:3917–21
100. Salmons, S., Sreter, F. A. 1976. Significance of impulse activity in the transformation of skeletal muscle type. *Nature* 263:30–34
101. Sarzala, M. G., Michalak, M. 1978. Studies on the heterogeneity of sarcoplasmic reticulum vesicles. *Biochim. Biophys. Acta* 513:221–35
102. Sarzala, M. G., Pilarska, M., Zubrzycka, E., Michalak, M. 1975. Changes in the structure, composition and function of sarcoplasmic reticulum membrane during development. *Eur. J. Biochem.* 57:25–34
103. Sarzala, M. G., Pilarska, M. 1976. Phospholipid biosynthesis in sarcoplasmic reticulum membrane during development. *Biochim. Biophys. Acta* 441:81–92
104. Sarzala, M. G., Zubrzycka, E., Michalak, M. 1975. Comparison of some features of undeveloped and mature sarcoplasmic reticulum vesicles. In *Calcium Transport in Contraction and Secretion*, ed. E. Carafoli, F. Clementi, W. Drabikowski, A. Margreth, pp. 329–38. Amsterdam: North Holland

105. Scales, D., Inesi, G. 1976. Assembly of ATPase protein in sarcoplasmic reticulum membranes. *Biophys. J.* 16:735–51
106. Scarpa, A., Carafoli, E. 1978. Calcium transport and cell function. *Ann. NY Acad. Sci.* 307:1–655
107. Schiaffino, S., Cantini, M., Sartore, S. 1977. T-system formation in cultured rat skeletal tissue. *Tissue Cell.* 9:437–46
108. Schiaffino, S., Margreth, A. 1969. Coordinated development of the sarcoplasmic reticulum and T system during postnatal differentiation of rat skeletal muscle. *J. Cell Biol.* 41:855–75
109. Schibeci, A., Martonosi, A. 1980. Ca^{2+}-binding proteins in nuclei. *Eur. J. Biochem.* 113:5–14
110. Schneider, M. F. 1981. Membrane charge movement and depolarization-contraction coupling. *Ann. Rev. Physiol.* 43:507–17
111. Schuurmans Stekhoven, F., Bonting, S. L. 1981. Transport adenosine triphosphatases: properties and functions. *Physiol. Rev.* 61:1–76
112. Shainberg, A., Yagil, G., Yaffe, D. 1969. Control of myogenesis in vitro by Ca^{2+} concentration in nutritional medium. *Exp. Cell Res.* 58:163–67
113. Somlyo, A. P., Shuman, H., Somlyo, A. V. 1978. Mitochondrial and sarcoplasmic reticulum contents *in situ:* electron probe analysis. In *Frontiers in Biological Energetics,* ed. P. L. Dutton, J. S. Leigh, A. Scarpa, pp. 742–51. NY: Academic
114. Somlyo, A. P., Somlyo, A. V., Shuman, H., Sloane, B., Scarpa, A. 1978. Electron probe analysis of calcium compartments in cryo sections of smooth and striated muscles. *Ann. NY Acad. Sci.* 307:523–44
115. Somlyo, A. V., Shuman, H., Somlyo, A. P. 1977. Elemental distribution in striated muscle and the effects of hypertonicity. Electron probe analysis of cryo sections. *J. Cell Biol.* 74:828–57
116. Somlyo, A. V., Shuman, H., Somlyo, A. P. 1977. Composition of sarcoplasmic reticulum *in situ* by electron probe X-ray microanalysis. *Nature* 268:556–58
117. Spitzer, N. C. 1979. Ion channels in development. *Ann. Rev. Neurosci.* 2:363–97
118. Sreter, F. A. 1969. Temperature, pH and seasonal dependence of Ca-uptake and ATPase activity of white and red muscle microsomes. *Arch. Biochem. Biophys.* 134:25–33
119. Sreter, F. A. 1970. Effect of denervation on fragmented sarcoplasmic reticulum of white and red muscle. *Exp. Neurol.* 29:52–64
120. Sreter, F. A., Luff, A. R., Gergely, J. 1975. Effect of cross-reinnervation on physiological parameters and on properties of myosin and sarcoplasmic reticulum of fast and slow muscles of the rabbit. *J. Gen. Physiol.* 66:811–21
121. Tada, M., Yamamoto, T., Tonomura, Y. 1978. Molecular mechanism of active calcium transport by sarcoplasmic reticulum. *Physiol. Rev.* 58:1–79
122. Tillack, T. W., Boland, R., Martonosi, A. 1974. The ultrastructure of developing sarcoplasmic reticulum. *J. Biol. Chem.* 249:624–33
122a. Volpe, P., Damiani, E., Salviati, G., Margreth, A. 1982. Transitions in membrane composition during postnatal development of rabbit fast muscle. *J. Muscle Res. Cell Motil.* 3
123. Waku, K. 1977. Skeletal muscle. In *Lipid Metabolism in Mammals,* ed. F. Snyder, pp. 189–208. NY/London: Plenum
124. Wang, C.-T., Saito, A., Fleischer, S. 1979. Correlation of ultrastructure of reconstituted sarcoplasmic reticulum membrane vesicles with variation in phospholipid to protein ratio. *J. Biol. Chem.* 254:9209–19
125. Wang, T., Grassi de Gende, A. O., Schwartz, A. 1979. Kinetic properties of calcium adenosine triphosphatase of sarcoplasmic reticulum isolated from cat skeletal muscles. A comparison of caudofemoralis (fast), tibialis (mixed), and soleus (slow). *J. Biol. Chem.* 254:10675–78
126. Warren, G. B., Bennett, J. P., Hesketh, T. R., Houslay, M. D., Smith, G. A., Metcalfe, J. C. 1975. The lipids surrounding a calcium transport protein: their role in calcium transport and accumulation. *Proc. 10th FEBS Meet.* 41:3–15
127. Whitlock, J. P. Jr., Augustine, R., Schulman, H. 1980. Calcium-dependent phosphorylation of histone H3 in butyrate-treated HeLa cells. *Nature* 287:74–76
128. Winegrad, S. 1971. Studies of cardiac muscle with a high permeability to calcium produced by treatment with ethylenediaminetetraacetic acid. *J. Gen. Physiol.* 58:71–93
129. Wu, F. S., Park, Y. C., Roufa, D., Martonosi, A. 1981. Selective stimulation of the synthesis of an 80,000 dalton protein by calcium ionophores. *J. Biol. Chem.* 256:5309–12

130. Wuhrmann, P., Ineichen, H., Riesen-Willi, U., Lezzi, M. 1979. Change in nuclear potassium electrochemical activity and puffing of potassium-sensitive salivary chromosome regions during *Chironomus* development. *Proc. Natl. Acad. Sci. USA* 76:806–8
131. Yamamoto, T., Takisawa, H., Tonomura, Y. 1979. Reaction mechanisms for ATP hydrolysis and synthesis in the sarcoplasmic reticulum. *Curr. Top. Bioenerg.* 9:179–236
132. Yen, S.-S., Chen-Yen, S.-H., Klein, R. L. 1974. Ontogenesis of Ca^{++} and Mg^{++} contents of embryonic chick heart. *Differentiation* 2:351–55
133. Zubrzycka, E., MacLennan, D. H. 1976. Assembly of the sarcoplasmic reticulum. Biosynthesis of calsequestrin in rat skeletal muscle cell cultures. *J. Biol. Chem.* 251:7733–38
134. Zubrzycka, E., Michalak, M., Kosk-Kosicka, D., Sarzala, M. G. 1979. Properties of microsomal subfractions isolated from developing rabbit skeletal muscle. *Eur. J. Biochem.* 93:113–21

IONIC CHANNELS IN SKELETAL MUSCLE

E. Stefani

Departments of Physiology and Biophysics, Centro de Estudios Avanzados del Instituto Politécnico Nacional, Apartado Postal 14–740, México 14, D.F.

D. J. Chiarandini

Departments of Ophthalmology and Physiology and Biophysics, New York University Medical Center, New York, NY 10016

INTRODUCTION

There is now considerable evidence that ions can move across the cell membrane through voltage-gated aqueous pores called "ionic channels." Each channel has a characteristic permeability, selectivity, and kinetics (15, 66–68, 92, 125). Electrical excitation in skeletal muscle involves voltage- and time-dependent changes of the permeabilities to Na^+ and K^+ which induce a transient inflow of Na^+ into the fiber followed by an outflow of K^+. As a result of these ionic movements the action potential is generated. Besides these Na^+ and K^+ channels, a voltage-dependent Ca^{2+} channel has been described recently in frog skeletal muscle (21, 104, 115). At rest, Cl^- and other K^+ channels are responsible for the dominant conductance (4). Most of these K^+ channels rectify inward current (3–5, 80).

Here we review recent research on ionic channels in twitch and tonic skeletal muscle fibers. In these studies the technique of voltage clamp has been widely used. The application of this tool to skeletal muscle has been reviewed recently (27). Models and mechanisms of ionic permeation across biological membranes have been discussed in recent articles (15, 23, 64, 65, 68, 70).

Na^+ CHANNELS

The introduction of the three microelectrode voltage-clamp technique (2) allowed the first examination of the ionic currents underlying the action potential in frog muscle. Adrian et al demonstrated an early, transient inward current that was abolished by tetrodotoxin (TTX) and reversed at a membrane potential value close to the Na^+ equilibrium potential. They successfully applied the Hodgkin & Huxley model to describe the currents and compute the action potential, using a m^3 relation.

Action potentials were later computed using an equivalent circuit that included the tubular system (TS). This was represented as a radial cable of sixteen elements. The tubular membrane was assumed to have activable Na^+ channels similar to those of the fiber surface (7). The best fit was obtained by postulating a density of Na^+ channels in the tubular wall of about one twentieth of that in the fiber surface.

Several lines of evidence indicate the presence of Na^+ channels in the membrane of the TS. The radial spread of mechanical activation has a Q_{10} of 2, which is not easily explained by pure passive propagation (56), and is reduced by TTX or a Na^+-deficient medium (38, 39). Twitch tension is diminished when the luminal Na^+ concentration is reduced or when an action potential imposed to the surface membrane only spreads passively to the TS after TTX treatment (20, 26, 34). Detubulation of muscle fibers reduces TTX binding by about 50% (79). Tubular Na^+ currents have been measured directly by several authors (31, 69, 86). At 12–15°C an extra inward current with a slow time course and a long delay was found. This current was blocked by TTX more slowly than the major Na^+ current.

Very recently, Nakajima & Gilai (89, 90) elegantly demonstrated the occurrence of action potentials in the TS by measuring the radial propagation of action potentials along the TS with potential-sensitive dyes. They calculated a value of 6.4 cm sec^{-1} for the radial conduction velocity of the action potential along the tubules at 24.5°C, in fair agreement with the results of Gonzalez-Serratos (56).

Na^+ Channel Kinetics

The introduction of the double sucrose-gap clamp (77) and the potentiometric vaseline-gap method (69) have greatly improved the time resolution of early Na^+ currents.

Na^+ permeability rises steeply for small depolarizations, increasing e-fold for a 3.7 mV depolarization, and levels off above 0 mV. The Na^+ current can be described by a m^3 relationship. The half-activation potential ranges from –42 to –50 mV (2, 33, 78). The time constant of activation at 0 mV (τ_m) ranges from 0.5 to 1.25 msec (1–2°C) (2, 33). With a E_{Na} of +63 mV and a P_{Na} of $1.5–7 \times 10^{-3}$ cm sec^{-1} the calculated peak I_{Na} near -15 mV is 3.7 mA cm^{-2}. The instantaneous I-V relationship for Na$^+$ currents shows the rectification predicted by the Goldman-Hodgkin-Katz equation, but it deviates from it at high positive potentials (33).

Na$^+$ current inactivates. At a holding potential of -90 mV, 10–30% of the Na$^+$ channels are inactivated and the half-inactivation potential ranges from -70 to -80 mV. The time constant of inactivation (τ_h) at 0 mV is about 1.5 msec (1–5°C) (2, 6, 33).

Ionic Selectivity

The permeability sequence of Na$^+$ channels is: Na$^+$ $\simeq$ Li$^+$ $\simeq$ hydroxylammonium > hydrazinium > ammonium > guanidinium > K$^+$ > aminoguanidinium with a ratio of 1: 0.96: 0.94: 0.31: 0.11: 0.093: 0.048: 0.031, respectively (32). No detectable inward currents have been recorded for Ca^{2+}, methylammonium, methylguanidinium, tetraethylammonium and tetramethylammonium. It appears that the channel selects strongly against methyl groups. In fact, methylammonium is nearly identical in size and shape to both hydroxylammonium and hydrazinium, but is impermeant. In addition, aminoguanidinium and guanidinium are much larger than methylammonium but are permeant (32).

Hille (61, 62, 64) has explained such ionic selectivity in the node of Ranvier by postulating a 3.1×5.1 Å oxygen-lined channel. According to this model the formation of hydrogen bonds allows the hydroxyl and amino groups of permeating ions to approach the oxygens of the pore about 0.8–0.9 Å closer than if hydrogen bonds were not formed. As a consequence, cations with methyl groups that cannot form hydrogen bonds would appear larger to the selectivity filter. Recently, this model has been extended to explain deviations from the independence principle (24, 25, 70).

Chemical Studies of Na$^+$ Channels

The presence of carboxyl groups in the channels appears to be responsible for the ionic selectivity of the channels and for the binding sites of saxitoxin (STX) and TTX (16–18, 47, 106). However, the carboxyl groups that determine the toxin binding site and the selectivity filter appear to be separate, either in whole or in part. Trimethyloxonium ion (TMO), a reagent that esterifies carboxyl groups, makes the Na$^+$ channel resistant to TTX without modifying its ionic selectivity (108, 109).

The activation and inactivation processes can be dissociated by various chemicals. Inactivation is reduced by external application of 0.7 mM N-bromoacetamide, 40 mM formaldehyde, or 1.8 mM glutaraldehyde, or by reduction of internal pH to pH 4 with 48–60 mM biphtalate buffer. This effect would be produced by modifications of tyrosine and arginine groups or by reaction with dysyl-or-N-terminal amino groups on the cytoplasmic side of the membrane. The activation process is little affected perhaps because the gate molecules are deeply buried inside the membrane (94).

Single Channel Conductance

Based on the density of TTX or STX binding sites in frog muscle, which ranges from 195 to 380 sites per μm^2 (12, 79, 103), and on a Na^+ membrane conductance of 50–328 mS cm^{-2} (69, 78), a single channel conductance of 1.5–15 pS can be calculated. In mammalian EDL muscle with a peak Na^+ conductance of 40–50 mS cm^{-2} and a density of STX binding sites of 535 ± 43 μm^{-2}, a single channel conductance of 1.68 pS has been obtained (58, 99), assuming that half the Na^+ channels are opened at the time of the peak current (99).

The extracellular patch clamp technique has provided direct information on the properties of single channels (91). In cultured muscle cells of rat this technique has shown that single channel conductance changes are stochastic with rapid transitions between two states: open and closed. The single channel currents correspond to a single channel conductance of 18 pS (107).

Na^+ Current in Mammalian Muscle

Na^+ currents in innervated fast- and slow-twitch mammalian muscle and in frog twitch muscle are similar. The kinetic model (m^3h) applied to frog muscle can also be applied to mammalian Na^+ current with minor modifications. In rat muscle the activation of Na^+ current occurs at more negative potentials than in frog muscle and the rates of activation are significantly slower at potentials more negative than –40 mV. Whether the rates of inactivation are slower in mammalian muscle is unclear (6, 43, 99).

The values of I_{Na} and maximum Na^+ conductance ($\overline{G}_{Na}$) are similar in frog and rat when measured with the same technique. In the frog $\overline{G}_{Na} \simeq$ 300 mS cm^{-2} and $I_{Na} \simeq$ 4 mA cm^{-2}, in rat, $\overline{G}_{Na} \simeq$ 120 mS cm^{-2} and $I_{Na} \simeq$ 2.5–4.5 mA cm^{-2} (33, 99).

Fast- and slow-twitch mammalian fibers have similar Na^+ channels. The maximum amplitude and time course of the Na^+ current are similar in both types of fibers. The main differences are that in slow-twitch fibers the maximum Na^+ current occurs at less negative potentials and that the reversal potential is more negative (44).

Effect of Denervation

In the adult rat muscle after denervation the action potential becomes partially resistant to TTX (102) and the density of binding sites for STX, measured in purified sarcolemma, is reduced (18). In innervated muscles TTX blocks Na^+ channels in a manner predicted by the binding of the toxin to a single population of channels with a dissociation constant of 5 nM. After denervation a second population of Na⁺ channels with a dissociation constant in the micromolar range appears (99). This second type of channel is responsible for 25–30% of the total Na⁺ conductance. Denervation shifts activation and inactivation curves by 10 mV to more negative potentials without appreciably affecting $\overline{G}_{Na}$ and the rates of activation and inactivation.

K⁺ CHANNELS

Various types of K⁺ channels have been demonstrated in frog skeletal muscle: (*a*) delayed rectifier, which gives rise to a fast K⁺ current (delayed current) responsible for the repolarization phase of the action potential (2); this current reaches a maximum in about 0.1 sec at -30 mV (3°C); (*b*) slow K⁺ channel, which generates a slow current that reaches a maximum in about 3 sec at -30 mV (3°C) (3); (*c*) inward rectifier, which is responsible for a conductance increase when the membrane potential of the fiber is shifted to more negative levels (3, 4, 80); and (*d*) Ca^{2+}-dependent K⁺ channel, which opens following increases in intracellular $[Ca^{2+}]$, first described in nerve cells (87), has been postulated to exist in skeletal muscle (19, 48).

Delayed Rectifier

The delayed K⁺ current can be fitted by a n^4 relationship. The mean maximum K⁺ conductance ranges from 5.8 to 23.0 mS cm^{-2}. The K⁺ conductance shows an e-fold increase for a 3 mV depolarization. In the activation curve the value of membrane potential for half activation is -40 to -50 mV and the time constant of activation (τ_n) is 5–8 msec at 0 mV (3°C) (2, 10, 82, 85, 111, 113).

The delayed channel has an approximate linear instantaneous I-V relationship and a mean equilibrium potential of -85 mV. It inactivates exponentially with a time constant of 0.4–1.2 sec at $+10$ mV (20°C). The steady-state inactivation curve is less steep than that of Na⁺ current and the potential for half inactivation is 20 mV more positive (3). The delayed rectifier has a selectivity sequence of $K^+ \geq Rb^+ \geq Cs^+ \geq Na^+ \geq Li^+$ (52). This sequence is similar to that of the equivalent K⁺ channels in nerve (57, 63).

The delayed rectifier channel is located mainly in the surface membrane. The variation of the reversal potential of the delayed K⁺ current as a function of the amount of current injected to clamp the membrane indicates that K⁺ accumulates in a space equivalent to one sixth to one third of the fiber volume, which is much larger than the TS volume (2). The presence of a clear repolarization phase of the action potential and of the delayed K^+ current in detubulated fibers supports that location (50, 93). In addition, the delayed rectifier seems to be located also in the TS. This is suggested by the demonstration that the late after-potential that follows a train of spikes becomes slower as the fiber radius increases (81).

The density of the delayed K^+ channels would be 1.4×10^9 channels cm^{-2} of external surface, assuming a single open channel conductance of 4 pS. This density is about twenty times less than that postulated for Na^+ channels (10, 12).

Tetraethylammonium (TEA) blocks delayed rectification in muscle as it does in nerve, but the affinity of the muscle binding sites is lower by more than one order of magnitude. The dissociation constant is 8×10^{-3} M (111). TEA shifts the threshold for delayed rectification and the activation curve to more negative potentials. Furthermore, it slows the rate of onset of the current by about 80% (111).

Tetracaine (2 mM) shifts the voltage dependence of the delayed current to more positive potentials, delays the onset of the current and slows its kinetics about 3.7-fold (10).

The application of Zn^{2+} prolongs the action potentials by depressing the K^+ delayed current (112). Zn^{2+} (0.1 mM) reduces the maximum K^+ conductance by 60% and does not alter the reversal potential. The main effect of Zn^{2+} is to reduce about ten-fold the rate constants for opening and closing of the gating mechanism of the conductance. It has little effect on the rate of inactivation. Zn^{2+} does not alter the effective valency of the gating particle, but the activation curve is shifted toward more positive potentials (112).

4-Aminopyridine (1 mM) reduces maximal K^+ conductance by about 50%. This effect is associated with a shift of about 10 mV to more positive potentials in the threshold and activation curve of the delayed current. The blockade is voltage-dependent and the drug appears to act only on open channels (53, 54).

The delayed rectifier has similar characteristics in mammalian and frog muscle (99). The mean maximal conductance in mammals is 11.7 ± 1.8 mS cm^{-2}, within the range observed in frog muscle, and the rate constants of activation are comparable when allowance is made for temperature differences. The amplitude and time course of the delayed K^+ current are similar in fast- and slow-twitch fibers of rat. In both types of fibers the fast tail currents decay with a similar time constant of 10 msec (43–45). However, slow-twitch (soleus) fibers exhibit an additional large slow K^+ current which decays with a time constant about ten times larger than that of the delayed current (see below).

Slow K^+ Channels

In frog muscle (3) tail current measurements show the presence of a slowly developing K^+ current that reaches a maximum in about 3 sec (−30 mV, 3°C) and then slowly inactivates to reach a final steady level of about one third the maximum amplitude. For a given depolarization the maximum value of this current is about one sixth of the maximum delayed current (3). The slow K^+ current can be described by a n^2 relationship with a τ_n of about 90 msec at 0 mV (5°C) (85). The slow K^+ current has been measured during depolarizing pulses after blocking the delayed rectifier with TEA in the intact muscle (110) and in the cut fiber preparation (13). The slow K^+ current has a reversal potential of about 10 mV more negative than that of the delayed current (3). This difference has been explained by postulating that the slow K^+ channel is more selective toward K^+ than the delayed rectifier.

In addition to the different kinetics and reversal potential, pharmacological evidence suggests that the slow K^+ current arises from a distinct channel since TEA and Zn^{2+} block the delayed rectifier but have a small effect on the slow K^+ system (110, 113). Moreover, diethylpyrocarbonate (1 mM), a histidine reagent, selectively blocks the slow K^+ current (85).

In slow-twitch fibers of rat soleus a prominent slow K^+ current is present. In fast-twitch fibers (iliacus muscle) a much smaller slow component can be inferred from tail currents. However, this component has been explained by an accumulation of K^+ within the TS (44, 45).

Inward Rectifier

The resting K^+ conductance of frog skeletal muscle shows inward or "anomalous" rectification. This conductance decreases when the net flow of K^+ is outward and increases when it is inward (3–5, 72, 80). The conductance change is rapid and depends on the membrane potential and $[K^+]_o$ rather than on the difference between the membrane potential and the K^+ equilibrium potential (119).

Under voltage clamp conditions, during hyperpolarizing pulses, the inward K^+ current declines with an exponential time course (3). For instance, a negative pulse of 75 mV produces a mean initial current of 6.9 μA cm^{-2}, which declines to a steady level of 1.6 μA cm^{-2} with a time constant (τ_d) of 0.25 sec (3°C). The I V relationship measured at a steady level of current shows a region of negative slope conductance for potentials less negative than −150 mV (3, 8).

The decline of the current is primarily due to depletion of K^+ in the TS for small hyperpolarizing pulses and to a fall in K^+ conductance for large pulses (4, 8, 118). The location of the inward rectifier in the TS is consistent with the depletion hypothesis (128). Furthermore, Standen & Stanfield (118) found that the fall in K^+ conductance for large pulses is due to a potential-dependent blockade of the inward rectifier by the Na^+ of the external solution. This blockade increases with increasing hyperpolarizations. For moderate hyperpolarizations the decay of K^+ current is mainly due to K^+ depletion in the TS. In Na^+ and Na^+-free saline, τ_d is identical. For larger and increasing hyperpolarizations, as expected for a voltage-dependent Na^+ blockade, τ_d becomes progressively briefer in Na^+ saline than in Na^+-free saline. For instance, with a hyperpolarization to -230 mV, τ_d was 14.8 ± 1.6 msec in Na^+ saline and 58.1 ± 7.2 in Na^+-free saline (118). The inward rectifier is also blocked in a potential-dependent manner by Rb^+, Cs^+, Li^+, Ba^{2+} and Sr^{2+} (51, 116, 118). It has been suggested that the blocking cation does not interact with gates but that under the influence of the electric field it is driven partway into the channel, where it may bind to a site and block the channel (116).

External application of TEA blocks inward rectification (dissociation constant: 2×10^{-2} M), while Zn^{2+} does not affect it. This suggests that the inward and delayed rectifiers are two separate channels (110). In addition, a 30 min exposure to 10 mM formaldehyde greatly reduces inward rectification sparing the delayed rectifier (14, 76).

Thallium permeates through the inward rectifier (114). In the absence of K^+, when Tl^+ carries the current, the inward rectifier inactivates rapidly with a time constant of about 27 msec for a 100 mV hyperpolarizing pulse. The mechanism of this action is unknown, but clearly the gating process depends on the permeating ion species as was recently shown in nerve cells for the inactivation of Ca^{2+} channels (124). Models for inward rectification have been described (1, 37, 117).

CHLORIDE CHANNELS

Chloride ions contribute the major share to the resting membrane conductance in frog and mammalian muscle fibers (4, 97). In frog skeletal muscle Cl^- channels are mainly located on the surface membrane (46, 72). On the other hand, at least 60% of chloride conductance (G_{Cl}) is located in the TS in mammalian muscle fibers (diaphragm and sternomastoid muscle) (42, 97). The resting G_{Cl} at physiological pH is about 0.2 mS cm^{-2} in frog (4, 71) and 2 mS cm^{-2} in rat diaphragm (97).

G_{Cl} is markedly dependent on external pH. It increases at alkaline pH and it decreases at acid pH. G_{Cl} and pH are related by a sigmoid curve. The apparent pKs for the groups controlling G_{Cl} are about 7.0 in frog (74, 75) and 5.5 in mammals (97). In frog at pH 7.4, I-V curves show that the steady-state G_{Cl} decreases as the fiber is hyperpolarized. The voltage dependence of G_{Cl} is consistent with the predictions of the constant field theory (4, 75). In alkaline and acid pH the changes in the I-V relationship can be predicted only to a certain extent by the constant field relation modified to include a surface potential term (49). Assuming that the surface potential is −80 mV at pH 5.0 and +90 mV at pH 9.8, the modified constant field equation describes the tendency of the I-V relationship but does not provide a good fit for the experimental curves (75).

G_{Cl} also is time dependent. During a negative pulse of 85 mV from the holding potential, the current carried by Cl⁻ decays with a τ_d of 90 msec at pH 7.4 At alkaline pH (9.8) the decay is slower. In contrast, at pH 5.0 Cl⁻ current increases approximately exponentially throughout the pulse (127). In the three pHs tested, the instantaneous I-V relationship is linear. However, the steady-state I-V relationship suggests that Cl⁻ current reaches a limiting value for large hyperpolarizations. According to Warner (127) several features of the Cl⁻ conductance are consistent with the supposition that anions combine with a carrier molecule to an extent determined by the external concentration of H^+.

In *Xenopus* and rat muscle (97, 126), G_{Cl} has properties similar to that in frog muscle. G_{Cl} is time-dependent, the instantaneous I-V relationship is linear, G_{Cl} is modified by external pH, and the steady-state I-V relationship shows that G_{Cl} is reduced by large hyperpolarizing pulses. In addition, the steady-state I-V relationship shows that Cl⁻ current saturates at potentials about 20–40 mV more negative than the holding potential and decreases at more negative potentials, displaying a region of negative slope conductance, as found for the inward rectifier (97, 126). In frog and rat muscle the temperature dependence of G_{Cl} is similar with a Q_{10} of about 1.3 and 1.6, respectively (4, 97).

In *Xenopus* it has been demonstrated that the voltage dependence of G_{Cl} is similar in normally polarized and depolarized fibers, suggesting that G_{Cl} is a function of the difference between the resting potential and the membrane potential during a testing pulse rather than a function of the absolute membrane potential and of the $[Cl]_o/[Cl]_i$ (126).

The findings in *Xenopus* and rat have been explained by proposing the existence of aqueous channels for Cl⁻ (97, 126). To explain the deviation of the I-V relationship from the constant field equation at alkaline pH, a blocking moiety has been postulated in these channels. As the membrane potential becomes more negative, this particle would move to a site that must be occupied by Cl⁻ to permeate; thus the particle competes with Cl⁻ for occupancy. This type of model has been previously proposed for the inward rectifier (117).

More insight on the Cl⁻ channel has been obtained by studying the inhibition of G_{Cl} by aromatic carboxylic acids (29, 98). These compounds do not appear to inhibit G_{Cl} by altering membrane surface potential, and their inhibitory action is not voltage-dependent. Conductance to all anions is not uniformly altered by these compounds as would be expected from a steric occlusion of a common anionic channel. The permeability sequence at the normal Cl⁻ channel is $Cl^- > Br^- > NO_3^- > I^-$. In the presence of 5×10^{-9} M anthracene-9-COOH this sequence is reversed to $I^- > NO_3^- > Br^- > Cl^-$. Apparently, aromatic carboxylic acids inhibit G_{Cl} by binding to a specific intramembrane site and altering the selectivity sequence of the channel (98).

CA²⁺ CHANNELS

Twitch muscle fibers of frog generate Ca^{2+}-dependent action potentials with a slow time course when K⁺ outward currents are blocked with TEA and the Cl⁻ shunt is eliminated by replacing an impermeant anion for Cl⁻ (21). This ionic condition is necessary to eliminate the K⁺ and Cl⁻ currents which would tend to repolarize the fiber and eliminate the Ca^{2+} action potential. A similar response has been observed in rat EDL muscle under comparable ionic conditions (E. Stefani, D. J. Chiarandini, unpublished results).

Voltage-clamp studies (three microelectrode clamp) have demonstrated a slow inward current that is mainly carried by Ca^{2+} (22, 104, 115). The current is abolished by Co^{2+}, Cd^{2+}, D-600, or in the absence of external Ca^{2+}. It is not affected by the removal of external Na⁺ or addition of TTX. The I-V relationship of this current (I_{Ca}) shows that it becomes evident at about −40 mV, reaches a maximum amplitude at 0 mV, and reverses at about +40 mV. This reversal potential is more negative than that expected for a Ca^{2+} electrode. This difference is explained by the existence of a remaining K⁺ outward current not blocked by TEA (104). When K⁺ current is blocked to a greater degree with overnight incubation of the muscle in a K⁺-free saline with 60 mM Cs⁺ and 60 mM TEA, the reversal potential increases to about +100 mV (40). K⁺ currents also can be blocked by the addition of 5 mM, 3,4-diaminopyridine (121).

I_{Ca} can be fitted to the M^3 kinetics. The half-activation potential is −39 mV (20°C); τ_m at 0 mV is 0.11 sec, and τ_h is 1.1 sec (121). The mean maximum inward I_{Ca} is 80 μA cm⁻² and the maximum G_{Ca} is 2.0–5.0 mS cm⁻² (104, 115, 121). With maintained depolarization I_{Ca} inactivates completely. The half-inactivation potential is −42 mV. The decay of I_{Ca} is not due to a time-dependent superimposed outward K⁺ current since the amplitudes of I_{Ca} and of the tail currents elicited with pulses of different duration with a holding potential of −90 mV (i.e. about K^+ equilibrium potential) decay with a similar time constant (120).

A current of similar nature has been recorded recently using the cut fiber preparation (13). To block contraction the muscle fibers were internally equilibrated with isotonic EGTA (K^+ or TEA salt). With TEA outside and isotonic K_2EGTA inside, the inward I_{Ca} is followed by a prominent slow outward K^+ current that corresponds to the slow K^+ channel and that increases when external Ca^{2+} is replaced by Mg^{2+}. This current is not evident in intact fibers when Ca^{2+} is replaced by Mg^{2+} or Co^{2+} (104, 115). In this case the dominant current is the remaining outward current of the delayed K^+ channel not fully blocked by TEA. However, in recent experiments an outward K^+ current similar in nature to the one reported by Almers & Palade (13) has been recorded in intact fibers when large pulses were applied driving the membrane potentials to +20 to +40 mV (E. Stefani, G. Cota, unpublished observations).

Ca^{2+} channels are located mainly in the TS. In detubulated muscle fibers a linear correlation was found between the degree of electric continuity of the TS with the surface membrane and I_{Ca} (93). A similar localization for Ca^{2+} channels has been proposed on the basis of an analysis of the amplitude and rate of decline of the current (11).

In view of the tubular localization of G_{Ca} it was important to know whether the decline of I_{Ca} is due to an inactivation of the channels or to Ca^{2+} depletion in the TS. In the intact fiber preparation bathed in hypertonic medium to reduce contractility, the decay of I_{Ca} appears to be due mainly to inactivation. However, calculations of $[Ca^{2+}]$ in the TS during I_{Ca} suggest that depletion may occur (93). The following observations support the hypothesis that inactivation is the main mechanism. In two-pulse experiments I_{Ca} is reduced during the second pulse when the conditioning prepulse is unable to elicit detectable I_{Ca} (40, 104). The rate of decay of control and inactivated Ca^{2+} currents is identical, which indicates that the rate of decay is independent of the size of I_{Ca}. Finally, the Q_{10} of I_{Ca} decay between 10 and 20°C is 2.9, which is compatible with the value expected for a gating mechanism; a value close to 1.3 would be expected for a diffusional process (40).

In the cut fiber preparation there is conclusive evidence that the decay of I_{Ca} is mainly related to Ca^{2+} depletion in the TS. The rate of decline is directly proportional to I_{Ca} amplitude. Furthermore, in two pulse experiments I_{Ca} is reduced during the second pulse only when the prepulse evokes I_{Ca} (11). It is not clear why the decay mechanisms in intact and cut fiber preparations are different.

The permeability sequence of the Ca^{2+} channel is $Ba^{2+} \geq Sr^{2+} > Ca^{2+}$

> Mn^{2+} > Mg^{2+}. Ni^{2+} and Co^{2+} are not permeant. The Ca^{2+} channel is blocked by nifedipine, D-600, local anesthetics and barbiturates (13, 96). The functional role of Ca^{2+} channels is not known. During a normal action potential an insignificant amount of Ca^{2+} would enter into the muscle cell through these channels. However, during sustained depolarizations a sizable amount of Ca^{2+} or Ba^{2+} enough to trigger tension may enter into the cell through these channels (100, 101, 121).

IONIC CHANNELS IN SLOW (TONIC) MUSCLE FIBERS

In addition to focally innervated twitch fibers, skeletal muscle possess another type of muscle cells, the slow or tonic fibers, which have distinct properties. They display multiple nerve endings, generate a slowly developing tension under repetitive nerve stimulatior, and give rise to a maintained or tonic tension when continuously depolarized with K^+ or cholinergic drugs (35, 60, 83, 95).

Tonic fibers occur either intermingled with twitch fibers in amphibians and mammals or as a pure slow muscle, the anterior latissimus dorsi, in birds (59). This type of fiber is difficult to study electrophysiologically because the fibers are scarce or have a smaller diameter.

Ionic Channels at Rest

In tonic fibers there is no detectable resting G_{Cl}. The membrane resistances measured in Cl^- and Cl^--free saline are identical. The lack of G_{Cl} largely explains the recorded high values of effective resistance. The mean resting conductance (mS cm^{-2}), mostly due to K^+, is 0.008 in frog, 0.17 in chicken, and 0.25 in rat (28, 35, 36, 55, 73, 122). Although the inward rectifier has not been thoroughly studied, I-V curves indicate that this channel is present in chicken and rat tonic fibers (36, 73). The lack of inward rectifier in frog tonic fibers (55, 122) could explain their small resting conductance. In rat it appears that the resting conductance has a small G_{Na} since Na-free solution hyperpolarizes the fibers by 10–15 mV (28).

Na^+ Channels

In frog tonic fibers activable Na^+ channels appear to be absent (55). However, after denervation these cells are capable of generating propagated action potentials, which are Na^+-dependent and TTX-sensitive. The G_{Na} underlying these potentials is small (88). Apparently Na^+ channels are incorporated into the membrane in patches (84, 105).

In contrast, avian slow fibers can generate action potentials that are Na$^+$-dependent and blocked by TTX. These action potentials are smaller and slower than in chicken twitch fibers (41). Mammalian tonic fibers appear to be an intermediate case between amphibian and avian slow fibers. Depolarizing pulses evoke a graded depolarizing response that involves a G_{Na} since it disappears in Na$^+$-free saline, is blocked by TTX, and is not affected by D-600 (28, 36).

Delayed Rectifier

In frog, chicken, and rat tonic fibers the delayed rectifier is present (28, 30, 36, 41). In frog, voltage-clamp experiments (55) have shown K$^+$ delayed currents. The reversal potential of the corresponding tail currents is modified by changes in external [K] (55).

In frog tonic fibers the maximum G_K is 0.5 mS cm^{-2}, about ten times smaller than in twitch fibers. The steady-state G_K-voltage relation is less steep in tonic fibers than in twitch fibers, with an e-fold change for 15 mV. The rate constants of activation are at most 2–4 times smaller than those of twitch fibers and have smaller voltage dependence. Whether the delayed rectifier inactivates during a prolonged pulse is uncertain (55). TEA blocks the delayed rectifier in mammalian and frog tonic fibers (28, 55).

Slow K$^+$ Channel

Until now this channel has been demonstrated only in frog tonic fibers, in which it is often present (55).

Ca^{2+} Channel

This channel appears to be present in tonic muscle fibers of toad. The electrical response attributed to these channels is unmodified by external Na$^+$ removal or addition of TTX, and is abolished by Co^{2+} and by the removal of external Ca^{2+} (123).

ACKNOWLEDGMENTS

The authors wish to thank the many investigators who provided articles and manuscripts for this review. This work was supported by the U.S.-Latin American Cooperative Program (PCAIEUA 790057 and NSF INT-7920212 and, in part, by grants PCCBNAL 790022 (CONACyT, Mexico) to E.S. and USPHS EY-01297 to D.J.C.

Literature Cited

1. Adrian, R. H. 1969. *Prog. Biophys. Biophys. Chem.* 19:341–69
2. Adrian, R. H., Chandler, W. K., Hodgkin, A. L. 1970. *J. Physiol.* 208:607–44
3. Adrian, R. H., Chandler, W. K., Hodgkin, A. L. 1970. *J. Physiol.* 208:645–68
4. Adrian, R. H., Freygang, W. H. 1962. *J. Physiol.* 163:61–103
5. Adrian, R. H., Freygang, W. H. 1962. *J. Physiol.* 163:104–14
6. Adrian, R. H., Marshall, M. W. 1977. *J. Physiol.* 268:233–50
7. Adrian, R. H., Peachey, L. D. 1973. *J. Physiol.* 235:103–31
8. Almers, W. 1972. *J. Physiol.* 225:33–56
9. Almers, W. 1972. *J. Physiol.* 225:57–83
10. Almers, W. 1976. *J. Physiol.* 262:613–37
11. Almers, W., Fink, R., Palade, P. T. 1981. *J. Physiol.* In press
12. Almers, W., Levinson, S. R. 1975. *J. Physiol.* 247:483–509
13. Almers, W., Palade, P. T. 1981. *J. Physiol.* In press
14. Argibay, J. A., Hutter, O. F. 1973. *J. Physiol.* 232:41–43P
15. Armstrong, C. M. 1975. *Q. Rev. Biophys.* 7:179–210
16. Baker, P. F., Rubinson, K. A. 1975. *Nature* 257:412–14
17. Baker, P. F., Rubinson, K. A. 1977. *J. Physiol.* 206:3–4P
18. Barchi, R. L., Weigele, J. B. 1980. *J. Physiol.* 295:383–96
19. Barrett, J. N., Barrett, E. F., Dribin, L. B. 1981. *Devel. Biol.* In press
20. Bastian, J., Nakajima, S. 1974. *J. Gen. Physiol.* 63:257–78
21. Beaty, G. N., Stefani, E. 1976. *Proc. R. Soc. London Ser. B.* 194:141–50
22. Beaty, G. N., Stefani, E. 1976. *J. Physiol.* 260:27–28P
23. Begenisich, T., Cahalan, M. 1979. In *Membrane Transport Processes*, ed. C. F. Stevens, R. W. Tsien, 3:113–16. NY: Raven Press. 156 pp.
24. Begenisich, T., Cahalan, M. 1980. *J. Physiol.* 307:217–42
25. Begenisich, T., Cahalan, M. 1980. *J. Physiol.* 307:243–57
26. Bezanilla, F., Caputo, C., Gonzalez-Serratos, H., Venosa, H. 1972. *J. Physiol.* 223:507–23
27. Bezanilla, F., Vergara, J., Taylor, R. 1980. In *Methods of Experimental Physics*, ed. G. M. Ehrenstein, A. Lecar. In press
28. Bondi, A. Y., Chiarandini, D. J. 1979. *J. Physiol.* 295:273–81
29. Bryant, S. H., Morales-Aguilera, A. 1971. *J. Physiol.* 219:367–83
30. Burke, W., Ginsborg, B. L. 1956. *J. Physiol.* 132:586–98
31. Caille, J., Ildefonse, M., Rougier, O. 1978. *Pflügers Arch.* 374:167–77
32. Campbell, D. T. 1976. *J. Gen. Physiol.* 67:295–307
33. Campbell, D. T., Hille, B. 1976. *J. Gen. Physiol.* 67:309–23
34. Caputo, C., Dipolo, R. 1973. *J. Physiol.* 229:547–57
35. Chiarandini, D. J., Davidowitz, J. 1979. In *Current Topics in Eye Research*, ed. J. A. Zadunaisky, H. Davson, 1:91–142. NY: Academic. 243 pp.
36. Chiarandini, D. J., Stefani, E. 1979. *J. Physiol.* 290:453–65
37. Ciani, S., Krasne, S., Miyazaki, S., Hagiwara, S. 1978. *J. Membr. Biol.* 44:103–34
38. Costantin, L. L. 1970. *J. Gen. Physiol.* 55:703–15
39. Costantin, L. L., Taylor, S. R. 1971. *J. Physiol.* 218:13P
40. Cota, G., Nicola Siri, L., Stefani, E. 1981. *Int. Congr. Biophys., 7th, Mexico City.* In press
41. Cullen, M. J., Harris, J. B., Marshall, M. W., Ward, M. R. 1975. *J. Physiol.* 245:371–85
42. Dulhunty, A. F. 1979. *J. Membr. Biol.* 45:293–310
43. Duval, A., Léoty, C. 1978. *J. Physiol.* 278:403–23
44. Duval, A., Léoty, C. 1980. *J. Physiol.* 307:23–41
45. Duval, A., Léoty, C. 1980. *J. Physiol.* 307:43–57
46. Eisenberg, R. S., Gage, P. W. 1969. *J. Gen. Physiol.* 53:279–97
47. Eisenman, G. 1962. *Biophys. J.* 2:259–323
48. Fink, R., Lüttgau, H. C. 1976. *J. Physiol.* 263:215–38
49. Frankenhaeuser, B. 1960. *J. Physiol.* 151:491–501
50. Gage, P. W., Eisenberg, R. S. 1969. *J. Gen. Physiol.* 53:298–310
51. Gay, L. A., Stanfield, P. R. 1977. *Nature* 267:169–70
52. Gay, L. A., Stanfield, P. R. 1978. *Pflügers Arch.* 378:177–79
53. Gillespie, J. I. 1977. *J. Physiol.* 273:64P
54. Gillespie, J. I., Hutter, O. F. 1975. *J. Physiol.* 252:70P
55. Gilly, W. F., Hui, C. S. 1980. *J. Physiol.* 301:157–73
56. Gonzalez-Serratos, H. 1971. *J. Physiol.* 212:777–99
57. Hagiwara, S., Eaton, D. C., Stuart, A. E., Rosenthal, N. P. 1972. *J. Membr. Biol.* 9:373–84

58. Hansen Bay, C. M., Strichartz, C. R. 1980. *J. Physiol.* 300:89–103
59. Hess, A. 1970. *Physiol. Rev.* 50:40–62
60. Hess, A., Pilar, G. 1963. *J. Physiol.* 169:780–98
61. Hille, B. 1971. *J. Gen. Physiol.* 58:599–619
62. Hille, B. 1972. *J. Gen. Physiol.* 59:637–58
63. Hille, B. 1973. *J. Gen. Physiol.* 61:669–86
64. Hille, B. 1975. In *Membranes—A Series of Advances,* ed. G. Eisenman, 3:255–323. NY: Marcel Dekker. 538 pp.
65. Hille, B. 1975. *J. Gen. Physiol.* 66:535–60
66. Hille, B. 1976. *Ann. Rev. Physiol.* 38:139–52
67. Hille, B. 1978. *Biophys. J.* 22:283–94
68. Hille, B. 1979. In *Membrane Transport Processes,* ed. C. F. Stevens, R. W. Tsien, 3:5–16. NY: Raven Press. 156 pp.
69. Hille, B., Campbell, D. T. 1976. *J. Gen. Physiol.* 67:265–93
70. Hille, B., Schwarz, W. 1978. *J. Gen. Physiol.* 72:409–42
71. Hodgkin, A. L., Horowicz, P. 1959. *J. Physiol.* 148:127–60
72. Hodgkin, A. L., Horowicz, P. 1960. *J. Physiol.* 153:370–85
73. Huerta, M., Stefani, E. 1981. *J. Physiol.* In press
74. Hutter, O. F., Warner, A. E. 1967. *J. Physiol.* 189:403–25
75. Hutter, O. F., Warner, A. E. 1972. *J. Physiol.* 227:275–90
76. Hutter, O. F., Williams, T. L. 1979. *J. Physiol.* 286:591–606
77. Ildefonse, M., Rougier, O. 1972. *J. Physiol.* 222:373–95
78. Ildefonse, M., Roy, G. 1972. *J. Physiol.* 227:419–31
79. Jaimovich, E., Venosa, R. A., Schrager, P., Horowicz, P. 1976. *J. Gen. Physiol.* 67:399–416
80. Katz, B. 1949. *Archs. Sci. Physiol.* 3:285–300
81. Kirsch, G. E., Nichols, R. A., Nakajima, S. 1977. *J. Gen. Physiol.* 70:1–21
82. Kovács, L., Schneider, M. F. 1978. *J. Physiol.* 277:483–506
83. Kuffler, S. W., Vaughan Williams, E. M. 1953. *J. Physiol.* 121:289–317
84. Lehouelleur, J., Schmidt, H. 1980. *Proc. R. Soc. London Ser. B* 209:403–13
85. Lynch, C. 1978. *Biophys. J.* 21:55a
86. Mandrino, M. 1977. *J. Physiol.* 269:605–25
87. Meech, R. W., Standen, N. B. 1975. *J. Physiol.* 249:211–39
88. Miledi, R., Stefani, E., Steinbach, A. B. 1971. *J. Physiol.* 217:737–54
89. Nakajima, S., Gilai, A. 1980. *J. Gen. Physiol.* 76:729–50
90. Nakajima, S., Gilai, A. 1980. *J. Gen. Physiol.* 76:571–62
91. Neher, E., Sakmann, B., Steinbach, J. H. 1978. *Pflügers Arch.* 375:219–28
92. Neher, E., Stevens, C. F. 1977. *Ann. Rev. Biophys. Bioeng.* 6:345–81
93. Nicola Siri, L., Sánchez, J. A., Stefani, E. 1980. *J. Physiol.* 305:87–96
94. Nonner, W., Spalding, B. C., Hille, B. 1980. *Nature* 284:360–63
95. Page, S. G. 1969. *J. Physiol.* 205:131–45
96. Palade, P. T., Almers, W. 1981. *Biophys. J.* 33:151a
97. Palade, P. T., Barchi, R. L. 1977. *J. Gen. Physiol.* 69:325–42
98. Palade, P. T., Barchi, R. L. 1977. *J. Gen. Physiol.* 69:879–96
99. Pappone, P. A. 1980. *J. Physiol.* 306:377–410
100. Potreau, D., Raymond, G. 1980. *J. Physiol.* 303:91–109
101. Potreau, D., Raymond, G. 1980. *J. Physiol.* 307:9–22
102. Redfern, P., Thesleff, S. 1971. *Acta Physiol. Scand.* 82:70–78
103. Ritchie, J. M., Rogart, R. B. 1977. *J. Physiol.* 269:341–54
104. Sánchez, J. A., Stefani, E. 1978. *J. Physiol.* 283:197–209
105. Schalow, G., Schmidt, H. 1979. *Proc. R. Soc. London Ser. B* 203:445–57
106. Shrager, P., Profera, C. 1973. *Biochim. Biophys. Acta* 318:141–46
107. Sigworth, F. J., Neher, E. 1980. *Nature* 287:447–49
108. Sigworth, F. J., Spalding, B. C. 1980. *Nature* 283:293–95
109. Spalding, B. C. 1980. *J. Physiol.* 305:485–500
110. Stanfield, P. R. 1970. *J. Physiol.* 209:231–56
111. Stanfield, P. R. 1970. *J. Physiol.* 209:209–29
112. Stanfield, P. R. 1973. *J. Physiol.* 235:639–54
113. Stanfield, P. R. 1975. *J. Physiol.* 251:711–35
114. Stanfield, P. R., Ashcroft, F. M., Plant, T. D. 1981. *Nature* 289:509–11
115. Stanfield, P. R. 1977. *Pflügers Arch.* 368.267 70
116. Standen, N. B., Stanfield, P. R. 1978. *J. Physiol.* 280:169–91
117. Standen, N. B., Stanfield, P. R. 1978. *Pflügers Arch.* 378:173–76
118. Standen, N. B., Stanfield, P. R. 1979. *J. Physiol.* 294:497–520

119. Standen, N. B., Stanfield, P. R. 1980. *J. Physiol.* 304:415–35
120. Stefani, E. 1981. *Int. Congr. Biophys., 7th, Mexico City* In press
121. Stefani, E., Sánchez, J. A., Nicola Siri, L. 1981. In *Advances in the Physiological Sciences, Vol. 5. Molecular and Cellular Aspects of Muscle Function,* ed. E. Varga, A. Köver, T. Kovácz, L. Kovácz. Budapest: Pergamon. In press
122. Stefani, E., Steinbach, A. B. 1969. *J. Physiol.* 203:383–401
123. Stefani, E., Uchitel, O. D. 1976. *J. Physiol.* 255:435–48
124. Tillotson, D. 1979. *Proc. Natl. Acad. Sci. USA* 76:1497–1500
125. Ulbricht, W. 1977. *Ann. Rev. Biophys. Bioeng.* 6:7–31
126. Vaughn, P. C., McLarnon, J. G., Loo, D. D. F. 1980. *Can. J. Physiol.* 58:999–1010
127. Warner, A. E. 1972. *J. Physiol.* 227:291–312
128. Williams, T. L. 1976. *J. Physiol.* 256:125P

CARDIOVASCULAR PHYSIOLOGY

Introduction, Nick Sperelakis, *Guest Editor,* and Robert M. Berne, *Editor*

The theme of this section is the function of the plasma membrane and the sarcoplasmic reticulum (SR) membrane of the myocardial cell. Both membranes are involved in regulation of the force of contraction of the heart by exercising control over the myoplasmic calcium concentration ($[Ca]_i$); both exhibit permeability changes and voltage changes during excitation. Each chapter, written by an outstanding expert, focuses on a different function or malfunction of these membranes.

The cell membrane, about 70 Å thick, is the permeability barrier between the cell interior and the interstitial fluid that regulates the inflow and outflow of inorganic ions and metabolites. It consists of a thin phospholipid bilayer matrix in which many types of protein molecules float. Some of these proteins span the entire membrane—e.g. the (Na,K)-ATPase, Ca-ATPase, and the various types of voltage-dependent ionic channels. Others are inserted only into the inner or outer leaflet of the membrane. Enzymes located in the cell membrane are those required to carry out the functions of the cell membrane—e.g. ion-transporting enzymes, or those that are coupled to pharmacological receptors on the outer surface of the membrane [for example, adenylate cyclase (localized at the inner membrane surface) is controlled by such receptors, including the beta-adrenergic receptor and histamine (H_2) receptor]. Some small protein molecules may shuttle back and forth across the lipid bilayer to act as ion exchange carriers. The Ca-Na exchange reaction exchanges 1 internal Ca^{2+} ion for 2 or 3 external Na^+ ions, and thus acts as a Ca pump, the energy from the Na^+ electrochemical gradient (provided by the Na,K-ATPase) being used indirectly to transport Ca^{2+} uphill.

The Na-K pump, Ca pump, and Ca-Na exchange systems maintain the steady-state ion distributions across the cell membrane. These ion concentration gradients, coupled with different relative ionic permeabilities, give rise to a large net diffusion potential (E_{diff}). This store of potential energy is drawn upon for the propagation of action potentials during excitation. An electrogenic Na pump potential also makes a contribution to the transmembrane potential (E_m)—i.e. the resting E_m is a few millivolts greater (more negative) than E_{diff}. Changes in E_m have important physiological repercussions.

The action potentials, with their underlying ionic conductance changes and ionic fluxes, activate the contractile mechanism and control the force of contraction of the heart. In excitation-contraction coupling, Ca^{2+} ion is the key messenger. Ca^{2+} influx across the sarcolemma, through voltage-dependent slow Ca-Na channels, occurs during excitation and is part of the inward slow current (I_{si}) that flows during most of the action potential plateau. This Ca^{2+} influx directly helps to raise $[Ca]_i$ to the level required to activate the myofilaments, and in addition, acts to bring about further release of Ca^{2+} from the SR.

The amount of Ca releasable from the SR is dependent on the total Ca sequestered in the SR, which is regulated by a number of factors, including cyclic AMP level and previous activation (treppe). For example, activation of the beta-adrenergic receptor and adenylate cyclase by the neurotransmitter norepinephrine elevates the cyclic AMP level, which activates cyclic AMP–dependent protein kinases that phosphorylate a number of proteins. One such protein stimulates the Ca-ATPase and Ca^{2+} uptake into the SR. Greater loading of the SR is one mechanism whereby beta-adrenergic agonists exert a positive inotropic action.

A second important mechanism for their positive inotropic action is the potentiation of I_{si} (hence Ca^{2+} influx) during the action potential, due to an increase in the density of functional slow channels in the sarcolemma. Phosphorylation of the slow channel protein by means of a cyclic AMP–dependent protein kinase may be responsible. That is, the phosphorylated slow channel may be available for voltage activation, whereas the dephosphorylated form may be electrically silent. Functioning of the slow channels is related to cyclic AMP and is strongly dependent on metabolic energy, in contrast to the other types of channels. The slow channels also possess other unique properties, including a selective blockade by acidosis.

These special properties of the myocardial slow channels allow the extrinsic and intrinsic control of the force of contraction of the heart by regulating the amount of Ca^{2+} influx. Neurotransmitters increase (norepinephrine) or decrease (acetylcholine) I_{si}. Circulating hormones, such as angiotensin II, also affect I_{si}. Hypoxia and regional ischemia decrease I_{si}, decreasing contractile force and conserving ATP.

Various ionic currents and conductance changes underlie the normal automaticity of the heart, as well as of ectopic pacemaker foci and reentry, leading to arrhythmias. A number of therapeutic agents act on the cell membrane of the myocardial cell. These include drugs that inhibit the Na,K-ATPase, such as cardiac glycosides, and drugs that act on the ionic channels, such as local anesthetics and calcium antagonists (slow-channel blockers). In addition, a number of toxins from a variety of animals and plants, and of widely different chemical structures, affect the ionic channels in cardiac membranes.

These topics are considered in detail in the chapters that comprise the section on myocardial membranes.

MYOCARDIAL MEMBRANES: REGULATION AND FUNCTION OF THE SODIUM PUMP

Tai Akera and Theodore M. Brody

Department of Pharmacology and Toxicology, Michigan State University, East Lansing, Michigan 48824

INTRODUCTION

In 1957, Skou (73) demonstrated in crab nerves the presence of an ATPase activated by the simultaneous presence of Na^+ and K^+. Skou wrote: "Characteristics of the system suggest that the ATPase studied here may be involved in the active extrusion of sodium from the nerve fiber." Subsequently Post et al (64) suggested that a single entity performs both the transport and the hydrolase functions. Recently, using liposomes in which the purified enzyme has been incorporated, this ATPase has been shown to mediate the active transport of Na^+ and K^+ across membranes with the concomitant hydrolysis of ATP (see 12, 70, 77). It is now well-established that Na,K-ATPase is the sodium pump (70).

Na,K-ATPase and the sodium pump are the subjects of several recent reviews (33, 70, 77, 80). The present review focuses on topics not dealt with in these reviews and is limited to recent papers. The following abbreviations are used in the text: Na,K-ATPase (Na^+ and K^+ activated, Mg^{2+}-dependent adenosine triphosphatase); $[\ ]_e$ and $[\ ]_i$ (extracellular and intracellular concentration, respectively, of the substance shown in the bracket).

CONTROL IN INTACT CELLS

The activity of the sodium pump may be regulated by various factors. Overall sodium pumping is determined by the number of active pumping sites and their turnover rates. Additionally, the number of Na^+ and K^+ ions translocated during each cycle of pumping, and hence per each ATP molecule hydrolyzed, may be altered.

Enzyme Concentrations

In the guinea-pig heart, the concentration of the sodium pump is higher in ventricular than in atrial muscle (21, 84). In bovine heart, Na,K-ATPase activity is lower in Purkinje fibers than in papillary muscle homogenate (60). Significant differences in enzyme activity in the heart, however, were not observed in the dog (46) or the cat (57). Myocardial $^{86}Rb^+$ uptake and ouabain binding are higher in newborn guinea pigs and dogs than in the corresponding adult animals (55).

One possible explanation for regional differences, or the lack of differences, in enzyme concentrations is the regional differences in sodium influx rates. The "demand" for increased Na^+ or K^+ transport governs the number of sodium pump units in other tissues (30, 54, 72). Thus if the sodium influx rate is higher, the sodium pump concentration in a tissue may be increased. Since the balance between the sodium influx rate and the capacity of the sodium pump is physiologically important, whether the above difference in myocardial Na,K-ATPase is the consequence, or is independent, of an enhanced sodium influx should have a significant influence on myocardial performance when Na^+ influx is enhanced or the sodium pump is suppressed.

Other factors may also determine the tissue Na,K-ATPase concentrations. For example, differences in Na,K-ATPase concentrations in liver and skeletal muscle of mice (51–53) or in cultured Chinese hamster cells (17) are apparently determined genetically.

Enzyme Turnover Rates

Owing to the "sidedness" of Na,K-ATPase, effects of various ligands on the sodium pump in intact cells are substantially different from those observed with isolated enzyme preparations. Kinetic studies under conditions that retain "sidedness" are mostly performed with resealed erythrocyte membranes or purified enzyme preparations incorporated into liposomes. Although data obtained with cardiac Na,K-ATPase are not available, results with other tissues should be generally applicable to myocardial Na,K-ATPase, since the effects of various ligands on the enzyme are relatively independent of the source of enzyme except for the affinity for cardiac glycosides and related inhibitors.

The concentration of Na^+ to cause a half-maximal activation of isolated Na,K-ATPase is dependent on the K^+ concentration and is estimated to vary from 0.16 to 8.1 mM when extrapolated to zero K^+ concentration—i.e. in the absence of competition by K^+ (see 70). In intact cells with high $[K^+]_i$, the $[Na^+]_i$ required to cause a half-maximal activation of the sodium pump is apparently higher than the above values (13). Since the activity of intracellular Na^+ is reported to be 5.7 mM in resting cardiac muscle (49), an increase in $[Na^+]_i$ should increase sodium pump activity. External Na^+ slows the turnover of the sodium pump by inhibiting the dephosphorylation reaction (15). This action of Na^+ probably has a minimal regulatory effect, since $[Na^+]_e$ is unlikely to change significantly.

ATP concentration affects the affinity of the enzyme for K^+ (37). For the K^+-induced activation of the sodium pump in erythrocyte membranes, $[ATP]_i$ should be above 20 μM; otherwise, K^+ on both sides inhibits ATP hydrolysis (16). Despite the relatively high affinity of the enzyme for K^+ at external "loading" sites (see 67, 70), it is generally thought that $[K^+]_e$ influences the sodium pump activity of myocardial cells. $[K^+]_e$ apparently affects the rate of active K^+ uptake, but may not alter Na^+ extrusion if the amount of intracellular Na^+ available to the sodium pump is limited (11).

Sodium pump activity may also be affected by the concentration of available ATP, which has been shown to modulate affinities of Na^+ binding sites on the enzyme (13, 68, 74, 75) in addition to serving as the energy source for the transport. Cytoplasmic [ATP], however, is at least five times higher (above 1 mM) than the K_d value (less than 0.2 mM) for low-affinity ATP binding sites (74). Thus, slight changes in [ATP], which may occur in myocardial cells, is unlikely to affect sodium pump activity, unless ATP in a discrete small pool is the source available to the sodium pump. Preferential use of membrane-compartmentalized ATP by Na,K-ATPase in erythrocyte ghosts has been reported (65); however, whether similar pools exist in myocardial cells is presently unknown.

Changes in $[Mg^{2+}]$ also affect the affinity of the enzyme for Na^+ in a complex manner (27); however, $[Mg^{2+}]_i$ is unlikely to change to such an extent as to affect the sodium pump activity in intact myocardial cells. Inhibition of Na,K-ATPase and its partial reaction, K-stimulated p-nitrophenyl phosphatase activity, by inorganic phosphate have been reported, suggesting a product inhibition of the forward enzyme reaction (71). Again, the concentration of inorganic phosphate in cytoplasm is unlikely to increase above the 1 mM needed for a significant inhibition of the enzyme reaction.

A 50% inhibition of isolated Na,K-ATPase may be observed at about 0.5 mM Ca^{2+}. Since the sodium pump apparently functions in the presence of a higher $[Ca^{2+}]_e$, the inhibitory sites are not accessible to external Ca^{2+}. The ability of Ca^{2+} to inhibit Na,K-ATPase isolated from guinea-pig heart is influenced by H^+ and Na^+; however, a minimal inhibition requires at least 10 μM Ca^{2+} (35). Since $[Ca^{2+}]_i$ is substantially lower than this value in myocardial cells, it seems unlikely that $[Ca^{2+}]_i$ regulates sodium pump activity. Bentfeld et al (14), however, suggested that Ca^{2+}-overload results in an inhibition of the myocardial sodium pump. More recently, Yingst & Hoffman (85) reported that 1 μM $[Ca^{2+}]_i$ causes a 40% sodium pump inhibition in resealed erythrocyte ghosts. Thus despite the results obtained in isolated enzyme studies, $[Ca^{2+}]_i$ could have a role in the regulation of sodium pump activity.

Free fatty acids, either added to isolated enzyme in vitro (47) or released by heparin injection in vivo (43) have been reported to inhibit myocardial Na,K-ATPase. Liver and kidney obtained from rats chronically fed on essential fatty acid–deficient diets have higher concentrations of Na,K-ATPase (50). These results may be consistent with an earlier report (44) that the enzyme activity is highly dependent on membrane lipid fluidity. A partial removal of membrane lipids results in reversible inactivation of the enzyme, associated with a slowing of the conformational change in phosphoenzyme from an ADP-sensitive to a K^+-sensitive form (39). Although phosphatidylserine has been implicated by many as being important for maximal enzyme activity, phosphatidylserine or phosphatidylinositol is not specifically required for enzyme activity (22). Thus membrane fluidity, rather than a specific lipid, appears important for the turnover of the sodium pump.

Na^+/K^+-Counter-Transport Ratio

Earlier studies, performed with erythrocytes, squid axons, or isolated Na,K-ATPase incorporated into liposomes, generally indicated that three Na^+ and two K^+ are exchanged for each ATP hydrolyzed when $[Na^+]_i$ and $[K^+]_e$—i.e. concentrations of cations at the side from which they are transported—are relatively high. This concept is consistent with the finding that there are three Na^+ and two K^+ binding sites per phosphorylation site and that the affinities of these sites for Na^+ or K^+ are altered by phosphorylation of the enzyme (81). The reported transport ratio, however, is not always consistent. In erythrocyte ghosts, the presence of a Na^+-K^+-cotransport system (unrelated to Na^+,K^+-ATPase) may alter the observed net Na^+/K^+ exchange ratio (32). Moreover, Jorgensen & Anner (42) have shown that the Na^+/K^+ transport ratio can be altered by selective trypsin digestion of purified enzyme performed before incorporation into phospholipid vesicles.

The ratio of Na^+/K^+ transport in vesicles containing rabbit kidney Na,K-ATPase changes from 0.76 to 1.93 as the intravesicular K^+ concentration decreases (12). In these reconstituted vesicles, only the "inside-out" sodium pump is activated, since ATP is added to the incubation medium after the formation of the vesicles. Therefore, the intravesicular K^+ actually represents $[K^+]_e$ and hence the above results indicate that $[K^+]_e$ regulates the transport ratio when the $[Na^+]_i$ is 50 mM. Since only the net flux of labelled Na^+, K^+, or Rb^+ was assayed in the above study, the change in transport ratio was attributed to variation in the proportion of the sodium pump operating in Na^+/K^+ and Na^+/Na^+ exchange modes. Alternatively, the actual transport ratio might have been altered as the $[K^+]_e$ was reduced.

The transport ratio in the myocardium may also be regulated by $[Na^+]_i$. Is it an absolute requirement that all Na^+ sites must be occupied before transport occurs (31)? In functional cardiac muscle, one can assume that the time-averaged Na^+ efflux and K^+ influx rates are precisely equal to the Na^+ influx and K^+ efflux rates, respectively. In quiescent canine myocardium, the rates of Na^+ influx and K^+ efflux are approximately 0.3 and 0.7 mmol kg^{-1} min^{-1}, respectively (48). This means that the Na^+/K^+ exchange ratio of the sodium pump in quiescent muscle is approximately 0.4, assuming that other mechanisms that cause net Na^+ efflux and K^+ influx do not play a significant role. If a part of the Na^+ efflux is mediated by a Na^+/Ca^{2+} exchange reaction, then the transport ratio is even lower than 0.4, a value significantly below the optimal ratio of 1.5. In the beating myocardium, the Na^+ influx rate increases roughly in proportion to the frequency of membrane depolarization (48). For example, in the myocardium beating at 80 min^{-1}, the Na^+ influx is approximately 4.2 mmol kg^{-1} min^{-1}, whereas K^+ efflux is independent of membrane depolarization and is approximately 0.7 mmol kg^{-1} min^{-1}. Therefore, the net transport ratio should be about 6. The actual value may be lower because the increase in $[Na^+]_i$ may stimulate other mechanisms for Na^+ extrusion, such as the Na^+/Ca^{2+} exchange reaction or a Na^+-K^+-cotransport system. Indirect evidence to support the flexibility of the transport ratio of the sodium pump has been reported (11, 15).

The sodium pump is generally considered to be electrogenic because it exchanges three Na^+ for two K^+, translocating one net positive charge across the membrane with each pumping cycle. The above consideration, however, suggests that the sodium pump may or may not be electrogenic, or may even be electrogenic in the opposite direction. Evidence frequently presented to support the electrogenicity of the sodium pump—i.e. a slight membrane depolarization upon complete sodium pump inhibition—may rather indicate that the transmembrane K^+ gradient decreases rapidly after inhibition of the sodium pump, before a measurable change in average $[K^+]_i$ occurs. It is also interesting to note that the calculated contribution of the electrogenicity of the sodium pump to the membrane potential is relatively small in cardiac muscle.

Owning to the flexibility of the transport ratio, ouabain-sensitive $^{86}Rb^+$ uptake, which is often used as an index of sodium pump activity, may not precisely reflect changes in Na^+ transport (11). It should also be noted that the $^{86}Rb^+$ uptake assayed without "Na^+ loading" in quiescent muscle preparations is an index of Na^+ influx as well as of Na^+ efflux, as these values are equal in steady-state preparations. Without "Na^+ loading", $^{86}Rb^+$ uptake may be altered by changes in Na^+ influx rates (11, 82).

In cardiac muscle beating under normal conditions, an increase in Na^+ influx due to a moderate increase in the frequency of electrical stimulation fails to cause a marked elevation of $[Na^+]_i$, indicating that the sodium pump has a reserve capacity. This is also consistent with the observation that an elevation of Na^+ influx by monensin, a sodium ionophore, or moderate sodium pump inhibition by the cardiac glycosides also fails markedly to elevate $[Na^+]_i$ (see 2). In the presence of a reserve capacity of the sodium pump, changes in Na^+ influx rate, rather than sodium pump capacity, should determine the rate of ouabain-sensitive $^{86}Rb^+$ uptake.

SODIUM PUMP INHIBITION

As discussed above, the sodium pump in cardiac muscle has a reserve capacity when the heart is beating under normal conditions. When the degree of sodium pump inhibition is such that the remaining capacity is insufficient to match the ongoing sodium influx rate, then myocardial Na^+ accumulation, K^+ loss, and resultant changes are anticipated. The significance of a "moderate" sodium pump inhibition, however, is controversial.

Cardiac Glycosides

Cardiac glycosides are remarkably specific inhibitors of Na,K-ATPase. Because of the impressive parallel relationship between the binding of digitalis derivatives to Na,K-ATPase observed in vitro and their positive inotropic action, and because Na,K-ATPase and the sodium pump are moderately inhibited by therapeutic (positive inotropic) concentrations of the glycosides, this enzyme system has been regarded as the receptor for the therapeutic and toxic actions of the glycosides (see 1–3).

An apparent lack of a marked $[Na^+]_i$ elevation resulting from a moderate sodium pump inhibition caused by therapeutic concentrations of the glycoside is explained as follows: A reserve capacity of the sodium pump exists only during the later phase of each cycle of myocardial function. During the early phase, when the Na^+ influx rate is high, $[Na^+]_i$ at the inner surface of sarcolemma is elevated, maximally activating the sodium pump. A glycoside-induced reduction in sodium pump capacity thus delays the extrusion of Na^+ that entered the cell associated with membrane depolarization, and also allows $[Na^+]_i$ at the inner surface of the sarcolemma to reach a higher level. Such an enhancement of the "sodium transient" augments a calcium transient, presumably by triggering a Na^+/Ca^{2+} exchange reaction, and increases the force of contraction (1). This sequence of events, however, has not been experimentally demonstrated, and alternative explanations that relate the glycoside binding to inotropic effects are also possible (see 3). There are also opposing views to the concept that Na,K-ATPase is the receptor for the therapeutic action of the cardiac glycosides (59). Data continue to accumulate supporting the hypothesis that Na,K-ATPase is the receptor for the inotropic action of the glycosides (5, 6, 9, 10, 26, 57, 84), favoring opposing views (23, 41, 58), or suggesting that multiple mechanisms may exist for the inotropic action (34).

One way to test the hypothesis that Na,K-ATPase is the receptor for the therapeutic action of the cardiac glycosides is to compare the relationship between the glycoside binding to Na,K-ATPase and that to the inotropic receptor. However, a comparison of the affinity of isolated Na,K-ATPase for the glycoside with the drug's potency to increase force of contraction requires caution. Although a good correlation has generally been claimed (see 2), a close examination of reported data indicates that there is often as much as a several-fold difference in the concentration of a glycoside needed to inhibit Na,K-ATPase and that needed to increase the force of contraction. Since intracellular Na^+ stimulates, or is required for, the glycoside binding to Na,K-ATPase in intact cells (20, 82), the affinity of the sodium pump for a glycoside may be increased several-fold by increasing the frequency of stimulation in isolated heart muscle preparations (82). Similarly, the development of the positive inotropic action of cardiac glycosides and aglycones requires intracellular Na^+ (5, 6, 82, 83). These aspects cannot be reproduced in isolated enzyme studies.

In isolated enzyme, K^+ lowers its affinity for the cardiac glycosides. This is because K^+ reduces both association and dissociation rate constants for the glycoside-enzyme interaction, but reduces the former value to a greater degree (18). The degree of K^+-induced change in the affinity depends on the chemical structure or the physicochemical properties of the compound (8, 9, 18). The glycoside binding site is either hydrophobic or is viscous and shielded from water (29), and K^+ seems to enhance the effectiveness of the shield to stabilize the glycoside-enzyme complex (8, 18). Although the K^+-effect on the glycoside-enzyme interaction is reflected in the effect of K^+ to modify the inotropic action of these compounds (9), a precise comparison using preparations with and without "sidedness" is again impossible.

A subtle change in the glycoside sensitivity of Na,K-ATPase may be caused by enzyme modification due to acetic anhydride or trinitrobenzene sulfonate treatment, probably secondary to altered affinity for K^+ (69), or by purification procedures (19). These alterations in glycoside sensitivity, however, may be relatively limited in extent, and marked differences in glycoside sensitivity, such as seen in rat heart Na,K-ATPase, probably result from differences in enzyme protein (40, 62). Digitalis sensitivity of Na,K-ATPase in cultured cells is apparently genetically determined (17, 66).

It is generally accepted that an excessive sodium pump inhibition accounts for the myocardial toxicity of cardiac glycosides. It is also established that the arrhythmias (toxicity) are initiated in Purkinje fibers and not in ventricular muscle proper (24). Attempts to demonstrate that the glycoside sensitivity of Na,K-ATPase isolated from Purkinje fibers is higher than that obtained from ventricular muscle were unsuccessful (60). Again, the difference may be due to a higher $[Na^+]_i$ in Purkinje fiber, either enhancing the glycoside binding to Na,K-ATPase in intact cells, or reducing the reserve capacity of the sodium pump. Other explanations are also possible (60). Infants and young animals tolerate higher doses of digitalis glycosides than do adults when the glycoside is given on a body weight basis. This has been explained by a higher Na,K-ATPase activity and Rb uptake in the myocardium of young animals (55), although a more precise evaluation should probably be based on the glycoside sensitivity of Na,K-ATPase in situ and the reserve capacity of the sodium pump.

The ischemic heart is more sensitive to the toxic effects of the glycosides. Ku & Lucchesi (45) have shown that the ouabain sensitivity of the sodium pump is increased in coronary-occluded and re-perfused canine hearts, although the glycoside sensitivity of the isolated enzyme is unchanged. These results suggest that the higher glycoside sensitivity of the ischemic heart is not due to a reduced reserve capacity of the sodium pump. Nor is elevated $[Na^+]_i$ secondary to inhibition of sodium pump turnover, since inhibition of turnover should decrease glycoside binding to the sodium pump (20).

Other Inhibitors

If Na,K-ATPase inhibition by cardiac glycosides is the cause of their positive inotropic effects, other inhibitors of the enzyme may also produce positive inotropic effects. Thus the effects of known Na,K-ATPase inhibitors on myocardial sodium pump activity and on the force of contraction have been extensively studied. The results indicate that Na,K-ATPase inhibitors such as cassaine, prednisolone-bisguanylhydrazone, sulfhydryl blocking agents, and certain cations do produce positive inotropic effects associated with enzyme or sodium pump inhibition, but that other potent inhibitors of isolated Na,K-ATPase, such as dihydroxychlorpromazine and vanadate, fail to inhibit the sodium pump in intact cells, apparently because they are incapable of gaining access to inhibitory sites (see 4). The latter compounds produce positive inotropic effects by other mechanisms (78, 79). These results again emphasize the difficulty in comparing inhibition of isolated Na,K-ATPase with physiological effects observed in intact cells.

Of particular interest is vanadate. This essential trace nutrient has been proposed as an intrinsic regulator of the sodium pump. Vanadate is present in the tissue in such a quantity that Na,K-ATPase might be significantly inhibited if this metal were available in an inhibitory form. Since pentavalent vanadate is inhibitory whereas tetravalent vanadyl ion is not, the conversion between these two forms has been proposed as the potential mechanism for regulation of the sodium pump (63).

Endogenous Ligands of the Digitalis Receptor

The successful elucidation of the endogenous ligands of the opiate receptor has provoked a search for endogenous ligands of the glycoside binding sites on Na,K-ATPase. If such a ligand exists, it could conceivably have a wide range of physiological functions, such as modulation of cardiac contractility, vascular resistance, renal salt excretion, neurotransmission, or stimulus-secretion coupling. Toad skin and its glands contain a "digitalis-like" compound, bufotalin, and related steroid derivatives. Toad serum contains a substance that reacts with an antibody against digoxin and also inhibits ouabain binding to erythrocytes; however, similar compounds were not detected in human serum (28).

Despite the steroidal structure of the digitalis derivatives, corticosteroids that have a trans-fusion between C and D rings instead of the cis-fusion of cardiotonic steroids, are generally devoid of digitalis-like activity. Attachment of a guanylhydrazone group at the 3 and 20 positions, however, makes prednisolone capable of inhibiting Na,K-ATPase and producing positive inotropic effects (see 4). Therefore, it is possible that the endogenous ligand is a steroid derivative.

Heparinized dog plasma has been shown to contain trace peptides possessing a positive inotropic action (56). A low molecular weight peptide in plasma apparently produced from a precursor inhibits sodium transport in isolated toad bladder (36). Extracts from guinea-pig brain (25), human kidney (61), bovine hypothalamus (38) and serum and various tissues of the dog (7) have been shown to inhibit Na,K-ATPase, sodium transport, and/or ouabain binding. An extract of dog skeletal muscle has a positive inotropic effect and is inactivated by ashing at 625°C (7), whereas a rat brain extract is resistant to ashing and sensitive to beta-mercaptoethanol, suggesting the inorganic nature of the latter material (76). Some of the endogenous substances in the above extracts were acid stable (7, 38); however, the exact nature of these materials is presently unknown. Several reports also indicate the presence of natriuretic or vasoconstrictive factors or sodium pump inhibitors in the plasma of hypertensive animals.

The above findings must be interpreted cautiously. The ability of a crude extract to inhibit Na,K-ATPase or ouabain binding and to produce a positive inotropic effect or to inhibit sodium transport does not necessarily indicate that a single compound is responsible for these actions. When a ligand is identified by immunoreactivity, it should be appreciated that most digitalis antibodies do not "recognize" the same part of the digitalis molecule as does Na,K-ATPase, since the pattern of cross-reactivity of antibodies is different from the pattern of "recognition" by glycoside binding sites on Na,K-ATPase.

CONCLUSION

The primary factor that determines the sodium pump activity is $[Na^+]_i$; its elevation increases the turnover rate of the sodium pump, alters the Na^+/K^+ transport ratio and electrogenicity of the sodium pump, and also potentially increases the number of pumping sites if an elevation of $[Na^+]_i$ persists. Additionally, $[K^+]_i$, $[K^+]_e$, $[Na^+]_e$, and $[Ca^{2+}]_i$ may have secondary regulatory functions.

Cardiac glycosides are highly selective inhibitors of Na,K-ATPase. Although not universally supported, overwhelming evidence suggests that Na,K-ATPase is the receptor for the therapeutic as well as the toxic actions of the glycoside. Results obtained with other sodium pump inhibitors and also with agents and under conditions that enhance Na^+ influx indicate that a reduction in the reserve capacity of the sodium pump enhances Na^+ transients, Ca^{2+} transients, and myocardial contractility, whereas sodium pump inhibition beyond a complete depletion of the reserve capacity causes toxicity (see 2, 4). Animal tissues or plasma may contain an endogenous ligand for digitalis binding sites on Na,K-ATPase. Such a ligand may be a physiological regulator of the sodium pump.

ACKNOWLEDGEMENTS

This review was supported in part by U.S. Public Health Service grant HL-16052 from the National Heart, Lung and Blood Institute. We thank Ms. Diane K. Hummel for her help in preparing the manuscript.

Literature Cited

1. Akera, T. 1977. Membrane adenosinetriphosphatase: A digitalis receptor? *Science* 198:569–74
2. Akera, T. 1981. Effects of cardiac glycosides on Na^+,K^+-ATPase. In *Heffter's Handbook of Experimental Pharmacology, 56I: Cardiac Glycosides*, ed. K. Greeff; series ed. G. V. R. Born, O. Eichler, A. Farah, H. Herken, A. D. Welch, pp. 288–336. Heidelberg: Springer.
3. Akera, T., Brody, T M. 1977. The role of Na^+,K^+-ATPase in the inotropic action of digitalis. *Pharmacol. Rev.* 29:187–220
4. Akera, T., Fox, A. A. L. Greeff, K. 1981. Substances possessing inotropic properties similar to cardiac glycosides. See Ref. 2, pp. 459–86
5. Akera, T., Hirai, M., Oka, T. 1979. Sodium ions and the development of the inotropic action of ouabain in guinea-pig heart. *Europ. J. Pharmacol.* 60:189–98
6. Akera, T., Olgaard, M. K. Temma, K., Brody, T. M. 1977. Development of the positive inotropic action of ouabain: Effects of transmembrane sodium movement. *J. Pharmacol. Exp. Ther.* 203:675–84
7. Akera, T., Takeda, K., Temma, K., Brody, T. M. 1980. Presence of Na,K-ATPase inhibitors in blood and other

tissues of the dog. *Pharmacologist* 22:287
8. Akera, T., Temma, K., Wiest, S. A., Brody, T. M. 1978. Reduction of the equilibrium binding of cardiac glycosides and related compounds to Na^+, K^+-ATPase as a possible mechanism for the potassium-induced reversal of their toxicity. *Naunyn-Schmiedeberg's Arch. Pharmacol.* 304:157–65
9. Akera, T., Wiest, S. A., Brody, T. M. 1979. Differential effect of potassium on the action of digoxin and digoxigenin in guinea-pig heart. *Europ. J. Pharmacol.* 57:343–51
10. Akera, T., Yamamoto, S., Chubb, J., McNish, R., Brody, T. M. 1979. Biochemical basis for the low sensitivity of the rat heart to digitalis. *Naunyn-Schmiedeberg's Arch. Pharmacol.* 308:81–88
11. Akera, T., Yamamoto, S., Temma, K., Kim, D-H., Brody, T. M. 1981. Is ouabain-sensitive rubidium or potassium uptake a measure of sodium pump activity in isolated cardiac muscle? *Biochim. Biophys. Acta* 640:779–90
12. Anner, B. M. 1980. Ratio of Na:K transport in reconstituted sodium pump vesicles. *Biochem. Biophys. Res. Commun.* 94:1233–41
13. Beauge, L. A. DiPolo, R. 1979. Sidedness of the ATP-Na^+-K^+ interactions with the Na^+ pump in squid axons. *Biochim. Biophys. Acta* 553:495–500
14. Bentfeld, M., Lüllmann, H., Peters, T., Proppe, D. 1977. Interdependence of ion transport and the action of ouabain in heart muscle. *Brit. J. Pharmacol.* 61:19–27
15. Blostein, R. 1979. Side-specific effects of sodium on (Na,K)-ATPase. *J. Biol. Chem.* 254:6673–77
16. Blostein, R., Chu, L. 1977. Sidedness of (sodium, potassium)-adenosine triphosphatase of inside-out red cell membrane vesicles. *J. Biol. Chem.* 252:3035–43
17. Chang, C. C. Trosko, J. E. Akera, T. 1978. Characterization of ultraviolet light-induced ouabain-resistant mutations in Chinese hamster cells. *Mutation Res.* 51:85–98
18. Choi, Y. R., Akera, T. 1977. Kinetic studies on the interaction between ouabain and (Na^+,K^+)-ATPase. *Biochim. Biophys. Acta* 481:648–59
19. Choi, Y. R., Akera, T. 1978. Membrane (Na^++K^+)-ATPase of canine brain, heart and kidney. Tissue-dependent differences in kinetic properties and the influence of purification procedures. *Biochim. Biophys. Acta* 508:313–27

20. Clausen, Y., Hansen, O. 1977. Active Na-K transport and the rate of ouabain binding. The effect of insulin and other stimuli on skeletal muscle and adipocytes. *J. Physiol. London* 270:415–30
21. Curfman, G. D., Crowley, T. J., Smith, T. W. 1977. Thyroid-induced alterations in myocardial sodium- and potassium-activated adenosine triphosphatase, monovalent cation active transport, and cardiac glycoside binding. *J. Clin. Invest.* 59:586–90
22. De Pont, J. J. H. H. M., Van Eeden, A. V. P., Bonting, S. L. 1978. Role of negatively charged phospholipids in highly purified (Na^+K^+)-ATPase from rabbit kidney outer medulla. Studies on (Na^++K^+)-activated ATPase. *Biochim. Biophys. Acta* 508:464–77
23. Erdmann, E., Philipp, G., Scholz, H. 1980. Cardiac glycoside receptor, (Na^++K^+)-ATPase activity and force of contraction in rat heart. *Biochem. Pharmacol.* 29:3219–29
24. Ferrier, G. R., Saunders, J. H., Mendez, C. 1973. A cellular mechanism for the generation of ventricular arrhythmias by acetylstrophanthidin. *Circ. Res.* 32:600–9
25. Fishman, M. C. 1979. Endogenous digitalis-like activity in mammalian brain. *Proc. Natl. Acad. Sci. USA* 76:4661–63
26. Flasch, H., Heinz, N. 1978. Correlation between inhibition of (Na^+,K^+)-membrane-ATPase and positive inotropic activity of cardenolides in isolated papillary muscle of guinea pig. *Naunyn-Schmiedeberg's Arch. Pharmacol.* 304:37–44
27. Flashner, M. S., Robinson, J. D. 1979. Effects of Mg^{2+} on activation of the (Na^++K^+)-dependent ATPase by Na^+. *Arch. Biochem. Biophys.* 192:584–91
28. Flier, J. S., Maratos-Flier, E., Pallotta, J. A., McIssac, D. 1979. Endogenous digitalis-like activity in the plasma of the toad Bufo marinus. *Nature* 279:341–43
29. Fortes, P. A. G. 1977. Anthroyl-ouabain: A specific fluorescent probe for the cardiac glycoside receptor of the Na,K-ATPase. *Biochemistry* 16:531–40
30. Gallis, J-L., Lasserre, P., Belloc, F. 1979. Freshwater adaptation in the euryhaline teleost, Chelon labrosus. I. Effects of adaptation, prolactin, cortisol and actinomycin D on osmotic balance and (Na^+-K^+)ATPase in gill and kidney. *Gen. Comp. Endocrinol.* 38:1–10
31. Garay, R. P., Garrahan, P. J. 1973. The interaction of sodium and potassium

with the sodium pump in red cells. *J. Physiol. London* 231:297–325
32. Garay, R. P., Dagher, G., Pernollet, M-G., Devynck, M-A., Meyer, P. 1980. Inherited defect in a Na^+,K^+-co-transport system in erythrocytes from essential hypertensive patients. *Nature* 284:281–83
33. Glynn, I. M., Karlish, S. J. D. 1975. The sodium pump. *Ann. Rev. Physiol.* 37:13–55
34. Godfraind, T., Ghysel-Burton, J. 1980. Independence of the positive inotropic effect of ouabain from the inhibition of the heart Na^+/K^+ pump. *Proc. Natl. Acad. Sci. USA* 77:3067–69
35. Godfraind, T., De Pover, A., Verbeke, N. 1977. Influence of pH and sodium on the inhibition of guinea-pig heart $(Na^+ +K^+)$-ATPase by calcium. *Biochim. Biophys. Acta* 481:202–11
36. Gruber, K. A., Buckalew, V. M. Jr. 1978. Further characterization and evidence for a precursor in the formation of plasma antinatriuretic factor. *Proc. Soc. Exp. Biol. Med.* 159:463–67
37. Hastings, D., Skou, J. C. 1980. Potassium binding to the $(Na^+ +K^+)$-ATPase. *Biochim. Biophys. Acta* 601:380–85
38. Haupert, G. T. Jr., Sancho, J. M. 1979. Sodium transport inhibitor from bovine hypothalamus. *Proc. Natl. Acad. Sci. USA* 76:4658–60
39. Hegyvary, C., Chigurupati, R., Kang, K., Mahoney, D. 1980. Reversible alterations in the kinetics of cardiac sodium- and potassium-activated adenosine triphosphatase after partial removal of membrane lipids. *J. Biol. Chem.* 255:3068–74
40. Hegyvary, C., Chigurupati, R., Mahoney, D. 1981. Do membrane lipids modify the ouabain-sensitivity of cardiac (Na,K)ATPase? *Res. Commun. Chem. Path. Pharmacol.* 31:195–203
41. Huang, W., Rhee, H. M., Chiu, T. H., Askari, A. 1979. Re-evaluation of the relationship between the positive inotropic effect of ouabain and its inhibitory effect on $(Na^+ +K^+)$-dependent adenosine triphosphatase in rabbit and dog hearts. *J. Pharmacol. Exp. Ther.* 211:571–82
42. Jorgensen, P. L., Anner, B. M. 1979. Purification and characterization of $(Na^+ +K^+)$-ATPase. VIII. Altered $Na^+:K^+$ transport ratio in vesicles reconstituted with purified $(Na^+ +K^+)$-ATPase that has been selectively modified with trypsin in presence of NaCl. *Biochim. Biophys. Acta* 555:485–92
43. Karli, J. N., Karikas, G. A., Levis, G. M., Moulopoulos, S. N. 1978. Inhibition of Na^+ and K^+-stimulated ATPase of rabbit heart sarcolemma after administration of heparin. *Biochem. Biophys. Res. Commun.* 81:168–75
44. Kimelberg, H. K. 1975. Alterations in phospholipid-dependent $(Na^+ +K^+)$-ATPase activity due to lipid fluidity. Effects of cholesterol and Mg^{2+}. *Biochim. Biophys. Acta* 413:143–56
45. Ku, D. D., Lucchesi, B. R. 1979. Ischemic-induced alterations in cardiac sensitivity to digitalis. *Europ. J. Pharmacol.* 57:135–47
46. Kuhlmann, J., Erdmann, E., Rietbrock, N. 1979. Distribution of cardiac glycosides in heart and brain of dogs and their affinity to the $(Na^+ +K^+)$-ATPase. *Naunyn-Schmiedeberg's Arch. Pharmacol.* 307:65–71
47. Lamers, J. M. J., Hülsmann, W. C. 1977. Inhibition of $(Na^+ +K^+)$-stimulated ATPase of heart by fatty acids. *J. Mol. Cell. Cardiol.* 9:343–46
48. Langer, G. A. 1974. Ionic movements and the control of contraction. In *The Mammalian Myocardium*, ed. G. A. Langer, A. J. Brady, pp. 193–217. NY: Wiley
49. Lee, C. O., Fozzard, H. A. 1975. Activities of potassium and sodium ions in rabbit heart muscle. *J. Gen. Physiol.* 65:695–708
50. Lin, M. H., Romsos, D. R., Akera, T., Leveille, G. A. 1979. Increase in Na^+, K^+-ATPase enzyme units in liver and kidneys from essential fatty acid deficient rats. *Experientia* 35:735–36
51. Lin, M. H., Romsos, D. R., Akera, T., Leveille, G. A. 1979. Na^+,K^+-ATPase enzyme units in skeletal muscle and liver of 14-day-old lean and obese (ob/ob) mice. *Proc. Soc. Exp. Biol. Med.* 161:235–38
52. Lin, M. H., Vander Tuig, J. G., Romsos, D. R., Akera, T., Leveille, G. A. 1979. Na^+,K^+-ATPase enzyme units in lean and obese (ob/ob) thyroxine-injected mice. *Am. J. Physiol.* 237:E265–72
53. Lin, M. H., Vander Tuig, J. G., Romsos, D. R., Akera, T., Leveille, G. A. 1980. Heat production and Na^+-K^+-ATPase enzyme units in lean and obese (ob/ob) mice. *Am. J. Physiol.* 238: E193–99
54. Lingham, R. B., Stewart, D. J., Sen, A. K. 1980. The induction of $(Na^+ +K^+)$-ATPase in the salt gland of the duck. *Biochim. Biophys. Acta* 601:229–34

55. Marsh, A. J., Lloyd, B. L., Taylor, R. R. 1981. Age dependence of myocardial Na$^+$-K$^+$-ATPase activity and digitalis intoxication in the dog and guinea pig. *Circ. Res.* 48:329–33
56. Masiar, E., Masiar, P. 1974. Trace peptides with positive inotropic action in blood plasma of dogs. I. Chemistry and physiological significance of the peptide constituents of an intermediate molecular weight (1000–10,000 daltons) peptide fraction from heparinized dog blood plasma. *Comp. Biochem. Physiol.* 49A:65–80
57. Michael, L. H., Schwartz, A., Wallick, E. T. 1979. Nature of the transport adenosine triphosphatase-digitalis complex: XIV. Inotropy and cardiac glycoside interaction with Na$^+$,K$^+$-ATPase of isolated cat papillary muscles. *Mol. Pharmacol.* 16:135–46
58. Noack, E., Felgentrager, J., Zettner, B. 1979. Changes in myocardial Na and K content during the development of cardiac glycoside inotropy. *J. Mol. Cell. Cardiol.* 11:1189–94
59. Okita, G. T. 1977. Dissociation of Na$^+$, K$^+$-ATPase inhibition from digitalis inotropy. *Fed. Proc.* 36:2225–30
60. Palfi, F. J., Besch, H. R. Jr., Watanabe, A. M. 1978. Ouabain sensitivity of the Na$^+$,K$^+$-ATPase activity from single bovine cardiac Purkinje fiber and adjacent papillary muscle. *J. Mol. Cell. Cardiol.* 10:1149–55
61. Paraskevova, J., Alken, R. G. 1979. An extract from human kidneys with a cardiotonic effect. *Naunyn-Schmiedeberg's Arch. Pharmacol.* 307:R37
62. Periyasamy, S. M., Lane, L. K., Askari, A. 1979. Ouabain-insensitivity of highly active Na$^+$+K$^+$-dependent adenosinetriphosphatase from rat kidney. *Biochem. Biophys. Res. Commun.* 86: 742–47
63. Post, R. L. 1979. A model for regulation of vanadate inhibition of (Na,K)-ATPase by reduction. *Fed. Proc.* 38:242 (Abstr.)
64. Post, R. L., Merritt, C. R., Kinsolving, C. R., Albright, C. D. 1960. Membrane adenosine triphosphatase as a participant in the active transport of sodium and potassium in the human erythrocyte. *J. Biol. Chem.* 235:1796–802
65. Proverbio, F., Hoffmann, J. F. 1977. Membrane compartmentalized ATP and its preferential use by the Na,K-ATPase of human red cell ghosts. *J. Gen. Physiol.* 69:605–32
66. Robbins, A. R., Baker, R. M. 1977. (Na,K)ATPase activity in membrane preparations of ouabain-resistant HeLa cells. *Biochemistry* 16:5163–68
67. Robinson, J. D. 1976. Substrate sites of the (Na$^+$+K$^+$)-dependent ATPase. *Biochim. Biophys. Acta* 429:1006–19
68. Robinson, J. D. 1977. Na$^+$ sites of the (Na$^+$+K$^+$)-dependent ATPase. *Biochim. Biophys. Acta* 482:427–37
69. Robinson, J. D. 1980. Enzyme modifications that alter interactions of K$^+$ and cardioactive steroids with (Na$^+$+K$^+$)-dependent ATPase. *Biochem. Pharmacol.* 29:1995–2000
70. Robinson, J. D., Flashner, M. S. 1979. The (Na$^+$+K$^+$)-activated ATPase. *Biochim. Biophys. Acta* 549:145–76
71. Robinson, J. D., Flashner, M. S., Marin, G. K. 1978. Inhibition of the (Na$^+$+K$^+$)-dependent ATPase by inorganic phosphate. *Biochim. Biophys. Acta* 509:419–28
72. Rodriguez, H. J., Hogan, W. C., Hellman, R. N., Klahr, S. 1980. Mechanism of activation of renal Na$^+$-K$^+$ ATPase in the rat: Effects of potassium loading. *Am. J. Physiol.* 238:F315–23
73. Skou, J. C. 1957. The influence of some cations on an adenosine triphosphatase from peripheral nerves. *Biochim. Biophys. Acta* 23:394–401
74. Skou, J. C. 1979. Effects of ATP on the intermediary steps of the reaction of the (Na$^+$+K$^+$)-ATPase. IV. Effect of ATP on $K_{0.5}$ for Na$^+$ and on hydrolysis at different pH and temperature. *Biochim. Biophys. Acta* 567:421–35
75. Skou, J. C., Esmann, M. 1980. Effects of ATP and protons on the Na:K selectivity of the (Na$^+$+K$^+$)-ATPase studied by ligand effects on intrinsic and extrinsic fluorescence. *Biochim. Biophys. Acta* 601:386–402
76. Swedner, K. J. 1980. The two Na,K-ATPases of brain: Inhibition by cardiac glycosides and endogenous factors. *Fed. Proc.* 39:1704 (Abstr.)
77. Swedner, K. J., Goldin, S. M. 1980. Active transport of sodium and potassium ions: Mechanism, function, and regulation. *New Engl. J. Med.* 302: 777–83
78. Takeda, K., Akera, T., Yamamoto, S., Shieh, I.-S. 1980. Possible mechanisms for inotropic actions of vanadate in isolated guinea pig and rat heart preparations. *Naunyn-Schmiedeberg's Arch. Pharmacol.* 314:161–70
79. Temma, K., Akera, T., Brody, T. M. 1977. Hydroxylated chlorpromazine metabolites: Positive inotropic action and the release of catecholamines. *Mol. Pharmacol.* 13:1076–85

80. Wallick, E. T., Lane, L. K., Schwartz, A. 1979. Biochemical mechanism of the sodium pump. *Ann. Rev. Physiol.* 41: 397–411
81. Yamaguchi, M., Tonomura, Y. 1979. Simultaneous binding of three Na^+ and two K^+ ions to Na^+,K^+-dependent ATPase and changes in its affinities for the ions induced by the formation of a phosphorylated intermediate. *J. Biochem. (Tokyo)* 86:509–23
82. Yamamoto, S., Akera, T., Brody, T. M. 1979. Sodium influx rate and ouabain-sensitive rubidium uptake in isolated guinea pig atria. *Biochim. Biophys. Acta* 555:270–84
83. Yamamoto, S., Akera, T., Brody, T. M. 1980. Positive inotropic action of digoxigenin and sodium pump inhibition. Effects of enhanced sodium influx. *J. Pharmacol. Exp. Ther.* 213:105–9
84. Yamamoto, S., Akera, T., Kim, D.-H., Brody, T. M. 1981. Tissue concentration of Na^+,K^+-adenosine triphosphatase and the positive inotropic action of ouabain in guinea-pig heart. *J. Pharmacol. Exp. Ther.* 217:701–7
85. Yingst, D. R., Hoffman, J. F. 1981. Effect of intracellular Ca on inhibiting the Na-K pump and stimulating Ca-induced transport in resealed human red cell ghosts. *Fed. Proc.* 40:543

ELECTROGENIC NA PUMPING IN THE HEART

H. G. Glitsch[1]

Department of Cell Physiology, Ruhr University Bochum, 4630 Bochum, Federal Republic of Germany

INTRODUCTION

As in other animal cells, an active Na-K transport operates in cardiac fibers. The transport extrudes Na from the cell and takes up K into the cell. Because Na extrusion from cells is effected against an electrochemical gradient, the Na transport mechanism is often called the 'Na pump'. The Na pump maintains the Na and K gradients across the cell membrane at rest and restores them following excitation. The maintenance and restoration of these ionic gradients are essential for the production of action potentials. The active Na transport is a most important ion transport mechanism in cells like cardiac cells which display frequent spontaneous action potentials.

A Na pump that takes up one K ion into the cell for each Na extruded does not separate electrical charges during its activity. It is therefore called *electroneutral*. An electroneutral active Na transport affects only *indirectly* the membrane potential of a cell by maintaining the Na and K concentration gradients across the cell membrane. An *electrogenic* Na pump contributes *directly* to the membrane potential by net charge transport. During the last two decades evidence has accumulated that active Na transport in many cells including cardiac cells is electrogenic. The main observation favoring an electrogenic active Na transport is that changes in membrane potential (or current) occur that cannot be accounted for by alterations in membrane conductance or in the passive ionic distribution at the cell membrane. These changes in membrane potential (or current) share several characteristics with the Na pump. The present article appraises findings that suggest the

[1]Supported by the Deutsche Forschungsgemeinschaft (SFB 114 'Bionach')

existence of an electrogenic Na pump in cardiac cells. A general description of the relationship between electrogenic Na pumping and some passive electrical properties of cell membranes is given by Rapoport (39). For further information on cardiac electrogenic Na pumping the reader is referred to more detailed reviews (17, 22, 31).

REPORTS ON A NA PUMP CONTRIBUTION TO CARDIAC MEMBRANE POTENTIAL

Sinoatrial Node Cells

If perfusion with K-containing solution is resumed following a period of several minutes in a K-free medium, cardiac preparations display a transient hyperpolarization. This 'K-activated response' was studied (34, 35) in sinoatrial node cells of the rabbit's heart. The response is increased by prolongation of the K-free period, by augmenting $[K]_o$ after the conditioning period, or by reactivating the Na pump in Cl-deficient solution (Cl^- replaced by the less permeant propionate$^-$). The latter finding might indicate that an increase in membrane resistance induces a stronger K-activated response. If Na is replaced by Li in the K-free solution, no K-activated response is observed after the K-free period. The response is also inhibited if $1.3 \cdot 10^{-5}$ M ouabain is applied during the last minute of the K-free period. The authors conclude that an electrogenic Na pump causes the K-activated response; they discuss the pump as a factor that modulates the heart rate under physiological conditions.

Purkinje Fibers

Délèze (5) was the first to suggest the existence of an electrogenic Na pump in cardiac preparations. He reported that the resting potential of sheep and calf Purkinje fibers (and ventricular cells) displays a higher temperature sensitivity than that predicted for a K electrode. This marked sensitivity of membrane potential is not due to changes in the P_K/P_{Na} ratio or to K accumulation (or depletion) at the outside of the cell membrane. Only when the fibers are poisoned by metabolic inhibitors is the membrane potential proportional to the absolute temperature as expected for a K electrode. The author suggests that an electrogenic Na pump contributes directly to the cardiac resting potential. Purkinje fibers stop beating spontaneously after stimulation with an unphysiologically high stimulus frequency. This 'overdrive suppression' (e.g. 45) is due to a transient hyperpolarization that is not observed if extracellular Na is replaced by Li or if the preparation is exposed to 2,4-dinitrophenol. The input resistance of the fibers is much the same before and after the overdrive. According to the author an electrogenic Na pump is probably the major cause of overdrive suppression. Under physiological conditions overdrive by the sinoatrial node cells might suppress the spontaneous activity of Purkinje fibers. Hiraoka & Hecht (23) reported on a transient hyperpolarization in sheep Purkinje fibers rewarmed in K-containing media after prolonged (24 hr) hypothermia. The hyperpolarization does not occur if Na is replaced by Li in the bathing fluid during hypothermia. Similarly, the hyperpolarization is blocked in preparations perfused for 1 hr in a solution containing 10^{-5} M ouabain before rewarming. A change in $[K]_o$ from 5.4 to 2.7 mM during rewarming causes a decline in the transient hyperpolarization. However, an identical change in $[K]_o$ evokes a hyperpolarization under steady-state conditions. Clearly, these findings are consistent with the concept of an electrogenic active Na transport that is activated in rewarmed cardiac Purkinje fibers. Beating canine Purkinje fibers display a transient hyperpolarization if perfusion with a bathing solution containing a normal (5.4 mM) or elevated (10.8 mM) K concentration is resumed after a period in a low K medium (0.54 mM) (28). The authors suggest that this hyperpolarization is brought about by both an increase in the K conductance of the cell membrane and the activation of the Na pump. A transient hyperpolarization is also observed following reapplication of a Na-containing bathing fluid to Na-depleted canine Purkinje fibers. The hyperpolarization is inhibited at low temperature, in K-free media, or in solutions containing Li instead of Na. It is completely blocked by 10^{-3} M ouabain (47). Similar findings have been reported from experiments on guinea-pig auricles (20). The first studies on the pump contribution to membrane currents in Purkinje fibers by means of a voltage clamp technique have been carried out by Isenberg & Trautwein (26, 27), who describe a voltage-independent outward current with a high temperature sensitivity. The current is inhibited by cardiac glycosides (10^{-5} M dihydroouabain, 10^{-6} M ouabain). According to the authors the current is generated by an electrogenic active Na transport. Following perfusion with a K-free, Cl-deficient medium an outward current is measured in voltage-clamped canine Purkinje fibers if the Na pump is activated by an increase in $[K]_o$. The current is acetylstrophanthidin-sensitive. Its amplitude depends on the duration of the prior exposure to K-free solution. The decline of this transient outward current obeys first-order kinetics, suggesting a constant part of the augmented $[Na]_i$ to be extruded by an electrogenic mechanism probably identical with the Na pump (12). The pump current has been further analyzed (11). The rate constant of its single exponential decline depends on $[K]_o$. The rate constant increases with increasing $[K]_o$ toward a maximum value. Based on earlier considerations (38; cf. also 7), Gadsby (11) points out that the rate constants at various $[K]_o$ are related to the activation of the electrogenic Na pump by K_o. He concludes that the pump is half-maximally activated at $\sim$ 1 mM K_o, in good agreement with observations on guinea-pig auricular cells (18). Similar experiments were conducted (6, 7, 8) on voltage-clamped sheep Purkinje fibers. The authors describe a transient outward current activated by several external activator cations (e.g. K_o, Rb_o, Cs_o, Li_o) and inhibited by strophanthidin (10^{-5} M) in preparations previously perfused for several minutes with a medium without activator cations. The Rb_o- or Cs_o-activated transient outward current is nearly voltage-independent within a membrane potential range of -90 to $+10$ mV and displays Q_{10} values between 1.6 and 2.3 in a range between 26 and 46°C. Again, an increase in $[Rb]_o$ or $[Cs]_o$ causes a faster decline of the outward current. The rate constants of the decline tend to a maximum value at high $[Rb]_o$ or $[Cs]_o$. Half-maximal activation of the current occurs at 6.3 mM Rb_o or 14.2 mM Cs_o. With respect to the activation of the Na pump, 2 mM K_o, 2 mM Rb_o, 6 mM Cs_o, 6 mM NH_{4o}, and 22 mM Li_o are equipotent, while Tl_o is a stronger activator cation than K_o or Rb_o. The papers mentioned above strongly suggest that the transient hyperpolarization of the cardiac cell membrane or the transient outward current underlying this hyperpolarization is due to electrogenic Na pumping. However, until recently direct evidence was not available that changes in Na_i do in fact occur simultaneously with the observed changes in membrane potential or current. Eisner et al (9) have now shown that the time constant of decline in the outward current and in the intracellular Na activity is much the same during a Rb-activated response in sheep Purkinje fibers. Similarly, the time constants of decline in membrane potential and in the internal Na activity are identical during a K- or Rb-activated response in sheep Purkinje fibers (21). These reports lend further support to the concept that the Na pump is electrogenic and contributes directly to the membrane potential of cardiac cells.

Atrial and Ventricular Preparations

The membrane potential of cat papillary muscles is quickly restored in K-containing solution after hypothermia. However, the recovery of $[Na]_i$ and $[K]_i$ is much slower. Ten minutes after perfusion at 27–28°C is resumed, the membrane potential is recorded to be 6 mV more negative than the (calculated) K equilibrium potential (E_K) (36). Similar experiments on guinea-pig auricles have confirmed these results and shown that the cell membrane of atrial cells hyperpolarizes transiently during rewarming (14). There is a strong correlation between the magnitude of hyperpolarization and of active Na efflux from rewarmed guinea-pig atria. The hyperpolarization is only observed if $[Na]_i$ has been increased during the hypothermia. Choline ions cannot replace Na ions in this mechanism (15). The hyperpolarization is not seen in K-free media where active Na efflux is diminished (14). Half-maximal pump activation occurs at 1.5 mM K_o (18). A corresponding finding suggests a half-maximal activation of a K-activated outward current at 1.3 mM K_o in voltage-clamped bullfrog atrial fibers (1). This outward current is inhibited by cardiac glycosides known to be specific inhibitors of active Na transport (cf 41). Similarly, the transient hyperpolarization of rewarmed guinea-pig atrial cells in K-containing solutions is blocked by cardiac glycosides (16). Clearly, all these observations point to the existence of an electrogenic Na pump in atrial and ventricular cells. Sano et al (40) studied the transient hyperpolarization in rewarmed ventricular and Purkinje fibers of various species, including 6- and 15-day-old chicken ventricular cells. The authors suggest that the contribution of electrogenic Na pumping to the membrane potential increases during cardiac ontogenesis.

Cultured Chick Embryo Heart Cells

The morphological complexity of many cardiac preparations does not always permit an unequivocal interpretation of experimental data. Therefore, it is interesting to look for a preparation with a simple geometry. Tissue cultured chick embryo heart cells seem to be suitable objects for studies on cardiac ion transport. In these preparations diffusion distances are less than 30 μm if the cells are grown around a very thin monofilament of nylon. Horres et al (24) studied the membrane potential, the cellular Na and K concentrations, and the ^{42}K uptake of these cells in a K-containing medium following a period of several minutes in a K-free solution. Application of a bathing fluid containing 5.4 mM K causes the membrane potential to rise from −38 to −91 mV within 5 seconds, while $[Na]_i$ and $[K]_i$ do not reach their steady-state level during 2 minutes. The membrane potential is initially more negative than E_K. ^{42}K uptake during the K-activated response is enhanced compared to steady-state conditions. The increase in membrane potential is greatly reduced by 10^{-4} M ouabain. According to the authors these results strongly suggest the existence of a cardiac electrogenic Na pump.

DISCUSSION OF THE EXPERIMENTAL DATA

The experimental data presented above demonstrate that a mechanism that shares several characteristics with active Na transport contributes to the cardiac membrane potential under conditions known to activate the Na pump. These characteristics fall into two groups. The first comprises properties that agree qualitatively with characteristics of the Na pump. For example, both mechanisms are activated by an increase in $[Na]_i$ or $[K]_o$. Neither choline nor Li can replace Na at the intracellular activation sites, while several monovalent cations such as Rb, Cs or Tl can act as substitutes for K at the extracellular activation sites of the active Na transport system and of the mechanism under study. Lowering the temperature of the bathing solution or application of cardiac glycosides inhibits both the Na pump and this mechanism. Since cardiac glycosides are known to be specific inhibitors of active Na transport, the latter finding strongly suggests that the mechanism described and the Na pump are identical. This conclusion is further backed by the second group of characteristics, which are in quantitative agreement with properties of cardiac active Na transport. For example, the mechanism under investigation is half maximally activated by 1–1.5 mM K_o in the working myocardium (1, 18) and in the cardiac conducting system (11). A *Km* value of 2 mM K is reported from studies on the activation of the cardiac transport ATPase (10, 37). External Rb and K ions are equipotent activators of the mechanism (e.g. 6, 19), and both cations are potent activators of the cardiac Na^+,K^+-activated, Mg^{2+}-dependent ATPase (transport ATPase) (37). Moreover, direct measurements of changes in membrane potential or current and in the intracellular Na activity during activation of the Na pump reveal an identical time course (9, 21). Thus it seems reasonable to conclude that the observed changes in membrane potential (current) are due to Na pumping.

ELECTROGENIC VERSUS ELECTRONEUTRAL PUMP

How is the contribution of active Na transport to the membrane potential brought about? Two hypotheses demand discussion. Obviously, all findings so far presented can easily be accounted for if an electrogenic Na pump is assumed to exist, which extrudes more Na ions from the cells than it takes up K ions. The Na pump activity is accompanied by a Na outward current. The intriguing second hypothesis states that many of the data reported are equally well interpreted as being due to the activation of an electroneutral active Na transport. For example, the observation that the membrane potential reaches values more negative than E_K during activation of the Na pump (e.g. 14, 28, 36) might suggest at first glance an electrogenic Na transport because the Goldman equation predicts that the membrane potential will never exceed E_K if it is due to ionic gradients across the cell membrane and the respective ionic permeabilities. This conclusion is premature for the following reason. Activation of an electroneutral pump might reduce $[K]_o$ at the outside of the cell membrane well below the K concentration of the bathing medium. However, the latter value is normally used for the calculation of E_K. Thus it might well be that the 'true' E_K remains more negative than the membrane potential during pump activation. While it is likely that K depletion does occur during the activation (see 30), several observations demonstrate quite clearly that depletion cannot be the only cause of the hyperpolarization of the cell membrane under these conditions. For example, the membrane potential of rewarmed guinea-pig auricles is more negative in K-containing than in K-free solutions (18). It is not easily understood why K depletion should be stronger in K-containing than in K-free bathing fluids. Furthermore, it is possible to differentiate between effects of K depletion and pump outward current on the I-V relation in voltage-clamped Purkinje fibers (7). Gadsby & Cranefield (13) investigated the pump activation in canine Purkinje fibers under three conditions in which K depletion causes changes opposite to those expected from an increase in the outward pump current. Under all three conditions pump activation induces effects expected for an increase in pump current. The authors also discuss and reject the alternative possibility that the changes in membrane potential are not at all causally related to Na pumping but to simultaneous alterations in passive membrane characteristics, such as reduction of the background Na inward current, increase in the steady state K conductance due to an augmented $[Ca]_i$ (cf 25), or activation of a voltage- and time-dependent K conductance. Of course, the fact that specific inhibitors of active Na transport like cardiac glycosides inhibit the hyperpolarization during enhanced active Na efflux does not favor this view. In conclusion, our present knowledge strongly suggests that the Na pump is electrogenic and contributes directly to the cardiac membrane potential. Neither K depletion nor changes in membrane conductance alone or in combination can account for the effects on membrane potential during enhanced Na pumping.

COUPLING RATIO AND THE ELECTROGENIC MECHANISM

In erythrocytes $3Na^+$ are extruded and $2K^+$ are taken up by the Na-K pump per ATP molecule split. The coupling ratio, i.e. the number of Na^+ extruded divided by the number of K^+ taken up per pump cycle, amounts to 1.5 under physiological conditions. Similar coupling ratios are found in a variety of cells with electrically excitable membranes [for references see (44)]. For example, in neurones of *Helix aspersa* about 30% of the total Na charge injected is pumped as an outward current, suggesting a coupling ratio of 1.5 (43). Based on this observation it seems reasonable to suppose that cardiac active Na transport is electrogenic because part of the intracellular Na is extruded as a Na current. Neither the physiological coupling ratio nor the factors controlling the ratio are known for cardiac cells. Changes in $[Na]_i$ (7, 12, 21) or in $[K]_o$ (11) do not affect the ratio in cardiac Purkinje fibers. Similarly, alterations in $[Rb]_o$ or $[Cs]_o$ are ineffective (7). The electrogenic mechanism is not yet understood on a molecular level.

EFFECT OF ELECTROGENIC NA PUMPING ON CARDIAC RESTING POTENTIAL

The resting potential is the membrane potential of a quiescent cell under steady-state conditions. There can be no net flux of any ion across the cell membrane in the steady state. Thus any Na outward current generated by an electrogenic Na pump in a resting cell must be compensated by net passive ionic fluxes. Although it is impossible to measure the pump current in the steady state, it is quite possible to calculate the contribution of an electrogenic Na pump to the resting potential. If Cl ions are passively distributed at resting potential, the potential of a resting cell (E_{rest}) with an electrogenic Na transport obeys the following equation (32):

$$E_{rest} = \frac{R \cdot T}{F} \ln \frac{P_{Na} \cdot [Na]_o + r \cdot P_K \cdot [K]_o}{P_{Na} \cdot [Na]_i + r \cdot P_K \cdot [K]_i} \qquad 1.$$

where r denotes the coupling ratio and the other symbols have their usual meaning. Equation 1 can be extended to nonsteady-state conditions (see 42). If $r = 1$ (electroneutral pump), equation 1 corresponds to a simplified Goldman equation. If it is assumed that the Na pump operates purely electrogenically ($r \to \infty$), the resting potential should approach E_K. The observation that the resting potential of cardiac cells in physiological media is always less negative than E_K suggests some kind of chemical coupling between active Na efflux and active K influx, i.e. $r \neq \infty$. According to Ascher (cited in 44) the pump contribution (E_p) to the resting potential cannot exceed:

$$E_P = \frac{R \cdot T}{F} \cdot \ln \frac{1}{r} \qquad 2.$$

If $r = 1.5$ as in other cells under physiological conditions, active Na transport cannot contribute more than 10–11 mV to the cardiac resting potential. Experimental data suggest a pump contribution to the cardiac resting potential of up to 10 mV (4, 15, 26).

EFFECT OF ELECTROGENIC NA PUMPING ON ACTION POTENTIAL

Isenberg & Trautwein (26) and Eisner & Lederer (7) described a glycoside-sensitive outward current probably generated by the Na pump in voltage-clamped Purkinje fibers. The reported pump current densities (0.3–0.8 μA $\cdot$ cm^{-2}) are large enough to affect markedly the plateau and repolarization phase during the cardiac action potential (cf 33). Correspondingly, application of 10^{-5} M dihydro-ouabain lengthens the action potential duration within 1 min (26). In canine Purkinje fibers driven at physiological frequencies a small additional activation of the Na pump shortens the action potential duration by as much as 20% (13). As pointed out by these authors, neither K depletion nor changes in membrane conductance can account for the observed changes in action potential. In summary, there is little doubt that the electrogenic Na pump does affect the action potential, especially in frequently beating cardiac preparations. The quantitative contribution of active Na transport to the outward currents during the action potential remains to be clarified. First steps in this direction are the papers by Chapman et al (2) and Johnson et al (29) in which the interaction between membrane potential and electrogenic Na pumping in beating cardiac fibers is studied by computer simulation.

EFFECT OF ELECTROGENIC NA PUMPING ON CARDIAC AUTOMATIC ACTIVITY

Noma & Irisawa (34) have suggested that the active electrogenic Na transport is one of the factors that modulate the heart rate under physiological conditions. Similarly, 'overdrive suppression,' which is probably due to electrogenic Na pumping, might be a physiological mechanism (cf 45, 46). In vivo the Purkinje fibers are overdriven by the faster rhythm of the sinus node. This overdrive suppression prevents an escape of the pacemaker activity in the fibers from the control by the dominant pacemaker. A transient inhibition of the spontaneous activity of cardiac Purkinje fibers is also observed if the Na pump is activated after a K-free period. This inhibition occurs regardless of whether the pacemaker activity arises from the lower or higher level of membrane potential (13).

POSSIBLE ANTIARRHYTHMIC EFFECTS OF CARDIAC ELECTROGENIC NA PUMPING

Spontaneous activity at any level of membrane potential, reentry of excitation due to circus movements, and action potentials triggered repetitively by delayed after-depolarizations are considered possible causes of arrhythmias (cf 3). Electrogenic Na pumping might have an antiarrhythmic effect for the following reasons (13). Activation of the Na pump generates a Na outward current, which reduces the slow diastolic depolarization and thereby the pacemaker activity at all membrane potentials. The current is able to repolarize a fiber from the lower to the higher level of membrane potential and to abolish the spontaneous activity at the lower level [Figure 8 in (13)]. Hyperpolarization of the cardiac cell membrane by Na pump activation also counteracts the slow conduction of excitation, a prerequisite of reentry. Finally, the pump (outward) current reduces the amplitude of delayed afterdepolarizations and thus the risk of triggered activity.

MEASUREMENTS OF PUMP CURRENT AS A TOOL FOR STUDIES IN CARDIAC MEMBRANE PHYSIOLOGY AND PHARMACOLOGY

Measurements of the Na outward current generated by the Na pump in voltage-clamped cardiac preparations can be used to study the interaction of active Na transport with other cellular activities. This approach has been pioneered by Eisner & Lederer (6, 8), who have studied the relationship between Na pump and contraction in sheep Purkinje fibers. The Na pump activity has been measured by examining the pump outward current. The same method is useful for studies on the cardiac Na pump as affected by drugs (e.g.1).

Literature Cited

1. Akasu, T., Ohta, Y., Koketsu, K. 1978. The effect of adrenaline on the electrogenic Na^+ pump in cardiac muscle cells. *Experientia* 34:488–90
2. Chapman, J. B., Kootsey, J. M., Johnson, E. A. 1979. A kinetic model for determining the consequences of electrogenic active transport in cardiac muscle. *J. Theor. Biol.* 80:405–24
3. Cranefield, P. F. 1977. Action potentials, afterpotentials, and arrhythmias. *Circ. Res.* 41:415–23
4. Daut, J., Rüdel, R. 1980. The electrogenic pump current in guinea-pig myocardium. *J. Physiol. London* 305:22P (Abstr.)
5. Délèze, J. 1960. Possible reasons for drop of resting potential of mammalian heart preparations during hypothermia. *Circ. Res.* 8:553–57
6. Eisner, D. A., Lederer, W. J. 1979. The role of the sodium pump in the effects of potassium-depleted solutions on mammalian cardiac muscle. *J. Physiol. London* 294:279–301
7. Eisner, D. A., Lederer, W. J. 1980. Characterization of the electrogenic sodium pump in cardiac Purkinje fibres. *J. Physiol. London* 303:441–74
8. Eisner, D. A., Lederer, W. J. 1980. The relationship between sodium pump activity and twitch tension in cardiac Purkinje fibres. *J. Physiol. London* 303: 475–94
9. Eisner, D. A., Lederer, W. J., Vaughan-Jones, R. D. 1980. Electrogenic sodium pumping in cardiac muscle: simultaneous measurement of intracellular sodium activity, membrane current and tension. *J. Physiol. London* 300:42–43P (Abstr.)
10. Erdmann, E., Bolte, H.-D., Lüderitz, B. 1971. The (Na^++K^+)-ATPase activity of guinea pig heart muscle in potassium deficiency. *Arch. Biochem. Biophys.* 145:121–25
11. Gadsby, D. C. 1980. Activation of electrogenic Na^+/K^+ exchange by extracellular K^+ in canine cardiac Purkinje fibers. *Proc. Natl. Acad. Sci. USA* 77:4035–39
12. Gadsby, D. C., Cranefield, P. F. 1979. Direct measurement of changes in sodium pump current in canine cardiac Purkinje fibers. *Proc. Natl. Acad. Sci. USA* 76:1783–87
13. Gadsby, D. C., Cranefield, P. F. 1979. Electrogenic sodium extrusion in cardiac Purkinje fibers. *J. Gen. Physiol.* 73:819–37
14. Glitsch, H. G. 1969. Über das Membranpotential des Meerschweinchenvorhofes nach Hypothermie. *Pfluegers Arch.* 307:29–46
15. Glitsch, H. G. 1972. Activation of the electrogenic sodium pump in guinea-pig auricles by internal sodium ions. *J. Physiol. London* 220:565–82

16. Glitsch, H. G. 1972. Hemmung der elektrogenen Na-Pumpe am Meerschweinchenvorhof durch Digitoxigenin. *Pfluegers Arch.* 335:243–51
17. Glitsch, H. G. 1979. Characteristics of active Na transport in intact cardiac cells. *Am. J. Physiol.* 236:H189–99
18. Glitsch, H. G., Grabowski, W., Thielen, J. 1978. Activation of the electrogenic sodium pump in guinea-pig atria by external potassium ions. *J. Physiol. London* 276:515–24
19. Glitsch, H. G., Kampmann, W., Pusch, H. 1981. Activation of the Na pump in sheep Purkinje fibres by extracellular K or Rb ions. *Pfluegers Arch.* 389:Suppl. R8 (Abstr.)
20. Glitsch, H. G., Klare, J. 1977. On the membrane potential of Na depleted guinea-pig auricles after addition of Na to the extracellular solution. *Pfluegers Arch.* 368:Suppl. R3 (Abstr.)
21. Glitsch, H. G., Pusch, H. 1980. Correlation between changes in membrane potential and intracellular Na activity during K activated response in sheep Purkinje fibres. *Pfluegers Arch.* 384:189–91
22. Haas, H. G. 1972. Active transport in heart muscle. In *Electrical Phenomena in the Heart,* ed. W. C. de Mello, pp. 163–89. NY/London: Academic. 415 pp.
23. Hiraoka, M., Hecht, H. H. 1973. Recovery from hypothermia in cardiac Purkinje fibers: considerations for an electrogenic mechanism. *Pfluegers Arch.* 339:25–36
24. Horres, C. R., Aiton, J. F., Lieberman, M., Johnson, E. A. 1979. Electrogenic transport in tissue cultured heart cells. *J. Mol. Cell. Cardiol.* 11:1201–5
25. Isenberg, G. 1977. Cardiac Purkinje fibres. [Ca^{2+}]$_i$ controls steady state potassium conductance. *Pfluegers Arch.* 371:71–76
26. Isenberg, G., Trautwein, W. 1974. The effect of dihydro-ouabain and lithiumions on the outward current in cardiac Purkinje fibers. Evidence for electrogenicity of active transport. *Pfluegers Arch.* 350:41–54
27. Isenberg, G., Trautwein, W. 1975. Temperature sensitivity of outward current in cardiac Purkinje fibers. Evidence for electrogenicity of active transport. *Pfluegers Arch.* 358:225–34
28. Ito, S., Surawicz, B. 1977. Transient, "paradoxical" effects of increasing extracellular K^+ concentration on transmembrane potential in canine cardiac Purkinje fibers. *Circ. Res.* 41:799–807
29. Johnson, E. A., Chapman, J. B., Kootsey, J. M. 1980. Some electrophysiological consequences of electrogenic sodium and potassium transport in cardiac muscle: a theoretical study. *J. Theor. Biol.* 87:737–56
30. Kunze, D. L. 1977. Rate-dependent changes in extracellular potassium in the rabbit atrium. *Circ. Res.* 41:122–27
31. Lüttgau, H. C., Glitsch, H. G. 1976. Membrane physiology of nerve and muscle fibers. *Fortschr. Zool.* 24:1–132
32. Mullins, L. J., Noda, K. 1963. The influence of sodium-free solutions on the membrane potential of frog muscle fibers. *J. Gen. Physiol.* 47:117–32
33. Noble, D. 1975. *The Initiation of the Heartbeat.* Oxford: Clarendon Press. 156 pp.
34. Noma, A., Irisawa, H. 1974. Electrogenic sodium pump in rabbit sinoatrial node cell. *Pfluegers Arch.* 351:177–82
35. Noma, A., Irisawa, H. 1975. Contribution of an electrogenic sodium pump to the membrane potential in rabbit sinoatrial node cells. *Pfluegers Arch.* 358:289–301
36. Page, E., Storm, S. R. 1965. Cat heart muscle *in vitro.* VIII. Active transport of sodium in papillary muscles. *J. Gen. Physiol.* 48:957–72
37. Portius, H. J., Repke, K. R. H. 1967. Eigenschaften und Funktion des Na^+ +K^+-aktivierten, Mg^{++}-abhängigen Adenosintriphosphat Phosphohydrolase-Systems des Herzmuskels. *Acta Biol. Med. Ger.* 19:907–38
38. Rang, H. P., Ritchie, J. M. 1968. On the electrogenic sodium pump in mammalian non-myelinated nerve fibres and its activation by various external cations. *J. Physiol. London* 196:183–221
39. Rapoport, S. I. 1970. The sodium-potassium exchange pump: relation of metabolism to electrical properties of the cell. *Biophys. J.* 10:246–59
40. Sano, T., Hiraoka, M., Sawanobori, T. 1975. Electrogenic contribution to resting potential in different cardiac tissues. In *Developmental and Physiological Correlates of Cardiac Muscle,* ed. M. Lieberman, T. Sano, pp. 299–310. NY: Raven Press. 336 pp.
41. Schatzmann, H.-J. 1953. Herzglykoside als Hemmstoffe tur den aktiven Kalium- und Natriumtransport durch die Erythrocytenmembran. *Helv. Physiol. Pharmacol. Acta* 11:346–54
42. Sjodin, R. A., Ortiz, O. 1975. Resolution of the potassium ion pump in muscle fibers using barium ions. *J. Gen. Physiol.* 66:269–86

43. Thomas, R. C. 1969. Membrane current and intracellular sodium changes in a snail neurone during extrusion of injected sodium. *J. Physiol. London* 201:495–514
44. Thomas, R. C. 1972. Electrogenic sodium pump in nerve and muscle cells. *Physiol. Rev.* 52:563–94
45. Vassalle, M. 1970. Electrogenic suppression of automaticity in sheep and dog Purkinje fibers. *Circ. Res.* 27:361–77
46. Vassalle, M. 1977. The relationship among cardiac pacemakers. *Circ. Res.* 41:269–77
47. Wiggins, J. R., Cranefield, P. F. 1974. Effect on membrane potential and electrical activity of adding sodium to sodium-depleted cardiac Purkinje fibers. *J. Gen. Physiol.* 64:473–93

PHOSPHORYLATION OF THE SARCOPLASMIC RETICULUM AND SARCOLEMMA

Michihiko Tada

First Department of Medicine, Osaka University School of Medicine, Osaka 553, Japan

Arnold M. Katz

Department of Medicine, Division of Cardiology, University of Connecticut School of Medicine, Farmington, Connecticut 06032

INTRODUCTION

The β-adrenergic actions of catecholamines represent a key control mechanism that regulates the metabolic, electrical, and mechanical performance of the myocardium. Cyclic AMP(cAMP) serves as the intracellular second messenger for these regulatory processes in that β-adrenergic activation of a sarcolemmal adenylate cyclase causes intracellular cAMP levels to rise (75). This second messenger appears to exert most of its cellular effects by activating cAMP-dependent protein kinases (cAMP-PKs) (42). These enzymes catalyze phosphorylation of phosphorylase kinase and glycogen synthetase, resulting in a marked increase in glycolysis (7), although these metabolic changes are not essential for the positive inotropic response of the well-oxygenated heart to catecholamines (87).

Catecholamines and their intracellular messenger cAMP alter several aspects of the excitation-contraction process in the myocardium (26, 27, 87), including increases in the flux of ions across the sarcolemma (SL) (60, 65) and sarcoplasmic reticulum (SR) (26, 80, 83). The present review describes findings that define the mechanism by which cAMP-PKs react to induce these changes in membrane function.

Membranes of Sarcoplasmic Reticulum and Sarcolemma

Cardiac SR vesicles, isolated in microsomal fractions by differential centrifugation of a myocardial homogenate, retain their major physiological characteristic of actively transporting Ca^{2+} in an ATP-dependent manner (85). These membrane vesicles are composed of several proteins and lipids organized such that the various proteins are situated within the basic matrix of a lipid bilayer (85). The major protein of the cardiac SR, the *Ca pump ATPase enzyme* (mw $\sim$100,000), which represents up to 40% of the total protein (82a, 85) and is responsible for the active transport of Ca from the cytoplasm into the lumen of the SR during relaxation, has been partially purified (89). *Phospholamban,* a 22,000-dalton protein in the cardiac SR membrane, is a PK substrate that can regulate the Ca transport system. *Calsequestrin,* a protein believed to bind Ca within the SR, has been purified from skeletal SR (85); its molecular weight has been reported to be in the range 44,000–65,000. It is not yet clear whether cardiac SR contains calsequestrin.

Cardiac SL vesicles have been purified from 1.5- to 16.3-fold by differential centrifugation (20, 68). Greater enrichment has been achieved by centrifugation on sucrose density gradients (22, 88). More than 75% of the SL vesicles prepared by Besch et al (3) were inside out, while those of Bers et al (2) were about 80% right-side out. Weglicki et al (92) obtained SL fractions from isolated myocytes that contained six major protein components, including proteins of 50,000, 91,000, and 140,000 daltons. SL vesicles prepared by conventional methods contain about 20 polypeptides ranging from 18,000 to 200,000 daltons (22, 70). A protein of 105,000 daltons may represent both the Ca^{2+}, Mg^{2+}-dependent ATPase and Na^+, K^+-ATPase enzymes, major ATPase activities exhibited by SL vesicles (19, 20, 57, 88). Recently, however, Caroni & Carafoli (103) reported a molecular weight of 150,000 for the Ca^{2+}-ATPase of cardiac sarcolemma.

PHOSPHORYLATION OF THE SR

Cardiac SR can form two classes of phosphoprotein (summarized in Figure 1) that differ in both molecular weight and chemical characteristics (80, 82). The Ca pump ATPase of the cardiac SR, which serves as an energy transducer that translocates Ca across the membrane, forms a phosphoprotein intermediate by incorporating the terminal phosphate of ATP into a hydroxylamine-labile acylphosphate (64a, 82, 85). In skeletal muscle SR, phosphorylation occurs at the carboxyl group of an aspartic acid residue (85). The amount of phosphorylation in cardiac SR ranges between 0.6 and 1.3 nmoles P/mg (64a, 82a), and probably occurs within the tryptic subfragment of 30,000–33,000 daltons (85).

Cardiac SR is able to form a second type of phosphoprotein when reacted with cAMP-PKs. In this phosphoprotein, the terminal phosphate of ATP is incorporated mainly into a serine residue of a 22,000-dalton protein that exhibits the stability characteristics of a phosphoester (35). This phosphorylatable protein, which represents one of the sites at which cAMP-PK exerts its action, appears to function as a modulator of the Ca^{2+}-ATPase and Ca pump of cardiac SR.

Phosphorylation of 22,000-dalton Protein (Phospholamban)

The formation of a phosphoester phosphoprotein when cardiac SR is incubated with cAMP in the absence (45, 101) and presence (35) of exogenous cAMP-PK is independent of Ca^{2+} up to 0.1 mM, but is markedly dependent on cAMP between 0.1 and 10 μM. Half-maximal phosphorylation occurs at 0.2 μM cAMP (35). Tada et al (77) found that a single protein of $\sim$ 22,000 daltons incorporated most of the ^{32}P label when cardiac microsomes incubated with [γ-^{32}P]ATP and cAMP-PK were solubilized in sodium dodecyl sulfate (SDS) and electrophoresed on SDS-polyacrylamide gels. In the presence of NaF, an inhibitor of phosphoprotein phosphatase, phosphorylation of this protein increased several fold, and small amounts of two other phosphoproteins (55,000 and 11,000 daltons) appeared. The 55,000-dalton protein could be attributed to autophosphorylation of the regulatory subunits of PK. LaRaia & Morkin (45) also found that phosphorylation of 20,000-dalton SR protein was catalyzed by an endogenous PK. These reports, which were subsequently confirmed by a number of groups (Table 1), indicate that the 20,000–24,000-dalton protein is the major phosphorylation site at which cAMP-PK acts on cardiac SR. We named this 22,000-dalton protein *phospholamban* ($\lambda\alpha\mu\beta\alpha\nu\epsilon\iota\nu$ = receive), meaning "phosphate receptor" (29, 77).

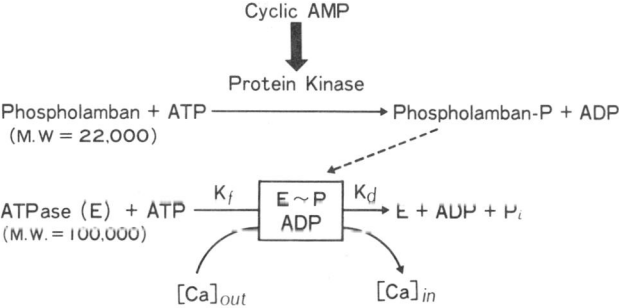

Figure 1 Control of the Ca^{2+}-dependent ATPase of the calcium pump of the cardiac SR by cAMP-dependent phosphorylation of phospholamban.

Phospholamban phosphorylation occurred in cardiac SR preparations that were virtually freed from SL vesicles by means of density gradient centrifugation (22, 43, 94). Other lower molecular weight proteins are also phosphorylated by cAMP-PKs, including a phosphoprotein of 6000 (5, 47) or 7000 (22, 48) daltons. The chemical characteristics of these smaller proteins are distinct from those of phospholamban (5). Another 15,000 (94) or 16,000 (71) dalton phosphoprotein has also been noted: This protein is phosphorylated at a threonine residue (94), while phosphorylation of phospholamban is at a serine residue (35, 48, 94). An 11,000-dalton phospho-

Table 1 Molecular weights of proteins phosphorylated by cAMP-dependent protein kinase in sarcoplasmic reticulum and sarcolemmal vesicles from cardiac muscle

Investigators year (ref.)	Sources (animals)	Sarcoplasmic reticulum Molecular weight (1000 dalton)	Sarcolemma Molecular weight (1000 dalton)
LaRaia & Morkin 1974 (45)	rabbit	20K	
Tada et al 1975 (77)	Dog	22K, 11K	
Kirchberger & Tada 1976 (34)	Dog Cat Guinea Pig Rabbit	22K, 11K	
Schwartz et al 1976 (64)	Dog	20K	
Wray & Gray 1977 (100)	Dog	20K	
Will et al 1978 (94)	Pigeon	22K, 15K	11.5K
St. Louis & Sulakhe 1979 (71)	Pig	56K, 22K, 16K	125K, 86K, 58K, 48K, 22K, 16K
Jones et al 1979 (22)	Dog	20K, 7K	165K, 90K, 56K, 24K, 11K, 7K
Lamers & Stinis 1980 (43)	Rat	24K, 9K	24K, 9K, 7K
Bidlack & Shamoo 1980 (5)	Dog	22K, 6K	
Le Peuch et al 1980 (48)	Dog	22K, 11K, 7K	
Tada et al 1981 (76)	Dog	22K, 11K	

protein that appears under some conditions may be a monomer of phospholamban (see below).

STRUCTURAL CHARACTERISTICS OF PHOSPHOLAMBAN Phospholamban is intimately associated with the SR membrane, and appears to be an acidic proteolipid (5, 48). A portion of this molecule is exposed at the surface of SR vesicles, as evidenced by the observations that it can be phosphorylated by exogenous cAMP-PK and that proteolytic digestion prevents its phosphorylation (5, 48, 77). The failure to iodinate phospholamban (54) probably reflects the location of the tyrosine residues (48) within the membrane interior.

Phospholamban was initially thought to represent a single peptide. However, addition of Triton X-100 and boiling of the phosphorylated SR solubilized in SDS resulted in replacement of the 22,000-dalton phosphoprotein by an 11,000-dalton phosphoprotein (43, 47, 48, 76). While these findings suggest the existence of a dimer-monomer transition, they could also be explained if the boiling in Triton X-100 hydrolyzed a peptide bond.

The phospholamban content of cardiac SR was estimated to be about 4% (83) or 6% (47) of the total protein. Phospholamban purified by two-dimensional gel electrophoresis contained 45% hydrophobic and 22% acidic amino acids (48).

Cyclic AMP-mediated Control of Ca Transport

ACTIVE CA TRANSPORT Marked stimulation of active Ca transport in cardiac SR accompanies the phosphorylation of phospholamban by cAMP-PK (18, 32, 35, 36, 77, 79, 81), the rates of oxalate- or phosphate-supported Ca uptake being more than doubled. Employing a rapid quenching apparatus, Will et al (93) reported that PK-catalyzed phosphorylation was capable of stimulating Ca transport in the absence of a Ca precipitating anion. Under these conditions, the stimulatory effect was apparent only during the initial phase of Ca transport (up to 300 msec after the start of the reaction). Phosphorylase kinase also was reported to stimulate the rate of Ca uptake (64), although the underlying mechanism has not been clarified.

Ca^{2+}-DEPENDENT ATPase The hydrolysis of ATP by Ca^{2+}-ATPase is tightly coupled to active Ca transport, two molecules of Ca being taken up for each molecule of ATP hydrolyzed (85). Thus the observed stimulation of Ca uptake by phosphorylation of phospholamban could reflect either enhanced turnover of the ATPase or increased efficiency of Ca transport. Tada et al (79, 82) found that cAMP-PK stimulated Ca^{2+}-ATPase activity as well as Ca uptake with the expected coupling stoichiometry of 2. Wray & Gray (100) also found stimulation of Ca^{2+}-ATPase by endogenous PK. Thus the turnover rate of a normally coupled transport ATPase is enhanced when phospholamban is phosphorylated.

CA EFFLUX Katz et al (29) reported that when cardiac SR was preincubated with cAMP-PK, the rate of EGTA-induced Ca efflux from SR loaded with ^{45}Ca was significantly increased, and the Ca^{2+} concentration needed to attain half-maximal activation of Ca efflux is reduced (37). These results indicate that the rate at which the Ca pump can run in reverse is enhanced by phospholamban phosphorylation. The relevance of these observations to the physiological increase in Ca release from the SR of the intact heart caused by catecholamines (27), however, remains uncertain.

Mechanism by Which Ca-ATPase is Regulated

ELEMENTARY STEPS OF ATPase During Ca^{2+} transport into the SR, the ATPase undergoes a complex series of reactions in which several phosphorylated intermediates (EP) are sequentially formed and degraded. Steady-state EP levels were not altered by phospholamban phosphorylation at saturating concentrations of Ca^{2+} and ATP (80, 82). However, phosphorylation of phospholamban reduced EP levels at $[Ca^{2+}]$ below 10 μM while increasing the rate of P_i liberation (v) (82). Thus, the ratio $v/[EP]$ is stimulated by phospholamban phosphorylation (82), suggesting that the rate-limiting step of EP decomposition (40, 85) is markedly enhanced. These results are consistent with the finding that the rate constant, k_d, of EP decay was enhanced (40, 82). Arrhenius plots for k_d indicated that the activation energy remained constant, with marked increase in free energy (82). At [ATP] below 5 μM, where EP formation was slowed, EP levels were also lowered by phosphorylation of phospholamban (84). These results indicate that phospholamban phosphorylation can enhance the rate of EP decomposition relative to that of EP formation when the latter is inhibited at low ATP concentrations.

The formation of EP is extremely rapid, taking place within tens of milliseconds (85). A rapid mixing device (16) allowed the rate of EP formation to be determined for reactions initiated at two different states of the ATPase. E_1 represents the state of enzyme that binds Ca (Equation 1) when SR vesicles are incubated at micromolar concentrations of ionized Ca^{2+}, while E_2 represents the state of the enzyme that does not bind Ca (Equation 1) produced when the SR is incubated with EGTA (84). In reactions initiated during the E_1 state, the initial rates of EP formation were virtually unaltered by phospholamban phosphorylation, whereas EP levels were slightly increased (40, 84). In reactions initiated during the E_2 state, where EP formation requires the initial conversion of E_2 to E_1 (see Equation 1), the initial rate of EP formation was markedly enhanced by phospholamban phosphorylation (84). Under the latter conditions, initial EP formation was much slower than when reactions were initiated at the E_1 state because conversion of E_2 to E_1 is rate-determining (84, 9a). As the conversion of E_2 to E_1 is independent of Ca^{2+} concentration, the values of $t_{1/2}$ (the time at which half of the maximal EP is formed) were shortened (from 43 to 22 msec) after phosphorylation of phospholamban (84), indicating that the conversion of E_2 to E_1 is accelerated by phospholamban phosphorylation. In accord with these findings, kinetic analysis of Ca transport indicates that the rapid initial rate of transport is markedly enhanced by phosphorylation of phospholamban when reactions were initiated at E_2, but not at E_1 (84).

The observed alterations in two distinct steps of the ATPase reaction can be interpreted in light of a current scheme of the Ca^{2+}-ATPase reaction (9a, 85):

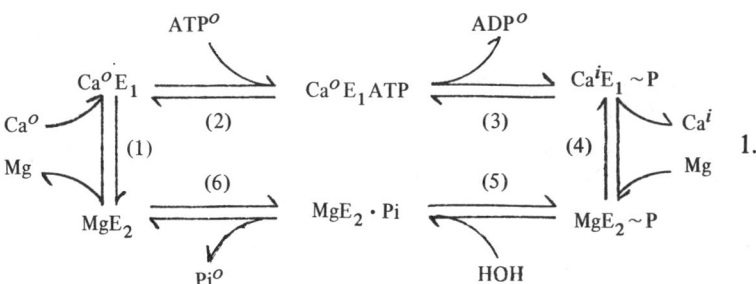

1.

E_1 and E_2 represent two different states of the ATPase (see above); i and o indicate the inside and outside of SR membranes, respectively. E_1P is the phosphorylated intermediate that has high affinity for Ca^{2+}, while E_2P has low affinity for Ca^{2+}. E_1P can donate its phosphate to ADP, thereby forming ATP, and is thus termed an ADP-sensitive phosphoenzyme. As usually measured, however, EP represents the sum of E_1P and E_2P. Under conditions where concentrations of Ca^{2+}, Mg^{2+} and KCl are saturating, E_1P usually predominates.

As pointed out above, the rate of conversion from E_2 to E_1 (Step 1), the rate-limiting step during EP formation, is accelerated when phospholamban is phosphorylated (84). The reversal rate of Step 1 ($E_1 \to E_2$) is also enhanced by phospholamban phosphorylation (82a). No change was found when the reaction was started at E_1 (40, 84) unless pH was lowered (39); however, lowering of pH can affect the control of ionized Ca^{2+} by the Ca/EGTA buffer. The finding that the rate of EP decomposition can be enhanced by phospholamban phosphorylation (40, 85) suggests that Step 4 is also accelerated, and the rate of decay of the ADP-sensitive E_1P is accelerated by phospholamban phosphorylation (82a).

Major changes in Ca^{2+}-affinity occur in two of the rate-determining steps (1 and 4) during the turnover of the ATPase that are enhanced by phospholamban phosphorylation, suggesting that phospholamban may exert its action by regulating Ca^{2+}-mediated conformational changes of the ATPase (see below). This interpretation is supported by the finding of Hicks et al (18) that phospholamban phosphorylation alters the Ca^{2+} sensitivity of the Ca pump and influences an interaction between the two Ca^{2+}-binding sites of the ATPase enzyme in a manner that reduces a positive cooperativity for Ca^{2+} found when phospholamban is in its dephospho-form (see below).

CALMODULIN-DEPENDENT PHOSPHORYLATION It was reported (38a, 47) that phosphate could be incorporated into a 22,000-dalton protein, presumably phospholamban, in the presence of Ca^{2+} and calmodulin, and that formation of this phosphoprotein (with stability characteristics of a phosphoester) was dependent on calmodulin but independent of cAMP-PK. It was postulated that phospholamban could be phosphorylated by an endogenous PK, associated with the SR, that was activated by calmodulin and Ca^{2+}. The maximal amount of phospholamban phosphorylation catalyzed by calmodulin-dependent PK at the optimal $[Ca^{2+}]$ of 5–10 μM was about the same as that within the cAMP-PK (76). While phosphorylation catalyzed by the two different PKs occurred independently, the amounts of phosphorylation were additive when SR was serially incubated under conditions favorable for each kinase (38a, 47, 76).

It was reported that preincubation of cardiac SR with calmodulin enhances the rate of calcium uptake (30) and that calmodulin-dependent phosphorylation of phospholamban is accompanied by an enhanced rate of Ca transport (38a, 47, 53). However, it is not clear how such enhancement is related to cAMP-mediated augmentation of active Ca transport. It was suggested (47,100) that cAMP-mediated control of active Ca transport is not seen unless calmodulin-dependent phosphorylation also occurs. In contrast, Tada et al (76, 82a) and Hicks et al (18) found that cAMP-mediated enhancement of Ca transport could be seen even when phosphorylation was carried out under Ca^{2+}-free conditions (pCa > 8). The Ca^{2+} concentrations employed in the Ca uptake assay by Le Peuch et al (47) were high (0.1 mM) and might mask the cAMP-mediated enhancement of Ca uptake by SR seen at $[Ca^{2+}]$ below 10 μM (18, 76, 79). Thus, at the physiological $[Ca^{2+}]$ (0.1–10 μM), stimulation of Ca transport by either cAMP-PK or calmodulin-PK systems appears to occur independently. It has been reported that these effects on Ca transport can be additive (29a, 76, 82a), in accord with the finding that phosphorylation by the two kinases occurred independently; but it is not known whether this effect is fully additive (38a, 76).

Le Peuch et al (47) suggested that calmodulin-dependent phosphorylation of phospholamban does not affect Ca^{2+}-ATPase activity at high Ca^{2+} concentrations, whereas others found that at physiological intracellular Ca^{2+} concentrations (0.1–50 μM), such phosphorylation stimulated the Ca^{2+}-dependent ATPase (29a, 53, 76). S. Katz (29a) has observed effects of calmodulin-dependent phosphorylation on the Ca^{2+}-dependence of Ca transport by cardiac SR similar to those of the cAMP-PK (18). The lack of effects of the calmodulin-dependent phosphorylation on EP level, and the resulting enhancement of $v/[EP]$, along with an increased rate of EP decomposition (29a) are also similar to findings obtained after cAMP-dependent phosphorylation (see above).

MOLECULAR MODEL   It is generally agreed that cAMP-dependent phosphorylation of phospholamban profoundly alters Ca transport by cardiac SR. Since this effect is accompanied by significant changes in both the Ca^{2+}-dependent (18, 79) and kinetic (40, 77, 82, 84) properties of the Ca^{2+}, Mg^{2+}-dependent ATPase, it can be assumed that phospholamban is closely associated with the Ca pump ATPase protein. Molecular models have been presented in which phosphorylated phospholamban was suggested to serve as an activator (80) or a derepressor (18, 28) of this ATPase. In light of more recent information, we now present a newer working model to describe the interaction between these two membrane proteins (Figure 2).

The ATPase is an amphipathic single polypeptide with a molecular weight of ~100,000 that functions, either as a monomer or an oligomer, to affect the ATP-dependent translocation of Ca^{2+} across the SR membrane (85, 102). A protein–protein interaction has been suggested to occur between phospholamban and the ATPase (18, 82, 84), although an alternative mode of interaction such as that involving the membrane lipids is also possible. This association may be increased when phospholamban is in its phosphorylated state (5, 83). A direct molecular interaction between these proteins is supported by the finding that the ATPase enzyme and phospholamban remain associated with each other after solubilization of SR with detergents (5, 48, 83). Although part of phospholamban is located at the cytoplasmic surface of SR membrane (48, 77), a large portion of the molecule appears to be imbedded within the membrane (5, 48, 54). After phosphorylation, the reactive serine group may be translocated towards the interior of the membrane (5, 77), although this site remains exposed to phosphoprotein phosphatases (33, 78). A 1:1 stoichiometry between phospholamban and the ATPase is suggested by the approximately 1:1 ratio

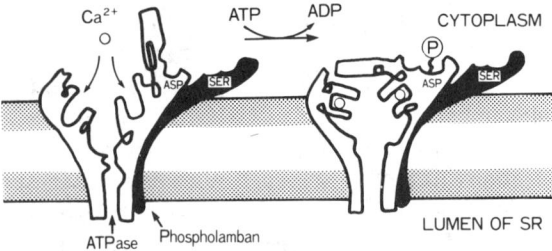
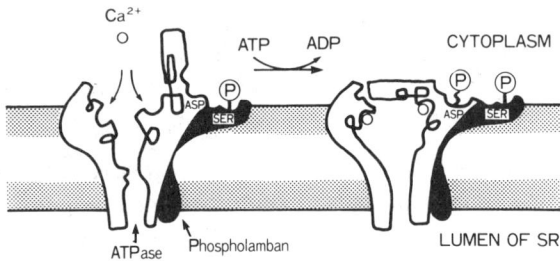

Figure 2 Diagrammatic representation of the action of phospholamban on the calcium pump of the cardiac SR. *ASP* represents an aspartic acid residue in the ATPase and *SER* a serine residue in phospholamban. The phosphates attached to these amino acids are $\sim$ P, the acyl phosphate ATPase intermediate, and -P, the regulatory phosphoester. See text for further explanation.

between the amounts of the two phosphoproteins formed within the membrane (27, 47, 82, 83).

During Ca^{2+} translocation, the region of the ATPase molecule containing the active ATPase site may undergo a conformational change that causes the Ca-binding moiety to be translocated from the outer to the inner surface of the membrane (85, 102). Alternatively, the environment around the bound Ca^{2+} may change from that of the outer surface of the SR membrane to that of the vesicle interior. Since both formation (84) and decomposition (40, 82) of EP are enhanced by phosphorylation of phospholamban, the conformational state of the ATPase region of the molecule appears to be under the direct control of phospholamban. Thus, when phosphorylated, phospholamban enhances the rate of the cation-mediated conformational changes of the ATPase region that results in an increased rate of translocation of Ca^{2+}. These changes appear to be accompanied by changes in both the Ca^{2+}-affinity and cooperativity of the two Ca^{2+}-binding sites of the ATPase (18).

A clue to whether dephospho-phospholamban acts as an inhibitor of the cardiac SR, or whether phosphorylated phospholamban is an activator, can be found in the study by Hicks et al (18), who compared the Ca^{2+} dependence of cardiac SR vesicles before and after cAMP-PK phosphorylation with that of a comparable preparation from rabbit fast skeletal muscle. The latter tissue appears to lack phospholamban (34) and cannot be stimulated by cAMP-PK catalyzed phosphorylation. Hicks et al (18) confirmed earlier findings (64a) that the Ca^{2+} sensitivity of the nonphosphorylated cardiac SR was significantly less than that of catecholamine-insensitive skeletal muscle, but found that the Ca^{2+} dependence of Ca transport by the cardiac SR after phospholamban phosphorylation closely resembled that of the skeletal preparations. The finding that the Ca^{2+} sensitivity of phosphorylated cardiac SR vesicles is similar to that of skeletal SR vesicles that appear to lack phospholamban suggests that the dephospho-form of phospholamban is an inhibitor of the cardiac Ca pump (18). However, the possibility that the skeletal preparations contain an activator similar to phosphorylated phospholamban cannot be ruled out.

Physiological Relevance of Phospholamban

We first proposed that cAMP-mediated acceleration of calcium uptake by cardiac SR may explain two of the major mechanical effects of catecholamines on heart muscle: abbreviation of systole and increased contractility (28a, 29, 77, 79). A number of findings support this view [see (27, 71a, 82a and 87) for review]. Thus the cAMP-mediated increase in the rate of Ca accumulation by SR could explain the faster rate of relaxation because of the increased rate at which Ca^{2+} would be removed from troponin in the catecholamine-stimulated myocardium. Enhanced accumulation of Ca^{2+} could increase the amount of Ca stored within the SR by retaining within the cell some of the Ca that would otherwise be lost during diastole (28a). This could increase the amount of Ca^{2+} available for delivery to the contractile proteins in subsequent contractions, thereby producing the inotropic effect of catecholamines. A third mechanical effect of catecholamines, namely acceleration of tension rise, could also be explained by the acceleration of Ca release from the SR when phospholamban is phosphorylated (29, 37). The abbreviation of systole seen in the catecholamine-stimulated heart might be explained by an action of phosphorylated phospholamban to accelerate spontaneous transitions from Ca uptake to Ca release in the SR (27). This proposed relationship between the cAMP-mediated changes in Ca transients is supported by experiments in intact hearts, in which catecholamines rapidly accelerate relaxation followed by augmentation of contraction (56).

The onset of increased tension development after exposure of the heart to catecholamines is gradual, reaching steady state after about 20 beats (62). Allen & Blinks (1), who used aequorin, a Ca-sensitive bioluminescent protein, to measure cytosolic Ca^{2+}, found that catecholamines augmented both the rate of Ca release into the cytosol during the initial phase of contraction and the rate of reduction of Ca^{2+} during relaxation. Employing a skinned cardiac cell preparation that exhibited cycles of phasic contractions upon addition of appropriate amounts of Ca^{2+} (14), Fabiato & Fabiato (13) demonstrated that brief preincubation with cAMP increased both the amplitude of contraction and the rates of tension development and relaxation while at the same time shortening overall contraction duration. These findings provide convincing evidence that changes in the SR associated with phospholamban phosphorylation can underlie these important mechanical effects of catecholamines.

A role for cAMP in mediating the inotropic effect of catecholamines is indicated by an increase in cAMP levels prior to the increase in contractility [see (87) for review]. It has been reported, however, that contractility could be increased with no detectable increase in cAMP by catecholamines covalently attached to glass beads (21, 90). These latter findings might reflect either a very slight increase in cAMP; or the action of a cAMP–PK system that is intimately related to the SR membrane. Immunocytochemical studies have shown that cAMP (59) and possibly PK (67) may be concentrated near the SR and SL. That phospholamban can be phosphorylated by catecholamines in vivo is indicated by the findings [(46, 52); A. Schwartz, E. G. Kranias, personal communication] that the addition of isoproterenol to ^{32}P-perfused hearts increased ^{32}P incorporation into a 22,000-dalton microsomal protein at the same time that the rates of tension development and relaxation were increased under conditions where a calmodulin inhibitor (fluphenazine) reduced Ca^{2+}-dependent phosphorylation of phospholamban (46). In hyperthyroid animals, the amount of phospholamban in cardiac microsomes was increased (50), as was the rate of Ca uptake (50, 51). Stimulation of Ca uptake in heart microsomes produced by the cAMP-phospholamban system was not altered in aged rat hearts, although aging reduced the rate of unstimulated microsomal Ca uptake (23).

There is growing evidence that proteins similar to the phospholamban system play a regulatory role in tissues other than cardiac muscle. Microsomes isolated from slow-contracting skeletal muscle contain a protein similar to phospholamban that, when phosphorylated by cAMP-PK, is associated with stimulated Ca uptake; neither a phosphorylatable 22,000-dalton protein nor Ca uptake stimulation was seen in fast-contracting skeletal muscle (29, 34). These findings are paralleled by a weak relaxation-promoting effect of catecholamines in the slow twitch muscle, but no such effect in fast muscle (see 27, 29), supporting the existence of a causal relationship between phospholamban phosphorylation and the acceleration of relaxation by catecholamines. However, Schwartz et al (64), who confirmed these findings with cardiac and slow skeletal muscles, found that a cAMP-PK increased the rate of Ca uptake by SR from fast-contracting skeletal muscle without accompanying membrane phosphorylation. This discrepancy has not been resolved.

A regulatory mechanism involving phosphorylation of a protein similar to phospholamban may also operate in platelets (17, 25) and in vascular (4, 15a) and visceral (1a, 31, 58a) smooth muscles. In platelet microsomes, a phospholamban-like protein was suggested to regulate Ca accumulation (24).

PHOSPHORYLATION OF THE SARCOLEMMA

Protein Kinase–Catalyzed Phosphorylation of Cardiac Sarcolemma

The status and physiological significance of sarcolemmal (SL) phosphorylation are more controversial than the studies of the SR reviewed above. The discrepancies that have arisen in studies of the SL are due primarily to difficulties in purifying these membranes, and to the relatively low concentrations of individual functional and regulatory proteins in the SL membrane.

A NaI-treated microsomal preparation from pig heart enriched in Na^+,K^+-ATPase was phosphorylated by both endogenous and exogenous cAMP-PKs, two membrane proteins of 24,000 and 90,000 daltons being phosphorylated (41). Phosphorylation of the lower molecular weight protein depended on cAMP. Phosphorylation of a different protein (11,700 daltons) was observed in a similar preparation that was treated with NaI and deoxycholate (11). A 93,000-dalton protein in this preparation (11) and a similar preparation from kidney (12) were also phosphorylated, though to a much less extent.

SL vesicles isolated from guinea-pig ventricles were found to be phosphorylated by both endogenous and exogenous cAMP-PKs (20, 68). This phosphoprotein exhibited the characteristics of a phosphoester, and was hydrolyzed by alkaline phosphatase (41, 71, 94). Incubation of SL from a variety of species with cAMP-PKs demonstrated a number of phosphorylated protein bands of different molecular weights (Table 1). Although lower molecular weight proteins (9000–24,000 daltons) were preferentially phosphorylated (22, 43, 71, 94), other proteins also could be phosphorylated.

Will et al (94) demonstrated that cardiac microsomes containing SR and SL vesicles exhibited cAMP phosphorylation of three proteins with molecular weights of 22,000, 15,000, and 11,500. When SL was separated from SR by density gradient centrifugation, the 11,500-dalton phosphoprotein was enriched in SL vesicles, whereas SR vesicles contained the 22,000- and

15,000-dalton phosphoproteins (94). Phosphorylation of the 11,500-dalton protein in SL occurred mainly at serine residues, and was catalyzed by an endogenous PK (22, 71, 94) in contrast to phosphorylation of the SR that was markedly enhanced by exogenous PK. Jones et al (22) and St. Louis & Sulakhe (71) reported that other SL proteins of 24,000 (22) or 22,000 (71) daltons were phosphorylated to a similar extent by endogenous PK (Table 1), and that addition of exogenous PK augmented their phosphorylation. Together, these findings suggest that the two membrane-bound peptides of 9000–16,000 daltons (22, 71, 94) and 22,000–24,000 daltons (22, 43, 94) serve as the major substrates for PK in the SL, although other proteins with higher molecular weights are also phosphorylated (see below). It is possible that the SL protein of 9000–11,500 daltons is identical with a lipoprotein (12,300 daltons) that binds Ca^{2+} (15).

Lamers & Stinis (43) found that 24,000- and 9000-dalton proteins could be phosphorylated by cAMP-PKs, and that the 24,000-dalton phosphoprotein was completely dissociated into 9000-dalton peptides after phosphorylated SL solubilized in SDS was heated to 95° C. These results, which suggest that the 24,000-dalton SL protein may be composed of two or more subunits and that the 24,000-dalton and 9000-dalton proteins are interconvertible (43), are analogous to recent findings suggesting that phospholamban in the SR may also be a dimer (see above). Endogenous phosphorylation of SL proteins with higher molecular weights, between 110,000 and 165,000 (22, 71), has also been observed. The phosphorylatable protein of 90,000–110,000 daltons (22, 71) may correspond to one of the two peptides of the Na^+,K^+-ATPase (11, 12), while the 56,000–58,000 and 86,000–90,000 dalton phosphoproteins (22, 71) may reflect autophosphorylation of the regulatory subunits of 55,000 daltons (monomer) and 98,000 daltons (dimer) of an endogenous PK.

Phosphorylase kinase has also been reported to catalyze a Ca^{2+}-dependent phosphorylation of cardiac SL (73). The molecular weights of these phosphoproteins, which have the characteristics of a phosphoester, are approximately 95,000, 70,000, 48,000 and 35,000. At least some of these phosphoproteins may arise from contaminant glycogen-protein particles (55a).

Cardiac SL preparations contain an endogenous phosphoprotein phosphatase activity capable of catalyzing the dephosphorylation of phosphoproteins (20, 71), raising the possibility that phosphoester phosphoproteins can be hydrolyzed in vivo by a membrane-bound phosphatase. The mechanisms that regulate this phosphatase activity are unclear, and it is not known whether different phosphatases are required to hydrolyze each of the phosphoproteins.

Walsh et al (91) observed that two SL proteins of 36,000 and 27,000 daltons were phosphorylated in SL vesicles isolated from rat hearts perfused with inorganic ^{32}P, and that phosphorylation of these membranes increased when the hearts were exposed to epinephrine. A 27,000-dalton protein in isolated SL also was phosphorylated by a cAMP-PK. These studies indicate that epinephrine leads to PK-catalyzed phosphorylation of an SL protein that may play a role in the inotropic response to catecholamines. The relationship between this phosphoprotein and phospholamban of the SR remains to be clarified.

Calcium Movements Across Sarcolemmal Membranes

CALCIUM PUMP AND Ca^{2+}, Mg^{2+}-DEPENDENT ATPase   The cardiac SL appears to contain two mechanisms that can pump Ca^{2+} out of the cell and thereby maintain a low cytosolic Ca^{2+} concentration in the presence of the transsarcolemmal Ca^{2+} influx that occurs during excitation-contraction coupling (44). While it appears that the major portion of this active Ca^{2+} efflux is effected by exchange with Na^+ (Na-Ca exchange) (44, 61), a number of studies indicate that the cardiac SL also contains an ATP-energized Ca pump.

Preparations of cardiac SL have long been known to exhibit Ca^{2+},Mg^{2+}-ATPase activity (66, 75a), but possible contamination of these membranes with SR fragments precluded concluding that a Ca pump existed in the SL. Recently, improved procedures for the isolation of SL membranes have led to a renewed search for an ATP-energized Ca pump (74, 103). Cardiac SL vesicles, isolated by these more elaborate procedures (68), exhibit ATP-dependent Ca^{2+} binding in the presence of Mg^{2+} and micromolar concentrations of Ca^{2+}. Since this process requires ATP, which is hydrolyzed during Ca binding, SL vesicles were suggested to exhibit active Ca^{2+} transport (20, 68, 103). However, other investigators (10, 88) failed to detect Ca^{2+}, Mg^{2+}-dependent ATPase activity.

SL vesicles exhibit ATP-dependent Ca accumulation in the presence of oxalate (20, 68, 72, 96), indicating the existence of a transmembrane Ca^{2+} flux that leads to Ca oxalate precipitation within the vesicles. While these data suggest that the SL contains an ATP-energized Ca pump, others (6, 8, 9, 22, 86) have reported that oxalate was ineffective in facilitating ATP-dependent Ca transport. These discrepancies could be explained, at least in part, by differences in the amount of SR contamination.

Recent reports that Ca binding accompanies Ca^{2+},Mg^{2+}-ATP hydrolysis (55, 58) did not always distinguish actively transported Ca^{2+} from Ca^{2+} that remained associated with (bound to) the outer surface of these membranes (see below). More convincing evidence that Ca^{2+} can be actively transported into SL vesicles was presented by studies in which Ca transport into the vesicles was operationally distinguished from Ca bound to their external surface by means of washing with La^{3+} (8) or EGTA (86). Under these conditions, the calcium ionophore A21387 completely released the Ca taken up into the vesicles. The properties of the Ca transport and its associated Ca^{2+},Mg^{2+}-dependent ATP hydrolysis differed from the Na-Ca exchange and Na^+,K^+-ATPase (8, 86). The Ca^{2+},Mg^{2+}-ATPase, however, was found to be co-purified with Na^+,K^+-ATPase after solubilization and revesiculation of SL-enriched membranes (57). Future work will be needed to determine whether the Ca^{2+}, Mg^{2+}-ATPase thought to mediate translocation of Ca^{2+} can form phosphorylated intermediates such as are seen in the Ca pump ATPase of SR (see above) and erythrocyte membranes (38), and whether a Ca^{2+}-dependent protein activator (calmodulin) found in erythrocyte membranes (38) can modulate the Ca^{2+}-dependency of the Ca pump ATPase in cardiac SL (9).

ATP-INDEPENDENT CALCIUM BINDING   Two classes of ATP-independent Ca-binding sites have been described in purified preparations of cardiac SL: high-affinity sites ($K_D = 0.1$-5 μM) and a greater number of low-affinity sites ($K_D = 0.3$-2.0 mM) (10, 49, 95, 97). High-affinity Ca binding was suggested to occur at protein component(s) of the SL because pretreatment with trypsin decreased the amount of bound Ca (41, 95). This activity was suggested to reside in a protein of approximately 100,000 daltons (49). Feldman & Weinhold (15) reported that a Ca-binding fraction contained a lipoprotein complex of about 71,400 daltons with a K_D for Ca^{2+} of 74 μM and a 12,300-dalton protein moiety. They suggested that this lipoprotein complex might function as a membrane storage site for Ca^{2+} prior to its transport into the muscle cells. While these Ca-binding activities cannot yet be assigned with confidence to any membrane functions, it is intriguing to speculate that the low-affinity Ca-binding sites may represent a point of entry for Ca^{2+} at the outer surface of the SL, while the high-affinity sites may represent Ca-binding sites at the inner surface.

Functional Implication of Sarcolemmal Phosphorylation

CALCIUM PUMP   The rate of Ca^{2+}-dependent ATP hydrolysis by SL preparations was augmented after phosphorylation by cAMP-PK (20, 104), while ATP-dependent Ca binding was unaffected (20, 72). Wollenberger et al (99), however, found an increase in Ca binding to a detergent-treated SL preparation. An early study attempted to correlate the increased Ca binding with cAMP-dependent phosphorylation of a 24,000-dalton protein (99), but the phosphorylation of other SL proteins (Table 1) could participate in the observed effect.

Enhancement of oxalate-dependent Ca uptake by cardiac SL vesicles after phosphorylation by PK remains controversial (see above), and it is possible that this transport activity may not arise from SL. The existence of a calmodulin-activatable SL Ca pump (ATPase) was recently reported (103). It remains to be seen whether calmodulin could exert its action directly (103) or via protein phosphorylation (104).

Phosphorylase kinase–catalyzed phosphorylation of the SL was also reported to increase the rate of Ca accumulation (73) without an increase in Ca^{2+}-ATPase activity, although ouabain-sensitive Na^+,K^+-ATPase activity was enhanced (69). It is thus possible that stimulation of the Na^+,K^+-ATPase may increase Ca accumulation indirectly, possibly by increasing Na-Ca exchange (98).

ATP-INDEPENDENT CALCIUM BINDING   Phosphorylation of SL proteins by cAMP and endogenous PKs has been reported to enhance ATP-independent Ca binding (41, 95, 99), but the physiological significance of these findings is not yet clear. These effects may reflect changes in a SL Ca pump, or in Ca storage sites at the outer surface of the SL membrane.

CALCIUM CHANNEL   The slow inward current, which contributes to the plateau phase of the cardiac action potential, is due largely to a Ca-selective ion channel. Cyclic AMP derivatives, as well as β-adrenergic agonists, increase both the plateau height and duration of the action potential (62), and catecholamines have been shown to increase the number of available Ca channels (60, 63, 63a). Sperelakis & Schneider (65) proposed a model of the cardiac SL in which cAMP-dependent phosphorylation of a membrane protein increased the number of Ca channels that could open during each action potential; Reuter (63) suggested that the dephospho-form of a voltage-independent gate inactivated Ca channels, and that phosphorylation of this gate enabled the channel to allow Ca entry. One or more of the phosphorylatable peptides found in isolated cardiac SL membranes may represent this putative modulator of slow channel activity. The 11,500-dalton protein, which resembles the 12,300-dalton protein moiety of a lipoprotein complex with high Ca-binding capacity (15), was suggested by Wollenberger & Will (98) to play a functional role because its phosphorylation could increase the Ca^{2+} affinity of the SL (41, 95, 99). However, further work will be needed to define the relationship between these proteins in isolated SL preparations and the mechanism by which cAMP regulates the electrophysiological properties of the Ca channels of the SL.

ACKNOWLEDGMENTS

The writing of this article and recent work of the authors were supported by the Ministry of Education, Science and Culture, Japan, and the Muscular Dystrophy Association (to M. T.) and by National Institutes of Health Grants HL-21812 and HL-22135 and the Connecticut and American Heart Associations (to A.M.K.).

Literature Cited

1. Allen, D. G., Blinks, J. R. 1978. Calcium transients in aequorin-injected frog cardiac muscle. *Nature* 273:509–13
1a. Andersson, R., Nilsson, K., Wikberg, J., Johansson, S., Mohme-Lundholm, E., Lundholm, L. 1975. Cyclic nucleotides and the contraction of smooth muscle. *Adv. Cyclic Nucl. Res.* 5:491–518
2. Bers, D. M., Philipson, K. D., Nishimoto, A. Y. 1980. Sodium-calcium exchange and sidedness of isolated cardiac sarcolemmal vesicles. *Biochim. Biophys. Acta* 601:358–71
3. Besch, H. R. Jr., Jones, L. R., Watanabe, A. M. 1976. Intact vesicles of canine cardiac sarcolemma: evidence from vectorial properties of Na^+,K^+-ATPase. *Circ. Res.* 39:586–95
4. Bhalla, R. C., Webb, R. C., Singh, D., Brock, T. 1978. Role of cyclic AMP in rat aortic microsomal phosphorylation and calcium uptake. *Am. J. Physiol.* 234:H508–14
5. Bidlack, J. M., Shamoo, A. E. 1980. Adenosine 3',5'-monophosphate-dependent phosphorylation of a 6000 and a 22,000 dalton protein from cardiac sarcoplasmic reticulum. *Biochim. Biophys. Acta* 632:310–25
6. Brandt, N. R., Caswell, A. H. 1979. ATP-energized Ca^{2+} pump in transverse tubules isolated from rabbit skeletal muscle. *Fed. Proc.* 38:1442 (Abstr.)
7. Brostrom, M. A., Reimann, E. M., Walsh, D. A., Krebs, E. G. 1970. A cyclic 3',5'-AMP-stimulated protein kinase from cardiac muscle. *Adv. Enzyme Regul.* 8:191–203
8. Caroni, P., Carafoli, E. 1980. An ATP-dependent Ca^{2+}-pumping system in dog heart sarcolemma. *Nature* 283:765–67
9. Caroni, P., Malmstroem, K., Carafoli, E. 1980. An ATP-dependent Ca^{2+} transporting system in heart sarcolemma. In *Calcium-Binding Proteins: Structure and Function*, ed. F. L. Siegel, E. Carafoli, R. H. Kretsinger, D. H. MacLennan, R. H. Wasserman, pp. 145–46. Amsterdam & NY: Elsevier North Holland
9a. De Meis, L., Vianna, A. L. 1979. Energy interconversion by the Ca^{2+}-dependent ATPase of the sarcoplasmic reticulum. *Ann. Rev. Biochem.* 48:275–92
10. Dhalla, N. S., Anand, M. B., Harrow, J. A. C. 1976. Calcium binding and ATPase activities of heart sarcolemma. *J. Biochem. Tokyo* 79:1345–50
11. Dowd, F. J. Jr., Pitts, B. J. R., Schwartz, A. 1976. Phosphorylation of a low molecular weight polypeptide in beef heart Na^+,K^+-ATPase preparations. *Arch. Biochem. Biophys.* 175:321–31
12. Dowd, F., Schwartz, A. 1975. The presence of cyclic AMP-stimulated protein kinase substrates and evidence for endogenous protein kinase activity in various Na^+,K^+-ATPase preparations from brain, heart and kidney. *J. Mol. Cell. Cardiol.* 7:483–97
13. Fabiato, A., Fabiato, F. 1975. Relaxing and inotropic effects of cyclic AMP on skinned cardiac cells. *Nature* 253:556–58
14. Fabiato, A., Fabiato, F. 1977. Calcium release from sarcoplasmic reticulum. *Circ. Res.* 40:119–29
15. Feldman, D. A., Weinhold, P. A. 1977. Calcium binding to rat heart plasma membranes: isolation and purification of a lipoprotein component with a high calcium binding capacity. *Biochemistry* 16:3470–75
15a. Fitzpatrick, D. F., Szentivanyi, A. 1977. Stimulation of calcium uptake into aortic microsomes by cyclic AMP and cyclic AMP-dependent protein kinase. *Naunyn-Schmiedeberg's Arch. Pharmacol.* 298:255–57
16. Froehlich, J. P., Sullivan, J. V., Berger, R. L. 1976. A chemical quenching apparatus for studying rapid reactions. *Anal. Biochem.* 73:331–41
17. Haslam, R. J., Davidson, M. M. L., Davies, T., Lynham, J. A., McClenaghan, M. D. 1978. Regulation of blood platelet function by cyclic nucleotides. *Adv. Cyclic Nuc. Res.* 9:533–52
18. Hicks, M. J., Shigekawa, M., Katz, A. M. 1979. Mechanism by which cyclic adenosine 3':5'-monophosphate-dependent protein kinase stimulates calcium transport in cardiac sarcoplasmic reticulum. *Circ. Res.* 44:384–91
19. Hobbs, A. S., Albers, R. W. 1980. The structure of proteins involved in active membrane transport. *Ann. Rev. Biophys. Bioeng.* 9:259–91
20. Hui, C.-W., Drummond, M., Drummond, G. I. 1976. Calcium accumulation and cyclic AMP-stimulated phosphorylation in plasma membrane-enriched preparations of myocardium. *Arch. Biochem. Biophys.* 173:415–27
21. Ingebretsen, W. R. Jr., Becker, E., Friedman, W. F., Mayer, S. E. 1977. Contractile and biochemical responses of cardiac and skeletal muscle to iso-

proterenol covalently linked to glass beads. *Circ. Res.* 40:474–84
22. Jones, L. R., Besch, H. R. Jr., Fleming, J. W., McConnaughey, M. M., Watanabe, A. M. 1979. Separation of vesicles of cardiac sarcolemma from vesicles of cardiac sarcoplasmic reticulum. *J. Biol. Chem.* 254:530–39
23. Kadoma, M., Sacktor, B., Froehlich, J. P. 1980. Stimulation by cAMP and protein kinase of calcium transport in sarcoplasmic reticulum from senescent rat myocardium. *Fed. Proc.* 39:2040 (Abstr.)
24. Käser-Glanzmann, R., Gerber, E., Lüscher, E. F. 1979. Regulation of the intracellular calcium level in human blood platelets: cyclic adenosine 3′,5′-monophosphate dependent phosphorylation of a 22,000 dalton component in isolated Ca^{2+}-accumulating vesicles. *Biochim. Biophys. Acta* 558:344–47
25. Käser-Glanzmann, R., Jakábová, M., George, J. N., Lüscher, E. F. 1977. Stimulation of calcium uptake in platelet membrane vesicles by adenosine 3′,5′-cyclic monophosphate and protein kinase. *Biochim. Biophys. Acta* 466:429–40
26. Katz, A. M. 1977. Excitation-contraction coupling. In *Physiology of the Heart*, pp. 137–59. NY: Raven. 450 pp.
27. Katz, A. M. 1979. Role of the contractile proteins and sarcoplasmic reticulum in the response of the heart to catecholamines: an historical review. *Adv. Cyclic. Nuc. Res.* 11:303–43
28. Katz, A. M. 1980. Relaxing effects of catecholamines in the heart. *Trends Pharmacol. Sci.* 1:434–36
28a. Katz, A. M., Repke, D. I. 1973. Calcium-membrane interactions in the myocardium: Effects of ouabain, epinephrine and 3′,5′-cyclic adenosine monophosphate. *Am. J. Cardiol.* 31:193–201
29. Katz, A. M., Tada, M., Kirchberger, M. A. 1975. Control of calcium transport in the myocardium by the cyclic AMP-protein kinase system. *Adv. Cyclic Nuc. Res.* 5:453–72
29a. Katz, S. 1980. Mechanism of stimulation of calcium transport in cardiac sarcoplasmic reticulum preparations by calmodulin. *Ann. NY Acad. Sci.* 356:267–78
30. Katz, S., Remtulla, M. A. 1978. Phosphodiesterase protein activator stimulates calcium transport in cardiac microsomal preparations enriched in sarcoplasmic reticulum. *Biochem. Biophys. Res. Commun.* 83:1373–79
31. Kimura, M., Kimura, I., Kobayashi, S. 1977. The activation of cyclic 3′,5′-adenosine monophosphate-dependent protein kinase on sarcoplasmic reticulum fractions of various smooth muscles and its related novel relaxants. *Biochem. Pharmacol.* 26:994–96
32. Kirchberger, M. A., Chu, G. 1976. Correlation between protein kinase-mediated stimulation of calcium transport by cardiac sarcoplasmic reticulum and phosphorylation of a 22,000 dalton protein. *Biochim. Biophys. Acta* 419:559–62
33. Kirchberger, M. A., Raffo, A. 1977. Decrease in calcium transport associated with phosphoprotein phosphatase-catalyzed dephosphorylation of cardiac sarcoplasmic reticulum. *J. Cyclic Nuc. Res.* 3:45–53
34. Kirchberger, M. A., Tada, M. 1976. Effects of adenosine 3′:5′-monophosphate-dependent protein kinase on sarcoplasmic reticulum isolated from cardiac and slow and fast contracting skeletal muscles. *J. Biol. Chem.* 251:725–29
35. Kirchberger, M. A., Tada, M., Katz, A. M. 1974. Adenosine 3′:5′-monophosphate-dependent protein kinase–catalyzed phosphorylation reaction and its relationship to calcium transport in cardiac sarcoplasmic reticulum. *J. Biol. Chem.* 249:6166–73
36. Kirchberger, M. A., Tada, M., Repke, D. I., Katz, A. M. 1972. Cyclic adenosine 3′,5′-monophosphate-dependent protein kinase stimulation of calcium uptake by canine cardiac microsomes. *J. Mol. Cell. Cardiol.* 4:673–80
37. Kirchberger, M. A., Wong, D. 1978. Calcium efflux from isolated cardiac sarcoplasmic reticulum. *J. Biol. Chem.* 253:6941–45
38. Klee, C. B., Crouch, T. H., Richman, P. G. 1980. Calmodulin. *Ann. Rev. Biochem.* 49:489–515
38a. Kranias, E. G., Bilezikjian, L. M., Potter, J. D., Piasick, M. T., Schwartz, A. 1980. The role of calmodulin in regulation of cardiac sarcoplasmic reticulum phosphorylation. *Ann. NY Acad. Sci.* 356:279–90
39. Kranias, E. G., Mandel, F., Schwartz, A. 1980. Involvement of cAMP-dependent protein kinase and pH on the regulation of cardiac sarcoplasmic reticulum. *Biochem. Biophys. Res. Commun.* 92:1370–76
40. Kranias, E. G., Mandel, F., Wang, T., Schwartz, A. 1980. Mechanism of the stimulation of calcium ion dependent adenosine triphosphatase of cardiac sar-

coplasmic reticulum by adenosine 3',5'-monophosphate dependent protein kinase. *Biochemistry* 19:5434–39
41. Krause, E.-G., Will, H., Schirpke, B., Wollenberger, A. 1975. Cyclic AMP-enhanced protein phosphorylation and calcium binding in a cell membrane-enriched fraction from myocardium. *Adv. Cyclic Nuc. Res.* 5:473–90
42. Krebs, E. G. 1972. Protein kinases. *Curr. Top. Cell. Regul.* 5:99–133
43. Lamers, J. M. H., Stinis, H. T. 1980. Phosphorylation of low molecular weight proteins in purified preparations of rat heart sarcolemma and sarcoplasmic reticulum. *Biochim. Biophys. Acta* 624:443–59
44. Langer, G. A. 1973. Heart: excitation-contraction coupling. *Ann. Rev. Physiol.* 35:55–86
45. LaRaia, P. J., Morkin, E. 1974. Adenosine 3',5'-monophosphate-dependent membrane phosphorylation. A possible mechanism for the control of microsomal calcium transport in heart muscle. *Circ. Res.* 35:298–306
46. Le Peuch, C. J., Guilleux, J.-C., Demaille, J. G. 1980. Phospholamban phosphorylation in the perfused rat heart is not solely dependent on β-adrenergic stimulation. *FEBS Lett.* 114:165–68
47. Le Peuch, C. J., Haiech, J., Demaille, J. G. 1979. Concerted regulation of cardiac sarcoplasmic reticulum calcium transport by cyclic adenosine monophosphate dependent and calcium-calmodulin-dependent phosphorylations. *Biochemistry* 18:5150–57
48. Le Peuch, C. J., Le Peuch, D. A. M., Demaille, J. G. 1980. Phospholamban, activator of the cardiac sarcoplasmic reticulum calcium pump. Physicochemical properties and diagonal purification. *Biochemistry* 19:3368–73
49. Limas, C. J. 1977. Calcium-binding sites in rat myocardial sarcolemma. *Arch. Biochem. Biophys.* 179:302–9
50. Limas, C. J. 1978. Enhanced phosphorylation of myocardial sarcoplasmic reticulum in experimental hyperthyroidism. *Am. J. Physiol.* 234:H426–31
51. Limas, C. J. 1978. Calcium transport ATPase of cardiac sarcoplasmic reticulum in experimental hyperthyroidism. *Am. J. Physiol.* 235:H745–51
52. Lindemann, J. P., Jones, L. R., Watanabe, A. M. 1980. Effects of muscarinic cholinergic agonists on beta receptor-induced increases in membrane protein phosphorylation in intact ventricles. *Clin. Res.* 28:471A (Abstr.)
53. Lopaschuk, G., Richter, B., Katz, S. 1980. Characterization of calmodulin effects on calcium transport in cardiac microsomes enriched in sarcoplasmic reticulum. *Biochemistry* 19:5603–7
54. Louis, C. F., Katz, A. M. 1977. Lactoperoxidase-coupled iodination of cardiac sarcoplasmic reticulum proteins. *Biochim. Biophys. Acta* 494:255–65
55. Mas-Oliver, J., Williams, A. J., Nayler, W. G. 1979. ATP-induced stimulation of calcium binding to cardiac sarcolemma. *Biochem. Biophys. Res. Commun.* 87:441–47
55a. Michalak, M., Sarzala, M. G., Drabikowski, W. 1977. Sarcoplasmic reticulum vesicles and glycogen-protein particles in microsomal fraction of skeletal muscle. *Acta Biochem. Pol.* 24:105–16
56. Morad, M., Rolett, E. L. 1972. Relaxing effects of catecholamines on mammalian heart. *J. Physiol. London* 224:537–58
57. Morcos, N. C., Drummond, G. I. 1980. ($Ca^{2+}+Mg^{2+}$)-ATPase in enriched sarcolemma from dog heart. *Biochim. Biophys. Acta* 598:27–39
58. Morcos, N. C., Jacobson, A. L. 1978. Interaction of sodium with the sarcolemmal calcium system. *Can. J. Biochem.* 56:1–6
58a. Nishikori, K., Maeno, H. 1979. Close relationship between adenosine 3':5'-monophosphate-dependent endogenous phosphorylation of a specific protein and stimulation of calcium uptake in rat uterine microsomes. *J. Biol. Chem.* 254:6099–106
59. Ong, S. M., Steiner, A. L. 1977. Localization of cyclic GMP and cyclic AMP in cardiac and skeletal muscle: immunocytochemical demonstration. *Science* 195:183–85
60. Reuter, H. 1973. Divalent cations as charge carriers in excitable membranes. *Prog. Biophys. Mol. Biol.* 26:1–43
61. Reuter, H. 1974. Exchange of calcium ions in the mammalian myocardium. Mechanisms and physiological significance. *Circ. Res.* 34:599–605
62. Reuter, H. 1974. Localization of beta adrenergic receptors, and effects of noradrenaline and cyclic nucleotides on action potentials, ionic currents and tension in mammalian cardiac muscle. *J. Physiol. London* 242:429–51
63. Reuter, H. 1979. Properties of two inward membrane currents in the heart. *Ann. Rev. Physiol.* 41:413–24
63a. Reuter, H., Scholz, H. 1977. A study of the ion selectivity and the kinetic properties of the calcium-dependent

slow inward current in mammalian cardiac muscle. *J. Physiol. London* 264:17–47
64. Schwartz, A., Entman, M. L., Kaniike, K., Lane, L. K., Van Winkle, W. B., Bornet, E. P. 1976. The rate of calcium uptake in sarcoplasmic reticulum of cardiac muscle and skeletal muscle. Effects of cyclic AMP-dependent protein kinase and phosphorylase b kinase. *Biochim. Biophys. Acta* 426:57–72
64a. Shigekawa, M., Finegan, J. -A. M., Katz, A. M. 1976. Calcium transport ATPase of canine cardiac sarcoplasmic reticulum. A comparison with that of rabbit fast skeletal muscle sarcoplasmic reticulum. *J. Biol. Chem.* 251:6894–900
65. Sperelakis, N., Schneider, J. A. 1976. A metabolic control mechanism for calcium ion influx that may protect the ventricular myocardial cell. *Am. J. Cardiol.* 37:1079–85
66. Stam, A. C. Jr., Weglicki, W. B., Gertz, E. W., Sonnenblick, E. H. 1973. A calcium-stimulated, ouabain-inhibited ATPase in a myocardial fraction enriched with sarcolemma. *Biochim. Biophys. Acta* 298:927–31
67. Steiner, A. L., Koide, Y., Earp, H. S., Bechtel, P. J., Beavo, J. A. 1978. Compartmentalization of cyclic nucleotides and cyclic AMP-dependent protein kinases in rat liver: Immunocytochemical demonstration. *Adv. Cyclic Nuc. Res.* 9:691–705
68. St. Louis, P. J., Sulakhe, P. V. 1976. Adenosine triphosphate-dependent calcium binding and accumulation by guinea pig cardiac sarcolemma. *Can. J. Biochem.* 54:946–56
69. St. Louis, P. J., Sulakhe, P. V. 1977. Stimulation of cardiac sarcolemmal (Na^+-K^+) ATPase activity by phosphorylase kinase. *Eur. J. Pharmacol.* 43:277–80
70. St. Louis, P. J., Sulakhe, P. V. 1978. Protein analysis of cardiac sarcolemma: effects of membrane-perturbing agents on membrane proteins and calcium transport. *Biochemistry* 17:4540–50
71. St. Louis, P. J., Sulakhe, P. V. 1979. Phosphorylation of cardiac sarcolemma by endogenous and exogenous protein kinases. *Arch. Biochem. Biophys.* 198:227–40
71a. Stull, J. T., Mayer, S. E. 1979. Biochemical mechanisms of adrenergic and cholinergic regulation of myocardial contractility. In *Handbook of Physiology. II. The Cardiovascular System. Vol. 1,* ed. R. M. Berne, N. Sperelakis, pp. 741–74. Bethesda: Am. Physiol. Soc. 970 pp.
72. Sulakhe, P. V., Leung, N. L.-K., St. Louis, P. J. 1976. Stimulation of calcium accumulation in cardiac sarcolemma by protein kinase. *Can. J. Biochem.* 54:438–45
73. Sulakhe, P. V., St. Louis, P. J. 1977. Stimulation of calcium accumulation in cardiac sarcolemma by phosphorylase kinase. *Biochem. J.* 164:457–59
74. Sulakhe, P. V., St. Louis, P. J. 1980. Passive and active calcium fluxes across plasma membranes. *Prog. Biophys. Mol. Biol.* 35:135–95
75. Sutherland, E. W., Rall, T. W. 1960. The relation of adenosine-3',5'-phosphate and phosphorylase to the actions of catecholamines and other hormones. *Pharmacol. Rev.* 12:265–99
75a. Tada, M., Finney, J. O. Jr., Swartz, M. H., Katz, A. M. 1972. Preparation and properties of plasma membranes from guinea pig hearts. *J. Mol. Cell. Cardiol.* 4:417–26
76. Tada, M., Inui, M., Yamada, M., Kadoma, M., Kuzuya, T., Abe, H., Kakiuchi, S. 1981. Effects of phospholamban phosphorylation catalyzed by adenosine 3':5'-monophosphate- and calmodulin-dependent protein kinases on calcium transport ATPase of cardiac sarcoplasmic reticulum. *J. Molec. Cell. Cardiol.* In press
77. Tada, M., Kirchberger, M. A., Katz, A. M. 1975. Phosphorylation of a 22,000–dalton component of the cardiac sarcoplasmic reticulum by adenosine 3':5'-monophosphate-dependent protein kinase. *J. Biol. Chem.* 250:2640–47
78. Tada, M., Kirchberger, M. A., Li, H.-C. 1975. Phosphoprotein phosphatase-catalyzed dephosphorylation of the 22,000-dalton phosphoprotein of cardiac sarcoplasmic reticulum. *J. Cyclic Nuc. Res.* 1:329–38
79. Tada, M., Kirchberger, M. A., Repke, D. I., Katz, A. M. 1974. The stimulation of calcium transport in cardiac sarcoplasmic reticulum by adenosine 3': 5'-monophosphate-dependent protein kinase. *J. Biol. Chem.* 249:6174–80
80. Tada, M., Ohmori, F., Kinoshita, N., Abe, H. 1978. Cyclic AMP regulation of active calcium transport across membranes of sarcoplasmic reticulum. Role of the 22,000-dalton protein phospholamban. *Adv. Cyclic Nuc. Res.* 9:355–69
81. Tada, M., Ohmori, F., Nimura, Y., Abe, H. 1977. Effect of myocardial protein kinase modulator on adenosine 3':

5'-monophosphate-dependent protein kinase-induced stimulation of calcium transport by cardiac sarcoplasmic reticulum. *J. Biochem. Tokyo* 82: 885–92
82. Tada, M., Ohmori, F., Yamada, M., Abe, H. 1979. Mechanism of the stimulation of Ca^{2+}-dependent ATPase of cardiac sarcoplasmic reticulum by adenosine 3',5'-monophosphate-dependent protein kinase. Role of the 22,000-dalton protein. *J. Biol. Chem.* 254:319–26
82a. Tada, M., Yamada, M., Kadoma, M., Inui, M., Ohmori, F. 1981. Calcium transport by sarcoplasmic reticulum and phosphorylation of phospholamban. *Mol. Cell. Biochem.* In press
83. Tada, M., Yamada, M., Ohmori, F., Kuzuya, T., Abe, H. 1979. Mechanism of cyclic AMP regulation of active calcium transport by cardiac sarcoplasmic reticulum. In *Cation Flux Across Biomembranes,* ed. Y. Mukohata, L. Packer, pp. 179–90. NY/San Francisco/London: Academic. 444 pp.
84. Tada, M., Yamada, M., Ohmori, F., Kuzuya, T., Inui, M., Abe, H. 1980. Transient state kinetic studies of Ca^{2+}-dependent ATPase and calcium transport by cardiac sarcoplasmic reticulum. Effect of cyclic AMP-dependent protein kinase-catalyzed phosphorylation of phospholamban. *J. Biol. Chem.* 255:1985–92
85. Tada, M., Yamamoto, T., Tonomura, Y. 1978. Molecular mechanism of active calcium transport by sarcoplasmic reticulum. *Physiol. Rev.* 58:1-79
86. Trumble, W. R., Sutko, J. L., Reeves, J. P. 1980. ATP-dependent calcium transport in cardiac sarcolemmal membrane vesicles. *Life Sci.* 27:207–14
87. Tsien, R. W. 1977. Cyclic AMP and contractile activity in heart. *Adv. Cyclic Nuc. Res.* 8:363–420
88. Van Alstyne, E., Bartschat, D. K., Wellsmith, N. V., Poe, S. L., Schilling, W. P., Lindenmayer, G. E. 1979. Isolation of highly enriched sarcolemma membrane fraction from canine heart. *Biochim. Biophys. Acta* 553:388–95
89. Van Winkle, W. B., Pitts, B. J. R., Entman, M. L. 1978. Rapid purification of canine cardiac sarcoplasmic reticulum Ca^{2+}-ATPase. *J. Biol. Chem.* 253: 8671–73
90. Venter, J. C., Ross, J. Jr., Kaplan, N. O. 1975. Lack of detectable change in cyclic AMP during the cardiac inotropic response to isoproterenol immobilized on glass beads. *Proc. Natl. Acad. Sci. USA* 72:824–28
91. Walsh, D. A., Clippinger, M. S., Sivaramakrishnan, S., McCullough, T. E. 1979. Cyclic adenosine monophosphate dependent and independent phosphorylation of sarcolemma membrane proteins in perfused rat heart. *Biochemistry* 18:871-77
92. Weglicki, W. B., Owens, K., Kennett, F. F., Kessner, A., Harris, L., Wise, R. M., Vahouny, G. V. 1980. Preparation and properties of highly enriched cardiac sarcolemma from isolated adult myocytes. *J. Biol. Chem.* 255:3605–9
93. Will, H., Blanck, J., Smettan, G., Wollenberger, A. 1976. A quench-flow kinetic investigation of calcium ion accumulation by isolated cardiac sarcoplasmic reticulum. Dependence of initial velocity on free calcium ion concentration and influence of preincubation with a protein kinase, MgATP and cyclic AMP. *Biochim. Biophys. Acta* 449:295–303
94. Will, H., Levchenko, T. S., Levitsky, D. O., Smirnov, V. N., Wollenberger, A. 1978. Partial characterization of protein kinase-catalyzed phosphorylation of low molecular weight proteins in purified preparations of pigeon heart sarcolemma and sarcoplasmic reticulum. *Biochim. Biophys. Acta* 543:175–93
95. Will, H., Schirpke, B., Wollenberger, A. 1973. Binding of calcium to cell membrane-enriched preparation from pig myocardium: increase in calcium affinity upon membrane protein phosphorylation enhanced by membrane-bound cyclic AMP-dependent protein kinase. *Acta Biol. Med. Ger.* 31:K45–52
96. Will, H., Schirpke, B., Wollenberger, A. 1976. Stimulation of Ca^{2+}-uptake by cyclic AMP and protein kinase in sarcoplasmic reticulum-rich and sarcolemma-rich microsomal fractions from rabbit heart. *Acta Biol. Med. Ger.* 35: 529–41
97. Williamson, J. R., Woodrow, M. L., Scarpa, A. 1975. Calcium binding to cardiac sarcolemma. In *Recent Advances in Studies on Cardiac Structure and Metabolism,* ed. A. Fleckenstein, N. S. Dhalla, 5:61–71. Baltimore: University Park Press
98. Wollenberger, A., Will, H. 1978. Protein kinase-catalyzed membrane phosphorylation and its possible relationship to the role of calcium in the adrenergic regulation of cardiac contraction. *Life Sci.* 22:1159–78

99. Wollenberger, A., Will, H., Krause, E.-G. 1975. Adenosine 3',5'-monophosphate, the myocardial cell membrane, and calcium. See Ref. 97, 5:81–93
100. Wray, H. L., Gray, R. R. 1977. Cyclic AMP stimulation of membrane phosphorylation and Ca^{2+}-activated, Mg^{2+}-dependent ATPase in cardiac sarcoplasmic reticulum. *Biochim. Biophys. Acta* 461:441–59
101. Wray, H. L., Gray, R. R., Olsson, R. A. 1973. Cyclic adenosine 3',5'-monophosphate-stimulated protein kinase and a substrate associated with cardiac sarcoplasmic reticulum. *J. Biol. Chem.* 248:1496–98
102. Yamamoto, T., Takisawa, H., Tonomura, Y. 1979. Reaction mechanisms for ATP hydrolysis and synthesis in the sarcoplasmic reticulum. *Curr. Top. Bioenerg.* 9:179–236
103. Caroni, P., Carafoli, E. 1981. The Ca^{2+}–pumping ATPase of heart sarcolemma. Characterization, calmodulin dependence, and partial purification. *J. Biol. Chem.* 256:3263–70
104. Lamers, J. M. J., Stinis, H. T., De Jonge, H. R. 1981. On the role of cyclic AMP and Ca^{2+}–calmodulin-dependent phosphorylation in the control of (Ca^{2+}+Mg^{2+})-ATPase of cardiac sarcolemma. *FEBS Lett.* 127:139–43

THE SLOW INWARD CALCIUM CURRENT IN THE HEART

Terence F. McDonald

Department of Physiology and Biophysics, Dalhousie University, Halifax, Nova Scotia, Canada B3H 4H7

INTRODUCTION

Ionic current flow through two distinct voltage-gated membrane channels underlies excitation in the heart: the fast inward Na current and the Ca-dependent "secondary" or "slow" inward current (I_{si}). These currents differ in their voltage dependencies, kinetics, ionic selectivity and responses to chemical agents (reviewed in 65).

This review focuses on I_{si} in the heart with an emphasis on voltage clamp studies. The first section deals with important properties of I_{si}, including selectivity, saturation, blocking, voltage and time dependencies, steady-state I_{si}, and inactivation mechanisms. The second section concerns Ca-activated potassium conductance, especially as linked to I_{si}, and the third discusses selected physiological and pharmacological interventions. I_{si}-contraction coupling is not specifically covered (for reviews see 14, 19, 22).

PROPERTIES OF I_{si}

I_{si} can be described by $I_{si} = g_{si} \cdot d(V,t) \cdot f(V,t) \cdot (V_m - V_{si})$ where d is the activation variable and f the inactivation variable. The widespread use of this formulation (64) has provided a framework for the investigation of I_{si} and its role in the healthy, diseased, or drug-treated heart. In addition, this description of I_{si} has been employed in computer models that successfully simulate electrical activity in Purkinje fibers (44), ventricular muscle (6), and S-A node (74). Despite this, much of the present information on I_{si} may turn out to be less than definitive. Quantitative measurements are hindered by problems associated with (*a*) voltage clamping cardiac preparations (5, 65), (*b*) the separation of I_{si} from other overlapping currents (49, 67), (*c*) external ion substitution (5, 37, 64), and (*d*) the determination of I_{si} reversal potentials [T. McDonald, unpublished observations; (31, 41, 55)]. These same problems plague the investigation of Ca channels in non-cardiac cell membranes (see 27).

Selectivity, Saturation, and Blocking

The difficulty of measuring reversal potentials and the consequences of varying Ca_o and Na_o on intracellular ion concentrations have hampered studies on selectivity and saturation of Ca channels in the heart. In the only methodical study on selectivity, P_{Ca}/P_{Na} and P_{Ca}/P_K were about 0.01 in bovine ventricular muscle, and probably higher in dog, cat, and sheep ventricle (66). Divalent cation selectivity has not been determined, but indirect evidence suggests Ba > Sr > Ca >> Mg (66; also see 39, 42, 55, 64). The amplitude of I_{si} increases with Ca_o but a quantitative description of this relation is lacking. There are indications of saturation with high Ca_o (55, 60) or Sr_o (71), behavior that is in keeping with the saturation of I_{Ca} (see 27) and I_{Na} (28) in other excitable membranes.

Inorganic blockers of I_{si} include Cd, Co, La, Mn, and Ni (25, 29, 43, 48, 58, 60, 64), although Mn also has a limited ability to carry I_{si} (58). These cations may compete with Ca for external sites important for Ca channel permeation (60). The most widely studied organic blockers are verapamil and D600; micromolar concentrations block I_{si} in cardiac tissue (10, 18, 36, 38, 47, 54, 56, 57, 60), embryonic heart cell aggregates (53), and isolated ventricular cells (34). Their well-known frequency-dependent effects on the action potential and force of contraction appear to be related to a frequency- and voltage-dependent block of the Ca channel (18, 47). This type of block is similar in many ways to that of the Na channel by local anesthetics (see 30) in that block requires depolarization (channel opening) and the rate of unblock increases with more negative diastolic potentials. Similar behavior has been found in cardiac tissue treated with a new agent, AQA 39 (70). The implications are that Ca channels in heart cells having a low diastolic potential and rapid rate of firing will be most susceptible to block by these agents.

Activation, Inactivation, and Recovery from Inactivation

I_{si} activation is not well described in cardiac cells but it is assumed to follow a monoexponential time course (64). Between -40 and $+30$ mV, the time constant τ_d ranges from 5 to 20–30 msec in ventricular muscle (6, 64, 69) but it is much shorter in the S-A node (57). The threshold of I_{si} lies positive to -50 mV and the steady-state activation variable, d_∞, has a sigmoidal shape with $V_{0.5}$ at about -20 mV and saturation at $+10$ mV (34, 64, 66, 69). In the S-A node, the threshold for I_{si} may be as negative as -60 mV (57).

The relation between voltage and the steady-state inactivation variable, f_∞, has a sigmoidal shape and is near 1.0 at −60 mV, 0.5 at −25 mV, and 0 at +10 mV (31, 34, 57, 64, 66, 69). There is general agreement that inactivation develops with an exponential time course, but less concurrence on estimates of τ_f and its dependence on voltage. In mammalian ventricular tissue (6, 24, 49, 66), inactivation at 0 mV appears to be a slow process (τ_f of 80–200 msec) compared to that (τ_f of 10–30 msec) observed in isolated rat ventricular cells (34), rabbit S-A (57) and A-V node (56), embryonic heart cell aggregates (53), and, with allowance for temperature, frog atrium (31). In the latter study, the dependence of τ_f on voltage was bell-shaped with the maximum near −15 mV. The relation in all other preparations was either U-shaped with a minimum near −25 mV (34, 40, 53, 57) or it curved upwards with positive potential (6, 24, 49, 64, 66, 69). [Note discussion in (69); compare (57) with (74).] The time constants of recovery from inactivation are of similar magnitude (24) or several times larger (40, 66, 69) than the time constants describing the development of inactivation.

There are at least two reasons for the divergent results noted in the survey above: (*a*) measurement problems (separation of I_{si}; resolution of just-detectable I_{si} near threshold potentials), and (*b*) genuine differences between preparations and species. Compared to the simple excitatory role of I_{Na} and (correspondingly?) rather similar properties in a variety of excitable cells, I_{si} in the heart must fulfill several different functions under widely varying conditions (6, 22, 65, 74). The addition of the species factor (e.g. rat versus cow) increases the complexity.

Steady-State I_{si}

There is a significant overlap of d_∞ and f_∞ such that the product, $d_\infty f_\infty$, is bell-shaped between −40 and 0 mV and predicts that 10–20% of the maximum conductance ought to remain activated around −20 mV (34, 64, 66, 69). Other than one indirect piece of evidence, a bell-shaped dependence of steady-state tension on voltage (69), proof for the existence and shape of the "window" rests solely on the overlap of d_∞ and f_∞. Little encouragement is offered by the results of one study that might have detected the overlap. In calf Purkinje fibers, the application of 9.6 μM D600 abolished time-dependent current attributed to I_{si} and induced an outward shift in the late (500 msec) membrane current (35; see also 67). The D600-sensitive "displacement" current was attributed to steady-state I_{si} since it increased with Ca_o and epinephrine. However, its dependence on voltage discounted a connection with $d_\infty f_\infty$. Instead, the conclusion was that some Ca channels that activate normally somehow fail to inactivate. Nearly 50% of the channels may belong to this category (35, 67).

Is residual I_{si} unmasked when other tissues are treated with Ca channel blockers? There is no evidence of an outward shift in the late current when peak I_{si} is depressed by 50–90% in guinea-pig or cat ventricular muscle, rabbit S-A node, or isolated rat ventricular cells treated with Co (29, 48), D600 (34, 54, 57), or verapamil (18). A slight outward shift has been recorded in D600-treated rabbit A-V node (56) and embryonic heart cell aggregates (53), indicating that perhaps 5–10% of channels remain open. Finally, when isolated bovine ventricular cells were treated with D600 or Ni, an outward shift in the late current suggests that 15–25% of the Ca channels do not inactivate at –5 mV (G. Isenberg, personal communication). One explanation is that the proportion of Ca channels that does not inactivate is dependent on species and/or tissue. A second is that a simultaneous block of outward currents in some preparations effectively masks the block of time-independent I_{si}.

The D600 experiments on Purkinje fibers (35, 67) are also interesting from two other points of view. (a) There is a strong similarity in the use- and voltage-dependent block of Ca channels by D600 (47) and that of Na channels by local anesthetics (30). Since use-dependent block by many local anesthetics depends on a functional, inactivation-gating mechanism (11), the block of inactivation-resistant Ca channels in D600-treated Purkinje fibers suggests that either the inactivation machinery is present or the drug-channel interaction differs from that envisaged in the Na channel. (b) The amount of Ca carried by I_{si} in voltage clamp experiments may have been seriously underestimated (e.g. in studies on I_{si}-contraction).

Mechanism of Inactivation

Nearly all studies of I_{si} in cardiac cells show that τ_f increases at membrane potentials more positive than –20 mV (see above); this behavior is more compatible with Ca-induced inactivation than with voltage-dependent inactivation gates. The concept of Ca-induced inactivation evolved from studies on Ca currents in *Paramecium* and *Aplysia* (see 8): during the flow of I_{Ca} there is an accumulation of Ca near the inner mouth of the channel and this local build-up of Ca somehow leads to functional inactivation. More recently, there is evidence that Ca-dependent inactivation is also present in stick insect skeletal muscle (3) and *Helix* neurons (9). Since the depression of Ca conductance by small increases in Ca_i has been observed in different cell types (see 27), there is some support for the suggestion (3) that this may be a universal characteristic of Ca channels.

The demonstration of Ca-dependent inactivation involves four basic experiments (8, 21). (a) A two-pulse clamp protocol in which pulse I is imposed to voltages between the holding potential and E_{Ca} (to inactivate I_{Ca}), and pulse II, following closely thereafter, is to a test potential that normally elicits I_{Ca}. The plot of test I_{Ca} versus pulse I voltage is expected to be sigmoid for Hodgkin-Huxley type inactivation. If, instead, it assumes a U-shape by increasing at positive voltages, it suggests Ca-dependent inactivation, since Ca influx will decrease as pulse I nears E_{Ca}. (*b*) An increased rate of inactivation when Ca_o is increased. (*c*) A greatly reduced rate and degree of inactivation when Ba or Sr are substituted for external Ca, these ions being less effective inactivators. (*d*) A reduced rate and degree of I_{Ca} inactivation, and a shorter recovery time, with high intracellular concentrations of EGTA. Application of these criteria indicates that a pure form of the mechanism may not be universal. For example, Ba and Sr currents decay at the same rate as Ca currents in many egg cell membranes (see 27). In addition, the steady-state inactivation curve in *Neanthes* egg cell membrane is sigmoid with no demonstrable recovery at positive potentials (the two-pulse protocol), Ba and Sr currents relax at the same rate and to the same degree as I_{Ca}, and a 5-fold increase in I_{Ca} amplitude by raising Ca_o has no effect on the time course of inactivation (21). Finally, the depletion of external Ca in the T-tubules of frog skeletal muscle is thought to be the cause of I_{Ca} relaxation in that tissue (2).

There is evidence favoring the operation of Ca-induced inactivation of I_{si} in the heart. (*a*) In Cs-loaded calf Purkinje fibers, double pulse experiments resulted in a U-shaped inactivation relation, and the rate and degree of inactivation decreased when Ca_o was replaced by Ba (42). (*b*) Raising Ca_o from 1.8 to 8.2 mM increased the rate of inactivation in cat ventricular muscle (40). (*c*) Sr currents inactivated more slowly than I_{si} in cow Purkinje fibers (71) and cat papillary muscle (39). (*d*) The disparity between the time constants of development and recovery from inactivation (see above) favor Ca-dependent inactivation, since the build-up and removal of Ca at internal blocking sites are unlikely to occur at the same rate. (*e*) There is little or no steady-state inactivation at voltages more negative than I_{si}-activating voltages.

Other observations seem to be inconsistent with this type of inactivation. (*a*) Catecholamines increased the amplitude of I_{si} but not the rate of inactivation in ventricular muscle (65) and S-A node (57). (*b*) The injection of Ca into sheep Purkinje fibers increased I_{si}, whereas the injection of EGTA depressed the current (32; see also 67). (*c*) In cat ventricular muscle, an increase in Ca_o increased I_{si} but shortened the restoration time (40).

Ca ACTIVATED POTASSIUM CONDUCTANCE

Ca-activated potassium currents have been identified in a wide variety of cells (see 50). In most cases, only one of several available voltage-dependent membrane potassium conductances is activated by Ca_i and abolished upon the removal of external Ca or block of I_{Ca} (26, 27, 50).

In cardiac Purkinje fibers, the injection of Ca ions increased both I_{K1} and I_{K2}; the converse effects were obtained following the injection of EGTA (33). The modification of I_{si} (amplitude; stimulation rate) has also been ultilized to alter Ca_i and gauge the effect on potassium currents in the heart. The results are inconclusive. (a) Increasing Ca_o slightly reduced I_x and increased I_{K1} (36) or had no direct effect on I_{K1}, I_{K2}, or I_x (17) in Purkinje fibers. (b) An increase in stimulation rate increased the late outward current in sheep and calf ventricular muscle (4) but produced no increase in rat ventricular muscle despite a large increase in I_{si} per unit time (61). In frog atrial trabeculae, large frequency-dependent changes in I_{si} (and tension) had no consistent effect on the late outward current (55). (c) Epinephrine increases I_{si} and time-dependent outward plateau current in Purkinje fibers and in frog atrial muscle (see 59). In rabbit S-A node, epinephrine also increased both I_{si} and I_K (10, 56). However, a block of the epinephrine effect on I_{si} (pretreatment with D600) blocked the effect on I_K in one study (10) but not in the other (56).

The most recent nominee for a Ca-activated current in heart is the transient outward current (previously known as the positive dynamic current, I_{Cl} or I_{qr}). It overlaps I_{si} in Purkinje fibers, is largely responsible for phase 1 repolarization, and has not (as yet?) been observed in other cardiac tissues. The transient outward current is carried mainly by potassium ions (37, 67) and is thought to be activated by Ca_i (35, 67): (a) It is sensitive to Ca_o and D600, (b) it is diminished after EGTA injection, and (c) it varies directly with peak force. A Ca-activated, voltage-dependent g_K in Purkinje fibers is consistent with the increase in ^{42}K efflux from driven (but not quiescent) fibers when Ca_o was raised from 0.9 to 8.1 mM (51). The results of a similar challenge with Ca_o, and two additional protocols designed to alter transmembrane Ca movement, did not support Ca_i-linked ^{42}K efflux in paced, turtle pacemaker tissue (20).

PHYSIOLOGICAL AND PHARMACOLOGICAL INTERVENTIONS

Post-rest stimulation and changes in stimulation rate produce complex, species-dependent changes in the action potential configuration, I_{si}, and tension (see 7, 12). In rat ventricular muscle, the relation between I_{si} amplitude and stimulation rate was bell-shaped with maximum I_{si} at 75 min^{-1}, whereas the relation between τ_f and rate was U-shaped with a minimum at 100 min^{-1} (61). In frog atrial muscle, post-rest stimulation at rates up to 60 min^{-1} was accompanied by a staircase increase in I_{si} and tension (55). The conclusion preferred by these authors was that the Ca influx increased membrane Ca conductance. Since, in addition, the injection of Ca increased I_{si} and shortened τ_f (32), and yet high Ca_i seems to depress calcium current (see above), there may be a bell-shaped relation between Ca_i and Ca channel conductance. If so, an action of cardiac glycosides on the cell membrane or on intracellular sites may alter Ca_i and increase, decrease, or have little effect on I_{si} (cf 46, 73).

Ca_i also features in two other interventions affecting I_{si}: pH and hypoxia. I_{si} is depressed in frog atrium superfused with acidic (pH 5.6) solution (15), slow action potentials in guinea-pig ventricular muscles are abolished at pH 6.4 (68), lowering pH from 7.4 to 6.6 reduces peak tension and ^{47}Ca uptake in rabbit ventricle (63), and low pH blocks Ca channels in nerve terminals (apparent pKa of binding site $\sim$ 6.3) (52). Assuming that these effects are a sequel to intracellular acidification, there is pertinent evidence from heart (16, 23) and nerve cells (1) that a decrease in pH_i increases Ca_i, and that an influx of Ca decreases pH_i. During hypoxia, there is probably an increase in Ca_i and a decrease in cell pH, both of which ought to depress I_{si}. There is voltage-clamp evidence for (62) and against (72) the proposition that I_{si} is reduced by hypoxia (45; see also 68).

The stimulation of I_{si} by β-adrenergic agonists, the reduction of I_{si} by acetylcholine, and the possibility that these effects may be related to cAMP and cGMP actions on Ca channels, has recently been reviewed (65). A new observation is that in addition to increasing I_{si} and I_K, epinephrine also speeds the development of an inward current activated negative to -50 mV in rabbit S-A node [i_f(10); i_h(57)]. One group concluded that the positive chronotropic action of epinephrine is mainly due to the effect on i_f (10); the other thought the key effect was on I_{si} because epinephrine did not increase the slope of diastolic depolarization when I_{si} was blocked by D600 (57). Both may be correct in that (a) the I_{si} mechanism is likely to dominate in preparations having a maximum diastolic potential of -65 mV or less, where the activation of i_f-i_h is slow, and (b) i_f-i_h should become more important when the maximum diastolic potential reaches -70 mV or more where the activation of i_f-i_h is considerably quicker (also see 43).

Recent studies have demonstrated that acetylcholine blocks I_{si} in sheep Purkinje fibers (13) and guinea-pig ventricular muscle (29). Since acetylcholine also depresses I_{si} in frog and mammalian atria (see 13), it would be surprising if I_{si} in S-A and A-V nodal preparations proved insensitive to that drug.

CONCLUDING REMARKS

The discovery and investigation of I_{si} over the past 12 years have reshaped concepts on cellular processes in healthy and diseased hearts. Despite this, information on the properties of Ca channels in the heart remains scant in some areas and confusing or contradictory in others. Technical improvements and further characterization of genuine tissue and species differences ought to improve the situation. The advent of the isolated adult heart cell opens up the possibility of controlling the ionic environment on both sides of the membrane and of observing gating currents and single channel events. Information gleaned from this model, and from systematic studies of chemical blockade and modification of Ca channels, should lead to an improved understanding of the properties and role of I_{si}.

ACKNOWLEDGMENTS

I am grateful to L. Delbridge, M. Vohra Jr., and T. Watanabe for their assistance, and to the Medical Research Council and the Nova Scotia Heart Foundation for support.

Literature Cited

1. Ahmed, Z., Connor, J. A. 1980. Intracellular pH changes induced by calcium influx during electrical activity in molluscan neurons. *J. Gen. Physiol.* 75:403–26
2. Almers, W., Fink, R., Palade, P. T. 1981. Calcium depletion in frog muscle tubules: The decline of calcium current under maintained depolarization. *J. Physiol.* 312:177–207
3. Ashcroft, F. M., Stanfield, P. R. 1980. Inactivation of calcium currents in skeletal muscle fibres of an insect depends on calcium entry. *J. Physiol.* 308:36P
4. Bassingthwaighte, J. B., Fry, C. H., McGuigan, J. A. S. 1976. Relationship between internal calcium and outward current in mammalian ventricular muscle; a mechanism for the control of the action potential duration? *J. Physiol.* 262:15–37
5. Beeler, G. W., McGuigan, J. A. S. 1978. Voltage clamping of multicellular myocardial preparations: Capabilities and limitations of existing methods. *Prog. Biophys. Mol. Biol.* 34:219–54
6. Beeler, G. W., Reuter, H. 1977. Reconstruction of the action potential of ventricular myocardial fibres. *J. Physiol.* 268:177–210
7. Boyett, M. R., Jewell, B. R. 1980. Analysis of the effects of changes in rate and rhythm upon electrical activity in the heart. *Prog. Biophys. Mol. Biol.* 36:1–52
8. Brehm, P., Eckert, R., Tillotson, D. 1980. Calcium-mediated inactivation of calcium current in *Paramecium*. *J. Physiol.* 306:193–203
9. Brown, A. M., Akaike, N., Tsuda, Y., Morimoto, K. 1980. Ion migration and inactivation in the calcium channel. *J. Physiol. Paris* 76:395–402
10. Brown, H., DiFrancesco, D. 1980. Voltage-clamp investigations of membrane currents underlying pace-maker activity in rabbit sino-atrial node. *J. Physiol.* 308:331–51
11. Cahalan, M. D. 1978. Local anesthetic block of sodium channels in normal and pronase-treated squid giant axons. *Biophys. J.* 23:285–311
12. Carmeliet, E. 1977. Repolarisation and frequency in cardiac cells. *J. Physiol. Paris* 73:903–23
13. Carmeliet, E., Ramon, J. 1980. Effects of acetylcholine on time-dependent currents in sheep cardiac Purkinje fibers. *Pfluegers. Arch.* 387:217–23
14. Chapman, R. A. 1979. Excitation-contraction coupling in cardiac muscle. *Prog. Biophys. Mol. Biol.* 35:1–52
15. Chesnais, J. M., Coraboeuf, E., Sauviat, M. P., Vassas, J. M. 1975. Sensitivity to H, Li and Mg Ions of the slow inward sodium current in frog atrial fibres. *J. Mol. Cell. Cardiol.* 7:627–42
16. Deitmer, J. W., Ellis, D. 1980. Interactions between the regulation of the intracellular pH and sodium activity of sheep cardiac Purkinje fibres. *J. Physiol.* 304:471–88
17. DiFrancesco, D., McNaughton, P. A. 1979. The effects of calcium on outward membrane currents in the cardiac Purkinje fibre. *J. Physiol.* 289:347–73
18. Ehara, T., Kaufmann, R. 1978. The voltage and time-dependent effects of (-)-verapamil on the slow inward current in isolated cat ventricular myocardium. *J. Pharmacol. Exp. Ther.* 207:49–55
19. Fabiato, A., Fabiato, F. 1979. Calcium and cardiac excitation-contraction coupling. *Ann. Rev. Physiol.* 41:473–84

20. Fleming, B. P., Giles, W. 1981. Changes in ^{42}K efflux produced by alterations in transmembrane calcium movements in turtle cardiac pace-maker tissue. *J. Physiol.* 314:65–77
21. Fox, A. P. 1981. Voltage-dependent inactivation of a calcium channel. *Proc. Natl. Acad. Sci. USA* 78:953–56
22. Fozzard, H. A. 1977. Heart: excitation-contraction coupling. *Ann. Rev. Physiol.* 39:201–20
23. Fry, C. H., Poole-Wilson, P. A. 1981. Effects of acid-base changes on excitation-contraction coupling in guinea-pig and rabbit cardiac ventricular muscle. *J. Physiol.* 313:141–60
24. Gettes, L. S., Reuter, H. 1974. Slow recovery from inactivation of inward currents in mammalian myocardial fibres. *J. Physiol.* 240:703–24
25. Giles, W. R., Hume, J. R. 1981. Inward current underlying the 'slow response' in single bullfrog atrial cells. *J. Physiol.* In press
26. Gorman, A. L. F., Thomas, M. V. 1980. Potassium conductance and internal calcium accumulation in a molluscan neurone. *J. Physiol.* 308:287–313
27. Hagiwara, S. 1981. Calcium channel. *Ann. Rev. Neurosci.* 4:69–125
28. Hille, B. 1975. Ionic selectivity, saturation, and block in sodium channels, a four-barrier model. *J. Gen. Physiol.* 66: 535–60
29. Hino, N., Ochi, R. 1980. Effect of acetylcholine on membrane currents in guinea-pig papillary muscle. *J. Physiol.* 307:183–97
30. Hondeghem, L. M., Katzung, B. G. 1977. Time- and voltage-dependent interactions of antiarrhythmic drugs with cardiac sodium channels. *Biochim. Biophys. Acta* 472:373–98
31. Horackova, M., Vassort, G. 1976. Calcium conductance in relation to contractility in frog myocardium. *J. Physiol.* 259:597–616
32. Isenberg, G. 1977. Cardiac Purkinje fibres. The slow inward current component under the influence of modified $[Ca^{2+}]_i$. *Pfluegers Arch.* 371:61–69
33. Isenberg, G. 1977. Cardiac Purkinje fibres. $[Ca^{2+}]_i$ controls the potassium permeability via the conductance components g_{k1} and g_{k2}. *Pfluegers Arch.* 371:77–85
34. Isenberg, G., Klöckner, U. 1980. Glycocalyx is not required for slow inward calcium current in isolated rat heart myocytes. *Nature* 284:358–60
35. Kass, R. S., Siegelbaum, S., Tsien, R. W. 1976. Incomplete inactivation of the slow inward current in cardiac Purkinje fibres. *J. Physiol.* 263:127–28P
36. Kass, R. S., Tsien, R. W. 1975. Multiple effects of calcium antagonists on plateau currents in cardiac Purkinje fibers. *J. Gen. Physiol.* 66:169–92
37. Kenyon, J. L., Gibbons, W. R. 1979. 4-Aminopyridine and the early outward current of sheep cardiac Purkinje fibers. *J. Gen. Physiol.* 73:139–57
38. Kohlhardt, M., Bauer, B., Krause, H., Fleckenstein, A. 1972. Differentiation of the transmembrane Na and Ca channels in mammalian cardiac fibers by the use of specific inhibitors. *Pfluegers Arch.* 335:309–22
39. Kohlhardt, M., Herdey, A., Kubler, M. 1973. Interchangeability of Ca ions and Sr ions as charge carriers of the slow inward current in mammalian myocardial fibres. *Pfluegers Arch.* 344:149–58
40. Kohlhardt, M., Krause, H., Kubler, M., Herdey, A. 1975. Kinetics of inactivation and recovery of the slow inward current in the mammalian ventricular myocardium. *Pfluegers Arch.* 355:1–17
41. Lee, K. S., Lee, E. W., Tsien, R. W. 1981. Slow inward current carried by Ca^{2+} or Ba^{2+} in single isolated heart cells. *Biophys. J.* 33:143a
42. Marban, E., Tsien, R. W. 1981. Is the slow inward calcium current of heart muscle inactivated by calcium? *Biophys. J.* 33:143a
43. Maylie, J., Morad, M. 1981. Ionic characterization of pacemaker current in voltage clamped rabbit SA node. *Biophys. J.* 33:11a
44. McAllister, R. E., Noble, D., Tsien, R. W. 1975. Reconstruction of the electrical activity of cardiac Purkinje fibres. *J. Physiol.* 251:1–59
45. McDonald, T. F., MacLeod, D. P. 1973. Metabolism and the electrical activity of anoxic ventricular muscle. *J. Physiol.* 229:559–82
46. McDonald, T. F., Nawrath, H., Trautwein, W. 1975. Membrane currents and tension in cat ventricular muscle treated with cardiac glycosides. *Circ. Res.* 37: 674–83
47. McDonald, T. F., Pelzer, D., Trautwein, W. 1980. On the mechanism of slow calcium channel block in heart. *Pfluegers Arch.* 385.175–79
48. McDonald, T. F., Pelzer, D., Trautwein, W. 1981. Does the calcium current modulate the contraction of the accompanying beat? A study of E-C coupling in mammalian ventricular muscle using cobalt ions. *Circ. Res.* 49:576–83

49. McDonald, T. F., Trautwein, W. 1978. Membrane currents in cat myocardium: Separation of inward and outward components. *J. Physiol.* 274:193–216
50. Meech, R. W. 1978. Calcium dependent potassium activation in nervous tissues. *Ann. Rev. Biophys. Bioeng.* 7:1–18
51. Musso, E., Vassalle, M. 1978. Effects of norepinephrine, calcium, and rate of discharge on ^{42}K movements in canine cardiac Purkinje fibers. *Circ. Res.* 42:276–84
52. Nachshen, D. A., Blaustein, M. P. 1979. Regulation of nerve terminal calcium channel selectivity by a weak acid site. *Biophys. J.* 26:329–34
53. Nathan, R. D., DeHaan, R. L. 1979. Voltage clamp analysis of embryonic heart cell aggregates. *J. Gen. Physiol.* 73:175–98
54. Nawrath, H., Ten Eick, R. E., McDonald, T. F., Trautwein, W. 1977. On the mechanism underlying the action of D-600 on slow inward current and tension in mammalian myocardium. *Circ. Res.* 40:408–14
55. Noble, S., Shimoni, Y. 1981. The calcium and frequency dependence of the slow inward current 'staircase' in frog atrium. *J. Physiol.* 310:57–75
56. Noma, A., Irisawa, H., Kokobun, S., Kotake, H., Nishimura, M., Watanabe, Y. 1980. Slow current systems in the A-V node of the rabbit heart. *Nature* 285:228–29
57. Noma, A., Kotake, H., Irisawa, H. 1980. Slow inward current and its role mediating the chronotropic effect of epinephrine in the rabbit sinoatrial node. *Pfluegers Arch.* 388:1–9
58. Ochi, R. 1976. Manganese-dependent propagated action potentials and their depression by electrical stimulation in guinea-pig myocardium perfused by sodium-free media. *J. Physiol.* 263:139–56
59. Pappano, A. J., Carmeliet, E. E. 1979. Epinephrine and the pacemaking mechanism at plateau potentials in sheep cardiac Purkinje fibers. *Pfluegers Arch.* 382:17–26
60. Payet, M. D., Schanne, O. F., Ruiz-Ceretti, E. 1980. Competition for slow channel of Ca^{2+}, Mn^{2+}, verapamil, and D-600 in rat ventricular muscle? *J. Mol. Cell. Cardiol.* 12:635–38
61. Payet, M. D., Schanne, O. F., Ruiz-Ceretti, E. 1981. Frequency dependence of the ionic currents determining the action potential repolarization in rat ventricular muscle. *J. Mol. Cell. Cardiol.* 13:207–15
62. Payet, M. D., Schanne, O. F., Ruiz-Ceretti, E., Demers, J. M. 1978. Slow inward and outward currents of rat ventricular fibers under anoxia. *J. Physiol. Paris* 74:31–35
63. Poole-Wilson, P. A., Langer, G. A. 1979. Effects of acidosis on mechanical function and Ca^{2+} exchange in rabbit myocardium. *Am. J. Physiol.* 236:H525–33
64. Reuter, H. 1973. Divalent cations as charge carriers in excitable membranes. *Prog. Biophys. Mol. Biol.* 26:1–43
65. Reuter, H. 1979. Properties of two inward membrane currents in the heart. *Ann. Rev. Physiol.* 41:413–24
66. Reuter, H., Scholz, H. 1977. A study of the ion selectivity and the kinetic properties of the calcium dependent slow inward current in mammalian cardiac muscle. *J. Physiol.* 264:17–47
67. Siegelbaum, S. A., Tsien, R. W. 1980. Calcium-activated transient outward current in calf cardiac Purkinje fibres. *J. Physiol.* 299:485–506
68. Sperelakis, N., Schneider, J. A. 1976. Metabolic control mechanism for calcium ion influx that may protect the ventricular myocardial cell. *Am. J. Cardiol.* 37:1079–85
69. Trautwein, W., McDonald, T. F., Tripathi, O. 1975. Calcium conductance and tension in mammalian ventricular muscle. *Pfluegers Arch.* 354:55–74
70. Trautwein, W., Pelzer, D., McDonald, T. F., Osterrieder, W. 1981. AQA 39, a new bradycardic agent which blocks myocardial calcium channels in a frequency- and voltage-dependent manner. *Naunyn-Schmiedeberg's Arch. Pharmacol.* 317:228–32
71. Vereecke, J., Carmeliet, E. 1971. Sr action potentials in cardiac Purkinje fibres. I. Evidence for a regenerative increase in Sr conductance. *Pfluegers Arch.* 322:60–72
72. Vleugels, A., Vereecke, J., Carmeliet, E. 1980. Ionic currents during hypoxia in voltage-clamped cat ventricular muscle. *Circ. Res.* 47:501–8
73. Weingart, R., Kass, R. S., Tsien, R. W. 1978. Is digitalis inotropy associated with enhanced slow inward calcium current? *Nature* 273:389–91
74. Yanagihara, K., Noma, A., Irisawa, H. 1980. Reconstruction of sino-atrial node pacemaker potential based on the voltage clamp experiments. *Jpn. J. Physiol.* 30:841–57

SODIUM-CALCIUM EXCHANGE IN THE HEART

G. A. Langer

University of California School of Medicine, Los Angeles, California 90024

INTRODUCTION

A short historical perspective of the subject provides a basis for division of the chapter into two major components: (*a*) a discussion of the relation of sodium (Na) to calcium (Ca) in *binding* at the sarcolemmal membrane, and (*b*) a discussion of the relation as it pertains to transsarcolemmal *transport*.

The first indications that the Na ion might affect Ca movement in the heart appeared in 1921. "The effect of lack of sodium shows a striking resemblance to the effect of strophanthidin" (14). The prophetic nature of this conclusion is discussed below. Von Wilbrandt & Koller (56) almost 30 years later developed the concept that the "site of action of calcium in the heart is most probably the cell membrane," and proposed a Ca_o/Na_o^2 relationship as being of importance in the control of the contractile response. In 1958 Lüttgau & Niedergerke (32) proposed that Na and Ca ions compete for anionic groups (R) at the cell surface (frog heart) on the basis of charge such that force was determined according to a Ca_o/Na_o^2 relationship. The idea was that the Ca, once bound to an anionic component (R^{2-}) according to $2Na + CaR \leftrightarrows Na_2R + Ca$, would move across the membrane and activate contraction. The control of force, via Ca, was placed at the cellular surface with little attention paid to possible events at the intracellular side of the sarcolemma.

In 1964, Repke in a consideration of the mechanism of action of digitalis (48) and Langer in a consideration of the staircase response (24) raised the possibility that intracellular accumulation of Na might be linked to augmented Ca influx, but no definitive mechanisms were offered. In 1968 Reuter & Seitz (51) found evidence for a membrane carrier system in the heart involving Na and Ca and demonstrated that Ca efflux rate was sensitive to the concentration gradient for Na across the sarcolemma. Next Baker et al (2) demonstrated, in the internally perfused squid axon, that in addition to inward Na movement coupled to outward Ca movement, exchange also occurred in the opposite direction—i.e. elevation of internal Na increased Ca influx. All of these studies indicated a system more complex than binding to an anionic sarcolemmal component and suggested that Na-Ca exchange might involve a carrier system operative in the control of the contractile response of the heart.

The results of work over the dozen years since the Baker et al study confirm that at least two aspects of the Na-Ca system exist in heart muscle: (a) interaction of the ions at *binding* sites, and (b) interaction of the ions in a transsarcolemmal *transport* system.

It is not clear, at present, the relative role played by the Na-Ca system in total ionic exchange in the heart.

BINDING

Using thin strips of frog atrium stimulated slowly (4/min), Chapman & Tunstall (12) noted a rapid development ($t_{1/2} = 2.4$–8.8 sec) of contracture upon removal of Na and an even more rapid decline ($t_{1/2} = 1.7$–5.5 sec) when Na was returned. These rapid responses of the strips beating at low rate indicated that the "Na/Ca antagonism occurs in a region of the muscle fibres accessible to the extracellular fluid and less accessible to the sarcoplasm. A structure fulfilling these requirements is the muscle cell membrane."

A recent study of Na-Ca exchange in the same tissue, frog atrium, also suggested that binding might be distinguished from a transport process. Horackova & Vassort (22) studied tonic tension responses elicited by depolarizing pulses as affected by alteration of $[Na]_o$ and $[Ca]_o$. As $[Na]_o$ was decreased at fixed $[Ca]_o$ peak tonic tension increased transiently by as much as 50% before it declined toward zero. The peak of the positive transient occurred within 30–60 seconds and became greater as $[Na]_o$ was decreased short of zero. At zero $[Na]_o$ the initial positive tension transient was eliminated. This study is consistent with the idea that as $[Na]_o$ falls it first increases the Ca/Na binding ratio at the cellular surface before $[Na]_i$ begins to fall significantly. The sequence is supported by comparison of the value for the exchange rate of the extracellular space with the rate of decline in $[Na]_i$. Horackova & Vassort measured a $t_{1/2}$ for washout of the extracellular space of 6.2 sec. Using a Na-sensitive microelectrode in Purkinje fibers, Ellis (17) noted that $[Na]_i$ declined with a $t_{1/2}$ of 2.3 min upon reduction of $[Na]_o$ from 140 to 14 mM, a rate less than 5% of the extracellular rate. Therefore it might be expected that as $[Na]_o$ fell additional Ca would bind competitively at surface sites, and if this Ca participates in the excitation contraction coupling (EC) process, positive inotropism would result.

Further evidence for surface binding and interaction of these sites with a Na-Ca membrane carrier comes from functional studies on mammalian heart. In rabbit ventricle Tillisch et al (54, 55) demonstrated a positive force transient upon lowering $[Na]_o$ from 142 mM and a negative transient upon raising $[Na]_o$. After the initial transients, force returned to near control levels (force at 142 mM $[Na]_o$) despite maintenance of the changed $[Na]_o$. The transient initial response had a $t_{1/2}$ of 1.2 min as compared to a $t_{1/2}$ of 5.6 min for the final return to control levels of force (muscles beating at 30/min). The rate of development of the initial response was independent of stimulation rate whereas the rate of return toward control levels increased as stimulation rate increased.

These functional responses are again consistent with the idea that decrements or increments in $[Na]_o$ increase or decrease Ca binding, respectively, at the cellular surface and that this competition affects the amount of Ca that, in some way, feeds the carrier system. Secondarily $[Na]_i$ responds in a frequency-dependent manner to the changed $[Na]_o$ which subsequently changes the rate of carrier movement (see below). For example, lowering of $[Na]_o$ increases Ca bound at sites that are in rapid equilibrium with the extracellular space. These sites are of importance in supplying the carrier. $[Na]_i$ begins to decline, but much more slowly than the effects on binding. Therefore carrier movement remains relatively unchanged but Ca influx is increased owing to augmented loading of the carrier system. As $[Na]_i$ falls, carrier movement slows and, though Ca binding remains high, influx falls toward control levels.

If the above interpretation is correct, the following should be experimentally demonstrable upon an abrupt decrease in $[Na]_o$: 1. a rapid and sustained increase in tissue Ca content; 2. no alteration in the rate of Ca efflux from the tissue; 3. an increase in sarcolemmal binding of Ca proportional to the transient increase in force; 4. no evidence for increased Ca influx via the slow Ca channel in the sarcolemma; and 5. progressive decline of $[Na]_i$ such that $[Na]_i/[Na]_o$ returns toward control values. All of these conditions have been demonstrated:

1.Increased Ca content: Wendt & Langer (57) demonstrated a rapid increase in tissue ^{45}Ca uptake in rabbit ventricle upon decrease of $[Na]_o$ to 36 mM. The increase was sustained until $[Na]_o$ was returned to 142 mM. The net increase was 125 μmoles/kg wet tissue. A similar response was defined in myocardial tissue culture with a net Ca increase amounting to 450 μmoles/kg wet tissue (29) upon reduction of $[Na]_o$ to 33 mM.

2.Unchanged Ca efflux rate: Neither of the studies above (29, 57) demonstrated any alteration in the rate of Ca efflux when $[Na]_o$ was reduced to

33–36 mM. Reuter & Seitz (51) demonstrated a marked decrease in Ca efflux, but this occurred when [Na]$_o$ was decreased to zero. Current results (29, 57) indicate that Na$_o$-dependent Ca efflux is fully saturated with respect to [Na]$_o$ at levels above 30 mM but is reduced when Na$_o$ is reduced below this level.

3. Sarcolemmal Ca binding: Philipson et al (41) have studied the effect of variation of Na concentration on Ca binding (non-energy dependent) to isolated sarcolemmal vesicles from rabbit heart. The percent change in binding was compared to the peak percent change in contractile force recorded by Tillisch (54) using the same Na concentrations in the same tissue. Through a range of [Na]$_o$ from 75 to 200 mM the percent changes of binding and force were virtually identical.

4. Other routes of influx: In at least 3 voltage-clamp studies, slow inward current was measured as [Na]$_o$ was decreased. No alteration in current was observed in either dog ventricle (50), cat ventricle (37), or frog atrium (4). Therefore there is no evidence that the augmented Ca influx secondary to decreased [Na]$_o$ occurs via the slow Ca channel. This implies that the Na-Ca carrier can function "in parallel" with Ca exchange through the slow channel.

5. Progressive decline of [Na]$_i$: Ellis (17) demonstrated that as [Na]$_o$ is reduced from 140 mM the new level of [Na]$_i$ is described by a linear relationship with a slope of 0.6. The slope indicates that [Na]$_i$/[Na]$_o$ will stablize at somewhat higher levels as [Na]$_o$ declines but apparently not enough to affect Ca efflux at [Na]$_o$ levels above 30 mM (see 2 above).

The relative roles of [Ca]$_o$ and [Na]$_o$ in the determination of Ca binding to the cellular surface (and therefore, in the determination of contractile force) are illustrated in Figure 1. The relationship is derived from experiments on rabbit ventricle, including studies of the relation of [Ca]$_o$ and [Na]$_o$ to Ca binding to sarcolemma (41), the relation of [Ca]$_o$ to force development (40), and the relation of the [Ca]$_o$/[Na]$_o$ ratio to force development (54). As can be seen force development is essentially interchangeable with Ca bound (41). Half-saturation occurs at Ca$_o$ ~ 1.3 mM. The effect of changing [Na]$_o$ between 75 and 175 mM is shown and indicates that Ca bound or the peak force transient increases 65% as [Na]$_o$ is decreased from 175 to 75 mM. The [Na]$_o$ reduction of 100 mM is equivalent to an increase of 2.5 mM [Ca]$_o$ at [Na]$_o$ = 142 mM. The relation will vary with diverse tissues from other species but serves to emphasize the interaction of Na and Ca with respect to binding at the cellular surface. When studies directed at the operation of transmembranous Na-Ca exchange employ various [Na]$_o$ levels it cannot be assumed that the effective [Ca]$_o$ remains unchanged. Such an assumption is seriously misleading, particularly with respect to evaluation of the stoichiometry of the system.

TRANSPORT

Transsarcolemmal movement via Na-Ca exchange has attracted a great deal more attention than Ca binding since Reuter & Seitz (51) and Baker et al (2) showed clear evidence for coupled Na-Ca movement. Work has proceeded concurrently on tissue preprations and on isolated sarcolemmal vesicles.

Tissue Preparations

Glitsch et al (21) found evidence in guinea-pig auricles not only for an effect of $[Na]_o$ on Ca efflux but, as in squid axon, an effect of $[Na]_i$ on Ca influx. Their discussion continued to assume that the exchange was an electroneut-

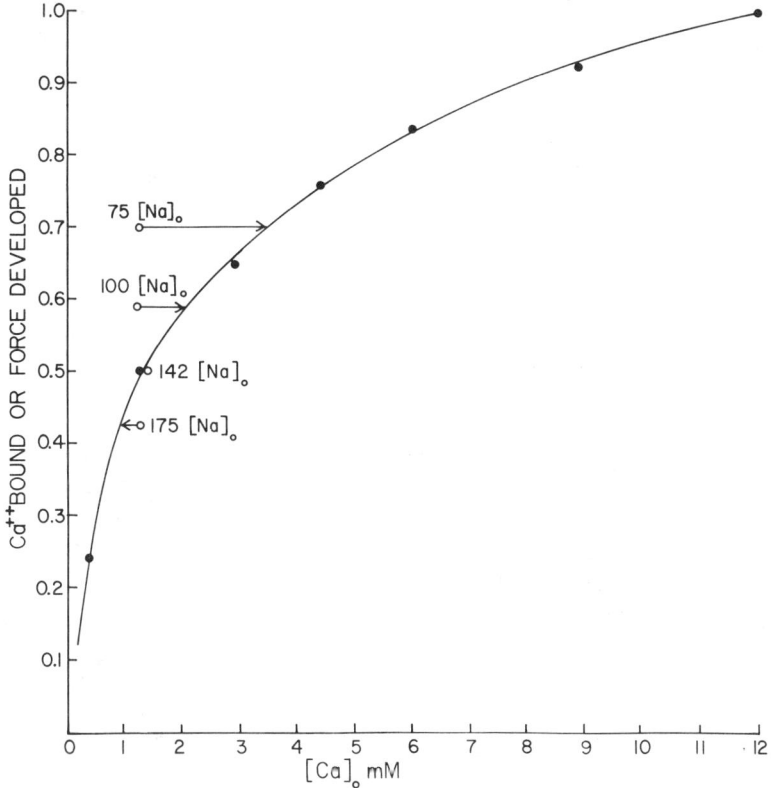

Figure 1 The relation between $[Ca]_o$ and Ca bound to sarcolemma or force development at 142 mM $[Na]_0$ (●). Ca bound or force is in relative units. The effect of various $[Na]_0$ (○) on binding/force is indicated. For example, a decrease in $[Na]_0$ from 175 to 75 mM is equivalent to increasing $[Ca]_0$ from 1.0 to 3.5 mM in the presence of 142 mM $[Na]_0$. Data are from rabbit ventricle.

ral one, i.e. 2 Na for 1 Ca transported in either direction. This 2:1 coupling relative to Ca extrusion ran into thermodynamic difficulty (4). The free energy change, ΔF (= RT ln concentration gradient for ion), required for movement of one Ca ion outward during diastole in mammalian ventricle (assuming $aCa_o = 0.7$ mM and $aCa_i = 10^{-4}$ mM) is -5.2 kcal. The free energy for movement of one Na ion inward [assuming $aNa_o = 115$ mM and $aNa_i = 7.2$ mM (17)] is 1.6 kcal. Therefore exchange of 2 Na for 1 Ca leaves a deficiency of 2.0 kcal. Unless coupling is greater than 2 to 1 or there is another source of energy, the system as visualized won't work.

Mullins & Brinley (36) demonstrated in squid axon that calcium efflux was sensitive to the membrane potential. Hyperpolarization increased and depolarization decreased Ca efflux and the sensitivity to membrane potential was lost if $[Na]_o$ was removed. This provided strong evidence for generation of current as the carrier moved Na and Ca—i.e. the exchange was electrogenic. Further direct evidence for electrogenicity, developed from study of sarcolemmal vesicles, is discussed below.

Studies in whole tissues have been significantly advanced by the development of ion sensitive microelectrodes (ISME). Using these electrodes in rabbit ventricular papillary muscles, Lee & Fozzard (30) found internal Na activity (aNa_i) to be 5.7 mM. This value compares well with that found by Ellis (17) in sheep heart Purkinje fibers of 7.7 mM. These values are much lower than measurements of total intracellular Na and would indicate that the intracellular activity coefficient for Na is about 0.175 (30). This suggests that over 80% of intracellular Na is bound or otherwise compartmentalized. Deitmer & Ellis (16) followed the aNa_i after high doses of cardioactive steroids and found that, despite complete inhibition of the Na-K pump, the aNa_i leveled off at values much below those expected at electrochemical equilibrium for Na. However, Deitmer & Ellis had previously demonstrated (15) that the presence of Ca in the extracellular fluid was required for stabilization of aNa_i and that raising the concentration of Ca (or Sr or Ba) could produce a decrease in aNa_i in the presence of Na-K pump inhibition. Conversely, decrease of $[Ca]_o$ in the presence of Na-K pump inhibition produced a further increase in aNa_i. Using flux techniques, Bridge et al (8) confirmed a marked stimulation of Na efflux upon raising $[Ca]_o$ in the presence of a nonfunctional Na-K pump. Thus it is likely that the "aNa_i plateau" can be attributed to exchange of Ca_o for Na_i via a Na-Ca exchange system.

Recent developments in Simon's laboratory (39) have resulted in the production of a neutral ligand quite selective for Ca that can be incorporated into microelectrodes sufficiently small to enable intracellular recording of aCa_i. It should be cautioned that the Ca microelectrodes are subject to instability, require careful and frequent calibration, and tend to demonstrate significant interference in their response in the range between 10^{-8} and 10^{-6} M aCa$_i$. In addition there is a possibility that they may damage components of the sarcotubular system and produce local increases in Ca concentration. Nevertheless these electrodes have provided significant further qualitative and semiquantitative information with respect to Na-Ca exchange.

Working with right ventricular papillary muscles from the ferret, Marban et al (33) found a mean resting aCa$_i$ of 0.26 μM which increased by 40% upon removal of Na from the perfusate. The rise in aCa$_i$ and the contracture response to Na$_o$ removal could be greatly augmented by partial membrane depolarization with 30 mM [K]$_o$ and introduction of caffeine. In rabbit ventricular muscle, Lee at al (31) found a resting aCa$_i$ of 38 nM or about 15% that found by Marban. However, the two studies assumed different stability constants for Ca-EGTA used in calibration, correction for which made the values comparable. This fact again emphasizes that absolute values for aCa$_i$ must be accepted with caution. In the Lee et al study, reduction of [Na]$_o$ to 20 mM increased aCa$_i$ almost 3-fold and was associated with a marked increase in twitch and a small increase in rest tension. The 20 mM [Na]$_o$ perfusion was maintained for less than 3 min in the record shown so that the late decline in tension noted by Tillisch et al (54) in rabbit ventricle would not have been evident. Ellis et al (18) also found, in sheep Purkinje tissue, a small increase in aCa$_i$ upon reduction of [Na]$_o$ to 14 mM. If high-dose acetylstrophanthidin was present, the [Na]$_o$ reduction produced a large increase in aCa$_i$ and a transient contracture response. Finally, Shing-Sheu & Fozzard (52) have succeeded in simultaneous impalement of sheep ventricular muscle with Na and Ca sensitive microelectrodes. They found aNa$_i$ to be 7.5 mM and aCa$_i$ to average 96 nM [using a Ca-EGTA stability constant for calibration intermediate to the values used by Marban (33) and Lee (31)]. With the ability to monitor aNa$_i$ and aCa$_i$ the equilibrium potentials, V_{Na} and V_{Ca}, can be calculated. The membrane potential (V_m) is also measured so the coupling ratio (n) of Na/Ca can be calculated according to (35):

$$n = \frac{2(V_{Ca} - V_m)}{(V_{Na} - V_m)} \qquad\qquad 1.$$

The n value was calculated after perfusion of the tissue with different concentrations of Na from control (151 mM) to 40% control. Despite the fact that aNa$_i$ decreased by as much as 25% at the low [Na]$_o$, the coupling ratio calculated by Shing-Sheu and Fozzard was 2.5–2.6 throughout the entire range. Finally Bers & Ellis (7) have also succeeded, with sheep Purkinje fibers, in measurement of aNa$_i$ and aCa$_i$ simultaneously. Using their values for V_m (−75 mV), aCa$_i$ (79–316 nM), and aNa$_i$ (5–7 mM), the coupling ratio, n, is 2.5–2.9. It should be stressed that the application of Equation 1 assumes that Na and Ca are the only ions exchanged, active transport of the ions by other mechanisms is small, and the passive leak of Ca is small. With all of these assumptions and the problems that affect quantitation previously noted above, the values derived should not be taken as absolute. But the fact that the coupling ratio in these studies was significantly above 2.0 is strong indication in these whole tissue experiments that the exchange is not electroneutral.

Other approaches in whole tissue support the operation of a Na-Ca exchange system. Two studies (22, 45) clearly indicate a role for Na in the maintenance of contracture states. In frog atrium, Horackova & Vassort (22) showed that voltage clamp-induced ionic tension was proportional to the level of $[Na]_o$. In rabbit ventricular muscle Ponce-Hornos & Langer (45) demonstrated that the contracture induced by zero $[K]_o$ and low $[Na]_o$ was markedly reduced if lithium (Li) instead of sucrose was used for Na_o replacement. Li substitution also abruptly halted the augmented Ca influx associated with contracture development. The study demonstrates that Li cannot substitute for Na in exchange for influxing Ca in the system. Once inside the cell it seems that Li displaces Na from the carrier but is incapable of supporting carrier movement.

Finally, in embryonic cardiac cell culture, Fosset et al (20) used veratridine to slow the inactivation of Na conductance and thus produce Na loading of the cell. As Na loading occurred, Ca influx was markedly stimulated and was dependent on the presence of Na. Augmented Ca influx did not occur when Li replaced Na in the presence of veratridine. This is in agreement with the study of Ponce-Hornos & Langer (45) reviewed above. It was also of interest that verapamil or D-600 (blockers of the slow channel) had no effect on the veratridine-dependent Na or Ca uptakes. This is consistent with the fact that augmentation of Ca binding and influx via the carrier by reduction of $[Na]_o$ produces no changes in slow inward current (4, 37, 50), as reviewed above.

In summary, the application of various techniques to whole, functional tissue provides clear evidence for the operation of a Na-Ca carrier. Its role in contributing to Ca influx can be distinguished from the role of the slow channel and there is good reason to believe that the ratio of Na carried to Ca carried is greater than 2—i.e. the exchange is electrogenic.

Sarcolemmal Vesicles

The first demonstration of the operation of Na-Ca exchange in vitro was that of Reeves & Sutko (46) using a relatively crude sarcolemmal preparation. Membrane vesicles isolated from rabbit ventricle accumulated Ca when an outwardly directed Na gradient was established in the absence of another energy source, e.g. ATP. A K gradient had little effect on Ca exchange and if the Na gradient was depleted by addition of Na_o the accumulated Ca was rapidly discharged. Pitts (44) also demonstrated Na-Ca exchange in vesicles from dog ventricle and estimated a coupling ratio of 3 Na for 1 Ca from his measurements of initial rates of ^{45}Ca efflux and ^{22}Na influx in previously Ca-loaded vesicles. Again, the stoichiometry has to be viewed with caution since the flux curves are nonlinear and small differences in estimated slope will make for large differences in the coupling ratio.

Four subsequent studies demonstrated that the exchange was, indeed, electrogenic. Bers et al (6) and Philipson & Nishimoto (42) used the K ionophore, valinomycin, to induce a potential across sarcolemmal vesicles purified approximately 30-fold with respect to classical membrane markers. The movement of K^+ into the vesicles produces an inside positive potential which, if the Na-Ca exchange results in net outward movement of positive charge, should augment the exchange. Such was clearly demonstrable and proportional to the membrane potential calculated from the Nernst relation. Reeves & Sutko (47) showed that the vesicular uptake of tetraphenylphosphonium (TPP^+), a positively charged highly lipid soluble ion, was enhanced by operation of the Na-Ca exchange system. Finally, Caroni et al (11) also found that K^+ in the presence of valinomycin significantly increased the exchange, and found that under conditions of charge neutralization provided by the counter K^+ movement, the K_m(Ca) of the exchanger was 1.5 μM. Such a K_m would be physiologically meaningful if it represented the affinity of the carrier for Ca at the inside of the sarcolemma in the intact cell. This value is compared to that found by Caroni & Carafoli (9, 10) for an ATP-dependent Ca transport system of the sarcolemma which is 0.3 μM (Ca). They have also measured the transport velocities of the Na-Ca exchange and the ATP-dependent Ca pump and found that the exchanger has a much greater velocity than the pump.

It is apparent that the in vitro studies support those of intact tissue and provide direct evidence for the electrogenic nature of Na-Ca exchange. The electrogenicity will make the movements of the exchanger sensitive to membrane potential during the course of excitation. The reversal potential (V_R) for the exchanger is defined by (35):

$$V_R = \frac{nV_{Na} - 2V_{Ca}}{n - 2} \qquad 2$$

If the values for V_{Na} (73 mV) and V_{Ca} (116 mV) calculated by Shing-Sheu & Fozzard (52) are taken and $n = 3$, then $V_R = -13$ mV. A similar calculation using Bers & Ellis' (7) data indicates $V_R = -28$ mV. Therefore the exchanger would provide net inward movement of Ca during the plateau of the action potential and net outward movement during rest. The exchanger would then contribute to net Ca influx during systole and net Ca efflux during diastole.

STOICHIOMETRY

The previous sections support electrogenic Na-Ca exchange as an operative mechanism in the heart. At this point one would like to have evidence that would provide an unequivocal value for the Na/Ca coupling ratio. A useful relationship at equilibrium is:

$$aCa_i = aCa_o \, (aNa_i/aNa_o)^n \, e^{(n-2)\frac{V_mF}{RT}} \qquad 3.$$

where the symbols are as previously given and F = Faraday (96,500 coulombs/mole), R = gas constant (8.2 joules/mol degree), and T = absolute temperature (310° at 37°C). ($V_mF/RT = -3.04$ at $V_m = -80$ mV). A great deal of information indicates that aCa_i during diastole is $< 10^{-7}$M. If this is used as a general guide we can test some values. If nominal $[Ca]_o$ is 1.5 mM the study of Bers & Langer (5) indicates that Ca bound to the low-affinity membrane sites that seem important in regulation of force is about 0.7 mmolar. This value will be used as an "effective $[Ca]_o$." A reasonable value for aNa_i seems to be 7.5 mM and Shing-Sheu & Fozzard's data (52) indicate aNa_o of 115 mM. Assuming $n = 3$ and $V_m = -80$ mV in diastole, $aCa_i = 9.3 \times 10^{-9}$M. This value is reasonable and fits the condition required during the diastolic state. Given the assumptions required, however, such analysis is little more than a "numbers game." First it has to be assumed that the level of aCa_i during diastole is solely dependent upon the operation of the Na-Ca exchanger. This is obviously not the case. The sarcotubular system and its active pumping contributes to the maintenance of aCa_i. In addition, the recently well-documented sarcolemmal Ca pump (9, 10, 34) with a K_m (Ca) in the range of 0.3 μM would also contribute. Second, testing of the stoichiometry of the system by the common practice of manipulation of aNa_o is fraught with problems. Even though aNa_i is monitored, the "effective $[Ca]_o$" cannot be assumed to remain unchanged. The binding of Ca at the sarcolemma is greatly affected by Na_o, as indicated in Figure 1. For example, a change of Na_o from 142 mM to 75 mM increases Ca bound by a factor of 1.4. If bound Ca "feeds" the carrier (as seems likely —see below), the effects would be significant. In addition there is good evidence that the exchanger may participate in Ca-Ca exchange (3, 43) and that Ca and Na compete for sites on the carrier itself. This would have obvious effects on the stoichiometry since the coupling ratio Na/Ca would then be affected by Na_o/Ca_o. Third, there is evidence that ATP may have an allosteric effect on Na-Ca exchange. Jundt & Reuter (23) found in guinea-pig auricles that the apparent affinity and maximal velocity of the exchange system were significantly reduced as ATP levels were depressed. In squid axon, Requena & Mullins (49) found that as Na_i increases there is an increase of the effect of ATP on Ca efflux. This, again, is compatible with an effect of ATP on the affinity of the Na-Ca exchanger, and this would be expected to affect the stoichiometry.

In summary, experimental systems have not yet been developed that can determine the coupling ratio of the system. However, it can be stated with assurance that more than 2 Na are carried for each 1 Ca and that the system is, therefore, electrogenic.

ROLE OF Na-Ca EXCHANGE IN CONTRACTILE CONTROL

Finally it is appropriate to attempt to place Na-Ca exchange in the context of its contribution to contractile control in the intact, functioning myocardium. The *net* amounts of Na and Ca cellular flux (exchange between cell and extracellular environment) during the course of a single contraction are markedly different. On the basis of membrane capacitance and magnitude of the action potential spike Na influx can be calculated to be approximately 13 μmoles/kg tissue/beat. There is evidence from electrophysiological studies (13) and flux studies (25) that N conductance remains significantly above resting levels through the action potential plateau with the addition of some 50 μmoles/kg more Na influx. In addition, resting Na flux is at least 3 μmoles/kg/sec (16). Thus in the course of one contraction-relaxation cycle in a ventricle beating 60 times/min with action potentials of 300 msec duration the total Na flux is about 65 μmoles/kg. By contrast, net Ca influx per cycle is probably between 1 and 2 μmoles/kg (26, 58). Therefore *net* Na:Ca flux is at least 30:1 from the point of view of beat-to-beat net gain and loss by the cell. The Ca requirement at the myofilaments is much greater than the net flux. In mammalian ventricle approximately 30 μmoles/kg wet wt/beat is needed for 90% activation of the myofilaments (53). Therefore if Na-Ca exchange is concerned only with net fluxes there is a large excess of Na—no matter what the coupling ratio. However, it is possible that a component of the total Ca flux involved in contraction coupling cycles (without leaving the cell) from sarcolemmal sites to myofilaments and back to sarcolemma. This seems to be the case for cultured myocardial cells (27). A fraction of such a component may be coupled to transsarcolemmal Na exchange.

Consideration of three experimental facts places an important condition upon operation of the Na-Ca exchanger: (a) Na-Ca exchange is electrogenic; (b) 30 µmoles Ca/kg wet wt/beat are required for near maximal force activation; and (c) net Ca flux is only 1–2 µmoles/kg wet wt/beat. Since the exchange is electrogenic, the carrier must cross the membrane potential barrier with each cycle. Since net Ca flux is 1–2 µmoles/beat the carrier cannot transport more than this amount between the extracellular space and the cell. It then follows that if the carrier is responsible for transport of more than 1–2 µmoles of the $\sim$ 30 required for near-maximal activation, it is necessary that the Ca be derived from storage sites on the cell surface between the membrane bilayer and the interstitial space. At present it is not possible accurately to quantitate and apportion that Ca which contributes to contraction coupling through the slow channel, that which contributes through the Na-Ca exchanger, and that derived directly or via regenerative release (19) from the sarcotubular system. Thus the relative importance of the Na-Ca exchanger in the control of beat-to-beat contraction is presently unknown.

There is, however, a pharmacological intervention that almost certainly involves the Na-Ca exchanger. This is the administration of digitalis. Though there remains controversy concerning the response of the myocardium to very low doses of the drug (38), there is strong evidence that higher doses and toxic doses depend upon stimulation of Na-Ca exchange (1, 28). It is proposed that Na_i increases as the drug inhibits the Na-K pump. The increased Na_i then provides the stimulus for augmented Ca influx via the Na-Ca exchanger.

CONCLUSIONS

In consideration of Na-Ca exchange in the myocardium there are two systems to be considered: (a) binding of the ions within the sarcolemmal-glycocalyx complex, (b) transsarcolemmal transport of the ions via an exchanger or carrier. The former is particularly important when variation of $[Na]_o$ is part of an experimental protocol. The transport system is electrogenic—i.e. more than 2 Na are exchanged for each 1 Ca. In most physiological circumstances $[Na]_i$ is the controller of transport activity. Increased $[Na]_i$ leads to enhanced Ca uptake by slowing Ca efflux, increasing Ca influx, or both. The exact stoichiometry of the coupling is not known and, indeed, may vary with such things as relative cation concentrations, level of cellular ATP, and activity of other Ca transport systems. The relative contribution of the Na-Ca exchange system to total cellular exchange of both ions in the heart under steady-state conditions is not known. If its contribution to the supply of coupling Ca is substantial its Ca supply must come from a store at the cell surface external to the lipid potential barrier.

The response to digitalis or conditions that involve Ca-dependent contracture seem to depend upon participation of the Na-Ca exchanger.

Literature Cited

1. Akera, T., Brody, T. M. 1978. The role of Na,K-ATPase in the inotropic action of digitalis. *Pharmacol. Rev.* 29:187–220
2. Baker, P. F., Blaustein, M. P., Hodgkin, A. L., Steinhardt, R. A. 1969. The influence of calcium on sodium efflux in squid axons. *J. Physiol. London* 200:431–58
3. Bartschat, D. K., Lindenmayer, G. E. 1980. Calcium movements promoted by vesicles in highly enriched sarcolemma preparation from canine ventricle. Calcium-calcium transport. *J. Biol. Chem.* 255:9626–34
4. Benninger, C., Einwachter, H. M., Haas, H. G., Kern, R. 1976. Calcium-sodium antagonism on the frog's heart: a voltage clamp study. *J. Physiol. London.* 259:617–45
5. Bers, D. M., Langer, G. A. 1979. Uncoupling cation effects on cardiac contractility and sarcolemmal Ca^{2+} binding. *Am. J. Physiol.* 273:H332–41
6. Bers, D. M., Philipson, K. D., Nishimoto, A. Y. 1980. Sodium-calcium exchange and sidedness of isolated cardiac sarcolemmal vesicles. *Bioch. Biophys. Acta.* 601:358–71
7. Bers, D. M., Ellis, D. 1981. Changes in intracellular calcium and sodium activity in sheep heart Purkinje fibers produced by changes of external sodium and internal pH. *J. Physiol. London.* In press
8. Bridge, J. H. B., Cabeen, W. R. Jr., Langer, G. A., Reeder, S. 1981. Sodium efflux in rabbit myocardium: relationship to sodium-calcium exchange. *J. Physiol. London.* 316:555–74
9. Caroni, P., Carafoli, E. 1980. An ATP-dependent Ca^{2+}-pumping system in dog heart sarcolemma. *Nature* 283:765–67
10. Caroni, P., Carafoli, E. 1981. The Ca^{2+}-pumping ATPase of heart sarcolemma: characterisation, calmodulin dependence and partial purification. *J. Biol. Chem.* 256:3263–70
11. Caroni, P., Reinlib, L., Carafoli, E. 1980. Charge movements during the Na^+/Ca^{2+} exchange in heart sarcolemmal vesicles. *Proc. Natl. Acad. Sci. USA* 77:6354-58
12. Chapman, R. A., Tunstall, J. 1969. Evidence for the site of Na/Ca antagonism in cardiac muscle of frog, *Rana pipiens. J. Physiol. London.* 201:9–11P (Abstr.)
13. Coraboeuf, E. 1978. Ionic basis of electrical activity in cardiac tissues. *Am. J. Physiol.* 234:H101–16
14. Daly, I. de Burgh, Clark, A. J. 1921. The action of ions upon the frog's heart. *J. Physiol. London.* 54:367–83
15. Deitmer, J. W., Ellis, D. 1978. The intracellular sodium activity of cardiac Purkinje fibres during inhibition and reactivation of the Na-K pump. *J. Physiol. London.* 284:241–59
16. Deitmer, J. W., Ellis, D. 1978. Changes in the intracellular sodium activity of sheep heart Purkinje fibres produced by calcium and other divalent cations. *J. Physiol. London.* 277:437–53
17. Ellis, D. 1977. The effects of external cations and ouabain on the intracellular sodium activity of sheep heart Purkinje fibres. *J. Physiol. London.* 273:211–40
18. Ellis, D., Deitmer, J. W., Bers, D. M. 1981. Intracellular pH, Na^+ and Ca^{++} activity measurements in mammalian heart muscle. In *Progress in Enzyme- and Ion-selective Electrodes,* ed. D. W. Lubbers, H. Acker, R. Buck, G. Eisenman, M. Kessler, W. Simon, pp. 148–55. NY: Springer
19. Fabiato, A., Fabiato, F. 1978. Calcium-induced release of calcium from the sarcoplasmic reticulum of skinned cells from adult human, dog, cat, rabbit, rat and frog hearts and from fetal and new born rat ventricles. *Ann. N.Y. Acad. Sci.* 307:491–522
20. Fosset, M., DeBarry, J., Lenoir, M-C., Lazdunski, M. 1977. Analysis of molecular aspects of Na^+ and Ca^{2+} uptakes by embryonic cardiac cells in culture. *J. Biol. Chem.* 252:6112–17
21. Glitsch, H. G., Reuter, H., Scholz, H. 1970. The effect of the internal sodium concentration on calcium fluxes in isolated guinea pig auricles. *J. Physiol. London.* 209:25–43

22. Horackova, M., Vassort, G. 1979. Sodium-calcium exchange in regulation of cardiac contractility. Evidence for an electrogenic, voltage-dependent mechanism. *J. Gen. Physiol.* 73:403–24
23. Jundt, H., Reuter, H. 1977. Is sodium-activated calcium efflux from mammalian cardiac muscle dependent on metabolic energy? *J. Physiol. London.* 266:78–79P (Abstr.)
24. Langer, G. A. 1964. Kinetic studies of calcium distribution in ventricular muscle of the dog. *Circ. Res.* 15:393–405
25. Langer, G. A. 1967. Sodium exchange in dog ventricular muscle. Relation to frequency of contraction and its possible role in the control of myocardial contractility. *J. Gen. Physiol.* 50:1221–39
26. Langer, G. A. 1974. Ionic movements and the control of contraction. In *The Mammalian Myocardium*, ed. G. A. Langer, A. J. Brady, pp. 193–217. NY: Wiley Interscience
27. Langer, G. A., Frank, J. S., Nudd, L. M. 1979. Correlation of calcium exchange, structure and function in myocardial tissue culture. *Am. J. Physiol.* 237:H239–46
28. Langer, G. A. 1977. Relationship between myocardial contractility and the effects of digitalis on ionic exchange. *Fed. Proc.* 36:2231–34
29. Langer, G. A., Nudd, L. M., Ricchiuti, N. V. 1976. The effect of sodium deficient perfusion on calcium exchange in cardiac tissue culture. *J. Mol. Cell. Cardiol.* 8:321–28
30. Lee, C. O., Fozzard, H. A. 1975. Activities of potassium and sodium ions in rabbit heart muscle. *J. Gen. Physiol.* 65:695–708
31. Lee, C. O., Uhm, D. Y., Dresdner, K. 1980. Sodium-calcium exchange in rabbit heart muscle cells: direct measurement of sarcoplasmic Ca^{2+} activity. *Science* 209:699–701
32. Lüttgau, H. C., Niedergerke, R. 1958. The antagonism between Ca and Na ions on the frog's heart. *J. Physiol. London.* 143:486–505
33. Marban, E., Rink, T. J., Tsien, R. W., Tsien, R. Y. 1980. Free calcium in heart muscle at rest and during contraction measured with Ca^{2+}-sensitive microelectrodes. *Nature* 286:845–50
34. Morcos, N. C. 1981. Isolation of frog heart sarcolemma possessing (Ca^{2+}+Mg^{2+})-ATPase and Ca^{2+}-pump activities. *Biochim. Biophys. Acta* 643:55–62
35. Mullins, L. J. 1979. The generation of electric currents in cardiac fibers by Na/Ca exchange. *Am. J. Physiol.* 236:C103–10
36. Mullins, L. J., Brinley, F. J. Jr. 1975. Sensitivity of calcium efflux from squid axons to changes in membrane potential. *J. Gen. Physiol.* 65:135–52
37. New, W., Trautwein, W. 1972. The ionic nature of slow inward current and its relation to contraction. *Pflügers Arch.* 334:24–38
38. Noble, D. 1980. Mechanism of action of therapeutic levels of cardiac glycosides. *Cardiovasc. Res.* 14:495–514
39. Oehme, M., Kessler, M., Simon, W. 1976. Neutral carrier Ca^{2+} microelectrode. *Chimica* 30:204–6
40. Philipson, K. D., Langer, G. A. 1979. Sarcolemmal-bound calcium and contractility in the mammalian myocardium. *J. Mol. Cell. Cardiol.* 11:857–75
41. Philipson, K. D., Bers, D. M., Nishimoto, A. Y., Langer, G. A. 1980. Binding of Ca^{2+} and Na^+ to sarcolemmal membranes: relation to control of myocardial contractility. *Am. J. Physiol.* 238:H373–78
42. Philipson, K. D., Nishimoto, A. Y. 1980. Na^+-Ca^{2+} exchange is affected by membrane potential in cardiac sarcolemmal vesicles. *J. Biol. Chem.* 255:6880–82
43. Philipson, K. D., Nishimoto, A. Y. 1981. Efflux of Ca^{2+} from cardiac sarcolemmal vesicles: influence of external Ca^{2+} and Na^+. *J. Biol. Chem.* 256:3698–702
44. Pitts, B. J. R. 1979. Stoichiometry of sodium-calcium exchange in cardiac sarcolemmal vesicles. *J. Biol. Chem.* 254:6232–35
45. Ponce-Hornos, J. E., Langer, G. A. 1980. Sodium-calcium exchange in mammalian myocardium: the effects of lithium. *J. Mol. Cell. Cardiol.* 12:1367–82
46. Reeves, J. P., Sutko, J. L. 1979. Sodium-calcium exchange in cardiac membrane vesicles. *Proc. Natl. Acad. Sci. USA* 76:590–94
47. Reeves, J. P., Sutko, J. L. 1980. Sodium-calcium exchange activity generates a current in cardiac membrane vesicles. *Science* 208:1461–64
48. Repke, K. 1964. Über den biochemischen Wirkungsmodus von Digitalis. *Klin. Wochenschrift.* 42:157–65
49. Requena, J., Mullins, L. J. 1979. Calcium involvement in nerve fibers. *Q. Rev. Biophys.* 12:371–460

50. Reuter, H., Beeler, G. W. Jr. 1969. Calcium current and activation of contraction in ventricular myocardial fibers. *Science* 162:399–401
51. Reuter, H., Seitz, N. 1968. The dependence of calcium efflux from cardiac muscle on temperature and external ion composition. *J. Physiol. London.* 195:45-70
52. Shing-Sheu, S., Fozzard, H. A. 1981. The stoichiometry of Na-Ca exchange in the mammalian myocardium. *Biophys. J.* 33:11 (Abstr.)
53. Solaro, R. J., Wise, R. M., Shiner, J. S., Briggs, F. N. 1974. Calcium requirements for cardiac myofibrillar activation. *Circ. Res.* 34:525–30
54. Tillisch, J. H., Fung, L. K., Hom, P. M., Langer, G. A. 1979. Transient and steady-state effects of sodium and calcium on myocardial contractile response. *J. Mol. Cell. Cardiol.* 11:137–48
55. Tillisch, J. H., Langer, G. A. 1974. Myocardial mechanical responses and ionic exchange in high-sodium perfusate. *Circ. Res.* 34:40–50
56. von Wilbrandt, W., Koller, H. 1948. Die Calciumwirkung am Froschherzen als Funktion des Ionengleichgewichts zwischen Zellmembran und Umgebung. *Helv. Physiol. Pharmacol. Acta* 6:208–21
57. Wendt, I. R., Langer, G. A. 1977. The sodium-calcium relationship in mammalian myocardium: effect of sodium deficient perfusion on calcium fluxes. *J. Mol. Cell. Cardiol.* 9:551–64
58. Winegrad, S. 1979. Electromechanical coupling in heart muscle. In *Handbook of Physiology,* ed. R. M. Berne, N. Sperelakis, pp. 393–428. Bethesda, Md: Am. Physiol. Soc.

CALCIUM RELEASE FROM CARDIAC SARCOPLASMIC RETICULUM

Saul Winegrad

Department of Physiology, School of Medicine, University of Pennsylvania, Philadelphia, Pennsylvania 19104

Several reviews (18, 27–29, 35, 60) published in the last few years have dealt with cardiac sarcoplasmic reticulum, and elsewhere in this section (61a) the molecular events of Ca transport in the reticulum are considered. Here attention is focused on specific properties of the reticulum that are important to its integrated function within the cardiac cell, in particular the triggering mechanism for release of calcium from the reticulum, changes in the properties of the reticulum during the cardiac cycle, and the modulation of sarcoplasmic Ca during a normal contraction of a myocardial cell.

Triggered Release of Ca from the Sarcoplasmic Reticulum

One of the major undetermined factors in the calcium activation of the contractile proteins in the heart is the mechanism by which the depolarization of the sarcolemma and transverse tubules causes a release of calcium from the sarcoplasmic reticulum. There have been two basic approaches to the problem. The first involves measuring the force or free Ca in skinned fiber preparations and studying procedures that produce phasic contractions, which have been assumed to result from transient releases of Ca^{2+} from an intracellular store (20, 23, 24, 26, 39). In the second, controlled depolarization of the sarcolemma and transverse tubules is used to excite the intact cell, and a correlation is evaluated between depolarization and some parameter such as birefringence (6), intracellular Ca concentration (11), or charge movement (14) that is thought to be related to the excitation-contraction coupling steps.

Four different procedures can produce phasic contractions in skinned striated muscle fibers: (a) application of a subthreshold concentration of Ca

(Ca-induced Ca release) (24, 26); (*b*) abrupt increase in the concentration of Cl or decrease in the concentration of K^+ in the bathing solution (depolarization-induced Ca release) (16, 20, 23); (*c*) osmotic shock (47); and (*d*) electrical stimulation (16, 17). The last two have been systematically examined only in skeletal and not in cardiac muscle fibers.

Because the mechanism of Ca-induced Ca release has been reviewed recently, it is not discussed here in detail (28). There is one important feature of this mechanism, in which the application of a subthreshold concentration of Ca^{2+} causes an abrupt release of a large amount of Ca from a Ca-loaded sarcoplasmic reticulum, that must be fundamental and yet is totally unexplained. In spite of the potential for positive feedback, Ca-induced Ca releases do not produce maximal contractions. Force generation is graded with the concentration of the trigger Ca, the Ca content of the reticulum, muscle length, and cyclic nucleotide concentration. While the appropriate mix of rate constants could produce a mechanism that is graded, this important characteristic has not been carefully studied. Although the concentrations of Ca to induce a Ca release vary to some extent, in three different models of the intact cardiac cells—namely the mechanically skinned, the saponin skinned and the hyperpermeable cardiac cell—release is definitely induced at sarcoplasmic concentrations of Ca that could be produced during the action potential by the slow inward calcium current. Skeletal muscle is different inasmuch as extracellular Ca is not required for electrically stimulated contractions.

More controversy surrounds the phenomenon of "depolarization-induced Ca release," a term used to describe the sudden increase in the concentration of calcium around the myofibrils that occurs when the relatively impermeant anion propionate or methylsulfonate is replaced in the bathing solution by Cl^-. Fabiato & Fabiato (27) have been unable to demonstrate this type of release in their truly mechanically skinned single cardiac cell preparation, whereas Kerrick & Best (39) find the mechanical response to elevated Cl in their multicellular preparation with disrupted membranes. Fabiato & Fabiato attribute this difference to the presence of cells with intact membranes and resting potentials in the center of the multifiber preparation. Little if any Cl-induced Ca release has been observed in the saponin skinned cardiac cell although K replacement by choline was effective in producing force (22).

The question has been raised whether Cl-induced Ca release is due to a depolarization of the membranes of the sarcoplasmic reticulum by Cl or to osmotic shock that results as Cl entry into the sarcoplasmic reticulum drags a large volume of water with it (47). Activation by Cl was initially observed when a droplet of solution containing a high concentration of Cl was applied to a skinned fiber immersed in oil (16). Endo and his colleagues (20,

23, 24), using 25 mM caffeine to empty the reticulum of Ca and tension as a measure of the amount of Ca release, observed that the amplitude of mechanical responses increased with the concentration of Cl used. Since certain local anesthetics produce similar effects on the Ca-induced contraction in skinned fibers and the electrically induced contraction in intact cells, they suggested that Cl was initiating the normal steps in the contractile process. Because the responses to K withdrawal and Cl addition were also similar they inferred that a change in potential across the reticular membranes initiated the Ca release.

Ca release in response to Cl elevation but not to K removal can be produced in isolated vesicles from skeletal sarcoplasmic reticulum. However when Meissner & McKinley (47) reduced water flow into the isolated vesicles by using a slowly permeating cation or anion with the Cl or K, they were unable to produce Ca release by rapid changes in the ionic composition of the bath that should have produced changes in potential in both directions. On the other hand, it is possible to release additional Ca with KCl after an osmotically induced release, and this has led to the suggestion (37) of two different pools of Ca in the sarcoplasmic reticulum, one released by a change in ionic composition and the other by water flow. In an attempt to prevent water flow across the reticulum membrane when the concentrations of K and Cl were changed in solutions bathing skinned fibers, Mobley (48) maintained a constant [K] · [Cl] product, which should have prevented the movement of water and eliminated any osmotic component to the Cl-induced Ca release (32). Cl contractures still occurred when Cl was increased but they were substantially weaker than those produced when [K] · [Cl] was not held constant.

When the initial Ca released by Cl in a skinned fiber is trapped by EGTA, little additional release occurs and the total amount is much smaller (25). The elevation of Cl may result in a small release of Ca followed by a Ca-induced Ca release of larger proportion. This model is incomplete, even if correct, because it does not account for the graded nature of the Cl-induced release. Stephenson (60) has suggested that Cl may contribute to a net increase in sarcoplasmic Ca concentration by inhibiting the reticulum, as she finds a 96% reduction of Ca^{45} uptake in the presence of Cl.

The sarcoplasmic reticulum can be separated into light and heavy fractions that both contain the Ca pump (46). Calsequestrin, the major Ca binding protein, however, is restricted to the heavy fraction. After vesicles have been loaded, Ca can be released from either fraction by substitution of Cl for methylsulfonate, but osmotic swelling of the vesicles should occur as the Cl enters the vesicles. Osmotically buffering the bathing solution with 200 mM sucrose before replacing methylsulfonate with Cl markedly reduces the amount of Ca released from the light fraction of the reticulum but only slightly reduces the release from the heavy fraction (12). The release from the heavy fraction is relatively specific for Ca, whereas the light fraction releases a large amount of Na^{22} as well. Na dantrolene inhibits the release from the heavy fraction but not from the light fraction. These results support the existence of a non-osmotic, Cl-induced release of Ca, but more direct measurements are necessary to prove it.

Using fluorescent oxonal and oxacarbocyanine dyes to detect changes in the membranes in isolated sarcoplasmic reticular vesicles that might be related to differences in electrical potential, Beeler et al (8) have found a clear separation between change in the optical signal and rate of release of Ca. The accelerated release when a methylsulfonate medium was diluted with a chloride medium was not accompanied by any large change in optical signal, whereas dilution with methylsulfonate and valinomycin produced a large change in optical signal with only a small change in Ca release. Other dissociations of potential and Ca release were effected with the ionophore A23187. In support of the suggestion of Stephenson (60), they found a slower Ca uptake in Cl than in methylsulfonate medium.

Further evidence against a critical role for depolarization of a resting potential across the reticulum membrane in releasing stored Ca comes from analysis of the concentrations of Cl, Na, and K in the reticulum and in the cytoplasm (57, 58). With electron probe microanalysis, Somlyo et al (57, 58) have found only a small gradient of Cl and K across the reticulum membrane, and the gradient is outward, the wrong direction to explain the Cl-induced Ca release on the basis of depolarization. The concentrations are, however, quite different from those of the extracellular fluid, indicating that the lumen of the reticulum is not continuous with the extracellular space. This technique measures total amount and not ionic concentration; nevertheless, these data argue forcefully against a resting potential across the reticulum membrane since the reticulum membrane, at least in isolated vesicles, is permeable to Cl, Na, and K. These data do not rule out the production of a signal from a transient polarization of the sarcoplasmic reticulum membrane by the high concentration of Cl^-.

Further problems become apparent when one examines the change in the content of Na, K, and Cl in the sarcoplasmic reticulum and in the cytoplasm as a result of 1.2 sec tetanus. In agreement with the autoradiographic data, electron probe analysis shows a loss of 59% of the Ca in the terminal cisternae (62 m mol/kg dry TC). However, even though Mg and K in the TC increase, they are insufficient to balance the loss of Ca, and there is a deficit of 62 mEq/kg dry wt in the terminal cisternae. Since this net movement of charge would produce a prohibitive potential difference, the calcium loss must be balanced by a combination of proton and organic ion movements that cannot be detected by electron probe analysis.

Changes in Intracellular Membranes during Contraction

Using a skeletal muscle preparation in which ionic current was reduced to a minimum, Chandler et al (14) measured a nonlinear charge movement that had a voltage dependence and time course similar to those of the activation of contraction. This charge movement also resembles the inactivation and repriming process in time course and appearance. They attributed it to some crucial step in the excitation-contraction coupling process. Although the size of the charge movement decreased markedly when the transverse tubules were partially disrupted by glycerol shock treatment (15), the disparity between the decline of charge movement and capacitance—the latter declined only half as much as the former—precludes any definite conclusion about the importance of the transverse tubules. The amount of charge movement associated with activation as determined by the minimum observable contraction remained the same even when it was elicited by depolarizations of different amplitudes under different conditions.

Charge movement and Ca concentration have been measured simultaneously with the cut end preparation of muscle (41, 42). The rise in intracellular Ca appeared at about −50 to −40 mV. Nonlinear charge movement generally began at somewhat more negative voltages than the rise in intracellular Ca and the onset of contraction, indicating that a certain amount of charge movement occurs before the contractile filaments are activated. The calcium transient, after allowance of a certain transition time from off to on, can be described by the sum of a constant and two exponential voltage-dependent functions that have time-independent coefficients. The faster of the two rises while the slower falls with depolarization. Only a single exponential is required for the decay of the transient. These data have been interpreted in terms of a closed, three-compartment system connected by time-independent fractional transfer coefficients. The compartments have been tentatively identified as the cytoplasm, the SR uptake sites, and SR release sites. Since the off rate for the calcium transient depends on the voltage of the preceding pulse, the performance of the reticulum is not governed only by the instantaneous membrane depolarization. The establishment of these time-independent Ca movements comes only after charge movement is 98% complete, and their decay begins only after a certain fraction of charge has returned to its resting position.

The relation between activation of contraction and the nonlinear charge movement has been disputed by Adrian et al (1) because much of the charge movement disappears in hypertonic solution at long sarcomere lengths, but Hui & Gilly (34) have found that both the strength-duration relation and the amount of charge moved are substantially the same over a range of sarcomere lengths of 2.4 μm to about 3.3 μm.

In an attempt to detect changes in the reticular membrane that are associated with depolarization of the sarcolemma and the rise in cytoplasmic Ca^{2+}, changes that have so far been inaccessible to microelectrode recording, optical techniques have been used (5, 6, 9, 50, 51). Either an intrinsic optical property of the muscle cell, such as birefringence, or the fluorescence of dyes that concentrate at membranes and are sensitive to changes in the electrical field have been measured during the contractile cycle. If the dye does not penetrate the cell, it can be used as an indicator of the potential across the surface membrane—e.g. the use of merocyanine to measure the action potential in frog hearts (3). Penetrating dyes, such as Nile Blue, offer the hope of detecting voltage signals from the sarcoplasmic reticulum. Since its surface area is much greater than that of the surface membrane, simultaneous signals from the sarcolemma or the transverse tubules would be swamped by a signal from the reticulum. In frog semitendinosus muscle the intensity of Nile Blue fluorescence increases transiently following a single electrical stimulus. The signal has several properties that have led to its association with the process of excitation-contraction coupling (9). It increases after substituting nitrate for chloride to prolong the action potential. Addition of signals to make a greater amplitude does not occur during trains of stimuli. Its threshold voltage is the same as that for active tension. It begins during the declining phase of the action potential, continues after it, and declines during maintained depolarization. Heavy water decreases its amplitude. In the voltage-clamped, cut-end muscle preparation (63) the signal displayed a nonlinear relation to membrane potential. Local anesthetics like procaine and tetracaine that are thought to inhibit Ca release from the sarcoplasmic reticulum (10) without inhibiting charge movement (4) block both the development of tension and the change in Nile Blue fluorescence normally produced by depolarization (13). One problem however in the execution of these optical studies is the need to use nonphysiological conditions such as hypertonic solutions or excessive stretch to prevent movement artifacts.

When single skeletal muscle fibers are depolarized by a propagated action potential two changes in birefringence occur that may reflect steps in the coupling process (5, 6). The first component is coincident with the action potential but is normally obscured by a much larger second component that begins just after stimulation and is almost complete by the onset of tension. The second component propagates along the fiber at the same velocity as the action potential and is enhanced by nitrate and paired stimulation, both of which increase tension. When a slit of light is moved across the fiber the amplitude of the signal changes more closely in relation to fiber volume than to surface area. Hypertonicity and stretch reduce tension much more than the signal. The time course of the second component of the birefringence signal is similar to that produced by Nile Blue (51). Injection of EGTA into the cell abolishes both the birefringence signal and the calcium transient measured with intracellular arsenazo III (61), leading to the suggestion that the birefringence signal is due not to a change in the sarcoplasmic reticulum itself but to some event associated with the released Ca. This is not the only interpretation since Stephenson has shown that the total amount of calcium released can be substantially reduced when EGTA chelates the initially released Ca (59).

Measurements of the Concentration of Intracellular Ca

The photoprotein aequorin (11, 21), dyes such as arsenazo III and antipyrolazo III (41) that have Ca sensitive absorption properties, and Ca-sensitive microelectrodes (43) have all been used to measure intracellular concentration of Ca. Aequorin has been the most useful in providing information about the function of the sarcoplasmic reticulum in heart muscle. Aequorin was first used in vertebrate muscle by microinjecting it into single twitch muscle fibers from frog and toad (11). Resting cells showed no luminescence, but light was emitted following electrical stimulation. The amplitude of light was greatest at about the time of the maximum rate of rise of tension, and it decayed considerably before tension. This temporal relation has been interperated as indicating that cytoplasmic Ca rises first; then, as troponin binds Ca to reduce its concentration in the cytosol, tension rises. Non-uniform distribution of aequorin in the cell, such as a higher concentration between than within the myofibrils, could also contribute to the phase difference between cytoplasmic Ca and tension.

In skeletal muscle the amplitude and shape of the aequorin signal are influenced by the stimulation frequency and the length of the muscle fiber. Repeated isometric contractions cause a decrease in the amplitude and the rate of decay of the aequorin signal as one might predict on the basis of the redistribution of Ca that occurs within the reticulum (66, 67). During a tetanus, the aequorin signals fuse, the amplitude rises, and the concentration of cytoplasmic Ca clearly become supramaximal for activation of contraction. The sensitivity of the aequorin signal to muscle length—it is maximum at about 2.5–2.7 μm sarcomere length—is consistent with the apparent length dependence of Ca release with electrical but not caffeine stimulation that has been inferred from Ca^{45} flux studies (30).

For studies with the smaller frog cardiac cells that would produce weaker signals, the aequorin technique has been modified by using multiple injections into many cells in a trabecula or papillary muscle (2, 3, 64). As in skeletal muscle, the amplitude of light rises before tension, peaks at about the time of the fastest rate of rise of tension, and decays before tension. Certain inotropic interventions modify the signal. The amplitude of both tension and the light flash increases with frequency of stimulation and with increasing concentrations of extracellular Ca. The response to a β-adrenergic agonist is particularly interesting but more complex. Tension and the aequorin signal rise to a greater amplitude faster and decay more quickly, but the upstroke of the aequorin signal has two phases. Catecholamines presumably produce a larger influx of Ca from the extracellular space and an additional release from an internal store likely to be the sarcoplasmic reticulum. Though sparser in frog than in mammalian hearts, the sarcoplasmic reticulum can still accumulate and release Ca when exposed to an adequate loading concentration (52, 68). Acetyl strophanthidin, on the other hand, increases the aequorin flash without producing an obvious second phase. Less easy to explain is the increase in contractile force with a decrease in aequorin flash that occurs at lengths up to 140% of slack length (2). Apparently either Ca sensitivity changes with length or (less likely) aequorin is redistributed (19, 25).

There are interesting similarities and differences in mammalian hearts (3). The amplitude of the aequorin flash changes in parallel with tension as extracellular Ca is increased but with little change in the time course of the light. The amplitude of the light also increases with force as the frequency of stimulation is increased, but in this case the flash and not tension decays more rapidly at higher frequencies. Time to peak tension decreases without a change in time to peak light. Epinephrine produces a large increase in light intensity and the appearance of a second slower and smaller signal after the first has been completed. The second increase is not associated with a rise in tension although its amplitude is as large as flashes that were associated with tension before epinephrine was administered. The amount of tension associated with a given amplitude of light is reduced by epinephrine in agreement with the known decrease in Ca sensitivity of cardiac contractile protein produced by a cAMP-regulated phosphorylation of TNI (33, 44, 49, 54). Interestingly, the decrease in sensitivity is complete at a lower concentration of the drug than the increase in amplitude of the aequorin flash. Caffeine increases contractile force and as the contraction becomes stronger, the aequorin signal increases and shows a prominent second phase that temporally coincides more closely than the fast phase with the time course of tension.

The two phases of the upstroke of the aequorin signal have been studied in greater detail in injected canine cardiac Purkinje fibers (64). As the interval between stimuli is reduced, the height of the aequorin flash and contractile tension decreases, but the two phases do not change proportionally. For increasing intervals from 0.6–2.5 sec between a conditioning impulse and a second test stimulus, the shape of the aequorin flash associated with the second stimulus changes. At 0.6 sec interval there is only a fast phase in the light signal and tension is small. The fast phase is the same with a 2.5 sec separation between conditioning and test stimulus, but a slower phase begins to appear that is much smaller than the one associated with the conditioning impulse. The amplitude of the contraction following test stimulus is also smaller than that following the conditioning stimulus. Force correlates better with the second phase of the aequorin flash. The rapid recovery of the fast phase following a stimulus resembles the reversal of inactivation of the slow inward Ca current (31), while both direct and indirect evidence indicate that complete recovery of the reticulum after a release of Ca occurs more slowly (66, 67, 69). For these reasons the rapid and the slow phases of light are attributed respectively to a transmembrane flux and an intracellular release of Ca.

Integration of Function of Sarcoplasmic Reticulum into the Contractile Cycle

For an appreciation of the physiological function of the sarcoplasmic reticulum it is necessary to place it in the structural context of the cell and the functional context of the cardiac cycle. In mammalian hearts, the sarcoplasmic reticulum provides the major portion of the calcium for activation of the contractile system, although there is a substantial influx of calcium during the plateau of the action potential (7). The sarcoplasmic reticulum of amphibian cardiac muscle, on the other hand, probably becomes important in excitation-contraction coupling only when the inotropic state of the tissue has been raised considerably by catecholamine stimulation or elevated extracellular Ca (2, 68). Within the reticulum of cardiac muscle there may be separation of release and uptake sites that can influence the mechanism of activation during normal repeated contractions, but this has only been shown in skeletal and not in heart muscle (36, 46, 66, 67).

Although several calcium binding sites might compete with the sarcoplasmic reticulum and troponin for cytoplasmic Ca, especially when the concentration is elevated following an action potential, only the mitochondria have been seriously considered because of their large capacity for Ca. Several different kinds of studies, however, indicate that mitochondria do not play a significant role in the Ca cycling during contraction of cardiac cells. When Ca accumulation by the sarcoplasmic reticulum in skinned cardiac fibers has been blocked by removal of ATP, uptake of calcium by the mitochondria can still occur in the presence of substrate; yet under these conditions slow loss of calcium from the sarcoplasmic reticulum still increases tension (68). Inhibition of mitochondrial uptake of calcium in intact hearts does not alter the time course of relaxation (65). According to the results of electron probe analysis in skeletal muscle there is no calcium in the mitochondria either before or after a tetanus unless the cell has been damaged (57). The kinetics of calcium uptake by isolated mitochondria is too slow for the organelle to be important in normal Ca cycling (40, 56). Consequently, intracellular Ca cycling during activation and relaxation of the contractile system involves only the sarcoplasmic reticulum and the sarcolemma.

The function of the sarcoplasmic reticulum and of the sarcolemma and contractile proteins changes when the inotropic state of the cells has been increased with catecholamines. The influx of calcium during the plateau of the action potential rises (55), the rate of calcium uptake by the reticulum increases (62), the concentration of Ca necessary to produce a given percentage of maximal activation increases (33, 44, 49, 54) because of an increase in the rate at which Ca comes off troponin (53), and the maximum force generating capacity of the contractile proteins increases (45). These changes are coordinated to produce a faster and stronger contraction. The combination of the faster release of calcium from troponin and the more rapid uptake by the reticulum accelerates relaxation. The need for a larger concentration of Ca to produce a given degree of activation is met by the combined effects of an increase in the slow inward calcium current and a greater release of Ca from the reticulum, which contains more calcium as a result of the increase in both influx and uptake. The effect of these two mechanisms for increasing cytoplasmic Ca may exceed the additional calcium requirements imposed by a decrease in Ca sensitivity and contribute to the increase in the amount of force. By requiring less time the stronger contraction that results from coordinated modification of the properties of the sarcolemma, contractile proteins, and sarcoplasmic reticulum facilitates the filling of the heart at the higher rates of contraction that normally accompany the enhanced inotropic state produced by adrenergic stimulation.

Literature Cited

1. Adrian, R., Caputo, C., Huang, C. 1978. Effect of stretch on intramembrane charge movement in striated muscle. *J. Physiol. London* 284:151P
2. Allen, D., Blinks, J. 1978. Calcium transients in aequorin-injected frog cardiac muscle. *Nature* 273:509–13
3. Allen, D., Kurihara, S. 1980. Calcium transients in mammalian ventricular muscle. *Europ. Heart J.* 1:(Suppl A) 5–15
4. Almers, W., Best, P. 1976. Effects of tetracaine on displacement currents and contraction in frog skeletal muscle. *J. Physiol. London* 263:582–611
5. Baylor, S., Oetliker, H. 1977. A large birefringence signal preceding contraction in single twitch fibres of the frog. *J. Physiol. London* 264:141–62
6. Baylor, S., Oetliker, H. 1977. Birefringence signals from surface and T system membranes of frog single muscle fibres. *J. Physiol. London* 264:199–213
7. Beeler, W., Reuter, H. 1970. Membrane calcium current in ventricular myocardial fibers. *J. Physiol. London* 207:191–209
8. Beeler, T., Russell, J., Martonosi, A. 1979. Optical probe responses on sarcoplasmic reticulum: oxacarbocyanines as probes of membrane potential. *Eur. J. Biochem.* 95:579–91
9. Benzanilla, F., Horowicz, P. 1975. Fluorescence intensity changes associated with contractile activation in frog muscle stained with Nile Blue A. *J. Physiol. London* 246:709–35
10. Bianchi, C. P., Bolton, T. 1967. Action of local anesthetics on coupling systems in muscle. *J. Pharmacol. Exp. Ther.* 153:388–405
11. Blinks, J., Rudel, R., Taylor, S. 1978. Calcium transients in isolated amphibian skeletal muscle fibres: detection with aequorin. *J. Physiol. London* 277:291–323
12. Campbell, K., Shamoo, A. 1980. Chloride induced release of actively loaded calcium from light and heavy sarcoplas-

mic reticulum vesicles. *J. Membr. Biol.* 54:73–80
13. Caputo, C., Vergara, J., Bezanilla, F. 1979. Local anesthetics inhibit tension development and Nile blue fluorescence signals in frog muscle fiber. *Nature* 277:401–2
14. Chandler, K., Rakowski, R., Schneider, M. 1976. A non-linear voltage dependent charge movement in frog skeletal muscle. *J. Physiol. London* 254:245–83
15. Chandler, K., Rakowski, R., Schneider, M. 1976. Effects of glycerol treatment and maintained depolarization on charge movement in skeletal muscle. *J. Physiol. London* 254:285–316
16. Costantin, L., Podolsky, R. 1967. Depolarization of the internal membrane system in the activation of frog skeletal muscle. *J. Gen. Physiol.* 50:1101–24
17. Costantin, L., Taylor, S. 1973. Graded activation in frog muscle fibers. *J. Gen. Physiol.* 61:424–43
18. Ebashi, S. 1976. Excitation-contraction coupling. *Ann. Rev. Physiol.* 38:293–313
19. Endo, M. 1972. Stretch induced-increase in activation of skinned muscle fibers by calcium. *Nature New Biol.* 237:211–13
20. Endo, M. 1977. Calcium release from the sarcoplasmic reticulum. *Physiol. Rev.* 57:71–108
21. Endo, J., Blinks, J. 1973. Inconstant association of aequorin luminescence with tension during calcium release in skinned muscle fibres. *Nature New Biol.* 246:218–20
22. Endo, M., Kitazawa, T. 1978. E-C coupling studies on skinned cardiac fibers. In *Biophysical Aspects of Cardiac Muscle*, ed. M. Morad, pp. 307–27. NY: Academic
23. Endo, M., Nakajima, Y. 1973. Release of calcium induced by depolarization of the sarcoplasmic reticulum membrane. *Nature New Biol.* 246:216–18
24. Endo, M., Tanaka, M., Ogawa, Y. 1970. Calcium induced calcium release of calcium from the sarcoplasmic reticulum of skinned skeletal muscle fibers. *Nature* 228:34–36
25. Fabiato, A., Fabiato, F. 1975. Dependence of contractile activation of skinned cardiac cells on the sarcomere length. *Nature* 256:54–56
26. Fabiato, A., Fabiato, F. 1975. Contractions induced by a calcium-triggered release of calcium from the sarcoplasmic reticulum of single skinned cardiac cell. *J. Physiol. London* 249:469–95
27. Fabiato, A., Fabiato, F. 1977. Calcium release from the sarcoplasmic reticulum. *Circ. Res.* 40:119–29
28. Fabiato, A., Fabiato, F. 1979. Calcium and cardiac excitation-contraction coupling. *Ann. Rev. Physiol.* 41:473–84
29. Fozzard, H. 1977. Heart: Excitation-contraction coupling. *Ann. Rev. Physiol.* 39:201–20
30. Frank, J., Winegrad, S. 1976. Effect of muscle length on ^{45}Ca efflux in resting and contracting skeletal muscle. *Am. J. Physiol.* 231:555–59
31. Gibbons, W. R., Fozzard, H. 1975. Slow inward current and contraction of sheep cardiac Purkinje Fibers. *J. Gen. Physiol.* 65:367–84
32. Hodgkin, A., Horowicz, P. 1959. The influence of potassium and chloride ions on the membrane potential of single muscle fibers. *J. Physiol. London* 148:126–60
33. Holroyde, M. J., Potter, J., Solaro, R. J. 1979. The calcium binding properties of phosphorylated and unphosphorylated cardiac and skeletal myosin. *J. Biol. Chem.* 254:6478–82
34. Hui, C., Gilly, W. 1979. Mechanical activation and voltage dependent charge movement in stretched muscle fibers. *Nature* 281:223–25
35. Inesi, G. 1979. Transport across single biological membranes. In *Membrane Transport in Biology*, ed. G. Giebiesch, D. Tosteson, H. Ussing, pp. 357–93. Heidelberg: Springer
36. Jorgensen, A., Kalnins, V., MacLennan, D. 1979. Localization of sarcoplasmic reticulum proteins in rat skeletal muscle by immunofluorescence. *J. Cell. Biol.* 80:372–84
37. Kasai, M., Miyamoto, H. 1976. Depolarization-induced calcium release from sarcoplasmic reticulum fragments . I. Release of calcium taken up upon using ATP. *J. Biochem. Tokyo* 79:1053–66
38. Deleted in proof
39. Kerrick, G., Best, P. 1974. Calcium ion release in mechanically disrupted heart cells. *Science* 183:435–37
40. Kitazawa, T. 1976. Physiological significance of Ca uptake by mitochondria in the heart in comparison with that by cardiac sarcoplasmic reticulum. *J. Biochem.* 80:1129–47
41. Kovacs, L., Rios, E., Schneider, M. 1979. Calcium transients and intramembrane charge movement in skeletal muscle fibers. *Nature* 279:391–96
42. Kovacs, L., Schneider, M. 1977. Increased optical transparency associated with excitation-contraction coupling in

voltage-clamped cut muscle fibres. *Nature* 265:556–60
43. Marban, E., Rink, T., Tsien, R., Tsien, R. 1980. Free calcium in heart muscle at rest and during contraction measured with Ca-sensitivity microelectrodes. *Nature* 286:845–50
44. McClellan, G., Winegrad, S. 1978. The regulation of calcium sensitivity of the contractile system in mammalian cardiac muscle. *J. Gen. Physiol.* 72:737–64
45. McClellan, G., Winegrad, S. 1980. Cyclic nucleotide regulation of the contractile proteins in mammalian cardiac muscle. *J. Gen. Physiol.* 75:283–95
46. Meissner, G., Conner, G., Fleischer, S. 1973. Isolation of the sarcoplasmic reticulum by zonal centrifugation and purification of Ca^{++} pump and Ca-binding protein. *Biochem. Biophys. Acta* 298:246–69
47. Meissner, G., McKinley, D. 1976. Permeability of sarcoplasmic reticulum membrane. The effect of changed ionic environment on Ca^{++} release. *J. Membr. Biol.* 30:79–88
48. Mobley, B. 1979. Chloride and osmotic contractures in skinned frog muscle fibers. *J. Membr. Biol.* 46:315–29
49. Mope, L., McClellan, G., Winegrad, S. 1980. Calcium sensitivity of the contractile system and phosphorylation of troponin in hyperpermeable cardiac cells. *J. Gen. Physiol.* 75:271–82
50. Morad, M., Salama, G. 1979. Optical probes of membrane potential in heart muscle. *J. Physiol. London* 292:267–95
51. Oetliker, H., Baylor, S., Chandler, K. 1975. Simultaneous changes in fluorescence and optical retardation in single muscle fibres during activity. *Nature* 257:693–96
52. Page, S., Niedergerke, R. 1972. Structure of physiological interests in frog heart ventricle. *J. Cell. Sci.* 11:179
53. Robertson, S. P., et al. 1982. *J. Biol. Chem.* In press
54. Ray, K., England, P. 1976. Phosphorylation of the inhibitory subunit of troponin and its effect on calcium dependence of cardiac myofibril ATPase. *FEBS Lett.* 70:11–17
55. Reuter, H. 1974. Localization of beta adrenergic receptors and effects of noradrenaline and cyclic nucleotides on action potentials, ionic currents and tension in mammalian cardiac muscle. *J. Physiol. London* 242:429–51
56. Schwartz, A., Entman, M., Kaniike, K., Lane, L., Van Winkle, W. B., Bornet, E. 1976. The rate of calcium uptake into sarcoplasmic reticulum of cardiac muscle and skeletal muscle. Effects of cyclic AMP-dependent protein kinase and phosphorlyase b kinase. *Biochem. Biophys. Acta* 426:57–72
57. Somlyo, A. V., Gonzalez-Serratos, H., Shuman, H., McClellan, G., Somlyo, A. P. 1981. Calcium release and ionic changes in the sarcoplasmic reticulum of tetanized muscle: an electron probe study. *J. Cell. Biol.* 90:577–94
58. Somlyo, A. V., Shuman, H., Somlyo, A. P. 1977. Elemental distributions in striated muscle and the effects of hypertonicity. *J. Cell. Biol.* 74:828–57
59. Stephenson, E. 1978. Properties of chloride-stimulated ^{45}Ca flux in skinned muscle fibers. *J. Gen. Physiol.* 71:411–30
60. Stephenson, E. 1981. Activation of fast skeletal muscle: contributions of studies on skinned fibers. *Am. J. Physiol.: Cell Physiol.* 9:C1–19
61. Suarez-Kurtz, G., Parker, I. 1977. Birefringence signals and calcium transients in skeletal muscle. *Nature* 270:746–48
61a. Tada, M., Katz, A. M. 1982. Phosphorylation of the sarcoplasmic reticulum and sarcolemma. *Ann. Rev. Physiol.* 44:401–22
62. Tada, M., Kirchberger, J., Repke, D., Katz, A. 1970. The stimulation of calcium transport in cardiac sarcoplasmic reticulum by adenosine 3':5' monophosphate dependent protein kinase. *J. Biol. Chem.* 249:6174–80
63. Vergera, J., Bezanilla, F., Salzberg, B. 1978. Nile blue fluorescence signals from cut single muscle fibers under voltage or current clamp condition. *J. Gen. Physiol.* 72:775–800
64. Wier, W. 1980. Calcium transient during excitation contraction coupling in mammalian heart: aequorin signals of canine Purkinje fibers. *Science* 207:1085–1087
65. Williamson, J., Schaffer, S., Scarpa, A., Safer, B. 1974. Investigations of the calcium cycle in perfused rat and frog hearts. In *Myocardial Biology*, ed. N. Dhalla, pp. 375–92. University Park Press
66. Winegrad, S. 1968. Intracellular calcium movements of frog skeletal muscle during recovery from tetanus. *J. Gen. Physiol.* 51:65–83
67. Winegrad, S. 1970. The intercellular site of calcium activation of contraction of frog skeletal muscle. *J. Gen. Physiol.* 55:77–88
68. Winegrad, S. 1973. Intracellular calcium binding and release in frog heart. *J. Gen. Physiol.* 62:693–706
69. Wood, E., Heppner, R., Weidmann, S. 1969. Inotropic effects of electric currents. *Circ. Res.* 24:409–45

THE ACTION OF CARDIOTOXINS ON CARDIAC PLASMA MEMBRANES

M. Lazdunski and J. F. Renaud

Centre de Biochimie du CNRS, Parc Valrose, Faculté des Sciences, 06034 Nice Cedex, France

During the last decade interest has increased in the many neurotoxins synthesized by microorganisms, plants, snakes, scorpions, spiders, and marine organisms. Each toxin is valuable because of its extreme specificity of action and its high affinity for its receptor, characteristics that allow its use in the pharmacological analysis of important physiological functions. Many neurotoxins are also cardiotoxins.

We here review natural substances that are toxic to cardiac cells through their interaction with ionic channels. Excluded from this discussion (reviewed in 26) are toxins that specifically act on nerve terminals to alter the storage and release of neurotransmitters such as α-latrotoxin, botulinum toxin, tetanus toxin, β-bungarotoxin, crotoxin, etc.

TOXINS SPECIFIC FOR THE FAST SODIUM CHANNEL

Tetrodotoxin and Saxitoxin

Tetrodotoxin (TTX) is a heterocyclic guanidine found in the puffer fish and in some species of newt, octopus, frog, and goby (37, 43). TTX produces a reversible inhibition of the inward fast Na^+ current in nerve, skeletal muscle, and cardiac muscle (37).

Saxitoxin (STX), like TTX, is a heterocyclic guanidine, but it has two guanidine groups. It is produced by *Gonyaulax* dinoflagellates and is concentrated in shellfishes that feed on these organisms. The mechanism of action of TTX and STX on the Na^+ channel is the same, and both are specific for this channel (37, 43). Half-maximum inhibition of both TTX and STX occurs at 1–5 nM (43, 53).

Biochemical studies of the binding of TTX and STX to their common receptor site were made possible by the availability of tritiated TTX and STX (43) and highly radioactive derivatives of TTX (10). Photoactivable TTX derivatives have been synthesized; they irreversibly block the fast Na^+ channel (9).

It was proposed that TTX and STX bind to the selectivity filter of the fast Na^+ channel (43). This model of interaction assumes that guanidinium moieties of TTX and STX enter the channel like the free guanidinium cation itself, and that the bulky part of the toxins occludes the mouth of the channel like a plug blocking the passage of Na^+ ions. However, recent chemical modifications of the Na^+ channel do not seem to support this mechanism of association. TTX binding is irreversibly blocked by treatment of nerve membranes with carboxyl-modifying reagents (4, 57), suggesting the presence of an essential carboxylate at the toxin receptor site. If TTX binds to the selectivity filter at the Na^+ site, such a chemical modification should also abolish the Na^+ channel conductance. Instead, Na^+ channels that have been made TTX-insensitive by chemical modifications are still active in the generation of action potentials (APs) with the same kinetics of activation and inactivation, with the same ionic selectivity, and with only a slightly reduced unit conductance (58).

The most specific aspect of TTX action on the heart is the marked insensitivity of cardiac muscle to TTX. A substantial reduction in the rate of rise of APs in ventricular trabeculae and Purkinje fibers does not occur until a TTX concentration of about 1 μM is reached (17). In experiments designed to determine whether the difference in TTX-sensitivity reflects basic differences in the TTX receptor of cardiac muscle, Baer et al (3) reported a voltage-dependence of TTX block in rat papillary muscle. The TTX sensitivity of the Na^+ channel was markedly increased when the membrane potential was more positive. Such an effect has never been reported for Na^+ channels in nerve or skeletal muscle. However, these conclusions have been criticized on the basis that upstroke velocity measurements are not a good measure of the true variations of the Na^+ conductance (11). Colatsky et al (12) found no voltage-dependence for steady-state TTX block of excitatory and background (leakage) Na^+ channels when peak I_{Na} was measured under voltage clamp. TTX block of background Na^+ channels in voltage-clamped dog Purkinje fibers is described by a 1:1 binding reaction with a voltage-independent dissociation constant (K_d) of 1.1 μM (12). This K_d is about three orders of magnitude higher than normally found for TTX-receptor interactions in nerve and skeletal muscle membranes.

The affinity of TTX for its cardiac receptor varies with the animal species. The K_d of the TTX-receptor complex measured in voltage-clamped frog atrial fibers is 3.4 nM and is not voltage-dependent (50). Moreover different families of cardiac TTX receptors corresponding to different classes of Na^+ channels also have been postulated (13).

The biochemical properties of the cardiac Na^+ channels have been studied using a TTX derivative in which tritiated ethylenediamine was grafted to tetrodotoxin ($[^3H]$en-TTX) (34). Saturable high-affinity binding of $[^3H]$en-TTX occurred in rat, rabbit, guinea pig, and embryonic chick heart membranes. The saturation isotherm showed a single population of TTX receptors with an equilibrium constant (K_d) of 0.5–1.0 nM. TTX and STX specifically displaced $[^3H]$en-TTX binding with K_d values of 2.0 nM and 3.4 nM, respectively. $[^3H]$en-TTX binding is inhibited at acidic pH, and half-inhibition occurs at pH 6.2. This pH value was ascribed to the pK of the carboxylate that can be modified by carbodiimides or trialkyloxonium salts (4, 40). It is close to values found for nerve and skeletal muscle membranes (8). Monovalent cations selectively displaced bound $[^3H]$en-TTX, indicating that the toxin binds to a specific ion coordination site. The displacement of $[^3H]$en-TTX by monovalent cations occurs in the sequence: guanidinium $> Tl^+ > NH_4^+ > Li^+ > Na^+ > K^+ > Rb^+ > Cs^+$. An interesting feature of inorganic monovalent cation binding to the Na^+ channel is a positive cooperativity that suggests multiple and coupled monovalent ion binding sites at the mouth of the channel. Divalent cations also displaced $[^3H]$en-TTX binding in the sequence Mg^{2+} ($K_d = 1.3$ mM) $> Ca^{2+}$ ($K_d = 3.2$ mM). Because cations like Na^+, K^+, or Ca^{2+} compete with TTX for association to the TTX receptor, K_d values measured in a Ringer solution are six times higher than in a choline medium (34).

Biochemical studies indicate that the interaction of TTX with Na^+ channels in cardiac cells has properties similar to those in nerve and skeletal muscle membranes (43). That binding occurs in unpolarized membranes suggests that there may be no voltage-dependence in the TTX inhibition of the heart Na^+ conductance. The biochemical approach used (34) could not have detected low-affinity TTX binding sites if they existed on the cardiac cell membrane.

Ceveratrum Alkaloids, Batrachotoxin, Aconitine and Grayanotoxins

These compounds represent lipid-soluble molecules that have dramatic effects on cardiac and other excitable cells. Veratridine and other veratrum alkaloids are steroids that occur in liliaceous plants belonging to the genera *Veratrum, Zygadenus, Stenanthium,* and *Schoenocaulon* (31, 37). Batrachotoxin (BTX) (reviewed in 1) is a steroidal alkaloid extracted from the skin of a Columbian frog *Phyllobates aurotaenia*. It is a most toxic substance, its lethal effect on mammals being due mainly to its cardiac arrhythmogenic action. Aconitine is the main alkaloid of the plant *Aconitum napellus*. Its structure is not closely related to that of veratridine and batrachotoxin (64). Grayanotoxins (GTXs) are toxic diterpenoids found in various species of *Rhododendron, Kalmia,* and *Leucothoe* (*Ericaceae*) (28).

Veratridine, BTX, and aconitine depolarize the nerve or muscle membrane by causing the voltage-sensitive fast Na^+ channels to become persistently active at the resting membrane potential (29, 51, 62). The hyperexcitability and depolarization of excitable cells induced by these toxins is caused by two effects: (*a*) a shift of the voltage dependence of activation of Na^+ channels towards more negative potentials and (*b*) a block of the inactivation process of these channels. In addition, these three toxins not only alter the gating mechanism of the fast Na^+ channel, they also change its ion selectivity and increase its permeability to larger cations (32). These three types of toxins may bind preferentially to an open state of the Na^+ channel since repetitive stimulation to activate the Na^+ channels enhances their action (27, 29, 62). Depolarization of excitable cells by veratridine, BTX, and aconitine are blocked by TTX (22, 37).

GTXs, although they are very different in structure from veratrum alkaloids, aconitine, and BTX, seem to have essentially the same mechanism of action on Na^+ channels (8, 37, 38).

Two approaches have been used to analyze the mechanism of action of these neurotoxic compounds. The first is the voltage-clamp technique, which provides information on the kinetics and voltage-dependence of the activation and inactivation steps as well as on the conductance properties of the open form of the channel. The second is radioactive monovalent cation flux. Veratridine, BTX, and GTXs all increase the initial rate of $^{22}Na^+$ entry into neuroblastoma or skeletal muscle cells (8). This stimulation, due to their chemical activation of Na^+ channels, is inhibited by TTX. BTX is the most powerful of this class of toxins.

A considerable literature exists on the cardiac action of veratridine, BTX, aconitine, and GTXs. A critical analysis of the results has recently been made (22).

Voltage-clamp experiments with veratrine[1] on frog atrial fibers and with aconitine on sheep Purkinje fibers (24, 39) showed that these molecules block the inactivation mechanism of the Na^+ channel. Ceveratrum alkaloids, aconitine, and BTX at low concentrations prolong the repolariza-

---

[1]Veratrine is a mixture of toxic alkaloids containing the pure compound veratridine.

tion phase of the cardiac AP; their effect is both dose- and frequency-dependent and is inhibited by TTX (22). These effects are seen at concentrations in the nanomolar range for BTX (23, 52) and in the micromolar range for other compounds (22, 39, 60). At higher concentrations, all these toxins depolarize the myocardial cell membrane, and this depolarization is blocked by TTX. Moreover, veratridine at 0.1 mM maximally increases the initial rate of $^{22}Na^+$ influx in embryonic chick cardiac cells in culture (18) by a factor of about 5; half-maximum effect occurred at 22 μM. Veratridine stimulation of Na^+ influx is inhibited by TTX [$K_{0.5\ (TTX)}$ of 6.6 nM at 22 μM veratridine].

GTXs activate Na^+ channels in cardiac preparations (30). The effect of GTX I has even been observed on the sinoatrial node, though this cell type is characterized by TTX-resistant electrical activity (55). It is one example in which the presence of Na^+ channels can be demonstrated by using specific toxins.

Veratridine, BTX, and GTXs all produce a dose-dependent and reversible positive inotropic effect (22). Aconitine is the only compound for which no inotropic effect has been reported. Comparisons of the concentration-effect relationships have been made for 14 ceveratrum ester alkaloids (22), and for 18 structurally related GTXs (25). Honerjäger (22) has shown that the maximum increase in force of contraction is nearly the same for the different ceveratrum alkaloids, but that these compounds have different half-maximum effects ranging between 10 nM and 100 μM. Ceveratrum alkaloids like germitetrine, or protoveratrines, produce positive inotropic effects at concentrations more than ten-fold lower than those necessary with veratridine.

Two Ca^{2+} entry systems are involved in the coupling between excitation and contraction in the heart: the slow Ca^{2+} channel and the Na^+/Ca^{2+} exchange system (42). The positive inotropic effect of veratridine, BTX, and GTXs is the indirect result of their action on the Na^+ channel. By triggering a persistent activation of the Na^+ channel, these toxins increase the intracellular Na^+ concentration that provokes an influx of Ca^{2+} through the Na^+/Ca^{2+} exchange system. The implication of the Na^+/Ca^{2+} exchange system in the mechanism of the inotropic effect produced by these toxins has been demonstrated directly both by voltage-clamp studies on frog atrial trabeculae (24) and by ^{22}Na and ^{45}Ca flux studies with cultured heart cells (18). When veratridine stimulates ^{22}Na influx by a factor of 5, it increases the initial rate of ^{45}Ca entry by a factor of 25. As expected for an indirect effect of veratridine on the Na^+/Ca^{2+} exchange system, veratridine-stimulated Ca^{2+} influx is inhibited by TTX, is Na^+-dependent, and Na^+ cannot be replaced by Li^+.

Scorpion and Sea Anemone Toxins

Toxins from both American and African scorpions are all single-chain polypeptides (7,000 daltons) extensively cross-linked by disulfide bonds (44). Scorpion venom contains a mixture of multiple toxins, some being more specific to mammals, others to insects or crustaceans. A lysine residue has been identified as one of the essential amino-acids at the active site (48). Voltage-clamp experiments on nerve preparations with pure toxins from *Androctonus australis* and *Buthus eupeus* (36, 46) showed that their main effect is to slow down considerably the inactivation of the Na^+ channel, which may even remain incomplete. These toxins have no effect on the activation of the channel. All scorpion toxins may not have the same mechanism of action. The venom of the American scorpion *Centruroides sculpturatus* has no effect on the inactivation of the Na^+ channel, but it shifts the voltage dependence of activation to more negative membrane potentials (6).

Toxins active on different animal species have been isolated from a variety of sea anemone species (54). They are single-chain basic polypeptides of 2,500–5,000 daltons cross-linked with disulfide bridges. An arginine residue is the essential amino-acid at their active site (5). Sea anemone and scorpion toxins have no obvious sequence homology. Like scorpion toxins, sea anemone toxins specifically slow the inactivation of Na^+ channels in nerve membranes without altering the activation step (45). This effect is inhibited by TTX. Both scorpion and sea anemone toxins enhance the action of veratridine, BTX, aconitine, and GTXs on the Na permeability of excitable cells (8, 44).

Scorpion and sea anemone toxins provide a tool for analyzing the biochemical properties of the Na^+ channel inactivation gate. The main properties of the binding of [^{125}I]-labeled toxins are the following: (*a*) Both toxins bind to sites that are different but interdependent, the stoichiometry of binding being higher for sea anemone toxins (63); and (*b*) the binding of scorpion toxins to their specific sites is voltage-dependent and does not occur on a depolarized membrane (7, 40, 44) whereas the binding of sea anemone toxins does not change with the membrane potential (63).

Scorpion toxin II from *Androctonus australis Hector* dramatically prolongs the repolarization phase of the cardiac AP. This prolongation is inhibited by TTX and is not seen in low-Na^+ solutions (14). Pure anemone toxins from *Anemonia sulcata* and *Anthopleura xanthogrammica* display the same type of effect on various cardiac preparations (47, 56). It was concluded that these two classes of polypeptide toxins slow the inactivation of the Na^+ channel.

Low concentrations of both types of polypeptide toxins produce a positive inotropic effect on a variety of cardiac preparations (2, 47, 56). The analysis of the mechanism of action of these toxins on the heart has been carried out with cultured embryonic chick cardiac cells using (a) electrophysiological measurements with the simultaneous recording of contraction and (b) measurements of ^{22}Na and ^{45}Ca fluxes. Results obtained with sea anemone toxins indicate that the toxin (a) induces APs of long duration, slows the beating rate, and simultaneously increase the amplitude and duration of contractions; (b) maximally increases the rate of ^{45}Ca entry in ventricular cells by a factor of about 12, this effect being blocked by TTX and dependent on $[Na^+]_o$ (Li^+ cannot replace Na^+); (c) works in synergy with veratridine to stimulate both ^{22}Na and ^{45}Ca influx; and (d) still induces contractions when the slow Ca^{2+} channel is blocked by D_{600}, verapamil, or Mn^{2+}. The toxin-induced contractions are blocked when the external Na^+ is replaced by Li^+. Thus the increase in amplitude and duration of cardiac contractions caused by the toxin is most likely due to an indirect activation of the Na^+-Ca^{2+} exchange system. *Androctonus australis Hector* scorpion toxins probably have the same mechanism of action since they also stimulate the rates of both ^{22}Na and ^{45}Ca influx into cardiac cells and since this stimulation is inhibited by TTX (15) and is synergetically increased by veratridine (16).

A polypeptide toxin isolated from a coral (*Goniopora* spp.) evokes positive inotropic effects that are suppressed by TTX (20). The mechanism of the cardiostimulant action of this peptide is probably the same as that of scorpion and sea anemone toxins.

Toxins as Tools in the Study of the Differentiation of Cardiac Na^+ Channels

Most groups working on embryonic heart ontogenesis have observed that the maximum upstroke velocity of the cardiac AP and the susceptibility of both AP and contraction to blockade by TTX increase during the embryonic development of the chick heart (61). At the early stage of development (both in vivo and in vitro), the inward current responsible for the rising phase of the AP is a slow Na^+ current sensitive to D_{600} but not to TTX. The TTX-sensitive fast Na^+ channel appears at later stages of development. Electrophysiological and biochemical studies using specific neurotoxins (41) have shown that (a) veratridine, BTX, sea anemone toxin, and TTX receptors are present at the early stage when APs are insensitive to TTX; (b) the fast Na^+ channel at this stage is in a nonfunctional (or silent) form that cannot be electrically activated but can be activated by the gating system toxins (veratridine and sea anemone toxin); and (c) the transition from the early to the late stage of development corresponds both to the transformation of silent into functional channels and to a 4–5-fold increase in the density of channels.

OTHER TOXINS

Ervatamine is a toxic alkaloid isolated from an Australian tree, *Ervatamia orientalis*. Studies on the mechanism of action of ervatamine and epiervatamine (a structural analog 4–8 times as potent as ervatamine) on nerve cells (19) have shown that (a) they provoke a frequency-dependent block of the fast Na^+ channel; (b) they inhibit the effects of veratridine, BTX, scorpion toxin, and sea anemone toxin on the Na^+ channel in a manner similar to that of TTX; (c) they do not bind to receptors specific for TTX, scorpion toxin, or sea anemone toxin; (d) they associate with the gating system of the channel and are competitive inhibitors of BTX action, suggesting that the two classes of toxins have a common receptor site; and (e) they have affinities similar to that of BTX for the Na^+ channel. Ervatamine blocks the cardiac AP without altering the resting potential (49). It inhibits the fast Na^+ current by binding to a single class of receptors ($K_d = 20$ μM). The inhibition is frequency-dependent and ervatamine prolongs the rate of reactivation of the channel. Ervatamine also inhibits the slow Na^+ channel seen in Ca^{2+}-free solutions containing TTX as well as the slow Ca^{2+} channel (49).

The effect of the so-called snake venom cardiotoxins (21, 33) on ionic permeabilities is due to their disrupting action on membrane architecture.

The mechanism of the cardiotoxicity of palytoxin, the most potent toxin (35), is still unknown.

SUMMARY AND CONCLUSIONS

Most of the known natural cardiotoxins are specific for the fast Na^+ channel of the heart cell. They may be classified into three groups based on their mechanism of action and the available information concerning their receptor sites: (a) TTX and STX, (b) veratrum alkaloids, aconitine, BTX and GTXs, and (c) polypeptide toxins from scorpion, sea anemone, and coral venoms. Through their action on the fast Na^+ channel, these toxins can either block contraction, produce positive inotropic effects and spontaneous contractions or induce arrhythmias and fibrillation. Pyrethroids, another class of molecules acting on the fast Na^+ channel (32) deserve study as possible cardiotoxins.

No natural toxic compound available is specific for other ionic channels important for the cardiac AP and excitation-contraction coupling. The present pharmacology of these channels includes compounds like TEA, D_{600}, and verapamil (59) that are not natural substances, are not absolutely specific for a given channel and do not act at very low concentrations. The discovery of natural toxins specific for the slow Ca^{2+} channel, for the Na^+/Ca^{2+} exchange system, or for other channels involved in the pacemaker mechanisms would be of a considerable interest.

Literature Cited

1. Albuquerque, E. X., Daly, J. W. 1976. Batrachotoxin a selective probe for channels modulating sodium conductances in electrogenic membranes. In *The Specificity and Action of Animal Bacterial and Plant Toxins.* London: Chapman and Hall
2. Alsen, C., Beress, L., Fischer, K., Proppe, O., Reinberg, T., Sattler, R. W. 1976. The action of a toxin from the sea anemone *Anemonia sulcata* upon mammalian heart muscles. *Naunyn-Schmiedeberg's Arch. Pharmacol.* 295:55–62
3. Baer, H., Best, P. M., Reuter, H. 1976. Voltage-dependent action of tetrodotoxin in mammalian cardiac muscle. *Nature* 263:344–45
4. Baker, P. F., Rubinson, K. A. 1975. Chemical modification of crab nerves can make them insensitive to the local anaesthetics tetrodotoxin and saxitoxin. *Nature* 257:412–14
5. Barhanin, J., Hughes, M., Schweitz, H., Vincent, J. P., Lazdunski, M. 1981. Structure-function relationship of sea anemone toxin II from *Anemonia sulcata. J. Biol. Chem.* 256:5764–69
6. Cahalan, M. D. 1975. Modification of sodium channel gating in frog myelinated nerve fibers by *Centruroïdes sculpturatus* scorpion venom. *J. Physiol. London* 244:511–34
7. Catterall, W. A. 1977. Membrane potential dependent binding of scorpion toxin to the action potential sodium ionophore. Studies with a toxin derivative prepared by lactoperoxidase catalyzed iodination. *J. Biol. Chem.* 252:8660–68
8. Catterall, W. A. 1980. Neurotoxins that act on voltage-sensitive sodium channels in excitable membranes. *Ann. Rev. Pharmacol. Toxicol.* 20.15–43
9. Chicheportiche, R., Balerna, M., Lombet, A., Romey, G., Lazdunski, M. 1979. Synthesis and mode of action on axonal membranes of photoactivable derivatives of tetrodotoxin. *J. Biol. Chem.* 254:1552–57
10. Chicheportiche, R., Balerna, M., Lombet, A., Romey, G., Lazdunski, M. 1980. Synthesis of new, highly radioactive tetrodotoxin derivatives and their binding properties to the sodium channel. *Eur. J. Biochem.* 104:617–25
11. Cohen, I. S., Strichartz, G. 1977. On the voltage-dependent action of tetrodotoxin. *Biophysical J.* 17:275–79
12. Colatsky, T. J., Gadsby, D. C. 1980. Is tetrodotoxin block of background sodium channels in canine cardiac Purkinje fibers voltage-dependent? *J. Physiol. London* 306:20P
13. Coraboeuf, E., Deroubaix, E., Coulombe, A. 1979. Effect of tetrodotoxin on action potential of the conducting system in the dog heart. *Am. J. Physiol.* 236:561–67
14. Coraboeuf, E., Deroubaix, E., Tazieff-Depierre, F. 1975. Effect of toxin II isolated from scorpion venom on action potential and contraction of mammalian heart. *J. Mol. Cell Cardiol.* 7:643–53
15. Couraud, F., Rochat, H., Lissitzky, S. 1976. Stimulation of sodium and calcium uptake by scorpion toxin in chick embryo heart cells. *Biochim. Biophys. Acta* 433:90–100
16. Couraud, F., Rochat, H., Lissitzky, S. 1980. Binding of scorpion neurotoxins to chick embryonic heart cells in culture and relationship to calcium uptake and membrane potential. *Biochemistry* 19: 457–62
17. Dudel, J., Peper, K., Rüdel, R., Trautwein, W. 1967. The effect of tetrodotoxin on the membrane current in cardiac muscle (Purkinje fibers). *Pflügers Arch.* 295:213–26
18. Fosset, M., De Barry, J., Lenoir, M. C., Lazdunski, M., 1977. Analysis of molecular aspects of Na^+ and Ca^{2+} uptakes by embryonic cardiac cells in culture. *J. Biol. Chem.* 252:6112–17
19. Frelin, C., Vigne, P., Ponzio, G., Romey, G., Tourneur, Y., Husson, H. P., Lazdunski, M. 1981. The interaction of ervatamine and epiervatamine with

the action potential Na⁺ ionophore. *Mol. Pharmacol.* 20:107–12
20. Fujiwara, M., Muramatsu, I., Hidaka, H., Ikushima, S., Ashida, K. 1979. Effects of goniopora toxin, a polypeptide isolated from coral, on electromechanical properties of rabbit myocardium. *J. Pharmacol. Exp. Ther.* 210:153–57
21. Gulik-Krzywichi, T., Balerna, M., Vincent, J. P., Lazdunski, M. 1981. Freezefracture study of cardiotoxin action on axonal membrane and axonal membrane lipid vesicles. *Biochim. Biophys. Acta* 643:101–14
22. Honerjäger, P. 1981. Cardioactive substances that prolong the open state of sodium channel. *Physiol. Biochem. Pharmacol.* 92. In press
23. Honerjäger, P., Reiter, M. 1977. The cardiotoxic effect of batrachotoxin. *Naunyn-Schmiedeberg's Arch. Pharmacol.* 299:239–52
24. Horackova, M., Vassort, G. 1974. Excitation-contraction coupling in frog heart. Effect of veratrine. *Pflügers Arch.* 352:291–302
25. Hotta, Y., Takeya, K., Kobayashi, S., Harada, N., Sakakibara, J., Shirai, N. 1980. Relationship between structure, positive inotropic potency and lethal dose of grayanotoxins in guinea pig. *Arch. Toxicol.* 44:259–67
26. Howard, B. D., Gunderson, C. B., Jr., 1980. Effects and mechanisms of polypeptide neurotoxins that act presynaptically. *Ann. Rev. Pharmacol.* 20:307–36
27. Jacques, Y., Fosset, M., Lazdunski, M. 1978. Molecular properties of the action potential Na⁺ ionophore in neuroblastoma cells. *J. Biol. Chem.* 253:7383–92
28. Kakisawa, H., Kozima, T., Yanai, M., Nakanishi, K. 1965. Stereochemistry of grayanotoxins. *Tetrahed. Lett.* 21:3091–104
29. Khodorov, B. I., Revenko, S. V. 1979. Further analysis of the mechanisms of action of batrachotoxin on the membrane of myelinated nerve. *Neuroscience* 4:1315–30
30. Ku, D. D., Akera, T., Frank, M., Brody, T. M., Iwasa, J. 1977. The effects of grayanotoxin I and α-dihydrograyanotoxin II on guinea pig myocardium. *J. Pharmacol. Exp. Ther.* 200:363–72
31. Kupchan, S. M., By, A. W. 1968. Steroid alkaloids: the veratrum group. In *Alkaloids,* ed. P. H. F. Manske, 10:193–285. NY: Academic Press
32. Lazdunski, M., Balerna, M., Barhanin, J., Chicheportiche, R., Fosset, M., Frelin, C., Jacques, Y., Lombet, A., Pouyssegur, J., Renaud, J. F., Romey, G., Schweitz, H., Vincent, J. P. 1980. Molecular aspects of the structure and mechanism of the voltage-dependent sodium channel. *Ann. N.Y. Acad. Sci.* 358:169–82
33. Lee, C. Y. 1979. Recent advances in chemistry and pharmacology of snake toxins. In *Advance in Cytopharmacology,* ed. B. Ceccarelli, F. Clement, 3:1–16. NY: Raven Press
34. Lombet, A., Renaud, J. F., Chicheportiche, R., Lazdunski, M. 1981. A cardiac tetrodotoxin binding component: biochemical identification, characterization, and properties. *Biochemistry* 20:1279–85
35. Moore, R. E., Scheuer, P. J. 1971. Palytoxin: A new marine toxin from coelenterate. *Science* 172:495–98
36. Mozhayeva, G. N., Naumov, A. P., Nosyreva, E. D., Grishin, E. V. 1980. Potential-dependent interaction of toxin from venom of the scorpion *Buthus eupeus* with sodium channels in myelinated fiber. Voltage clamp experiments. *Biochim. Biophys. Acta* 597:587–602
37. Narahashi, T. 1974. Chemicals as tools in the study of excitable membranes. *Physiol. Rev.* 54:813–89
38. Narahashi, T. 1979. Modulation of nerve membrane sodium channels by neurotoxins. See Ref. 33, pp. 293–303
39. Peper, K., Trautwein, W. 1967. The effect of aconitine on the membrane current in cardiac muscle. *Pflügers Arch.* 296:328–36
40. Ray, R., Morrow, C. S., Catterall, W. 1978. Binding of scorpion toxin to receptor sites associated with voltage sensitive sodium channels in synaptic nerve ending particles. *J. Biol. Chem.* 253:7307–17
41. Renaud, J. F., Romey, G., Lombet, A., Lazdunski, M. 1981. Differentiation of the fast sodium channel in embryonic heart cells followed by its interaction with neurotoxins. *Proc. Natl. Acad. Sci. USA* 78:5348–52
42. Reuter, H. 1974. Exchange of calcium ions in the mammalian myocardium. Mechanisms and physiological significance. *Circ. Res.* 34:599–605
43. Ritchie, J. M., Rogart, R. B. 1977. The binding of saxitoxin and tetrodotoxin to excitable tissue. *Rev. Physiol. Biochem. Pharmacol.* 79:1–50
44. Rochat, H., Bernard, P., Couraud, F. 1979. Scorpion toxins: chemistry and mode of action. See Ref. 33, pp. 325–34

45. Romey, G., Abita, J. P., Schweitz, H., Wunderer, G., Lazdunski, M. 1976. Sea anemone toxin: a tool to study molecular mechanisms of nerve conductions and excitation-secretion coupling. *Proc. Natl. Acad. Sci. USA* 73:4055–59
46. Romey, G., Chicheportiche, R., Lazdunski, M., Rochat, H., Miranda, F., Lissitzky, S. 1975. Scorpion neurotoxin a presynaptic toxin which affects both Na^+ and K^+ channels in axons. *Biochem. Biophys. Res. Commun.* 64:115–21
47. Romey, G., Renaud, J. F., Fosset, M., Lazdunski, M. 1980. Pharmacological properties of the interaction of a sea anemone polypeptide toxin with cardiac cells in culture. *J. Pharmacol. Exp. Ther.* 213:607–15
48. Sampieri, F., Habersetzer-Rochat, C. 1978. Structure function relationships in scorpion neurotoxins. Identification of the super reactive lysine residue in toxin I of *Androctonus australis Hector*. *Biochim. Biophys. Acta* 535:100–9
49. Sauviat, M. P. 1980. Effects of ervatamine chlorhydrate on cardiac membrane currents in frog atrial fibers. *Br. J. Pharmacol.* 71:41–49
50. Sauviat, M. P. 1981. On the action of TTX on frog atrial trabeculae. *Pflügers Arch.* In press
51. Schmidt, H., Schmitt, O. 1974. Effect of aconitine on the sodium permeability of node of Ranvier. *Pflügers Arch.* 349:133–48
52. Shotzberger, G. S., Albuquerque, E. X., Daly, J. W. 1976. The effects of batrachotoxin on cat papillary muscle. *J. Pharmacol. Exp. Ther.* 196:433–44
53. Schwarz, J. R., Ulbricht, W., Wagner, H. H. 1973. The rate of action of tetrodotoxin on myelinated nerve fibers of *Xenopus laevis* and *Rana esculenta*. *J. Physiol. London.* 233:167–94
54. Schweitz, H., Vincent, J. P., Barhanin, J., Frelin, C., Linden, G., Hugues, M., Lazdunski, M. 1981. Purification of eight sea anemone toxins from *Anemonia sulcata, Anthopleura xantogrammica, Stoichactis giganteus* and *Actinodendron plumosum*. *Biochemistry* 20:5245–52
55. Seyama, I. 1978. Effect of grayanotoxin I on SA node and right atrial myocardia of the rabbit. *Am. J. Physiol.* 235:C136–C142
56. Shibata, S., Izumi, T., Seriguchi, D. G., Norton, T. R. 1978. Further studies on the positive inotropic effect of the polypeptide *anthopleurin-A* from a sea anemone. *J. Pharmacol. Exp. Ther.* 205:683–92
57. Shrager, P., Profera, C. 1973. Inhibition of the receptor for tetrodotoxin in nerve membranes by reagents modifying carboxyl groups. *Biochim. Biophys. Acta* 318:141–46
58. Sigworth, F. J., Spalding, B. C. 1980. Chemical modification reduces the conductance of sodium channels in nerve. *Nature* 283:293–95
59. Sperelakis, N. 1981. Effects of cardiotoxic agents on the electrical properties of myocardial cells. In *Cardiac Toxicology*, ed. T. Balazs, 1:39–108. Boca Raton: CRC
60. Sperelakis, N., Pappano, A. J. 1969. Increase in P_{Na} and P_K of cultured heart cells produced by veratridine. *J. Gen. Physiol.* 53:97–114
61. Sperelakis, N., Shigenobu, K., Mc Lean, M. J. 1975. Membranes cation channels. Changes in developing hearts, in cell culture and in organ culture. In *Developmental Physiological Correlates of Cardiac Muscle*, ed. M. Lieberman, T. Sano, pp. 209–34. NY: Raven Press
62. Ulbricht, W. 1969. The effect of veratridine on excitable membranes of nerve and muscle. *Erg. Physiol.* 61:18–71
63. Vincent, J. P., Balerna, M., Barhanin, J., Fosset, M., Lazdunski, M. 1980. Binding of sea anemone toxin to receptor sites associated with the gating system of the sodium channel in synaptic nerve endings in vitro. *Proc. Natl. Acad. Sci. USA* 77:1646–50
64. Wiesner, K., Kao, W., Jay, E. W. K. 1969. A rigorous, purely chemical structure proof for aconitine and delphinine. *Can. J. Chem.* 47:2734–37

ADRENERGIC RECEPTORS IN THE HEART[1]

Brian B. Hoffman* and Robert J. Lefkowitz

Howard Hughes Medical Institute Laboratory, Departments of Medicine (Cardiology) and Biochemistry, Duke University Medical Center Durham, North Carolina 27710.

Introduction

Catecholamines, acting through alpha- and beta-adrenergic receptors, modulate a variety of physiological responses in the heart. Most importantly catecholamines increase the rate and force of cardiac contraction. These actions occur mainly as a consequence of the binding of the endogenous substances norepinephrine and epinephrine to specific adrenergic receptors on the plasma membrane of cells in the heart. Whereas the effects of sympathetic nervous stimulation on the heart have been examined for many years (reviewed in 57), only recently has it become possible to measure directly the properties of the receptors for catecholamines in the heart. Here we discuss a few selected areas of active research where radioligand binding techinques have been applied to the study of adrenergic receptors in the heart.

Direct Demonstration of Adrenergic Receptors

Beta-adrenergic receptors were first identified in the dog heart using [^{3}H]DHA in 1975 (3). [^{3}H]DHA binding sites had the characteristics expected of interaction with beta-adrenergic receptors. For example, the sites recognized catecholamines with the appropriate stereospecific potency series. Subsequently, beta-adrenergic receptors have been identified and characterized in the hearts of many other species including rat, frog, cat, chick embryo, rabbit, guinea pig, and mouse (5, 10, 14, 16, 17, 25, 27, 28, 37, 45,

---

[1]Abbreviations: [^{3}H] DHA, (−)[^{3}H]dihydroalprenolol; [^{3}H]DHE, [^{3}H]dihydroergocryptine; SHR, spontaneously hypertensive rats.
*Current address: Division of Clinical Pharmacology (S–155), Department of Medicine, Stanford University School of Medicine, Stanford, CA 94305

49, 65, 67, 70). Radioligand binding has been used to measure directly the affinity of a variety of drugs for beta receptors. Also, regulation of the number of beta receptors in the heart by various influences has been extensively examined. For example, the effects of development and ageing (7, 9, 14, 56), ischemia (51), guanine nucleotides (29, 66), and altered thyroid state (reviewed in 32) have been extensively investigated.

In the hearts of several species there also appear to be alpha-adrenergic receptors that induce contractile responses (reviewed in 57). Alpha-adrenergic receptors have been demonstrated in the heart with [^{3}H]DHE in rat (24, 59, 62, 69), rabbit (58), and fetal lamb (15). Also, [^{3}H]WB4101 (55, 75), and [^{3}H]prazosin (35) have been used to label alpha receptors in the heart.

Beta-Adrenergic Receptor Subtypes in the Heart

Ahlquist initially distinguished alpha- and beta-adrenergic receptors in 1948 (2). Subsequently, in 1967, Lands et al (39) found two subtypes of beta receptors termed beta$_1$ and beta$_2$. Beta$_1$ and beta$_2$ receptors were operationally defined by their relative affinities for epinephrine and norepinephrine: Beta$_1$ receptors have approximately equal affinity for epinephrine and norepinephrine, whereas beta$_2$ receptors have a higher affinity for epinephrine than for norepinephrine. A wide variety of drugs are now available with different affinities for one or the other beta receptor subtype [for recent review see (53)].

Recent interest has been focused on the beta-adrenergic receptor subtypes mediating inotropic and chronotropic responses in the heart, particularly in light of the clinical implications. Both beta$_1$ and beta$_2$ receptors are found in the hearts of certain species. A number of studies have found differences in the chronotropic and inotropic effects of series of catecholamines (e.g. 11, 19). Using a number of selective beta-adrenergic agonists and antagonists, Carlsson et al (12) found that the chronotropic effects of these drugs seemed to be mediated by both beta$_1$ and beta$_2$ receptors in the cat heart. Subsequently, these workers found that the atria from dog and human hearts exhibited the same heterogeneous response (1). Carlsson et al have extended their physiological findings in cat heart to indicate that while beta$_1$ receptors are the predominant beta-adrenergic receptor in both cat atrium and ventricle, beta$_2$ receptors have a more significant role in the sinus node than in the ventricle (13). O'Donnell & Wanstall (52) also found that both beta$_1$ and beta$_2$ receptors mediated chronotropic responses in isolated cat atria but that chronotropic responses in isolated guinea pig atria were apparently mediated exclusively by beta$_1$ receptors. The study of Dreyer & Offermeier suggests differences in the beta-adrenergic receptors mediating chronotropic and inotropic responses in guinea-pig atria (19). Thus the cat atrium might serve as a good model for the human situation, whereas the guinea-pig atrium might be a better tissue for examining the potency of novel agonists and antagonists at $beta_1$-adrenergic receptors since chronotropic responses do not seem to involve functionally significant $beta_2$ receptors (52). Some workers have speculated that the $beta_1$ receptors respond to norepinephrine released from nerve terminals in the mammalian heart whereas the $beta_2$ receptors are activated by circulating epinephrine from the adrenal medulla. Interestingly, in the amphibian heart where epinephrine rather than norepinephrine is the neurotransmitter, the predominant receptor mediating atrial contraction is a $beta_2$ receptor (61).

Stimulated by such observations, extensive efforts have been made to determine quantitatively the number of $beta_1$ and $beta_2$ receptors in the heart. No radioligands presently available label exclusively $beta_1$ or $beta_2$ receptors; indeed, the two most commonly used beta-adrenergic receptor ligands, [^{3}H]DHA and [^{125}I]iodohydroxybenzylpindolol, do not distinguish between $beta_1$ and $beta_2$ receptors. Therefore in order to distinguish $beta_1$ from $beta_2$ receptors in the heart, several groups have used unlabelled drugs shown to have different affinities at $beta_1$ and $beta_2$ receptors in intact tissues. Competition curves of these drugs with the radioligand are constructed. The relative proportion of $beta_1$ and $beta_2$ receptors can then be determined by a number of analytical techniques. The affinities of the $beta_1$ and $beta_2$ receptors for the selective drugs may be calculated. Using a graphical method based on a modified Scatchard plot, Nahorski and collaborators (8) found that while rat lung contained both $beta_1$ and $beta_2$ receptors, rat heart contained predominantly, if not exclusively, $beta_1$ receptors. Minneman, Molinoff and their co-workers have investigated beta receptors in the hearts of several species in a similar fashion and analyzed their data using a computer-assisted iterative technique (47). A third approach for delineating receptor subtypes (25, 30) involves the computer modeling of untransformed binding data from competition curves using a nonlinear least-squares curve-fitting technique. In our view, the latter approach is the most appropriate; the relative merits of these techniques have been discussed elsewhere (41, 48). Using the curve-modelling technique Hancock et al (25) found that rat ventricle contains exclusively $beta_1$ receptors whereas frog ventricle contains predominantly $beta_2$ receptors as well as ~20% $beta_1$ receptors. Minneman et al (47) suggested that rat heart contains 83% $beta_1$ and 17% $beta_2$ receptors, although they did not demonstrate that the data were statistically better fit by two receptor subtypes than by $beta_1$ receptors exclusively. These studies (46) provide useful information about directly determined affinities of a variety of selective drugs for $beta_1$ and $beta_2$ receptors. The heart contains a variety of cell types. Since fibroblast-like cells from the heart grown in culture contain $beta_2$ receptors (40), results of such studies from cardiac homogenates must be interpreted with caution. By these techniques it has also been shown that the atria from cats and guinea pigs contain both $beta_1$ and $beta_2$ receptors whereas the ventricles contain essentially only $beta_1$ receptors (28). These results are in good general agreement with the physiological findings discussed above.

Regulation of Adrenergic Receptors by Catecholamines

Exposure of a wide variety of cells to agonist drugs or hormones for a period of time leads to a state of decreased responsiveness to further stimulation by that agonist. This diminished responsiveness, occurring as a result of prolonged exposure to elevated concentrations of, for example, catecholamines, has been referred to as desensitization, tachyphylaxis, or refractoriness. Desensitization of beta-adrenergic receptor stimulation of adenylate cyclase has been characterized extensively in a number of model systems —e.g. frog erythrocytes (50, 68) and cultured cells (60, 63).

The mechanisms by which catecholamines may produce desensitization of beta-adrenergic receptors coupled to adenylate cyclase are multiple and complex. The phenomenon of desensitization is quite general; however, the molecular mechanisms by which this occurs may vary considerably from one cell type to another. Radioligand binding techniques have demonstrated that alterations in the beta-adrenergic receptor itself may contribute to at least some forms of catecholamine-induced refractoriness. However, other forms of desensitization appear to involve lesions occurring mainly or perhaps exclusively distal to the receptors. Intracellular cAMP concentration is modulated by its synthesis by adenylate cyclase and degradation by phosphodiesterases. Either of these processes may be modified in a desensitized cell. The activation of adenylate cyclase by catecholamines is dependent on the functioning of beta-adrenergic receptors, adenylate cyclase molecules, and guanine nucleotide regulatory components which appear to be essential intermediates between beta receptor stimulation and adenylate cyclase activation. Modification of one or more of these components may be the mechanism for desensitization in a particular cell. The role of cyclic nucleotides in the heart has recently been reviewed (20).

The results of direct examination and characterization of some of these components for the desensitized beta-adrenergic receptor-adenylate cyclase system in the heart have recently been reported. Chronic injection of rats with isoproterenol leads to a hypertrophied heart that exhibits a decreased magnitude and sensitivity of contractile response to in vitro stimulation by isoproterenol. Tse et al (64) extensively investigated the mechanisms by which this change occurred. Desensitization of the heart to isoproterenol was associated with a reduction in both sensitivity and maximal response of adenylate cyclase to activation by isoproterenol. A decrease in the number of beta-adrenergic receptors in the desensitized hearts was also noted. The beta receptors were measured with [^{3}H]DHA whose affinity for the receptors was unchanged after desensitization. Thus it appears that loss in beta-adrenergic receptors with isoproterenol injection plays a major role in desensitization. However, there was also a loss in NaF-stimulated adenylate cycle in the desensitized hearts which might suggest a lesion(s) in either the catalytic unit of adenylate cyclase or the guanine nucleotide regulatory component. The latter possibility may be reflected in the apparent decrease in affinity of the remaining receptors for isoproterenol (64). It has been suggested that high-affinity agonist binding reflects interaction of the beta receptor with the guanine nucleotide regulatory component (44). There appears to be an impaired ability of agonists to bind with high affinity to the beta receptors in this model of desensitization in the heart; this phenomenon has been extensively characterized in desensitized frog erythrocytes (36). Quite different findings have been reported in another model of cardiac desensitization. March et al (45) exposed chick embryo ventricles to 1 μm isoproterenol for 30 min; this led to a diminution in the subsequent inotropic response to isoproterenol. There was no loss, however, in beta-adrenergic receptor number measured with [^{3}H]DHA; also NaF-stimulated adenylate cyclase was unchanged in the densensitized hearts. The major defect appeared to be diminished maximal stimulation by isoproterenol of adenylate cyclase in membranes derived from the desensitized hearts. This was interpreted to indicate an uncoupling of the beta-adrenergic receptor —adenylate cyclase complex, though there was apparently no loss in the beta receptors's affinity for isoproterenol assessed in competition curves with [^{3}H]DHA. There are thus major phenomenological differences in the results of these two important studies. These could reflect differences in the species used (rat vs chick embryo); compare, for example, the mechanism of desensitization in frog erythrocytes (68) with that in turkey red cells (31). Also, the time point at which the mechanisms of desensitization were evaluated could be an important variable. In the astrocytoma cell line studied by Harden et al (26), a functional uncoupling of the beta-adrenergic receptors occurs prior to the loss of beta receptors from the membranes.

These models of desensitization may be relevant to changes that occur in the heart with various forms of hypertension (4). For example, the hearts of spontaneously hypertensive rats (SHR) show diminished sensitivity to the inotropic and chronotropic effects of isoproterenol (38). Limas & Limas (42) reported a diminished number of beta-adrenergic receptors (without change in affinity for [^{3}H]DHA) in SHR compared with normal controls. Interestingly, the diminished number of beta-adrenergic receptors was almost fully evident by 5 weeks of age, even though the rise in blood pressure was quite modest at that point. The authors conjectured that these early changes in beta receptors might reflect increased sympathetic drive in these animals (33, 34). However, Bhalla et al found similar numbers of beta-adrenergic receptors in SHR and controls with a reduced affinity of isoproterenol for the receptors in SHR (9). Woodcock et al (71–73) reported that in 3 different models of hypertension (one-kidney Goldblatt, desoxycorticosterone acetate/salt, and SHR) there were similar reductions in cardiac beta-adrenergic receptors. Adenylate cyclase activities were measured in cardiac membranes from the former two models of hypertension. It was found that maximal isoproterenol-activated adenylate cyclase was dimished whereas NaF-stimulated activities were not diminished compared to controls (72). The authors also reported decreased alpha-adrenergic receptor numbers in the hearts of hypertensive animals, whereas there was no change in the number of beta receptors in the lungs or kidneys of these animals (73).

The receptor changes in these models of hypertension in the rat are reminiscent of those in isoproterenol-induced desensitization of the heart (64) described above. Whether the desensitization seen in hypertension is induced by elevated catecholamines, increased sympathetic tone or transmitter release (18), or some other mechanism is unclear. While there seems to be some agreement that hypertension causes a reduced beta-adrenergic receptor number in the heart, one other study (21) found no change in receptor number.

Since catecholamines can desensitize beta-adrenergic receptors, it is possible that the presence of endogenous catecholamines modulates the number and/or properties of beta-adrenergic receptors in the normal heart. Indeed, several models have been developed wherein the heart is depleted of catecholamines leading to hypersensitivity. Increased responsiveness of adenylate cyclase to catecholamines is seen in rat hearts denervated by 6-hydroxydopamine or depleted of catecholamines by reserpine (54). Also, increased sensitivity is observed in the hearts of SHR after withdrawal of propranolol treatment (38). Treatment of rats with propranolol for two weeks was found to lead to an increase in beta-adrenergic receptor number in the heart (22). However, another study found no change (6). Whether the number of beta receptors in the human heart increases after propranolol treatment is not known. Also, denervation of the rat heart with guanethidine (23) or 6-hydroxydopamine (74) is associated with an increased number of beta-adrenergic receptors. Furthermore, depletion of catecholamines in the heart as a consequence of hypertrophy induced by aortic constriction apparently led to an increase in beta-adrenergic receptor number (43), though the relationship of this finding to the decrease in the number of beta-adrenergic receptors seen in certain models of hypertension is unclear. While beta-adrenergic receptors may increase in the heart, the relationship of these findings to the putative "propranolol withdrawal syndrome" in man is unknown.

ACKNOWLEDGMENTS

Ms. Tina Stephenson lent her gracious hands to the typing of this manuscript. Dr. Hoffman was a Fellow of the Medical Research Council of Canada.

Literature Cited

1. Åblad, B., Carlsson, B., Carlsson, E., Dahlöf, C., Ek, L., Hultberg, E. 1974. Cardiac effects of β-adrenergic receptor antagonists. *Adv. Cardiol.* 12:290–302
2. Ahlquist, R. P. 1948. Study of adrenotropic receptors. *Am. J. Physiol.* 153:586–600
3. Alexander, R. W., Williams, L. T., Lefkowitz, R. J. 1975. Identification of cardiac β-adrenergic receptors by (-)[^{3}H]alprenolol binding. *Proc. Natl. Acad. Sci. USA* 72:1564–68
4. Amer, S. M., Gomoll, A. W., Perhuch, J. L. Jr., Ferguson, H. C., McKinney, G. R. 1974. Aberrations of cyclic nucleotide metabolism in the hearts and vessels of hypertensive rats. *Proc. Natl. Acad. Sci. USA* 71:4930–34
5. Baker, S. P., Boyd, H. M., Potter, L. T. 1980. Distribution and function of β-adrenoceptors in different chambers of the canine heart. *Br. J. Pharmacol.* 68:57–63
6. Baker, S. P., Potter, L. T. 1980. Effect of propranolol on β-adrenoceptors in rat hearts. *Br. J. Pharmacol.* 68:8–10
7. Baker, S. P., Potter, L. T. 1980. Cardiac β-adrenoceptors during normal growth of male and female rats. *Br. J. Pharmacol.* 68:65–70
8. Barnett, D. B., Rugg, E. L., Nahorski, S. R. 1978. Direct evidence of two types of β-adrenoceptor binding site in lung tissue. *Nature* 273:166–68
9. Bhulla, R. C., Sharma, R. V., Ramanathan, S. 1980. Ontogenetic development of isoproterenol subsensitivity of myocardial adenylate cyclase and β-adrenergic receptors in spontaneously hypertensive rats. *Biochim. Biophys. Acta* 632:497–506
10. Bobic, A., Korner, P., Carson, V., Oliver, J. R. 1980. Cardiac β-adrenoceptors and adenylate cyclase activation in rabbit heart during conditions of altered sympathetic activity. *Circ. Res.* 46 (Suppl. I):43–44
11. Brittain, R. B., Jack, D., Ritchie, A. C. 1970. Recent β-adrenoreceptor stimulants. *Adv. Drug Res.* 5:157–253
12. Carlsson, E., Åblad, B., Brandstrom, A., Carlsson, B. 1972. Differentiated blockage of the chronotropic effects of various adrenergic stimuli in the cat heart. *Life Sci.* 11 (Part I):953–58
13. Carlsson, E., Dahlof, C-G., Hedberg, A., Persson, H., Tangstrand, B. 1977. Differentiation of cardiac chronotropic and inotropic effects of β-adrenoceptor agonists. *Naunyn-Schmiedeberg's Arch. Pharmacol.* 300:101–5
14. Chen, F.-C. M., Yamamura, H. I., Roeske, W. R. 1979. Ontogeny of mammalian myocardial β-adrenergic receptors. *Eur. J. Pharmacol.* 58:255–64
15. Cheng, J. B., Cornett, L. E., Goldfien, A., Roberts, J. M. 1980. Decreased concentration of myocardial α-adrenoceptors with increasing age in fetal lambs. *Br. J. Pharmacol.* 70:515–17
16. Chenieux-Guicherey, P., Dausse, J. P., Meyer, P., Schmitt, H. 1978. Inhibition of [^{3}H]dihydroalprenolol binding to rat cardiac membranes by various β-blocking agents. *Br. J. Pharmacol.* 63:177–82
17. Ciaraldi, T., Marinetti, G. V. 1977. Thyroxine and propylthiouracil effects *in vivo* on alpha and beta adrenergic receptors in rat heart. *Biochem. Biophys. Res. Commun.* 74:984–91
18. De Champlain, J. 1977. The sympathetic system in hypertension. *Clin. Endocrinol. Metab.* 6:633–55
19. Dreyer, A. C., Offermeier, J. 1975. Indications for the existence of two types of cardiac β-adrenergic receptors. *Pharmacol. Res. Commun.* 7:151 61
20. Drummond, G. I., Severson, D. L. 1979. Cyclic nucleotides and cardiac function. *Circ. Res.* 44:145–53
21. Giachetti, A., Clark, T. L., Berti, F. 1979. Subsensitivity of cardiac β-adrenoceptors in renal hypertensive

rats. *J. Cardiovasc. Pharmacol.* 1: 467–71
22. Glaubiger, G., Lefkowitz, R. J. 1977. Elevated beta-adrenergic receptor number after chronic propranolol treatment. *Biochem. Biophys. Res. Commun.* 78:720–25
23. Glaubiger, G., Tsai, B. S., Lefkowitz, R. J., Weiss, B., Johnson, E. M. Jr. 1978. Chronic guanethidine treatment increases cardiac β-adrenergic receptors. *Nature* 273:240–42
24. Guicheney, P., Garay, R. P., Levy-Marchal, C., Meyer, P. 1978. Biochemical evidence for presynaptic and postsynaptic α-adrenoceptors in rat heart membranes: positive homotropic cooperativity of presynaptic binding. *Proc. Natl. Acad. Sci. USA* 75:6285–89
25. Hancock, A. A., De Lean, A. L., Lefkowitz, R. J. 1979. Quantitative resolution of beta-adrenergic receptor subtypes by selective ligand binding: application of a computerized model fitting technique. *Mol. Pharmacol.* 16:1–9
26. Harden, T. K., Su, Y.-F., Perkins, J. P. 1979. Catecholamine-induced desensitization involves an uncoupling of beta-adrenergic receptors and adenylate cyclase. *J. Cyclic Nucleotide Res.* 5:99–106
27. Harden, T. K., Wolfe, B. B., Molinoff, P. B. 1976. Binding of iodinated beta-adrenergic antagonist to protein derived from rat heart. *Mol. Pharmacol.* 12:1–15
28. Hedberg, A., Minneman, K. P., Molinoff, P. B. 1980. Differential distribution of $beta_1$ and $beta_2$ adrenergic receptors in cat and guinea pig heart. *J. Pharmacol. Exp. Ther.* 213:503–8
29. Hegstrand, L. R., Minneman, K. P., Molinoff, P. B. 1979. Multiple effects of guanosine triphosphate on beta adrenergic receptors and adenylate cyclase activity in rat heart, lung and brain. *J. Pharmacol. Exp. Ther.* 210:215–21
30. Hoffman, B. B., De Lean, A., Wood, C. L., Schocken, D. D., Lefkowitz, R. J. 1979. Alpha-adrenergic receptor subtypes: quantitative assessment by ligand binding. *Life Sci.* 24:1736–46
31. Hoffman, B. B., Mullikin-Kilpatrick, D., Lefkowitz, R. J. 1979. Desensitization of beta-adrenergic stimulated adenylate cyclase in turkey erythrocytes. *J. Cyclic Nucleotide Res.* 5:355–66
32. Hoffman, B. B., Lefkowitz, R. J. 1980. Radioligand binding studies of adrenergic receptors: new insights into molecular and physiological regulation. *Ann. Rev. Pharmacol. Toxicol.* 20:581–608

33. Iriuchijima, J. 1973. Sympathetic discharge rate in spontaneously hypertensive rat. *Jpn. Heart J.* 14:350–56
34. Judy, W. V., Watanabe, A. M., Henry, D. P., Besch, H. R. Jr., Murphy, W. R., Hockel, G. M. 1976. Sympathetic nerve activity: role in regulation of blood pressure in the spontaneously hypertensive rat. *Circ. Res.* 38:Suppl. II, pp. 21–29
35. Karliner, J. S., Barnes, P., Hamilton, C. A., Dollery, C. T. 1979. $Alpha_1$-adrenergic receptors in guinea pig myocardium: Identification by binding of a new radioligand, [^{3}H]prazosin. *Biochem. Biophys. Res. Commun.* 90:142–49
36. Kent, R. S., De Lean, A., Lefkowitz, R. J. 1980. A quantitative analysis of beta-adrenergic receptor interactions: resolution of high and low affinity states of the receptor by computer modeling of ligand binding data. *Mol. Pharmacol.* 17:14–23
37. Kravietz, W., Poppert, D., Erdmann, E., Glossmann, H., Struck, C. J., Konrad, C. 1976. β-adrenergic receptors in guinea-pig myocardial tissue. *Naunyn-Schmiedebergs Arch. Pharmacol.* 295:215–24
38. Kunos, G., Robertson, B., Kan, W. H., Preiksaitis, H., Mucci, L. 1978. Adrenergic reactivity of the myocardium in hypertension. *Life Sci.* 22:847–54
39. Lands, A. M., Arnold, A., McAuliff, J. P., Luduena, F. P., Brown, T. G. 1967. Differentiation of receptor systems activated by sympathomimetic ammis. *Nature* 214:597–98
40. Lau, Y. H., Robinson, R. B., Rosen, M. R., Bilezikian, J. P. 1980. Subclassification of β-adrenergic receptors in cultured rat cardiac myoblasts and fibroblasts. *Circ. Res.* 47:41–48
41. Lefkowitz, R. J., Hoffman, B. B. 1980. New directions in adrenergic receptor research part I. *Trends Pharmacol. Sci.* 1:314–318
42. Limas, C., Limas, C. J. 1978. Reduced number of β-adrenergic receptors in the myocardium of spontaneously hypertensive rats. *Biochem. Biophys. Res. Commun.* 83:710-14
43. Limas, C. J. 1979. Increased number of β-adrenergic receptors in the hypertrophied myocardium. *Biochim. Biophys. Acta.* 588:174–78
44. Limbird, L. E., Gill, D. M., Lefkowitz, R. J. 1980. Agonist-promoted coupling of the β-adrenergic receptors with the guanine nucleotide regulatory protein of the adenylate cyclase system. *Proc. Natl. Acad. Sci. USA* 77:775–79

45. Marsh, J. D., Barry, W. H., Neer, E. J., Alexander, R. W., Smith, T. W. 1980. Desensitization of chick embryo ventricle to the physiological and biochemical effects of isoproterenol: evidence for uncoupling of the β-receptro–adenylate cyclase complex. *Circ. Res.* 47:493–501
46. Minneman, K. P., Hegstrand, L. R., Molinoff, P. B. 1979. Pharmacological specificity of beta$_1$ and beta$_2$ adrenergic receptors in rat heart and lung *in vitro*. *Mol. Pharmacol.* 16:21–33
47. Minneman, K. P., Hegstrand, L. R., Molinoff, P. B. 1979. Simultaneous determination of beta$_1$ and beta$_2$ adrenergic receptors in tissues containing both receptor subtypes. *Mol. Pharmacol.* 16:34–46
48. Minneman, K. P., Molinoff, P. B. 1980. Classification and quantitation of β-adrenergic receptor subtypes. *Biochem. Pharmacol.* 29:1317–23
49. Moustafa, E., Giachetti, A., Downey, H. F., Bashour, F. A. 1978. Binding of [^{3}H]dihydroalprenolol to beta adrenoceptors of cells isolated from adult rat heart. *Naunyn-Schmiedeberg's Arch. Pharmacol.* 303:107–9
50. Mukherjee, C., Caron, M. G., Lefkowitz, R. J. 1975. Catecholamine-induced subsensitivity of adenylate cyclase associated with loss of beta-adrenergic receptor binding sites. *Proc. Natl. Acad. Sci. USA* 72:1945–49
51. Mukherjee, A., Wong, T. M., Buja, L. M., Lefkowitz, R. J., Willerson, J. T. 1979. Beta-adrenergic and muscarinic cholinergic receptors in canine myocardium: effects of ischemia. *J. Clin. Invest.* 64:1423–38
52. O'Donnell, S. R., Wanstall, J. C. 1979. pA$_2$ values of selective β-adrenoceptor antagonists on isolated atria demonstrate a species difference in the β-adrenoceptor population mediating chronotropic responses in the cat and guinea-pig. *J. Pharm. Pharmacol.* 31:686–90
53. Phillips, D. K. 1980. Chemistry of alpha and beta-adrenergic agonists and antagonists. In *Adrenergic Activators and Inhibitors Part I*, ed. L. Szekenes, pp. 3–61: Berlin/Heidelberg/NY: Springer. 1210 pp.
54. Pik, K., Wollemann, M. 1977. Catecholamine hypersensitivity of adenylate cyclase after chemical denervation in rat heart. *Biochem. Pharmacol.* 26:1448–49
55. Raisman, R., Briley, M., Langer, S. Z. 1979. Specific labelling of postsynaptic α^1 adrenoceptors in rat heart ventricle by [^{3}H]WB4101. *Naunyn-Schmiedeberg's Arch. Pharmacol.* 307:223–26
56. Rockson, S. G., Homcy, C. J., Quinn, P., Manders, W. T., Haber, E., Vatner, S. F. 1981. Cellular mechanisms of impaired adrenergic responsiveness in neonatal dogs. *J. Clin Invest.* 67:319–27
57. Scholtz, H. 1980. Effects of beta- and alpha-adrenoceptor activators and adrenergic transmitter releasing agents in the mechanical activity of the heart. See Ref. 53, pp. 651–712
58. Schumann, H. J., Brodde, O.-E. 1979. Demonstration of α-adrenoceptors in rabbit heart by [^{3}H]-dihydroeryocryptine binding. *Naunyn-Schmiedeberg's Arch. Pharmacol.* 308:191–98
59. Sharma, V. K., Banerjee, S. P. 1978. Alpha-adrenergic receptor in rat heart: effects of thyroidectomy. *J. Biol. Chem.* 253:5277–79
60. Shear, M., Insel, P. A., Melmon, K. L., Coffino, P. 1976. Agonist-specific refractoriness induced by isoproterenol. *J. Biol. Chem.* 251:7572–76
61. Stene-Larsen, G., Helle, K. B. 1978. Cardiac β$_2$-adrenoceptor in the frog. *Comp. Biochem. Physiol. C.* 60:165–73
62. Story, B. D., Briley, M. S., Langer, S. Z. 1979. The effects of chemical sympathectomy with 6-hydroxydopamine on α-adrenoceptor and muscarinic cholinergic binding in rat heart ventricle. *Eur. J. Pharmacol.* 57:423–26
63. Su, Y.-F., Johnson, G. L., Cubeddu-Ximenez, L., Leichtling, B. H., Ortmann, R., Rerkins, J. P. 1976. Regulation of adenosine 3':5'-monophosphate content of human astrocytoma cells: mechanism of agonist-specific desensitization. *J. Cyclic Nucleotide Res.* 2:271–85
64. Tse, J., Powell, J. R., Baste, C. A., Priest, R. E., Kuo, J. F. 1979. Isoproterenol-induced cardiac hypertrophy: modifications in characteristics of β-adrenergic receptor, adenylate cyclase, and ventricular contration. *Endocrinology* 105:246–55
65. U'Prichard, D. C., Bylund, D. B., Snyder, S. H. 1978. (±)[^{3}H]Epinephrine and (-)[^{3}H]dihydroalprenolol binding to β$_1$ and β$_2$ noradrenergic receptors in brain, heart, and lung membranes. *J. Biol. Chem.* 253:5090–102
66. Watanabe, A. M., McConnaughey, M. M., Strawbridge, R. A., Fleming, J. W., Jones, L. R., Besch, H. R. Jr. 1978. Muscarinic cholinergic receptor modulation of β-adrenergic receptor affinity

for catecholamines. *J. Biol. Chem.* 253:4833–36
67. Wei, J.-W., Sulakhe, P. V. 1979. Regional and subcellular distribution of β- and α-adrenergic receptors in the myocardium of different species. *Gen. Pharmacol.* 10:263–67
68. Wessels, M. R., Mullikin, D., Lefkowitz, R. J. 1979. Selective alteration in high affinity agonist binding: a mechanism of beta-adrenergic receptor desensitization. *Mol. Pharmacol.* 16:10–20
69. Williams, R. S., Lefkowitz, R. J. 1978. Alpa-adrenergic receptors in rat myocardium: identification by binding of [^{3}H]dihydroergocryptine. *Circ. Res.* 44:72–27
70. Winek, R., Bhalla, R. 1979. [^{3}H]Dihydroalprenolol binding sites in the rat myocardium: relationship between a single binding site population and the concentration of radioligand. *Biochem. Biophys. Res. Commun.* 91:200–6
71. Woodcock, E. A., Funder, J. W., Johnston, C. I. 1978. Decreased cardiac β-adrenoceptors in hypertensive rats. *Clin. Exp. Pharmacol. Physiol.* 5:545–50
72. Woodcock, E. A., Funder, J. W., Johnston, C. I. 1979. Decreased cardiac β-adrenergic receptors in deoxycorticosterone salt and renal hypertensive rats. *Circ. Res.* 45:560–65
73. Woodcock, E., Johnston, C. I. 1980. Changes in tissue alpha- and beta-adrenergic receptors in renal hypertension in the rat. *Hypertension* 2:156–61
74. Yamada, S., Yamamura, H. I., Roeske, W. R. 1980. Alterations in cardiac autonomic receptors following 6-hydroxydopamine treatment in rats. *Mol. Pharmacol.* 18:185–92
75. Yamada, S., Yamamura, H. I., Roeske, W. R. 1980. Characterization of alpha$_1$-adrenergic receptors in the heart using [^{3}H]WB4101: effect of 6-hydroxydopamine treatment. *J. Pharmacol. Exp. Ther.* 215:176–85

MECHANISMS OF CARDIAC ARRHYTHMIAS

Joseph F. Spear and E. Neil Moore[1]

Department of Animal Biology, School of Veterinary Medicine, University of Pennsylvania, Philadelphia, Pennsylvania 19104

The mechanisms involved in the disruption of the rate and rhythm of the heart have classically been categorized as those due to abnormalities in impulse formation (automaticity) or those due to abnormalities in impulse conduction (59, 66). Current knowledge has expanded these concepts to include additional mechanisms of arrhythmogenesis that are just beginning to be understood. Many of these new concepts have recently been reviewed (6, 10, 21, 23, 24, 29, 30, 38, 39, 52, 64, 65, 69, 74, 78, 89, 95, 99, 101, 106, 107, 110). In this review we consider some of the newer concepts; we restrict discussion to membrane electrophysiologic phenomena.

ABNORMAL AUTOMATICITY

The sinoatrial node is the site of the normal dominant pacemaker of the heart. Normally the sinus node maintains dominance of the heart's rhythm through overdrive suppression of latent pacemaker tissues in other regions of the heart (94, 95). Regions capable of automaticity include atrial fibers exhibiting plateau-type action potentials (44), the atrioventricular junction (40), the bundle of His (9), and the bundle branch–Purkinje systems (40). Automatic fibers have also been reported in the canine atrioventricular valves (14) and in the ostium of the coronary sinus (105). Normal atrial and ventricular myocardial fibers do not exhibit automaticity (40). Ectopic rhythms may originate if the rate of automaticity in one of these subsidiary pacemakers increases sufficiently to capture control of the heart from the

---

[1]Supported in part by grants HL-25213, HL-23071, and HL-16076 from the National Heart, Lung and Blood Institute.

sinoatrial node. Also, depression in the rate of automaticity of the sinoatrial node can allow the passive escape of residual pacemaker activity in other areas of the heart.

Sites of spontaneous activity may become isolated owing to conduction block between the pacemaker site and the rest of the heart. Impulses arising in a pacemaker site may intermittently not exit from the site (exit block) (34) or the site of impulse formation may be protected from overdrive suppression from dominant pacemaker sites owing to entrance block (102). In the latter situation, parasystolic rhythms may arise in which two pacemakers are competing for dominance in controlling the rhythm of the heart, one of which is protected from suppression by entrance block. The diagnosis of a parasystolic rhythm relies on the behavior of the protected focus; ectopic impulses capture the rhythm of the heart in an intermittent but predictable way based on their undisturbed spontaneous cycle length. Sustained fixed coupling of an ectopic beat to the dominant normal beat or fixed patterns of ectopic rhythms such as bigeminy or trigeminy have been traditionally thought to indicate a nonparasystolic mechanism for ectopic impulse formation (19, 63). Recently, it has been shown experimentally that a protected ectopic focus could nonetheless be influenced by the dominant rhythm acting electrotonically across the area of partial block (50, 72, 73). This electrotonic influence may either accelerate or delay the time of occurrence of the parasystolic discharge depending on the time within the cycle at which it occurred (reset its phase). In this analysis, it was shown using a computer simulation that under certain circumstances, rhythms such as bigeminy and trigeminy would be expected to appear over certain ranges of dominant heart rates (73). Therefore a parasystolic rhythm can manifest itself as sustained fixed coupling of ectopic beats, and this characteristic will not always indicate a nonautomatic mechanism. A recent clinical study has provided support of this hypothesis (70).

Acceleration of subsidiary pacemakers to capture the heart may occur in pathological situations. Catecholamines accelerate pacemaker activity (5, 40, 41, 91) and digitalis intoxication is known to increase the rate of automaticity (58, 96, 97). Twenty-four hours following experimental myocardial infarction, fibers of the Purkinje system exhibit accelerated rates of automaticity (35, 42, 45, 87). In such cases, cells may not be greatly depolarized and the rapid inward sodium current may not be inactivated.

In tissues that have been depolarized to resting potentials within the range of −40 to −50 millivolts, the rapid inward sodium current is inactivated. These depolarized cells, however, may maintain the ability to generate action potentials owing to the activation of a slow inward current (7,

8, 20). In addition, spontaneous rhythmicity can occur in Purkinje fibers in this range of resting potentials (7, 8, 20), and even in working myocardial cells rhythmic activity has been reported under these conditions (46, 60, 61). This type of rhythmic activity is characteristically sensitive to agents that block the slow inward current. "Slow-response" automaticity has also been reported in atrial muscle from diseased human hearts (43, 84) and in infarcted ventricular tissue removed from patients at the time of aneurysmectomy (85, 90). It is interesting that the automaticity that normally occurs in the sinoatrial and atrioventricular nodes occurs in tissues that naturally have low resting potentials in the range of −60 to −50 millivolts. The action potentials and spontaneous automaticity in these tissues can be eliminated by agents that block the slow inward current, suggesting that they may also rely on similar mechanisms (47, 62, 69, 109). The role of slow-response automaticity in clinical arrhythmias has still not been conclusively verified.

TRIGGERED ACTIVITY

The response of an ectopic rhythm to electrical pacing has often been used to distinguish those rhythms that rely on an automatic focus for their mechanism from those that rely on abnormal conduction (reentry). Recently, rhythms have been described in experimental situations in which the precipitating factor is dependent on activity from a distant site triggering an automatic rhythm. This appearance of triggered activity has been linked to the occurrence of afterdepolarizations (21). Triggered activity can be distinguished from the abnormal enhanced automaticity that occurs in depolarized fibers since triggerable fibers will remain quiescent unless activity is evoked by an exogenous source. Triggered activity has been described in both depolarized tissue and tissues that exhibit relatively normal resting potentials (18, 21, 22, 25, 31, 33, 44, 68, 77–79, 104, 105). In cardiac Purkinje fibers, acetylstrophanthidin and ouabain can cause transient depolarizations that appear in the diastolic period following repolarization of an action potential (25, 33, 77). The amplitudes of these depolarizations are dependent on the rate of driving of the fibers. Following appropriate programmed electrical stimulation these transient depolarizations may reach threshold, causing a run of nondriven triggered impulses (33). These transient depolarizations can be enhanced by increasing the concentration of calcium in the bathing medium; they are depressed by manganese and slow-channel blocking agents (32). The relationship between the acetylstrophanthidin-induced transient depolarization and the slow inward current that appears only at more depolarized resting potentials has not been determined. While calcium probably plays a role in the transient depolarization, this role may be indirect; calcium may modify the inward current carried by other ions (100).

Triggered activity has been reported in diseased human atrium (44) and in atrial tissues removed from cats with spontaneous cardiomyopathy (18). Triggered activity may also complicate the automatic rhythms that dominate hearts at 24 hours following coronary occlusion (26). Although it has not been definitely proven, triggered activity may cause some clinical arrhythmias (76, 78, 108).

In cat sinus node cells and in isolated Purkinje fibers with maximum diastolic potentials around −60 mV a subthreshold depolarizing pulse early in the cycle or hyperpolarizing pulse later in the cycle can "annihilate" pacemaker activity (48, 49). A computer model of a sheet of excitable cells indicates that the electrotonic influence of adjacent nonpacemaker tissue can suppress pacemaker activity in automatic fibers (93). A propagated response in this system may induce subthreshold oscillations in the pacemaker fibers or may bring out sustained rhythmic activity. These studies emphasize that given the appropriate conditions, triggering the initiation and termination of rhythmic activity by exogenous activity may be a basic characteristic of all automaticity.

ECTOPIC IMPULSES INDUCED BY INJURY CURRENT

During the acute phases of myocardial ischemia, cells within the ischemic zone become significantly depolarized and their action potential durations are reduced compared to normal. In this period cells remain electrically coupled to adjacent nonischemic tissue, and therefore significant current may flow between injured depolarized cells and normal tissue. The flow of injury current may induce reexcitation of nonischemic cells; this mechanism may be involved in early ischemic arrhythmias (52, 53, 61). A focal origin of ectopic activity has been observed on the normal side of the ischemic border possibly induced by injury currents (53). For this mechanism to operate, a region of inexcitability must be imposed between the normal tissues and the ischemic border. In well-coupled cells, no abrupt transition in action potential duration or resting potential can occur since current flow would tend to equalize adjacent potentials. Local injury currents, however, have yet to be proven to cause the arrhythmias associated with acute ischemia.

SLOW CONDUCTION

The tissues in the normal heart exhibit a considerable range of conduction velocities. In the bundle branch–Purkinje system, conduction velocity may be as high as 3–4 m sec^{-1} while within the central region of the atrioventricular node it drops as low as 0.02 m sec^{-1}. Disruptions in impulse conduction may lead to bradyarrhythmias associated with atrioventricular block or with exit block of pacemaker activity. On the other hand, disruption in impulse conduction leading to reentry can produce tachyarrhythmias.

The conduction of the cardiac impulse relies on local current circuits generated by active regions of the cardiac membrane. These local currents act to depolarize adjacent membrane to threshold. The effectiveness of transmission of the cardiac impulse therefore depends on the magnitude of the current source (the amplitude and the rate of depolarization of the action potential), the effectiveness of electrotonic transmission of the current (fiber diameter, geometric arrangement, and cable properties), and the excitability of the resting membrane in the pathway of conduction.

The rate of action potential depolarization is an important determinant of conduction velocity. In the heart, areas that normally exhibit very slow conduction such as the sinoatrial and atrioventricular nodes also have action potentials with very slow maximum rates of depolarization (40). In addition, depolarized tissues exhibiting slow response action potentials have very slow rates of depolarization and slow conduction velocities (20).

Using the assumption of uniform conduction, the velocity of a propagated action potential can be modeled mathematically using cable equations and cable properties derived from experimental measurements. The importance of the rate of depolarization versus other factors such as cell-to-cell coupling can be estimated by varying the parameters. It was found using these calculations that a 100-fold increase in internal resistance or coupling resistance between cells should produce only a 10-fold decrease in conduction velocity (23). Based on this cable analysis, the 100-fold difference in conduction velocity between Purkinje fiber and atrioventricular node seems more likely to be due to the reduced rate of depolarization found in AV node as compared to Purkinje fibers. In contrast, other investigators have shown in synthetically grown strands of cardiac muscle that recorded conduction velocities as low as 0.002 m sec^{-1} could only be accounted for by an increased internal resistance between cells (67).

Cell coupling is an important determinant of conduction velocity when one is considering conduction through a region of local block. In regions of local conduction block (induced by a sucrose gap, a local decrease in external sodium, an imposed voltage, or a focal elevation of extracellular potassium concentration), transmission of an impulse across the inexcitable gap to a distal excitable region may still occur electrotonically (4, 11–13, 51, 103). The effectiveness of this transmission will be determined by the efficiency of the electrotonic spread of current from the proximal excitable region through the inexcitable region to the distal excitable region. For transmission to occur, current generated in the proximal excitable region by an action potential must depolarize the distal region by discharging the membrane through local current circuits. Depending on the cable properties of the system, it may require considerable time for the distal excitable region to be brought to threshold. More than a hundred milliseconds may elapse following activity in the proximal segment. The action potential in the distal segment is preceded by a pre-potential that reflects the electrotonic transmission of current from the proximal segment. The conduction velocity through a tissue under these conditions can be very slow owing to the step delay in transmission of the cardiac impulse through the blocked region. Small increases in the resistance to current flow between cells across the inexcitable gap produce large increments in conduction time in this form of transmission delay.

There is evidence that this mechanism may play a role in the normal transmission of the cardiac impulse through the atrioventricular node. Studies using microelectrodes to map the sequence of activation through rabbit AV node have shown regions of the central node where it appears there is a step delay in transmission between closely approximated regions (16, 80). Step-like delays in transmission have also been reported in human infarcted myocardium (86).

Cell excitability, too, influences conduction velocity. The local currents generated by an active region of membrane must depolarize the adjacent membrane in the conduction pathway to threshold potential in order for an active response to occur. If the difference between the threshold potential and the resting potential is increased or decreased, a corresponding increase or decrease in conduction velocity results since it requires more or less time to charge the adjacent membrane to threshold. During the period of supernormal excitability in Purkinje fibers conduction velocity is faster than that which occurs when a beat is evoked later in the diastolic period. This acceleration in conduction velocity occurs even though the rate of depolarization of the evoked action potentials may be reduced by 30% compared to beats conducted later during diastole. (88). Correlations between changes in membrane excitability and conduction velocity independent of changes in action potential rate of depolarization were also found during phase 4 depolarization in Purkinje fibers and following variations in potassium and calcium ion concentration in the external medium (75). Beats evoked during the peak of the transient depolarization induced by digitalis toxicity also conduct faster despite the presence of a reduced rate of depolarization (75).

Changes in any of the above factors may result in slow conduction leading to cardiac arrhythmias. Each factor has been shown to be an important determinant of conduction velocity in various experimental models under specific conditions.

REENTRY

As originally described, reentry involved basically a circus movement in which a region of the heart previously excited was reexcited by the same conducted beat after it traversed a circuitous route (81). The elements in such a model require unidirectional block in one limb of the circular pathway. Conduction in the other limb must be so slow, the pathway so long, and/or the refractory period so brief that when the impulse arrives back at its origin it may reexcite the previously excited area. Such reexcitation has been demonstrated in the Wolff-Parkinson-White syndrome, in which an anomalous bypass tract between atrial and ventricular tissue can circumvent the normal atrioventricular conduction pathway (36).

Reentrant mechanisms, however, may operate in the absence of a fixed obstacle around which the beat must conduct. Sustained reentrant activity has been induced in relatively homogeneous sheets of atrial tissue by imposing a premature beat (1–3). The impulse proceeding in one direction around the tissue sets up a vortex of conduction in which a central area of the vortex is not part of the pathway of conduction. The central functional obstacle results from electrotonic depolarization of the focus of the vortex by the encircling conducted activity (3, 93). A characteristic of vortex reentry is that the cycle time of the reentrant circuit is determined primarily by the duration of the refractory period of the tissue since the vortex shrinks in diameter until the cycling beat impinges upon the tail of its refractory wake. The frequency of the tachycardia therefore can be increased within limits by decreasing the refractory period of the tissue. This is in contrast to the fixed obstacle circus movement in which the cycle time is not usually determined by the refractory period of the tissue but rather by the conduction velocity around the circuit.

Separate pathways are not required for reentry (4, 51). In an experimental preparation using a sucrose gap to produce an inexcitable segment in a preparation of Purkinje fibers, a driven impulse on the proximal side of the gap will be electrotonically transmitted after a delay to activate the tissue on the distal side. When this delay is long enough, electrotonic transmission in the reverse direction over the same blocked segment can reexcite the proximal segment.

Dissociated conduction of a wavefront under some circumstances may produce an extra excitation. In this model, an impulse originating in one region of the heart may be conducted to another region via two or more pathways. If the transmission time in one pathway lags sufficiently behind the transmission time in the other, the two impulses may arrive at the common distal site sufficiently separated in time so that the delayed impulse can produce a second response. This mechanism has been described in a case of Wolff-Parkinson-White syndrome where one atrial response produced two ventricular responses owing to dissociated conduction over the normal and anomalous pathways (56). This mechanism however does not lead to sustained reentrant activity.

A reentrant circuit can be very small. In circumstances where conduction is slow and the reentrant pathway is serpiginous, the reentrant circuit may be small enough to behave as a focus providing ectopic impulses to the rest of the heart. It has been estimated in a computer model of a sheet of excitable cells that a reentrant vortex can be established in which the length of the pathway is approximately 20 space constants (93). In the "leading circle" model (vortex reentry) described previously, the reentrant circuit diameter can be 6–8 mm (3). In the intact heart it would be difficult to distinguish microreentry of this sort from other possible mechanisms such as triggered activity. Both could behave as a small focus, and the ectopic activity could be initiated and terminated using similar stimulation techniques.

Localized reentry has been postulated as a mechanism for the sustained ventricular tachycardia observed in experimental animals and patients after myocardial infarction (17, 27, 28, 37, 54, 55, 57, 71, 98). The ability to record continuous and fractionated electrical activity from regions of the infarct during ventricular tachycardia has been interpreted as indicating delayed conduction and reentrant activity. Recently, such activity has been detected on the surface of the body using specialized signal processing of the electrocardiogram (15, 83). Low amplitude, high frequency signals at the termination of the QRS complex and extending into the ST segment appear to be manifestations of the slowly conducting activity within infarcted tissue. A high correlation between the appearance of these body surface potentials and the propensity for patients to have ventricular tachyarrhythmias has also been demonstrated (82, 92).

ACKNOWLEDGMENTS

The authors wish to thank Drs. Eric L. Michelson and Michael B. Simson for their helpful comments and Marianne Adams for preparation of the manuscript.

Literature Cited

1. Allessie, M. A., Bonke, F. I. M., Schopman, F. J. G. 1973. Circus movement in rabbit atrial muscle as a mechanism of tachycardia. *Circ. Res.* 33:54–62
2. Allessie, M. A., Bonke, F. I. M., Schopman, F. J. G. 1976. Circus movement in rabbit atrial muscle as a mechanism of tachycardia. II. The role of nonuniform recovery of excitability in the occurrence of unidirectional block, as studied with multiple microelectrodes. *Circ. Res.* 39:168–77
3. Allessie, M. A., Bonke, F. I. M., Schopman, F. J. G. 1977. Circus movement in rabbit atrial muscle as a mechanism of tachycardia. III. The "leading circle" concept: A new model of circus movement in cardiac tissue without the involvement of an anatomical obstacle. *Circ. Res.* 41:9–18
4. Antzelevitch, C., Jalife, J., Moe, G. K. 1980. Characteristics of reflection as a mechanism of reentrant arrhythmias and its relationship to parasystole. *Circulation* 61:182–91
5. Armour, J. A., Hageman, G. R., Randall, W. C. 1972. Arrhythmias induced by local cardiac nerve stimulation. *Am. J. Physiol.* 223:1068–75
6. Arnsdorf, M. F. 1977. Membrane factors in arrhythmogenesis: Concepts and definitions. *Prog. Cardiovasc. Dis.* 19:413–29
7. Aronson, R. S., Cranefield, P. F. 1973. The electrical activity of canine cardiac Purkinje fibers in sodium-free, calcium-rich solutions. *J. Gen. Physiol.* 61:786–808
8. Aronson, R. S., Cranefield, P. F. 1974. The effect of resting potential on the electrical activity of canine cardiac Purkinje fibers exposed to Na-free solution or to ouabain. *Pfluegers Arch.* 347:101–16
9. Bailey, J. C., Greenspan, K., Elizari, M. V., Anderson, G. J., Fisch, C. 1972. Effects of acetylcholine on automaticity and conduction in the proximal portion of the His-Purkinje specialized conduction system of the dog. *Circ. Res.* 30:210–16
10. Bandura, J. P. 1980. The role of electrotonus in slow potential development and conduction in canine Purkinje tissue. In *The Slow Inward Current and Cardiac Arrhythmias*, ed. D. P. Zipes, J. C. Bailey, V. Elharrar. The Hague: Martinus Nijhoff
11. Bandura, J. P., Brody, D. A. 1975. Electrotonic transmission through blocked canine Purkinje tissue: Role of calcium? *Circulation* 52(4):11–18
12. Bandura, J. P., Brody, D. A. 1976. The role of resting membrane potential and slow diastolic depolarization in restoration of transmission through blocked segments of canine Purkinje tissue. *Am. J. Cardiol.* 37:119
13. Bandura, J. P., Wennemark, J. R., Brody, D. A. 1975. Microelectrode study of block mechanisms in Purkinje tissue segments exposed to a sodium-free milieu. *Am. J. Cardiol.* 35(1):121
14. Bassett, A. L., Fenoglio, J. J. Jr., Wit, A. L., Myerburg, R. J., Gelband, H. 1976. Electrophysiological and ultrastructural characteristics of the canine tricuspid valve. *Am. J. Physiol.* 230:1366–73
15. Berbari, E. J., Scherlag, B. J., Hope, R. R., Lazzara, R. 1978. Recording from the body surface of arrhythmogenic ventricular activity during the ST segment. *Am. J. Cardiol.* 41:697–702
16. Billette, J., Janse, M. J., Van Capelle, F. J. L., Anderson, R. H., Touboul, P., Durrer, D. 1976. Cycle-length-dependent properties of AV nodal activation in rabbit hearts. *Am. J. Physiol.* 231:1129–39
17. Boineau, J. P., Cox, J. L. 1973. Slow ventricular activation in acute myocardial infarction: A source of reentrant premature ventricular contraction. *Circulation* 48:702–13
18. Boyden, P. A., Tilley, L. P., Liu, S. K., Wit, A. L. 1977. Effects of atrial dilatation on atrial cellular electrophysiology: Studies on cats with spontaneous cardiomyopathy. *Circulation* 56(4):48
19. Chung, E. K. 1971. *Principles of Cardiac Arrhythmias*, Ch. 10. Baltimore: Williams & Wilkins
20. Cranefield, P. F. 1975. *The Conduction of the Cardiac Impulse: The Slow Response and Cardiac Arrhythmias*. Mount Kisco, NY: Futura
21. Cranefield, P. F. 1977. Action potentials, afterpotentials and arrhythmias. *Circ. Res.* 41:415–23
22. Cranefield, P. F., Aronson, R. S. 1974. Initiation of sustained rhythmic activity by single propagated action potentials in canine cardiac Purkinje fibers exposed to sodium-free solution or to ouabain. *Circ. Res.* 34:477–81
23. Cranefield, P. F., Dodge, F. A. 1980. Slow conduction in the heart. See Ref. 10
24. Cranefield, P. F., Wit, A. L. 1979. Car-

diac arrhythmias. *Ann. Rev. Physiol.* 41:459–72
25. Davis, L. D. 1973. Effect of changes in cycle length on diastolic depolarization produced by ouabain in canine Purkinje fibers. *Circ. Res.* 32:206–14
26. El-Sherif, N., Gough, W. B., Zeiler, R., Mehra, R. 1981. Endocardial mapping of triggered automaticity in canine ischemic Purkinje fibers. *Am. J. Cardiol.* 47:489
27. El-Sherif, N., Sherlag, B. J., Lazzara, R. et al. 1977. Reentrant ventricular arrhythmias in the late myocardial infarction period. I. Conduction characteristics in the infarction zone. *Circulation* 55:686–702
28. El-Sherif, N., Smith, R. A., Evans, K. 1981. Canine ventricular arrhythmias in the late myocardial infarction period in the dog. 8. Epicardial mapping of reentrant circuits. *Circ. Res.* 49:255–65
29. Elharrar, V., Zipes, D. P. 1977. Cardiac electrophysiologic alterations during myocardial ischemia. *Am. J. Physiol.* 233(3):H329–45
30. Elharrar, V., Zipes, D. P. 1980. Voltage modulation of automaticity in cardiac Purkinje fibers. See Ref. 10
31. Ferrier, G. R. 1978. Effects of transmembrane potential on oscillatory afterpotentials induced by acetylstrophanthidin in canine ventricular tissues. *Fed. Proc.* 37:419 (Abstr.)
32. Ferrier, G. R., Moe, G. K. 1973. Effect of calcium on acetylstrophanthidin-induced transient depolarization in canine Purkinje tissue. *Circ. Res.* 33:508–15
33. Ferrier, G. R., Saunders, J. H., Mendez, C. 1973. A cellular mechanism for the generation of ventricular arrhythmias by acetylstrophanthidin. *Circ. Res.* 32:600–9
34. Fisch, C., Greenspan, K., Anderson, G. J. 1971. Exit block. *Am. J. Cardiol.* 28:402
35. Friedman, P. L., Stewart, J. F., Fenoglio, J. J., Wit, A. L. 1973. Survival of subendocardial Purkinje fibers after extensive myocardial infarction in dogs. In vitro and in vivo correlations. *Circ. Res.* 33:597–611
36. Gallagher, J. J., Gilbert, M., Svenson, R. H., Sealy, W. C., Kasell, J., Wallace, A. G. 1975. Wolff-Parkinson-White Syndrome: The problem, evaluation and surgical correction. *Circulation* 51:767–85
37. Garan, H., Fallon, J. T., Ruskin, J. N. 1980. Sustained ventricular tachycardia in recent canine myocardial infarction. *Circulation* 62:980–87
38. Gettes, L. S. 1976. Possible role of ionic changes in the appearance of arrhythmias. *Pharmacol. Ther. B* 2:787–810
39. Hauswirth, O., Singh, B. N. 1979. Ionic mechanisms in heart muscle in relation to the genesis and the pharmacological control of cardiac arrhythmias. *Pharmacol. Rev.* 30:5–63
40. Hoffman, B. F., Cranefield, P. F. 1960. *Electrophysiology of the Heart.* NY: McGraw-Hill
41. Hogan, P. M., Davis, L. D. 1968. Evidence for specialized fibers in the canine atrium. *Circ. Res.* 23:387–96
42. Hope, R. R., Scherlag, B. J., El-Sherif, N., Lazzara, R. 1976. Hierarchy of ventricular pacemakers. *Circ. Res.* 39:883–88
43. Hordof, A. J., Edie, R., Malm, J. R., Hoffman, B. F., Rosen, M. R. 1976. Electrophysiologic properties and response to pharmacologic agents of fibers from diseased atria. *Circulation* 54:774–79
44. Hordof, A. J., Spotnitz, A., Mary-Rabine, L., Edie, R., Rosen, M. R. 1978. The cellular electrophysiologic effects of digitalis on human atrial fibers. *Circulation* 57:223–29
45. Horowitz, L. N., Spear, J. F., Moore, E. N. 1975. Subendocardial origin of ventricular arrhythmias in 24-hour-old experimental myocardial infarction. *Circulation* 53:56–63
46. Imanishi, S., Surawicz, B. 1976. Automatic activity in depolarized guinea pig ventricular myocardium: Characteristics and mechanisms. *Circ. Res.* 39:751–59
47. Irisawa, H., Yanagihara, K. 1980. The slow inward current of the rabbit sinoatrial nodal cells. See Ref 10
48. Jalife, J., Antzelevitch, C. 1979. Phase resetting and annihilation of pacemaker activity in cardiac tissue. *Science* 206:695–97
49. Jalife, J., Antzelevitch, C. 1980. Pacemaker annihilation: Diagnostic and therapeutic implications. *Am. Heart J.* 100:128–30
50. Jalife, J., Moe, G. K. 1976. Effect of electrotonic potentials on pacemaker activity of canine Purkinje fibers in relation to parasystole. *Circ. Res.* 39:801
51. Jalife, J., Moe, G. K. 1981. Excitation, conduction and reflection of impulses in isolated bovine and canine cardiac Purkinje fibers. *Circ. Res.* 49:233–47
52. Janse, M. J., Kleber, A. G. 1981. Electrophysiologic changes and ventricular

arrhythmias in the early phase of regional myocardial ischemia. *Circ. Res.* 49:1069–81
53. Janse, M. J., Van Cappelle, F. J. L., Morsink, H., Kleber, A. G., Wilms-Schopman, F., Cardinal, R., D'Alnoncourt, N., Durrer, D. 1980. Flow of "injury" current and patterns of excitation during early ventricular arrhythmias in acute regional myocardial ischemia in isolated porcine and canine hearts. *Circ. Res.* 47(2):151–65
54. Josephson, M. E., Horowitz, L. N., Farshidi, A. 1978. Continuous local electrical activity: A mechanism of recurrent ventricular tachycardia. *Circulation* 57:659–65
55. Josephson, M. E., Horowitz, L. N., Farshidi, A., Kastor, J. A. 1978. Recurrent sustained ventricular tachycardia. 1. Mechanisms. *Circulation* 57:431–40
56. Josephson, M. E., Seides, S. F., Damato, A. N. 1976. Wolff-Parkinson-White Syndrome with 1:2 atrioventricular conduction. *Am. J. Cardiol.* 37:1094–96
57. Karagueuzian, H. S., Fenoglio, J. J., Weiss, M. B., Wit, A. L. 1979. Protracted ventricular tachycardia induced by premature stimulation of the canine heart after coronary artery occlusion and reperfusion. *Circ. Res.* 44:833–46
58. Kastor, J. A., Spear, J. F., Moore, E. N. 1972. Localization of ventricular irritability by epicardial mapping. *Circulation* 45:952–64
59. Katz, L. N., Pick, A. 1956. *Clinical Electrocardiography. Part I: The Arrhythmias With an Atlas of Electrocardiograms.* Philadelphia: Lea and Ferbiger
60. Katzung, B. G. 1975. Effects of extracellular calcium and sodium on depolarization-induced automaticity in guinea pig papillary muscle. *Circ. Res.* 37:118–27
61. Katzung, B. G., Hondeghem, L. M., Grant, A. O. 1975. Cardiac ventricular automaticity induced by current of injury. *Pfluegers Arch.* 360:193–97
62. Kokubun, S., Nishimura, M., Noma, A., Irisawa, H. 1980. The spontaneous action potential of rabbit atrioventricular node cells. *Jpn. J. Physiol.* 30:529–40
63. Langendorf, R., Pick, A. 1967. Parasystole with fixed coupling. *Circulation* 35:304
64. Lazzara, R., El-Sherif, N., Hope, R. R., Scherlag, B. J. 1978. Ventricular arrhythmias and electrophysiological consequences of myocardial ischemia and infarction. *Circ. Res.* 42:740–49
65. Lazzara, R., Scherlag, B. J. 1980. Role of the slow current in the generation of arrhythmias in ischemic myocardium. See Ref. 10
66. Lewis, T. 1925. *The Mechanism and Graphic Registration of the Heart Beat.* London: Shaw
67. Lieberman, M., Kootsey, J. M., Johnson, E. A., Swanobori, T. 1973. Slow conduction in cardiac muscle: A biophysical model. *Biophys. J.* 13:37–55
68. Mary-Rabine, L., Rosen, M. R. 1977. Sustained rhythmic activity in human atria. *Circulation* 56:III-48
69. Mendez, C. 1980. The slow inward current and AV nodal propagation. See Ref. 10
70. Michelson, E. L., Morganroth, J., Spear, J. F., Kastor, J. A., Josephson, M. E. 1978. Fixed coupling: Different mechanisms revealed by exercise-induced changes in cycle length. *Circulation* 58:1002–9
71. Michelson, E. L., Spear, J. F., Moore, E. N. 1980. Electrophysiologic and anatomic correlates of sustained ventricular tachyarrhythmias in a model of chronic myocardial infarction. *Am. J. Cardiol.* 45:583–90
72. Moe, G. K., Jalife, J., Mueller, W. J. 1977. Reciprocation between pacemaker sites: reentrant parasystole? In *Reentrant Arrhythmias,* ed. H. E. Kulbertus, p. 271. Lancaster: MTP Press, Ltd.
73. Moe, G. K., Jalife, J., Mueller, W. J., Moe, B. 1977. A mathematical model of parasystole and its application to clinical arrhythmias. *Circulation* 56:968–79
74. Opie, L. H., Nathan, D., Lubbe, W. F. 1979. Biochemical aspects of arrhythmogenesis and ventricular fibrillation. *Am. J. Cardiol.* 43:131–48
75. Peon, J., Ferrier, G. R., Moe, G. K. 1978. The relationship of excitability to conduction velocity in canine Purkinje tissue. *Circ. Res.* 43:125–35
76. Rosen, M. R., Fisch, C., Hoffman, B. F., Danilo, P. Jr., Lovelace, D. E., Knoebel, S. B. 1980. Can accelerated atrioventricular junctional escape rhythms be explained by delayed after depolarizations? *Am. J. Cardiol.* 45:1272–84
77. Rosen, M. R., Gelband, H., Hoffman, B. F. 1973. Correlation between effects of ouabain on the canine electrocardiogram and transmembrane potentials of isolated Purkinje fibers. *Circulation* 47:65–72

78. Rosen, M. R., Reder, R. F. 1981. Does triggered activity have a role in the genesis of cardiac arrhythmias? *Ann. Int. Med.* 94:794–801
79. Saito, T., Otoguro, M., Matsubara, T. 1978. Electrophysiological studies of the mechanism of electrically induced sustained rhythmic activity in the rabbit right atrium. *Circ. Res.* 42:199–206
80. Sano, T., Suzuki, F., Takigawa, S. 1964. The genesis of the step or notch on the upstroke of the action potential obtained from the atrioventricular node. *Jpn. J. Physiol.* 14:659–68
81. Schmitt, F. O., Erlanger, J. 1928. Directional differences in the conduction of the impulse through heart muscle and their possible relation to extrasystolic and fibrillatory contractions. *Am. J. Physiol.* 87:326–47
82. Simson, M. B. 1981. Identification of patients with ventricular tachycardia after myocardial infarction from signals in the terminal QRS complex. *Circulation.* 64:235–42
83. Simson, M. B., Euler, D. E., Michelson, E. L., Falcone, R. A., Spear, J. F., Moore, E. N. 1981. Detection of delayed ventricular activation on the body surface in dogs. *Am. J. Physiol.* 241:H363–69
84. Singer, D. H., Ten Eick, R. E., DeBoer, A. 1973. Electrophysiological correlates of human atrial tachyarrhythmias. In *Cardiac Arrhythmias*, ed. L. S. Dreifus, W. Likoff, pp. 97–111. NY: Grune & Stratton
85. Spear, J. F., Horowitz, L. N., Hodess, A. B., MacVaugh, H., Moore, E. N. 1979. Cellular electrophysiology of human myocardial infarction. I. Abnormalities of cellular activation. *Circulation* 59:247–56
86. Spear, J. F., Horowitz, L. N., Josephson, M. E., Harken, A., Moore, E. N. 1982. Electrophysiologic characteristics of automaticity and conduction in isolated human infarcted myocardium. In *Ventricular Tachycardia: Mechanism and Management*, ed. M. E. Josephson. Mt. Kisco, NY: Futura. In press
87. Spear, J. F., Michelson, E. L., Spielman, S. R., Moore, E. N. 1977. The origin of ventricular arrhythmias 24 hours following experimental anterior septal coronary artery occlusion. *Circulation* 55: 844–52
88. Spear, J. F., Moore, E. N. 1974. Supernormal excitability and conduction in the His-Purkinje system of the dog. *Circ. Res.* 35:782–92
89. Sperelakis, N. 1979. Propagation mechanisms in heart. *Ann. Rev. Physiol.* 41:441–57
90. Talano, J. V., Singer, D. H., Loeb, H. S., Ten Eick, R. E., Elson, J., Euler, D. E., Randall, W. C., Moran, J. M., Gunnar, R. M. 1976. Intractable ventricular tachyarrhythmia in post-infarction aneurysm: Clinical, electrophysiologic and electropharmacologic studies. *Clin. Res.* 24:242A
91. Tsien, R. W. 1974. Effects of epinephrine on the pacemaker potassium current of cardiac Purkinje fibers. *J. Gen. Physiol.* 64:293–319
92. Uther, J. B., Dennett, C. J., Tan, A. 1978. The detection of delayed activation signals of low amplitude in the vectorcardiogram of patients with recurrent ventricular tachycardia by signal averaging. In *Management of Ventricular Tachycardia—Role of Mexiletine*, ed. E. Sandoe, D. G. Julian, J. W. Bell, pp. 80–82. Amsterdam: Excerpta Medica
93. Van Capelle, F. J. L., Durrer, D. 1980. Computer simulation of arrhythmias in a network of coupled excitable elements. *Circ. Res.* 47:454–66
94. Vassalle, M. 1977. The relationship among cardiac pacemakers: Overdrive suppression. *Circ. Res.* 41:269–77
95. Vassalle, M. 1977. Generation and conduction of impulses in the heart under physiological and pathological conditions. *Pharmacol. Ther. B.* 3:1–39
96. Vassalle, M., Karis, J., Hoffman, B. F. 1962. Toxic effects of ouabain on Purkinje fibers and ventricular muscle fibers. *Am. J. Physiol.* 203:433
97. Vassalle, M., Musso, E. 1976. On the mechanisms underlying digitalis toxicity in cardiac Purkinje fibers. In *Recent Advances in Studies on Cardiac Structure and Metabolism*, ed. P. E. Roy, N. S. Dhalla, 9:355–76. Baltimore: University Park Press
98. Waldo, A. L., Kaiser, G. A. 1973. A study of ventricular arrhythmias associated with acute myocardial infarction in the canine heart. *Circulation* 47: 1222–28
99. Weidmann, S. 1974. Heart: electrophysiology. *Ann. Rev. Physiol.* 36:155–70
100. Weingart, R., Kass, R. S., Tsien, R. W. 1977. Roles of calcium and sodium ions in the transient inward current induced by strophanthidin in cardiac Purkinje fibers. *Biophys. J.* 17:3A (Abstr.)
101. Wellens, H. J. J., Farre, J., Bar, F. W. 1980. The role of the slow inward cur-

rent in the genesis of ventricular tachyarrhythmias in man. See Ref. 10
102. Wennemark, J. R., Bandura, J. P. 1974. Microelectrode study of Wenckebach periodicity in canine Purkinje fibers. *Am. J. Cardiol.* 33:390–98
103. Wennemark, J. R., Bandura, J. P., Brody, D. A., Ruesta, V. J. 1975. Microelectrode study of high grade block in canine Purkinje fibers. *J. Electrocardiol.* 8(4):299–306
104. Wit, A. L., Cranefield, P. F. 1976. Triggered activity in cardiac muscle fibers of the simian mitral valve. *Circ. Res.* 85–98
105. Wit, A. L., Cranefield, P. F. 1977. Triggered and automatic activity in the canine coronary sinus. *Circ. Res.* 41:435–45
106. Wit, A. L., Cranefield, P. F. 1978. Reentrant excitation as a cause of cardiac arrhythmias. *Am. J. Physiol.* 235:H1–17
107. Wit, A. L., Cranefield, P. F., Gadsby, D. C. 1980. Triggered activity. See Ref. 10
108. Zipes, D. P., Foster, P. R., Troup, P. J., Pedersen, D. H. 1979. Atrial induction of ventricular tachycardia: Reentry versus triggered automaticity. *Am. J. Cardiol.* 44:1–8
109. Zipes, D. P., Mendez, C. 1973. Action of manganese ions and tetrodotoxin on atrioventricular nodal transmembrane potentials in isolated rabbit hearts. *Circ. Res.* 32:447–54
110. Zipes, D. P., Rinkenberger, R. L., Heger, J. J., Prystowsky, E. N. 1980. The role of the slow inward current in the genesis and maintenance of supraventricular tachyarrhythmias in man. See Ref. 10

SPECIAL TOPIC: CHEMOTAXIS AND MOTILITY

Introduction, D. E. Koshland, Jr., *Section Editor*

Chemotaxis, defined as the movement towards or away from chemicals, is a universal biological response. Organisms chemotax towards nutrients, sexual attractants, etc, and individual cells within an organism also migrate along chemotactic gradients. In addition to its biological importance, chemotaxis has assumed a special role in the study of behavior since it is a response that can be studied objectively and analyzed quantitatively. In the following pages, four systems that have been particularly prominent in the study of chemotaxis are examined.

In the first of these chapters, Boyd & Simon examine the bacterial system, possibly the simplest system and the most amenable to biochemical and genetic analysis. Bacteria migrate towards chemicals by a temporal sensing mechanism that utilizes a reversible methylation system. The combination of genetic engineering tools and the ease of biochemical isolations have made it possible to delineate a number of features that give insight into sensory transduction processes in general.

In the second chapter, Kung & Saimi examine the sensing system of paramecia, a system that, like the bacteria, is susceptible to genetic analysis. The genetics are more complex and the biochemistry more difficult, but because paramecia are much larger they are susceptible to electrophysiological microinjection experiments. Studies on this system are intriguing since the chemotactic signals seem to be transmitted through an action potential and then through calcium concentrations that influence the reversal of the cilia. Because the paramecium, like the bacterium, is a single cell, selection and biochemical analyses are relatively easy.

In the next chapter, Gerisch examines *Dictyostelium,* which at one stage in development comprises single amoeboid cells that chemotax toward cyclic AMP. This system has additional unusual features. The single cells can aggregate to form a structure and hence are a favorite object for the study of developmental organization and differentiation. Moreover, it has been found that the cellular response depends on an oscillating concentration of cyclic AMP and is not generated by a constant level of this important chemical. The system not only responds to cyclic AMP but the cyclic AMP stimulates its own production and influences the aggregation of cells, their differentiation, and the internal biochemistry.

Finally, Schiffmann analyses the response of leukocytes to chemotactic stimuli. These cells protect mammals against infection; they chemotax toward invaders and destroy them by phagocytosis. These cells seem to be attracted by parts of the complement fixation system, C5A, formylated peptides, and arachidonic acid derivatives. As in the bacterial system, there seems to be evidence that methylation is involved in the leukocyte chemotactic response, and the formyl peptides that can be synthesized have led to the identification of receptor sites on the surface of the cell. Thus leukocyte chemotaxis offers the opportunity in a mammalian system for studying the intercellular and intracellular events in signal processing.

These four articles, therefore, bring together four of the most advanced systems for relating the processing of signals to the ultimate behavioral response of the cell. Each system has advantages and disadvantages as a model, but the study of each has advanced to a stage where significant similarities and provocative differences are emerging. Already they are providing insight into important processes.

BACTERIAL CHEMOTAXIS

Alan Boyd and Melvin Simon

Department of Biology, University of California, San Diego, La Jolla, California 92093

INTRODUCTION

Recent reviews have treated the physiology and molecular biology of bacterial motility and chemotaxis (29, 34, 37, 47, 51, 64, 74, 78, 81, 84). We concentrate here upon the most recent advances in the field and attempt to define areas for further investigation. To place this discussion in context, we first provide an overview of the physiological aspects of bacterial behavior. We do not cite all primary references (for these, the reader is referred the reviews listed above).

Our understanding of bacterial chemotaxis has come from work with two closely related gram-negative species *Escherichia coli* and *Salmonella typhimurium*, which seem to have almost identical sensory systems. We use data from both interchangeably as is now traditional.

PHYSIOLOGY OF BACTERIAL BEHAVIOR—AN OVERVIEW

Bacterial chemotaxis is the process by which bacterial cells migrate through concentration gradients of attractants and repellents. Typical attractants include serine, aspartate, maltose, ribose, galactose, fructose, sorbitol and N-acetylglucosamine; repellents include leucine, indole, weak acids, cobalt, and nickel (1, 57, 69). Bacteria also respond to gradients of oxygen (50), pH (41), and temperature (54). Evidence indicates that a single basic mechanism regulating flagellar activity is involved in the mediation of all of these tactic responses.

Motility and Chemotaxis

Bacterial behavior in an isotropic environment is characterized by periods of smooth swimming interspersed with brief episodes of tumbling (11). Since a tumble leads to random reorientation, an individual cell moves in a three-dimensional random walk. In gradients of chemoeffectors this random walk is biased to achieve net migration. For example, a cell swimming toward higher concentrations of attractant has a reduced probability of tumbling relative to a cell swimming in the opposite direction (11).

Observations of cells with latex beads adhering to their flagella, and of cells tethered to microscope slides by a single flagellum, have revealed the mechanical basis of bacterial motility (10, 76). A motile cell of *E. coli* possesses several flagella, which arise from points distributed over the entire cell surface. These flagella are rigid helical filaments composed of repeating subunits of a single protein species, flagellin (36). The stiff filament is connected by a flexible structure termed the hook (36) to the flagellar basal body embedded in the cell surface layers (24). The helical filaments are rotated at their base; when cells are tethered to a microscope slide, this is observable as a rotation of the cell body relative to the immobilized flagellar filament. In a free swimming cell, all of the flagella come together to form a synchronously rotating bundle of filaments that drives the cell through the medium (53). During smooth swimming, the flagella are all rotated counter-clockwise, as viewed looking down the filament from tip to base (76). A reversal of rotation of one or more filaments disrupts the flagellar bundle and leads to a tumbling episode (53). Thus chemotactic behavior, which stems from the regulation of tumble frequency, arises as a result of the regulation of flagellar reversal (9, 49).

Flagellar rotation is thought to be driven by a rotary motor located in the basal structure of the flagellum. We know little of the mechanisms of energy transduction and reversal of rotation in this motor. The basal structure is complex (24) and contains 10 major polypeptides (37, 74, 78). However, the isolated basal structure does not contain all of the components necessary for flagellar activity. A case in point is that of the *mot*A and *mot*B gene products: Mutations in either of these genes lead to paralyzed cells that nevertheless possess complete flagellar structures (6, 77). These two proteins, which have been identified (56), are thus implicated directly in the energization of flagellar rotation; yet they are not recovered in preparations of basal structures.

The Bacterial Sensory Strategy

Bacterial chemotaxis is readily observed when cells are placed in an environment in which there is a graded change in the concentration of an attractant or repellent through space. In such a spatial gradient, the bacteria migrate to regions of higher attractant or lower repellent concentrations. Since bacteria are small and are subject to the effects of Brownian motion, it is advantageous for them to compare concentrations over distances substantially greater than their own length. This enables them to detect very shallow gradients. Swimming bacteria are able to do this by detecting changes in the concentrations of chemoeffectors as a function of time. Thus spatial gradients of chemoeffector concentration are sensed as changes of concentration through time, and this sensory strategy is termed *temporal sensing*.

These conclusions are based upon observations of bacteria subjected to instantaneous changes in chemoeffector concentration, commonly referred to as *temporal stimuli* (16, 52). In these experiments, cells respond to an increase in attractant concentration with a prolonged cessation of tumbling, followed eventually by a resumption of random swimming. It is convenient to think of this as a two-phase response, with an immediate excitation response (suppression of tumbling) and a subsequent slower adaptation process (return to random swimming). When attractant is removed, or repellent is added, the cells immediately tumble but rapidly adapt once more and return to random swimming.

To account for the ability of bacteria to compare concentrations of chemoeffectors through time, we may envisage a sensing mechanism consisting of five elements: (*a*) a measure of the current environment; (*b*) a measure of the past environment, continually adjusted according to recent experience; (*c*) a comparator to measure the relative values of (*a*) and (*b*); (*d*) a signal from the comparator that influences the probability of reversal of flagellar rotation; and (*e*) a switch that influences motor reversal according to the input from the comparator. We discuss below the known or potential molecular identities of each of these components of the sensory mechanism.

MOLECULAR COMPONENTS OF THE SENSING SYSTEM

Genetic studies of bacterial chemotaxis have revealed three classes of mutants that fail to exhibit chemotactic responses under some circumstance, even though the cells remain actively motile. These mutants define three corresponding classes of molecular components of the sensory system in bacteria.

Chemoreceptors

Chemoreceptor mutants are defective in a specific response to a single chemoeffector. The mutations result in the loss of activity of the corresponding chemoreceptor protein. For sugars such as maltose, ribose, and galactose the chemoreceptor is a small soluble protein located in the periplasmic space between the cell membrane and the outer cell evelope (3, 28, 30). These proteins are also active in the uptake of the corresponding sugar, although it is clear that uptake is not necessary for taxis (34). The chemoreceptors for amino acids appear to be integral membrane proteins that specifically bind serine or aspartate (20, 35, 90), as discussed below. Integral membrane proteins may also act as chemoreceptors for those sugars transported into the cell by the phosphotransferase system (2, 57).

Transducers

Mutations in three genes, *tsr*, *tar*, and *trg*, abolish chemotaxis toward a subset of effector substances (31, 33, 46, 61, 69, 79, 83). Thus the *tsr* product is necessary for taxis to the attractant serine and from repellents such as weak acids, indole, and leucine; the *tar* product is necessary for taxis to the attractants aspartate and maltose, and from the repellents cobalt and nickel; and the *trg* product is necessary for taxis to the attractants ribose and glactose. A new gene, *tap*, has recently been found that may encode a transducer for another as yet unidentified subset of effectors (14). The products of the transducer genes are all integral membrane proteins (71).

Central Components

Nonchemotactic mutants are actively motile yet fail to migrate through gradients of any effector. In general, such mutations identify central components of the sensory system. There are eight loci for which this phenotype is observed: *che*A, *che*B, *che*C, *che*D, *che*W, *che*R, *che*Y, and *che*Z (62, 63, 64, 65, 80, 81). Nonchemotactic mutants fall into two broad phenotypic classes; *che*B and *che*Z mutants exhibit continuous tumbling while mutations in the *che*A, *che*W, *che*Y, *che*R, and *che*D loci all lead to a reduced frequency of tumbling (62, 63, 65, 66). *che*C mutants may exhibit either of these phenotypes (40, 63).

Many of the chemotaxis genes form contranscribed units: the *mot*A, *mot*B, *che*A, and *che*W genes constitute the "mocha" operon (77), which is directly adjacent to an operon consisting of the *tar*, *tap*, *che*R, *che*B, *che*Y, and *che*Z genes (14, 65, 80). All of these genes have been cloned, either on lambda transducing phage vehicles or onto plasmid vehicles (56, 75). This has facilitated the identification and characterization of many of the corresponding gene products.

Specific functions have been ascribed to several chemotaxis gene products (considered in detail below). However, the *che*A, *che*W, and *che*Y products have not been associated with any specific activity. The *che*A gene is complex, encoding two distinct polypeptides (molecular weights 76K and 66K) that correspond to different genetic complementation groups (77, 82). The two polypeptides are both required for chemotaxis and differ only in that the larger has an additional 90–100 amino acids at the amino terminal end. It has been suggested that the smaller *che*A product has a cytoplasmic function while the larger product may have a membrane-associated function. The *che*W and *che*Y gene are both small cytoplasmic polypeptides composed of approximately 120 and 80 amino acids, respectively (80).

THE TRANSDUCER PROTEINS

The transducer proteins are thought to play an important role in information processing, acting as the comparator in the sensory system. An important operational difference between transducer mutants and the various *che* mutants that helps to define transducer function is that many of the latter class lose chemotactic ability due to a failure to adapt to stimuli. However, most *che* mutants can be shown to undergo the excitation phase of the behavioral response (26, 65, 68). In contrast, a transducer mutant loses even the excitation phase of the behavioral response to one or more chemoeffectors. This observation strongly supports the view that the transducer proteins are transmembrane receptors that relay information to the flagellar apparatus about changes in the concentration of chemoeffectors. In fact, it is now clear that two distinct states of the transducers embody the cell's information about present and past effector concentrations.

1. The *degree of occupancy* of the transducers as chemoreceptors is a measure of the cell's present environment. The *tsr* and *tar* transducers themselves bind the amino acid attractants serine and aspartate (20, 35, 90) and thus act as the primary chemoreceptors for these compounds. The periplasmic sugar receptors undergo a conformational change when they bind ligand (92), and it is thought that this triggers an interaction between the occupied chemoreceptor and the corresponding transducer (44). In a sense the transducer acts as a secondary chemoreceptor in this case.

2. The transducers are post-translationally methylated and their *degree of methylation* reflects the cell's past environment (84). For this reason the transducers are often referred to as methyl-accepting chemotaxis proteins (MCPs). During adaptation to a specific stimulus the degree of methylation of the corresponding transducer changes. Adaptation to increased attractant levels is associated with increased transducer methylation; removal of attractant leads to decreased methylation (83).

Adaptation and de-adaptation are so closely correlated with the processes of methylation and de-methylation that these processes have been intensively studied. The transducers are methylated at glutamyl residues to form carboxymethyl esters (43, 89). Methyl groups donated by S-adenosylmethionine (4, 5, 7, 8) are added by a methyltransferase (molecular weight 28K) encoded by the *che*R gene (85); the *che*B gene product (molecular weight 38K) is the methylesterase that removes methyl groups from the transducers (86). The functions ascribed to these gene products correlate well with the mutant phenotypes. During adaptation to a temporal attractant stimulus, a lower level of transducer methylation than that demanded by the environment is associated with suppression of tumbling; *che*R mutants are unable to methylate and have a smooth-swimming phenotype. Conversely, a high rate of tumbling is associated with a transient demand for demethylation during adaptation to attractant removal; *che*B mutants have anomalously high levels of transducer methylation and a tumbly phenotype.

When the transducers are labelled with radioactive methyl groups and analyzed by sodium dodecyl sulfate polyacrylamide gel electrophoresis (SDS-PAGE) they form characteristics sets of multiple bands of apparent molecular weight 60–70K (83). This multiplicity of electrophoretic forms reflects, at least in part, the multiple methylation of single gene products (13, 18, 22, 25). Multiple methylation is seen clearly when cloned *tsr* and *tar* genes carried on lambda transducing phages are expressed in UV-irradiated bacteria. Under these circumstances, it is possible to label the transducers with a radioactive amino acid independently of the methylation process. Each gene encodes multiple species that are resolved in SDS-PAGE. The addition of stimuli results in the appearance of new faster-migrating forms of the polypeptides, which persist only as long as the stimulus is present. Methylation at multiple sites on a transducer molecule appears to cause progressive quantal anodic shifts in electrophoretic mobility. These mobility shifts may be due to the enhanced SDS binding resulting from each charge-neutralizing carboxymethylation. Similar effects of covalent modification upon electrophoretic mobility of polypeptides have been proposed as the basis for an analytical method for counting integral numbers of amino acid residues per polypeptide chain (17).

In two-dimensional electrophoretic separations (isoelectric focusing + SDS-PAGE), the transducers form complex diagonal arrays of spots since methylation also leads to changes in the isoelectric point of the polypeptides [(27, 32); C. Rollins, F. W. Dahlquist, personal communication.] The gel patterns have been interpreted as being consistent with the presence of 3–4 methylated sites per transducer. Structural studies indicate that all of these sites are present on a single tryptic fragment in the *tsr* and *tar* transducers, suggesting that the methyl-accepting function is associated with a single domain of the proteins (19). However, the most clearly resolved gel patterns also reveal complexities inexplicable in terms of multiple methylation alone (32, 72a, 72b). Furthermore bacteria defective in *che*R function might be expected to possess only a single (unmethylated) form of each transducer, but this is not the case. Although several of the faster migrating forms do not appear in these cells, at least two forms are seen (13, 22). The only circumstance in which single forms are observed is in cells lacking both *che* R and *che*B function, indicating that the *che*B product performs a second modification unrelated to demethylation (72a, 72b). This modification results in a decrease in gel mobility and may correspond to the unmasking of a carboxyl group through the deamination of a glutamine residue in the primary gene product. Thus methylation and further modification by the methylesterase of the *tsr, tar, tap,* and *trg* gene products can account for the appearance of approximately twenty bands in SDS-PAGE analysis of methyl-labelled protein.

The Comparator Function of Transducers

All available evidence is consistent with the view that the transducers constitute the comparator element of the sensory system defined above. Transducer function may be envisaged in terms of three functional domains: an extracellular domain responsible for chemoreception, an intracellular methylated domain, and an intracellular domain that generates the signal to the flagellar apparatus (15, 90). Evidence for this domain structure comes from the finding that the genes encoding the *tar, tsr,* and *tap* transducers possess nucleotide sequence homology (14). The region of homology is located toward the carboxy terminal segment of these genes, extending over approximately one third of each gene. These homologous segments probably encode the methylated domains of the proteins while the amino terminal regions of the proteins may form the chemoreceptor domains; the gene segments encoding chemoreceptor function have presumably diverged considerably or may have evolved from unrelated nucleotide sequences. Proteolytic fragments of transducers have been reported to retain subfunctions such as capacity to be methylated (90).

Comparator function could arise from the regulation of signaling by the two other domains. Several models have been proposed in which effector binding (primary or secondary chemoreception) induces a conformational change in the transducer molecule, which activates the signaling domain; the level of signaling is then regulated by the methylation of an intracellular domain (15, 84, 90). The number of activated transducers at any moment may be determined solely by the prevailing chemoeffector concentration whereas the summed signaling of these activated molecules may be determined by their mean level of methylation, which is adjusted during sensory adaptation (15). In all models, the basis for temporal sensing lies in the mechanism that relates transducer methylation to the environment of the recent past.

CENTRAL COMPONENTS AND CONTROL OF METHYLATION

The nature of the bacterial behavioral response indicates a fundamental bifurcation of the sensory information relating to a stimulus. The transducer output has two effects: an immediate effect upon flagellar rotation, and a more gradual effect upon transducer methylation. We now consider this bifurcation in terms of the possible identity of the signal from the transducer to the flagellar structure, the nature of the flagellar switch, and the evidence of feedback from the switch to the transducers.

The Nature of the Signal

The transducers are thought to convey information somehow to the flagellar apparatus about changes in their degree of occupancy as primary or secondary chemoreceptors. The molecular nature of this signal remains obscure. The following is a list of formal possibilities.

1. A direct interaction between the transducer and a switch component. There is no direct evidence relating to this possibility, but the great excess of transducers over flagella makes this an unlikely candidate (51).

2. A change in the cytoplasmic level of an ionic species associated with the transducer functioning as either a gate pore or a binding protein. One report suggests a role for calcium in chemotactic signaling in *B. subtilis* (60), and it has been pointed out that methylation of the transducers could affect an ion binding site or ion channel (19).

3. Changes in membrane potential. While cells exhibit behavioral responses to changes in membrane potential, it seems that stimulation by attractants and repellents is not necessarily accompanied by changes in membrane potential (58).

4. Generation of a low molecular weight compound. A recent report has described rapid transient changes in cytoplasmic cGMP levels following attractant stimulation (12).

5. A cascade of protein-protein interactions, in which information is relayed via interactions between chemotaxis gene products. The fact that mutations in the *che*R, *che*B, *che*Y, and *che*Z genes do not block the excitation phase of the behavioral response argues against any obligatory role in signal transmission. On the other hand, the *che*A and *che*W gene products are not ruled out as participants in such a mechanism, since the corresponding mutants show no behavioral response (66a).

The Flagellar Switch

The switch is the postulated component of the flagellar motor that regulates the sense of flagellar rotation according to the information received from the transducers. Genetic studies have shown that the *che*C locus encodes a switch function. This locus seems to identify the interface between the genes determining flagellar structure (*fla* genes) and those encoding chemotaxis functions, since *che*C mutations are actually rare alleles of the *fla*A locus (63).

One well-studied mutant of this type is a *che*C mutant of *S. typhimurium* that exhibits inverted chemotactic responses (40). These bacteria swim as a result of prolonged stable clockwise rotation of their flagella. Addition of attractant, which favors counterclockwise rotation as in normal cells, leads in the mutant to reversal of flagella rotation which causes tumbling. Consequently, increases in attractant concentration transiently increase tumble frequency in the mutant, just as repellents do in normal cells. An alteration in the *fla*A/*che*C gene product is thus able to reset the intrinsic rotational behavior of the flagellar motor yet leave it responsive to input from the transducers.

Mutations mapping in the *fla*B gene have also been found to cause che⁻ phenotypes, and these have been designated *che*V mutations (23, 66a). This suggests that the *fla*A and *fla*B gene products are both flagellar components that are active in the regulation of the direction of flagellar rotation. The observation that *che*W and *che*A mutants never tumble (91) has led to the suggestion that the *che*W and *che*A products may be cytoplasmic components of the switch mechanism essential for flagellar reversal (66a).

Little is known about the relationship among the polypeptides controlling motility, the switch, and the flagellar basal structure. Flagellar rotation is energized by the protonmotive force (39, 48, 55, 56a, 56b), and this energy is apparently channeled through the *mot*A and *mot*B products since mutations in these genes cause flagellar paralysis (6). The *mot*A and *mot*B products are membrane proteins (71); but since they are not recovered with the basal structure (36), it is not clear whether they are physically coupled to the flagellar apparatus or widely distributed in the membrane and only chemically coupled to the motor.

The switch is also sensitive to changes in the physiological state of the cell. When the protonmotive force falls below the level necessary to saturate the motors and give maximal swimming velocity, rotation becomes heavily biased in favor of the counterclockwise direction (i.e. no tumbling). This may constitute a dispersal mechanism in terms of dire emergency (38).

Regulation of Transducer Methylation

Whatever the exact function of the multiple covalent modifications of the transducers may be, we can, to a first approximation, consider sensory adaptation in terms of the overall level of methylation that results from the relative activities of the *che*R methyltransferase and the *che*B methylesterase. Two observations are central to an understanding of the mechanism of regulation of transducer methylation: (*a*) The adapted state reached following a specific temporal stimulus is correlated with the acquisition of a new steady-state level of methylation of the corresponding transducer, while other transducers undergo no net change in methylation (83). (*b*) During adaptation to an attractant stimulus the dynamic equilibrium between methylation and demethylation is disturbed by a block to methylesterase activity that causes a transient rise in the level of methylation of all transducers (12).

Clearly, a nonspecific block to demethylation cannot create a change in the level of methylation of a single transducer. Thus a change in the susceptibility of a transducer molecule to the *che*R and *che*B products probably results when the protein binds chemoeffector. Effects of this type have been reported from studies of reconstituted systems containing partially purified components (42, 90). The block to methylesterase activity is presumably a kinetic device that accelerates the attainment of the adapted state.

How is the block accomplished and at what point in the pathway of information flow does the signal generated by the transducers feed back to regulate methylesterase activity? More specifically, does this bifurcation occur before or after the signal has reached the switch? There is evidence that it occurs after—that in a sense the switch signals back to the transducers: Genetic studies strongly suggest that the *che*Z gene product interacts directly with both the *fla*A/*che*C product (i.e. a switch component) and the *che*B methylesterase. Thus in studies of intergeneric complementation (23), *che*B or *che*Z defects in *S. typhimurium* was remedied by supplying the corresponding *E. coli* gene, but only if both *che*Z and *che*B were supplied, even though the recipient was only defective in one of them. Furthermore, some *che*Z mutants exhibit a defect in methylesterase function like that in *che*B mutants (86). The inference is that the *che*Z product modifies methylesterase activity through a direct protein-protein interaction not possible between gene products from different species. Genetic studies have also revealed allele-specific intergenic suppression of *che*C defects by mutations in the *che*Z gene (67); this, too, suggests a direct interaction between the gene products concerned. This evidence indicates a role for the *che*Z product as a regulator of methylesterase activity, which is itself influenced by the state of the flagellar switch.

OTHER CLASSES OF BEHAVIORAL RESPONSE

As a paradigm for information processing the methylated transducer is central to our current understanding of the bacterial behavioral response. Above, we have surveyed the evidence for this paradigm, with specific reference to the major attractant responses that appear to be mediated via the three known methylated transducers, products of the genes *tsr* (serine), *tar* (aspartate), and *trg* (ribose and galactose). Bacteria respond to a wide range of other effectors—e.g. they are attracted to sugars transported via the phosphotransferase system (PTS) and are repelled by weak acids (such as acetate and benzoate) and by cobalt and nickel ions. They migrate through gradients of pH, oxygen concentration (aerotaxis), and temperature (thermotaxis). These responses have been attributed to the activity of methylated transducers, either know or not yet identified. However, such environmental parameters differ from the major chemoattractants. While the evidence is good that the major chemoattractants are detected outside the cell via specific ligand-chemoreceptor interactions, and that transmembrane transduction follows, the other effectors listed above could also change some intracellular parameter and thus affect the behavioral regulatory network at several points.

The assignment of a behavioral response to the activity of an individual methylated transducer often involves experiments with mutants defective in one or more of the major transducers. However, the sensory input from each transducer is not completely independent of other channels, and the information processing system is not well-buffered against the loss of a major transducer. The *tsr* transducer in particular seems to be associated with such effects: Wild-type cells do not adapt fully to a saturating concentration of serine, but exhibit a permanent reduction in tumble frequency in the continued presence of the attractant (11); *tsr* mutants show prolonged behavioral responses to other classes of stimuli (31); and certain mutations in *tsr*, mapping at the *che*D locus (66), create a dominant smooth-swimming phenotype associated with overmethylation of all transducers (our unpublished data). Furthermore, the *tsr-tar* double mutants are predominantly smooth-swimming and exhibit severe defects in adaptation to stimuli mediated via the *trg* transducer (31). Triple transducer mutants, *tsr-tar-trg*, show no alternation of flagellar rotation, but are permanently smooth-swimming (31). This catalog of phenotypes indicates that the transducers are interdependent components of a finely tuned mechanism and suggests that conclusions drawn from the pleiotropic effects of transducer mutations should be regarded with caution. In examining cases of a behavioral response where the transducer has not been completely defined, we indicate several instances where general effects upon the physiological state of the cell might also be involved.

Weak Acid Repellents and pH Taxis

Bacteria accumulate at the neutral point in pH gradients and exhibit migratory and temporal responses to weak acids such as acetate or benzoate. Recent evidence suggests that these two responses are manifestations of a single phenomenon, a behavioral response to intracellular pH (41, 70) in which the weak acid response is attributable to the acidification of the cytoplasm caused by protonated organic acids. Macnab (41) has suggested that this indicates a pH-sensitive device in the system that regulates the flagellar motor.

The assignment of weak acid repellent responses to an activity of the *tsr* transducer is based partly upon the pleiotropic effects of *tsr* null mutations (69). However, since such mutants respond to weak acids as attractants (59) the role of an intact *tsr* transducer in this repellent response is clearly not straightforward. During the behavioral response to weak acid repellents, the level of transducer methylation falls. This decrease has been ascribed to the loss of methyl label from the *tsr* transducer (83), although the lack of an effect upon *tar* transducer methylation was demonstrated in a *tsr* mutant, which in any case fails to exhibit the behavioral response. These effects might be fruitfully reexamined now that sensitive methods are available for analysis of changes in specific transducer methylation (13).

Following weak acid stimulus, the fall in the methylation level of the transducers and the behavioral adaptation are both extremely rapid, so that it is not possible to infer a fixed sequence of events like that associated with an attractant response (83). We have summarized above the evidence that the transducers and the flagellar switch are connected with a closed loop of interactions whereby a signal generated by the transducer influences switch activity and interactions between the switch and the *che*Z gene product modulate activity of the *che*B methylesterase. Clearly, any change in an intracellular parameter impinging upon any one of these components would disturb the equilibrium of the adapted state and would lead to changes in transducer methylation and switch activity.

Behavioral Responses to Changes in Protonmotive Force

Bacteria migrate in gradients of oxygen concentration (50). This is probably the most physiologically significant example of the phenomenon termed energy taxis or protonmotive force (PMF) taxis (87).

The PMF is the energy source for flagellar rotation (55) but also appears to be the parameter by which changes in cellular energy are sensed by the flagellar switch. Sudden decreases in PMF transiently enhance tumbling; sudden increases transiently suppress tumbling (58, 87). In the case of oxygen taxis, bacteria with lesions in any one of the three transducer genes show normal responses (50), ruling out a direct obligatory role for any one of these transducers in the behavioral response. These responses may be mediated via another transducer species, or may perhaps represent direct effects of an intracellular state upon one or more components of the sensory system.

The divalent cation repellents cobalt and nickel apparently act via the Mg^{2+}, Ca^{2+}ATPase of the bacterial membrane (93). Such effects could easily be mediated via the PMF-sensing mechanism. Nevertheless, *tar* mutants appear to be specifically defective in this class of repellent response, suggesting a role for the *tar* transducer in the response (69).

The Thermotactic Response

The thermoresponse in bacteria is blocked by the presence of a high concentration of serine, although it is also moderately sensitive to the presence of high concentrations of aspartate (54). This finding suggests that the thermoresponse might be mediated via the *tsr* transducer as a thermosensory device. Alternatively, other components of the system regulating the direction of flagellar rotation may be responsive to changes in temperature and may transmit information in a manner dependent upon the presence of an intact *tsr* transducer.

Response to PTS Sugars

Sugars such as glucose, fructose, mannitol, and N-acetyl glucosamine are taken into the cell via the phosphotransferase (PTS) system, which operates by a group translocation mechanism. For each sugar transported there is a specific membrane protein receptor, the enzyme II component. The remaining components of the PTS are shared by all of the sugars. *E. coli* and *S. typhimurium* exhibit attractant response to PTS sugars (2, 57). For the non-PTS sugars maltose, ribose, and galactose, the evidence is good that transport is not necessary for chemotaxis. For PTS sugars, the evidence is somewhat less convincing since it is technically difficult to separate the processes of transport and taxis by genetic or physiological manipulations.

There is no evidence that any of the methylated transducers are required for PTS sugar responses; accordingly, the existence of a further methylated transducer has been postulated to account for these responses (31). Alternatively, PTS sugar sensing may be related in some way to changes in an intracellular state accompanying uptake.

How Many Methylated Transducers Are There?

Repellent responses, pH taxis, and the thermoresponse are thought to be mediated via the *tsr* and *tar* transducers; a fourth transducer has been invoked for the PTS sugar response (31) and a fifth for the PMF response (50). Several candidates for a fourth transducer have been identified in gel analysis of methyl-labelled proteins (33, 45).

An alternative approach to counting transducer genes has recently been tried, based upon the finding that extensive sequence homology between *tsr* and *tar* is detectable by Southern blot hybridization (14). A new gene, *tap*, was found immediately adjacent to *tar* that also possesses homology with *tsr* and *tar* and encodes multiple gel bands, properties suggestive of a transducer function for this gene product. A study of the effect of a specific lesion in this gene upon chemotactic responses should determine the role of the *tap* product. Surprisingly, no other genes related to the *tsr-tar-tap* family were detected by preliminary hybridization experiments with chromosomal DNA; this suggests that even the *trg* gene may be only distantly related to the major transducer genes.

Implications for the Evolution of Chemotaxis

A hierarchy of regulatory variables may be envisaged as affecting the switch. This, in turn, could reflect the evolution of the bacterial sensory repertoire. Primitive responses may have been mediated via direct effects of transient changes in intracellular states such as PMF or pH. This behavioral system would of necessity have operated within narrow limits since it would have relied upon cellular mechanisms of homeostasis to restore the intracellular state and thus provide for behavioral adaptation.

The evolution of responses to a broader range of environmental factors would entail the appearance of a regulatory mechanism specific for the behavioral response allowing behavioral adaptation independent of physiological homeostasis. Such a mechanism might be formed from a circuit containing a methylated protein and ancillary enzymes. In this mechanism, the methylated protein would merely act as a regulator of switching, not as a sensory transducer. Finally, in the development of the modern repertoire, a new sensory input would be developed in the form of responsiveness of the methylatd protein to extracellular concentrations of attractants such as amino acids.

CONCLUSIONS

We have learned much about the components of the chemotaxis system but little about how they interact. Part of the difficulty may derive from a failure to identify all the components of the motility-chemotaxis system. Furthermore, even where the components have been identified we often have no clear idea of their detailed function. For example, we have no idea how the *mot*A and *mot*B gene products are related to the structural components of the flagellar basal structure.

Further reconstruction experiments involving chemotaxis and motility protein-membrane complexes and assays for intermediates in the energy transduction and signaling processes will lead us to a clearer picture of how the system works. Such experiments, for example, can help define the role of the transducer proteins. One family of transducers, the *tar, tsr,* and *tap* group, appear to behave like classical cell-surface receptors. As in the case of the insulin receptor, (21) one portion of the bacterial chemoreceptor is on the outer surface of the plasma membrane and binds specific ligands. Another domain in the protein is intracellular; again as in some receptors in mammalian systems [e.g. rhodopsin (73)], it undergoes covalent modification in response to specific ligand binding. The most recent suggestion that the level of a cyclic nucleotide varies in response to changes in receptor occupancy may carry the analogy even further since it is clear that such effects mediate many hormone responses (72). These considerations have led to the suggestion that the transducer molecules in the chemotaxis system may be composed of a number of discrete domains, each responsible for a function such as ligand binding, signal transmission, or adaptation. Further genetic and structural studies of these proteins may provide general information about the function and evolution of complex cell surface receptors.

There appear to be ways, other than through the known transducer molecules, of affecting and intervening in the process of signal transmission. Such factors are difficult to define since little is known about the biochemical nature of the regulation of flagellar reversal, and still less is known about the feedback mechanisms that inform the adaptation system about the state of the flagellar switch.

Sophisticated genetic tools for the analysis of this system are being developed. Biochemical techniques are emerging that will allow further exploration of the system. The motility-chemotaxis system in prokaryotes remains an important model. It embodies many fundamental biological problems concerning the mechanisms involved in energy transduction into mechanical force, the structure and function of cell-surface receptors, the nature of information transmission, and the evolution of mechanisms for sensory reception.

Literature Cited

1. Adler, J. 1969. *Science* 166:588–97
2. Adler, J., Epstein, W. 1974. *Proc. Natl. Acad. Sci. USA* 71:2895–99
3. Aksamit, R., Koshland, D. E. 1974. *Biochemistry* 13:4473–78
4. Armstrong, J. B. 1972. *Can. J. Microbiol.* 18:591–94
5. Armstrong, J. B. 1972. *Can. J. Microbiol.* 18:1695–701
6. Armstrong, J. B., Adler, J. 1967. *Genetics* 56:363–73
7. Aswad, D., Koshland, D. E. 1974. *J. Bacteriol.* 118:640–45
8. Aswad, D., Koshland, D. E. 1975. *J. Molec. Biol.* 97:207–23
9. Berg, H. C. 1974. *Nature* 249:77–79
10. Berg, H. C., Anderson, R. A. 1973. *Nature* 245:380–82
11. Berg, H. C., Brown, D. A. 1972. *Nature* 239:500–4
12. Black, R. A., Hobson, A. C., Adler, J. 1980. *Proc. Natl. Acad. Sci. USA* 77:3879–83
13. Boyd, A., Simon, M. 1980. *J. Bacteriol.* 143:809–15
14. Boyd, A., Krikos, A., Simon, M. 1981. *Cell.* In press
15. Boyd, A., Mandel, G., Simon, M. 1981. *Symp. Soc. Exp. Biol.* In press
16. Brown, D. A., Berg, H. C. 1974. *Proc. Natl. Acad. Sci. USA* 71:1388–92
17. Creighton, T. E. 1980. *Nature* 284:487–89
18. Chelsky, D., Dahlquist, F. W. 1980. *Proc. Natl. Acad. Sci. USA* 77:2434–38
19. Chelsky, D., Dahlquist, F. W. 1981. *Biochemistry.* In press
20. Clarke, S., Koshland, D. E. 1979. *J. Biol. Chem.* 254:9695–702
21. Czech, M. P. 1977. *Ann. Rev. Biochem.* 46:359–84
22. DeFranco, A. L., Koshland, D. E. 1980. *Proc. Natl. Acad. Sci. USA* 77:2439–43
23. DeFranco, A. L., Parkinson, J. S., Koshland, D. E. 1979. *J. Bacteriol.* 139:107–14
24. DePamphilis, M. L., Adler, J. 1971. *J. Bacteriol.* 105:396–407
25. Engstrom, P., Hazelbauer, G. L. 1980. *Cell* 20:165–71

26. Goy, M. F., Springer, M. S., Adler, J. 1978. *Cell* 15:1230–40
27. Hayashi, H., Koiwai, O., Kozuka, M. 1979. *J. Biochem.* 85:1213–23
28. Hazelbauer, G. L. 1975. *Bacteriol.* 122:206–14
29. Hazelbauer, G. L. 1980. *Endeavour, N.S.* 4:67–73
30. Hazelbauer, G. L., Adler, J. 1971. *Nature New Biol.* 230:101–4
31. Hazelbauer, G. L., Engstrom, P. 1980. *Nature* 283:98–100
32. Hazelbauer, G. L., Engstrom, P. 1981. *J. Bacteriol.* 145:35–42
33. Hazelbauer, G. L., Engstrom, P., Harayama, S. 1981. *J. Bacteriol.* 145:43–49
34. Hazelbauer, G. L., Parkinson, J. S. 1977. In *Receptors and Recognition: Microbiol Interactions,* Ser. B. Vol. 3, pp. 59–98. ed. J. Reissing. London: Chapman and Hall
35. Hedblom, M. L., Adler, J. 1980. *J. Bacteriol.* 144:1048–60
36. Hilmen, M., Simon, M. 1976. In *Cell Motility,* ed. R. Goldman, T. Pollard Rosenbaum, pp. 35–45. NY: Cold Spring Harbor Press
37. Iino, T. 1977. *Ann. Rev. Genet* 11:161–82
38. Khan, S., Macnab, R. M. 1980. *J. Molec. Biol.* 138:563–97
39. Khan, S., Macnab, R. M. 1980. *J. Molec. Biol.* 138:599–614
40. Khan, S., Macnab, R. M., DeFranco, A. L., Koshland, D. E. 1978. *Proc. Natl. Acad. Sci. USA* 75:4150–54
41. Kihara, M., Macnab, R. M. 1981. *J. Bacteriol.* 145:1209–21
42. Kleene, S. J., Hobson, A. C., Adler, J. 1979. *Proc. Natl. Acad. Sci. USA* 76:6309–13
43. Kleene, S. J., Toews, M. L., Adler, J. 1977. *J. Biol. Chem.* 252:3214–18
44. Koiwai, O., Hayashi, H. 1979. *J. Biochem.* 86:27–34
45. Koiwai, O., Minoshia, S. Hayashi, H., 1980. *J. Biochem.* 87:1365–70
46. Kondoh, H., Ball, C. B., Adler, J. 1979. *Proc. Natl. Acad. Sci. USA* 76:260–64
47. Koshland, D. E. 1979. *Physiol. Rev.* 59:811–62
48. Larsen, S. H., Adler, J., Gargus, J. J., Hogg, R. W. 1974. *Proc. Natl. Acad. Sci. USA* 71:1239–43
49. Larsen, S. H., Reader, R. W., Kort, E. M., Tso, W. W., Adler, J. 1974. *Nature* 249:74–77
50. Laszlo, D. J., Taylor, B. L. 1981. *J. Bacteriol.* 145:990–1001
51. Macnab, R. M. 1980. In *Bacterial Chemotaxis in Biological Regulation and Development,* ed. R. F. Goldberger, pp. 377–411 London: Plenum
52. Macnab, R. W., Koshland, D. E. 1972. *Proc. Natl. Acad. Sci. USA* 69:2509–12
53. Macnab, R. W., Ornston, M. K. 1977. *J. Molec. Biol.* 112:1–30
54. Maeda, K., Imae, Y. 1979. *Proc. Natl. Acad. Sci. USA* 72:3939–43
55. Manson, M. D., Tedesco, P., Berg, H. C., Harold, F. M., Van der Drift, C. 1977. *Proc. Natl. Acad. Sci. USA* 74:3060–64
56. Matsumura, P., Silverman, M., Simon, M. 1977. *J. Bacteriol.* 132:996–1002
56a. Matsuura, S., Shioi, J., Imae, Y. 1977. *FEBS Lett.* 82:187–90
56b. Matsuura, S., Shioi, J., Imae, Y., Iida, S. 1979. *J. Bacteriol.* 140:28–36
57. Melton, T., Hartman, P. E., Stratis, J. P., Lee, T. L., Davis, A. T. 1978. *J. Bacteriol.* 133:708–16
58. Miller, J. B., Koshland, D. E. 1977. *Proc. Natl. Acad. Sci. USA* 74:4752–56
59. Muskavitch, M. A., Kort, E. M., Springer, M. S. Goy, M. F., Adler, J. 1978. *Science* 201:63–65
60. Ordal, G. W. 1977. *Nature* 270:66–67
61. Ordal, G. W., Adler, J. 1974. *J. Bacteriol.* 117:517–26
62. Parkinson, J. S. 1974. *Nature* 252:317–19
63. Parkinson, J. S. 1976. *J. Bacteriol.* 126:758–70
64. Parkinson, J. S. 1977. *Ann. Rev. Genet.* 11:397–414
65. Parkinson, J. S. 1978. *J. Bacteriol.* 135:45–53
66. Parkinson, J. S. 1980. *J. Bacteriol.* 142:953–61
66a. Parkinson, J. S. 1981. In *31st Symp. Soc. Gen. Microbiol.,* pp. 265–90
67. Parkinson, J. S., Parker, S. R. 1979. *Proc. Natl. Acad. Sci. USA.* 76:2390–94
68. Parkinson, J. S., Revello, P. T. 1978. *E. coli. Cell* 15:1221–30
69. Reader, R. W., Tso, W. W., Springer, M. S., Goy, M. F., Adler, J. 1979. *J. Gen. Microbiol.* 111:363–74
70. Repaske, D. R., Adler, J. 1981. *J. Bacteriol.* 145:1196–208
71. Ridgway, H. F., Silverman, M., Simon, M. 1977. *J. Bacteriol.* 132:657–65
72. Rodbell, M. 1980. *Nature* 284:17–22
72a. Rollins, C., Dahlquist, F. W. 1981. *Cell* 25:333–40
72b. Sherris, D., Parkinson, J. S. 1981. *Proc. Natl. Acad. Sci. USA* 78:6051–55
73. Shichi, H., Somers, R. L. 1978. *J. Biol. Chem.* 253:7040–46
74. Silverman, M. 1980. *Q. Rev. Biol.* 55:395–407

75. Silverman, M., Matsumura, P., Draper, R., Edwards, S., Simon, M. 1976. *Nature* 261:248–50
76. Silverman, M., Simon, M. 1974. *Nature* 249:73–74
77. Silverman, M., Simon, M. I. 1976. *Nature* 284:477–479
78. Silverman, M., Simon, M. I. 1977. *Ann. Rev. Microbiol.* 31:397–419
79. Silverman, M., Simon, M. 1977. *Proc. Natl. Acad. Sci. USA* 74:3317–21
80. Silverman, M., Simon, M. 1977. *J. Bacteriol.* 130:1317–25
81. Simon, M., Silverman, M., Matsumura, P., Ridgway, H., Komeda, Y., Hilmen, M. 1978. In *28th Symp. Soc. Gen. Microbiol.*, pp. 271–84
82. Smith, R. A., Parkinson, J. S. 1980. *Proc. Natl. Acad. Sci. USA* 77:5370–74
83. Springer, M. S., Goy, M. F., Adler, J. 1977. *Proc. Natl. Acad. Sci. USA* 74:3312–16
84. Springer, M. S., Goy, M. F., Adler, J. 1979. *Nature* 264:577–79
85. Springer, W. R., Koshland, D. E. 1977. *Proc. Natl. Acad. Sci. USA* 74:533–37
86. Stock, J. R., Koshland, D. E. 1978. *Proc. Natl. Acad. Sci. USA* 75:3659–63
87. Taylor, B. L., Miller, J. B., Warrick, H. M., Koshland, D. E. 1979. *J. Bacteriol.* 140:567–73
88. Toews, M. L., Goy, M. F., Springer, M. S., Adler, J. 1979. *Proc. Natl. Acad. Sci. USA* 76:5544–48
89. Van der Werf, P., Koshland, D. E. 1977. *J. Biol. Chem.* 252:2793–95
90. Wang, E. A., Koshland, D. E. 1980. *Proc. Natl. Acad. Sci. USA* 77:7157–61
91. Warrick, H. N., Taylor, B. L., Koshland, D. E. 1977. *J. Bacteriol.* 130:223–31
92. Zukin, R. S., Hartig, P. R., Koshland, D. E. 1977. *Proc. Natl. Acad. Sci. USA* 74:1932–36
93. Zukin, R. S., Koshland, D. E. 1976. *Science* 193:405–8.

THE PHYSIOLOGICAL BASIS OF TAXES IN *PARAMECIUM*

Ching Kung and Yoshiro Saimi

Laboratory of Molecular Biology and Department of Genetics, University of Wisconsin, Madison, Wisconsin 53706

INTRODUCTION

The electric and ionic control of behavior is better understood in *Paramecium* and related ciliated protozoa than in the other three systems reviewed in this section. We therefore focus on our knowledge of such control. Reviews with different emphases are available (1, 20, 21, 29, 52, 62, 79, 119).

Various stimuli increase the probability that a paramecium will give the "avoiding reaction"—i.e. will back away from the stimulus and then swim forward in a randomly chosen direction. Other stimuli decrease that probability. Jennings (48) and his contemporaries described such behavior in detail. The phenomenon is formally equivalent to the modulation of "tumbling" frequency by stimuli in bacteria (see Simon's review in this Section). Although the receptors and the ciliary force regulators are not well understood in the *Paramecium* system, we know that a Ca action potential connects the receptor to the ciliary motion. This situation may be contrasted with the study of bacterial taxes, where a great deal is known about the receptors and the adaptation process but where it is not clear how the occupied receptors pass on their information nor, indeed, what the physical nature of that information is.

The present article emphasizes several concepts that may not be evident to readers: (*a*) the regulation of membrane potential as a means of regulating behavior; (*b*) the use of mutations in electrophysiological research; (*c*) the local control of Ca^{2+} concentration; (*d*) the importance of Na^+ in the natural behavior of *Paramecium;* (*e*) the presence of at least six different types of ion conductances on the membrane, whose interaction leads to the "spontaneous" electric discharges and the "spontaneous" avoiding reactions; (f) ciliate chemotaxis to inorganic and organic substances as it relates to the chemoreception of these substances in the taste and smell receptors of higher forms; (g) complications in the unicellular systems where sensory transduction, excitation, and other functions occur on the same membrane; and (h) the prospects of a multidisciplinary study of the *Paramecium* membrane.

THE BASIC SCHEMA

The basic schema of *Paramecium* behavior can be summarized as follows: stimulus → receptor potential → action potential → $[Ca^{2+}]_{in}$ increase → ciliary reversal.

The concepts here originate in metazoan neurophysiology and have been developed by Naitoh, Eckert and their co-workers. The apparent validity of this scheme for a unicell argues that the molecular mechanisms for sensory transduction, membrane excitation, and cell motility predate cellular differentiation and synaptic communication. One of the hopes in this field has been to reach an understanding of some of these mechanisms by using unicells, of which a combined study involving electrophysiology, genetics, and biochemistry is feasible (29, 52, 79).

This scheme, applicable to other ciliates [(22, 41, 123); Y. Naitoh, personal communication], is considered basic because the paramecia have little tactic behavior without it. Mutants without action potentials, for example, though capable of locomotion and some velocity modulation, are grossly deficient in their responses to heat (42) and to chemicals (115, 116).

The individual steps of the basic schema are documented and discussed below.

Stimulus → Receptor Potential

Since a paramecium can respond to chemicals, heat, touch, and, in some cases, light (48), one would expect the presence of a variety of sensory mechanisms to transduce the stimuli of different modalities. Although we know little about the transducing mechanisms and the nature of the receptors, a functionally important consequence is clearly the generation of a receptor potential—a change in the membrane potential graded to the strength of the stimuli. The best-documented receptor potential in *Paramecium* is generated by touch. Naitoh & Eckert (32, 72) showed that a mechanical prod at the anterior end of *P. caudatum* generates a transient depolarizing receptor potential on which the depolarization-triggered action potential rides. The channels responsible for the first depolarization (the receptor potential) are different from those for the second depolarization (the action potential). The former, presumably consisting of a sensor and an ionophore, are located on the body membrane instead of the ciliary membrane (83), have less ion specificity [(24); C. Kung, A. D. Murphy, Y. Satow, in preparation] and are not blocked by mutations that abolish the action potential (52, 107).

Van Houten (113, 115–117) showed that certain organic repellents, such as quinidine hydrochloride, cause a rise of the action potential level (depolarization) that is sustained throughout the presence of the repellent and triggers frequent action potentials in *Paramecium.* Inorganic cations, such as Ba^{2+}, which cause depolarization and action potentials, can be repellents. Kobatake and co-workers (2, 108) found that the concentration thresholds of various cations and "odorants" needed to effect a depolarization, indirectly measured by rhodamine-fluorescence increase, are the same as the thresholds for repulsion in *Tetrahymena.*

Receptor Potential → Action Potential

Any suprathreshold depolarization, such as a receptor potential, opens the Ca channel (Ca activation) and, after a delay, the K channel (delayed rectification), resulting in the Ca action potential (70, 75). The action potential is usually graded to the stimulus and is not of the all-or-none type, except in the presence of K-channel blockers such as Ba^{2+}, or tetraethylammonium (TEA^+) (71, 97). The advent of the voltage-clamp technique in this field has made possible a much deeper understanding of the function of these ion channels (15, 23, 63, 74, 80, 92, 100). A step depolarization (e.g. from a resting level of −35 mV to −10 mV within 0.5 msec) induces a transient current that flows into the cell and is followed by an outward current sustained as long as the depolarization. A region of negative resistance in the I-V plot explains the regenerative nature at the upstroke of the action potential of the unclamped membrane. Since the electrode reports the total current, which can have an inward and an outward current at any instant, it is important to separate these components. Such a separation can be achieved using a mutant that blocks its Ca-channel function or by suppressing the K-channel function through injected Cs^+ and TEA^+. Using the mutant, Oertel et al (80) showed that the Ca inward current peaks about 2 msec after the step depolarization and recedes rapidly thereafter within 5 msec (the Ca inactivation). Brehm & Eckert (15, 16, 29) showed that this recession is due to the increase in the internal concentration of Ca^{2+} (Ca-dependent Ca inactivation) and is not simply a voltage- and time-dependent event, as is the Na inactivation of axons. Whether this recession in Ca^{2+} current is due to a loss in the electromotive force of Ca^{2+} or to a genuine change in the state of the Ca channel is not clearly resolved, because the local concentration of Ca^{2+} at the vicinity of the Ca channel is not known; how this Ca^{2+} (or its analogs) are handled by the pump is also unknown.

The Ca channel passes Ca^{2+}, Sr^{2+}, and Ba^{2+}, but not Mg^{2+} and monovalent cations (15, 24, 71, 80). The threshold depolarization that activates the Ca current is close to the resting level as is evident from the current-voltage plots (29, 100).

Current through the delayed rectifying K channel is the main cause of the downstroke of the action potential. Under the voltage clamp, the threshold depolarization that activates the K outward current is about 20 mV from the resting level (15, 74, 80, 101). The channel passes K^+, and probably Rb^+, but not Na^+ and divalent cations. It is blocked by Cs^+ and TEA^+ (15).

The Ca channel is located on the ciliary membrane and not on the surface membrane covering the cell body, since deciliated bodies lose the ability to generate the action potential but regain it upon reciliation (28, 63, 84). Because the cilia do not regrow in unison (82), it is not possible to decide whether the Ca channels are evenly distributed through the ciliary membrane or further restricted to the basal region. Deciliated bodies retain the mechanosensory receptors and the delayed rectifying K current (63, 83).

Action Potential → $[Ca^{2+}]_{in}$ Increase

The entry of Ca^{2+} into the paramecium during excitation has been demonstrated by ^{45}Ca influx (17, 19). This is assayed at a low temperature to inhibit a powerful Ca-extrusion mechanism. The Ca^{2+} influx is also measured by a flow-through technique (B. Martinac, E. Hildebrand, submitted). ^{133}Ba influx through the Ca channel can be conveniently measured at physiological temperatures since there is no significant efflux of the preloaded Ba^{2+} (57). Recently, J. Thiele & J. E. Schultz (108a) have demonstrated the entry of Ca^{2+} into ciliary membrane vesicles by using arsenazo III absorbance change and employing inexcitable mutant membrane as control. This cell-free assay promises a biochemical dissection of the elements in the excitation process.

Ca^{2+}, in general, does not diffuse freely in the cytoplasm because of local sequestration (5, 91, 110). In *Paramecium*, a wild type and an inexcitable mutant forming a conjugating pair do not coordinate their behavior for a period of hours even though the two cytoplasmic chambers are confluent. This suggests that the Ca^{2+} entering the wild-type conjugant is not accessible to the mutant partner (10, 45). Inexcitable mutants of *P. caudatum* without Ca action potential nevertheless can generate a Ca-dependent receptor potential when touched at the anterior end. Only the cilia at or near the touched region show ciliary reversal (107), indicating the supply of Ca^{2+} through the receptor channel reaches only the local cilia. The same conclusion can be drawn from a different type of experiment with *Stylonichia* (23). The usual action potential and ciliary reversal are not accompanied by trichocyst discharge, a process also triggered by Ca^{2+} (6, 88). Since the trichocysts dock near the base of the cilia, the simplest explanation would be either that there is a large difference in the threshold Ca^{2+} concentrations for the two functions or that the Ca^{2+} entering through the voltage-sensitive Ca channel on the ciliary membrane stays near the ciliary lumen. Taken together, these observations suggest that rise and fall of Ca^{2+} concentration related to excitation may occur not in the general cytoplasm but in or near the ciliary space.

Note that the geometry of the paramecium soma is such that the cytoplasmic chamber is isopotential at any point (31). Propagation of the action potential is not necessary and does not occur in *Paramecium*. If the "cilioplasm" is as conductive as the cytoplasm, the geometry of the cilium also makes its lumen nearly isopotential at all points (16, 28). Thus any changes in electric potential are probably shared by all cilia at once. In sum, it is highly likely that the thousands of cilia on the paramecium body share the same electric information, but not their Ca^{2+}.

$[Ca^{2+}]_{in}$ Increase → Ciliary Reversal

The beat cycle of a cilium can be divided into a power stroke and a recovery stroke. The power stroke of paramecium cilia is normally toward the paramecium's posterior right. Increase in the intraciliary Ca^{2+} concentration causes the cilia to beat with the power stroke toward the cell's anterior right with a concomitant increase in beat frequency (58).

The axoneme responds to the rise of the intraciliary Ca^{2+} concentration and *not* directly to the electric signal (76, 77). After Triton treatment has destroyed the membrane barrier, the beat of the cilia can be reactivated with Mg^{2+} and ATP. Addition of Ca^{2+} above 10^{-6} M causes the reorientation of the beat direction of the reactivated cilia. A calculation shows that the Ca^{2+} concentration rises to 4×10^{-5} M in the cilium if the action current under the voltage clamp is similar to the natural action current and the whole ciliary lumen is available for the diffusion (15).

The mechanism by which the axoneme responds to the entered Ca^{2+} is not known. The beat form of the reactivated axonemes from *Chlamydomonas* can be altered by addition of Ca^{2+} (11). Since these axonemes are free of the basal bodies, the rootlet fibers, and the ciliary membrane, this observation argues against the involvement of these structures in the response to Ca^{2+}. Physiological concentrations of Ca^{2+} have no effect on the dynein arms in their ability to slide the peripheral tubules in cilia (120). It has been proposed that an interaction of the radial spokes and the central microtubule pairs is needed for bend formation (122) or in the regulation of the activity of the peripheral tubules (85, 86), and Ca^{2+} may affect this interaction (11, 86).

Several Ca-ATPases associated with cilia have been described, but it is not yet possible to assign any function to these ATPases. At least two functions may employ Ca-ATPases in *Paramecium:* the ciliary reversal mechanism and the Ca^{2+} extrusion mechanism [(25, 90); D. L. Nelson, J. Rauh, A. E. Levin, L. Riddle, personal communication]. Calmodulin or its analogs have been purified from ciliated or flagellated protozoa (37, 47, 50, 64, 90, 106, 112, 121); calmodulin activates dynein ATPase in *Tetrahymena* (12). Phenothiazines (calmodulin blockers) have been reported to inhibit a Ca-ATPase and affect paramecium behavior (90).

OTHER COMPONENTS

The locomotor behavior of *Paramecium* includes other components besides those in the basic schema reviewed above.

Hyperpolarization and Speed Control

Membrane hyperpolarization is correlated with an increase in the beat frequency of the cilia (14, 33, 59). This "ciliary augmentation" is accompanied by a slight shift of the power stroke of the beat even more toward the posterior end; this increases the cell's speed in its forward swimming direction. Transferring paramecia from one solution to another of a lower ionic strength usually causes a hyperpolarization and an acceleration. Mutants that are hyperpolarized upon transfer to certain solutions speed up after the transfer (96, 98). Under certain conditions, hyperpolarization can become regenerative, due to K^+ efflux through the anomalous rectifying K channel (81, 99).

A touch at the posterior end causes the paramecium to speed forward (69). A hyperpolarizing receptor potential due to an efflux of K^+ has been recorded from cells that are mechanically stimulated at the posterior end (22, 72, 73, 83). This receptor current, under a voltage clamp, survives deciliation. Thus both the anterior and posterior mechanoreceptors are in the soma.

What controls the ciliary beat frequency remains unclear. Machemer & Eckert (30, 60) proposed that "ciliary augmentation" is due to a decrease in the internal concentration of Ca^{2+}. It is difficult to evaluate this decrease since such factors as the resting Ca conductance, Ca-pump rate and local Ca distribution cannot yet be quantitatively estimated. Brehm & Eckert (14) favor the idea that electrophoretic effect of inward current through the cilium causes the higher beat frequency, but how it is caused is unclear and the inward current through the cilium upon hyperpolarization has not been demonstrated. It is agreed, however, that Ca^{2+} is the factor that causes the cilia to beat faster in the reverse direction (14, 30, 33, 60, 61).

Frequency of Avoiding Reactions vs Swimming Speed

A depolarization increases the frequency of the avoiding reaction (F_{AR}), and a hyperpolarization decreases that frequency, possibly to zero. On the other hand, hyperpolarization increases the speed (v) and encourages translational movement, while depolarization discourages translational movement, possibly to zero, if the avoiding reactons are so frequent that the paramecium jerks in place. There are two mechanisms for attraction or repulsion: orthokinesis through a modulation of speed for translation (Δv) and klinokinesis through a modulation of the rate in directional changes (ΔF_{AR}) (35). Van Houten (115, 116, 117) proposed that the membrane potential is the key controlling parameter in paramecium chemotaxis. Four situations can occur: Repellent type I causes moderate depolarizations that increase F_{AR} and disperse the paramecia through klinokinesis, while the concomitant v decrease is not enough to counteract that dispersal effect. Attractant type II causes strong depolarizations leading to very high F_{AR}, forcing v toward zero, and the paramecia are trapped in place through orthokinesis. Attractant type I causes moderate hyperpolarizations that decrease F_{AR} and attract the paramecia through klinokinesis, while the comcomitant v increase is not enough to counteract that attraction. Repellent type II causes strong hyperpolarizations resulting in very high v with no avoiding reaction and thus repulsion by orthokinesis. Van Houten has correlated the attraction or repulsion [quantified by a population behavioral test (118)] with the membrane-potential changes (recorded directly from individual specimens) upon the addition of various attractants or repellents (116). Data from mutants defective in sets of behavior and those defective in more specific chemotaxes further support her theory (114, 117).

Continuous Backing

A typical avoiding reaction lasts less than a second. The modulation of such avoiding reactions is the klinokinesis described above. There are many situations in which the paramecia swim backward continuously for seconds or minutes. There is a long tradition of timing this backward swimming when the paramecia are transferred from a solution with a low $[K^+]/[Ca^{2+}]^{1/2}$ ratio to one of a higher ratio (46, 68). While the Donnan ratio suggests ion exchange (78), presumably on the negatively charged surface, there is not yet a completely satisfactory theory to explain how the duration of backing is determined. Intracellular recording shows that the membrane depolarizes in solutions of high $[K^+]/[Ca^{2+}]^{1/2}$, but no change in either the membrane potential or the membrane resistance is seen to be correlated with the renormalization of the ciliary beat (D. Oertel, C. Kung, unpublished). Prolonged backing is also observed when the paramecia are confronted with a solution of high $[Ba^{2+}]^{1/2}/[Ca^{2+}]^{1/2}$ (57, 68).

A more natural environment for paramecia (freshwater, culture media, and different paramecium "salines") contains Na^+ besides K^+, Ca^{2+}, and Mg^{2+}. In such an environment, excessive stimulations or insults to the membrane induce continuous backward swimming, which could be a means of direct escape from the source of attack. Using a mutant, "paranoiac," which effects such backward swimming bouts without the harmful stimuli or membrane damage, Saimi & Kung (92) have discovered a Ca-induced Na current in *Paramecium*. Under a voltage clamp, this current carried by Na^+ (or Li^+) rises long after the Ca transient to maximum in some seconds and has an N-shaped I-V plot. They have suggested that the prolonged presence of internal Ca^{2+} in the wild type (due to cell damage or successive avoiding reactions, for example) induces the Na current that sustains a depolarization and therefore the backward swimming for seconds or minutes. Prolonged presence of internal Ca^{2+} also induces a slow K^+ outward current (13, 94, 102). The Na^+ gain and K^+ loss correlated with such prolonged depolarizations have been measured with isotope-flux and flame-photometric methods (38, 40, 95).

"Adaptation"

Paramecium shows a variety of slow changes in behavioral pattern that we loosely call "adaptation." For example, the frequency of avoiding reaction decays over tens of seconds to a ground rate after the organism is placed in a new solution. Paramecia also lose their avoiding reactions after prolonged incubation in K^+-rich solutions (39, 43, 104, 105). Raising the growth temperature increases the critical temperatures that trigger the thermo-avoidance (42), and changes the voltage sensitivity of channels, as seen under voltage clamp (103). Clearly, no single mechanism accounts for the various "adaptations."

KINDS OF ION CHANNELS AND THE ORIGIN OF "SPONTANEOUS" DISCHARGES

The paramecium surface membrane contains a variety of ion channels whose conductances can be distinguished by their ion specificities, triggering mechanisms, activation and inactivation kinetics, and in some cases by their locations (Table 1). The functions of some of these conductances in the locomotor behavior have been described above.

One of the problems in understanding the behavior of paramecium is the nature and origin of the so-called "spontaneous" avoiding reactions occurring at 0.1–1 Hz in the common culture conditions. The noise of the membrane potential, which apparently originates mostly from certain K channels (65, 66) showing thermal fluctuation between the opened and closed states, may serve to trigger the action potential. However, in culture media, fresh water and paramecium "salines," all having Na^+, besides K^+, Mg^{2+} and Ca^{2+}, the electric discharges are not the simple exemplary action potentials triggered from cells bathed in the Ca-K or Ca-Ba buffers. Such discharges begin with an action potential but have plateaus lasting 500–1000 msec (96). From the characteristics of ion channels summarized in Table 1, not only the fast, voltage-dependent ion channels but also the slow, Ca-dependent channels must be involved in such electric discharges. A series of channel interactions can be constructed to explain them as follows. Certain noise fluctuations or a rebound from a previous episode triggers the action potential by opening the fast channels. The Ca^{2+} that has entered then opens first the Ca-induced Na channel and then the Ca-induced K channel to establish the quasi-plateau. The Ca-induced K current eventually grows, repolarizing then hyperpolarizing the membrane. This hyperpolarization and the rebound from it would involve the channels activated by hyperpolarization. Though speculative, this or a similar series of channel interactions is more realistic than a model involving the fast, voltage-sensitive channels alone in the generation of the "spontaneous" discharges in Na^+-containing media. The modulation of these discharges may well be the basis of taxes in the natural settings.

RESPONSES TO CHEMICALS

In relation to the reception and transduction of taste response to cations, Beidler (7, 9) proposed that specific binding of ions to protein receptors on the membrane changes their conformations. This change affects neighboring structures and somehow leads to an increase of membrane conductance and therefore to the receptor potential. This scheme is generally accepted also for sugar- and pheromone-reception (49, 55, 67, 89) and is consistent with recent biochemical findings in bacterial chemotaxis, neurotransmitter reception, and hormone reception. Because it is cumbersome to do research on taste or smell cells, model systems including the ciliates may provide better opportunities to test definitively Beidler's proposal.

Although receptor proteins that bind various cations have not been sought, *Paramecium* reacts to three natural cations, Ca^{2+}, K^+, and Na^+, very differently, suggesting specific effects of these ions. Adding Ca^{2+} to the solution usually suppresses avoiding reactions and accelerates the paramecia (57, 68). Transferring paramecia to a K^+-rich medium induces continuous backward swimming (68). Transferring them to a Na^+-rich medium induces repetitive avoiding reactions (51). Nevertheless, all cations depolarize the *Paramecium* membrane; thus potential changes are necessary but not sufficient in determining behavior. The curious fact that all cations

Table 1 Characteristics of ion currents under voltage clamp in *Paramecium*

Current	Triggering mechanism (threshold)	Natural permeant ions (others)	Kinetics during voltage stimulation		After stimulation		
			Activation (typical peak time)	Inactivation (typical kinetics)	Relaxation (typical tail decay at $\Delta V = 0$ mV)[a]	Location	Reference[d]
Ca current	depol. ($\Delta V > 5$ mV)[a]	Ca^{2+} (Sr^{2+}, Ba^{2+})	fast (~ 2 ms at $\Delta V \cong 30$ mV)	fast (< 5 ms at $\Delta V \cong 30$ mV)	fast ($\tau < 2$ ms)[e]	ciliary membrane	28, 63, 80, 84, 101, f, g
Delayed rectifying K current	depol. ($\Delta V > 20$ mV)	K^+ (Rb^+?)	fast (20–50 ms at $\Delta V \cong 40$ mV)	slow ($\tau > 1$ s at $\Delta V \cong 40$ mV)	fast ($\tau < 3$ ms)	soma membrane	63, 102, f, g
Anomalous rectifying K current	hyperpol. ($\Delta V < -20$ mV)	K^+ (Rb^+?)	intermediate (> 50 ms at $\Delta V \cong -40$ mV)	?	fast ($\tau < 10$ ms)	soma membrane	63, 81
Ca-induced[b] K current	$[Ca^{2+}]_{in}$ ↑	K^+ (Rb^+?)	slow (> 100 ms variable with ΔV)	slow (half decay > 10 s)	slow ($\tau \cong 70$ ms)	?	102
Ca-induced[b] Na current	$[Ca^{2+}]_{in}$ ↑ (lower than above?)	Na^+ (Li^+)	slow (4–9 s at $\Delta V \cong 10$ mV)	slow (half decay > 30 s at $\Delta V \cong 10$ mV)	slow ($\tau \cong 400$ ms)	?	92
Na current upon hyperpol.	hyperpol.? ($\Delta V < -10$ mV)	Na^+ (Li^+?)	slow (> 1 s at $\Delta V \cong -30$ mV)	?	slow ($\tau > 200$ ms)	soma membrane	g

TAXES IN PARAMECIUM

		Kinetics after mechano-stimulation[c]			
		Rise	Fall		
Anterior mechano-receptor current	mechanical stimulation at the anterior	fast (10–20 ms)	intermediate (half decay = 20–70 ms)	soma membrane (anterior?)	24, 83, h
Posterior mechano-receptor current	mechanical stimulation at the posterior	fast (10–20 ms)	fast (half decay ≅ 5 ms)	soma membrane (posterior?)	83

Ion selectivity column: Ca^{2+}, Mg^{2+} (Sr^{2+}, Ba^{2+}) for anterior; K^+ (Rb^{+}?) for posterior.

[a] ΔV = membrane potential – holding potential; holding potential ≅ resting potential.
[b] Ca-induced currents are only indirectly triggered by voltage so that the kinetics should reflect the channel responses to the local changes in internal Ca concentration.
[c] mechanical stimulation lasted < 7 ms, at resting-potential level.
[d] References: 23, 63, 83, 84, f in *P. caudatum*; 80, 81, 92, 101, 102, g, h in *P. tetraurelia*; 24 in *Stylonychia mytilus*.
[e] τ is time constant of the exponential function.
[f] Y. Naitoh, personal communication.
[g] Y. Saimi, C. Kung, unpublished.
[h] From unclamped conditions, C. Kung, A. D. Murphy, Y. Satow, in preparation.

depolarize this membrane was first explained by a nonselective permeability to cations (70, 74, 75) but more recently by the neutralization of surface negative charges, which triggers a readjustment of the membrane potential (29, 100, 103). Changes in surface potential have been suggested as part of the sensory transduction in other systems (36).

Among the more natural organic compounds, acetate, proprionate, butyrate, lactate, and folate attract certain ciliates. These are known to hyperpolarize the membrane at millimolar concentrations [(111, 113, 115, 117, 118); J. Van Houten, M. DiNallo, M. Wohlford, submitted)]. Various "odorants," alcohols, aldehydes, and "bitter substances" are repellents to ciliates and are found to depolarize their membrane (4, 26, 108, 111). Tanabe et al (108) observed changes in the fluorescence and its polarization of a probe upon the application of "odorants" to *Tetrahymena* and suggested that gross structural changes in the membrane and its fluidity accompany the odorant reception. However, Ca^{2+}, K^+, and Na^+ are repellents by the same tests (2, 108) but do not change these parameters. A variety of physical changes are expected when the cells are responding to the addition of chemicals, and it is important to separate the *cause* of the taxis from the *physiological consequence* of the taxis or in some cases from the *pathological consequence* of application of chemicals. While *membrane depolarization* is the biological signal of the sensory cells, a *collapse* of the resting potential can easily be due to membrane damages. It is also important to monitor the actual behavior and conditions of the cells besides getting certain population test scores.

We think that Beidler's scheme is reasonable and to be expected, although it has not been proven applicable to ciliate chemotaxis. It remains possible, however, that the natural chemotaxes to hydrophilic and hydrophobic substances use very different transduction pathways. The involvement of the lipid domain of the membrane in excitation, though not necessarily sensory reception, has been implicated in the action of local anesthetics (18, 44, 54, 87), in the studies of thermotaxis (42), and in a mutant whose phospholipid defect is correlated with malfunctions of channels (34a).

COMPLICATIONS AND PROSPECTS

Many types of sensory cells bear cilia and much of the ultrastructure is conserved in these sensory cilia. If some of the sensory mechanisms are also conserved throughout evolution, the ciliated protozoa would be logical models for these sense cells. Experimentally, such models allow an integration of electrophysiological analysis with genetic and chemical analyses.

We should be reminded, however, that a protozoan is more complex (having the whole behavioral machinery in one cell) than a specialized receptor cell, neuron, or effector cell in the metazoan. Because sensory transduction and excitation occur on the same continuous membrane, triggering action potentials (and therefore avoiding reactions) with agents that affect the excitation process directly may strengthen our understanding of how action potential is produced, but may not help us understand the sensory processes that may (unless the ion channels are the receptors) have been bypassed. Ba^{2+}, TEA^+, and even quinine or quinidine (34, 56), for example, have known effects on the ion channels. As described above, many harmful agents that leak Ca^{2+} into the cell can cause the activation of the slow Na and K channels and lead to continuous backward swimming, which may show us how they leak Ca^{2+} but would not elucidate the biologically meaningful sensory processes.

Although we believe that ciliates are valid models for the study of chemoreception, a judicious choice of relevant stimuli is important. Thus the survey of relevant chemotactic stimuli by Jennings (48), Dryl (27), and Van Houten (113, 119) is invaluable. The detection mechanisms for the natural inorganic cations, such as Ca^{2+}, Na^+, and K^+, and for natural organic materials, such as folate and acetate, very likely evolved in ciliates and might be expected to be similar, in terms of their functional principles, to the detection mechanisms for salts and acids in the taste receptors of higher forms. Proton and alkali cations are the natural stimuli for sour and salty tastes, and the taste cells on frog and rat tongues respond to these ions by depolarizations (3, 93). Thus both the stimuli and the output are similar in protozoa and metazoa.

The study of chemotaxis to the "bitter substances" and "odorants" requires more justification. These are not substances natural to these ciliates and it is therefore doubtful that *specific* mechanisms exist in them to handle these substances. Since the issue of whether these substances are received by specific receptors (proteins) or by the general hydrophobic domain (mainly the lipids) has not been resolved by study of taste and olfaction (8, 53), it remains possible that the ciliates are valid model systems in the study of the chemoreception of these substances.

As model systems ciliates offer several advantages: (*a*) They have easily observable and quantifiable taxes to a variety of natural stimuli; (*b*) their large size allows direct electrophysiological measurement, and the appearance of receptor potentials upon stimulation has been established; (*c*) mutants are available that are blocked at the stimulus-response pathway at different points; (*d*) cell-biological techniques such as microinjection and deciliation have been developed; and (*e*) a large amount of biochemical information concerning the membrane proteins and lipids is now available (1, 79, 109).

The ciliate systems remain unique in providing such an opportunity for multidisciplinary research and may well lead us to a better understanding of the events between the reception of a stimulus, chemical or otherwise, and the generation of the receptor potential.

ACKNOWLEDGMENTS

We wish to thank many of our colleagues for providing unpublished results of their recent work, and Drs. M. Forte, J. Kung, and A. D. Murphy for critical reading of the manuscript. The preparation of this review was supported in part by NSF grant BNS79-18544 and NIH grant GM22714.

Literature Cited

1. Adouette, A., Ling, K.-Y., Forte, M., Ramanathan, R., Nelson, D. L., Kung, C. 1981. *J. Physiol. Paris.* In press
2. Aiuchi, T., Tanabe, H., Kurihara, K., Kobatake, Y. 1980. *Biochim. Biophys. Acta* 628:355–64
3. Akaike, N., Noma, A., Sato, M. 1976. *J. Physiol. London* 254:87–107
4. Ataka, M., Tsuchii, A., Ueda, T., Kurihara, K., Kobatake, Y. 1978. *Comp. Biochem. Physiol.* 61A:109–15
5. Baker, P. F. 1976. In *Calcium in Biological Systems,* ed. C. J. Duncan, pp. 67–88. Cambridge/London/NY: Cambridge Univ. Press
6. Balinski, M., Plattner, H., Matt, H. 1981. *J. Cell Biol.* 88:179–88
7. Beidler, L. M. 1954. *J. Gen. Physiol.* 38:133–39
8. Beidler, L. M., ed. 1971. *Handbook of Sensory Physiology, Vol. IV, Chemical Senses, Pts. 1, 2.* Berlin/Heidelberg/NY: Springer. 518 pp., 410 pp.
9. Beidler, L. M. 1971. See Ref. 8, Pt. 2, pp. 200–70
10. Berger, J. D. 1976. *Genet. Res.* 27:123–34
11. Bessen, M., Fay, R. S., Witman, G. B. 1980. *J. Cell Biol.* 86:446–55
12. Blum, J. J., Hayes, A., Jamieson, G. A. Jr., Vanaman, T. C. 1980. *J. Cell Biol.* 87:386–97
13. Brehm, P., Dunlap, K., Eckert, R. 1978. *J. Physiol. London* 274:639–54
14. Brehm, P., Eckert, R. 1978. *J. Physiol. London* 283:557–68
15. Brehm, P., Eckert, R. 1978. *Science* 202:1203–6
16. Brehm, P., Eckert, R., Tillotson, D. 1980. *J. Physiol. London* 306:193–203
17. Browning, J. L., Nelson, D. L. 1976. *Biochim. Biophys. Acta* 448:338–51
18. Browning, J. L., Nelson, D. L. 1976. *Proc. Natl. Acad. Sci. USA* 73:452–56
19. Browning, J. L., Nelson, D. L., Hansma, H. G. 1976. *Nature* 259:491–94
20. Byrne, B. J., Byrne, B. C. 1978. *CRC Crit. Rev. Microbiol.* 6:53–108
21. Cronkite, D. L. 1979. In *Biochemistry and Physiology of Protozoa,* ed. S. H. Hutner, M. Levandowsky, 2:222–75. NY: Academic
22. de Peyer, J., Machemer, H. 1978. *J. Comp. Physiol.* 127A:255–66
23. de Peyer, J., Machemer, H. 1978. *Nature* 276:285–87
24. de Peyer, J. E., Deitmer, J. W. 1980. *J. Exp. Biol.* 88:73–89
25. Doughty, M. J. 1978. *Comp. Biochem. Physiol.* 60B:339–45
26. Dryl, S. 1959. *Acta Biol. Exp.* 19:95–104
27. Dryl, S. 1973. In *Behavior of Microorganisms,* ed. A. Pérez-Miravete, pp. 16–30. NY: Plenum
28. Dunlap, K. 1977. *J. Physiol. London* 271:119–34
29. Eckert, R., Brehm, P. 1979. *Ann. Rev. Biophys. Bioeng.* 8:353–83
30. Eckert, R., Machemer, H. 1975. In *Molecules and Cell Movement,* ed. S. Inoué, R. E. Stephens, pp. 151–64. NY: Raven
31. Eckert, R., Naitoh, Y. 1970. *J. Gen. Physiol.* 55:467–83
32. Eckert, R., Naitoh, Y., Friedman, K. 1972. *J. Exp. Biol.* 56:683–94
33. Epstein, M., Eckert, R. 1973. *J. Exp. Biol.* 58:437–62
34. Falk, G. 1961. In *Biophysics of Physiological and Pharmacological Actions,* ed. A. M. Shanes, pp. 259–79. Washington DC: Am. Assoc. Adv. Sci.
34a. Forte, M., Satow, Y., Nelson, D. L., Kung, C. 1981. *Proc. Natl. Acad. Sci. USA.* In press

35. Fraenkel, G. S., Gunn, D. L. 1961. *The Orientation of Animals*. NY: Dover. 376 pp.
36. Gingell, D. 1971. In *Membranes and Ion Transport*, ed. E. E. Bittar, pp. 317–57. NY: Wiley
37. Gitelman, S. E., Witman, G. B. 1980. *J. Cell Biol.* 87:764–70
38. Hansma, H. G. 1979. *J. Cell Biol.* 81:374–81
39. Hansma, H. G. 1981. *J. Membr. Biol.* In press
40. Hansma, H. G., Kung, C. 1976. *Biochim. Biophys. Acta* 463:128–39
41. Hara, R., Asai, H. 1980. *Nature* 283:869–70
42. Hennessey, T., Nelson, D. L. 1979. *J. Gen. Microbiol.* 112:337–47
43. Hildebrand, E., Dryl, S. 1976. *Bioelectrochem. Bioenerg.* 3:543–44
44. Hille, B. 1977. *J. Gen. Physiol.* 69:497–515
45. Hiwatashi, K., Haga, N., Takahashi, M. 1980. *J. Cell Biol.* 84:476–80
46. Jahn, T. L. 1962. *J. Cell. Comp. Physiol.* 60:217–88
47. Jamieson, G. A. Jr., Vanaman, T. C., Blum, J. J. 1979. *Proc. Natl. Acad. Sci. USA* 76:6471–75
48. Jennings, H. S. 1906. *Behavior of Lower Animals*. Bloomington, Ind: Indiana Univ. Press. 366 pp.
49. Kaissling, K. E. 1971. See Ref. 8, Pt. 1, pp. 351–431
50. Kumagai, H., Nishida, E., Ishiguro, K., Murofushi, H. 1980. *J. Biochem.* 87:667–70
51. Kung, C. 1971. *Z. Vergl. Physiol.* 71:142–64
52. Kung, C. 1979. In *Topics in Neurogenetics*, ed. X. O. Breakfield, pp. 1–26. NY: Elsevier North-Holland
53. Kurihara, K., Kamo, N., Kobatake, Y. 1978. *Adv. Biophys.* 10:27–95
54. Lee, A. G. 1976. *Nature* 262:545–48
55. Le Magnen, J., MacLeod, P., eds. 1977. *Olfaction and Taste,* VI. London/Washington DC: IRL. 527 pp.
56. Lew, V. L., Ferreira, H. G. 1978. *Curr. Top. Membr. Transp.* 10:217–77
57. Ling, K.-Y., Kung, C. 1980. *J. Exp. Biol.* 84:73–87
58. Machemer, H. 1974. In *Cilia and Flagella*, ed. M. A. Sleigh, pp. 199–286. NY: Academic
59. Machemer, H. 1974. *J. Comp. Physiol.* 92:293–316
60. Machemer, H. 1976. *J. Exp. Biol.* 65:427–48
61. Machemer, H., Eckert, R. 1973. *J. Gen. Physiol.* 61:572–87
62. Machemer, H., de Peyer, J. 1977. *Verh. Dtsch. Zool. Ges.* 1977:86–110
63. Machemer, H., Ogura, A. 1979. *J. Physiol. London* 296:49–60
64. Maihle, N. J., Garofalo, R. S., Satir, B. H. 1981. *J. Cell Biol.* 89:695–99
65. Majima, T. 1980. *Biophys. Chem.* 11:101–8
66. Moolenaar, W. H., DeGoede, J., Verveen, A. A. 1976. *Nature* 260:344–46
67. Morita, H. 1972. *Adv. Biophys.* 3:161–98
68. Naitoh, Y. 1968. *J. Gen. Physiol.* 51:85–103
69. Naitoh, Y. 1974. *Am. Zool.* 14:883–93
70. Naitoh, Y., Eckert, R. 1968. *Z. Vergl. Physiol.* 61:427–52
71. Naitoh, Y., Eckert, R. 1968. *Z. Vergl. Physiol.* 61:453–72
72. Naitoh, Y., Eckert, R. 1969. *Science* 164:963–65
73. Naitoh, Y., Eckert, R. 1973. *J. Exp. Biol.* 59:53–65
74. Naitoh, Y., Eckert, R. 1974. In *Cilia and Flagella*, ed. M. A. Sleigh, pp. 305–52. NY: Academic
75. Naitoh, Y., Eckert, R., Friedman, K. 1972. *J. Exp. Biol.* 56:667–81
76. Naitoh, Y., Kaneko, H. 1972. *Science* 176:523–24
77. Naitoh, Y., Kaneko, H. 1973. *J. Exp. Biol.* 58:657–76
78. Naitoh, Y., Yasumasu, I. 1967. *J. Gen. Physiol.* 50:1303–10
79. Nelson, D. L., Kung, C. 1978. In *Taxis and Behavior (Receptors and Recognition, Ser. B, 5,)* ed. G. L. Haselbauer, pp. 75–100. London: Chapman & Hall
80. Oertel, D., Schein, S. J., Kung, C. 1977. *Nature* 268:120–24
81. Oertel, D., Schein, S. J., Kung, C. 1978. *J. Membr. Biol.* 43:165–85
82. Ogura, A. 1981. *Cell Struct. Funct.* 6:43–50
83. Ogura, A., Machemer, H. 1980. *J. Comp. Physiol.* 135(A):233–42
84. Ogura, A., Takahashi, K. 1976. *Nature* 264:170–72
85. Omoto, C. K., Kung, C. 1979. *Nature* 279:532–34
86. Omoto, C. K., Kung, C. 1980. *J. Cell Biol.* 87:33–46
87. Papahadjopoulos, D. 1972. *Biochim. Biophys. Acta* 265:169–86
88. Plattner, H., Reichel, K., Matt, H. 1977. *Nature* 267:702–4
89. Prosser, C. L. 1973. In *Comparative Animal Physiology*, ed. C. L. Prosser, pp. 553–76. Philadelphia: Saunders
90. Rauh, J., Levin, A. E., Nelson, D. L. 1980. In *Calcium-Binding Proteins: Structure and Function,* ed. F. L. Siegel,

E. Carafoli, R. H. Kretsinger, D. H. MacLennan, R. H. Wasserman, pp. 231–32. NY: Elsevier North-Holland
91. Rose, B., Loewenstain, W. R. 1975. *Science* 190:1204–6
92. Saimi, Y., Kung, C. 1980. *J. Exp. Biol.* 88:305–25
93. Sato, M. 1973. *Adv. Biophys.* 4:103–52
94. Satow, Y. 1978. *J. Neurobiol.* 9:81–91
95. Satow, Y., Hansma, H. G., Kung, C. 1976. *Comp. Biochem. Physiol.* 54A:323–29
96. Satow, Y., Kung, C. 1974. *Nature* 247:69–71
97. Satow, Y., Kung, C. 1976. *J. Exp. Biol.* 65:51–63
98. Satow, Y., Kung, C. 1976. *J. Neurobiol.* 7:325–38
99. Satow, Y., Kung, C. 1977. *J. Comp. Physiol.* 119:99–110
100. Satow, Y., Kung, C. 1979. *J. Exp. Biol.* 78:149–61
101. Satow, Y., Kung, C. 1980. *J. Exp. Biol.* 84:57–71
102. Satow, Y., Kung, C. 1980. *J. Exp. Biol.* 88:293–303
103. Satow, Y., Kung, C. 1981. *J. Membr. Biol.* 59:179–90
104. Shusterman, C. L. 1981. PhD thesis. Univ. of Wisconsin, Madison. 143 pp.
105. Shusterman, C. L., Thiede, E. W., Kung, C. 1978. *Proc. Natl. Acad. Sci. USA* 75:5645–49
106. Suzuki, Y. Hirabayashi, T., Watanabe, Y. 1980. *Biochim. Biophys. Res. Commun.* 90:253–60
107. Takahashi, M., Naitoh, Y. 1978. *Nature* 271:656–58
108. Tanabe, H., Kurihara, K., Kobatake, Y. 1980. *Biochemistry* 19:5339
108a. Thiele, J., Schultz, J. E. 1981. *Proc. Natl. Acad. Sci. USA.* In press
109. Thompson, G. A., Nozawa, Y. 1977. *Biochim. Biophys. Acta* 472:55–92
110. Tillotson, D., Gorman, A. L. F. 1980. *Nature* 286:816–17
111. Ueda, T., Kobatake, Y. 1977. *J. Membr. Biol.* 34:351–68
112. van Eldik, L. J., Piperno, G., Watterson, D. M. 1980. *Proc. Natl. Acad. Sci. USA* 77:4779–83
113. Van Houten, J. 1976. PhD thesis. Univ. California, Santa Barbara. 193 pp.
114. Van Houten, J. 1977. *Science* 198:746–48
115. Van Houten, J. 1978. *J. Comp. Physiol.* 127:167–74
116. Van Houten, J. 1979. *Science* 204:1100–3
117. Van Houten, J. 1981. *Olfact. Taste* 7:53–56
118. Van Houten, J., Hansma, H. G., Kung, C. 1975. *J. Comp. Physiol.* 104:211–23
119. Van Houten, J., Hauser, D. C. R., Levandowsky, M. 1981. In *Biochemistry and Physiology of Protozoan*, ed. M. Levandowsky, S. H. Hutner, 4:67–124. NY: Academic
120. Walter, M. F., Satir, P. 1979. *Nature* 278:69–70
121. Walter, M. F., Schultz, J. E. 1981. *Eur. J. Cell Biol.* 24:97–100
122. Warner, F. D., Satir, P. 1974. *J. Cell Biol.* 63:35–63
123. Wood, D. C. 1980. *Ann. Meet. Soc. Neurosci., 10th, Cincinnati,* 6:575 (Abstr.)

CHEMOTAXIS IN *DICTYOSTELIUM*

Günther Gerisch

Max-Planck-Institut für Biochemie, 8033 Martinsried bei München, Germany

FUNCTIONS OF CHEMOTAXIS IN *DICTYOSTELIUM* DEVELOPMENT

Dictyostelium discoideum exists as single amoeboid cells during the first phase of its developmental cycle. These cells phagocytose bacteria. Chemotaxis to compounds released from the bacteria (e.g. folic acid) is probably involved in food seeking (95). After the growth phase the amoebae differentiate into cells capable of aggregating into a multicellular organism. A breakthrough in the investigation of *Dictyostelium* chemotaxis was the discovery in 1967 that aggregating cells of *D. discoideum* are attracted by cyclic AMP (cAMP) (64). Normally in the laboratory *D. discoideum* develops asexually, but by combining cells of different mating types sexual development can be induced. Chemotaxis is also involved in sexual development as the cells are attracted by immature macrocysts (93). These are cell groups in which zygotes are formed. In the multicellular stage the cells differentiate into spores and cells forming a stalk and basal disk. These two types of cells constitute the fruiting body formed at the end of development. Evidence has been presented for an action of cAMP in the sorting out of pre-spore and pre-stalk cells, suggesting that the spatial pattern of these cells is determined by chemotaxis (88). Chemotaxis to oxygen seems also to be involved in aligning the pre-spore/pre-stalk pattern in the stage preceding fruiting body formation (114).

The discovery that cAMP is a chemoattractant established cAMP as a primary, extracellular messenger. Cyclic AMP exerts its function as a chemoattractant by binding to cell-surface receptors (47, 49, 70, 78). In addition, cAMP affects cell differentiation (16, 18, 35) and stimulates its own production (20, 37, 107, 113). cAMP-stimulated synthesis of cAMP is the basis of a relay system by which chemotactic stimuli are propagated as waves from cell to cell (103, 104, 112).

The multiplicity of cAMP actions indicates that the signal processing pathways connected to cAMP receptors diverge. But signal processing pathways also converge, as it is shown by the similarity of responses obtained with cAMP and folic acid (7, 131, 132). Folic acid, a chemoattractant of pre-aggregation cells of *D. discoideum* (94), also interacts with cell-surface receptors (121, 129) and, like cAMP, has multiple effects. In this review emphasis is put on the signal processing pathways involved in the chemotactic response. Previous reviews cover similar topics (17, 39, 67, 80, 91). The genetics of cell aggregation in *Dictyostelium* has also been reviewed (53, 92) and the biochemistry of *Dictyostelium* development in general is discussed in (66) and (11).

CHEMOATTRACTANTS IN VARIOUS SPECIES

Cyclic AMP is the chemoattractant of aggregating cells of *Dictyostelium discoideum* and a number of other, but not all, *Dictyostelium* species. Aggregating cells of *D. minutum* do not respond to cAMP but to a factor with a molecular weight of less than 700 that has been purified from yeast extract. The purified material contained glycine and a heterocyclic compound (56). Its action as an attractant is specific for *D. minutum*. Another factor purified from yeast extract proved to be specific for *D. lacteum* (85). This chemoattractant seems to consist of an aromatic moiety bound to the amino terminal of glycine (56). In cellular slime molds of the genus *Polysphondylium* other chemotactic agents of low molecular weight are produced that have tentatively been identified as oligopeptides (130).

Cells of all cellular slime mold species tested are sensitive to folic acid (94). In *D. discoideum* the chemotactic activity of folic acid is strongest for cells at or shortly after the end of growth. Whereas the sensitivity to folic acid declines during development, the sensitivity of *D. discoideum* cells to cAMP increases until the cells reach, after several hours, the aggregation-competent stage (9).

Studies on the interaction of chemoattractants with cell-surface receptors have concentrated on cAMP (47, 49, 70, 78, 89) and folic acid (121, 129). A report on chemotaxis to oxygen (114) suggests that *D. discoideum* cells possess, in addition to cAMP and folic acid receptors, an O_2-binding molecule, possibly an autoxidable redox system in the plasma membrane. In this context a cytochrome b found in plasma membranes of *D. discoideum* is of interest (111). Changes in the redox state of a cytochrome b of unknown location are correlated with cAMP production (36). A function of the plasma membrane cytochrome b in chemotaxis to O_2 and possibly other chemoattractants remains to be established.

CELL SURFACE BINDING SITES FOR CHEMOATTRACTANTS

Cyclic AMP Receptors

Specific, reversible, and saturable binding of cAMP to the surface of living cells can be measured under conditions where cAMP phosphodiesterase, also present on the cell surface, is inhibited. Because of the high specificity of the receptors for cAMP, the phosphodiesterase can be blocked by cGMP without inhibiting binding of cAMP to the receptors. An alternative is phosphodiesterase inhibition by dithiothreitol. Between 2 and 5×10^5 cAMP binding sites have been found on the surface of aggregation-competent cells (47, 70).

Scatchard plots of binding data are nonlinear and may indicate either multiple binding sites or negative cooperativity. The precocious appearance of low-affinity binding during development as compared to high-affinity binding argues for multiple sites (47). The increase of the dissociation rate of the (^{3}H)-cAMP receptor complex observed after the addition of excess unlabeled cAMP can be more easily explained by negative cooperativity (89). Assuming two types of binding sites, the K_D of high-affinity binding sites would be 9 nM and their number during aggregation 1.5×10^4 per cell; the K_D of low-affinity binding sites would be 160 nM and their number per cell 2×10^5 (47).

Cyclic AMP interaction with cell-surface receptors has been studied by the use of 50 cAMP analogs (77). The conclusions are: (a) cAMP binds to the receptors at the purine base by two hydrogen bonds at the 6-amino and the 7-N-position, at the ribose by one hydrogen bond at the 3'-oxygen position, and possibly at the phosphate group by ionic interaction; (b) the purine moiety also binds by interaction of its π-electron system with an acceptor at the receptor binding site; and (c) the receptor involved in chemotaxis differs in its specificity from the regulatory subunit of protein kinases of higher animals, although there are some binding sites in common on both cAMP binding proteins. The receptor of *D. discoideum* seems to interact with the anti-conformation of cAMP, whereas the protein kinase appears to possess a syn-type binding site.

With the possible exception of 9-(tetrahydro-2-furyl)adenine (61), no antagonist of cAMP is known for the chemotaxis receptors.

Cyclic AMP binds to membrane preparations of aggregation-competent cells, but binding to solubilized receptors has not yet been reported. Attempts to identify cell-surface receptors for cAMP by affinity labeling of living cells with 8-azido-cAMP have led to controversial results. Cooper et al (14) found no labeling of living cells when two intracellular, soluble cAMP-binding proteins of mol wt 42,000 and 39,000 were labeled in cell homogenates of strain NC4 (in strain V12, the molecular weights were 39,000 and 38,000, respectively). Incorporation of label into intact cells was reported by Wallace & Frazier (123). The only specifically labeled protein formed a band on SDS-polyacrylamide gels below actin. The estimated mol wt was 40,000. The protein labeled exhibited developmental regulation similar to that of cAMP receptors. Using careful controls, Juliani & Klein (54) studied photoaffinity labeling of living cells by 8-azido-(^{32}P) cAMP and found a developmentally regulated cAMP-binding protein of mol wt 45,000, slightly larger than actin. When cells were treated with high concentrations of cAMP, the receptors were down-regulated to about 25% of the binding activity of untreated cells. Under these conditions the 45,000-dalton component was no longer labeled. The label was found instead in a protein of mol wt 47,000. It is unknown whether this protein is a precursor, or a product of conversion, of the 45,000-dalton protein.

During normal development the number of cAMP binding sites increases by one order of magnitude between the end of the growth phase and the aggregation stage (47, 49, 70). This increase is promoted by cAMP applied in pulses of 5×10^{-9}M (106). Down-regulation of cell-surface binding sites requires application of cAMP at concentrations as high as 10^{-5} to 10^{-3}M (63). The same effect is obtained in a phosphodiesterase-deficient mutant with a cAMP concentration as low as 10^{-8}M (62). During down-regulation of binding sites, cAMP already bound becomes inaccessible to the extracellular space. A possible mechanism of down-regulation of cell-surface sites is internalization. Recycling of internalized membrane components has been demonstrated and quantitatively investigated in *D. discoideum* (115).

Folic Acid Receptors

Binding of folic acid to the cell surface of *D. discoideum* shows characteristics similar to cAMP binding. Curvilinear Scatchard plots indicate either negative cooperativity or multiple binding sites and a $K_{0.5}$ of about 2×10^{-7}M (129). The number of binding sites exposed to extracellular folic acid was 6×10^4 per cell shortly after the end of growth, and 3×10^4 at 10 hr later. This small decline contrasts remarkably to the inability of cells to respond chemotactically to folic acid at 6 hr after the end of growth or later, and has led to the suggestion that during cell development the folic acid receptors are disconnected from the signal processing pathways. However, if one measures not chemotaxis but outward spreading of cells placed in droplets on agar containing 10^{-5}M folic acid, the recognition and processing of folic acid signals can be demonstrated in cells that have developed for up to 9 hr (122).

In addition to folic acid binding sites, folic acid deaminase has been detected at the cell surface. The affinity of the deaminase catalytic site appears to be different from that of the folic acid binding sites thought to be involved in chemotaxis (129). The product of the deaminase-catalyzed reaction, 2-deamino-2-hydroxy-folic acid (55, 97), is chemotactically inactive but acts as a competitor with about the same affinity to the binding sites as the folic acid (129). Cyclic AMP does not interfere with folic acid binding in accord with the assumption that cAMP and folic acid receptors are independent entities.

The finding that the affinity of folic acid to chemotaxis receptors is primarily determined by the pteridine ring (95) opened the possibility of purifying the receptors by chromatography on folic acid coupled through its glutamic acid moiety to Sepharose (121). Seven polypeptides present in detergent extracts from a membrane fraction were eluted by folic acid from an affinity column. Some of these folic acid binding proteins were abundant in the soluble fraction of cell homogenates. However, one of the proteins appeared to be an integral membrane protein. The identity of this protein with the receptor acting in chemotaxis remains to be established.

REACTIONS PROBABLY INVOLVED IN THE CHEMOTACTIC RESPONSE

Short-Term Responses to Cyclic AMP

Since the beginning of chemotactic orientation of a cell is already observed 5 sec after stimulation by cAMP (40), only short-term biochemical changes are potential mediators of the chemotactic response. Short-term effects include (*a*) an increased influx of calcium, (*b*) an increase of intracellular cGMP concentration apparently due to the activation of guanylate cyclase, (*c*) a decrease in the phosphorylation of myosin heavy chains, (*d*) changes in the methylation of phospholipids, (*e*) the activation of adenylate cyclase, and (*f*) a decrease of extracellular pH (72, 73). The activation of adenylate cyclase is the basis of cAMP-stimulated cAMP production and thus a step in the relay of cAMP signals. The pH change is, at least in part, due to increased CO_2 production (38), indicating a transient increase of the respiration rate in stimulated cells. The other responses are possibly more directly related to the chemotactic response.

Cyclic GMP Regulation

After the addition of cAMP to a suspension of chemotactically responsive cells the intracellular cGMP concentration increases rapidly by a factor of about 10. This increase starts at less than 5 sec, reaches its peak after about 10 sec, and sharply declines thereafter (81, 133). Other species of cellular slime molds respond to their chemotactic factors with a similar, transient rise of cGMP: *Polysphondylium violaceum* (128) to a factor, possibly an oligopeptide (130), produced by aggregating cells of this organism; *Dictyostelium lacteum* to an unidentified, low molecular weight compound from yeast extract (79). No change in cAMP concentration was observed in chemotactically stimulated cells of these species. The same was true for folic acid stimulated preaggregative cells of *D. discoideum* in which adenylate cyclase activity was extremely low (133). Thus cAMP is apparently not involved in the processing of chemotactic signals.

The increase of cGMP appears to be due to the activation of guanylate cyclase (83). Most of the activity of this enzyme has been found to be associated with a particulate fraction in one laboratory (83) and by other workers to reside in the 100,000 $\times$ g supernatant (124). The first authors reported a 2–3-fold activation of guanylate cyclase by ATP which decreases the K_m for the substrate, GTP (76). The fall of cGMP concentration is due to phosphodiesterases. The major cyclic nucleotide phosphodiesterases of *D. discoideum* hydrolyze both cAMP and cGMP. In a mutant, HPX 235, defective in these phosphodiesterases, Dicou & Brachet (23) discovered an intracellular, soluble cGMP-specific phosphodiesterase that might play a key role in chemotaxis. Cyclic GMP has also been found in the extracellular medium of cAMP-stimulated cells where it is inactivated by extracellular and cell-surface phosphodiesterases (125).

Cyclic GMP probably exerts its action by binding to soluble intracellular proteins that specifically interact with this nucleotide (86, 87, 100). Cofractionation of these proteins with the major protein kinases was not observed. The apparent dissociation constant of soluble cGMP binding protein(s) in *D. discoideum* was estimated on millipore filters after washing of the samples, a method that does not always give reliable results. The K_D was reported to be 5×10^{-10}M (86, 87). Since the cGMP concentration in unstimulated cells is about 1×10^{-7}M (81, 133), it remains an open question whether proteins with such a high affinity for cGMP mediate chemotaxis. If the cGMP is located in the same intracellular compartment as the binding protein, the latter would already be almost completely saturated with cGMP before chemotactic stimulation of the cells. Another unexpected result is the localization of cGMP in *P. violaceum* cells as revealed by histochemistry using anti-cGMP antibodies (96). Whereas in preaggregative cells the cGMP appeared to be uniformly distributed in the cytoplasm, aggregating cells showed the antibody label preferentially in the nuclei. It seems reasonable to assume that the nuclear cGMP binding material is involved in cell functions other than chemotaxis.

Myosin Heavy Chain Dephosphorylation

If *D. discoideum* cells are stimulated by cAMP and then rapidly opened in a detergent solution containing (γ–^{32}P)-ATP, the incorporation of ^{32}P into myosin heavy chains is substantially increased (101). The reason is obviously that stimulation of whole cells with cAMP causes in vivo dephosphorylation of the myosin heavy chains (69). As has been first shown in another amoeba, *Acanthamoeba castellanii* (13, 75), heavy chain phosphorylation changes the actin-activated Mg^{2+}-ATPase activity of myosin, and thus has a regulatory function in the amoebae similar to that of light chain phosphorylation in mammalian smooth muscle and nonmuscle cells. In *D. discoideum* phosphorylation occurs at the tail portion of myosin heavy chains (98) and decreases the actin-activated Mg^{2+}-ATPase activity of the myosin (65). Another effect of myosin heavy chain phosphorylation is inhibition of self-assembly of myosin to thick filaments (65).

Stimulation of the cells with cAMP causes an increased influx of calcium (126). *D. discoideum* contains a calmodulin not identical with beef brain calmodulin but similar in function (5, 12). Myosin heavy chain kinase is inhibited by calcium and calmodulin in a slow, temperature-dependent reaction that does not require ATP (74). A tentative scheme of signal processing from cAMP receptors at the cell surface to the contractile system thus involves opening of calcium channels, inactivation of myosin heavy chain kinase in a calmodulin-mediated reaction, decrease of steady-state phosphorylation of myosin, increase of the actin-activated Mg^{2+}-ATPase activity, and polymerization of myosin.

Phospholipid Methylation

In cells labeled with (^{3}H-methyl)L-methionine, cAMP was reported to cause a transient decrease in methylation of the total methanol/chloroform extractable phospholipids (84). A two-fold increase in cAMP-stimulated cells of ^{3}H-methyl groups incorporated into a membrane protein of mol wt $\sim$ 120,000 has also been reported. These changes in the methylation of membrane components were thought to be the basis of an inhibitory effect of S-adenosyl-L-methionine on calcium uptake into vesicles, as was observed in cell homogenates.

In contrast to the decrease of total phospholipid methylation a fast, transient increase in (^{3}H-methyl)phosphatidylcholine, followed by a decrease in (^{3}H-methyl)lysophosphatidylcholine, was observed in cAMP-stimulated cells (2). This change in phospholipid methylation seems to be mediated by cGMP which, when added to cell homogenates together with S-adenosyl (^{3}H-methyl) methionine, stimulates the formation of mono- and dimethylated phosphatidyl-ethanolamine and of phosphatidylcholine (2). Calcium also seems to play a role. The lipid methyltransferase was stimulated two-fold by calcium plus calmodulin as compared to its activity in the presence of chlorpromazine, an inhibitor of calmodulin-mediated reactions (43).

CELL BEHAVIOR DURING CHEMOTAXIS

The movement of *Dictyostelium* cells in gradients of chemoattractants differs markedly from that of bacteria. *Dictyostelium* cells move in more or less straight lines, rather than in tracks reminiscent of the smooth-swimming and tumbling behavior of bacteria. Nevertheless, *Dictyostelium* cells show a behavior analogous to tumbling: Slight changes in the direction of movement (99) are caused by the retraction of a front of pseudopods and the formation of a new front from another area of the cell surface (31). This behavior can be simulated by local application of cAMP from a micropipette. Aggregating cells of *D. discoideum* are normally elongated and equipped with one front of pseudopods. Locally applied cAMP induces new pseudopods from any area of the cell surface and results in the retraction of the previous front (40). Thus the polarity of a cell can be changed by local chemotactic stimulation. Often the new pseudopods break abruptly through the cell surface as blebs with hyaline cytoplasm.

Stimulation of a cell causes not only local extension of pseudopods at the area close to the source of cAMP, but also a fast general contraction of the cell. The contraction, also called "cringing," is transient, lasting about 25 sec after the addition of cAMP (29). It is tempting to speculate that the contraction is related to dephosphorylation of myosin heavy chains, and the local extension of pseudopods to changes in phospholipid methylation which, in other cells, has been shown to influence membrane fluidity (51, 52).

The contraction of a cell is separated in certain mutants from cAMP-induced pseudopod extension (34). These are mutants whose cells are unable to coordinate pseudopod formation. The cells are not clearly polarized and form pseudopods simultaneously in all directions. The mutant cells are able to contract in response to cAMP, but the ability to extend pseudopods preferentially towards the source of the attractant is almost completely abolished. The mutant behavior makes it obvious that in wild type cells pseudopod formation is inhibited at any part of the cell surface except at the established front, and that this inhibition is essential for the directionality of movement in gradients of attractant.

MECHANISM OF GRADIENT DETECTION

The mechansism of gradient detection is controversial. Two extreme views are (a) cells sense a spatial gradient over their surface, so that the distribution of receptors occupied at a given time determines the direction of movement, (b) cells sense a gradient during the extension of pseudopods. In the latter case pseudopods would act as sensors that detect concentration differences in time by carrying, on their surface, receptors into various directions relative to a gradient (3, 37).

Mato et al (82) have calculated that the minimal concentration difference between the two ends of a cell that elicits a chemotactic response is 1% when the average concentration of cAMP at the position of the cells is 4×10^{-9}M. The response was judged as positive when in half of the samples about one third of all cells had moved towards the source of cAMP within 30 min. The authors assumed that the amoebae sense a gradient only once during a period of 1 msec, thought to be the cAMP receptor equilibration time. They concluded that an amoeba would be unable reliably to sense a gradient by the extension of a single pseudopod. If one assumes less stringent conditions for sensing of a gradient by an amoeboid cell, it can be calculated that spatial or temporal mechanisms are equally effective (6).

Adaptation is essential in bacteria for the temporal mechanism of gradient sensing to function (see review in this volume). The study of adaptation requires the controlled modulation of the attractant concentration. Such experiments are difficult to perform in *D. discoideum* because the cells produce attractant-destroying enzymes. One possibility of overcoming the effects of phosphodiesterases on the extracellular cAMP concentration is to expose the cells to attractant in a flow chamber (19). Another possibility is to use a slowly degradable analog of the attractant. In *D. discoideum* the cGMP increase caused by the slowly hydrolysed 3',5'-cyclic adenosine phosphorothioate as an agonist proved to be transient (109). Thus the activation of guanylate cyclase by chemotaxis receptors appears to underlie adaptation. Influx of calcium is another transient response to cAMP, but it has not yet been studied in the continued presence of attractant. Further studies are required to clarify whether *Dictyostelium* cells sense a gradient by a spatial or temporal mechanism, or by a combination of both.

PERIODIC SYNTHESIS AND RELEASE OF CYCLIC AMP

Aggregating cells of *D. discoideum* not only respond to cAMP, they also produce the attractant. Since the cells are capable of synthesizing cAMP periodically in the form of pulses, phases of cAMP production can alternate with phases in which the same cells respond chemotactically. In an agitated suspension, cells synchronize their activities (36). As a result, pulses of cAMP are produced every 6–8 min (41, 42). The generation of cAMP pulses is due to periodic activity changes of adenylate cyclase (108). The cAMP is released into the extracellular space shortly after synthesis, possibly by exocytosis (68), and hydrolyzed by cell-surface bound and extracellular phosphodiesterases. The intercellular signal that synchronizes the activities of the cells is cAMP which, by binding to cell-surface receptors, leads to the activation of adenylate cyclase (105, 107, 113). Control of the cellular oscillator by extracellular cAMP is shown by the induction of phase shifts. A pulse of 5–10 nM cAMP applied to a cell suspension causes either precocious cAMP production, a phase delay, or no response, depending on the phase of the oscillations in which the pulse is applied (72).

The mechanism of the oscillations is unknown. Theoretical work has suggested that adenylate cyclase is an intrinsic part of the oscillator and cell-surface cAMP receptors act as allosteric sites of the cyclase (44). Mathematical modeling indicates that under these conditions adenylate cyclase activity would oscillate, provided the cellular ATP concentration declines in each period as a result of cAMP synthesis, or for other reasons (45, 46). Such a decline was not detectable by measuring total cellular ATP (30, 108). It is not excluded, however, that the concentration changes of ATP were restricted to a cellular compartment in which adenylate cyclase is localized. It is of importance in this connection that adenylate cyclase has been found in a vesicle fraction that can be separated from plasma membranes (50).

The pulse shape of the cAMP signals is determined by three factors: (a) autocatalytic stimulation of cAMP synthesis via cell surface receptors; (b) adaptation of the cells to extracellular cAMP signals, which results in the deactivation of adenylate cyclase in the continued presence of extracellular cAMP; and (c) destruction of cAMP by cell-surface and extracellular phosphodiesterases. The importance of phosphodiesterase is emphasized by the behavior of a mutant deficient in phosphodiesterase production (4, 15). This mutant slowly accumulates cAMP nonperiodically but forms pulses periodically if phosphodiesterase is added to the medium (10). Adaptation also plays an essential role in the oscillatory control of cAMP production. The cells adapt to a constant concentration of cAMP within minutes and remain in this stage as long as cAMP or another agonist is present (38, 41). Quantitative data on the relationship between changes of extracellular cAMP concentrations on one side and cAMP synthesis and release on the other have been obtained (19, 20, 24, 25, 26). In a series of elegant studies labeled cells were exposed in a flow chamber to a defined program of concentration changes in time. By this technique it was possible to stimulate cells by the increase of extracellular cAMP concentrations from 10^{-12} to 10^{-5} M in the form of small successive steps (20). As a result, cells released cAMP continuously over a period of more than 30 min. Very slow and very rapid jumps from a low cAMP concentration to a higher one induced the same total amount of cAMP release.

The oscillations discussed above proved strongly temperature dependent (48, 90, 127). The length of period changed from 8 min at 22°C to 27 min at 8.5°C. A completely temperature compensated type of oscillation has been reported (59) in living cells and isolated membranes. These oscillations were characterized by changes in the binding capacity for cAMP which, in the living cells, was inversely related to oscillatory changes in the phosphorylation of membrane proteins (58).

It would be of great interest if these results could be confirmed by other laboratories. In an aggregation field, where the cells are coupled to each other by diffusion of cAMP, signaling cells form a spatio-temporal pattern. A pattern often observed in populations of aggregating *D. discoideum* cells is a spiral wave connecting all cells which, at a given time, release cAMP (1, 32, 116). The wave propagates from a central area to the periphery of a territory, thus extending the area from which cells are moving towards an aggregation center. Wave propagation can be initiated by an artificial, pulsating source of cAMP (103, 104). The necessity of newly synthesizing the transmitter, cAMP, in response to each stimulus distinguishes the transmitter system of *Dictyostelium* from neuronal systems and is the reason for its slowness. The signals propagate with velocities of less than a millimeter per minute (1, 32, 112).

REGULATION OF EXTRACELLULAR CHEMOATTRACTANT-DESTROYING ENZYMES

The concentrations of the two chemoattractants, cAMP and folic acid, in the extracellular medium are regulated by enzymes produced by *D. discoideum* cells. Production of these enzymes is regulated by the chemoattractants. The regulation represents part of a negative feedback loop that keeps extracellular attractant concentrations within certain limits.

Some of the phosphodiesterases that hydrolyze extracellular cAMP are bound to the cell surface (71), while some are soluble and are released into the extracellular medium (117, 119, 21). These enzymes also hydrolyze cGMP. In this respect the catalytic sites of the phosphodiesterases are less specific than the cell-surface receptors acting in chemotaxis (70, 77).

After the end of growth an inhibitor of extracellular phosphodiesterase is released into the extracellular medium (102). This glycoprotein forms a tight complex with the enzyme which has a much higher K_m than free phosphodiesterase (57). The inhibitor can be purified on a phosphodiesterase affinity column and forms a single band on SDS-polyacrylamide gels corresponding to a mol wt of about 43,000–47,000 (22, 28).

The relationship of membrane bound and extracellular phosphodiesterase is unclear. A mutant in which both types of phosphodiesterase are deficient suggests a common gene product or a common mechanism of regulation (15, 4). Both types of enzyme are concanavalin A binding glycoproteins (27). The phosphodiesterase at the surface of intact cells is not inactivated by the inhibitor of extracellular phosphodiesterase. However, the membrane bound phosphodiesterase becomes sensitive to the inhibitor after solubilization, indicating that this enzyme shares the inhibitor binding site with extracellular phosphodiesterase (71).

Folic acid is inactivated by deaminases at the cell surface and in the extracellular medium (97, 55, 8). The extracellular folate deaminases seem to be glycoproteins. They bind to lentil lectin and concanavalin A (8).

Folic acid stimulates production of extracellular deaminase (7) and also cAMP phosphodiesterase (109, 7). Cyclic AMP does the same (7, 60, 118, 120). Furthermore, both cAMP (33, 120) and folic acid suppress the production of phosphodiesterase inhibitor (7, 109). Thus the cAMP and folic acid receptors that control the chemoattractant-destroying extracellular enzymes feed into common pathways of signal transduction. These pathways of signal processing acting in extracellular phosphodiesterase and deaminase regulation apparently circumvent a step in which adaptation occurs. In contrast to other responses of *D. discoideum* cells, control of the extracellular enzymes is sustained as long as either cAMP or folic acid is present in the medium (7, 33, 109, 134).

The importance of attractant destruction for chemotaxis is questionable. Certainly in soil, the natural habitat of *Dictyostelium* species, cAMP and folic acid are liberated from other microorganisms. Under these conditions the phosphodiesterases and deaminases may prevent saturation of the chemotaxis receptors and thus enable the cells to detect concentration differences (9). However, hydrolysis of the attractant is not essential for the chemotactic response of a cell. A very slowly hydrolyzable compound like cyclic-3',5'-adenosine phosphorothioate is an active attractant (110).

Literature Cited

1. Alcantara, F., Monk, M. 1974. Signal propagation during aggregation in the slime mould *Dictyostelium discoideum*. *J. Gen. Microbiol.* 85:321–334
2. Alemany, S., Gil, M. G., Mato, J. M. 1980. Regulation by guanosine 3':5'-cyclic monophosphate of phospholipid methylation during chemotaxis in *Dictyostelium discoideum*. *Proc. Natl. Acad. Sci. USA* 77:6996–99
3. Alt, W. 1980. Biased random walk models for chemotaxis and related diffusion approximations. *J. Math. Biol.* 9:147–77
4. Barra, J., Barrand, P., Blondelet, M.-H., Brachet, P. 1980. *pdsA*, a gene involved in the production of active phosphodiesterase during starvation of *Dictyostelium discoideum* amoebae. *Molec. Gen. Genet.* 177:607–13
5. Bazari, W. L., Clarke, M. 1981. Characterization of a novel calmodulin from *Dictyostelium discoideum*. *J. Biol. Chem.* 256:3598–603

6. Berg, H. C., Purcell, E. M. 1977. Physics of chemoreception. *Biophys. J.* 20: 193–219
7. Bernstein, R. L., Rossier, C., van Driel, R., Brunner, M., Gerisch, G. 1981. Folate deaminase and cyclic AMP phosphodiesterase in *Dictyostelium discoideum*: their regulation by extracellular cyclic AMP and folic acid. *Cell Diff.* 10:79–86
8. Bernstein, R. L., Tabler, M., Vestweber, D., van Driel, R. 1981. Extracellular folate deaminase of *Dictyostelium discoideum. Biochem. Biophys. Acta* 677:295–302
9. Bonner, J. T., Barkley, D. S., Hall, E. M., Konijn, T. M., Mason, J. W., O'-Keefe, III G., Wolfe, P. B. 1969. Acrasin, acrasinase, and the sensitivity to acrasin in *Dictyostelium discoideum. Devel. Biol.* 20:72–87
10. Brachet, P., Dicou, E. L., Klein, C. 1979. Inhibition of cell differentiation in a phosphodiesterase defective mutant of *Dictyostelium discoideum. Cell Diff.* 8:255–65
11. Cappuccinelli, P., Ashworth, J. M., eds. 1977. *Development and Differentiation in the Cellular Slime Moulds.* Amsterdam/NY:Elsevier/North Holland Biomedical Press. 317 pp.
12. Clarke, M., Bazari, W. L., Kayman, S. C. 1980. Isolation and properties of calmodulin from *Dictyostelium discoideum. J. Bacteriol.* 141:397–400
13. Collins, J. H., Korn, E. D. 1980. Actin-activation of Ca^{2+}- sensitive Mg^{2+}-ATPase activity of *Acanthamoeba* myosin II is enhanced by dephosphorylation of its heavy chains. *J. Biol. Chem.* 255:8011–14
14. Cooper, S., Chambers, D. A., Scanlon, S. 1980. Identification and characterization of the adenosine 3',5'-cyclic monophosphate binding proteins appearing during the development of *Dictyostelium discoideum. Biochim. Biophys. Acta* 629:235–42
15. Darmon, M., Barra, J., Brachet, P. 1978. The role of phosphodiesterase in aggregation of *Dictyostelium discoideum. J. Cell Sci.* 31:233–43
16. Darmon, M., Barrand, P., Brachet, P., Klein, C., Pereira da Silva, L. 1977. Phenotypic suppression of morphogenetic mutants of *Dictyostelium discoideum. Devel. Biol.* 58:174–84
17. Darmon, M., Brachet, P. 1978. Chemotaxis and differentiation during the aggregation of *Dictyostelium discoideum* amoebae. In *Taxis and Behavior, Receptors and Recognition Ser. B*, ed. G. L. Hazelbauer, 5:101–39. London:Chapman & Hall. 341 pp.
18. Darmon, M., Brachet, P., Pereira da Silva, L. 1975. Chemotactic signals induce cell differentiation in *Dictyostelium discoideum. Proc. Natl. Acad. Sci. USA* 72:3163–66
19. Devreotes, P. N., Derstine, P. L., Steck, T. L. 1979. Cyclic 3',5'-AMP relay in *Dictyostelium discoideum*. I. A technique to monitor responses to controlled stimuli. *J. Cell Biol.* 80:291–99
20. Devreotes, P. N., Steck, T. L. 1979. Cyclic 3',5'-AMP relay in *Dictyostelium discoideum*. II. Requirements for the initiation and termination of the response. *J. Cell Biol.* 80:300–9
21. Dicou, E. L., Brachet, P. 1979. Multiple forms of an extracellular cyclic-AMP phosphodiesterase from *Dictyostelium discoideum. Biochem. Biophys. Acta* 578:232–42
22. Dicou, E., Brachet, P. 1979. Purification of the inhibitor of the 3'-5' cyclic AMP phosphodiesterase of *Dictyostelium discoideum* by affinity chromatography. *Biochem. Biophys. Res. Commun.* 90:1321–27
23. Dicou, E., Brachet, P. 1980. A separate phosphodiesterase for the hydrolysis of cyclic guanosine 3',5'-monophosphate in growing *Dictyostelium discoideum* amoebae. *Eur. J. Biochem.* 109:507–14
24. Dinauer, M. C., Mackay, S. A., Devreotes, P. N. 1980. Cyclic 3',5'-AMP relay in *Dictyostelium discoideum*. III. The relationship of cAMP synthesis and secretion during the cAMP signaling response. *J. Cell Biol.* 86:537–44
25. Dinauer, M. C., Steck, T. L., Devreotes, P. N. 1980. Cyclic 3',5'-AMP relay in *Dictyostelium discoideum*. IV. Recovery of the cAMP signaling response after adaptation to cAMP. *J. Cell Biol.* 86:545–53
26. Dinauer, M. C., Steck, T. L., Devreotes, P. N. 1980. Cyclic 3',5'-AMP relay in *Dictyostelium discoideum*. V. Adaptation of the cAMP signaling response during cAMP stimulation. *J. Cell Biol.* 86:554–61
27. Eitle, E., Gerisch, G. 1977. Implication of developmentally regulated concanavalin A binding proteins of *Dictyostelium* in cell adhesion and cyclic AMP regulation. *Cell Diff.* 6:339–46
28. Franke, J., Kessin, R. H. 1981. The cyclic nucleotide phosphodiesterase inhibitory protein of *Dictyostelium discoideum. J. Biol. Chem.* 256:7628–37
29. Futrelle, R. P., McKee, W. G., Traut, J. 1980. Response of *Dictyostelium dis-

coideum to localized cAMP stimuli; computer analysis of cell motion. *J. Cell Biol.* 87:CI415 (Abstr.)
30. Geller, J. S., Brenner, M. 1978. Measurements of metabolites during cAMP oscillations of *Dictyostelium discoideum*. *J. Cell. Physiol.* 97:413–19
31. Gerisch, G. 1964. *Dictyostelium minutum (Acrasina).* Aggregation. *Encyclopaedia Cinematographica E673/1964.* Göttingen: Inst. Wissenschaft. Film
32. Gerisch, G. 1965. Stadienspezifische Aggregationsmuster bei *Dictyostelium discoideum*. *Roux'Arch. Entwicklungsmech.* 156:127–144
33. Gerisch, G. 1979. Control circuits in cell aggregation and differentiation of *Dictyostelium discoideum*. In *Mechanisms of Cell Change*, ed. J. D. Ebert, T. Okada, pp. 225–239. VIIIth Congr. Int. Soc. Devel. Biol., Tokyo, 1977. NY:Wiley. 343 pp.
34. Gerisch, G. 1980. *Periodische Enzymaktivierung als Kontrollfaktor Multizellulärer Entwicklung.* Rheinisch-Westfälische Akad. Wissenschaft. Opladen: Westdeutscher Verlag. pp. 7–38
35. Gerisch, G., Fromm, H., Huesgen, A., Wick, U. 1975. Control of cell-contact sites by cyclic AMP pulses in differentiating *Dictyostelium* cells. *Nature* 255:547–49
36. Gerisch, G., Hess, B. 1974. Cyclic-AMP-controlled oscillations in suspended *Dictyostelium* cells: their relation to morphogenetic cell interactions. *Proc. Natl. Acad. Sci. USA* 71:2118–22
37. Gerisch, G., Hülser, D., Malchow, D., Wick, U. 1975. Cell communication by periodic cyclic-AMP pulses. *Philos. Trans. R. Soc. Lond. Ser. B.* 272:181–92
38. Gerisch, G., Maeda, Y., Malchow, D., Roos, W., Wick, U., Wurster, B. 1977. Cyclic AMP signals and the control of cell aggregation in *Dictyostelium discoideum*. See Ref. 11 pp. 105–24
39. Gerisch, G., Malchow, D. 1976. Cyclic AMP receptors and the control of cell aggregation in *Dictyostelium*. *Adv. Cyc. Nucl. Res.* 7:49–68
40. Gerisch, G., Malchow, D., Huesgen, A., Nanjundiah, V., Roos, W., Wick, U., Hülser, D. 1975. Cyclic AMP reception and cell recognition in *Dictyostelium discoideum*. In *ICN-UCLA Symp. Devel. Biol.*, ed. D. McMahon, C. F. Fox, pp. 76–88. Menlo Park, Calif:Benjamin. 604 pp.
41. Gerisch, G., Malchow, D., Roos, W., Wick, U. 1979. Oscillations of cyclic nucleotide concentrations in relation to the excitability of *Dictyostelium discoideum* cells. *J. Exp. Biol.* 81:33–47
42. Gerisch, G., Wick, U. 1975. Intracellular oscillations and release of cyclic AMP from *Dictyostelium* cells. *Biochem. Biophys. Res. Commun.* 65:364–70
43. Gil, M. G., Alemany, S., Cao, D. M., Castano, J. G., Mato, J. M. 1980. Calmodulin modulates phospholipid methylation in *Dictyostelium discoideum*. *Biochem. Biophys. Res. Commun.* 94:1325–30
44. Goldbeter, A. 1975. Mechanism for oscillatory synthesis of cyclic AMP in *Dictyostelium discoideum*. *Nature* 253:540–42
45. Goldbeter, A., Segel, L. A. 1977. Unified mechanism for relay and oscillation of cyclic AMP in *Dictyostelium discoideum*. *Proc. Natl. Acad. Sci. USA* 74:1543–47
46. Goldbeter, A., Segel, L. A. 1980. Control of developmental transitions in the cyclic AMP signalling system of *Dictyostelium discoideum*. *Differentiation* 17:127–35
47. Green, A. A., Newell, P. C. 1975. Evidence for the existence of two types of cAMP-binding sites in aggregating cells of *Dictyostelium discoideum*. *Cell* 6:129–36
48. Gross, J. D., Peacey, M. J., Trevan, D. J. 1976. Signal emission and signal propagation during early aggregation in *Dictyostelium discoideum*. *J. Cell Sci.* 22:645–56
49. Henderson, E. J. 1975. The cyclic adenosine 3',5'-monophosphate receptor of *Dictyostelium discoideum*. *J. Biol. Chem.* 250:4730–36
50. Hintermann, R., Parish, R. W. 1979. The intracellular location of adenylyl cyclase in the cellular slime molds *Dictyostelium discoideum* and *Polysphondylium pallidum*. *Exp. Cell Res.* 123:429–34
51. Hirata, F., Axelrod, J. 1978. Enzymatic methylation of phosphatidylethanolamine increases erythrocyte membrane fluidity. *Nature* 275:219–20
52. Hirata, F., Strittmatter, W. J., Axelrod, J. 1979. β-Adrenergic receptor agonists increase phospholipid methylation, membrane fluidity, and β-adrenergic receptor-adenylate cyclase coupling. *Proc. Natl. Acad. Sci. USA* 76:368–72
53. Jacobson, A., Lodish, H. F. 1975. Genetic control of development of the cellular slime mold *Dictyostelium discoideum*. *Ann. Rev. Genet.* 9:145–85

54. Juliani, M. H., Klein, C. 1981. Photoaffinity labeling of the cell surface adenosine 3':5'-monophosphate receptor of *Dictyostelium discoideum* and its modification in down-regulated cells. *J. Biol. Chem.* 256:613–19
55. Kakebeeke, P. I. J., De Witt, R. J. W., Konijn, T. M. 1980. Folic acid deaminase activity during development in *Dictyostelium discoideum*. *J. Bacteriol.* 143:307–12
56. Kakebeeke, P. I. J., Mato, J. M., Konijn, T. M. 1978. Purification and preliminary characterization of an aggregation-sensitive chemoattractant of *Dictyostelium minutum*. *J. Bacteriol.* 133:403–5
57. Kessin, R. H., Orlow, S. J., Shapiro, R. I., Franke, J. 1979. Binding of inhibitor alters kinetic and physical properties of extracellular cyclic AMP phosphodiesterase from *Dictyostelium discoideum*. *Proc. Natl. Acad. Sci. USA* 76:5450–54
58. King, A. C., Frazier, W. A. 1977. Reciprocal periodicity in cyclic AMP binding and phosphorylation of differentiating *Dictyostelium discoideum* cells. *Biochem. Biophys. Res. Commun.* 78:1093–99
59. King, A. C., Frazier, W. A. 1979. Properties of the oscillatory cAMP binding component of *Dictyostelium discoideum* cells and isolated plasma membranes. *J. Biol. Chem.* 254:7168–76
60. Klein, C. 1975. Induction of phosphodiesterase by cyclic adenosine 3',5'-monophosphate in differentiating *Dictyostelium discoideum*. *J. Biol. Chem.* 250:7134–38
61. Klein, C. 1979. The effects of inhibitors of adenylate cyclase and phosphodiesterase on *D. discoideum* aggregation. *Exp. Cell Res.* 124:205–13
62. Klein, C. 1979. A slowly dissociating form of the cell surface cyclic adenosine 3':5'-monophosphate receptor of *Dictyostelium discoideum. amoeba. J. Biol. Chem.* 254:12573–78
63. Klein, C., Juliani, M. H. 1977. cAMP-induced changes in cAMP-binding sites on *D. discoideum* amoebae. *Cell* 10:329–35
64. Konijn, T. M., van de Meene, J. G. C., Bonner, J. T., Barkley, D. S. 1967. The acrasin activity of adenosine-3',5'-cyclic phosphate. *Proc. Natl. Acad. Sci. USA* 58:1152–54
65. Kuczmarski, E. R., Spudich, J. A. 1980. Regulation of myosin self-assembly: Phosphorylation of *Dictyostelium* heavy chain inhibits formation of thick filaments. *Proc. Natl. Acad. Sci. USA* 77:7292–96
66. Loomis, W. F. 1975. *Dictyostelium discoideum. A Developmental System.* NY: Academic. 214 pp.
67. Loomis, W. F. 1979. Biochemistry of aggregation in *Dictyostelium*. A review. *Devel. Biol.* 70:1–12
68. Maeda, Y., Gerisch, G. 1977. Vesicle formation in *Dictyostelium discoideum* cells during oscillations of cAMP synthesis and release. *Exp. Cell Res.* 110:119–26
69. Malchow, D., Böhme, R., Rahmsdorf, H.-J. 1981. Regulation of phosphorylation of myosin heavy chain during the chemotactic response of *Dictyostelium* cells. *Eur. J. Biochem.* 117:213–18
70. Malchow, D., Gerisch, G. 1974. Short-term binding and hydrolysis of cyclic 3',5'-adenosine monophosphate by aggregating *Dictyostelium* cells. *Proc. Natl. Acad. Sci. USA* 71:2423–27
71. Malchow, D., Nägele, B., Schwarz, H., Gerisch, G. 1972. Membrane-bound cyclic AMP phosphodiesterase in chemotactically responding cells of *Dictyostelium discoideum*. *Eur. J. Biochem.* 28:136–42
72. Malchow, D., Nanjundiah, V., Gerisch, G. 1978. pH oscillations in cell suspensions of *Dictyostelium discoideum:* their relation to cyclic-AMP signals. *J. Cell Sci.* 30:319–30
73. Malchow, D., Nanjundiah, V., Wurster, B., Eckstein, F., Gerisch, G. 1978. Cyclic AMP-induced pH changes in *Dictyostelium discoideum* and their control by calcium. *Biochim. Biophys. Acta* 538:473–80
74. Maruta, H., Baltes, W., Gerisch, G., Dieter, P., Marmé, D. 1981. Signal transduction in chemotaxis of *Dictyostelium discoideum:* Role of Ca^{2+} and calmodulin in the regulation of myosin heavy chain kinases and other protein kinases. In *Higher Plant Membranes*, ed. D. Marmé. NY:Elsevier, North-Holland, Biochemical Press
75. Maruta, H., Korn, E. D. 1977. *Acanthamoeba* cofactor protein is a heavy chain kinase required for actin activation of the Mg^{2+}-ATPase activity of *Acanthamoeba* myosin I. *J. Biol. Chem.* 252:8329–32
76. Mato, J. M. 1979. Activation of *Dictyostelium discoideum* guanylate cyclase by ATP. *Biochem. Biophys. Res. Commun.* 88:569–74
77. Mato, J. M., Jastorff, B., Morr, M., Konijn, T. M. 1978. A model for cyclic AMP-chemoreceptor interaction in

78. Mato, J. M., Konijn, T. M. 1975. Chemotaxis and binding of cyclic AMP in cellular slime molds. *Biochim. Biophys. Acta* 385:173–79
79. Mato, J. M., Konijn, T. M. 1977. Chemotactic signal and cyclic GMP accumulation in *Dictyostelium*. See Ref. 11, pp. 93–103
80. Mato, J. M., Konijn, T. M. 1979. Chemosensory transduction in *Dictyostelium discoideum*. In *Biochemistry and Physiology of Protozoa*, ed. M. Levandowski, S. H. Hutner, pp. 181–219. NY:Academic
81. Mato, J. M., Krens, F. A., van Haastert, P. J. M., Konijn, T. M. 1977. 3':5'-cyclic AMP-dependent 3':5'-cyclic GMP accumulation in *Dictyostelium discoideum*. *Proc. Natl. Acad. Sci. USA* 74:2348–51
82. Mato, J. M., Losada, A., Nanjundiah, V., Konijn, T. M. 1975. Signal input for a chemotactic response in the cellular slime mold *Dictyostelium discoideum*. *Proc. Natl. Acad. Sci. USA* 72:4991–93
83. Mato, J. M., Malchow, D. 1978. Guanylate cyclase activation in response to chemotactic stimulation in *Dictyostelium discoideum*. *FEBS Lett.* 90:119–22
84. Mato, J. M., Marín-Cao, D. 1979. Protein and phospholipid methylation during chemotaxis in *Dictyostelium discoideum* and its relationship to calcium movements. *Proc. Natl. Acad. Sci. USA* 76:6106–9
85. Mato, J. M., van Haastert, P. J. M., Krens, F. A., Konijn, T. M. 1977. An acrasin-like attractant from yeast extract specific for *Dictyostelium lacteum*. *Devel. Biol.* 57:450–53
86. Mato, J. M., Woelders, H., Konijn, T. M. 1979. Intracellular cyclic GMP-binding proteins in cellular slime molds. *J. Bacteriol.* 137:169–72
87. Mato, J. M., Woelders, H., van Haastert, P. J. M., Konijn, T. M. 1978. Cyclic GMP binding activity in *Dictyostelium discoideum*. *FEBS Lett.* 90:261–64
88. Matsukuma, S., Durston, A. J. 1979. Chemotactic cell sorting in *Dictyostelium discoideum*. *J. Embryol. Exp. Morphol.* 50:243–51
89. Mullens, I. A., Newell, P. C. 1978. cAMP binding to cell surface receptors of *Dictyostelium*. *Differentiation* 10:171–76
90. Nanjundiah, V., Hara, K., Konijn, T. M. 1976. The effect of temperature on morphogenetic oscillations in *Dictyostelium discoideum*. *Nature* 260:705
91. Newell, P. C. 1977. Aggregation and cell surface receptors in cellular slime molds. In *Microbial Interactions, Receptors and Recognition Ser. B*, ed. J. L. Reissig, 3:1–57. London:Chapman and Hall. 436 pp.
92. Newell, P. C. 1978. Genetics of the cellular slime molds. *Ann. Rev. Genet.* 12:69–93
93. O'Day, D. H., Durston, A. J. 1979. Evidence for chemotaxis during sexual development in *Dictyostelium discoideum*. *Can. J. Microbiol.* 25:542–44
94. Pan, P., Hall, E. M., Bonner, J. T. 1972. Folic acid as second chemotactic substance in the cellular slime moulds. *Nature* 237:181–182
95. Pan, P., Hall, E. M., Bonner, J. T. 1975. Determination of the active portion of the folic acid molecule in cellular slime mold chemotaxis. *J. Bacteriol.* 122:185–91
96. Pan, P., Wedner, H. J. 1979. Immunohistochemical localization of cyclic GMP in aggregating *Polysphondylium violaceum*. *Differentiation* 14:113–18
97. Pan, P., Wurster, B. 1978. Inactivation of the chemoattractant folic acid by cellular slime molds and identification of the reaction product. *J. Bacteriol.* 136:955–59
98. Peltz, G., Kuczmarski, E. R., Spudich, J. A. 1981. *Dictyostelium* myosin: characterization of chymotryptic fragments and localization of the heavy-chain phosphorylation site. *J. Cell Biol.* 89:104–8
99. Potel, M. J., Mackay, S. A. 1979. Preaggregative cell motion in *Dictyostelium*. *J. Cell Sci.* 36:281–309
100. Rahmsdorf, H. J., Gerisch, G. 1978. Specific binding proteins for cyclic AMP and cyclic GMP in *Dictyostelium discoideum*. *Cell Diff.* 7:249–57
101. Rahmsdorf, H. J., Malchow, D., Gerisch, G. 1978. Cyclic AMP-induced phosphorylation in *Dictyostelium* of a polypeptide comigrating with myosin heavy chains. *FEBS Lett.* 88:322–26
102. Riedel, V., Gerisch, G., Müller, E., Beug, H. 1973. Defective cyclic adenosine-3',5'-phosphate-phosphodiesterase regulation in morphogenetic mutants of *Dictyostelium discoideum*. *J. Mol. Biol.* 74:573–85
103. Robertson, A., Drage, D. J. 1975. Stimulation of late interphase *Dictyostelium discoideum* amoebae with an ex-

ternal cyclic AMP signal. *Biophys. J.* 15:765–75
104. Robertson, A., Drage, D. J., Cohen, M. H. 1972. Control of aggregation in *Dictyostelium discoideum* by an external periodic pulse of cyclic adenosine monophosphate. *Science* 175:333–35
105. Roos, W., Gerisch, G. 1976. Receptor-mediated adenylate cyclase activation in *Dictyostelium discoideum. FEBS Lett.* 68:170–72
106. Roos, W., Malchow, D., Gerisch, G. 1977. Adenylyl cyclase and the control of cell differentiation in *Dictyostelium discoideum. Cell Diff.* 6:229–39
107. Roos, W., Nanjundiah, V., Malchow, D., Gerisch, G. 1975. Amplification of cyclic-AMP signals in aggregating cells of *Dictyostelium discoideum. FEBS Lett.* 53:139–42
108. Roos, W., Scheidegger, C., Gerisch, G. 1977. Adenylate cyclase activity oscillations as signals for cell aggregation in *Dictyostelium discoideum. Nature* 266: 259–61
109. Rossier, C., Eitle, E., van Driel, R., Gerisch, G. 1980. Biochemical regulation of cell development and aggregation in *Dictyostelium discoideum.* In *The Eucaryotic Microbial Cell,* ed. G. W. Gooday, D. Lloyd, A. P. J. Trinci, pp. 405–24. Soc. Gen. Microbiol. Symp. 30. London:Cambridge Univ. Press. 439 pp.
110. Rossier, C., Gerisch, G., Malchow, D., Eckstein, F. 1978. Action of a slowly hydrolysable cyclic AMP analogue on developing cells of *Dictyostelium discoideum. J. Cell Sci.* 35:321–38
111. Schmidt, W., Thomson, K., Butler, W. L. 1977. Cytochrome b in plasma membrane enriched fractions from several photoresponsive organisms. *Photochem. Photobiol.* 26:407–11
112. Shaffer, B. M. 1962. The Acrasina. *Adv. Morphogen.* 2:109–82
113. Shaffer, B. M. 1975. Secretion of cyclic AMP induced by cyclic AMP in the cellular slime mould *Dictyostelium discoideum. Nature* 255:549–52
114. Sternfeld, J., David, C. N. 1981. Oxygen gradients cause pattern orientation in *Dictyostelium* cell clumps. *J. Cell Sci.* 50:9–17
115. Thilo, L., Vogel, G. 1980. Kinetics of membrane internalization and recycling during pinocytosis in *Dictyostelium discoideum. Proc. Natl. Acad. Sci. USA* 77:1015–19
116. Tomchik, K. J., Devreotes, P. N. 1981. Adenosine 3',5'-monophosphate waves in *Dictyostelium discoideum:* A demonstration by isotope dilution-fluorography. *Science* 212:443–446
117. Toorchen, D., Henderson, E. J. 1979. Characterization of multiple extracellular cAMP-phosphodiesterase forms in *Dictyostelium discoideum. Biochem. Biophys. Res. Commun.* 87:1168–75
118. Tsang, A. S., Coukell, M. B. 1977. The regulation of cyclic AMP-phosphodiesterase and its specific inhibitor by cyclic AMP in *Dictyostelium. Cell Diff.* 6:75–84
119. Tsang, A. S., Coukell, M. B. 1979. Biochemical and genetic evidence for two extracellular adenosine 3':5'-monophosphate phosphodiesterases in *Dictyostelium purpureum. Eur. J. Biochem.* 95:407–17
120. Tsang, A. S., Coukell, M. B. 1979. Direct evidence for extracellular adenosine 3':5'-monophosphate phosphodiesterase induction and phosphodiesterase inhibitor repression by exogenous adenosine 3':5'-monophosphate in *Dictyostelium purpureum. Eur. J. Biochem.* 95:419–25
121. Van Driel, R. 1981. Binding of the chemoattractant folic acid by *Dictyostelium discoideum* cells. *Eur. J. Biochem.* 115: 391–95
122. Varnum, B., Soll, D. R. 1981. Chemoresponsiveness to cAMP and folic acid during growth, development, and dedifferentiation in *Dictyostelium discoideum. Differentiation* 18:151–60
123. Wallace, L. J., Frazier, W. A. 1979. Photoaffinity labeling of cyclic-AMP- and AMP-binding proteins of differentiating *Dictyostelium discoideum* cells. *Proc. Natl. Acad. Sci. USA* 76:4250–54
124. Ward, A., Brenner, M. 1977. Guanylate cyclase from *Dictyostelium discoideum. Life Sci.* 21:997–1008
125. Wick, U. 1978. Änderungen der intra- und extrazellulären Konzentrationen von zyklischem AMP, zyklischem GMP, sowie von Calcium während der Bildung von Aggregationssignalen bei Dictyostelium discoideum. PhD thesis. Univ. Tübingen, Germany. 93 pp.
126. Wick, U., Malchow, D., Gerisch, G. 1978. Cyclic-AMP stimulated calcium influx into aggregating cells of *Dictyostelium discoideum. Cell Biol. Int. Rep.* 2:71–79
127. Wurster, D. 1976. Temperature dependence of biochemical oscillations in cell suspensions of *Dictyostelium discoideum. Nature* 260:703–704
128. Wurster, B., Bozzaro, S., Gerisch, G. 1978. Cyclic GMP regulation and responses of *Polysphondylium violaceum*

to chemoattractants. *Cell Biol. Int. Rep.* 2:61–69
129. Wurster, B., Butz, U. 1980. Reversible binding of the chemoattractant folic acid to cells of *Dictyostelium discoideum*. *Eur. J. Biochem.* 109:613–18
130. Wurster, B., Pan, P., Tyan, G.-G., Bonner, J. T. 1976. Preliminary characterization of the acrasin of the cellular slime mold *Polysphondylium violaceum*. *Proc. Natl. Acad. Sci. USA* 73:795–99
131. Wurster, B., Schubiger, K. 1977. Oscillations and cell development in *Dictyostelium discoideum* stimulated by folic acid pulses. *J. Cell Sci.* 27:105–114
132. Wurster, B., Schubiger, K., Brachet, P. 1979. Cyclic GMP and cyclic AMP changes in response to folic acid pulses during cell development of *Dictyostelium discoideum*. *Cell Diff.* 8:235–42
133. Wurster, B., Schubiger, K., Wick, U., Gerisch, G. 1977. Cyclic GMP in *Dictyostelium discoideum*. Oscillations and pulses in response to folic acid and cyclic AMP signals. *FEBS Lett.* 76:141–44
134. Yeh, R. P., Chan, F. K., Coukell, M. B. 1978. Independent regulation of the extracellular cyclic AMP phosphodiesterase-inhibitor system and membrane differentiation by exogenous cyclic AMP in *Dictyostelium discoideum*. *Devel. Biol.* 66:361–74

LEUKOCYTE CHEMOTAXIS

Elliott Schiffmann

Laboratory of Developmental Biology and Anomalies, National Institute of Dental Research, National Institute of Health, Bethesda, Maryland 20205

Dedication: The author would like to dedicate this review to Professor David Shemin on the occasion of his 70th birthday.

INTRODUCTION

Chemotaxis, the directional migration of cells along a chemical gradient, is displayed by a variety of cells, including bacteria, the cellular slime molds, leukocytes, tumor cells, and neuronal cells (48). Chemotaxis has been demonstrated recently in differentiated cells such as fibroblasts (18, 45), endothelial cells (19), and smooth muscle cells (23). This motile response has been studied extensively in leukocytes, the major inflammatory cells, and shown to be a receptor-mediated process (2, 63), as in bacteria (4) and cellular slime molds (31). Studies using defined attractants have led to the discovery of biochemical events that may be obligatory for chemotaxis in phagocytic cells. I review some of these reactions and then indicate a possible sequence of reactions for the leukotactic response.

GENERAL CHARACTERISTICS OF LEUKOCYTE CHEMOTAXIS

I consider here a major phagocytic cell, the neutrophil, in which molecular events accompanying chemotaxis have been studied extensively.

The chemotactic response in leukocytes takes place on a substratum where the cells migrate toward a source of the attractant. This involves orientation of the cell, assumption of an asymmetric morphology, adherence, spreading, and then locomotion. Ultrastructural studies of migrating neutrophils have implicated both microtubules and microfilaments in chemotaxis (16). Chemoattractants stimulate assembly and organization of microtubules as well as the localization of microfilaments in the advancing cell. Colchicine, an agent that inhibits assembly of microtubules, causes a loss of these structures accompanied by decreased cell orientation and chemotaxis. Cytochalasin B, which inhibits assembly of microfilaments, also blocks migration but does not affect the chemoattractant-induced orientation. These observations indicate that microtubules may be necessary for orientation and direction of locomotion while microfilaments may be required for the contractile events.

The ability of the cell to orient by the use of its cytoskeletal elements may be involved in sensing a gradient of attractant. That is, the leukocyte appears to employ a *spatial* mechanism whereby it detects differences in concentrations of attractant across its dimensions (64). This implies a functional asymmetry of the cell, characterized, perhaps, by localization of receptors. This mode of sensing differs from that employed by bacteria, which use a *temporal* mechanism (30). These cells do not depend upon an existing gradient, but compare the level of attractant from a prior exposure to the present level. The detection of this difference induces a chemotactic response until adaptation to the present level occurs. However, a recent study (66) has suggested that the leukocyte may use a temporal sensing mechanism in addition to the spatial one. Locomoting cells in the presence of a uniform concentration of attractant show an asymmetric morphology. Upon addition of an increased level of attractant, these cells within seconds begin to round up and to exhibit surface ruffling with cessation of locomotion. After a delay that depends upon the magnitude of the change in concentration of attractant, the cells resume random locomotion. This transient behavior is characteristic of adaptation to a stimulus by a cell using a temporal sensing process.

In addition to displaying adaptive behavior, the cell becomes deactivated or desensitized when exposed continuously to levels of attractant that normally elicit a migratory response (61). After removal of the stimulus, the cell can recover, but this requires a longer period of time than the resumption of locomotion seen in adaptation (66). The two phenomena, however, may have a common feature: Changes in the availability of receptors may be involved (discussed below).

The migration of cells exposed to attractants is accompanied by a variety of other receptor-mediated events (48). These include chemiluminescence, active oxygen production, release of lysosomal enzymes, changes in membrane potential, decreased cell surface charge, increases in cationic fluxes, and internal redistribution of Ca^{2+}. The role of these in chemotaxis is not yet clear; I consider events likely to be involved in cell migration.

The interaction between attractant and surface receptor thus initiates a series of rapid, profound cellular alterations. These are accompanied by a large expenditure of energy known as the respiratory burst. The cell, therefore, appears to be 'primed' to carry out its major function of defending the host; first, by detecting a chemical signal from invading organisms such as bacteria, then by migrating to the infected area, and finally by phagocytizing the pathogens.

MEASUREMENT OF CHEMOTAXIS

Leukocyte chemotaxis can be measured in two ways: the tracking of single cells as they migrate toward a source of attractant, and observation of changes in the distribution of a population of cells as they migrate in a chemical gradient through a filter. The former assay is tedious, but modifications of it have been developed in which the orientation of cells to a stimulus, prior to migration, can be readily determined (64). The second method, such as the micropore filter assay involving the Boyden chamber, is more versatile and allows one to distinguish between chemokinetic, or nondirected, and chemotactic motility. This is accomplished by observing cellular migration in both the presence and absence of various gradients of an attractant (65). This technique has been improved for speed and accuracy by the use of computer-assisted image analysis of microscopic fields of migrated cells (24).

BIOCHEMICAL CHARACTERISTICS OF LEUKOCYTE CHEMOTAXIS

Chemotaxis may be characterized by three phases: an initial, or sensory, phase, in which a signal is generated by the interaction of the attractant and its receptor; an intermediate phase, in which the signal is processed to the cell's motility elements; and a terminal, or effector, phase in which the motility apparatus (both microtubules and microfilaments) is activated to produce directional migration.

The Sensory, or Recognition, Phase

THE CHEMOATTRACTANTS A great variety of substances attract leukocytes. The well-defined attractants include the complement-derived C5a (13), formylated peptides (47, 51), and derivatives of arachidonic acid (58). C5a and a formylated peptide each deliver a signal to the cell through a specific receptor (2, 11, 64).

C5a, a peptide of 74 amino acid residues, binds to a high-affinity receptor ($K_d \sim 5 \times 10^{-9}M$) (11). Removal of the C-terminal arginine resulted in diminished activity, but loss of the C-terminal pentapeptide (-Met-Gln-Leu-Gly-Arg) abolished activity. Neither this pentapeptide by itself nor it in noncovalent combination with the C5a [1–69] fragment was active. C5a [1–69] was, however, inhibitory. Furthermore, the affinities in binding to receptor of C5a and its derivatives corresponded to their efficacies in eliciting a chemotactic response: C5a > C5a des Arg > C5a [1–69]. The pentapeptide [res. 70–74] does not bind to the receptor. It was concluded that C5a possessed both a 'recognition' site within the C5a [1–69] portion of the molecule and an 'initiation site' in the C-terminal pentapeptide portion.

The formylated peptides have been widely used as probes to study molecular events in leukotaxis. They are related to bacterially derived leukoattractants (48), which themselves might be signal peptides that are released from N-terminal regions of newly synthesized proteins as these are transferred across or inserted into the cell membrane (10).

Structure-activity studies (15, 51) have demonstrated great specificity in the most active compounds. In a series of formylated tripeptides, exemplified by FMet-Leu-Phe, the presence of an aromatic residue in the terminal position confers much greater potency ($\sim 10^{-10}$M) on the molecule than a polar residue ($\sim 10^{-7}$M). Shifting the aromatic residue to the second position also results in a marked decrease in activity ($\sim 10^{-8}$M). The formyl group is uniquely responsible for high potency since either the absence of the formyl group or the neutralization of the charge of the α-amino group of N-terminal Met in alternative ways (acetylation, replacement with either a desamino residue or the isosteric 2-ethylhexanoyl moiety) produces much weaker agonists ($\sim 10^{-6}$M). It has recently been found that FMet-Leu-Phe has a preferred pleated sheet conformation in solution, as determined by nuclear magnetic resonance (5). This property may allow a unique fit of this highly potent formylated tripeptide with the receptor compared to the receptor complex with the nonacylated, weak agonist.

A study of variations in the side chain of the N-terminal parent residue norleucine, which can substitute for methionine, indicates that the most active compound is the one possessing a four carbon side chain (norleucine) while there is a corresponding decrease in activity with progressively smaller side chains. Branching of the four carbon side chain (isoleucine) does not reduce activity.

Substitution of the formyl group with the bulky t-butoxy-carbonyl (tBoc) resulted in weak ($\sim 10^{-7}$M) antagonists to the formylated peptides. Steric hindrance by the tBoc group may contribute to the inhibitory activity of these compounds.

The great specificity of the formylated peptides suggested that their effects were receptor-mediated. A high-affinity binding site ($K_d \sim 10^{-9}$M) was demonstrated on rabbit neutrophils and an estimate made of 10^5 sites per cell (2). The bacterially derived attractant, but not C5a, competed with formylated peptides for this receptor. The binding affinities and chemotactic potencies of a series of peptides were found to be closely correlated and in the same rank, indicating a physiologic role for the receptor. Similar receptors have been shown to exist on human neutrophils (63) and monocytes

(53). The use of affinity labeling procedures has led to isolation of a protein (∼70 kilodaltons) with characteristics of the human neutrophil receptor (38). A study (46) of the properties of the rabbit neutrophil receptor has shown a requirement for free sulfhydryl groups in binding peptides. Binding (4°) does not require energy or the presence of cations such as Na^+, K^+, Ca^{2+}, or Mg^{2+}. The pH maximum for binding is 6.5–7.0. In addition, brief exposure to proteases does not appreciably affect binding. These properties might well be an advantage to the cell that must operate under adverse conditions in inflammatory states.

Another class of well-defined potent leukoattractants consists of oxidized derivatives of arachidonic acid generated through the lipoxygenase pathway (Figure 1). These are the hydroxy-eicosatetraenoic acids, or HETEs. 5,12-di-HETE is active at about $10^{-9}M$, while 5-HETE and 11- and 12-HETE are much less potent (∼$10^{-6}M$). These compounds are synthesized in response to neutrophil stimulation by external attractants such as FMet-Leu-Phe (21). Their release from the cell may, therefore, serve to amplify the chemotactic response to the initial stimulus.

REGULATION OF ATTRACTANT BINDING TO RECEPTOR Formylated peptides were shown to be hydrolyzed by neutrophils at rates that correlated with their chemotactic potencies (3). These findings suggested that protease activity might be involved in the dissociation of peptide attractant from receptor after the chemical signal had been generated. In this way, the binding sites would be freed to detect new molecules of attractant. The fate of the receptor-ligand complex is clearly a more complicated

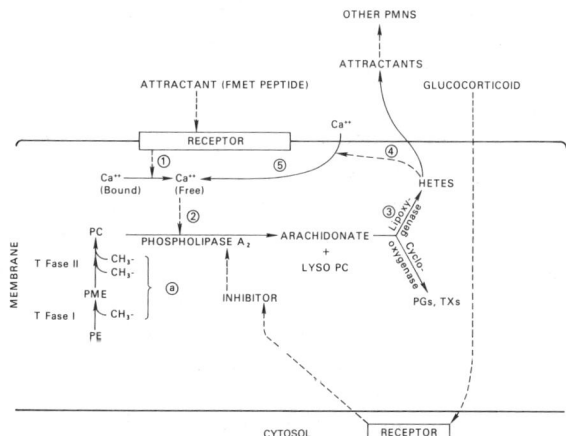

Figure 1 Molecular events in the leukocyte membrane that may be involved in chemotaxis (see text). TFase is methyl transferase. PE is phosphatidylethanolamine; PME is phosphatidylmonomethylethanolamine; PC is phosphatidylcholine; HETEs are hydroxy-eicosatetraenoic acids; PGs are prostaglandins; and TXs are thromboxanes.

process. Receptor-mediated endocytosis of radiolabeled formylated peptide occurs in a time- and temperature-dependent manner (41, 57, 60). The process is quite rapid at 24° and 37°. Within two minutes, more than half the labeled peptide taken up is no longer dissociable from the cell. After a longer interval, appreciable hydrolysis occurs, probably within the cell since cleavage products were found in the lysosomal fraction (60). In addition, visual evidence of internalization has been obtained with the aid of a fluorescent peptide (40).

Receptor-ligand internalization may reflect the phenomenon of "down-regulation," as measured by the decreased binding of labeled peptide after exposure of the cell to unlabeled peptide. This has been studied in both human and rabbit neutrophils, with reports of conflicting findings. For example, in one study (60) down-regulation occurred that could be reversed if the cells were allowed to recover by removing the chemotactic peptide and then incubating them at 37°. In another study (37), subsequent regain of binding capacity in down-regulated cells occurred poorly. Finally, in still another report (1), down-regulation was not observed at all. This could have resulted from the lengthy processing of the cells between their exposure to attractant and their use in the binding assay. Recovery may have occurred, obscuring an initial loss of receptors. The weight of evidence seems to support the conclusion that neutrophils can undergo down-regulation.

The progressive deactivation of cells during exposure to an attractant might, in part, be a consequence of down-regulation. Studies to test this point have indicated that no clear relationship exists between the number of available receptors and chemotactic responsiveness. For example, exposure of cells to complement-derived attractant caused a nonpreferential chemotactic deactivation (12); the cells lost responsiveness both to this stimulus and to the formylated peptide. Yet, the number of binding sites to the peptide was increased ("up-regulation"). Prior exposure of cells to peptide, however, produce a loss of binding sites to the peptide and a more marked depression of chemotaxis to the peptide than to the complement factor (preferential deactivation). "Up-regulation" also appears to occur during exposure of neutrophils to low concentrations of degranulating stimuli (ionophore, phorbol ester). These agents almost doubled the number of receptors and reduced apparent affinity of binding by about 20% (14).

If down-regulation, or ligand-receptor internalization, and chemotaxis are closely linked, the effects of inhibitors upon these two functions might be well correlated. In a test of this hypothesis, Niedel et al (39) found that internalization of peptide could be blocked by chymotryptic inhibitors. Yet, the levels of these reagents needed for this effect were much greater than those necessary to inhibit chemotaxis. In a related study, Warabi et al (in preparation) found that dansylcadaverine, a transglutaminase substrate shown to prevent internalization of ligands (epidermal growth factor) in fibroblasts, also inhibited internalization of formylated peptides in neutrophils, but, again, at a concentration greatly it excess of that required to block chemotaxis.

It appears, therefore, that "down-" or "up-regulation" of a receptor by a chemoattractant and the chemotactic response itself are not tightly coupled. Endocytotic and exocytotic events might play a role, however, in the adaptation of the cell to increasing levels of attractant as it moves along a gradient.

Processing the Chemotactic Signal

The interaction between chemoattractant and receptor generates a signal that is transmitted to the cytoskeleton, resulting in the directional movement of the cell. The molecular events that process this signal are not yet defined. However, certain receptor-mediated changes do occur that are rapid and transient and, therefore, might have a direct role in amplifying the chemical signal.

CHANGES IN MEMBRANE POTENTIAL   Rapid alterations in membrane potential are stimulated by chemoattractants (16). An initial depolarization is followed by a hyperpolarization. The latter, observed within one minute after stimulation, is accompanied by increased permeability to K^+. The presence of Ca^{2+} is required for an optimal response. These changes in membrane potential may not be required for chemotaxis since the cells from patients with chronic granulomatous disease (CGD) upon chemotactic stimulation exhibit little change in potential yet possess appreciable chemotactic responsiveness (60% of normal) (50). CGD cells are deficient in active oxygen production used in killing phagocytized bacteria. Normal changes in membrane potential may, therefore, be more related to efficient phagocytosis and its accompanying secretory events than to chemotaxis.

IONIC EVENTS   Chemotaxis, as measured in the micropore system, is optimal in the presence of Na^+, K^+, Ca^{2+}, and Mg^{2+}. Cells exposed to attractants undergo rapid alterations of monovalent cation fluxes and activation of a membrane-bound Na^+,K^+-ATPase (7). This reaction may function as a cation pump. The role of Ca^{2+} is not yet defined despite extensive studies. There appear to be at least two types of Ca^{2+} movement stimulated by attractants. These are an increased permeability to Ca^{2+}, as shown with $^{45}Ca^{2+}$ uptake studies (32), and a release of internally bound (perhaps to the membrane) cation to the cytoplasm (34), as shown by changes in fluorescence in cells treated with tetracycline. The changes in permeability could be blocked by inhibitors of arachidonic acid metabolism via the lipoxygenase pathway (33) (Figure 1), but the release of cellular Ca^{2+} internally could not. These changes are believed to have a significant role in chemotaxis. Becker has suggested (6) that an early event in processing the signal is the release of Ca^{2+} from the membrane. The rise in free Ca^{2+} may then activate phospholipase A_2, liberating arachidonate (Figure 1). Metabolites of arachidonate via the lipoxygenase pathway could then promote influx of Ca^{2+}, which is required for sustained locomotion in a manner not yet understood.

CHANGES IN CYCLIC NUCLEOTIDES A variety of studies have been performed to define a role in chemotaxis for cyclic nucleotides, the "second messengers" for many hormonally initiated responses. Agents that stimulate cyclic AMP (cAMP) formation inhibit chemotaxis while agents stimulating increases in cyclic GMP (cGMP) enhance the chemotactic response (25). The chemoattractants themselves, however, do not produce the expected changes in these nucleotides. For example, attractants induced greater than two-fold increases in cAMP. These changes were quite rapid and transient, occurring within 10 seconds after stimulation and disappearing after 5 minutes (52). No changes in cGMP were detected.

A possible role for cAMP was suggested in studies (62) showing a chemoattractant-induced phosphorylation of a 90-kilodalton protein, occurring within 5 seconds and decaying after 10 minutes. The phosphorylation step was blocked by inhibitors of the lipoxygenase pathway for arachidonate metabolism. The operation of the latter pathway may be required for chemotaxis (discussed below), but its activation by attractants occurs later than the stimulated changes in cAMP.

Certain characteristics of the increase in cAMP—i.e. it is rapid, transient, mediated by a receptor—make it an attractive candidate for an early event in processing the chemical signal. However, exposure of the cell to non-chemotactic ligands (isoproterenol, prostaglandin E) that activate adenyl cyclase depress chemotaxis as well as other attractant-induced functions (active oxygen production and lysosomal enzyme release). To explain this discrepancy it has been speculated (52) that the attractant-stimulated increases in cAMP are compartmentalized in the cell membrane, whereas increases in cAMP in the presence of certain pharmacological agents, for example, are "global." The former may be essential for chemotaxis while the latter may modulate it. However, it is still not clear whether either cyclic nucleotide changes or the formation of a phosphorylated protein are required for chemotaxis.

PROTEIN METHYLESTER METABOLISM The formation and hydrolysis of protein methyl esters play a role in processing the signal in bacterial chemotaxis (54). These reactions may also be involved in leukocyte chemotaxis. O'Dea et al (42) reported a rapid (30 sec to 1 min), transient stimulation of protein methylester formation in rabbit neutrophils exposed to formylated peptide in the presence of (^{3}H-methyl)methionine. The response appeared to be mediated by a receptor since a specific antagonist of formylated peptides abolished it. Also, the shape of the dose response curve for the methylation reaction correlated well with that for chemotaxis. In addition, chemoattractants stimulated hydrolysis of preformed methylesters (59). These findings suggested that a turnover of protein methylesters might be involved in chemotaxis, perhaps as a "second messenger." However, attractant-stimulated changes in protein methylesters have not yet been demonstrated in human neutrophils or macrophages. The great lability of these esters has hampered efforts to identify proteins that may participate in chemotaxis.

REACTIONS OF MEMBRANE PHOSPHOLIPIDS

Turnover of phospholipids Extensive studies have been performed on the role of lipid components in leukotaxis. Using agents that raise intracellular levels of S-adenosyl-homocysteine, a potent methyltransferase inhibitor, it was found that such compounds markedly reduced chemotaxis, phospholipid methylation, and protein methylester formation in monocytes (43). Similar results were obtained for neutrophils (49). In addition, these cells upon exposure to chemoattractants showed changes in phospholipid methylation (27). There occurred an apparent decrease in the transfer of [^{3}H]methyl group from labeled methionine to phospholipids. However, chemoattractants produced no changes in the levels of lipids synthesized through the CDP-choline pathway. Furthermore, this decreased methylation was well correlated with a rapid release of arachidonic acid, probably via the action of phospholipase A_2 upon phosphatidyl choline. Both this reaction and chemotaxis could be blocked by mepacrine, an inhibitor of phospholipase A_2. More recently, it has been shown (Bareis et al, submitted) that inhibition of methyltransferase reactions in neutrophils blocked the attractant-stimulated phospholipase A_2 activity but did not affect the stimulation of this enzymatic activity by the calcium ionophore A23187. Other receptor-mediated processes, such as release of histamine in mast cells and basophils, have similar properties (26). These findings suggest that a turnover of methylated phospholipids may play a significant role in leukocyte chemotaxis. Therefore, the relatively minor pathway in phospholipid formation, the successive methylation of phosphatidylethanolamine to produce phosphatidylcholine, appears to be considerably more important in certain cell functions than the major pathway of phospholipid synthesis via CDP-choline.

In macrophages, chemoattractants inhibit methylation of phospholipids but, unlike their effect in neutrophils, do not stimulate their degradation (44). Although there may be differences in lipid reactions between these two types of phagocytes, the findings with both cells are in accord with the suggestion that local changes in lipid components could alter membrane fluidity and contribute in a manner not yet clear to motile behavior.

Modulation of phospholipase A_2 Studies on the mechanism of the anti-inflammatory effects of glucocorticoids have provided further support for the role of phospholipase A_2 in leukocyte chemotaxis. These steroids induce the synthesis of an inhibitor of phospholipase in leukocytes and other tissues (8, 29). A 40-kilodalton protein has been isolated from neutrophils treated with glucocorticoids that inhibits both chemotaxis and phospholipase A_2 (30). A chemotactic response may depend in part upon the cell rendering its endogenous inhibitor inactive. Evidence to support this possibility lies in the finding that addition of Ca^{2+} to a mixture of the lipase and the inhibitor, lipomodulin, substantially restores the enzymatic activity (F. Hirata, submitted). In the intact cell, attractants stimulate the influx of Ca^{2+}, which might then "release" the enzyme as in the in vitro system. There are indications (25a) that chemoattractants stimulate the cell to phosphorylate lipomodulin in a Ca^{2+}-dependent reaction. There is evidence that the modified inhibitory protein is inactive. Therefore, processing the chemical signal may involve the inactivation of the phospholipase inhibitor, contributing to liberation of arachidonic acid and subsequent generation of the oxidized derivatives of this compound. Some of these derivatives may, in turn, be required for the chemotactic response.

Arachidonate metabolism Arachidonate liberated in stimulated cells triggers the energy-producing hexose monophosphate shunt and is converted to prostaglandins, and oxidized derivatives (9) such as 5-hydroxy-eicosatetraenoic acid, 5-HETE. This compound, formed via the lipoxygenase pathway, is itself an attractant. Inhibition of this pathway blocked chemotaxis. These findings suggested that oxidized derivatives of arachidonate may be involved in chemotaxis. The role of such intermediates in leukocyte chemotaxis has been clarified in part (20, 21). 5,12-di-HETE as well as 5-HETE were found to be potent attractants whose formation was induced by exposure of the cells to peptide attractants. Inhibitors of the lipoxygenase pathway blocked both chemotaxis and the formation of the HETEs. However, the inhibition of chemotaxis could be overcome by addition of 5- and 11-HETE. An indication of the role of these derivatives may be in findings (56) that ionophore-stimulated neutrophils produced 5-HETE and incorporated the latter into membrane phospholipids. Furthermore, 5-HETE stimulated a degranulation that was not blocked by lipoxygenase inhibitors. It was inferred that membrane changes induced by esterification of 5-HETE into lipids might be required for leukocyte functions. In addition, it was found that attractants stimulated a rapid, transient incorporation of labeled arachidonate into a neutrophil protein fraction (22). This was inhibited by blocking the lipoxygenase pathway. The results suggested that a derivative of arachidonate was used to form a membrane component involved in the chemotactic response. These findings suggest that lipoxygenase-produced metabolites of arachidonate may have two functions in chemotaxis: They could be released from the cell and as chemoattractants amplify the response induced by the external stimulus, and they may aid in signal processing at the membrane. The latter function is supported by the finding that lipoxygenase inhibitors suppressed chemoattractant-induced Ca^{2+} influx (33).

POSSIBLE REACTIONS IN PROCESSING THE CHEMICAL SIGNAL Figure 1 illustrates reactions at the membrane that may play a role in chemotaxis. The interaction of an attractant with a receptor generates a signal, step 1, which liberates bound Ca^{2+}. The free Ca^{2+} activates phospholipase A_2, step 2, which yields free arachidonate. This can be inhibited by agents such as mepacrine. The substrate for phospholipase A_2, phosphatidylcholine, is generated by methyltransferase reactions, step a. These can be inhibited by agents that cause a build-up of S-adenosylhomocysteine. The liberated arachidonate is converted to HETEs in step 3. The HETEs appear to have two functions: diffusion out of the cell to act as attractants for other cells in an amplification of chemotaxis, and action at the outer membrane, step 4, to increase permeability to Ca^{2+}. This influx of Ca^{2+} may enhance the phospholipase reaction, and may take part in cytoskeletal interactions (described below).

In addition, the increase in free Ca^{2+} (steps 1 and 4) may have another role: the inactivation of the endogenous steroid-induced inhibitor of phospholipase A_2. Such inactivation might occur via a Ca^{2+}-dependent phosphorylation.

Modulation of chemotaxis may occur through an endocytotic process, the internalization of the receptor-ligand complex, or "down regulation." The peptide attractant may be degraded and the receptor itself might undergo a similar fate or be reutilized. Alternatively, conditions leading to vigorous exocytosis, such as high levels of attractant, have been shown to cause a marked loss of receptors (17).

Effector Phase: The Cytoskeleton

Chemoattractants induce assembly of both actin-containing microfilaments and microtubules in cells. The studies of Stossel and co-workers have provided imaginative model systems for in vitro contractile events involving leukocyte actin and myosin (55). Polymerized actin interacting with myosin can stimulate the Mg^{2+}-dependent ATPase of the myosin. A number of cofactors appear to be involved. One of these, the actin-binding protein, cross-links actin, creating a "gel" lattice. The ATPase reaction is stimulated during gel formation, leading to contraction. Another factor, the protein gelsolin, in the presence of Ca^{2+}, breaks cross-links, favoring the "sol" transition. It has been suggested that a local influx of Ca^{2+}, stimulated by receptor-mediated events, initiates a focused dissolution of the actin lattice. The cytoplasm and the membrane of the cell move toward zones of the more highly cross-linked lattice as a result of the contraction generated by the myosin ATPase. This could lead to extension of a pseudopod by the cell. Thus a directional migration of the cell toward a source of attractant might be initiated. While these events have not yet been shown in vivo, the model provides an attractive role for Ca^{2+} at the membrane level.

The attractant-stimulated assembly of microtubules may be required for the vector of the cell's locomotion (16). Free sulfhydryls and Ca^{2+} may play a role in this assembly (36), but the reactions that transmit the chemical signal to these cytoskeletal elements are unknown. Recently it has been reported that chemoattractants stimulated in leukocytes a post-translational addition of tyrosine to the C-terminal residue of the α-tubulin chain (35). This was blocked by inhibitors at the receptor level and by agents that depress Ca^{2+} influx, methylation, and phospholipase A_2. The pattern of inhibition was similar to that observed in blockade of chemotaxis. Thus the tyrosylation reaction may prepare tubulin for proper assembly into microtubules during the motile response.

CONCLUDING REMARKS

The chemotactic response somewhat resembles hormonally induced cell behavior. A high-affinity receptor combines with a ligand, the attractant, to generate a chemical signal. It seems reasonable to expect that such an event may be localized in an area on the cell membrane since the response is a vectorial one. The chemical signal is amplified by mobilizing large amounts of energy accompanied by profound changes in the cell. A key reaction in this process may be the release of arachidonate by way of phospholipase A_2 in the membrane. The replenishment of the substrate for this reaction appears to occur through the "minor" path of phospholipid methylation. Therefore, the turnover of phospholipids may have a role as a "second messenger" in chemotaxis. It is apparently involved in a number of other receptor-mediated activities (26). An optimal chemotactic response requires the presence of Ca^{2+}. The receptor-mediated turnover of methylated phospholipids may prime the cell for its Ca^{2+}-initiated functions.

The neutrophil may be a good model for the study of rapid receptor-mediated processes. The profound changes it undergoes during chemotactic activation appear to be independent of de novo synthesis of macromolecules. The elucidation of the biochemical events that underlie chemotactic behavior in this relatively simple system may be applicable to motile responses in a variety of other animal cells as well as to chemosensory events generally.

Phagocytic cells are prime elements in host resitance to disease. However, in certain conditions such as allergy, the inflammatory response is inappropriate, leading to local derangement of homeostasis. The presence of lipomodulin, the steroid-induced phospholipase A_2 inhibitor, may play a crucial role here. Its inactivation may serve to "activate" the leukocyte, whether by an "appropriate" stimulus (bacterial infection) or an "inappropriate" one (allergen). In this regard, it has recently been shown (28) that high levels of autoantibodies to lipomodulin exist in the serum of patients with severe arthritis and lupus conditions. Their leukocytes may, therefore, be in a much more "inflammatory" state than those of normals. This suggests that homeostasis may in part depend upon the presence of normal levels of natural anti-inflammatory agents such as lipomodulin. It may be expected that further studies on the mechanism of leukocyte activation will contribute to understanding in this area of physiology.

ACKNOWLEDGMENT

The author would like to express his gratitude to Ms. Kathleen L. Moore for typing and editing the manuscript.

Literature Cited

1. Abita, J. P., Morgat, J. L. 1980. On the mechanism of polymorphonuclear leukocyte deactivation of chemotaxis by the peptide formylmethionyl-leucyl-phenylalanine. *FEBS Lett.* 111:14–17
2. Aswanikumar, S., Corcoran, B. A., Schiffmann, E., Day, A. R., Freer, R. J., Showell, H. J., Becker, E. L., Pert, C. B. 1977. Demonstration of a receptor on rabbit neutrophils for chemotactic peptides. *Biochem. Biophys. Res. Commun.* 74:810–17
3. Aswanikumar, S., Schiffmann, E., Corcoran, B. A., Wahl, S. M. 1976. Role of a peptidase in phagocyte chemotaxis. *Proc. Natl. Acad. Sci. USA* 73:2439–42
4. Adler, J. 1975. Chemotaxis in bacteria. *Ann. Rev. Biochem.* 44:341–56
5. Becker, E. L., Bleich, H. D., Day, A. R., Freer, R. J., Glasel, J. A., Visintainer, J. 1979. Nuclear magnetic resonance conformational studies on the chemotactic tripeptide formyl-L-methionyl-L-leucyl-L-phenylalanine. A small beta sheet. *Biochemistry* 18:4656–68
6. Becker, E. L., Stossel, T. P. 1980. Chemotaxis. *Fed. Proc.* 39:2949–52
7. Becker, E. L., Talley, J. V., Showell, H. J., Naccache, P. H., Sha'afi, R. I. 1978. Activation of rabbit PMN leukocyte membrane Na^+,K^+-ATPase by chemotactic factor. *J. Cell Biol.* 77:329–33
8. Blackwell, G. J., Flower, R. J. 1979. Anti-inflammatory steroids induce biosynthesis of a phospholipase A_2 inhibitor which prevents prostaglandin generation. *Nature* 278:456–59
9. Bokoch, G. M., Reed, P. W. 1980. Stimulation of arachidonic acid metabolism in the PMN leukocyte by an N-formylated peptide. *J. Biol. Chem.* 255:10223–26

10. Chang, C. N., Blobel, G., Model, P. 1978. Detection of prokaryotic signal peptidase in an *Escherichia coli* membrane fraction: Endoproteolytic cleavage of nascent fI pre-coat protein. *Proc. Natl. Acad. Sci. USA* 75:361-65
11. Chenoweth, D. E., Hugli, T. E. 1978. Demonstration of a specific C5a receptor on intact human polymorphonuclear leukocytes. *Proc. Natl. Acad. Sci. USA* 75:3943-47
12. Donabedian, H., Gallin, J. I. 1981. Deactivation of human neutrophil chemotaxis by chemoattractants. *J. Immunol.* 127:839-44
13. Fernandez, H. N., Hugli, T. E. 1978. Primary structural analysis of the polypeptide portion of human C5a anaphylatoxin. *J. Biol. Chem.* 253:6955-64
14. Fletcher, M. P., Gallin, J. I. 1980. Degranulating stimuli increase the availability of receptors on human neutrophils for the chemoattractant fMet-Leu-Phe. *J. Immunol.* 124:1585-88
15. Freer, R. J., Day, A. R., Radding, J. A., Schiffmann, E., Aswanikumar, S., Showell, H. J., Becker, E. L. 1980. Further studies on the structural requirements for synthetic peptide chemoattractants. *Biochemistry* 19:2404-10
16. Gallin, J. I., Gallin, E. K., Malech, H. L., Cramer, E. B. 1978. Structural and ionic events during leukocyte chemotaxis. In *Leukocyte Chemotaxis*, ed. J. I. Gallin, P. G. Quie, pp. 123-41. NY: Raven Press
17. Gallin, J. I., Wright, D. G., Schiffmann, E. 1978. Role of secretory events in modulating human neutrophil chemotaxis. *J. Clin. Invest.* 62:1364-74
18. Gauss-Miller, V., Kleinman, H. K., Martin, G. R., Schiffmann, E. 1980. Role of attachment factors and attractants in fibroblast chemotaxis. *J. Lab. Clin. Med.* 96:1071-80
19. Glaser, B. M., D'Amore, P. A., Seppa, H., Seppa, S., Schiffmann, E. 1980. Adult tissues contain chemoattractants for vascular endothelial cells. *Nature* 288:483-84
20. Goetzl, E. J. 1980. Role for endogenous monohydroxy-eicosatetraenoic acids (HETEs) in regulation of human neutrophil migration. *Immunology* 40:709-19
21. Goetzl, E. J., Pickett, W. C. 1980. Human PMN leukocyte chemotactic activity of complex hydroxy-eicosatetraenoic acids (HETEs). *J. Immunol.* 125:1789-91
22. Goldman, D. W., Goetzl, E. J. 1981. Novel metabolites of arachidonic acid in human neutrophils. *Fed. Proc.* 40:1004 (Abstr.)
23. Grotendorst, G. R., Seppa, H. E. J., Kleinman, H. K., Martin, G. R. 1981. Attachment of smooth muscle cells to collagen and their migration toward platelet-derived growth factor. *Proc. Natl. Acad. Sci. USA* 78:3669-72
24. Harvath, L., Falk, W., Leonard, E. J. 1980. Rapid quantitation of neutrophil chemotaxis: Use of a polyvinylpyrrolidone-free polycarbonate membrane in a multiwell assembly. *J. Immunol. Meth.* 37:39-45
25. Hill, H. R. 1978. Cyclic nucleotides as modulators of leukocyte chemotaxis. See Ref. 16, pp. 179-93
25a. Hirata, F. 1981. The regulation of lipomodulin, a phospholipase inhibitory protein, in rabbit neutrophils by phosphorylation. *J. Biol. Chem.* 256:7730-33
26. Hirata, F., Axelrod, J. 1980. Phospholipid methylation and biological signal transmission. *Science* 209:1082-90
27. Hirata, F., Corcoran, B. A., Venkatasubramanian, K., Schiffmann, E., Axelrod, J. 1979. Chemoattractants stimulate degradation of methylated phospholipids and release of arachidonic acid in rabbit leukocytes. *Proc. Natl. Acad. Sci. USA* 76:2640-43
28. Hirata, F., del Carmine, R., Nelson, C. A., Axelrod, J., Schiffmann, E., Warabi, H., DeBlas, A. L., Nirenberg, M., Manganiello, V., Vaughan, M., Kumagai, S., Green, E., Decker, J. L., Steinberg, A. D. 1981. Presence of autoantibody for phospholipase inhibitory protein, lipomodulin, in patients with rheumatic diseases. *Proc. Natl. Acad. Sci. USA* 78:3190-94
29. Hirata, F., Schiffmann, E., Venkatasubramanian, K., Salomon, D., Axelrod, J. 1980. A phospholipase A_2 inhibitory protein in rabbit neutrophils induced by glucocorticoids. *Proc. Natl. Acad. Sci. USA* 77:2533-36
30. MacNab, R. M., Koshland, D. E. Jr. 1972. Gradient-sensing mechanism in bacterial chemotaxis. *Proc. Natl. Acad. Sci. USA* 69:2509-12
31. Malchow, D., Gerisch, G. 1974. Short-term binding and hydrolysis of cyclic 3':5'-adenosine monophosphate by aggregating *Dictyostelium* cells. *Proc. Natl. Acad. Sci. USA* 71:2423-27
32. Naccache, P. H., Showell, H. J., Becker, E. L., Sha'afi, R. I. 1977. Transport of sodium, potassium, and calcium across rabbit polymorphonuclear leukocyte membranes. *J. Cell Biol.* 73:428-44

33. Naccache, P. H., Showell, H. J., Becker, E. L., Sha'afi, R. I. 1979. Pharmacological differentiation between chemotactic factor-induced calcium redistribution and transmembrane flux in rabbit neutrophils. *Biochem. Biophys. Res. Commun.* 89:1224–30
34. Naccache, P. H., Volpi, M., Showell, H. J., Becker, E. L., Sha'afi, R. I. 1979. Chemotactic factor-induced release of membrane calcium in rabbit neutrophils. *Science* 203:461–63
35. Nath, J., Flavin, M., Corcoran, B., Schiffmann, E. 1981. Stimulation of tubulin-tyrosylation in rabbit leukocytes evoked by the chemoattractant Met-Leu-Phe. *J. Cell Biol.* 91:232–39
36. Nath, J., Rebhun, L. I. 1976. Effects of caffeine and methylxanthines on the development and metabolism of sea urchin eggs. *J. Cell Biol.* 68:440–50
37. Nelson, R. D., Fiegel, V. D., Herron, M. J., Simmons, R. L. 1980. Chemotactic deactivation of human neutrophils. *J. Reticuloendothel. Soc.* 28:285–94
38. Niedel, J., Davis, J., Cuatrecasas, P. 1980. Covalent affinity labeling of the formyl peptide chemotactic receptor. *J. Biol. Chem.* 255:7063–66
39. Niedel, J., Frothingham, R., Cuatrecasas, P. 1980. Inhibition of ^{125}I-chemotactic peptide uptake by protease inhibitors. *Biochem. Biophys. Res. Commun.* 94:667–73
40. Niedel, J. E., Kahane, I., Cuatrecasas, P. 1979. Receptor-mediated internalization of fluorescent chemotactic peptide by human neutrophils. *Science* 205:1412–14
41. Niedel, J., Wilkinson, S., Cuatrecasas, P. 1979. Receptor-mediated uptake and degradation of ^{125}I-chemotactic peptide by human neutrophils. *J. Biol. Chem.* 254:10700–6
42. O'Dea, R. F., Viveros, O. H., Axelrod, J., Aswanikumar, S., Schiffmann, E., Corcoran, B. A. 1978. Rapid stimulation of protein carboxy-methylation in leukocytes by a chemotactic peptide. *Nature* 272:462–64
43. Pike, M. C., Kredich, N. M., Snyderman, R. 1978. Requirement of S-adenosylmethionine-mediated methylation for human monocyte chemotaxis. *Proc. Natl. Acad. Sci. USA* 75:3928–32
44. Pike, M. C., Kredich, N. M., Snyderman, R. 1979. Phospholipid methylation in macrophages is inhibited by chemotactic factors. *Proc. Natl. Acad. Sci. USA* 76:2922–26
45. Postlethwaite, A. E., Snyderman, R., Kang, A. H. 1976. The chemotactic attraction of human fibroblasts to a lymphocyte-derived factor. *J. Exp. Med.* 144:1188–1203
46. Schiffmann, E., Aswanikumar, S., Venkatasubramanian, K., Corcoran, B. A. Pert, C. B., Brown, J., Gross, E., Day, A. R., Freer, R. J., Showell, H. J., Becker, E. L. 1980. Some characteristics of the neutrophil receptor for chemotactic peptides. *FEBS Lett.* 117:1–7
47. Schiffmann, E., Corcoran, B. A., Wahl, S. M. 1975. N-formylmethionyl peptides as chemoattractants for leukocytes. *Proc. Natl. Acad. Sci. USA* 72:1059–62
48. Schiffmann, E., Gallin, J. I. 1979. Biochemistry of phagocyte chemotaxis. *Curr. Top. Cell. Regul.* 15:203–61
49. Schiffmann, E., O'Dea, R. F., Chiang, P. K., Venkatasubramanian, K., Corcoran, B., Hirata, F., Axelrod, J. 1979. Role for methylation in leukocyte chemotaxis. In *Modulation of Protein Function, ICN-UCLA Symposium,* ed. D. E. Atkinson, C. F. Fox, 13: 299–313. NY: Academic
50. Seligmann, B. E., Gallin, J. I. 1980. Use of lipophilic probes of membrane potential to assess human neutrophil activation. *J. Clin. Invest.* 66:493–503
51. Showell, H. J., Freer, R. J., Zigmond, S. H., Schiffmann, E., Aswanikumar, S., Corcoran, B., Becker, E. L. 1976. Structure-activity relations of synthetic peptides as chemoattractants and inducers of lysosomal enzyme secretion for neutrophils. *J. Exp. Med.* 143:1154–69
52. Simchowitz, L., Fischbein, L. C., Spilberg, I., Atkinson, J. P. 1980. Transient elevation in intracellular cyclic AMP by chemotactic factors. *J. Immunol.* 124:1482–91
53. Snyderman, R., Fudman, E. J. 1980. Demonstration of a chemotactic factor receptor on macrophages. *J. Immunol.* 124:2754–57
54. Springer, W. R., Koshland, D. E. Jr. 1977. Identification of a protein methyltransferase as the *cheR* gene product in the bacterial sensing system. *Proc. Natl. Acad. Sci. USA* 74:533–37
55. Stendahl, O. I., Stossel, T. P. 1980. Actin-binding protein amplifies actomyosin concentration, and gelsolin confers calcium control on direction of contraction. *Biochem. Biophys. Res. Commun.* 92:675–81
56. Stenson, W. F., Parker, C. W. 1980. Mono-eicosatetraenoic acids (HETEs) induce degranulation of human neutrophils. *J. Immunol.* 124:2100–4

57. Sullivan, S. J., Zigmond, S. H. 1980. Chemotactic peptide receptor modulation in PMN leukocytes. *J. Cell Biol.* 85:703–11
58. Turner, S. R., Tainer, J. A., Lynn, W. S. 1975. Biogenesis of chemotactic molecules by the arachidonate lipoxygenase system of platelets. *Nature* 257:680–1
59. Venkatasubramanian, K., Hirata, F., Gagnon, C., Corcoran, B. A., O'Dea, R. F., Axelrod, J., Schiffmann, E. 1980. Protein methylesterase and leukocyte chemotaxis. *Molec. Immunol.* 17:201–8
60. Vitkauskas, G., Showell, H. J., Becker, E. L. 1980. Specific binding of synthetic chemotactic peptides to rabbit peritoneal neutrophils. *Molec. Immunol.* 17:171–80
61. Ward, P. A., Becker, E. L. 1968. The deactivation of rabbit neutrophils by chemotactic factor and the nature of the activatable esterase. *J. Exp. Med.* 127:693–709
62. Wedner, H. J., Simchowitz, L., Atkinson, J., Stenson, W. 1980. Chemotactic factors induce rapid phosphorylation of a 90 kilodalton protein in human PMN leukocytes. *Fed. Proc.* 39:1950 (Abstr.)
63. Williams, L. T., Snyderman, R., Pike, M. C., Lefkowitz, R. J. 1977. Specific receptor sites for chemotactic peptides on human polymorphonuclear leukocytes. *Proc. Natl. Acad. Sci. USA* 74:1204–8
64. Zigmond, S. H. 1977. The ability of polymorphonuclear cells to orient in gradients of chemotactic factors. *J. Cell Biol.* 75:606–16
65. Zigmond, S. H., Hirsch, J. G. 1973. Leukocyte locomotion and chemotaxis: New Methods for evaluation and demonstration of cell-derived chemotactic factor. *J. Exp. Med.* 137:387–410
66. Zigmond, S. H., Sullivan, S. J. 1979. Sensory adaptation of leukocytes to chemotactic peptides. *J. Cell Biol.* 82:517–27

ENDOCRINOLOGY AND METABOLISM

Introduction, Dorothy T. Krieger, *Section Editor*

In the past decade, approximately thirty peptides have been described in brain. Many of them had previously been described in other loci—i.e. in the pituitary and in the gastrointestinal tract. It is increasingly evident that these peptides are synthesized independently in the various tissues in which they are found, and that their regulation, processing, and consequently their functions differ in their different sites of origin. Although progress has been made in the identification, characterization, and localization of these peptides in brain, much less is known about their function. However, a number of peptides evidently interact with neurotransmitters in the regulation of basic homeostatic systems.

In part one of this year's section on endocrinology and metabolism, two aspects of central nervous system regulation of such systems are reviewed. The first concerns those peptides that have been most *critically* assessed for a possible role in learning and behavior. Two other articles deal with central nervous system regulation of reproductive function. This field has become more complex since the discovery of gonadotropin-releasing hormone. Although the importance of this peptide in pituitary regulation is undeniable, it has become evident that changes in pituitary function in various phases of the reproductive cycle do reflect changes not in the level of gonadotropin-releasing hormone secretion alone but rather in pituitary responsiveness thereto, which, in turn, is governed by changes in gonadal secretion. Such effects of gonadal hormones are mediated at both the hypothalamic and pituitary levels. The frequency of secretory pulses of gonadotropin-releasing hormone governs the nature of the gonadotropin secreted by the pituitary —i.e. luteinizing hormone or follicle-stimulating hormone. Furthermore, the mechanisms involved in positive and negative feedback are not static but rather vary throughout the development of the organism. This raises the question of what other factors (presumably neural) are involved in the regulation of such feedback. Since the alterations in such feedback appear to be responsible for changes associated with puberty, future research in this field will further illuminate this basic biological stage.

Part two, the section on peptide hormones, is multi-faceted. We are learning that mammalian peptides are present not only in invertebrates but possibly in unicellular organisms as well. Taken together with the concept that peptides are synthesized in the form of large precursor molecules that are subsequently processed to their component peptide(s), this raises the possibility that the processing and functions of such peptides have differed in the course of evolution. New information has also accumulated on the mode of peptide hormone action. The valuable concept of membrane receptor binding of peptides, subsequent activation of adenylate cyclase, and eventual translation of this activation to a specific cellular response (mediated by the cAMP "second messenger" activation of a dependent protein kinase) does not explain certain peptide hormonal events. Evidence shows that hormone plasma membrane receptors form oligomeric complexes with nucleotide regulatory units that bind GTP and mediate either the stimulation or inhibition of adenylate cyclase by GTP. There is also evidence that peptide hormones and their receptors may be internalized, although the role of this mechanism in intracellular function is unclear. Additional evidence suggests that calcium can serve as a second messenger; the chapter on its cytosolic receptor (calmodulin) summarizes its known intracellular actions and its role in protein secretion and receptor endocytosis. Thus a number of loci may be responsible for a lack of cellular translation of peptide messengers. This lack can result in diseases, some of which have already been characterized.

BEHAVIORAL EFFECTS OF NEUROPEPTIDES: Endorphins and Vasopressin

George F. Koob and Floyd E. Bloom

A. V. Davis Center for Behavioral Neurobiology, Salk Institute, La Jolla, California 92037

INTRODUCTION

In the process of uncovering the full biological activity of almost every newly discovered neurotransmitter candidate, reports appear that the substance produces alterations of animal behavior after cerebroventricular or intracerebral injection. Neuropeptides are not an exception to this rule, and in fact appear to be rather favorite targets of such investigative thrusts. Generally, such early behavioral changes are accomplished with amounts of the material often larger than those contained in the entire nervous system. The resultant behaviors, generally a bizarre posture or abnormal movement pattern, have little obvious physiological significance. In contrast, subsequent behavioral tests often employ extremely small amounts of the materials to elicit behavioral actions after subcutaneous injection or, in virtually miniscule amounts, after intracerebroventricular injections. In these cases, the extreme potency of peptides in eliciting rather subtle behavioral phenomena tends to perpetuate the curiosity value of the effect without establishing its physiological value. In this review, we examine critically the reported behavioral actions of two families of neuropeptides—the endorphins (6) and the oxytocin-vasopressin peptides (66)—whose chemical, cytological, and cellular physiological actions have already been studied extensively (see 57). Other peptides have also been reported to have subtle, low-dose behavioral actions, especially non-opioid fragments of the pro-opiomelanocortin peptides [e.g. alpha MSH and corticotropin (see 24, 40)]. However, we focus here on the two series of studies with the most comprehensive documentation as of early 1981.

Two approaches are used to study the role of peptides in behavior. The first employs central or peripheral administration of the natural or synthetic peptide agonist to act directly on receptors whose location is unknown but generally regarded as intracranial. The second approach infers that an endogenous peptide has an effect on learning or some other behavior because the behavior is missing when the peptide system is destroyed, congenitally deficient, or susceptible to selective pharmacological antagonism. Both of these approaches require the same assumption: that the natural agonists and synthetic antagonists both act directly and specifically at the same peptide receptors.

Endorphins

Several different opioid receptor classifications have been reported in central and peripheral nervous systems based on relative effects of synthetic agonists and antagonists (19, 33, 51, 53, 56, 78). Although generally valid, the definition of opioid activity based on naloxone antagonism may not hold at high naloxone doses that also antagonize GABA (30). Behavioral effects of peripheral injections of peptides are generally not reversible by naloxone (23, 60).

Nevertheless, opiate alkaloids such as morphine and heroin share a variety of behavioral effects with the endogenous opioid peptides—e.g. analgesia, behavioral activation at low doses, sedation at high doses, and euphoria. Comparison of the behavioral actions of peptides with the stereospecific actions of opiate alkaloids and reversal by the specific antagonists naloxone or naltrexone could be used to establish response specificity.

LEARNING AND MEMORY

The role of endorphin peptides in altering behaviors thought to reflect memory and learning has been studied with respect to behaviors motivated by either aversive or appetitive conditions.

Endorphins and Aversively Motivated Behaviors

The role of opiates in pain perception has been known for centuries and it is therefore not surprising that opiates would disrupt the retention of pain-motivated tasks (4, 20, 73). The effect is least severe in highly overtrained animals and most severe when the drug is injected prior to each training session (4), presumably because opiates reduce the alarm response associated with a fear- or anxiety-motivated condition (32, 38). The suppression of the alarm or fear response by morphine is probably independent of its analgesic effects (4).

Because opiates can also produce memory deficits immediately after training (consolidation), the effect is consistent with a more cognitive-emotional effect (17). Retrograde amnesia after morphine has also been seen with inhibitory avoidance tests given parenterally (37) or intracerebroventricularly (68). Supporting the physiological role of an endorphin is the observation that naloxone increases apparent rentention both in an inhibitory avoidance task and in an active avoidance task (55). Post-training injections of the opiate agonist, levorphanol, into the amygdala of rats, an enkephalin-rich region (see 6, 56), produced a time-dependent and dose-dependent decrease in the latency to reenter the shock compartment in an inhibitory avoidance test (32). Post-training administration of naloxone into the amygdala significantly increased the latency to reenter the shock compartment, and the combination of levorphanol and naloxone produced no effect.

The effects of opiate-like peptides on aversive conditioning are even less clear, and to a large extent, the endorphin peptides produce effects opposite to those of opiate alkaloids. Thus endorphins attenuate amnesia in rats after systemic injection (59, 60, 62) when injected at low doses (10 μg/rat) subcutaneously at least 1 hr before the retention test. This dosage of beta-endorphin would be about 10-fold suprathreshold for analgesia after intracerebroventricular injections (7, 50), making its peripheral effectiveness in this test astounding. Furthermore, extraordinarily low doses of Met- and Leu-enkephalin (0.0003–30.0 μg/rat)—doses far below any detectable central effect (see 7)—produced results identical to those of beta-endorphin when injected subcutaneously 1 hr before either the acquisition or the retention test. This effect is not reversible by naloxone (60).

Although these results are consistent with an interpretation of an antiamnesia action for endorphins, broader hypotheses regarding changes in arousal, fear-motivation, or response to stress were not explored. Similar results and a similarly difficult interpretation can be found in the results of recent studies by De Wied and his colleagues (26, 27 29). Here peripheral injections of Met-enkephalin, beta-endorphin, and alpha-endorphin all delayed the extinction of a pole jump avoidance task at doses of less than 3.0 μg/rat when injected peripherally and at doses less than 0.1 μg/rat when injected intraventricularly (26); these peptides produce little lasting overt behavior in doses of 500 μg/rat (10). The receptor mechanism here remains unknown because although the opiate antagonist naltrexone produces an effect (i.e. facilitating extinction of the pole jump avoidance task) the antagonist does not block the delay in extinction produced by alpha-endorphin. Furthermore, this effect of naltrexone appears to be in direct contrast to the memory-enhancing effects of the closely related antagonist naloxone, above (55).

Even more curiously, gamma endorphin, only one amino acid longer than alpha-endorphin (see 65), also produces effects opposite to those of alpha endorphin [i.e. facililates the extinction of pole jump avoidance (29)]. This facilitation of extinction and the fact that Des-Tyr-γ-endorphin appeared to have "cataleptogenic" properties similar to those of haloperidol prompted De Wied to speculate that Des-Tyr-γ-endorphin could be an endogenous neuroleptic (26, 28). The differential effects of alpha- and gamma-endorphin on extinction of pole jump avoidance after peripheral injection have been replicated (47), but to date no one has observed cataleptogenic action of Des-Tyr-γ-endorphin (75) nor has Des-Tyr-γ-endorphin been isolated from rat brain in significant quantities.

However, in several other studies, the opposing effects of alpha- and gamma-endorphins on retention of inhibitory avoidance has been confirmed. Post-training or pre-retention treatment (1 hr) with alpha-endorphin significantly increased the response latency in a one-trial inhibitory avoidance task, whereas a similar treatment regime with Des-Tyr-γ-endorphin decreased the latency (28). Similar results in our laboratory showed an increase in latency in a one-trial inhibitory avoidance task after a 1 hr pre-retention treatment with alpha-endorphin, while gamma-endorphin produced the opposite effects (43). However, the time course of the training-testing protocol must be an important factor in determining (a) the mechanism of action of peripherally injected endorphins, and (b) the possibly independent sites of action of the enkephalins and the β-endorphin-derived peptides (62).

Endorphins and Appetitively Motivated Learning

Before concluding that endorphins or other peptides may function in memory operations, it is essential to generalize their effects from avoidance learning to appetitively motivated tasks. If the actions of the peptide were similar for both the aversive and appetitive tasks, then a primary learning or memory substrate might be more seriously proposed as a mechanism of action. Alternatively, if the effects of the peptides are task-specific, other mechanisms of action such as motivational variables may be more important.

In the first reported effect of endorphins on any learning task (the only study using appetitively motivated learning), Met^5-enkephalin and a potent analgesic analog [D-Ala^2]-Met^5-enkephalin-NH$_2$, facilitated performance of hungry rats in a complex 12-choice maze (42). Here, the peptide was injected at a dose of 80.0 μg/kg intraperitoneally 15 min before the test each day. Interestingly, a Met-enkephalin analog with virtually no opiate activity, [D-Phe]-Met^5-enkephalin, also facilitated maze performance whereas morphine produced the opposite results (see 39, 40).

More recently, rats were trained for 10 days in a continuous reinforcement lever-press task for food reward and tested with peptides on the eleventh day for extinction. Alpha-endorphin (10 µg/rat) delayed and gamma-endorphin (10 µg/rat) slightly facilitated extinction but only at the first 2-hr extinction trial (43).

Different results, were obtained in a test using water as a reward. Here, rats deprived of water for 23 hrs/day were trained for 7 days (one trial per day) to run down an angular alley to get 30 sec of access to water. On the eighth day a series of extinction trials were conducted every 2 hr, and saline or a peptide was injected (10.0 µg/rat s.c.) after the first extinction trial. In contrast to their effects on aversive tasks and on the food-reinforced appetitive task, both alpha- and gamma-endorphin delayed extinction (48). The rats receiving alpha- and gamma-endorphin continued to run to the dry tube even at the fourth extinction trial (6 hr). In a replication of this water task experiment with gamma-endorphin (48), naloxone (5 mg/kg) not only failed to block the delay in extinction, it also produced similar effects by itself. The naloxone plus gamma-endorphin group actually ran to the dry tube faster than the gamma-endorphin plus saline group (43). In both series of experiments, peptide actions on extinction were detectable for 4–6 hr after injection.

To examine the prolonged time course of apparent peptide action relative to the disappearance of the endorphin following peripheral administration, endorphins were measured in blood, hypothalamus, and brain by radioimmunoassay following subcutaneous injection. Alpha-endorphin levels in the blood peak 5 min after injection and virtually disappear by 20 min postinjection. No detectable levels of alpha-endorphin were ever seen in hypothalamus or the rest of the brain. Yet animals given such injections show significant alterations of behavior at time points several hours after injection (48).

The finding of rapid disappearance of alpha-endorphin in the blood after peripheral injection is in accord with previous work showing a rapid clearance of peptide from the blood. The half-time disappearance of radioactivity after the injection of labeled Met^5-enkephalin is less than one minute (41). Thus the behavioral effects of the endorphins appear not to require persistence of the peptide in the blood. In fact, there is evidence that behavioral effects immediately after injection (62) may be different from those produced by behavioral tests 1 hr post-injection (59–61, 63). Determining the actual active compound or mechanism by which these peptides produce this delayed effect is thus of major importance for future work.

At least one group has argued that radiolabeled Met-enkephalin (^{3}H-Tyrosine) crosses the blood-brain barrier, albeit to a very limited degree— 0.84% of the radioactivity injected into the carotid (41, 42). If peripherally derived peptides could actually act directly on brain receptors to produce their behavioral effects, the amounts reaching brain sites would be incredibly small.

There is as yet no obvious solution to this apparent mystery, though several tentative explanations have been proffered. Because peptides have been reported to have similar qualitative effects on extinction of aversive behaviors after intracerebroventricular injections at 1000-fold lower doses (23, 44), it would at first glance seem that the peripherally injected peptides could elicit their effects eventually in the brain, even though no net change in level is detected immediately in the brain. Similar dose relationships are reported for the drinking behaviors induced by peripheral and cerebroventricular injections of angiotensin (31). In this case, the site of the peptide action would appear to be in the specialized ventricular lining cells that form the circumventricular organ system (16, 67). These nonciliated neuronal cells (subfornical organ and organum vasulosum laminae terminalis) exist within the ependymal lining, yet retain their afferent and efferent circuits to adjacent brain nuclei. Furthermore, these cell systems appear able to concentrate circulating peptide hormones in physiological amounts (45, 69) and could, therefore, be regarded as neuronal monitors of circulating hormonal messages. Thus their position on the blood side of the blood-brain barrier, within the cerebrospinal fluid spaces, places them propitiously for such a monitoring function, and could explain how peripherally injected angiotensin can work at 1000 times smaller doses via the ventricles than after subcutaneous injection.

In the case of angiotensin-induced drinking behavior, one of the signals known to be elaborated by the brain during this behavior is vasopressin (14), even though vasopressin itself does not lead to drinking (64). Such an output might seem logically related to drinking as a corollary signal to conserve body water during hypovolemic conditions. Furthermore, opiates and endorphins have been reported to evoke vasopressin secretion (77), and vasopressin is itself an extremely potent peptide in analogous memory-like aversive conditioning tests. Although this sequential connection between endorphins and vasopressin, explored more fully below, might be regarded as a potential mechanistic explanation of the means by which endorphins affect learning performance, it would by no means explain how vasopressin alters learning. Furthermore, Martinez and co-workers (54) have proposed that another peripheral hormone source, the adrenal medulla, may mediate some of the effects of peripheral enkephalins on active avoidance behavioral.

VASOPRESSIN

The primary physiological action of vasopressin (or antidiuretic hormone) is to conserve water, primarily by enhancing renal tubular permeability to water (reviewed in 34, 57, 66). Vasopressin release is triggered by dehydration, enhanced plasma osmolarity, or reduction in effective plasma volume. After pathological, experimental, or congenital loss of vasopressin-secreting neurons, vasopressin deficiency results in a diuresis of dilute urine, a condition known as diabetes insipidus (see 57). The most important nonrenal action of vasopressin is its pressor effect, mediated directly on the smooth muscles of the vascular system. This vasoconstrictor effect may be physiologically significant during hypovolemic or hypotensive crises, and requires doses of vasopressin considerably higher than for maximal antidiuresis (see 34).

Effects of Vasopressin on Aversively Motivated Behavior

In early work De Wied (21, 22) found hypophysectomized rats to be deficient in a number of behavioral situations, especially the acquisition and extinction of aversively motivated tasks. These deficiencies were reversible by administration of a crude pituitary extract, Pitressin, and, in later work, by arginine vasopressin (AVP, the natural vasopressin of all mammals except swine) in microgram amounts injected subcutaneously (25). Vasopressin also reversed the behavioral deficits observed in Brattleboro strain rats with congenital diabetes insipidus (12, 25).

Further, lysine vasopressin (the natural vasopressin of swine) delays extinction in an active avoidance task in intact animals (21) and in an inhibitory avoidance task (1), as does intracerebroventricular injection of nanogram quantities of AVP. Anti-vasopressin immune serum injected intraventricularly inhibits retention in the passive avoidance test (71, 72). More recently, the same investigators reported that AVP enhanced retention of the passive avoidance response when injected subcutaneously either just after the training test (shock) or just before the retention test, but not at times in between (11), suggesting to these investigators that AVP enhances both consolidation and retrieval of "memory."

Although attempts to replicate these effects in other laboratories have met with mixed success (3, 18, 36), our recent studies on active avoidance have successfully reproduced and extended some of the findings of De Wied and associates. Here both subcutaneous and intracerebroventricular injection of vasopressin delayed the extinction of an active avoidance response (44). A pressor antagonist analog of arginine vasopressin, 1-deaminopenicillamine, 2-(o-methyl)tyrosine AVP (dPTyr (Me)AVP), which prevented the pressor response (5), also abolished the effects of subcutaneously injected AVP on prolongation of extinction of active avoidance (49). This blockade may indicate either that signals from peripheral visceral sources play an important role in the subsequent behavioral changes or that the receptors at which AVP elicits its pressor effect are similar to those leading to its behavioral action.

Evidence to support the hypothesis that vasopressin produces its behavioral effects independently of its classical renal or pressor effects comes from the work of De Wied and associates with different analogs of vasopressin. For example, desglycinamide-lysine vasopressin, a vasopressin analog with minimal renal activity, will nevertheless reverse the behavioral deficits associated with diabetes insipidus (25), and will have effects similar to those of vasopressin in tests with active avoidance (74) and in counteracting several drug-induced disruptions of consolidation, including diethyl dithiocarbamate; pentylenetetrazol, puromycin and CO_2-induced amnesia (2, 15, 46, 61). The relative behavioral potencies of these analogs suggests that the ring structure of vasopressin may be most important for the "consolidation" of acquired avoidance responses, while the C-terminal appears to be more important for reversing the effects of amnesia treatments (70).

Vasopressin and Appetitively Motivated Learning

With few exceptions, the data regarding a role for vasopressin in "learning" come from studies employing aversively motivated tasks. Obviously this limits the nature of the *behavioral* conclusions that can be drawn from such data. If vasopressin does have "memory" enhancing properties, this effect should also be demonstrable using positively motivated tasks. Two reports in the literature show some prolongation of extinction of appetitive learning with vasopressin. Rats trained to discriminate sides of a T-maze using a sex reward showed a facilitation of retention with desglycinamide-lysine vasopressin (13). Rats receiving vasopressin during acquisition training in a black-white discrimination T-maze task showed a prolongation of extinction, but only on the side using the black discriminative stimulus (35). Also, preliminary experiments in our laboratory (A. Ettenberg et al, in preparation) have shown that post-training administration of vasopressin can facilitate subsequent test performance in a one-trial water-finding task initially described by Major & White (52).

Clinical Studies

Recent clinical observations have demonstrated that lysine vasopressin can improve performance in tests involving attention, concentration, and memory in a double-blind study involving aged subjects (47). Others have reported improvements in retrograde and anteriograde amnesia in a preliminary study with lysine vasopressin (58). More recently, in double-blind studies, a synthetic analog of vasopressin, 1-desamino-8-D-arginine vasopressin (DDAVP) produced consistent improvements in tests designed to measure long-term memory in patients suffering from affective illness; DDAVP also produced improvements in serial learning and prompted free recall and recall of semantically related words in both cognitively impaired and unimpaired adults (76). Curiously, in animal studies, the replacement of the L-arginine in arginine vasopressin by D-arginine results in a marked decrease in the AVP's ability to alter extinction of pole jump avoidance (74). Whether this reflects species differences in response to vasopressin or the difference in the nature of the behavioral tests remains to be determined.

CONCLUSIONS

A critical set of questions must be confronted that center on three groups of unknowns: (a) molecular interactions: the essential sites and mechanisms of action of these peptides in the central nervous system; (b) cellular interactions: the result of such molecular interactions on the activity of receptive neurones; (c) behavioral interactions: the classes of central cellular systems that operate in specific behavioral operations. In the absence of experimental data on these interactions, innumerable possible interpretations can flourish without demonstrating that a specific peptide is either "involved with" or "mediates" a specific behavior. Nevertheless, approaches to these fundamental demonstrations are essential before we can attempt to extrapolate from animal behavior experiments to the human behavioral syndromes that may involve these peptides [i.e. substance abuse syndromes or mental illnesses (e.g. 8, 76)]. Although there is evidence for an effect of vasopressin independent of its physiological effects, this effect is reversible in some cases by blockade of the pressor receptor for vasopressin. Determining how and where the various peptide receptors interact to produce behavioral changes is the challenge of current research.

Literature Cited

1. Ader, R., De Wied, D. 1972. Effects of lysine vasopressin on passive avoidance learning. *Psychon. Sci.* 29:46–48
2. Asin, K. E. 1980. Lysine vasopressin attenuation of diethylidithiocarbamate-induced amnesia. *Pharmacol. Biochem. Behav.* 12:343–46
3. Bailey, W. H., Weiss, J. M. 1979. Evaluation of a 'memory deficit' in vasopressin-deficient rats. *Brain Res.* 162:174–78
4. Banerjee, U. 1971. Acquisitions of conditioned avoidance response in rats under the influence of addicting drugs. *Psychopharmacologia* 22:133–34
5. Bankowski, K., Manning, M., Haldar, J., Sawyer, W. H. 1978. Design of potent antagonists of the vasopressor response to arginine-vasopressin. *J. Med. Chem.* 21:850–53
6. Beaumont, A., Hughes, J. 1979. Biology of opioid peptides. *Ann. Rev. Pharmacol. Toxicol.* 19:245–67
7. Belluzzi, J. D., Grant, N., Garsky, V., Sarantakis, D. Wise, C. D., Stein, L. 1976. Analgesia induced *in vitro* by central administration of enkephalin in rat. *Nature* 260:625–26
8. Belluzzi, J. D., Stein, L. 1977. Enkephalin may mediate euphoria and drive-reduction reward. *Nature* 266:566–68
9. Bloom, F. E., Segal, D. S. 1980. Endorphins in cerebrospinal fluid. In *Neurobiology of Cerebrospinal Fluid,* ed. J. H. Wood, pp. 651–64. NY: Plenum
10. Bloom, F., Segal, D., Ling, N., Guillemin, R. 1976. Endorphins: profound behavioral effects in rats suggest new etiological factors in mental illness. *Science* 194:630–32
11. Bohus, B., Kovacs, G. L., De Wied, D. 1978. Oxytocin, vasopressin and memory: opposite effects on consolidation

and retrieval processes. *Brain Res.* 157:414–17
12. Bohus, B., Van Wimersma Greidanus, T. J. B., De Wied, D. 1975. Behavioral and endocrine responses of rats with hereditary hypothalamic diabetes insipidus (Brattleboro Strain). *Physiol. Behav.* 14:609–15
13. Bohus, B. 1977. Effect of desglycinamide-lysine vasopressin (DG-LVP) on sexually motivated T-maze behavior of the male rat. *Hormones Behav.* 8:52–61
14. Bonjour, J. P., Malvin, R. L. 1970. Stimulation of ADH release by the renin-angiotensin system. *Am. J. Physiol.* 218:1555–59
15. Bookin, H. B., Pfeifer, W. D. 1977. Effect of lysine vasopressin on pentylenetetrazol-induced retrogade amnesia in rats. *Pharmacol. Biochem. Behav.* 7:51–54
16. Buggy, J., Johnson, A. K. 1978. Angiotensin-induced thirst: effects of third ventricle obstruction and periventricular ablation. *Brain Res.* 149:117–28
17. Castellano, C. 1975. Effects of morphine and heroin on discrimination learning and consolidation in mice. *Psychopharmacologia* 42:235–42
18. Celestian, J. F., Carey, R. J., Miller, M. 1975. Unimpaired maintenance of a conditioned avoidance response in rats with diabetes insipidus. *Physiol. Behav.* 15:707–11
19. Chang, K. J., Cuatrecasas, P. 1979. Multiple opiate receptors. *J. Biol. Chem.* 254:2610–18
20. Cook, L., Weidley, E. 1957. Behavioral effects of some psychopharmacological agents. *NY Acad. Sci.* 66:740–52
21. De Wied, D. 1965. The influence of the posterior and intermediate lobe of the pituitary and pituitary peptides on the maintenance of a conditioned avoidance response in rats. *Int. J. Neuropharmacol.* 4:157–67
22. De Wied, D. 1971. Long term effect of vasopressin on the maintenance of a conditioned avoidance response in rats. *Nature* 232:58–60
23. De Wied, D. 1976. Behavioral effects of intraventricularly administered vasopressin and vasopressin fragments. *Life Sci.* 46:27–29
24. De Wied, D. 1977. Peptides and behavior. *Life Sci.* 20:195–204
25. De Wied, D., Bohus, B., Van Wimersma Greidanus, T. J. B. 1975. Memory deficit in rats with heriditary diabetes insipidus. *Brain Res.* 85:152–56
26. De Wied, D., Bohus, B., Van Ree, J. M., Kovacs, G. L., Greven, H. M. 1978. Neuroleptic-like activity of [des-tyr]-γ-endorphin in rats. *Lancet* 1:1046
27. De Wied, D., Bohus, B., Van Ree, J. M., Urban, I. 1978. Behavioral and electrophysiological effects of peptides related to lipotropin (β-LPH). *J. Pharmacol. Exp. Ther.* 204:570–80
28. De Wied, D., Kovacs, G. L., Bohus, B., Van Ree, J. M., Greven, H. M. 1978. Neuroleptic activity of the neuropeptide β-LPH$_{62-77}$ ([Des-Tyr]-endorphin; DY γe). *Eur. J. of Pharmacol.* 49:427–36
29. De Wied, D., Van Ree, J. M., Greven, H. M. 1980. Neuroleptic-like activity of peptides related to [Des-tyrγ]-γ-endorphin: Structure activity studies. *Life Sci.* 26:1575–79
30. Dingledine, R., Iversen, L. L., Breuker, E. 1978. Naloxone as a GABA antagonist; evidence from iontophoretic, receptor binding, and convulsant studies. *Eur. J. Pharmacol.* 47:19–27
31. Fitzsimons, J. T. 1972. Thirst. *Physiol. Rev.* 52:468–561
32. Gallagher, M., Kapp, B. S. 1978. Opiate administration into the amygdala: Effects of memory processes. *Life Sci.* 23:1973–78
33. Gilbert, P. E., Martin, W. R. 1976. The effects of morpine and nalorpine-like drugs in the nondependent, morphine dependent and cyclazocine-dependent chronic spinal dog. *J. Pharmacol. Exp. Ther.* 198:66–82
34. Hays, R. M. 1980. Agents affecting the renal conservation of water. In *The Pharmacologic Basis of Therapeutics*, ed. A. G. Gilman, L. S. Goodman, pp. 916–28. NY: MacMillan
35. Hostetter, G., Jubb, S. L., Kozlowski, G. P. 1977. Vasopressin affects the behavior of rats in a positively-rewarded discrimination task. *Life Sci.* 21:1323–28
36. Hostetter, G., Jubb, S. L., Kozlowski, G. P. 1980. An inability of subcutaneous vasopressin to affect passive avoidance behavior. *Neuroendocrinology* 30:174–77
37. Jensen, R. A., Martinez, J. L., Messing, R. B., Speihler, V., Vasquez, B. J., Soumireu-Mourat, B., Liang, K. C., McGaugh, J. L. 1978. Morphine and naloxone alter memory in the rat. *Soc. Neurosci. Abstr.* 4:260
38. Kapp, B. S., Gallagher, M. 1979. Opiates and memory. *Trends Neurosci.* 2:177–80
39. Kastin, A. J., Coy, D. H., Olson, R. D., Panksepp, J., Schally, A. V., Sandman, C. A. 1979. Behavioral effects of the brain opiates enkephalin and endor-

phin. In *Central Nervous System Effects of Hypothalamic and Other Peptides*, ed. R. Collu, pp. 273–81. NY: Raven
40. Kastin, A. J., Olson, R. D., Schally, A. V., Coy, D. H. 1979. CNS effects of peripherally administered brain peptides. *Life Sci.* 25:401–14
41. Kastin, A. J., Nissen, M. C., Schally, A. V., Coy, D. H. 1976. Blood brain barrier, half-time disappearance, and brain distribution for labeled enkephalin and a potent analog. *Brain Res. Bull.* 1:583–89
42. Kastin, A. J., Scollan, E. L., King, M. G., Schally, A. V., Coy, D. H. 1976. Enkephalin and a potent analog facilitate maze performance after intraperitoneal administration in rats. *Pharmacol. Biochem. Behav.* 5:691–95
43. Koob, G. F., Le Moal, M., Bloom, F. E. 1981. Enkephalin and endorphin influences on appetitive and aversive conditioning. In *Endogenous Peptides and Learning and Memory Processes*, ed. J. Martinez, J. McGaugh. NY: Academic. pp. 249–65
44. Koob, G. F., Le Moal, M., Gaffori, O., Manning, M., Sawyer, W. H., Rivier, J., Bloom, F. E. 1981. Arginine vasopressin and a vasopressin antagonist peptide: opposite effects on extinction of active avoidance in rats. *Regul. Peptides.* 2:153–64
45. Landas, S., Phillips, M. I., Stamler, J. F., Raizada, M. K. 1980. Visualization of specific angiotensin II binding sites in the brain by fluorescent microscopy. *Science* 210:791–93
46. Lande, S., Flexner, T. B., Flexner, L. L. 1972. Effects of corticotrophin and desglycinamide lysine vasopressin on suppression of memory by puromycin. *Proc. Natl. Acad. Sci. USA* 69:558–60
47. Legros, J. J., Gilot, P., Seron, X., Claessens, J., Adam, A., Moeglen, J. M., Audibert, A., Berchier, P. 1978. Influence of vasopressin on learning and memory. *Lancet* 1:41–42
48. Le Moal, M., Koob, G. F., Koda, L. Y., Bloom, F. E., Manning, M., Sawyer, W. H., Rivier, J. 1981. Vasopressin antagonist peptide: blockade of pressor receptor prevents behavioral action of vasopressin. *Nature* 291:491–93
49. Le Moal, M., Koob, G. F., Bloom, F. E. 1979. Endorphins and extinction: differential actions on appetitive and aversive tasks. *Life Sci.* 24:1631–36
50. Loh, H. H., Tseng, L. F., Wei, E., Li, C. H. 1976. β-Endorphin is a potent analgesic agent. *Proc. Natl. Acad. Sci. USA* 73:2895–98
51. Lord, J. A. H., Waterfield, A. A., Hughes, J., Kosterlitz, H. W. 1977. Endogeneous opioid peptides: multiple agonists and receptors. *Nature* 267:495–99
52. Major, R., White, N. 1978. Memory facilitation by self-stimulation reinforcement mediated by the nigro-neostriatal bundle. *Physiol. Behav.* 20:723–33
53. Martin, W. R., Eades, C. G., Thompson, J. A., Huppler, R. E., Gilbert, P. E. 1976. The effects of morphine- and nalorpine-like drugs in the nondependent and morphine-dependent chronic spinal dog. *J. Pharmacol. Exp. Ther.* 197:517–32
54. Martinez, J. L. Jr., Rigter, H. 1980. Enkephalin and $ACTH_{4-10}$ effects on active avoidance conditioning are attenuated by adrenal demedullation. *Neurosci. Abstr.* 6:319
55. Messing, R. B., Jensen, R. A., Martinez, J. L. Jr., Spiehler, V. R., Vasquez, B. J., Soumireu-Mourat, B., Liang, D. C., McGaugh, J. L. 1979. Naloxone enhancement of memory. *Behav. Neural Biol.* 27:266–75
56. Miller, R. J., Cuatrecasas, P. 1979. Neurobiology and neurophamacology of the enkephalins. *Adv. Biochem. Psychopharmacol.* 20:187–226
57. Moses, A. M., Share, L., eds. 1977. *Neurohypophysis: International Conference on the Neurohypophysis.* Basel: S. Karger. pp. 1–318
58. Oliveros, J. C., Jandali, M. K., Timsit-Berthier, M., Remy, R., Benghezal, A., Audibert, A., Moeglen, J. M. 1978. Vasopressin in amnesia. *Lancet* 1:42
59. Rigter, H. 1978. Attenuation of amnesia in rats by systemically administered enkephalin. *Science* 200:83–85
60. Rigter, H., Greven, H., Van Riezen, H. 1977. Failure of naloxone to prevent reduction of amnesia by enkephalins. *Neuropharmacology* 16:545–47
61. Rigter, H., Van Riezen, H., De Wied, D. 1974. The effects of ACTH- and vasopressin-analogues on CO_2-induced retrograde amnesia in rats. *Physiol. Behav.* 13:381–88
62. Rigter, H., Hannan, T. J., Messing, R. B., Martinez, J. L. Jr., Vasquez, B. J., Jensen, R. A., Veliquette, J., McGaugh, J. L. 1980. Enkephalins interfere with acquisition of an active avoidance response. *Life Sci.* 26:337–46
63. Rigter, H., Shuster, S., Thody, A. J. 1977. ACTH, alpha-MSH and beta-LPH: Pituitary hormones with similar activity in an amnesia test in rats. *J. Pharmacol. Physiol.* 29:10–111

64. Rolls, B. J. 1971. The effect of intravenous infusion of antidiuretic hormone on water intake in the rat. *J. Physiol.* 219:331–19
65. Rossier, J., Bloom, F. 1979. Central neuropharmacology of endorphins. *Adv. Biochem. Psychopharmacol.* 20:165–86
66. Sawyer, W. H. 1964. Vertebrate neurohypophysial principles. *Endocrinology* 75:981–90
67. Simpson, J. R., Mangiapane, M. L., Dellman, H.-D. 1978. Central receptor sites for angiotensin-induced drinking: A critical review. *Fed. Proc.* 37:2676–82
68. Stein, L., Belluzzi, J. D. 1978. Brain endorphins and the sense of well-being: a psychobiological hypothesis. *Adv. Biochem. Psychopharmacol.* 18:299–311
69. van Houten, M., Posner, B. I., Kopriwa, B. M., Brawer, J. R. 1979. Insulin binding sites in the rat brain: In vivo localization to the circumventricular organs by quantitative autoradiography. *Endocrinology* 105:666–73
70. Van Ree, J. M., Bohus, B., Versteeg, D. H., De Wied, D. 1978. Neurohypophyseal principles and memory processes. *Biochem. Pharmacol.* 27:1793–1800
71. Van Wimersma Greidanus, T. B., Dogterom, J., De Wied, D. 1975. Intraventricular administration of antivasopressin serum inhibit memory consolidation in rats. *Life Sci.* 16:637–44
72. Van Wimersma Greidanus, T. B., De Wied, D. 1976. Modulation of passive-avoidance behavior of rats by intracerebroventricular administration of antivasopressin serum. *Behav. Biol.* 18:325–33
73. Verhave, T., Owen, J. E., Robbins, E. G. 1959. The effect of morphine sulfate on avoidance and escape behavior. *J. Pharmacol. Exp. Ther.* 125:248–51
74. Walter, R., Van Ree, J. M., De Wied, D. 1978. Modification of conditioned behavior of rats by neurohypophyseal hormones and analogues. *Proc. Natl. Acad. Sci. USA* 75:2493–96
75. Weinberger, S. B., Arnstein, A., Segal, D. C. 1979. Des-tyrosine—endorphin and haloperidol: Behavioral and biochemical differentiation. *Life Sci.* 24:1637–44
76. Weingartner, H., Gold, P., Ballenger, J. C., Smallberg, S. A., Summers, R., Rubinow, D. R., Post, R. M., Goodwin, F. K. 1981. Effects of vasopressin on human memory functions. *Science* 211:601–3
77. Weitzman, R., Fisher, D., Minnick, S., Ling, N., Guillemin, R. 1977. Beta-endorphin stimulates secretion of arginine vasopressin *in vivo*. *Endocrinology* 101:1643–46
78. Wuster, M., Schulz, R., Herz, A. 1980. The direction of opioid agonists towards μ-, δ-, and e-receptors in the vas deferens of the mouse and the rat. *Life Sci.* 27:163–70

THE ROLE OF THE CENTRAL NERVOUS SYSTEM IN THE CONTROL OF OVARIAN FUNCTION IN HIGHER PRIMATES

C. R. Pohl and E. Knobil

Department of Physiology, University of Pittsburgh School of Medicine, Pittsburgh, Pennsylvania 15261

INTRODUCTION

The Mammalia have evolved a remarkable diversity both of reproductive strategies and of the neuroendocrine systems that control them. The nature and extent of these variations have only recently come to light, and we expect that their continued study will facilitate the formulation of unifying concepts applicable to all vertebrates.

Here we review recent studies of the control of ovarian function in the rhesus monkey that have relevance to human reproduction.

THE ARCUATE NUCLEUS AND PULSATILE GONADOTROPIN SECRETION

That the secretion of luteinizing hormone (LH) and follicle-stimulating hormone (FSH) by the pituitary is a rhythmic, pulsatile phenomenon has been recognized in a number of species, including the rat (40), mouse (26), guinea pig (32), rabbit (103), cat (52), sheep (18, 57), pig (15, 39), cattle (56, 96), chicken (125), monkey (29, 88), and human (78, 131). LH pulses in ovariectomized rhesus monkeys occur approximately once every hour and, for this reason, were termed "circhoral" (29). In the human, pulses occur once every one to two hours (76, 102, 105, 119, 131, 132).

The pulsatile release of gonadotropins is undoubtedly the consequence of rhythmic discharges of gonadotropin releasing hormone (GnRH) into the pituitary portal circulation (19). Inactivation of this hypothalamic decapeptide by the intravenous administration of antisera against synthetic GnRH to ovariectomized rhesus monkeys causes an abrupt reduction in plasma gonadotropin concentrations (74). Furthermore, a variety of neuroleptic drugs and α-adrenergic blocking agents, presumably acting at the neural level rather than directly on the pituitary (92), promptly interrupt the pulsatile discharge of LH in the monkey (8).

The intermittent delivery of endogenous GnRH to the pituitary gland must be the consequence of a synchronous discharge of GnRH-containing neurons effected by some neuronal signal generator or oscillator (44, 62). Since complete neural disconnection of the medial basal hypothalamus (MBH) from the remainder of the central nervous system does not interfere with the circhoral pattern of gonadotropin secretion in ovariectomized monkeys (64), the circhoral oscillator as well as the cell bodies of GnRH-producing neurons must reside within the MBH. Immunocytochemical studies in primates have demonstrated GnRH cell bodies in this region, among other areas (5, 6, 16, 72, 109).

Within the MBH, bilateral destruction of the arcuate nucleus by radiofrequency current abolishes gonadotropin secretion by the pituitary without producing deficits in other adenohypophysial functions (90). On the other hand, gonadotropin secretion is largely unaffected by large MBH lesions that spare the region of the arcuate nucleus. It appears, therefore, that this nucleus constitutes the primary structure within the MBH that mediates the hypothalamic control of gonadotropin secretion in the rhesus monkey and encompasses the circhoral oscillator.

No physiologic significance could be ascribed to the rhythmic, pulsatile pattern of gonadotropin secretion until recently when it was recognized that the functioning of the hypophysiotropic control system that directs gonadotropin secretion is obligatorily intermittent. Continuous GnRH infusion into female monkeys with lesions of the arcuate nucleus that abolished endogenous GnRH secretion was incapable of reinitiating sustained gonadotropin secretion, whereas infusion of the decapeptide as six-minute pulses once every hour reestablished normal pituitary function (77). Furthermore, continuous infusion of GnRH profoundly inhibited gonadotropin secretion previously established by the pulsatile administration of the decapeptide to such animals (7). This inhibition of gonadotropin secretion by continuous GnRH infusion was reversed when the pulsatile replacement regimen was reinstituted. That this phenomenon is the consequence of the pattern of hypophysiotropic stimulation rather than of the total mass of decapeptide delivered to the gonadotrophs was suggested by the failure of a wide range of infusion rates of GnRH to restore gonadotropin secretion in monkeys with hypothalamic lesions when given in the continuous mode

(7). Constant infusions of GnRH have been shown equally ineffective in eliciting sustained gonadotropin secretion in normal monkeys (36), sheep (20), rats (87, 108), humans (46, 95), and rat pituitary cells in vitro (111). The inhibition of gonadotropin secretion by administration of long-acting GnRH agonists has been found clinically useful for the treatment of idiopathic precocious puberty (27). Furthermore, in the rhesus monkey, increasing the frequency of GnRH administration from once every hour to two, three, or five pulses per hour results in gradual declines in plasma gonadotropin levels (121). The cellular mechanisms underlying this pattern-dependent desensitization are presently unknown.

In rhesus monkeys with arcuate lesions, decreasing the frequency of exogenous GnRH pulses from one per hour to one every three hours leads to declines in plasma LH levels while causing elevations in circulating FSH (121). These results permit the important conclusion that relatively small changes in the frequency of GnRH stimulation not only alter the concentrations of LH and FSH in the circulation but also have major effects on the ratio of FSH to LH. In contrast, major changes in the amplitude (i.e. the infusion rate) of GnRH pulses have but minor effects on gonadotropin secretion (121).[1]

The electrophysiologic bases of the pulsatile release of GnRH are currently under investigation. Striking increases in multiunit activity have been recorded from the region of the arcuate nucleus coincident with the initiation of circhoral LH pulses in ovariectomized monkeys (33, 62). The continued study of this system from a neurophysiologic standpoint will surely lead to new insights into the functioning of neuroendocrine control systems and their modulation.

OVARIAN CONTROL OF GONADOTROPIN SECRETION

Pulsatile gonadotropin secretion in monkeys (126) and women (132) is controlled by a classical negative feedback action of estradiol (53, 115). Although progesterone alone does not inhibit gonadotropin secretion (55), it can synergize with estradiol in this regard under some experimental circumstances (55, 98). The physiologic significance of such a synergism, however, is not clear because, during the menstrual cycle of the rhesus monkey, gonadotropin levels in the follicular phase of the cycle, when

---

[1] In comparison to other species, the rhesus monkey is relatively insensitive to exogenous GnRH (1, 34, 65). For example, a dose of approximately 1.0 µg GnRH/kg/pulse is required to sustain gonadotropin secretion in the rhesus monkey (121), whereas only 0.025 µg GnRH/kg/pulse is needed to restore gonadotropin secretion in humans with anorexia nervosa or Kallman's syndrome (28, 73, 116).

progesterone is undetectable, are essentially the same as in the luteal phase, when progesterone concentrations are maximal (60).

The initiation of the preovulatory surge of gonadotropins is the consequence of a positive feedback or stimulatory action of estrogen. When the rising plasma estradiol concentration that accompanies follicular maturation late in the follicular phase of the menstrual cycle exceeds a threshold of approximately 200 pg/ml for a period of two days (54, 134), gonadotropin surges are elicited in monkeys (53, 127) and women (68, 75, 81, 129, 130, 132). The small amount of progesterone released during the rising phase of estradiol-induced preovulatory gonadotropin surges (51, 99, 118) appears to be necessary for their complete development in the human (21, 68, 71, 84). In the rhesus monkey, however, estradiol alone suffices for the full expression of LH and FSH surges (60). In any event, it appears that the characteristic patterns of LH and FSH secretion throughout the entire menstrual cycle, basal secretion interrupted once every 28 days, on the average, by a gonadotropin surge, can be accounted for by the waxing and waning of ovarian estradiol secretion.

In a pharmacologic context, progesterone can induce surges of LH and FSH in estrogen-primed women and rhesus monkeys (41, 68, 69, 80, 83, 97, 98, 113). Furthermore, when progesterone is administered in association with a gonadotropin-releasing dose of estrogen, it can either facilitate (i.e. advance in time) (21, 23, 31, 47, 48, 71) or block (23, 31, 48, 71, 112) these gonadotropin discharges. The response actually obtained depends upon the dose of progesterone (31, 48) and the time of its administration relative to estrogen (31, 48, 71).

Complete neural disconnection of the MBH from the remainder of the brain does not interfere with the negative and positive feedback actions of estradiol (35, 64). Although it has been reported that lesions in the rostral hypothalamus block estrogen-induced gonadotropin discharges in monkeys (82), these results have not been confirmed [(91); see (61) for discussion]. Moreover, gonadotropin surges can be elicited by estrogen administration following aspiration of the entire brain anterior and dorsal to the MBH (49) or following separation of the ventral MBH from the anterior hypothalamus by placement of a Silastic barrier (24). It would seem, therefore, that the loci of the negative and positive feedback actions of estradiol reside within the MBH-hypophysial unit.

To differentiate between the hypothalamus and pituitary as the sites of feedback actions of estradiol, lesions were placed in the arcuate region of ovariectomized rhesus monkeys. These lesions abolished gonadotropin secretion, including estrogen-induced surges. Circulating levels of gonadotropins were then reestablished by the chronic, intermittent administration of GnRH (one pulse per hour). The administration of estrogen to such animals first inhibited LH and FSH secretion and then elicited gonadotropin surges indistinguishable from those in nonlesioned controls (77). Negative and positive feedback actions of estrogen have also been reported in pituitary stalk–sectioned monkeys similarly treated with GnRH (85). It was concluded from these experiments, in which the pituitary gland was "clamped" by the exogenous GnRH, that both the negative and positive feedback actions of estradiol were exerted at the pituitary. Indeed, estrogen can elicit gonadotropin surges up to 48 hr after cessation of GnRH administration to arcuate-lesioned monkeys, suggesting that the steroid is itself a gonadotropin-releasing hormone (120). This is consonant with the report that a surge of gonadotropins can be induced by estrogen given shortly after pituitary stalk section (38). These findings explain the earlier observation that α-adrenergic blocking agents that acutely inhibit tonic gonadotropin secretion in the monkey (8), presumably by blocking endogenous GnRH release, fail to modify estrogen-induced surges (60). Similarly, the acute neutralization of endogenous GnRH by the administration of antisera against the decapeptide was unable to inhibit the positive feedback action of estrogen (74).

A pituitary site of the negative feedback action of estradiol was also suggested by earlier findings that estrogen administration inhibits gonadotropin release in response to exogenously administered GnRH (58, 65, 114, 133).

While the foregoing considerations clearly point to the pituitary as the primary site of the classical feedback actions of estradiol on gonadotropin secretion, they do not rule out such actions at the level of the hypothalamus.[2] Evidence for such a locus, however, is not compelling. The mean concentrations of GnRH in pituitary portal plasma of monkeys are not influenced by ovariectomy (19, 79), a procedure that leads to a ten-fold increase in LH levels (2). Although it has been reported that the implantation of estradiol into the MBH of rhesus monkeys is followed by a decline and then a discharge of gonadotropins, whereas implantation of the steroid into the pituitary gland produces no effect (117), these findings may be interpreted in terms of the "implantation paradox" of Bogdanove (9), which holds that an intrapituitary depot of steroid influences only a small surround of cells, the bulk of the remainder being unaffected, whereas implantation into the hypothalamus permits efficient distribution of the steroid by the pituitary portal circulation to the entire anterior lobe. The author himself concludes: "These experiments do not, however, permit a clear distinction between actions of estrogen in the medial basal hypothalamus

---

[2]The negative feedback action of testosterone in the male is at least partially attributable to a decrease in the frequency of pulsatile gonadotropin secretion (89, 104).

and in the pituitary" (117). The same considerations pertain to the interpretations of other studies that purport to demonstrate a hypothalamic site of action on the basis of a suppression of circulating gonadotropins following injections of estradiol into the hypothalamus (37) or infusions of the steroid into the third ventricle (22).

In contrast to the pituitary site of action of estradiol, the progesterone blockade of estradiol-induced gonadotropin discharges must be exerted in the central nervous system because hypothalamic lesions abolished this action of progesterone (122). On the other hand, the blocking action of progesterone could be demonstrated in monkeys with intact nervous systems that had been amenorrheic and were receiving a pulsatile GnRH replacement regimen (94). Because the GnRH regimen was continued throughout these experiments, it was concluded that the blocking action of progesterone is not effected by an interruption of GnRH release. Rather, these findings suggested that progesterone may cause the release of an inhibitory agent from the hypothalamus that blocks the positive feedback action of estradiol on the pituitary (94).

The facilitating action of progesterone on gonadotropin discharges, on the other hand, is exerted at the pituitary as demonstrated by the advancing of estrogen-induced gonadotropin surges during progesterone treatment of arcuate-lesioned monkeys receiving intermittent GnRH replacement (122).

HYPOTHALAMIC CONTROL OF THE PITUITARY-OVARIAN AXIS

While the secretory activity of the gonadotrophs, including their responses to the feedback actions of estradiol, has an absolute requirement for hypothalamic GnRH, the role of the decapeptide in this regard must be viewed as a permissive one. This is because the inhibitory and stimulatory actions of estradiol on gonadotropin secretion are demonstrable in the face of an unvarying, pulsatile GnRH replacement regimen when administered to rhesus monkeys in which endogenous GnRH production has been abolished (77). Predictably, such a regimen is all that is required to subserve complete, 28-day, ovulatory menstrual cycles as demonstrated in rhesus monkeys with hypothalamic lesions (61, 63). These findings have formed the basis for the treatment of women with primary and secondary hypothalamic amenorrhea using intermittent GnRH administration, resulting in ovulation (28, 67) and pregnancy (70, 107).

These studies have permitted the construction of a model of the neuroendocrine control system that governs the 28-day ovarian cycle (61). In brief, this model has three basic components: the arcuate nucleus of the hypothalamus, the gonadotrophs of the pituitary gland, and the ovary. The arcuate nucleus is the central component of the control system. Its basic *unmodulated* operation consists of generating a signal, once per hour in the monkey or once every one to two hours in the human, that eventuates in the release of a bolus of GnRH into the pituitary portal circulation to which the gonadotrophs respond by releasing a pulse of LH and FSH. Immature ovarian follicles respond to these pulses of gonadotropic hormones by increasing in size and secreting increasing quantities of estradiol, which achieve peak levels in the circulation near mid-cycle. This process occupies approximately 14 days. The magnitude of the pituitary response to each GnRH pulse is controlled by estradiol acting directly on the gonadotrophs (the negative feedback loop). When estradiol exceeds a threshold of approximately 200 pg/ml for at least two days, the negative feedback action of the steroid is interrupted and estradiol now initiates the discharge of the preovulatory gonadotropin surge. (It should be reemphasized that the surge does not require an increment in GnRH release by the hypothalamus.)

The Graafian follicle promptly responds to the gonadotropin surge by full maturation, massive estradiol secretion, follicular rupture, ovulation, corpus luteum formation, and progesterone secretion. Despite continued gonadotropin release during the luteal phase of the cycle, follicular development is inhibited by the presence of progesterone (42, 50). The functional life-span of the corpus luteum, which is inherent in this structure (59), is approximately 14 days; when progesterone inhibition is removed following luteolysis, a new follicle is selected for development, and the cycle is repeated. The ovary, therefore, times the events of the menstrual cycle. Its characteristic 28-day duration simply represents the sum of the durations of follicular development and of the functional life-span of the corpus luteum.

While an unvarying pulsatile GnRH stimulus is capable of subserving the entire menstrual cycle, there can be no doubt that, in a physiologic context, the neural component of the control system is influenced by the activities of higher centers and by the ovarian hormones themselves. That these influences probably impinge on the hypothalamic oscillator to vary its frequency was suggested by the finding, already mentioned, that small changes in the frequency of exogenous GnRH administration to rhesus monkeys with hypothalamic lesions had profound influences on gonadotropin secretion (61). The consequences of these on ovarian function have recently been described (100). A decrease in the frequency of GnRH pulses from the physiologic rate of once per hour to once every 90 minutes appeared to reduce the incidence of ovulatory menstrual cycles. A regimen of one pulse every two hours yielded only anovulatory cycles, whereas one pulse every three hours was incapable of inducing follicular development. It should be recalled in this connection that these reductions in GnRH pulse frequencies tend to reduce plasma LH levels while elevating those of FSH (121).

In a more physiologic context, the frequency of gonadotropic hormone secretion is reduced during the luteal phase of the ovarian cycles of women (105, 119, 131), cows (96), and ewes (3, 45), a phenomenon attributable to the action of progesterone (43, 126). However, the physiologic sequelae of this effect remain to be defined for the rhesus monkey because in the animals with hypothalamic lesions receiving an unvarying replacement regimen of pulsatile GnRH administration, the functional life-span of the corpus luteum or any other aspect of the menstrual cycle does not appear to differ from normal (63).

The sensory input associated with suckling appears to completely inhibit the arcuate pulse generator (4, 93, 106, 124). In time, FSH makes its appearance in the circulation followed by LH (93, 101, 106). This observation is consonant with the view that the activity of the neural oscillator is reinitiated with a low frequency, favoring an accumulation of FSH in the circulation (see above), followed by a gradual acceleration to the normal unrestrained frequency when LH and FSH concentrations converge.

The longitudinal pattern of the gonadotropic hormones in the course of puberty, FSH rising first in the circulation followed by LH with a gradual convergence of the two hormones (17, 86, 110), also suggests a gradual acceleration of the arcuate oscillator from total inactivity to nocturnal activation (11, 12) to the final adult frequency (25, 73, 116). Such a maturation of the neural control system appears to constitute the only limiting factor in the initiation of puberty, since the pituitary and ovaries are fully competent in prepubertal monkeys, which can be induced to have normal 28-day ovulatory menstrual cycles by the administration of a pulsatile GnRH replacement regimen that sustains normal menstrual cycles in adults with hypothalamic lesions (123). Furthermore, the low plasma concentrations of estradiol characteristic of the perimenarchial period in monkeys (30) are replicated in adults receiving pulses of GnRH once every three hours (100). Increasing this frequency to once every two hours produces anovulatory episodes of follicular development, similar to those in the postmenarchial period preceding the first ovulatory menstrual cycle (30).

A variety of circumstances that compromise the normal functioning of the pituitary-ovarian axis in women are characterized by the absence or infrequency of gonadotropin pulses (66). These disorders include anorexia nervosa (13, 119), hyperprolactinemic amenorrhea (10), idiopathic hypogonadotropic hypogonadism (14), and secondary hypothalamic amenorrhea (128). Pulsatile GnRH infusions have restored pituitary-ovarian function in patients with each of these derangements (28, 67, 70, 73, 107), suggesting disturbances in rhythmic activation of the hypothalamic control system that governs pulsatile gonadotropin secretion.

Literature Cited

1. Arimura, A., Spies, H. G., Schally, A. V. 1973. *J. Clin. Endocrinol. Metab.* 36:372–74
2. Atkinson, L. E., Bhattacharya, A. N., Monroe, S. E., Dierschke, D. J., Knobil, E. 1970. *Endocrinology* 87:847–49
3. Baird, D. T. 1978. *Biol. Reprod.* 18:359–64
4. Baird, D. T., McNeilly, A. S., Sawers, R. S., Sharpe, R. M. 1979. *J. Clin. Endocrinol. Metab.* 49:500–6
5. Barry, J. 1977. *Cell Tiss. Res.* 181:1–14
6. Barry, J., Carette, B. 1975. *Cell Tiss. Res.* 164:163–78
7. Belchetz, P. E., Plant, T. M., Nakai, Y., Keogh, E. J., Knobil, E. 1978. *Science* 202:631–33
8. Bhattacharya, A. N., Dierschke, D. J., Yamaji, T., Knobil, E. 1972. *Endocrinology* 90:778–86
9. Bogdanove, E. M. 1963. *Endocrinology* 73:696–712
10. Bohnet, H. G., Dahlén, H. G., Wuttke, W., Schneider, H. P. G. 1975. *J. Clin. Endocrinol. Metab.* 42:132–43
11. Boyar, R. M. 1978. *Ann. Rev. Med.* 29:509–20
12. Boyar, R., Finkelstein, J., Roffwarg, H., Kapen, S., Weitzman, E., Hellman, L. 1972. *N. Engl. J. Med.* 287:582–86
13. Boyar, R. M., Katz, J., Finkelstein, J. W., Kapen, S., Weiner, H., Weitzman, E. D., Hellman, L. 1974. *N. Engl. J. Med.* 291:861–65
14. Boyar, R. M., Wu, R. H. K., Kapen, S., Hellman, L., Weitzman, E. D., Finkelstein, J. W. 1976. *J. Clin. Endocrinol. Metab.* 43:1268–75
15. Brinkley, H. J. 1981. *Biol. Reprod.* 24:22–43
16. Bugnon, C., Bloch, B., Fellmann, D. 1977. *Brain Res.* 128:249–62
17. Burr, I. M., Sizonenko, P. C., Kaplan, S. L., Grumbach, M. M. 1970. *Pediatr. Res.* 4:25–35
18. Butler, W. R., Malven, P. V., Willett, L. B., Bolt, D. J. 1972. *Endocrinology* 91:793–801
19. Carmel, P. W., Araki, S., Ferin, M. 1976. *Endocrinology* 99:243–48
20. Chakraborty, P. K., Adams, T. E., Tarnavsky, G. K., Reeves, J. J. 1974. *J. Anim. Sci.* 39:1150–57
21. Chang, R. J., Jaffe, R. B. 1978. *J. Clin. Endocrinol. Metab.* 47:119–25
22. Chappel, S. C., Resko, J. A., Norman, R. L., Spies, H. G. 1981. *J. Clin. Endocrinol. Metab.* 52:1–8
23. Clifton, D. K., Steiner, R. A., Resko, J. A., Spies, H. G. 1975. *Biol. Reprod.* 13:190–94
24. Cogen, P. H., Antunes, J. L., Louis, K. M., Dyrenfurth, I., Ferin, M. 1980. *Endocrinology* 107:677–83
25. Conte, F. A., Grumbach, M. M., Kaplan, S. L., Reiter, E. O. 1980. *J. Clin. Endocrinol. Metab.* 50:163–68
26. Coquelin, A., Bronson, F. H. 1980. *Endocrinology* 106:1224–29
27. Crowley, W. F. Jr., Comite, F., Vale, W., Rivier, J., Loriaux, D. L., Cutler, G. B. Jr. 1981. *J. Clin. Endocrinol. Metab.* 52:370–72
28. Crowley, W. F., McArthur, J. W. 1980. *J. Clin. Endocrinol. Metab.* 51:173–75
29. Dierschke, D. J., Bhattacharya, A. N., Atkinson, L. E., Knobil, E. 1970. *Endocrinology* 87:850–53
30. Dierschke, D. J., Weiss, G., Knobil, E. 1974. *Endocrinology* 94:198–206
31. Dierschke, D. J., Yamaji, T., Karsch, F. J., Weick, R. F., Weiss, G., Knobil, E. 1973. *Endocrinology* 92:1496–1501
32. Donovan, B. T., ter Haar, M. B., Parvizi, N. 1977. *J. Physiol. London* 265:597–613
33. Dufy, B., Dufy-Barbe, L., Vincent, J. D., Knobil, E. 1979. *J. Physiol. Paris* 75:105–8
34. Ehara, Y., Ryan, K. J., Yen, S. S. C. 1972. *Contraception* 6:465–78
35. Ferin, M., Antunes, J. L., Zimmerman, E., Dyrenfurth, I., Frantz, A. G., Robinson, A., Carmel, P. W. 1977. *Endocrinology* 101:1611–20
36. Ferin, M., Bogumil, J., Drewes, J., Dyrenfurth, I., Jewelewicz, R., Vande Wiele, R. L. 1978. *Acta Endocrinol.* 89:48–59
37. Ferin, M., Carmel, P. W., Zimmerman, E. A., Warren, M., Perez, R., Vande Wiele, R. L. 1974. *Endocrinology* 95:1059–68
38. Ferin, M., Rosenblatt, H., Carmel, P. W., Antunes, J. L., Vande Wiele, R. L. 1979. *Endocrinology* 104:50–52
39. Foxcroft, G. R., Pomerantz, D. K., Nalbandov, A. V. 1975. *Endocrinology* 96:551–57
40. Gay, V. L., Sheth, N. A. 1972. *Endocrinology* 90:158–62
41. Goldenberg, R. L., Grodin, J. M., Vaitukaitis, J. L., Ross, G. T. 1973. *Am. J. Obstet. Gynecol.* 115:193–96
42. Goodman, A. L., Hodgen, G. D. 1977. *J. Clin. Endocrinol. Metab.* 45:837–40
43. Goodman, R. L., Karsch, F. J. 1980. *Endocrinology* 107:1286–90
44. Goodman, R. L., Karsch, F. J. 1981. In *Biological Clocks in Seasonal Reproductive Cycles*, ed. B. K. Follett, D. E. Fol-

lett, pp. 223–36. Bristol: John Wright and Sons Ltd. 292 pp.
45. Hauger, R. L., Karsch, F. J., Foster, D. L. 1977. *Endocrinology* 101:807–17
46. Heber, D., Swerdloff, R. S. 1981. *J. Clin. Endocrinol. Metab.* 52:171–72
47. Helmond, F. A., Simons, P. A., Hein, P. R. 1980. *Endocrinology* 107:478–85
48. Helmond, F. A., Simons, P. A., Hein, P. R. 1981. *Endocrinology* 108:1837–42
49. Hess, D. L., Wilkins, R. H., Moossy, J., Chang, J. L., Plant, T. M., McCormack, J. T., Nakai, Y., Knobil, E. 1977. *Endocrinology* 101:1264–71
50. Hoffmann, F. 1962. *Geburtsh. Frauenheilk.* 22:433–40
51. Johansson, E. D. B., Wide, L. 1969. *Acta Endocrinol.* 62:82–88
52. Johnson, L. M., Gay, V. L. 1981. *Endocrinology* 109:240–46
53. Karsch, F. J., Dierschke, D. J., Weick, R. F., Yamaji, T., Hotchkiss, J., Knobil, E. 1973. *Endocrinology* 92:799–804
54. Karsch, F. J., Weick, R. F., Butler, W. R., Dierschke, D. J., Krey, L. C., Weiss, G., Hotchkiss, J., Yamaji, T., Knobil, E. 1973. *Endocrinology* 92:1740–47
55. Karsch, F. J., Weick, R. F., Hotchkiss, J., Dierschke, D. J., Knobil, E. 1973. *Endocrinology* 93:478–86
56. Katongole, C. B., Naftolin, F., Short, R. V. 1971. *J. Endocrinol.* 50:457–66
57. Katongole, C. B., Naftolin, F., Short, R. V. 1974. *J. Endocrinol.* 60:101–6
58. Keye, W. R. Jr., Jaffe, R. B. 1974. *J. Clin. Endocrinol. Metab.* 38:805–10
59. Knobil, E. 1973. *Biol. Reprod.* 8:246–58
60. Knobil, E. 1974. *Rec. Prog. Horm. Res.* 30:1–36
61. Knobil, E. 1980. *Rec. Prog. Horm. Res.* 36:53–88
62. Knobil, E. 1981. *Biol. Reprod.* 24:44–49
63. Knobil, E., Plant, T. M., Wildt, L., Belchetz, P. E., Marshall, G. 1980. *Science* 207:1371–73
64. Krey, L. C., Butler, W. R., Knobil, E. 1975. *Endocrinology* 96:1073–87
65. Krey, L. C., Butler, W. R., Weiss, G., Weick, R. F., Dierschke, D. J., Knobil, E. 1973. In *Hypothalamic Hypophysiotropic Hormones,* ed. C. Gual, E. Rosemberg, pp. 39–47. Amsterdam: Excerpta Medica. 428 pp.
66. Leyendecker, G. 1979. *Eur. J. Obstet. Gynecol. Reprod. Biol.* 9:175–86
67. Leyendecker, G., Struve, T., Plotz, E. J. 1980. *Arch. Gynäkol.* 229:177–90
68. Leyendecker, G., Wardlaw, S., Nocke, W. 1972. *Acta Endocrinol.* 71:160–78
69. Leyendecker, G., Wildt, L., Gips, H., Nocke, W., Plotz, E. J. 1976. *Arch. Gynäkol.* 221:29–45
70. Leyendecker, G., Wildt, L., Hansmann, M. 1980. *J. Clin. Endocrinol. Metab.* 51:1214–16
71. March, C. M., Goebelsmann, U., Nakamura, R. M., Mishell, D. R. Jr. 1979. *J. Clin. Endocrinol. Metab.* 49:507–13
72. Marshall, P. E., Goldsmith, P. C. 1980. *Brain Res.* 193:353–72
73. Marshall, J. C., Kelch, R. P. 1979. *J. Clin. Endocrinol. Metab.* 49:712–18
74. McCormack, J. T., Plant, T. M., Hess, D. L., Knobil, E. 1977. *Endocrinology* 100:663–67
75. Monroe, S. E., Jaffe, R. B., Midgley, A. R. Jr. 1972. *J. Clin. Endocrinol. Metab.* 34:342–47
76. Naftolin, F., Yen, S. S. C., Perlman, D., Tsai, C. C., Parker, D. C., Vargo, T. 1973. *J. Clin. Endocrinol. Metab.* 37:6–10
77. Nakai, Y., Plant, T. M., Hess, D. L., Keogh, E. J., Knobil, E. 1978. *Endocrinology* 102:1008–14
78. Nankin, H. R., Troen, P. 1971. *J. Clin. Endocrinol. Metab.* 33:558–60
79. Neill, J. D., Patton, J. M., Dailey, R. A., Tsou, R. C., Tindall, G. T. 1977. *Endocrinology* 101:430–34
80. Nillius, S. J., Wide, L. 1971. *Acta Endocrinol.* 67:362–70
81. Nillius, S. J., Wide, L. 1971. *J. Obstet. Gynaecol. Br. Commonw.* 78:822–27
82. Norman, R. L., Resko, J. A., Spies, H. G. 1976. *Endocrinology* 99:59–71
83. Odell, W. D., Swerdloff, R. S. 1968. *Proc. Natl. Acad. Sci. USA* 61:529–36
84. Odell, W. D., Swerdloff, R. S. 1975. *Adv. Biosci.* 15:141–56
85. Pavasuthipaisit, K., Hess, D. L., Norman, R. L., Adams, T. E., Baughman, W. L., Spies, H. G. 1981. *Neuroendocrinology* 32:42–49
86. Penny, R., Guyda, H. J., Baghdassarian, A., Johanson, A. J., Blizzard, R. M. 1970. *J. Clin. Invest.* 49:1847–52
87. Piper, E. L., Perkins, J. L., Tugwell, D. R., Vaught, W. G. 1975. *Proc. Soc. Exp. Biol. Med.* 148:880–82
88. Plant, T. M. 1980. In *Testicular Development, Structure, and Function,* ed. A. Steinberger, E. Steinberger, pp. 419–23. NY: Raven Press. 536 pp.
89. Plant, T. M. 1980. *Proc. 10th Ann. Meet. Soc. Neurosci., Cincinnati,* Abstr. 140.1
90. Plant, T. M., Drey, L. C., Moossy, J., McCormack, J. T., Hess, D. L., Knobil, E. 1978. *Endocrinology* 102:52–62
91. Plant, T. M., Moossy, J., Hess, D. L.,

Nakai, Y., McCormack, J. T., Knobil, E. 1979. *Endocrinology* 105:465–73
92. Plant, T. M., Nakai, Y., Belchetz, P., Keogh, E., Knobil, E. 1978. *Endocrinology* 102:1015–18
93. Plant, T. M., Schallenberger, E., Hess, D. L., McCormack, J. T., Dufy-Barbe, L., Knobil, E. 1980. *Biol. Reprod.* 23:760–66
94. Pohl, C. R., Richardson, D. W., Marshall, G. R., Knobil, E. 1981. *Proc. 63rd Ann. Meet. Endocrine Soc., Cincinnati*, Abstr. 196.
95. Rabin, D., McNeil, L. W. 1980. *J. Clin. Endocrinol. Metab.* 51:873–76
96. Rahe, C. H., Owens, R. E., Fleeger, J. L., Newton, H. J., Harms, P. G. 1980. *Endocrinology* 107:498–503
97. Rakoff, J. S., Yen, S. S. C. 1978. *J. Clin. Endocrinol. Metab.* 47:918–21
98. Resko, J. A., Ellinwood, W. E., Knobil, E. 1981. *Am. J. Physiol.* 240:E489–92
99. Resko, J. A., Koering, M. J., Goy, R. W., Phoenix, C. H. 1975. *J. Clin. Endocrinol. Metab.* 41:120–25
100. Richardson, D. W., Hutchison, J. S., Pohl, C. R., Germak, J. A., Knobil, E. 1981. *Proc. 63rd Ann. Meet. Endocrine Soc., Cincinnati*, Abstr. 513
101. Rolland, R., Lequin, R. M., Schellekens, L. A., De Jong, F. H. 1975. *Clin. Endocrinol.* 4:15–25
102. Root, A., DeCherney, A., Russ, D., Duckett, G., Garcia, C.-R., Wallach E. 1972. *J. Clin. Endocrinol. Metab.* 35:700–4
103. Rowe, P. H., Hopkinson, C. R. N., Shenton, J. C., Glover, T. D. 1975. *Steroids* 25:313–21
104. Santen, R. J. 1975. *J. Clin. Invest.* 56:1555–63
105. Santen, R. J., Bardin, C. W. 1973. *J. Clin. Invest.* 52:2617–28
106. Schallenberger, E., Richardson, D. W., Knobil, E. 1981. *Biol. Reprod.* 25:370–74
107. Schoemaker, J., Simons, A. H. M., van Osnabrugge, G. J. C., Lugtenburg, C., van Kessel, H. 1981. *J. Clin. Endocrinol. Metab.* 52:882–85
108. Schuiling, G. A., De Koning, J., Zürcher, A. F., Gnodde, H. P., van Rees, G. P. 1976. *Neuroendocrinology* 20:151 56
109. Silverman, A. J., Antunes, J. L., Forin, M., Zimmerman, E. A. 1977. *Endocrinology* 101:134–42
110. Sizonenko, P. C., Burr, I. M., Kaplan, S. L., Grumbach, M. M. 1970. *Pediatr. Res.* 4:36–45
111. Smith, M. A., Vale, W. W. 1981. *Endocrinology* 108:752–59
112. Spies, H. G., Niswender, G. D. 1972. *Endocrinology* 90:257–61
113. Terasawa, E., Rodriguez-Sierra, J. F., Dierschke, D. J., Bridson, W. E., Goy, R. W. 1980. *J. Clin. Endocrinol. Metab.* 51:1245–50
114. Thompson, I. E., Arfania, J., Taymor, M. L. 1973. *J. Clin. Endocrinol. Metab.* 37:152–55
115. Tsai, C. C., Yen, S. S. C. 1971. *J. Clin. Endocrinol. Metab.* 32:766–71
116. Valk, T. W., Corley, K. P., Kelch, R. P., Marshall, J. C. 1980. *J. Clin. Endocrinol. Metab.* 51:730–38
117. Weick, R. F. 1981. *Biol. Reprod.* 24:415–22
118. Weick, R. F., Dierschke, D. J., Karsch, F. J., Butler, W. R., Hotchkiss, J., Knobil, E. 1973. *Endocrinology* 93:1140–47
119. Wentz, A. C., Jones, G. S., Sapp, K. 1976. *Obstet. Gynecol.* 47:309–18
120. Wildt, L., Hausler, A., Hutchison, J. S., Marshall, G., Knobil, E. 1981. *Endocrinology* 108:2011–13
121. Wildt, L., Hausler, A., Marshall, G., Hutchison, J. S., Plant, T. M., Belchetz, P. E., Knobil, E. 1981. *Endocrinology.* 109:376–85
122. Wildt, L., Hutchison, J. S., Marshall, G., Knobil, E., Pohl, C. R. 1981. *Endocrinology* 109:1293–94
123. Wildt, L., Marshall, G., Knobil, E. 1980. *Science* 207:1373–75
124. Williams, R. F., Johnson, D. K., Hodgen, G. D. 1979. *J. Clin. Endocrinol. Metab.* 49:422–28
125. Wilson, S. C., Sharp, P. J. 1975. *J. Endocrinol.* 64:77–86
126. Yamaji, T., Dierschke, D. J., Bhattacharya, A. N., Knobil, E. 1972. *Endocrinology* 90:771–77
127. Yamaji, T., Dierschke, D. J., Hotchkiss, J., Bhattacharya, A. N., Surve, A. H., Knobil, E. 1971. *Endocrinology* 89:1034–41
128. Yen, S. S. C., Rebar, R., Vandenberg, G., Judd, H. 1973. *J. Clin. Endocrinol. Metab.* 36:811–16
129. Yen, S. S. C., Tsai, C. C. 1971. *J. Clin. Endocrinol. Metab.* 33:882–87
130. Yen, S. S. C., Tsai, C. C. 1972. *J. Clin. Endocrinol. Metab.* 34:298–305
131. Yen, S. S. C., Tsai, C. C., Naftolin, F., Vandenberg, G., Ajabor, L. 1972. *J. Clin. Endocrinol. Metab.* 34:671–75
132. Yen, S. S. C., Tsai, C. C., Vandenberg, G., Rebar, R. 1972. *J. Clin. Endocrinol. Metab.* 35:897–904
133. Yen, S. S. C., Vandenberg, G., Siler, T. M. 1974. *J. Clin. Endocrinol. Metab.* 39:170–77
134. Young, J. R., Jaffe, R. B. 1976. *J. Clin. Endocrinol. Metab.* 42:432–42

NEUROENDOCRINE CONTROL MECHANISMS AND THE ONSET OF PUBERTY

Edward O. Reiter

University of Massachusetts, Baystate Medical Center, Springfield, Massachusetts 01107

Melvin M. Grumbach

University of California, San Francisco, California 94143

Puberty is the transitional period between the juvenile state and adulthood during which the adolescent growth spurt occurs, secondary sexual characteristics appear, fertility is achieved, and profound psychological changes take place. We propose that the events characterizing pubertal maturation of the reproductive endocrine system can be viewed as part of a continuum extending from sexual differentiation and the ontogeny of the hypothalamic-pituitary gonadotropin-gonadal system in the fetus to the attainment of full sexual maturation and fertility, and then ultimately to senescence (20, 21). Two independent processes, controlled by different mechanisms but closely linked temporally, are involved in the increase of sex steroid secretion in the peripubertal and pubertal periods. One process, *adrenarche,* involves the increase in adrenal androgen secretion, which precedes by about two years the second event, *gonadarche,* the activation of the hypothalamic-pituitary gonadotropin-gonadal apparatus that had been active at a low level during childhood.

The regulatory systems that control human male and female reproduction comprise the following fundamental components:

1. The arcuate nucleus of the medial basal hypothalamus and its transducer neurosecretory neurons. These translate neural signals into a periodic,

oscillatory chemical signal, gonadotropin-releasing hormone (GnRH). GnRH, synthesized by the neurosecretory peptidergic neurons and released from their axon terminals at the median eminence into the primary plexus of the hypothalamic-hypophyseal portal circulation, is transported by this private conduit to the anterior pituitary gland. A large body of evidence in rodents indicates that catecholaminergic and opioid neuronal networks, as well as sex steroids, modulate the release of GnRH. The importance of neurotransmitters and opioid neural pathways in man remains to be established.

2. The pituitary gonadotropes that, in response to the GnRH rhythmic signal, release LH and FSH in a pulsatile manner at periodic intervals.
3. The gonads.

This control mechanism, with its 3 principal components (the arcuate GnRH neurosecretory neurons, the pituitary gonadotropes, and the gonadotropin-responsive elements of the gonad), is common to all mammalian species. It is at each of these loci that modulating factors exercise their effect. Further, at each of the latter two levels, the target cells contain specific cell-surface receptors for the peptide hormones that mediate the cellular response to the signal.

We have proposed that the hypothalamic-pituitary-gonadal system in the human differentiates and functions during fetal life and early infancy, is suppressed to a low level of activity for almost a decade during childhood, and is reactivated during puberty (20, 28). In this light, puberty represents not the initiation of pulsatile secretion of GnRH and, thus, of pituitary gonadotropins, but the reactivation, after a protracted period of quiescent or absent activity, of the GnRH neurosecretory neurons in the arcuate nucleus and their endogenous, apparently self-sustaining oscillatory secretion. This system initially is operative in the fetus. Experimental and clinical studies support the hypothesis that the CNS, not the pituitary gland nor gonads, restrains the activation of the hypothalamic-pituitary gonadotropin-gonadal system in prepubertal children. This inhibition appears to be mediated through the suppression of GnRH synthesis and its pulsatile secretion (12a, 20).

TIMING OF THE ONSET OF PUBERTY

The specific mechanisms involved in the timing of puberty are complex and poorly understood. The average age at onset of puberty shows a secular trend over the past century toward earlier occurrence; the trend cuts across geographic and ethnic lines (65, 72). This progressive decline in the age at puberty is thought to be due to improvements in socioeconomic conditions, nutrition, and general health; in developed nations such a trend appears to have slowed or ceased over the last 20 years (65).

Influences other than socioeconomic also affect the age at puberty. The association of a minimal percent of body fat with menarche and the initiation of the pubertal growth spurt, and the suggestion of a "critical" factor in altering metabolic rate and hypothalamic function have been widely debated (19, 45). An effect of nutritional factors and body composition upon the time of onset of puberty is supported by the earlier age of menarche in moderately obese girls (72), by delayed maturation of the reproductive endocrine system in states of malnutrition and chronic illness (37) and following early athletic or ballet training (18, 68), and by the relationship of amenorrhea to such states of diminished body fat as anorexia nervosa (59), voluntary weight loss (37), and vigorous physical conditioning (18, 38, 67). When increments in the excretion of urinary gonadotropin were correlated with changes in body composition at puberty, however, both developmental events appeared to occur simultaneously rather than sequentially (49). Although menarche is a relatively late pubertal event and is removed from those neural factors that influence both the gonadotropin-gonadal sex steroid and physical changes at the initiation of puberty, the possibility exists that some alteration of body metabolism, including the ratio of fat to lean body mass, may affect the CNS restraints of pubertal onset. In the human, the pineal gland and melatonin do not have an important inhibitory influence on this control system (17, 34, 62) despite an unconfirmed report to the contrary (60).

THE PATTERNS OF GONADOTROPIN SECRETION

There are two patterns of gonadotropin secretion: tonic and cyclic (20, 64). Tonic or basal secretion is regulated by negative or inhibitory feedback mechanisms: Changes in the concentration of circulating sex steroids, and possibly "inhibins" (putative nonsteroidal regulators of FSH secretion produced by germinal tissue), result in reciprocal changes in secretion of pituitary gonadotropins. This is the general pattern of secretion in the male and one of the control mechanisms in the female. Cyclic secretion involves positive or stimulatory feedback mechanisms: An increment in circulating estrogens to a critical level for a sufficient duration initiates a synchronous, pulsatile burst of LH and FSH (the midcycle surge), which is characteristic of the pattern in normal adult females prior to menopause. FSH and LH are probably always secreted in a pulsatile or episodic manner at periodic intervals, irrespective of whether the secretion is tonic or cyclic.

In the adult male, pulsatile release of LH has a periodicity of about 90 min; the periodicity of the LH secretory episodes is similar in adult females except during the mid- and late-luteal phases of the menstrual cycle, when the LH pulses are diminished to every 3–4 hr. The pulsatile secretion of FSH in normal adults is less prominent; this has been attributed, in part, to the longer half-life of FSH than LH, to differences in the factors that modulate the action of GnRH on FSH and LH release, and to fundamental differences in the secretory pattern for the two gonadotropins.

The physiologic implications of this new concept of gonadotropin secretion was unclear until the classic observations of Knobil and his associates (31) in the rhesus monkey. They described the inhibition of gonadotropin secretion by the continuous infusion of GnRH via desensitization or down-regulation of gonadotrope GnRH receptors (10, 44). Discontinuous or pulsatile stimulation (e.g. 1 μg min^{-1} of GnRH for 6 min every hour) reverses or prevents refractoriness to GnRH and restores pulsatile release of gonadotropin in adult monkeys with hypothalamic lesions that had obliterated the arcuate nucleus and thus extinguished endogenous GnRH secretion. These studies provided evidence that the GnRH input to the pituitary gonadotropes is frequency-coded. The studies of Carmel et al (8) have shown that intermittent abrupt rises of GnRH levels in portal blood occur at intervals of 2–3.5 hr. In a single patient whose portal blood was sampled during surgery, over 100-fold variation in GnRH levels was demonstrated.

GONADOTROPIN LEVELS FROM FETAL LIFE THROUGH PUBERTY

The changing pattern of gonadotropin and sex steroid secretion according to age and level of maturation has been discussed (20, 45, 64). Kaplan and Grumbach and their associates [see (28) for review] have described development of pituitary gonadotropin secretion in the fetus. The human fetal pituitary gland can synthesize and store FSH and LH by 10 weeks of gestation and can secrete these hormones by 11 and 12 weeks. The pattern of changes of FSH and LH concentrations in both pituitary glands and serum of fetuses is consistent with a sequence of increased synthesis and secretion, in which peak serum concentrations reach adult castrate levels, followed by a decline after mid-gestation that persists to term.

After the fall in sex steroids, especially estrogens, during the first days after birth, the concentration of FSH and LH increases and exhibits wide perturbations during the first months of life; intermittent high gonadotropin concentrations are associated with increased testosterone values in male infants and estradiol levels in females (70). By about 6 months in the male and 1–2 years of age in the female, concentrations of gonadotropins decrease to low levels present during childhood until the onset of puberty.

In the peripubertal period, gonadotropin concentrations rise. In girls, FSH levels rise during the early stages of puberty and then plateau, whereas LH levels tend to rise in later stages; in boys, FSH concentrations rise progressively during puberty, whereas LH levels increase sharply in early pubertal development and then gradually rise throughout the remainder of pubertal maturation. In addition to these changes in immunoreactive gonadotropins, serum concentrations of biologically active LH (measured by a sensitive rat or mouse interstitial cell testosterone generation assay) have been quantitated throughout fetal life and at all stages of extrauterine life (15, 35, 55). In human studies, bioactive LH concentrations are usually undetectable during prepubertal years, then rise dramatically during pubertal maturation (35, 55). The increment in mean LH levels between pubertal and prepubertal states has generally been greater using bioassays than standard immunoassay techniques, perhaps because of methodologic problems, but possibly related to qualitative changes in the LH molecule (1, 35, 56).

EPISODIC RELEASE OF GONADOTROPINS AND THE DEVELOPMENT OF CIRCADIAN RHYTHMS

In adult men and in women during the follicular and early luteal phases, discrete episodic or pulsatile bursts of LH occur about once every 90 min (27, 53). In prepubertal children, some, but not all, investigators have found secretory episodic bursts of LH (27, 45, 48, 64); the pulses are of lower amplitude than in pubertal children or adults (48). Penny et al (48) pointed out that the increase in the plasma concentration of LH during puberty correlates with the increase in magnitude of the LH pulses rather than a change in frequency. In such studies, concentrations of gonadotropins are low, and it is often difficult to demonstrate episodic release for methodologic and statistical reasons. In pubertal subjects, Boyar et al (4) first described the mainly sleep-associated pulsatile release of LH in early and mid-puberty; only late in puberty were prominent LH pulses noted during the day. Additionally, pulsatile release of FSH occurs in pubertal individuals, though the spikes are smaller than those of LH (27).

In addition to ultradian (episodic) fluctuations of gonadotropin release, a circadian rhythm (with nocturnal peaking) develops during late childhood and adolescent years. In prepubertal boys, Parker et al (46), but not Boyar et al (4), reported significant LH increments in some sleeping children. Nocturnal urinary LH excretion in prepubertal children is greater than in daytime, although absolute differences are small (33). Judd et al (27) described a larger group of prepubertal subjects and appeared to resolve the discrepancies. In boys younger than 10.5 years, LH concentrations rose significantly in only 22% of nights, while in the older prepubertal boys, LH levels were greater during sleep in 78% of nights. Generally similar, though somewhat less striking nocturnal increments have been described in serum FSH levels in prepubertal and late prepubertal children. During puberty, there is further maturation of sleep-enhanced LH secretion, presumably related to alterations in CNS restraint of the hypothalamic GnRH pulse generator that leads to increased activity initially, mainly during sleep. In early and mid-pubertal subjects, pulsatile LH release occurs largely during sleep; in late puberty, pulsatile release is demonstrable throughout the whole day and simulates the adult pattern. The factors that lead to the initiation and development of this circadian rhythm remain unclear; patients with sexual maturation as in idiopathic precocious puberty, or glucocorticoid-treated patients with congenital virilizing adrenal hyperplasia who have advanced bone age and early onset of true puberty exhibit the same pattern of LH secretion as normal pubertal children (5). The pattern of enhanced sleep-associated LH secretion occurring in agonadal patients during the pubertal period suggests that this pattern does not depend upon gonadal function (6).

Such sleep-associated LH release appears to correlate with increased sensitivity of pituitary gonadotropes to administration of GnRH in the peripubertal period and during puberty (2, 13).

CENTRAL NERVOUS SYSTEM AND THE ONSET OF PUBERTY

In both human and subhuman primates, the increased LH and FSH secretion in the fetus and during infancy is followed by a long period, approximately one decade, in which the reproductive endocrine system is suppressed (20, 21, 28, 64). The factors involved in this restraint of the onset of puberty are not well understood. Two mechanisms have been invoked to explain the prepubertal restraint by the CNS of gonadotropin secretion. One is a sex steroid–dependent mechanism, a highly sensitive hypothalamic-pituitary-gonadal negative feedback system. The other is a sex steroid–independent mechanism that can be ascribed to "intrinsic" CNS inhibitory influences (12, 12a, 20) (Figure 1).

Negative Feedback Mechanism (Sex Steroid–Dependent)

Fifty years ago Hohlweg & Dohrn suggested that at puberty a CNS "Sexualzentrum" that regulates gonadotropin secretion changed its sensitivity to circulating sex steroids (24). Studies in the fetal sheep, which has a pattern of fetal pituitary gonadotropin secretion similar to that of the hu-

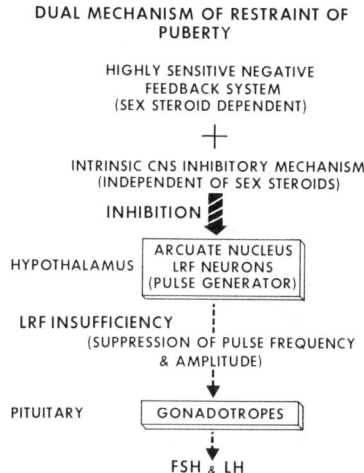

Figure 1 Dual mechanism of restraint of puberty.

man fetus, describe a decrement in GnRH-evoked gonadotropin release after midgestation with advancing gestational age (41). The inhibition of hypothalamic-GnRH release and lowered pituitary gonadotropin secretion appear to be consequences of the progressive acquisition of increased sensitivity of the hypothalamic "gonadostat" (the arcuate GnRH neurosecretory neurons), and probably the pituitary gland, to inhibitory effects of high concentrations of sex steroids in the fetal circulation (41). This hypothalamic-regulatory mechanism, which probably involves the maturation of sex steriod receptors on the GnRH neurons, is not fully developed at birth (28, 41). During childhood, this tonic control mechanism is exquisitely sensitive to the suppressive effect of small amounts of circulating sex steroids. Coincident with the onset of puberty, the hypothalamic-gonadostat (the arcuate nucleus), and possibly the pituitary gland, becomes progressively less sensitive to the inhibitory effects of sex steroids upon GnRH release, which results in the increased release of GnRH in a pulsatile pattern and enhanced secretion of gonadotropins. In adults, the hypothalamic-pituitary negative feedback mechanism is less sensitive to feedback by sex steroids, and adult levels of gonadotropins and sex steroids are present.

Evidence for an operative and highly sensitive negative feedback mechanism in prepubertal children has been summarized (21): (*a*) The pituitary gland in the prepubertal child secretes small amounts of FSH and LH, suggesting that the hypothalamic-pituitary gonadotropin-gonadal complex operates during childhood, but at a low level of activity. (*b*) The low level of gonadotropin secretion in childhood is rapidly shut off by administration of sex steroids. When small amounts of estrogen are administered to prepubertal children, a quick and significant decrease in gonadotropin secretion ensues (21, 29); in contrast, considerably higher levels of estrogen are required to suppress gonadotropin secretion in the adult (20, 32). The elevated gonadotropin concentrations in infancy and early childhood in patients with gonadal dysgenesis (12) are evidence that hormones secreted by the normal prepubertal gonad, despite their low level, inhibit gonadotropin secretion; they support the hypothesis that a highly sensitive negative feedback mechanism is operative in young prepubertal children.

"Intrinsic" CNS Inhibitory Mechanism (Sex Steroid–Independent)

The diphasic pattern of basal and GnRH-induced FSH and LH secretion from infancy to adulthood (i.e. higher in the first year of life than in the next 8–10 years) is qualitatively similar in normal individuals and in patients with gonadal dysgenesis, but in the latter, gonadotropin levels are strikingly higher except during the mid-childhood nadir (12). The striking fall in gonadotropin secretion and reserve (12a) in agonadal children 4–11 years old (mid-childhood nadir) suggests the presence of CNS inhibitory influences independent of gonadal sex steroid secretion that restrain gonadotropin production and delay the onset of puberty. Such a pattern cannot be explained by gonadal sex steroid feedback since functional gonads are lacking, or by increased secretion of adrenal sex steroids, since concentrations are low and dexamethasone suppression of the adrenal does not augment serum levels of gonadotropins. The nature of this postulated intrinsic CNS inhibitory system during infancy and childhood remains uncertain (Figure 2). Suppression of this neural inhibitory mechanism would lead to reactivation of gonadotropin secretion at puberty. In patients who develop true precocious puberty due to hypothalamic lesions, the intrinsic CNS inhibitory system is impaired and results in the premature appearance of the augmented, pulsatile gonadotropin secretion characteristic of puberty.

Whether the possible inhibitory effects of opioids (39, 40, 51, 52) and dopamine (25, 52) on gonadotropin secretion can be translated to the prepubertal period remains to be established.

LOCATION OF THE NEGATIVE FEEDBACK MECHANISM

Ferin et al found that micro-injections of estradiol into the hypothalamus limited pulsatile LH secretion in ovariectomized monkeys (16). Pituitary portal blood levels of immunoreactive GnRH are elevated in ovariectomized monkeys (43). Radiofrequency lesions in the arcuate region of the medial-basal hypothalamus of ovariectomized rhesus monkeys (42) [or ad-

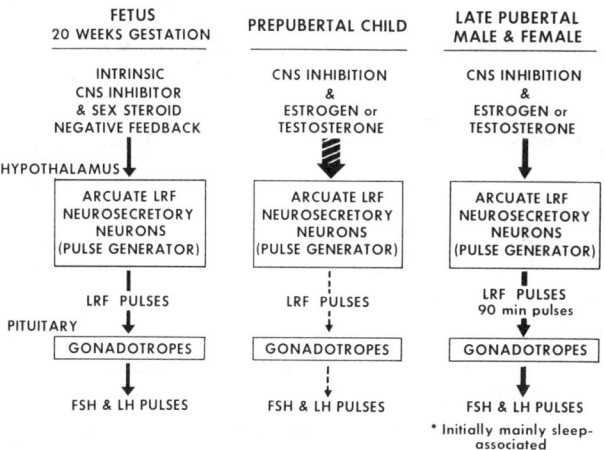

Figure 2 A scheme illustrating the changes in the activity of the arcuate GnRH pulse generator during development and the effect on pituitary gonadotropes and the hypothesis that the functional GnRH insufficiency of the prepubertal child is a consequence of CNS restraint by sex-steroid dependent and sex-steroid independent mechanisms.

ministration of anti-GnRH serum (30a)] dramatically lowered serum levels of LH and FSH. In the same animal model, pulsatile secretion of LH was reestablished by chronic, intermittent administration of GnRH (30a). Subsequent administration of estradiol quickly eliminated this episodic secretion of LH along with diminishing mean basal serum concentrations of LH and FSH. These experiments suggest an acute negative feedback effect of estradiol upon gonadotropin secretion in the pituitary gonadotropes. However, using a microinfusion system, Chappel et al delivered extremely small quantities of estrogen to the medial basal hypothalamus, primarily the arcuate nucleus or to the anterior pituitary (9). Estrogen injected into the arcuate nucleus decreased the rate of GnRH secretion, as reflected in lowered gonadotropin levels and inhibition of pulsatile LH release. Direct gonadotrope exposure to estrogen lowered responsivity to GnRH but did not diminish pulsatile LH secretion. The evidence thus suggests that estrogen-induced gonadotropin suppression occurs at both the hypothalamic and pituitary levels.

THE INTERACTION OF SYNTHETIC GnRH WITH PITUITARY GONADOTROPIN SECRETION

If increased secretion of gonadotropins with the approach of puberty is a consequence of a change in neural and hormonal restraints on synthesis and pulsatile secretion of GnRH, the disinhibition of the arcuate GnRH oscillator should lead to increased GnRH pulses (frequency and amplitude) initially, followed by increased gonadotropin secretion by the pituitary and, finally, to augmented output of sex steroids by the gonad. GnRH secretion is not readily measured directly in the human being; however, endogenous release can be assessed indirectly and qualitatively by the gonadotropin response to exogenous GnRH (21). With the availability of synthetic GnRH, the pituitary sensitivity to GnRH and the dynamic reserve or readily releasable pool of pituitary gonadotropins have been examined during different stages of pubertal maturation and in many disorders involving the hypothalamic-pituitary-gonadal system (20, 21, 57, 64). The results support the concept that the prepubertal state is characterized by functional GnRH insufficiency (20, 21).

The release of LH following the administration of GnRH is minimal in prepubertal children beyond infancy, increases strikingly during the peripubertal period and puberty, and is still greater in adult males and females (20, 21). The change in the maturity-related patterns of FSH release after administration of GnRH is quite different from that of LH and results in a striking reversal of the FSH/LH ratio of gonadotropin release. Prepubertal and pubertal females release much more FSH than males at all stages of sexual maturation. Prepubertal girls, in fact, have a larger readily releasable pool of pituitary FSH than pubertal girls or prepubertal or pubertal males. Because basal levels of serum FSH and LH are similar in prepubertal children, the dramatic difference in secretion of FSH and LH evoked by GnRH administration demonstrates clearly the difference between pituitary sensitivity and the actual basal secretory rate of FSH and LH.

The heightened LH response of pituitary gonadotropes to exogenous GnRH in peripubertal children who do not yet exhibit physical signs of sexual maturation provides further evidence that the self-priming effect (21, 57) of endogenous GnRH augments pituitary responsiveness to exogenous GnRH. The degree of previous exposure of gonadotropes to endogenous GnRH appears to affect both the magnitude and quality of LH responses to a single dose of GnRH. Before puberty, the low setpoint of the hypothalamic gonadostat to inhibitory feedback effects of low concentrations of plasma sex steroids and intrinsic CNS factors suppresses secretion of GnRH. As a result, the prepubertal pituitary gland has a small pool of releasable LH and, thus, decreased responsiveness to acute administration of synthetic GnRH. With the approach of puberty, increased release of endogenous GnRH increases the number of LH receptors on the gonadotrope (10), augments pituitary sensitivity to exogenous GnRH, and enlarges the reserve of LH in the gonadotrope. The explanation for the discordance of FSH and LH release during maturation is not clear, but recent studies by Knobil with differing modes of administration of GnRH suggest that the pulse frequency of the pattern of secretion of endogenous GnRH is a major factor (30). Reducing the frequency of the GnRH pulse from one every hour to one every 3 hours increased strikingly the ratio of FSH to LH. Further, endogenous sex steroids may affect this ratio—e.g. the inhibitory effects of testosterone and estradiol on FSH and LH release.

These findings and the above-noted phenomenon of gonadotrope desensitization by chronic GnRH administration led to the administration of GnRH by intermittent intravenous infusion by Belchetz et al (3). When GnRH was given intermittently, gonadotropin secretion did not show refractoriness and was fully reestablished in animals with lesions of the medial basal hypothalamus in which the arcuate nucleus (and its GnRH neurons) was obliterated (3, 30, 30a, 31). It has also been observed that when GnRH was administered to prepubertal monkeys in a pulsatile manner, spontaneous menstrual cycles were established (30, 68). These data dramatically demonstrated the importance of the pattern of GnRH administration and that puberty could be induced in prepubertal monkeys by a discontinuous GnRH pulse (1 per hour). The findings stimulated a series of investigations in humans.

When large doses of GnRH were administered by varying routes to patients with hypogonadotropic hypogonadism for approximately one week, immature responses to standard GnRH tests were normalized (71). Jacobson et al (26) reported that treatment of two males with Kallmann syndrome with hourly subcutaneous pulses of GnRH on 10 successive nights led to increased mean gonadotropin levels, the appearance of pulsatile LH secretion, and a greater augmentation of LH than FSH secretion. In three patients with anorexia nervosa (36), patterns of FSH and LH release were carefully studied during 5 days of low-dose intravenous GnRH pulse therapy (1 per 2 hours). Early in the treatment period, basal levels of FSH rose quickly and FSH responses to GnRH exceeded those of LH. At completion of the study, however, LH release exceeded that of FSH, as in normal pubertal maturation. In a series of patients with hypogonadotropic hypogonadism, the same protocol of intravenous administration of GnRH pulses achieved a similar alteration of basal FSH/LH ratio and maturation of gonadtrope responses to acute GnRH tests (14, 66). These studies suggest that administration of GnRH in a pulsatile manner, presumably comparable to that which occurs in the normal individual, leads to reversal of the FSH/LH ratio, as is commonly seen during spontaneous pubertal development. The data provide further indirect support for the reawakening of augmented pulsatile GnRH release by the arcuate nucleus oscillator (rather than basic changes at the level of the gonad or anterior pituitary) as the first hormonal change in the onset of puberty.

MATURATION OF THE POSITIVE FEEDBACK MECHANISM

In normal women the midcycle surge in LH and FSH secretion is attributed to the positive feedback effect of an increased, critical concentration of estradiol for a sufficient length of time during the latter part of the follicular phase. This stimulatory effect of estradiol upon gonadotropin secretion has not been demonstrated in prepubertal or early pubertal girls; it is a later maturational event and does not occur before mid-puberty (21, 29, 50, 54). Development of positive feedback action of estradiol requires: (*a*) ovarian follicles primed by FSH to secrete sufficient estradiol to reach and maintain a critical level in the circulation; (*b*) a pituitary gland that is sensitized by estrogen to amplify and augment the effect of GnRH and that contains a large enough pool of releasable LH to provide an LH surge; and (*c*) in addition to the usual adult pattern of pulsatile LH-RH secretion, quite likely sufficient GnRH stores for the GnRH neurosecretory neurons to respond to estradiol stimulation with an acute increase in GnRH release, though this last requirement has not been firmly established in primates (20).

Several studies suggest that estrogen-induced positive feedback occurs directly at the pituitary level so long as the gonadotropes are being or have been recently exposed to GnRH (11, 42, 69). Moreover, Knobil et al (30, 31) reestablished menstrual cyclicity in lesioned monkeys by the chronic intermittent administration of GnRH. Pulsatile GnRH treatment at a fixed dose led to sufficient gonadotropin secretion to stimulate estradiol synthesis and release by ovarian follicles and induced an ovulatory LH surge in the absence of an endogenous hypothalamic GnRH secretory surge. Hence an increase in neither the frequency nor the amplitude of the GnRH pulse is required to induce a midcycle LH surge. However, the data do not exclude the possibility that, in the normal adult female, estrogen also induces an increase in GnRH release and, thus, has a positive feedback effect on both the pituitary gland and the hypothalamic GnRH neurons.

Even though gonadotropin cyclicity (23) and estrogen-induced positive feedback have been demonstrated by mid-puberty and prior to menarche, the positive feedback loop does not appear to be complete at that time (20, 54, 64). Indeed, the modulating action of the pubertal ovary in its output of estradiol on the hypothalamic-pituitary-gonadotropin unit appears insufficient to induce an ovulatory LH surge even when there are adequate pituitary stores of readily releasable LH and FSH. The ovary, from either lack of sufficient gonadotropin stimulation or decreased responsivity, does not secrete estradiol in sufficient amount or duration to induce the ovulatory LH surge. Thus during the first two years of postmenarche as many as

50–90% of cycles are anovulatory, decreasing to less than 20% of cycles by 5 years after menarche (20, 64). Table I summarizes the proposed ontogeny of the hypothalamic pituitary gonadotropin-gonadal apparatus.

Table 1 Postulated ontogeny of hypothalamic-pituitary gonadotropin-gonadal circuit

A. Fetus
   1. Hypothalamic arcuate GnRH neurosecretory neurons (pulse generator) operative by 80 days gestation
   2. Episodic secretion of FSH and LH by 80 days gestation
   3. Initially unrestrained secretion of GnRH (100–150 days)
   4. Maturation of negative sex steroid feedback mechanisms after 150 days gestation—sex difference
   5. Low level of GnRH secretion at term
B. Early Infancy
   1. Arcuate GnRH pulse generator highly functional after 12 days of age
   2. Prominent RSH and LH episodic discharges until about age 6 months in males and 12 months in females with transient increase in plasma levels of testosterone and estradiol in males and females respectively.
C. Late Infancy and Childhood
   1. Negative feedback control of FSH and LH secretion becomes highly sensitive to sex steroids (low setpoint), maximum sensitivity attained by age 2–4 years
   2. Intrinsic CNS inhibition of arcuate GnRH pulse generator operative, maximum sensitivity by about age 4 years
   3. Arcuate GnRH pulse generator inhibited; amplitude and probably frequency of GnRH discharges low
   4. Secretion of FSH, LH, and sex steroids is low
D. Late Prepubertal Period
   1. Decreasing sensitivity of hypothalamic-pituitary unit to sex steroids (increased setpoint) and decreasing effectiveness of intrinsic CNS inhibitory influences
   2. Increased amplitude and frequency of GnRH pulses; initially most prominent with sleep (nocturnal)
   3. Increased sensitivity of gonadotropes to GnRH
   4. Increased secretion of FSH and LH
   5. Increased responsiveness of gonad to FSH and LH
   6. Increased secretion of gonadal hormones
E. Puberty
   1. Further decrease in CNS restraint of arcuate GnRH neurons including the sensitivity of negative feedback mechanism to sex steroids
   2. Prominent sleep-associated increase in episodic secretion of GnRH gradually changes to adult pattern of pulses every 90–120 minutes
   3. Pulsatile secretion of LH follows pattern of GnRH
   4. Progressive development of secondary sex characteristics
   5. Mid- to late puberty-maturation of *positive* feedback mechanism and capacity to exhibit an estrogen-induced LH surge
   6. Spermatogenesis in male; ovulation in female

CONTROL OF ADRENAL ANDROGENESIS (ADRENARCHE)

The role of the adrenal component of adolescent maturation, with its characteristic protopubertal increase in adrenal androgen secretion, is not clearly understood (14a, 22, 47). Both cross-sectional and longitudinal studies have defined a significant increment in adrenal androgen secretion by 8 years of age in both boys and girls [see (22) for review]. A progressive rise in plasma concentrations of adrenal androgens occurs during late childhood and adolescence, which correlates with the development and growth of the zona reticularis of the adrenal cortex, reaching adult levels in late adolescence. Gonadotropin and gonadal steroid concentrations do not begin to increase until 10–12 years of age, or approximately 2 years after the adrenarche. Thus it appears that activation of the adrenal androgen-secreting system (adrenarche) normally precedes activation of the hypothalamic-pituitary gonadotropin-gonadal axis (gonadarche). This temporal relationship between the adrenarche and maturation of gonadal function has suggested that adrenal androgens may play an important role in the onset of the gonadarche.

Considerable information, however, suggests that adrenarche and gonadarche are independent events, apparently controlled by separate mechanisms, and that adrenarche is not essential for the onset of gonadarche (22, 61). Supporting this hypothesis is the demonstration that the onset of puberty and the age of menarche occurred within the normal range in patients with primary adrenal insufficiency, as well as in patients with premature adrenarche, thus either in the absence of adrenal androgens or in the presence of modestly elevated levels (22). In further studies in patients with idiopathic precocious puberty, functional agonadism, and precocious adrenarche, Sklar et al (61) described a dissociation between adrenarche and gonadarche, additional evidence for separate modulation of adrenal androgen secretion.

The control of adrenal androgen secretion has not been well defined. Three different mechanisms have been discussed in detail (22): (a) an intrinsic alteration in the activity of the enzymes involved in adrenal androgenesis; (b) the secretion of an extra-pituitary stimulatory factor, such as estrogen; and (c) a pituitary adrenal androgen stimulating hormone, as yet uncharacterized.

Recent measurement of in vitro bioactivity of the 3-β-ol dehydrogenase-Δ4–5 isomerase, 17–20 lyase, and 17-hydroxylase enzyme systems suggests that a decrease of 3-β-ol dehydrogenase-Δ4–5 isomerase activity does not explain increased pubertal DHA and DHAS production: rather, there is increased 17–20 lyase and 17-hydroxylase activity (58). Such findings (63) appear to diminish further the validity of a hypothesis that estrogens directly enhance Δ_5-steroid production. The most compelling argument against the role of estrogens in the adrenarche is that levels of DHA and DHAS increase one to two years prior to the pubertal increment of estrogens in early adolescent girls. Although a wide variety of tissue growth-promoting peptides have been isolated over the past decade, the relation of any of these to adrenal androgenesis is not known. The involvement of endogenous opioid and other neuropeptides in adrenal androgen secretion has not been explored.

The third hypothesis for initiation of adrenal androgenesis relates to secretion of pituitary factors. Recent suggestive evidence (22, 47) favors the regulation of adrenal androgen secretion by a dual control mechanism: ACTH is obligatory for action of an unidentified pituitary adrenal androgen stimulating hormone (AASH) (22). The possibility that the postulated AASH is contained in the pro-opio-melanocortin molecule merits study. Neither prolactin, LH, nor FSH appears to be this putative factor. Parker & Odell (47) have infused a bovine pituitary extract into dexamethasone-suppressed adult orchiectomized male dogs and found a larger increment in the ratio of plasma androstenedione and DHA to cortisol than could be explained by the ACTH content of that extract (7).

In summary, the adrenal cortex does not appear to play a critical rate-limiting role in the onset of normal puberty, as there is a readily definable discordance between gonadarche and adrenarche in patients with primary adrenal insufficiency, gonadotropin deficiency, precocious adrenarche, and gonadal failure. Adrenal androgens do not appear to be essential for a normal adolescent growth spurt, though gonadal sex steroids and growth hormone are required. *Excess* adrenal androgens do affect gonadal function, as seen in young women with congenital adrenal hyperplasia due to 21-hydroxylase deficiency who develop amenorrhea or oligomenorrhea when receiving inadequate glucocorticoid replacement therapy. In such individuals, excess circulating androgens appear to alter the dynamics of gonadotropin secretion as well as intraovarian gonadotropin sex steroid interactions, resulting in augmented follicular atresia and polycystic ovaries (22). Finally, we suggest that a putative and elusive adrenal androgen-stimulating hormone in the presence of ACTH (14a, 22, 47) induces differentiation of the zona reticularis, and synthesis and secretion of adrenal androgens at adrenarche, an event that occurs independently of activation of the hypothalamic-pituitary gonadotropin-gonadal axis.

Literature Cited

1. Aggarwal, B. B., Papkoff, H. 1981. Relationship of sialic acid residues to in vitro biologic and immunological activities of equine gonadotropins. *Biol. Reprod.* 24:1082–87
2. Beck, W., Wuttke, W. 1980. Diurnal variations of plasma luteinizing hormone, follicle-stimulating hormone, and prolactin in boys and girls from birth to puberty. *J. Clin. Endorinol. Metab.* 50:635–39
3. Belchetz, P. E., Plant, T. M., Nakai, Y., Keogh, E. J., Knobil, E. 1978. Hypophysial responses to continuous and intermittent delivery of hypothalamic gonadotropin-releasing hormone. *Science* 202:631–32
4. Boyar, R., Finkelstein, J., Roffwarg, H., Kapen, S., Weitzman, E., Hellman, L. 1972. Synchronization of augmented luteinizing hormone secretion with sleep during puberty. *N. Engl. J. Med.* 287:582–86
5. Boyar, R., Finkelstein, J. W., David, R., Roffwarg, H., Kapen, S., Weitzman, E. D., Hellman, L. 1973. Twenty-four hour patterns of plasma luteinizing hormone and follicle stimulating hormone in sexual precocity. *N. Engl. J. Med.* 289:282–86
6. Boyar, R. M., Finkelstein, J. W., Roffwarg, H., Kapen, S., Weitzman, E. D., Hellman, L. 1973. Twenty four hour luteinizing hormone and follicle stimulating hormone secretory pattern in gonadal dysgenesis. *J. Clin. Endocrinol. Metab.* 37:521–25
7. Branchaud, C. T., Goodyer, C. G., Hall, C. S. G., Arato, J. S., Silman, R. E., Giroud, C. J. P. 1978. Steroidogenic activity of h-ACTH and related peptides on the human neocortex and fetal adrenal cortex in organ culture. *Steroids* 31:557–72
8. Carmel, P. W., Antunes, J. L., Ferin, M. 1979. Collection of blood from the pituitary stalk and portal veins in monkeys, and from the pituitary sinusoidal system of monkey and man. *J. Neurosurg.* 50:75–80
9. Chappel, S. C., Resko, J. A., Norman, R. L., Spies, H. G. 1981. Studies in rhesus monkeys on the site where estrogen inhibits gonadotropins: Delivery of 17-β-estradiol to the hypothalamus and pituitary gland. *J. Clin. Endocrinol. Metab.* 52:1–8
10. Clayton, R. N., Catt, K. J. 1981. Gonadotropin-releasing hormone receptors: characterization, physiological regulation and relationship to reproductive function. *Endocrine Rev.* 2:186–209
11. Cogen, P. H., Antunes, J. L., Louis, K. M., Dyrenfurth, I., Ferin, M. 1980. The effects of anterior hypothalamic disconnection on gonadotropin secretion in the female rhesus monkey. *Endocrinology* 107:677–83
12. Conte, F. A., Grumbach, M. M., Kaplan, S. L. 1975. A diphasic pattern of gonadotropin secretion in patients with the syndrome of gonadal dysgenesis. *J. Clin. Endocrinol. Metab.* 40:670–74
12a. Conte, F. A., Grumbach, M. M., Kaplan, S. L., Reiter, E. O. 1981. Correlation of LRF-induced LH and FSH release from infancy to 19 years with the changing pattern of gonadotropin secretion in agonadal patients. *J. Clin. Endocrinol. Metab.* 50:1163–68
13. Croley, K. P., Valk, T. W., Kelch, R. P., Marshall, J. C. 1981. Estimation of Gn-RH pulse amplitude during pubertal development. *Pediatr. Res.* 15:157–62
14. Crowley, W. F. Jr., McArthur, J. W. 1980. Stimulation of the normal menstrual cycle in Kallmann's syndrome by pulsatile administration of luteinizing hormone-releasing hormone (LHRH). *J. Clin. Endocrinol. Metab.* 51:173–75
14a. Cutler, G. B., Loriaux, D. L. 1980. Adrenarche and its relationship to the onset of puberty. *Fed. Proc.* 39:2384–90
15. Dufau, M. L., Beitins, I. Z., McArthur, J., Catt, K. 1977. Bioassay of serum LH concentrations in normal and LH-RH stimulated human subjects. In *The Testis in Normal and Infertile Men*, ed. P. Troen, H. R. Nankin, pp. 309–25. NY: Raven Press
16. Ferin, J., Rosenblatt, H., Carmel, P. W., Antunes, J. L., VandeWide, R. L. 1979. Estrogen-induced gonadotropin surges in female rhesus monkeys after pituitary stalk section. *Endocrinology* 104:50–52
17. Fevre, M., Boyar, R. M., Rollag, M. D. 1979. Dosage radioimmunologique de la mélatonine au cours du nycthemére chez le garcon pubère. *Ann. Endocrinol. Paris* 40:555–56
18. Frisch, R. E., Gotz-Welbergen, A. V., McArthur, J. W., Albright, T., Witschi, J., Bullen, B., Birnholz, J., Reed, R. B., Hermann, R. 1981. Delayed menarche and amenorrhea of college athletes in relation to age of onset of training. *J. Am. Med. Assoc.* 246:1559–63
19. Frisch, R. E., Revelle, R. 1970. Height and weight at menarche and a hypothe-

sis of critical body weights and adolescent events. *Science* 169:397–99
20. Grumbach, M. M. 1980. The neuroendocrinology of puberty. In *Neuroendocrinology,* ed. D. T. Krieger, J. C. Hughes, pp. 249–58. Sunderland, Mass: Sinauer Assoc.
21. Grumbach, M. M., Roth, J. C., Kaplan, S. L., Kelch, R. P. 1974. Hypothalamic-pituitary regulation of puberty in man: evidence and concepts derived from clinical research. In *The Control of the Onset of Puberty,* ed. M. M. Grumbach, G. D. Grave, F. E. Mayer, pp. 115–66. NY: Wiley
22. Grumbach, M. M., Richards, G. E., Conte, F. A., Kaplan, S. L. 1978. Clinical disorders of adrenal androgen function and puberty: An assessment of the role of the adrenal cortex in normal and abnormal puberty in man and evidence for an ACTH-like pituitary adrenal androgen stimulating hormone. In *The Endocrine Function of the Human Adrenal Cortex,* ed. V. H. T. James, M. Serio, G. Giusti, L. Martini, pp. 583–612. London/NY: Academic
23. Hansen, J. W., Hoffman, H. J., Ross, G. T. 1975. Monthly gonadotropin cycles in premenarchal girls. *Science* 190:161–63
24. Hohlweg, W., Dohrn, M. 1932. Über die Beziehungen zwischen Hypophysenvorderlap und Keimdrusen. *Klin. Wochenschr.* 11:233–35
25. Huseman, C. A., Kugler, J. A., Schneider, I. G. 1980. Mechanism of dopaminergic suppression of gonadotropin secretion in men. *J. Clin. Endocrinol. Metab.* 51:209–14
26. Jacobson, R. I., Seyler, L. E., Tamborlane, W. V., Gertner, J. M., Genel, M. 1979. Pulsatile subcutaneous nocturnal administration of Gn-RH by portable infusion pump in hypogonadotropic hypogonadism: initiation of gonadotropin responsiveness. *J. Clin. Endocrinol. Metab.* 49:652–54
27. Judd, H. L. 1979. Biorhythms of gonadotropins and testicular hormone secretion. In *Endocrine Rhythms,* ed. D T. Krieger, pp. 299–324. NY: Raven Press
28. Kaplan, S. L., Grumbach, M. M., Aubert, M. L. 1976. The ontogenesis of pituitary hormones and hypothalamic factors in the human fetus: Maturation of the central nervous system regulation of anterior pituitary function. *Rec. Prog. Horm. Res.* 32:161–243
29. Kelch, R. P., Kaplan, S. L., Grumbach, M. M. 1973. Suppression of urinary and plasma follicle-stimulating hormone by exogenous estrogens in prepubertal and pubertal children. *J. Clin. Invest.* 52:1122–28
30. Knobil, E. 1980. The neuroendocrine control of the menstrual cycle. *Rec. Prog. Horm. Res.* 36:53–88
30a. Knobil, E., Plant, T. M. 1978. The neuroendocrine control of gonadotropin secretion in the female rhesus monkey. In *Frontiers in Neuroendocrinology,* ed. W. F. Ganong, L. Martini, pp. 249–64. NY: Raven
31. Knobil, E., Plant, T. M., Wildt, L., Belchetz, P. E., Marshall, G. 1980. Control of the rhesus monkey menstrual cycle: permissive role of the hypothalamic gonadotropin-releasing hormone. *Science* 207:1371–73
32. Kulin, H. E., Reiter, E. O. 1972. Gonadotropin suppression by low dose estrogen: differential responses of FSH and LH. *J. Clin. Endocrinol. Metab.* 35:836–39
33. Kulin, H. E., Moore, R. C. Jr., Santner, S. J. 1976. Circadian rhythms in gonadotropin excretion in prepubertal and pubertal children. *J. Clin. Endocrinol. Metab.* 42:770–73
34. Lenko, H. L., Lang, U., Aubert, M. L., Paunier, L., Sizonenko, P. C. 1981. Melatonin in plasma and urine before and during puberty. *Pediat. Res.* 15:74 (Abstr.)
35. Lucky, A. W., Rich, B. H., Rosenfield, R. L., Fang, V. S., Roche-Bender, N. 1980. LH bioactivity increases more than immunoreactivity during puberty. *J. Pediatr* 97:205–13
36. Marshall, J. C., Kelch, R. P. 1979. Low dose pulsatile gonadotropin-releasing hormone in anorexia nervosa: A model of human pubertal development. *J. Clin. Endocrinol. Metab.* 49:712–18
37. McArthur, J. W., O'Loughlin, K. M., Beitins, I. Z., Johnson, L., Hourihan, J., Alonso, C. 1976. Endocrine studies during the refeeding of young women with nutritional amenorrhea and infertility. *Mayo Clin. Proc.* 51:607–16
38. McArthur, J. W., Bullen, B. A., Beitins, I. Z., Pagano, M., Badger, T. M., Klibanski, A. 1980. Hypothalamic amenorrhea in runners of normal body composition. *Endocrine Res. Comm.* 7: 13–25
39. Morley, J. E., Baranetsky, N. G., Wingert, T. D., Carlson, H. E., Hershman, J. M., Melmed, S., Levin, S. R., Jamison, K. R., Weitzman, R., Chang, R. J., Varner, A. A. 1980. Endocrine effects of naloxone-induced opiate receptor

blockade. *J. Clin. Endocrinol. Metab.* 50:251–57
40. Moult, P. J. A., Grossman, A., Evans, J. M., Rees, L. H., Besser, G. M. 1981. The effect of naloxone on pulsatile gonadotropin release in normal subjects. *Clin. Endocrinol.* 14:321–24
41. Mueller, P. L., Sklar, C. A., Gluckman, P. D., Kaplan, S. L., Grumbach, M. M. 1981. Hormone ontogeny in the ovine fetus. IX. Luteinizing hormone and follicle stimulating hormone response to luteinizing hormone releasing factor in mid- and late gestation and in the neonate. *Endocrinology* 108:881–86
42. Nakai, Y., Plant, T. M., Hess, D. L., Keogh, E. J., Knobil, E. 1978. On the site of the negative and positive feedback actions of estradiol in the control of gonadotropin secretion in the rhesus monkey. *Endocrinology* 102:1008–14
43. Neill, J. D., Patton, J. M., Dailey, R. A., Tsou, R. C., Tindall, G. T. 1977. Luteinizing hormone releasing hormone (LH-RH) in pituitary stalk blood of rhesus monkeys: relationship to level of LH release. *Endocrinology* 101: 430–34
44. Nett, T. M., Crowder, M. E., Moss, G. E., Duello, T. M. 1981. GnRH-receptor interaction. V. Down-regulation of pituitary receptors for GnRH in ovariectomized ewes by infusion of homolgous hormone. *Biol. Reprod.* 24: 1145–55
45. Ojeda, S. R., Andrews, W. W., Advis, J. P., White, S. S. 1980. Recent advances in the endocrinology of puberty. *Endocrine Rev.* 1:228–57
46. Parker, D. C., Judd, H. L., Rossman, L. G., Yen, S. S. C. 1975. Pubertal sleep-wake patterns of episodic LH, FSH and testosterone release in twin boys. *J. Clin. Endocrinol. Metab.* 40:1099–109
47. Parker, L. N., Odell, W. D. 1980. Control of adrenal androgen secretion. *Endocrine Rev.* 1:392–410
48. Penny, R., Olambiwonnu, N. O., Frasier, S. D. 1977. Episodic fluctuations of serum gonadotropins in pre- and post-pubertal girls and boys. *J. Clin. Endocrinol. Metab.* 45:307–11
49. Penny, R., Goldstein, I. P., Frasier, S. D. 1978. Gonadotropin excretion and body composition. *Pediatrics.* 61:294–300
50. Presl, J., Horejsi, J., Stroufova, A., Herzmann, J. 1976. Sexual maturation in girls and the development of estrogen-induced gonadotropic hormone release. *Ann. Biol. Anim. Biochim. Biophys.* 16:377–83

51. Quigley, M. E., Yen, S. S. C. 1980. The role of endogenous opiates on LH secretion during the menstrual cycle. *J. Clin. Endocrinol. Metab.* 51:179–81
52. Quigley, M. E., Sheehan, K. L., Caspar, R. F., Yen, S. S. C. 1980. Evidence for increased dopaminergic and opioid activity in patients with hypothalamic hypogonadotropic amenorrhea. *J. Clin. Endocrin. Metab.* 50:949–54
53. Rebar, R. W., Yen, S. S. C. 1979. Endocrine rhythms in gonadotropins and ovarian steroids with reference to reproductive processes. In *Endocrine Rhythms*, ed. D. T. Krieger, pp. 259–98. NY: Raven Press
54. Reiter, E. O., Kulin, H. E., Hamwood, S. M. 1974. The absence of positive feedback between estrogen and luteinizing hormone in sexually immature girls. *Pediatr. Res.* 8:740–45
55. Reiter, E. O., Beitens, I. Z., Optrea, T., Gutai, J. P. 1982. Bioassayable luteinizing hormone during childhood and adolescence and in patients with delayed pubertal development. *J. Clin. Endocrinol. Metab.* In press
56. Robertson, D. M., Puri, V., Lindberg, M., Diczfalusy, E. 1979. Biologically active luteinizing hormone (LH) in plasma. *Acta Endocrinol.* 92:615–26
57. Roth, J. C., Kelch, R. P., Kaplan, S. L., Grumbach, M. M. 1972. FSH and LH response to luteinizing hormone-releasing factor in prepubertal and pubertal children, adult males and patients with hypogonadotropic and hypergonadotropic hypogonadism. *J. Clin. Endocrinol. Metab.* 35:926–30
58. Schiebinger, R. J., Albertson, B. D., Cassorla, F. G., Bowyer, D. W., Geelhoed, G. W., Cutler, G. B., Loriaux, D. L. 1981. The developmental changes in plasma adrenal androgens during infancy and adrenarche are associated with changing activities of adrenal microsomal 17-hyroxylase and 17,20 desmolase. *J. Clin. Invest.* 67:1177–82
59. Sherman, B. M., Halmi, K. A., Zamudio, R. 1975. LH and FSH response to gonadotropin-releasing hormone in anorexia nervosa: effect of nutritional rehabilitation. *J. Clin. Endocrinol. Metab.* 41:135–42
60. Silman, R. E., Leone, R. M., Hooper, R. J. L., Preece, M. A. 1979. Melatonin, the pineal gland and human puberty. *Nature* 282:301–3
61. Sklar, C. A., Kaplan, S. L., Grumbach, M. M. 1980. Evidence for dissociation between adrenarche and gonadarche: studies in patients with idopathic preco-

cious puberty, gonadal dysgenesis, isolated gonadotroph deficiency, and constitutionally delayed growth and adolescence. *J. Clin. Endocrinol. Metab.* 51:548–56
62. Sklar, C. A., Conte, F. A., Kaplan, S. L., Grumbach, M. M. 1981. Human chorionic gonadotropin-secreting pineal tumor: relation to pathogenesis and sex limitation of sexual precocity. *J. Clin. Endocrinol. Metab.* 53:656–60
63. Sklar, C. A., Kaplan, S. L., Grumbach, M. M. 1981. Lack of effect of estrogens on adrenal androgen in children and adolescents with a comment on estrogens and pubic hair growth. *Clin. Endocrinol.* 14:311–20
64. Styne, D. M., Grumbach, M. M. 1978. Puberty in the male and female: its physiology and disorders. In *Reproductive Endocrinology,* ed. S. S. C. Yenn, R. Jaffe, pp. 189–240. Philadelphia: W. B. Saunders
65. Tanner, J. M. 1981. *A History of the Study of Human Growth.* Cambridge: Cambridge Univ. Press. pp. 286–98
66. Valk, T. W., Corley, K. P., Kelch, R. P., Marshall, J. C. 1980. Hypogonadotropic hypogonadism: Hormonal responses to low dose pulsatile administration of gonadotropin-releasing hormone. *J. Clin. Endocrinol. Metab.* 51:730–37
67. Warren, M. P. 1980. The effects of exercise on pubertal progression and reproductive function in girls. *J. Clin. Endocrinol. Metab.* 50:1150–57
68. Wildt, L., Marshall, G., Knobil, E. 1980. Experimental induction of puberty in the infantile female rhesus monkey. *Science* 207:1373–75
69. Wildt, L., Hausler, A., Hutchison, J. S., Marshall, G., Knobil, E. 1981. Estradiol as a gonadotropin releasing hormone in the rhesus monkey. *Endocrinology* 108:2011–13
70. Winter, J. S. D., Faiman, C., Hobson, W. C., Prasad, A. V., Reyes, F. I. 1975. Pituitary-gonadal regulations in infancy. I. Patterns of serum gonadotropin concentrations from birth to four years of age in man and chimpanzee. *J. Clin. Endocrinol. Metab.* 40:545–51
71. Yoshimoto, Y., Moridera, K., Imura, H. 1975. Restoration of normal pituitary gonadotropin reserve by administration of luteinizing hormone releasing hormone in patients with hypogonadotropic hypogonadism. *N. Engl. J. Med.* 292:242–45
72. Zacharias, L., Wurtman, R. J. 1969. Age at menarche. *N. Engl. J. Med.* 280:868–75

THE EVOLUTION OF PEPTIDE HORMONES

Hugh D. Niall

Howard Florey Institute of Experimental Physiology and Medicine, Parkville, Victoria 3052, Australia

INTRODUCTION

Our understanding of the evolutionary relationships of peptide hormones is itself still evolving. The present picture is incomplete partly because we know the structure of only a few peptide hormones from a restricted range of species, and we know nothing of the structure of hormone receptors. It is also incomplete because we can only study evolutionary successes; there is no "fossil-equivalent" for peptide hormones. Thus we are limited to an analysis of contemporary peptide hormones and can only guess at the structure of ancestral forms.

It is important to point out that a peptide hormone molecule is impotent or irrelevant without the means to reach a target cell and there to evoke a response. Thus the focus of this chapter is on the evolution of an intercellular communications system that includes effector (hormone), receptor, and feedback components, linked to mechanisms for processing, transport, storage, secretion, and metabolism. This system can only develop through a coordinated evolution of its components.

THE FIRST STEP

We generally think of hormones as molecules that modulate and coordinate function within relatively complex multicellular organisms. But at what point in evolution did hormones first appear? A plausible case can be made that hormone evolution started at the unicellular stage of life, which probably lasted some hundreds of millions of years.

It may seem fanciful to discuss unicellular organisms in terms of behavior and communication, but bacteria, for example, exhibit many of the characteristics of much "higher" organisms. They move towards nutrients and away from substances injurious to them (positive and negative chemotaxis). They are found singly, in pairs (diplococci), or in clusters (staphylococci). They readily exchange genetic material. They compete for food and lebensraum. When times get hard they go into the equivalent of hibernation (spore formation), which allows survival until living conditions improve. Above all, they operate as a community with a common cause (survival) to which individuals may be sacrificed. A single bacterium cannot survive where a colony may. Thus a particular bacterial genome (a selfish gene) would increase its chances of survival and propagation by promoting coordinated behavior relevant to the environmental circumstances (1, 11).

These comments in favor of a bacterial sociobiology have no point unless it can be shown that unicellular organisms exchange signals that regulate their function. Signal molecules (akin to pheromones in higher organisms) could be used by single cells to indicate the presence or direction of a food source or a toxic substance and perhaps to regulate rates of division. Support for this view comes from studies of cellular slime moulds. (*Dictyostelium discoideum*), a life form intermediate between unicellular and multicellular forms (4). When carbohydrate food sources are in short supply, the unicellular forms of the slime mould migrate towards one another and aggregate. This effect is mediated by release of an attractant identified as cyclic AMP (cAMP). The slime mould cells have surface receptors for cAMP that bind it with high affinity and that exhibit characteristics consistent with negative cooperativity and down-regulation as defined by criteria developed for vertebrate peptide hormones. Thus a system identical in general design to that of a hormone in a higher organism is found even in this primitive form of life.

A recent report (20) supports the presence of molecules closely resembling mammalian peptide hormones in unicellular eukaryotic organisms (e.g. *Tetrahymena*) and in bacteria. Small quantities of material cross-reacting with antisera to porcine insulin were isolated from large-scale cultures of these organisms. The material was shown to co-chromatograph with insulin on gel filtration columns, indicating a similar apparent molecular weight. Fractions immunoreactive with antisera to porcine insulin also contained insulin-like bioactivity as measured in a fat cell assay. Incubation with insulin antisera abolished this biological activity. Similar findings suggested the presence of ACTH and endorphin-like material in similar cell cultures (20a).

While these findings are discussed elsewhere in this section (see chapter by J. Roth), they require comment in the context of peptide hormone evolution. If this work is confirmed by further studies, it has major implications for our overall picture of evolution. If these contemporary unicellular organisms are representative of their remote ancestors, then the genes specifying at least some mammalian peptide hormones appeared very, very early in evolution and have been conserved virtually unchanged throughout the whole of invertebrate and vertebrate history. There is nothing inherently surprising about this. Insulin-like material has been found previously in insects and molluscs (7) and calcitonin-like immunoreactive material in primitive invertebrates. Chorionic gonadotropin-like material has been found in bacteria (8, 21). No doubt more detailed studies will soon elucidate the exact nature and origin of the insulin-like material in *Tetrahymena* and other unicellular organisms. These should include studies to demonstrate the presence of nucleotide sequences coding for insulin within the genome of the organism involved. This approach would eliminate any possibility of accidental contamination of cultures with porcine insulin. (Great care was taken to minimize the possibility of this in the studies cited above.)

A further possibility recognized by Roth and co-workers (20) is that the presence of the "insulin" results from a later recombination event. In other words, the unicellular organisms may have acquired the insulin gene recently (in evolutionary terms) by transfer of genetic material from a multicellular, more-evolved organism. A viral-type vector could be postulated. This sort of hypothesis would be difficult to prove. However an insulin gene transferred in this way would not survive in a stable form in its new host unless it provided a survival advantage of some kind. Whatever the explanation, it would certainly be ironic, in view of all the effort by biologists to clone genes for peptide hormones into bacteria, if such genes were there all the time!

PATTERNS OF PEPTIDE HORMONE EVOLUTION

Recurrent themes have become apparent in peptide hormone evolution. I have summarized these in four colloquial "rules," no one of which should be taken too literally.

Rule 1: "Gene Duplication is the name of the game"

Most evidence points to the generation of contemporary families of peptide hormones through successive events of gene duplication (23). Many peptide hormones are found in families with extensive amino acid sequence homology (e.g. the insulin like family, the prolactin-like family, the gut hormones, the glycoprotein hormones). This homology is most easily explained by a process of gene duplication. Through this mechanism a gene coding for a functional hormone is duplicated, giving a redundant copy (Figure 1). While the original hormone continues to carry out its functional role, the duplicate gene is free to mutate and to acquire the structural properties necessary to interact with a different set of receptors. Thus the organism gains a new hormone. Recombinant DNA techniques reveal that duplicate genes are much more common than had previously been recognized, whether functional (i.e. coding for a new hormone) or nonfunctional (i.e. a pseudogene). There are strong teleological arguments in favor of this mode of hormone evolution. Establishment of a hormone-receptor feedback loop requires a molecule that has the right properties to be processed, stored, and secreted from its cell of origin, to survive transport to its target, and to bind to and activate a receptor. Because it already possesses the required structural features, it is much easier for a preexisting hormone to give rise to a new one through rather limited mutation, than for example, a new hormone to evolve from a structural protein or from an enzyme lacking these molecular properties.

Similar arguments can be advanced for the evolution of peptide hormone receptors from preexisting receptors. Receptor molecules must possess several distinct structural domains: those that direct their insertion into the external cell membrane in the correct orientation, those that are responsible for binding the hormone, and those that interact with the post-receptor apparatus involved in the biological response of the target cell. If the receptor has more than one subunit, the structural features responsible for subunit interaction must also be conserved. For these reasons it seems likely that related receptors [e.g. receptors for insulin and for insulin-like growth factor (IGF), or for growth hormone and prolactin] have evolved from duplicated genes.

Processing enzymes provide a further example of gene duplication. The same hormone may at times be processed in a slightly different way in different cells, suggesting that more than one enzyme is involved. The 31K precursor to ACTH and β-endorphin, for example, is converted to different end-products in the anterior pituitary and the pars intermedia (14). The

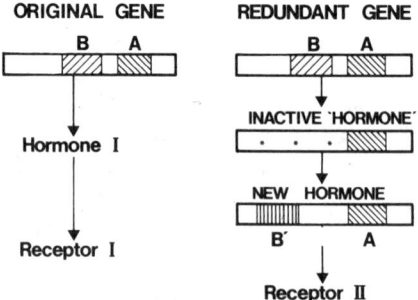

Figure 1 Schematic representation of the evolution of a new hormone by gene duplication. B, A: domains responsible for binding to and activation of receptors. B': structural features mediating binding to new receptors.

processing enzymes in different tissues usually (but not invariably) recognize local clusters of basic residues. Thus they have features in common with the family (19) of serine proteases (trypsin, tonin, NGF γ-subunit, EGF binding protein) and probably represent a group of structurally related enzymes associated with membranes of the endoplasmic reticulum and Golgi apparatus.

At present it is unclear when during peptide hormone evolution the main gene duplications occurred. Diagrams representing putative "family trees" for particular peptide hormones have been constructed by Dayhoff and co-workers (12) on the basis of amino acid sequence comparisons. The approximate time intervals between successive gene duplications can be estimated and the results compared with the fossil record. As more information accumulates, this kind of analysis looks less satisfactory. It assumes that mutations accumulate steadily over long evolutionary time periods, whereas there may in fact be sudden bursts of evolution. Moreover, it now appears that, as discussed above, gene duplications may have occurred early in evolution. For example, the large sequence differences between human and fish calcitonins have been used to estimate the time of duplication of an ancestral calcitonin gene. Human-type calcitonin was thought to be confined to primates and rats. However MacIntyre and co-workers (13) have recently reported the presence of "human-type" calcitonin in a variety of primitive species, including amphibia, protochordates, and cyclostomes. Other studies (5, 6) from the same group suggest that two calcitonin genes are present in fish, reptiles, and mammals. The proportions of each calcitonin differ greatly from one species to another. Though two calcitonin genes are typically conserved, one may be suppressed or only partially expressed. This illustrates the difficulty of studying peptide hormone evolution on the basis of the incomplete information available.

In summary, gene duplication may have been the major mechanism of peptide hormone evolution. Duplications of genes coding for hormones, receptors, and processing enzymes have all been involved. There is increasing reason to believe that many such gene duplications occurred very early, and some hormone genes may have existed at the unicellular stage of life.

Rule 2: "Everything is made everywhere"

While not literally true, this statement is not as exaggerated as it might seem. Several years ago the picture of hormone origin and distribution within the body was reassuringly simple. Each hormone had its endocrine gland of origin: Pituitary hormones came from the pituitary, insulin came from the pancreatic islets, and hypothalamic releasing factors exerted defined effects via the hypothalamic-hypophyseal portal vessels. The picture is now very different. "Pituitary" hormones are found in the brain and the placenta (17). Somatostatin is found in the gastrointestinal tract and pancreatic islets as well as in the hypothalamus (31). Molecules closely resembling gonadotropin-releasing hormone (GnRH) have been found in the gonads and the placenta (16, 29, 32). Gastrointestinal hormones (CCK, gastrin) are present in the brain. This confusing situation has been further compounded by a report that insulin is ubiquitous in extrapancreatic tissues of rats and humans, being found, for example, in brain, liver, cultured human lymphocytes, and cultured human fibroblasts (28). Moreover, the placenta seems to make almost any hormone one cares to think of, from ACTH to epidermal growth factor.

At present the role of these widespread hormones is not clear, but a few comments are possible. First, since all cells have a full complement of genes, there is no reason why a particular hormone cannot be made in any cell. That insulin may be found in all mammalian cells (or at least a wide variety) raises the possibility that all cells may make all hormones, albeit in very small quantities. Gene repression may not be an all-or-none phenomenon; perhaps all genes are at least expressed at a very low level, their products detectable with a sufficiently sensitive assay. This might explain the common phenomenon of peptide hormone secretion by tumors of nonendocrine organs (e.g. carcinoma of the lung).

Much further work is needed to evaluate the possibility of extensive low-level derepression of hormone-specific genes. Recombinant DNA techniques should be particularly useful here since direct demonstration of gene expression via mRNA production should be possible.

Rule 3: "Never make a new hormone if you can use an old one"

Nature is frugal. Zoologists and comparative biologists have noted for hundreds of years how certain mechanisms (e.g. protective coloration) are used in myriad different ways in different species. The molecular biologists can now make similar observations. I have noted above (Rule 1) that it makes sense for new hormones to evolve from preexisting hormones, thus taking advantage of a great deal of evolutionary groundwork. However, gene duplication (of hormone or receptor) is only one means of diversification. Another is the functioning of a single hormone in new or different roles. For example, receptors for an old hormone may evolve in a new target tissue. Prolactin is illustrative, since the hormone itself predated by millennia the evolution of mammals. Prolactin enables fish to adapt to water of varying salt concentration (25). With the mammalian development of a specialized tissue, the mammary gland, for nourishing the young, prolactin performed a new function. Antidiuretic hormone and calcitonin are two other hormones whose major function has change significantly with evolution. Medawar has summed up this situation well: "Endocrine evolution is not an evolution of hormones but an evolution of the uses to which they are put" (22).

A single hormone may also serve at different sites within the organism in the regulation of different local functions. A good example is the regulation of local events at nerve terminals by the same neurotransmitter (acetylcholine or norepinephrine, for example). Local release of neurotransmitters achieves a high local concentration of the effector in a circumscribed area. Deleterious effects at other sites are prevented partly by local destruction of the labile effector molecule and partly through its dilution in body fluids and its removal through re-uptake mechanisms. It has only recently become appreciated that many peptide hormones act as local regulatory agents in the manner of the classical neurotransmitters and local modulators such as the prostaglandins. Somatostatin acts in the hypothalamus, gastrointestinal tract, and pancreatic islets (and probably at several other sites) as such a local hormone. The other peptides originally thought to be dedicated hypothalamic-pituitary releasing factors likewise are found at multiple sites in the brain and elsewhere, where their roles are not yet defined. "Directed local hormones" represent a subclass; specific vascular channels (e.g. the pituitary portal system) or axons transport these regulatory peptides to small groups of localized target cells.

Measurement of the blood levels of the locally produced and locally active peptide hormones is unlikely to be rewarding (24). By definition, these must be below a threshold level that would interfere with the separate regulatory functions of the same hormone at different sites. Just as circulating levels of acetylcholine would reflect nothing but trivial spillover from its multiple sites of action, so the levels in the general circulation of hormones such as somatostatin or TRH or substance P seem unlikely to be physiologically meaningful.

A further circumstance that allows a single hormone to function in several different ways is the existence within the organism of biological barriers. In a pregnant woman, for example, ACTH (and presumably its companion endorphins) function in four compartments in addition to the placenta—i.e. in the general circulations of the mother and the fetus and in the brains of both. Here the placental barrier and the blood-brain barrier allow the same molecules to regulate different functions simultaneously. "Directed local transport" from pituitary cells into the brain substance may also play a role here (2).

Rule 4: "Conservation of structure = function"

Evolution has been possible only because the molecular apparatus for copying genes has a small but finite error rate, and because (presumably) random mutations can affect germ cells to produce new genotypes. As I have argued elsewhere (24), the error rate is probably itself subject to Darwinian evolution. It must be high enough to allow more complex and better adapted forms of life (or hormones) to evolve, yet low enough to preserve existing functionally important molecules. Gene duplication, which temporarily frees a segment of genome from selective pressure, has helped to establish such a rate. Regions of DNA that are free to mutate seem to do so rapidly. Thus in comparing amino acid sequences or nucleotide sequences coding for hormones the question to ask is not why there are differences but why certain regions have been conserved. For example, mammalian insulins are highly conserved, there being only one or two sequence differences between porcine, bovine, and human insulins (3). Relaxins and calcitonins, on the other hand, differ markedly: Fewer than half the residues are identical in corresponding sequence positions (15, 26). What can one conclude from this? Conservation of virtually the whole of the insulin molecule between related but not very close species (man and pig) indicates that virtually the whole structure is functional. Since available evidence (27) suggests that only one region of the insulin molecule interacts with receptors, what does the rest do? Part of the answer almost certainly lies in the molecular interactions of insulin molecules with one another to form dimers and Zn^{2+}-coordinated hexamers. The exact structural features involved here have been worked out through the elegant crystallographic studies of Hodgkin and associates (3). Other structural features may be important in the processing of the biosynthetic precursor to insulin and in the storage of insulin granules.

What of sequence variability in calcitonins, relaxins, and other hormones? Unfortunately, we cannot argue so strongly the converse proposition that regions of interspecific sequence variability in a hormone molecule demonstrate noncritical aspects of the structure. This is because we know neither all the functions of hormones nor to what extent a particular hormone is being conserved *within* a particular species—i.e. the degree of polymorphism. Recombinant DNA techniques now permit the rapid screening of chromosomal DNA from many individuals of a species. Synthesis of hormone variants is an alternate approach to this question.

Information is starting to accumulate on the disposition and nucleotide sequence of introns within genes coding for polypeptide hormones. It is too early to draw definite conclusions. However, whereas the location of introns seems to be fairly constant when the same hormone-specific gene is compared between species, the actual size and nucleotide sequences of such introns seem to vary greatly (30). Conservation of intron location, however, is probably a further indication of an evolutionary relationship between hormones. An important role for introns in the actual evolutionary process leading to new hormones has been suggested (9, 10) but is far from established.

SUMMARY

Despite limitations in our present knowledge it is already possible to discern the main features of peptide hormone evolution, since the same mechanisms (and indeed the same hormone molecules) function in many different ways. This underlying unity of organization has its basis in the tendency of biochemical networks, once established, to survive and diversify. The most surprising recent findings in endocrinology have been the discovery of vertebrate peptide hormones in multiple sites within the same organism, and the reports, persuasive but requiring confirmation, of vertebrate hormones in primitive unicellular organisms (20, 20a). Perhaps the major challenge for the future is to define the roles and interactions of the many peptide hormones identified in brain (18). The most primitive bacteria and the human brain, though an enormous evolutionary distance apart, may have more in common than we have recognized until now. As Axelrod & Hamilton have pointed out in a recent provocative article, "The Evolution of Cooperation" (1), bacteria, though lacking a brain, are capable of adaptive behavior that can be analysed in terms of game theory. It is clear that we can learn a great deal about the whole evolutionary process from a study of the versatile and durable peptide hormone molecules.

Literature Cited

1. Axelrod, R., Hamilton, W. D. 1981. The evolution of cooperation. *Science* 211:1390–96
2. Bergland, R. M., Page, R. B. 1978. Can the pituitary secrete directly to the brain? (affirmative anatomical evidence) *Endocrinology* 102:1325–38
3. Blundell, T. L., Dodson, G. G., Hodgkin, D. C., Mercola, D. A. 1972. Insulin: the structure in the crystal and its reflection in chemistry and biology. *Adv. Prot. Chem.* 26:280–402
4. Bonner, J. T. 1967. *The Cellular Slime Moulds.* Princeton, NJ: Princeton Univ. Press. 2nd ed.
5. Cano, R. P., Girgis, S. I., MacIntyre, I. 1981. Further evidence for calcitonin gene duplication: the identification of two different calcitonins in a fish, a reptile and two mammals. *Acta Endocrinol.* In press
6. Cano, R. P., Girgis, S. I., MacIntyre, I. 1981. Identification of both human and salmon calcitonin-like molecules in birds suggesting the existence of two calcitonin genes. *J. Endocrinol.* In press
7. Chan, S. J., Kwok, S. C. M., Steiner, D. F. 1981. The biosynthesis of insulin: some genetic evolutionary aspects. *Diabetes Care* 4:4–10
8. Cohen, H., Strampp, A. 1976. Bacterial synthesis of substance similar to human chorionic gonadotrophin. *Proc. Soc. Exp. Biol. Med.* 152:408–10
9. Crick, F. 1979. Split genes and RNA splicing. *Science* 204:264–71
10. Darnell, J. E. 1978. Implication of RNA–RNA splicing in evolution of eukaryotic cells. *Science* 202:1257–60
11. Dawkins, R. 1976. *The Selfish Gene.* Oxford: Oxford Univ. Press. 224 pp.
12. Dayhoff, M. O., Park, C. M., McLaughlin, P. J. 1972. *Atlas of Protein Sequence and Structure,* ed. M. O. Dayhoff, 5:7–28. Washington DC: Natl. Biochem. Res. Found.
13. Girgis, S. I., Galan Galan, F., Arnett, T. R., Rogers, R. M., Bone, Q., Ravazzola, M., MacIntyre, I. 1980. Immunoreactive human calcitonin-like molecule in the nervous systems of protochordates and a cyclostome, *Myxine. J. Endocrinol.* 87:375–82
14. Herbert, E. 1981. Discovery of pro-opiomelanocortin—a cellular polyprotein. *Trends Biochem. Sci.* 6:184–88
15. John, M. J., Borjesson, B. W., Walsh, J. R., Niall, H. D. 1981. Limited sequence homology between porcine and rat re-

laxins: implications for physiological studies. *Endocrinology* 108:726–29
16. Khodr, G. S., Siler-Khodr, T. M. 1980. Placental luteinizing hormone-releasing factor and its synthesis *Science* 207–315–17
17. Krieger, D. T., Liotta, A. D. 1979. Pituitary hormones in brain: where, how and why? *Science* 205:366–72
18. Krieger, D. T., Martin, J. B. 1981. Brain peptides *N. Engl. J. Med.* 304:876–85
19. Lazure, C., Seidah, N. G., Thibault, G., Boucher, R., Genest, J., Chretien, M. 1981. Sequence homologies between tonin, nerve growth factor γ-subunit, epidermal growth factor binding protein and serine proteases. *Nature* 292:383–84
20. Le Roith, D., Shiloach, J., Roth, J., Lesniak, M. A. 1980. Evolutionary origins of vertebrate hormones: Substances similar to mammalian insulins are native to unicellular eukaryotes. *Proc. Natl. Acad. Sci. USA* 77:6184–88
20a. Le Roith, D., Shiloach, J., Roth, J., Liotta, A. S., Krieger, D. T., Lewis, M., Pert, C. B. 1981. Evolutionary origins of vertebrate hormones: material very similar to ACTH, β-endorphin and dynorphin in protozoa. *Trans. Assoc. Am. Physiol.* In press
21. Maruo, T., Cohen, J., Segal, S. J., Kiode, S. S. 1979. Production of choriogonadotropin-like factor by a microorganism. *Proc. Natl. Acad. Sci. USA* 76:6622–26
22. Medawar, P. 1953. Some immunological and endocrinological problems raised by the evolution of viviparity in vertebrates. *Symp. Soc. Exp. Biol. Med.* 7:320–38
23. Niall, H. D. 1976. Peptide hormone homologies and evolution. In *Peptide Hormones*, ed. J. A. Parsons, pp. 8–32. London: MacMillan
24. Niall, H. D. 1980. The evolution of peptide hormones. In *Endocrinology 1980*, ed. I. A. Cumming, J. W. Funder, F. A. O. Mendelsohn, pp. 13–18. Canberra: Aust. Acad. Sci.
25. Nicoll, C. S., Bern, H. A. 1972. On the actions of prolactin among the vertebrates: is there a common denominator? In *Lactogenic Hormones*, ed. G. E. W. Wolstenholme, J. Knight, pp. 299–324. Edinburgh/London: Churchill Livingston
26. Potts, J. T. Jr., Deftos, L. J. 1969. Parathyroid hormone, thyrocalcitonin, Vitamin D, bone and bone mineral metabolism. In *Duncan's Diseases of Metabolism*, ed. P. K. Bondy, pp. 904–1082. Philadelphia/London/Toronto: W. B. Saunders
27. Pullen, R. A., Lindsay, D. G., Wood, S. P., Tickle, I. J., Blundell, T. L., Wollmer, A., Krail, G., Brandenburg, D., Zahn, H., Gliemann, J., Gammeltoft, S. 1976. Receptor-binding region of insulin. *Nature* 259:369–73
28. Rosenzweig, J. L., Havrankova, J., Lesniak, M. A., Brownstein, M., Roth, J. 1980. Insulin is ubiquitous in extrapancreatic tissues of rats and human. *Proc. Natl. Acad. Sci. USA* 77:572–76
29. Sharpe, R. M., Fraser, H. M., Cooper, I., Rommerts, F. F. G. 1981. Sertoli-Leydig cell communication via an LHRH-like factor. *Nature* 290:785–87
30. Ullrich, A., Dull, T. J., Gray, A., Brosius, J., Sures, I. 1980. Genetic variation in the human insulin gene. *Science* 209:612–15
31. Vale, W., Brazeau, P., Rivier, C., Brown, M., Boss, B., Rivier, J., Burgus, R., Ling, N., Guillemin, R. 1975. Somatostatin. *Rec. Prog. Horm. Res.* 31:365–98
32. Ying, S-Y., Ling, N., Böhlen, P., Guillemin, R. 1981. Gonadocrinins: peptides in ovarian follicular fluid stimulating the secretion of pituitary gonadotropins *Endocrinology* 108:1206–15

POST-TRANSLATIONAL PROTEOLYSIS IN POLYPEPTIDE HORMONE BIOSYNTHESIS

Kevin Docherty and Donald F. Steiner

Department of Biochemistry, University of Chicago, Chicago, Illinois 60637

Many cellular, viral, and hormonal polypeptides are initially synthesized as larger polypeptides that are processed proteolytically to yield their final active products. These precursors can be conveniently divided into groups that are processed either within their cells of origin or after their secretion into the blood or surrounding connective tissue matrix. The latter group includes the classical zymogen forms of various hydrolytic enzymes as well as such diverse proteins as the procollagens (1), vitellogenins (2), and promellitins (3). Those precursors that are processed at intracellular sites may be further conveniently subdivided into the *pre*proteins and *pro*proteins. These two large groups differ significantly in their processing kinetics, in the subcellular localizations of cleavage enzymes, and in their functions. This review focuses primarily on these forms, with special emphasis on recent progress in elucidating their primary enzymic cleavage mechanisms. At this stage it seems most likely that for each subgroup a single proteolytic mechanism accounts for the major processing events, with some specialized features requiring additional enzymes for several of the prohormonal systems.

The Preproteins

Since the discoveries of Milstein et al (4) of IgG light-chain precursors and of Kemper et al (5) of preproparathyroid hormone, many similar rapidly processed precursors have been studied in detail [for recent reviews see (6, 7)]. These precursors are all extended at their amino termini with regions that are approximately 20–30 amino acids long and have certain well-conserved physical properties and secondary structural features.[1] Their function, as defined in the Signal Hypothesis of Blobel & Dobberstein (8), is almost certainly to assist in the initial microsomal segregation of peptide materials destined for secretion. In bacteria similar precursors function to promote direct secretion of polypeptides into the periplasmic space or the extracellular medium (9). The bacterial presequences appear to be structurally analogous to their eukaryotic counterparts (6, 7, 9), and it has recently been shown that some eukaryotic presequences (10) as well as various hybrid (eukaryotic and prokaryotic) presequences can function to promote secretion when expressed in bacteria such as *E. coli* (10, 11).

While there seems to be general agreement that the nascent prepeptide regions of presecretory proteins interact with the membranes of the endoplasmic reticulum (ER), or the bacterial inner membrane, to promote the formation of membrane-bound polyribosomal complexes leading to peptide translocation, there are presently only theories to approach the possible mechanisms of this interaction. Three such theories are:

1. The Pore Model. As originally envisioned in the Signal Hypothesis, the nascent presequence causes the aggregation of protein subunits within the membrane to form a pore through which the growing polypeptide chain passes (8). Cleavage of the presequence then occurs cotranslationally within the cisternal space as the growing presecretory peptide chain enters this space.

2. The Loop Model. Several groups (7, 9) have based models on certain recurrent structural features of the presequences, particularly their strongly hydrophobic central regions (usually residues -7 to -17), as well as other regions indicative of greater flexibility of the peptide chain nearer the cleavage site (residues -3 to -7), and the nonrandom disposition of amino acids having charged side chains, especially within the more hydrophilic N-terminal region. According to these models the presequences enter the apolar membrane bilayer leaving the more hydrophilic N-terminal region behind on the ribosomal side and with their hydrophobic central regions extending across the membrane towards the cleavage site so as to form a loop in the entering peptide chain (Figure 1).

In these models the formation of the loop is assumed to result in the "threading" of the nascent presecretory peptide across the membrane. Cleavage on the inner membrane surface then creates a new N-terminus on the side of the membrane opposite the ribosome and the peptide chain continues to be transferred across the membrane as it folds progressively

---

[1]These extensions will be referred to as either *presequences, prepeptides,* or *signal sequences* in this review.

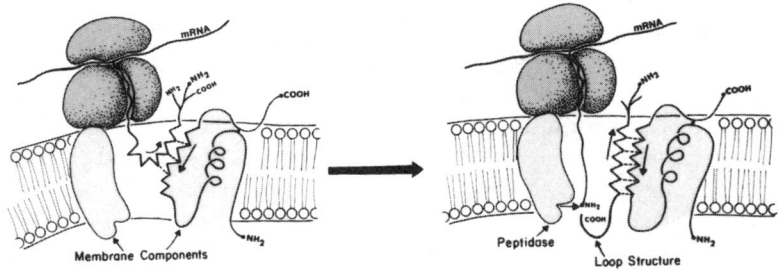

Figure 1. β Strand Loop Model for presecretory peptide interaction with membranes (see text for details).

to its mature conformation starting from the new N-terminus. Significant support for this kind of model has been provided by studies of the 3-dimensional structure of the influenza virus hemagglutinin protein (12). Although it has been proposed recently that the transmembrane portions of the presequence might be helical (13), this seems highly unlikely since in the case of several transmembrane proteins the sequences spanning the membrane in a helical arrangement contain approximately 22–26 strongly hydrophobic residues (14). This length can be shown to be required to form a stable α helix long enough to span the nonpolar core of the membrane. Inasmuch as the hydrophobic segment in many presequences is only 8–10 residues long, this segment could be expected to span the membrane only when in an extended chain configuration, as in a β sheet or β strand structure.

For this reason we have proposed (7, 15) that such structures are formed between the incoming hydrophobic signal sequence and one or more preexisting membrane proteins, which function essentially as receptors (16, 17). On further reflection it will be apparent that such a "receptor" functions by virtue of secondary structure interactions, and hence in theory could accept any sufficiently nonpolar peptide segment, thus accounting for two of the known characteristics of these segregative systems: (*a*) the presence in microsomes of saturable binding sites or "translocators" (16, 17), and (*b*) the ability of a wide variety of presecretory proteins having nonhomologous structures to compete for the same sites (15). On the other hand, a series of α helices formed from peptides with differing side chains would all be topographically unique and therefore incapable of competing for a common binding site. For these rather compelling reasons, and despite the reported tendency of the prepeptide of preproparathyroid hormone to form α helixes in organic solvents (18), the participation of helixes in prepeptide-mediated membrane translocation processes is likely to be transient at best (i.e. prior to the transition to the β strand form in interacting with the translocator molecules). Note that the coat protein of a filamentous bacteriophage, Pf3, exhibits a much higher amount of β structure in its membrane-associated state than in the virion itself (19).

Helical structures may, however, serve as a means of subsequently "rafting" segments or subdomains of nascent peptide chain across the membrane as synthesis and vectorial transfer proceed. Such partially folded structures might play an important, but passive, role in transfer completion, especially if the main driving force for effecting transfer after the initial loop has formed is further peptide chain folding (tertiation) on the inner surface of the microsomal or bacterial cytoplasmic membrane. If this is the case then downstream mutations that affect subdomain "raft" formation could disrupt this process, as has been noted in several cases (20, 21). That the role in transfer played by C-terminal regions of the presecretory proteins is largely *permissive* is further supported by the recent demonstration that its deletion from the maltose binding protein does not prevent segregation of the remaining segment (22).

3. The Trigger Hypothesis. According to this theory segregation is not a cotranslational event but instead occurs subsequently owing largely to information contained in the conformation of the newly synthesized (pre)-secretory polypeptide. Upon contact with the membrane it is presumed that a conformational change is induced that triggers the entry of the peptide into the membrane and its subsequent insertion or passage across the membrane (23). This seems to be an attractive hypothesis as it is based upon the known ability of proteins to undergo conformational transitions. Difficulties arise when one begins to consider the specifics of this theory. What are conformations A (outside) and B (inside) the membrane? How different are they? If state A proceeds spontaneously to state B on contacting the membrane, how can state B return to state A spontaneously as the protein exits the membrane on its other side? Why does it exit at all on the other side? If three conformations of the nascent protein are required to overcome this problem (A,B, and A' or C), how do these differ? It is likely, of course, that since conformational changes are fundamental to protein function some probably do occur during membrane segregation—e.g., such as folding on the luminal side of the membrane.[2] However, it is difficult to understand how conformational changes alone can account for either the specificity or

---

[2]Several studies have indicated that nascent presecretory proteins may have a nonnative conformation (24, 25). Such conformational instability is probably due to the presence of the hydrophobic prepeptide sequence which would be expected to interfere with other hydrophobic regions as these interact to form a stable tertiary structure (26). Removal of the prepeptide by cleavage, or by its partition into the membrane as in the loop model, would then allow normal folding to proceed (see also 12).

the vectorial characteristics of transmembrane protein segregation. Nonetheless, this theory seems to provide a relevant model for the incorporation of some proteins or precursors into biological membranes, viruses, or organelles (15, 19, 25).

The most satisfying theory for segregative transfer would combine elements of all the theories discussed above. It would entail the formation of a peptide loop involving the prepeptide in a beta structure at a special site on the membrane which might thus acquire some characteristics of a pore, and segregation might be driven in part by protein folding trans to the side of synthesis; completion of polypeptide transfer may require the formation of partially folded subdomains that allow more polar regions of the nascent secretory protein to raft across the membrane.

Preprotein Cleavage Mechanisms

Cleavage of presecretory proteins occurs rapidly after their synthesis (27, 28, 29), presumably owing to proteolytic enzymes associated with the rough ER or its counterpart after cell homogenization, the rough microsomes in eukaryotic cells (30), or the cell membrane in bacteria (31). Studies with reconstituted systems have confirmed this association and some efforts to identify and characterize a "Signal Peptidase" have been reported (32-36). The initial cleavage appears to be carried out by an endoprotease acting at the junction of the prepeptide with the secreted product peptide and acting preferentially on amino acids having small neutral side chains, such as Ala, Cys, Ser, or Thr. The presence of glycine at the cleavage sites in some prepeptides suggests that other factors, such as a β turn either just before or after the cleavage site, as described above, may also guide the enzyme. It is unlikely that highly specific converting enzymes exist for each presecretory protein since dog pancreas or ascites cell microsomes seem to be able correctly to cleave a wide variety of preproteins from unrelated organs or species (28, 32). Likewise, the ability of bacterial cells correctly to process eukaryotic presecretory proteins also supports the existence of a system having a relatively broad specificity. The microsomal enzyme exhibits latency, but can be unmasked by treatment with detergents (32, 37, 38). These findings have led to the assumption that the enzyme is located on the luminal side of the microsomal membrane. Some evidence has been reported that indicates the enzyme may be a metalloprotease (34, 36) Microsomal or bacterial membranes also appear to have the capacity completely to degrade signal peptides once these have been cleaved by the endoprotease (29, 35). Whether this is accomplished by exopeptidases, particularly amino peptidases, or by mixtures of membrane-associated proteases remains unclear.

Propolypeptide Processing

Propolypeptides occur predominantly in the biosynthesis of polypeptide hormones (Figure 2), although albumin is a notable exception (39, 40). With half-lives ranging from 20 min to 1 hr or longer (41, 42, 43, 44, 45), they are longer-lived than the presequences. Cleavage probably occurs at the stage of secretory processing near the Golgi apparatus and continues into the secretory granule (46, 47, 48, 49). The site of cleavage of the "pro" sequences is usually marked by a pair of basic amino acids. Apart from this highly characteristic feature, the "pro" sequences vary greatly in length and structure, and may occur in any part of the molecule.

The significance of the "pro" sequence remains obscure. It would appear that most of the propolypeptides have little if any biological activity. Possible roles for the "pro" region include ensuring the correct folding of the hormone (exemplified by proinsulin), providing a minimum critical length for segregation and transport through the secretory apparatus, or enabling the released peptides to act as signals (50). It has also been suggested that the "pro" regions reflect the evolutionary origin of the polypeptide, namely from the more primitive process of lysosomal digestion (51, 52).

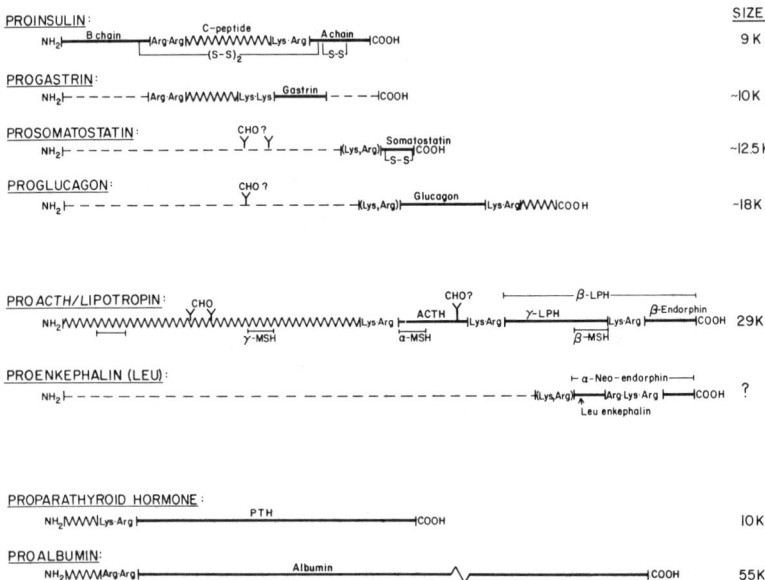

Figure 2. Diagrammatic representations of several polypeptide precursor forms [see (15) for sources].

Evidence suggests that most propolypeptides are processed by a special group of intracellular proteases located in the Golgi apparatus and developing secretory granules. There appears to be a requirement for dibasic residues as a signal for this converting activity, and Arg rather than Lys may be preferred on the carboxyl side of the pair. One exception to this is the Lys-Lys pair in the ACTH/endorphin precursor (Figure 2). However, this bond is cleaved in the pars intermedia during the maturation of the product (53, 54). Abnormal proinsulins and proalbumins have been described in which mutations have occurred in the basic residues at the site of cleavage, Arg to a neutral residue in the abnormal proinsulin (55, 56) and Arg to Gln in the abnormal proalbumin (57). In both situations the normal cleavage process is prevented or greatly slowed down. Biosynthetic studies on anglerfish islets in which the arginine and lysine residues of proinsulin, proglucagon, and prosomatostatin were replaced by the analogs canavanine and thialysine provided further evidence for the essential role of the dibasic amino acids (58). Substitution of the dibasic residues with the analogs prevented cleavage of the prohormones. The situation, however, is somewhat confused by chicken proalbumin, which unlike rat and bovine proalbumin has Ala-Arg at the C-terminal end of the propeptide and is readily cleaved (59).

The identification and characterization of the proteolytic enzymes involved in the conversion of prohormones to hormones, and in the further processing of the polypeptide hormones (see below), can only be achieved if rigid criteria are applied (summarized in Table 1). Although a great deal of information has accumulated on the enzyme activities involved, to date no enzyme has been identified that fulfills all these criteria.

Proinsulin was the first hormonal precursor to be discovered (60). Studies in vitro have shown that the combined effects of trypsin and carboxypeptidase B can readily reproduce the cleavage pattern in vivo to yield the known

Table 1 Criteria for identification of proprotein converting proteases

---

1. The protease must correctly cleave the precursor to generate *all* known products.
2. The enzyme must be resolved from other contaminating proteases before characterization.
3. Biochemical characterization should include studies of:
    (a) Cleavage specificity (i.e. whether trypsin-like, papain-like, etc)
    (b) pH characteristics (optimum, stability, etc)
    (c) Cleavage mechanism (serine, thiol, metallo-type)
    (d) Susceptibility to known protease inhibitors.
4. Cellular and subcellular localization must be appropriate for its putative role.
5. Final proof of normal participation in prohormone cleavage requires demonstration that inhibition, inactivation, or mutation of the enzyme prevents cleavage in the intact cell.

correct products, namely insulin, C-peptide, 3-Arg, and 1-Lys (61). Isolated granules will process prelabeled proinsulin contained in the granule in vitro. Studies on such lysed granules have failed to demonstrate clearly a trypsin-like activity, but have demonstrated the presence of carboxypeptidase B–like activity (62). Although in earlier experiments it was observed that only intact granules converted proinsulin to insulin (63), recent observations [K. Docherty, D. Steiner, unpublished; (64)] have shown that lysed granules will also convert the prohormone.

In the case of proinsulin it is apparent that trypsin-like and carboxypeptidase B–like enzymes must be involved, but whether these are more closely related to the serine and metalloproteases of the exocrine pancreas or to the catheptic thiol and metalloproteases of the lysosomes remains unresolved. Furthermore, it is not clear whether the enzymes are present in the soluble phase of the granule or are membrane-bound.

Both serine and thiol proteases have been implicated in prohormone conversion. A trypsin-like enzyme obtained from a by-product fraction of insulin production was shown to catalyze the conversion of bovine proinsulin to insulin (65). The effect of inhibitors indicated its trypsin-like characteristics, but its isoelectric point (4.82) and molecular size (70,500) clearly distinguished it from pancreatic trypsin. Two other serine proteases have been proposed as candidates for proinsulin converting activity: (a) pancreatic kallikrein, because of its presence in the β cell and its capacity to generate insulin-like intermediates from proinsulin that could be further modified by carboxypeptidase B (66); and (b) plasminogen activator, because it is produced by B cells, where its synthesis and secretion are coordinately regulated with those of insulin, and because plasmin transforms proinsulin to a substance resembling insulin and also fails to degrade insulin (67). Thiol proteases resembling lysosomal cathepsins have been implicated in the conversion of proinsulin to insulin on the basis (a) that they are localized in the secretion granule/mitochondrial fraction of pancreatic islets (68, 69), (b) that they can convert proinsulin to insulin in vitro (70, 71), and (c) that glucose injections given to rats over a 20-hr period increased islet cathepsin B activity (72). It has also been reported that cathepsin B will convert proalbumin to albumin in rat liver. The enzyme and propolypeptide are thought to be sequestered in separate populations of vesicles that fuse in the presence of Ca^{2+} at or near the intracellular concentration (73).

A membrane-bound proparathyroid hormone converting activity has been identified in bovine and porcine parathyroids (74). This activity, which was different from trypsin since it was not inhibited by pancreatic trypsin inhibitor, was inhibited 80% by chloroquine, a powerful inhibitor of cathepsin B_1. The idea that the proPTH converting enzyme was cathepsin B was discounted, however, when purified cathepsin B was tested on proPTH. The enzyme did not remove the terminal hexapeptide but instead cleaved in the central region of the molecule (75, 76). Further evidence against a general role for cathepsin B in the conversion of prohormones to hormones comes from the observation that cathepsin B degraded glucagon by a sequential cleavage of dipeptides from the C-terminal end of the molecule (77).

Recent experiments involving purified secretory granules from anglerfish islets (78, 79) have indicated that the prosomatostatin, proglucagon, and proinsulin converting enzyme may be a granule-membrane associated thiol proteinase (80, 64). This converting activity was inhibited by p-chloromecuribenzoate, dithiodipyridine, antipain, and leupeptin but not by inhibitors of serine or metalloproteases or by chloroquine or TLCK, thus implicating a thiol protease that is not cathepsin B.

Further proteolytic processing of hormones and neurosecretory peptides can occur to produce immunologically related peptides. Proglucagon, prosomatostatin, propancreatic polypeptide, and ProACTH/endorphin all undergo proteolysis at sites other than those that release the active hormone, to produce a variety of cosecreted products. Proglucagon is processed through several intermediates (81) of which a 10,000 mol wt peptide is a major cosecreted product (82). Propancreatic polypeptide is processed to pancreatic polypeptide and a 2500-3000 mol wt peptide (83), while prosomatostatin (84, 85) in the small intestine (86) and hypothalamus [(87, 88); see (89) for review] is processed to somatostatin and a large active form of somatostatin—somatostatin 28. A secreted 16K fragment of proACTH/endorphin also has recently been identified and sequenced (54, 90). Although the opiate peptide Met-enkephalin might be produced by cleavage of β-endorphin, recent evidence indicates the existence of separate precursors containing single or multiple copies of both Leu and Met enkephalin in bovine adrenal medulla (91). The specificity of cleavage involved in this processing, which may not necessarily be marked by Arg-Arg or Lys-Arg pairs of amino acids, is shown in Table 2.

Gastrin and cholecystokinin also provide excellent examples of this type of processing (92). Progastrin is cleaved at a dibasic site to gastrin 34 (93), which is then subject to proteolytic cleavage in the antrum to produce gastrins 17 and 14, and in the brain, tetrapeptide gastrin (94, 95, 96, 97). Cholecystokinin 39 (CCK 39) is cleaved to CCK 33 (98, 99), which in turn is processed in brain to CCK 12, CCK 8, and tetrapeptide cholecystokinin (100, 101). Other hormone-related peptides that undergo unusual cleavages (Table 2) include ACTH in the pars intermedia (53) and gastric tissue (102), proinsulin C-peptide in islets (103), prorelaxin in the corpus luteum (104), GIP in the small intestine (105), and prolactin in the pituitary (106, 107). Proteolysis of human growth hormone (hGH) in the pituitary has also been observed (108, 109, 110). Note that whereas dibasic amino acids may direct

Table 2 Some specialized cleavage enzymes for prohormones

Type	Cleavage specificity	Substrate	Source	Reference
1. Trypsin-like	Arg-Ile	CCK	brain	114
	Arg-Asp	CCK	brain	114
	Lys-Lys	γ LPH and β endorphin	pars intermedia	117
	Lys-Lys-Arg-Arg	ACTH	pars intermedia	53
2. Chymotrypsin-like	Leu-Ala	proinsulin	islets	103
	Leu-Ser	relaxin	corpus luteum	104
	Tyr-Ala	GIP	small intestine	105
3. Others	Pro-Trp	gastrin	antrum	92
	Gly-Trp	gastrin/CCK	brain	92

proteolysis in prohormone cleavage, amidation/deamidation of residues may direct points of attack of proteases to produce the various active fragments of hGH (111). It is not clear however whether all of this processing produces biologically active peptides, or whether some of it may be artifactual or related to degradative processes.

Little information is available on the enzymes responsible for the processing discussed above. A nontryptic arginine esterase that will process cholecystokinin to CCK 12 and CCK 8 has been identified in brain tissue (112). Two enzymes of molecular size greater than γ globulin and between albumin and γ globulin may be involved, but the evidence implicating these enzymes in CCK cleavage is not yet conclusive (113, 114). Arginine esteropeptidases responsible for processing of nerve growth factor (NGF) and epidermal growth factor (EGF) also have been identified (115). These appear to be more specific enzymes, since the EGF enzyme will not process NGF (116). Both unprocessed growth factors occur in multisubunit complexes that include the processing enzyme (115).

Conclusions

In this review we have focused mainly on the cleavage mechanisms and associated enzymes involved in polypeptide hormone precursor processing. This approach has necessitated the exclusion of more detailed considerations of some of the physiological and pathophysiological implications of these mechanisms. The recent findings that both the brain and the gastrointestinal tract produce similar or identical peptides for various regulatory purposes raises many intriguing questions that are only just beginning to be explored (92, 118, 119, 120, 121). Thus it is not yet clear whether the same or related genes are expressed in these various tissues of origin or whether variations in processing have a primary structural basis or are simply manifestations of differing levels and kinds of processing enzymes in different tissues expressing identical genes. The recognition that the products of cleavage are in many instances stored and cosecreted increases the possibilities for diversifying the regulatory functions of any given endocrine or neurosecretory cell. Thus until all such cosecretory products have been identified our knowledge of the physiology of many endocrine cells will remain incomplete. Evolutionary studies also will shed further light on the origins and more primitive functions of endocrine precursor proteins. Studies on preproinsulin in more primitive vertebrates already indicate that both pre- and prohormonal processing mechanisms are of relatively ancient origin, both probably having arisen before the evolution of the vertebrates (122, 123).

ACKNOWLEDGMENT

Work from the authors' laboratory has been supported by grants from the USPHS (AM 13914 and AM 20595), the Kroc Foundation, and the Coustan Memorial Fund. We thank Ms. Myrella Smith for able assistance in preparing this manuscript.

Literature Cited

1. Fessler, J. H., Fessler, L. J. 1978. *Ann. Rev. Biochem.* 47:129–62
2. Wiley, H. S., Wallace, R. A. 1981. *J. Biol. Chem.* 256:8626–34
3. Kreil, G., Molloy, C., Kaschnitz, R., Haiaml, L., Vilas, U. 1980. *Ann. N.Y. Acad. Sci.* 343:338–45
4. Milstein, C., Brownlee, G. G., Harrison, T. M., Mathews, M. B. 1972. *Nature New Biol.* 239:117–20
5. Kemper, B., Habener, J. F., Mulligan, R. C., Potts, J. T. Jr., Rich, A. 1974. *Proc. Natl. Acad. Sci. USA* 71: 3731–35
6. Kreil, G. 1981. *Ann. Rev. Biochem.* 50:317–48
7. Steiner, D. F., Quinn, P. S., Patzelt, C., Chan, S. J., Marsh, J., Tager, H. S. 1980. In *Cell Biology: A Comprehensive Treatise*, ed. L. Goldstein, D. M. Prescott, 4:175–201. NY: Academic
8. Blobel, G., Dobberstein, B. 1975. *J. Cell. Biol.* 67:835–51
9. DiRienzo, J. M., Nakamura, K., Inouye, M. 1978. *Ann Rev. Biochem.* 47:481–532
10. Talmadge, K., Kaufman, J., Gilbert, W. 1980. *Proc. Natl. Acad. Sci. USA* 77:3988–92
11. Chan, S. J., Weiss, J., Konrad, M., White, T., Bahl, C., Yu, S.-D., Marks, D., Steiner, D. F. 1981. *Proc. Natl. Acad. Sci. USA.* 78:5401–5
12. Wilson, I. A., Skehel, J. J., Wiley, D. C. 1981. *Nature* 289:366–73
13. Engelman, D. M., Steitz, T. A. 1981. *Cell* 23:411–22
14. Marchesi, V. T., Furthmayr, H., Tomitz, M. 1976. *Ann. Rev. Biochem.* 45:667–98
15. Steiner, D. F., Quinn, P. S., Chan, S. J., Marsh, J., Tager, H. S. 1980. *Ann. N.Y. Acad. Sci.* 343:1–16
16. Walter, P., Blobel, G. 1980. *Proc. Natl. Acad. Sci. USA* 77:7112–16
17. Meyer, D. I., Dobberstein, B. 1980. *J. Cell Biol.* 87:503–8
18. Rosenblatt, M., Beaudette, N. Y., Fasman, G. D. 1980. *Proc. Natl. Acad. Sci. USA* 77:3983–87
19. Thomas, G. J. Jr., Day, L. A. 1981. *Proc. Natl. Acad. Sci. USA* 78:2962–66

20. Bedouelle, H., Bassford, P. J. Jr., Fowler, A. V., Zabin, I., Beckwith, J., Hofnung, M. 1980. *Nature* 285:78–81
21. Moreno, F., Fowler, A. V., Hall, M., Silhavy, T. J., Zabin, I., Schwartz, M. 1980. *Nature* 286:356–59
22. Ito, K., Beckwith, J. R. 1981. *Cell* 25:143–50
23. Wickner, W. 1979. *Ann. Rev. Biochem.* 48:23–45
24. Lomedico, P. T., Chan, S. J., Steiner, D. F., Saunders, G. F. 1977. *J. Biol. Chem.* 252:7971–78
25. Oxender, D. L., Anderson, J. J., Daniels, C. J., Landick, R., Gunsalus, R. P., Zurawski, G., Yanofsky, C. 1980. *Proc. Natl. Acad. Sci. USA* 77:2005–9
26. Lesk, A. M., Rose, G. D. 1981. *Proc. Natl. Acad. Sci. USA* 78:4304–8
27. Dorner, A. J., Kemper, B. 1978. *Biochemistry* 17:5550–55
28. Shields, B., Blobel, G. 1978. *J. Biol. Chem.* 253:3753–56
29. Patzelt, C., Labrecque, A. D., Duguid, J. R., Carroll, R. J., Keim, P., Heinrikson, R. L., Steiner, D. F. 1978. *Proc. Natl. Acad. Sci. USA* 75:1260–64
30. Blobel, G., Dobberstein, B. 1975. *J. Cell Biol.* 67:852–62
31. Inouye, M., DiRienzo, J., Maeda, T., Movva, R., Nakamura, K., Lee, N., Pirtle, R., Pirtle, I. 1980. *Ann. N.Y. Acad. Sci.* 343:362–67
32. Jackson, R. C., Blobel, G. 1977. *Proc. Natl. Acad. Sci. USA* 74:5598–602
33. Chang, C. N., Blobel, G., Model, P. 1978. *Proc. Natl. Acad. Sci. USA* 75:361–65
34. Zimmerman, M., Ashe, B. M., Alberts, A. W., Pierzchala, P. A., Powers, J. C., Nishino, H., Strauss, A. W., Mumford, R. A. 1980. *Ann. N.Y. Acad. Sci.* 343:405–13
35. Zwizinski, C., Date, T., Wickner, W. 1981. *J. Biol. Chem.* 256:3593–97
36. Zwizinski, C., Wickner, W. 1980. *J. Biol. Chem.* 255:7973–77
37. Kaschnitz, R., Kreil, G. 1978. *Biochem. Biophys. Res. Commun.* 83:901–7
38. Jackson, R. C., Blobel, G. 1980. *Ann. N.Y. Acad. Sci.* 343:391–402
39. Judah, J. D., Gamble, M., Steadman, J. H. 1973. *Biochem. J.* 134:1088–91
40. Russell, J. H., Geller, D. M. 1975. *J. Biol. Chem.* 250:3409–13
41. Steiner, D. F., Cunningham, D. D., Spigelman, L., Aten, B. 1967. *Science* 157:697–700
42. Steiner, D. F. 1967. *Trans. N.Y. Acad. Sci.* 30:60–68
43. Judah, J. D., Nicholls, M. R. 1971. *Biochem. J.* 123:649–55
44. Mains, R. E., Eipper, B. A. 1976. *J. Biol. Chem.* 251:4115–20
45. Habener, J. F., Potts, J. R. Jr. 1978. *New Engl. J. Med.* 299:580–85
46. Sorensen, R. L., Shank, R. D., Lindall, A. W. 1972. *Proc. Soc. Exp. Biol. Med.* 139:652–55
47. Sun, A. M., Lin, B. J., Haist, R. E. 1973. *Can. J. Physiol. Pharmacol.* 51:175–82
48. Steiner, D. F., Kemmler, W., Tager, H. S., Peterson, J. D. 1974. *Fed. Proc.* 33:2105–15
49. Jamieson, J. D., Palade, G. E. 1977. In *International Cell Biology*, ed B. R. Brinkley, H. R. Porles, pp. 308–317. NY: Rockefeller Univ. Press
50. Steiner, D. F., Kemmler, W., Tager, H. S., Rubenstein, A. H., Lernmark, Å., Zühlke, H. 1975. In *Proteases and Biological Control*, ed. E. Reich, D. Rifkin, E. Shaw. Cold Spring Harbor, NY: Cold Spring Harbor Lab.
51. Hales, C. N. 1978. *FEBS. Lett.* 94:10–15
52. Hales, C. N., Docherty, K. 1979. In *Proteases and Hormones*, ed. M. K. Agarwal, pp. 19–46. Amsterdam: Elsevier/North Holland Biomedical Press
53. Scott, A. P., Ratcliffe, J. G., Rees, L. H., Landon, J., Bennett, H. P. J., Lowry, P. J., McMartin, C. 1973. *Nature* 244:65–67
54. Eipper, B. A., Mains, R. E. 1980. *Endocrine Rev.* 1:1–27
55. Gabbay, K. H., Bergenstal, R. M., Wolff, J., Mako, M. E., Rubenstein, A. H. 1979. *Proc. Natl. Acad. Sci. USA* 76:2881–85
56. Robbins, D. C., Blix, P. M., Rubenstein, A. H., Kanazawa, Y., Kosaha, K., Tager, H. S. 1981. *Nature* 291:679–81
57. Brennan, S. O., Carrell, R. W. 1978. *Nature* 274:908–9
58. Noe, B. D. 1981. *J. Biol. Chem.* 256:4940–46
59. Rosen, A. M., Geller, D. M. 1977. *Biochem. Biophys. Res. Commun.* 78:1060–66
60. Steiner, D. F., Oyer, P. E. 1967. *Proc. Natl. Acad. Sci. USA* 57:473–80
61. Kemmler, W., Peterson, J. D., Steiner, D. F. 1971. *J. Biol. Chem.* 246:6786–91
62. Kemmler, W., Steiner, D. F., Borg, J. 1973. *J. Biol. Chem.* 248:4544–51
63. Kemmler, W., Steiner, D. F. 1970. *Biochem. Biophys. Res. Commun.* 41:1223–30
64. Fletcher, D. J., Quigley, J. P., Bauer, G. E., Noe, B. D. 1981. *J. Cell. Biol.* 90:312–22

65. Yip, C. C. 1971. *Proc. Natl. Acad. Sci. USA* 68:1312–15
66. Ole-Moi Yoi, O., Seldin, D. C., Spragg, J., Pinkus, G. S., Austen, K. F. 1979. *Proc. Natl. Acad. Sci. USA* 76:3612–16
67. Virji, M. A. G., Vassalli, J.-D., Estensen, R. D., Reich, E. 1980. *Proc. Natl. Acad. Sci. USA* 77:875–79
68. Zühlke, H., Schmidt, J. S., Gottschling, D., Wilke, B. 1974. *Acta Biol. Med. Germ.* 33:407–18
69. Zühlke, H., Kohnert, K.-D., Jahr, H., Schmidt, S., Kirschke, H., Steiner, D. F. 1977. *Acta Biol. Med. Germ.* 36:1695–703
70. Ansorge, S., Kirschke, H., Friedrich, K. 1977. *Acta Biol. Med. Germ.* 36:1723–27
71. Puri, R. B., Anjaneyulu, K., Kidwai, J. R., Rao, V. K. M. 1978. *Acta Diabet. Lat.* 15:243–50
72. Puri, R. B., Kidwai, J. R., Sahig, M. K., Rao, V. K. M. 1976. *Indian J. Exp. Biol.* 14:567–73
73. Judah, J. D., Quinn, P. S. 1978. *Nature* 271:384–85
74. MacGregor, R. R., Chu, L. L. H., Cohn, D. V. 1976. *J. Biol. Chem.* 251:6711–16
75. MacGregor, R. R., Hamilton, J. W., Kent, G. N., Shofstall, R. E., Cohn, D. V. 1979. *J. Biol. Chem.* 254:4428–36
76. Cohn, D. V., MacGregor, R. R. 1981. *Endocrine Rev.* 2:1–26
77. Aronson, N. N. Jr., Barrett, A. J. 1978. *Biochem. J.* 171:759–65
78. Noe, B. D., Baste, C. A., Bauer, G. E. 1977. *J. Cell Biol.* 74:578–88
79. Noe, B. D., Baste, C. A., Bauer, G. E. 1977. *J. Cell Biol.* 74:589–604
80. Fletcher, D. J., Noe, B. D., Bauer, E., Quigley, J. P. 1980. *Diabetes* 29:593–99
81. Patzelt, C., Tager, H. S., Carroll, R. J., Steiner, D. F. 1979. *Nature* 282:260–66
82. Patzelt, C., Schug, G. 1981. *FEBS. Lett.* 129:127–30
83. Schwartz, T. W., Gingerich, R. L., Tager, H. S. 1980. *J. Biol. Chem.* 255:11494–98
84. Hobart, P., Crawford, R., Sher, L., Pictet, R., Rutter, W. J. 1980. *Nature* 288:137–41
85. Patzelt, C., Tager, H. S., Carroll, R. J., Steiner, D. F. 1980. *Proc. Natl. Acad. Sci. USA* 77:2410–14
86. Pradayrol, L., Jörnvall, H., Mutt, V., Ribet, A. 1980. *FEBS. Lett.* 109:55–58
87. Esch, F., Böhlen, P., Ling, N., Benoit, R., Brazeau, P., Guillemin, R. 1980. *Proc. Natl. Acad. Sci. USA* 77:6827–31
88. Schally, A. V., Huang, W.-Y., Chang, R. C. C., Arimura, A., Redding, T. W., Millar, R. P., Hunkapiller, M. W., Hood, L. E. 1980. *Proc. Natl. Acad. Sci. USA* 77:4489–93
89. Noe, B. D., Fletcher, D. J., Bauer, G. E. 1981. In *Biochemistry, Physiology and Pathology of the Islets of Langerhans*, ed. S. J. Cooperstein, D. T. Watkins. NY: Academic
90. Seidah, N. G., Chretien, M. 1981. *Proc. Natl. Acad. Sci. USA* 78:4236–40
91. Kimura, S., Lewis, R. V., Stern, A. S., Rossier, J., Stein, S., Udenfriend, S. 1980. *Proc. Natl. Acad. Sci. USA* 77:1681–85
92. Rehfeld, J. F. 1981. *Am. J. Physiol.* 240:G255–66
93. Noyes, B. A., Mevarech, M., Stein, R., Agarwal, K. L. 1979. *Proc. Natl. Acad. Sci. USA* 76:1770–74
94. Gregory, R. A., Tracy, H. J. 1964. *Gut* 5:103–17
95. Gregory, R. A., Tracy, H. J. 1974. *Gut* 15:683–85
96. Gregory, R. A., Tracy, H. J. 1975. In *Gastrointestinal Hormones*, ed. J. C. Thompson, pp. 13–24. Houston: Univ Texas Press
97. Rehfeld, J. F., Larsson, L.-I. 1979. In *Gastrins and the Vagus*, ed. J. F. Rehfeld, E. Amdrup, pp. 85–94. London/NY: Academic
98. Mutt, V., Jorpes, J. W. 1971. *Biochem. J.* 125:57p–58p
99. Mutt, V. 1976. *Clin. Endocrinol.* 5:1755–835
100. Rehfeld, J. F. 1978. *J. Biol. Chem.* 253:4022–30
101. Golterman, N. R., Rehfeld, J. F., Roigaard-Peterson, H. 1980. *J. Biol. Chem.* 255:6181–85
102. Larsson, L.-I. 1981. *Proc. Natl. Acad. Sci. USA* 78:2990–94
103. Tager, H. S., Emdin, S. O., Clark, J. L., Steiner, D. F. 1973. *J. Biol. Chem.* 248:3476–82
104. Hudson, P., Haley, J., Cronk, M., Shire, J., Niall, H. 1981. *Nature* 291:127–31
105. Jörnvall, H., Carlquist, M., Kwauk, S., Otte, S. C., McIntosh, C. H. S., Brown, J. C., Mutt, V. 1981. *FEBS. Lett.* 123:205–10
106. Mittra, I. 1980. *Biochem. Biophys. Res. Commun.* 95:1750–59
107. Mittra, I. 1980. *Biochem. Biophys. Res. Commun.* 95:1760–67
108. Singh, R. N. P., Seavy, B. K., Rice, V. P., Lindsey, T. T., Lewis, U. J. 1974. *Endocrinology* 94:883–91
109. Lewis, U. J., Singh, R. N. P., Vanderlaan, W. P., Tutwiler, G. F. 1977. *Endocrinology* 101:1587–603

110. Maciag, T., Forand, R., Ilsley, S., Cerundolo, J., Greenlee, R., Kelley, P. R., Canalis, E. 1980. *J. Biol. Chem.* 255:6064–70
111. Lewis, U. J., Singh, R. N. P., Bonewald, L. F., Seavy, B. K. 1981. *J. Biol. Chem.* In press
112. Straus, E., Malesci, A., Yalow, R. S. 1978. *Proc. Natl. Acad. Sci. USA* 75:5711–14
113. Malesci, A., Straus, E., Yalow, R. S. 1980. *Proc. Natl. Acad. Sci. USA* 77:597–99
114. Ryder, S. W., Straus, E., Yalow, R. S. 1980. *Proc. Natl. Acad. Sci. USA* 77:3669–71
115. Greene, L. A., Shooter, E. M. 1980. *Ann. Rev. Neurosci.* 3:353–402
116. Server, A. C., Shooter, E. M. 1976. *J. Biol. Chem.* 25:165–73
117. Smyth, D. G., Zakarian, S. 1980. *Nature* 288:613–15
118. Krieger, D. T., Martin, J. B. 1981. *New Engl. J. Med.* 304:876–85
119. Krieger, D. T., Martin, J. B. 1981. *New Engl. J. Med.* 304:944–51
120. Snyder, S. H. 1980. *Science* 209:976–83
121. Brownstein, M. J., Russell, J. T., Gainer, H. 1980. *Science* 207:373–78
122. Hobart, P. M., Shen, L-P., Crawford, R., Pictet, R. L., Rutter, W. J. 1980. *Science* 210:1360–63
123. Chan, S. J., Emdin, S. O., Kwok, S. C. M., Kramer, J. M., Falkmer, S., Steiner, D. F. 1981. *J. Biol. Chem.* 256:7595–602

RECEPTORS FOR PEPTIDE HORMONES: ALTERATIONS IN DISEASES OF HUMANS[1]

Jesse Roth and Simeon I. Taylor

Diabetes Branch, National Institute of Arthritis, Diabetes, Digestive and Kidney Diseases, National Institutes of Health, Bethesda, Maryland 20205

The receptor, the initial site of hormone interaction with the target cell, can be the focus of a disease process. As discovered in 1972, mice with testicular feminization and androgen resistance had an inborn deficiency of androgen receptors (9) and mice with genetic obesity and extreme insulin resistance had a severe but reversible deficiency of insulin receptors (21, 52). Since then it has become clear that the receptor is a target cell site that is frequently involved in human disease states. The receptors for peptide hormones that are involved in human disease are the focus of this review (37, 47–49).

RECEPTORS FOR PEPTIDE HORMONES

Peptide hormones, which represent 80% or more of all hormones, have specific receptors that are intrinsic proteins of the plasma membrane; typically these receptors have molecular weights in excess of 10^5 daltons and are composed of several subunits (in contrast to the hormones, which are all much smaller and simpler in structure). The binding of hormone to receptor on the surface of the target cell causes the receptor to express its own intrinsic program of activity, which initiates a series of steps that leads to production of a soluble intracellular ("second") messenger that leads to the panoply of biological events typical for that receptor on that target cell. Thus the receptor serves two fundamental functions: recognition of hormone and transfer of information. The strength of the signal transmitted to the cell by the receptor depends on the concentration of hormone-receptor complexes, which in vivo is a function of three co-equal variables: hormone

[1] The US Government has the right to retain a nonexclusive, royalty-free license in and to any copyright covering this paper.

concentration, receptor concentration, and receptor affinity. That the receptor itself (rather than the hormone) actually activates the target cell accounts for two observations. (*a*) When two or more hormones can bind to two or more receptor types, the nature of the biological response depends on which receptor is occupied and is independent of which hormone is bound (48). (*b*) Autoantibodies that bind to a specific receptor can mimic the action of the corresponding hormone even in the absence of the latter (24, 51).

METHODS FOR CHARACTERIZING RECEPTORS

The major method for the study of receptors is to measure in vitro the interaction of a radioactively labeled hormone with receptors on whole cells or membrane-rich fractions of cells. The interaction, when carried out under appropriate conditions over a range of concentrations of unlabeled hormone, can provide reasonable estimates of receptor number and receptor affinity. Selected unlabeled hormones and analogs as competitors for the labeled ligand are essential to define the specificity of the receptor, since each labeled hormone often binds to more than one type of receptor. Our ability to characterize the receptors as protein molecules or measure their biological activity is still inadequate. Therefore a receptor defect that is beyond the binding steps is likely to be overlooked with the present methods, and mistakenly classified as a "post-receptor defect." In effect, our review is limited to those diseases where hormone binding to receptors is affected.

For studies of receptors in humans, only a few tissues are readily available. Receptors on circulating cells have been the most popular. Fresh cells have been utilized to evaluate the status of a particular receptor as close as possible to the in vivo state. Cultured cells, especially fibroblasts, have been used to determine the genetic features or capabilities of a given receptor free of in vivo influences. More recently, lymphocytes that have been immortalized by Epstein-Barr virus have been used with substantial success.

The cells from humans that are available for study typically are not major target cells for that hormone. Hopefully the receptor on the available cell will reflect what is going on at the major target cells. This assumption needs to be verified as closely as possible in each situation. One approach is to measure the receptor on two unrelated cells; if the changes are identical in both cell types, it is likely that the defect is widespread. Another approach is to use animal models where receptors of multiple tissues can be measured. A third approach is to perturb the situation in vivo or in vitro to see if the receptor changes with alterations in the physiological state.

Since the introduction of the competitive binding method for studying cell surface receptors, numerous other in vitro methods have been employed. One other method of possibly great utility for studies in humans measures hormone binding to its receptors in the whole body in vivo (61). This approach avoids the bias introduced by sampling the receptors on only one or two cell types and studying them under arbitrary conditions in vitro.

INSULIN RECEPTORS

In some disturbances of glucose metabolism, events at the secretory cell dominate; in most disorders the target cell is dominant, and alterations in the target cell receptor appear to have an important role in the pathology. Later, it will be seen that even in those conditions where secretory defects dominate, changes in the target cell receptors are often important modifiers of the clinical state.

Moderate Insulin Resistance

The most common disorders of glucose and insulin metabolism are those that involve moderate insulin resistance—i.e. patients require or produce two or three times as much insulin as normal.

OBESITY   Overweight patients with normal glucose tolerance often have hyperinsulinemia and a reduction in sensitivity to insulin associated with a decrease in the number of insulin receptors throughout the body, including monocytes, erythrocytes, and adipocytes (2–4, 13, 14, 22, 36, 37).

DIABETES   Among diabetic patients, only 10–20% are totally lacking insulin and require insulin to sustain life (Type I). Most diabetic patients (Type II) have endogenous insulin secretion. Many are not treated with insulin; in those who are, treatment is to control symptoms rather than sustain life. Many obese and thin patients with Type II diabetes have decreased sensitivity to insulin, have elevated levels of insulin and a reduction in receptor concentrations. In the hyperglycemic patients, both thin and obese, as well as in the euglycemic obese patients, there is an excellent correlation between the severity of the insulin resistance and the severity of the receptor deficiency. In addition, therapy that is effective in relieving the insulin resistance and ameliorating the hyperinsulinemia also produces improvement at the level of the receptor. Such therapies include reduction in total calorie intake, redistribution of major classes of foodstuffs, an increase in the fiber content of the diet, muscular exercise, and administration of some of the oral anti-diabetic drugs (2, 4, 7, 36, 38, 44).

ACROMEGALY   Another condition characterized by moderate insulin resistance, with or without hyperglycemia, is acromegaly, which is due to excess growth hormone produced by the pituitary. In patients with acromegaly a reduction in the concentration of receptor correlates with the severity of the hyperinsulinemia and the severity of the insulin resistance, which is similar to findings in patients with obesity or Type II diabetes. In addition, in patients with acromegaly who retain normal glucose tolerance, there is a compensatory increase in receptor affinity, which is absent in the patients in whom glucose tolerance is impaired (34). Other conditions characterized by moderate insulin resistance in which insulin receptors have been studied but in which final conclusions are as yet unclear include glucocorticoid excess, myotonic dystrophy, Werner's syndrome, renal failure, and hepatic failure (16, 33, 35, 47–49, 55).

Extreme Insulin Resistance

In contrast to patients with moderate insulin resistance in whom insulin levels are elevated several fold, patients with extreme insulin resistance typically require 5–50 fold elevations of endogenous and exogenous insulin to achieve a normal response. The two types of biochemical mechanisms involved in these syndromes are presented below. Irrespective of the mechanism, patients with extreme insulin resistance often have two other abnormalities: (*a*) acanthosis nigricans, a skin lesion, and (*b*) masculinization in females due to excess circulating androgens produced by the ovaries. The high frequency of these two findings in patients with extreme insulin resistance of diverse causes suggests that extreme hyperinsulinemia itself may be responsible, but the biochemical pathways involved are at present obscure.

ANTI-RECEPTOR ANTIBODIES In some patients with extreme insulin resistance, designated Type B extreme insulin resistance with acanthosis nigricans, insulin binding to its receptors is markedly reduced and there are circulating polyclonal antibodies specific for insulin receptors (19, 20). Other findings typically suggest an autoimmune disorder, but only a minority of patients have a defined autoimmune disease such as lupus erythematosus. The antibodies bind to the insulin receptor, reduce the affinity of the receptor for the hormone, and thereby impair binding of insulin; conversely, insulin binding to receptor reduces antibody binding. The antibodies are also insulinomimetic but only acutely, not chronically. In addition they can desensitize the target cell to insulin by effects at a post-binding site. The dominant clinical effect is insulin resistance; in the one patient whose adipose tissue was studied in vitro, the antibodies shifted the dose response curve for insulin to the right without significantly altering the maximal effects of high concentrations of the hormone (39). In a minority of the patients, hypoglycemia that is difficult to control may sometimes become the dominant concern (18). It is not yet known whether a single immunoglobulin species can produce all of the effects observed with whole serum. With remission, either spontaneous or drug-induced, antibody disappears, and insulin levels, insulin sensitivity, and glucose metabolism become normal. Overall, as in the patients with moderate insulin resistance, there is excellent correlation between the receptor and the clinical state, for patients as a group, for individual patients, and during changes in the clinical course including the response to treatment.

PRIMARY TARGET CELL DISORDERS We have studied insulin binding to its receptor in four syndromes characterized by severe insulin resistance and hyperinsulinemia without autoimmunity: leprechaunism, Rabson Mendenhall syndrome (extreme insulin resistance associated with pineal hyperplasia and somatic abnormalities), Type A extreme insulin resistance, and lipoatrophic diabetes (5, 11, 28, 41, 50, 53, 54, 56). In some of the patients in each group (most patients with Type A but only a minority of patients with lipoatrophic diabetes), insulin binding to receptor is affected, as described below. In the remainder, insulin binding to receptors is normal, which suggests that the defect is beyond the receptor or in the receptor but beyond the binding step. Within each of the four syndromes, the patients with defects in binding are indistinguishable from their counterparts with these syndromes who have normal binding of insulin to receptor. This suggests that defects at several loci early in the path of insulin action yield similar net results. While the four syndromes share many features, they each have enough unique phenotypic features to suggest unique etiologies.

Quantitative defects With circulating cells and cultured lymphocytes there is a decrease in binding due to a decrease in the number of receptors per cell; the residual receptors are qualitatively normal (5, 53). Not only are they recognized normally by insulins, insulin derivatives, and by antibodies directed against the insulin receptor, but their subunit structure appears to be normal. The cause for the decrease in receptor number in these patients is not known.

Qualitative abnormalities in the insulin receptor We have studied in detail one patient with a qualitative abnormality in insulin binding (54). This patient has the clinical syndrome of leprechaunism and initially had been thought to have normal insulin receptors; the defect in insulin sensitivity had been assigned to a post-receptor locus (28). In fact, insulin binding to cultured cells under physiological conditions showed a modestly elevated level of binding due to an increase in receptor affinity. Detailed studies showed that the receptor binding of insulin was quite abnormal in response to alterations in pH and temperature (54). We think that this defect in insulin binding is not the cause of the insulin resistance but a marker for the fact the receptor is seriously disordered, possibly in its ability to transfer information to the cell. Preliminary studies with cultured fibroblasts from other patients have suggested that such qualitative abnormalities may be more widespread (25, 43).

Post-binding defects Many patients with extreme insulin resistance have normal insulin binding to their cells. Traditionally, these patients have been referred to as having "post-receptor defects." Because of the possibility that structural abnormalities in the receptor may not affect insulin binding yet give rise to insulin resistance, we suggest the term "post-binding defects." While many such patients have been described, in no case has the site or nature of the defect been identified.

Supersensitivity to Insulin

At the opposite end of the spectrum are anorexia nervosa, growth hormone deficiency, and glucocorticoid deficiency, disorders of the target cell in which blood glucose concentrations and insulin levels are low-normal or subnormal along with a heightened responsiveness to exogenously administered insulin. In patients with anorexia nervosa the concentration of insulin receptors is elevated; refeeding results in restoration to normal of insulin sensitivity, plasma insulin, and insulin receptors (57).

With hGH deficiency, binding of insulin to its receptor is normal, and chronic hGH replacement produces restoration to normal of plasma insulin and sensitivity to insulin but no changes in insulin binding (31). Thus in contrast to anorexia nervosa (and to acromegaly) the target cell changes that affect hormone sensitivity are beyond the level of hormone binding. Glucocorticoid deficiency, which in rats produces an elevation in the affinity of the insulin receptor, has not been characterized in detail in humans.

Insulin Deficiency

Insulin deficiency in experimental animals is associated with an elevation in the concentration of insulin receptors. A similar situation is observed in patients with insulin deficiency due to pancreatitis or pancreatectomy. Interestingly, patients with Type I diabetes, who have equally severe insulin deficiency, typically have insulin receptors in the normal range; the reason that these patients fail to increase receptor concentrations is not known. In addition, patients with pancreatic diabetes, who typically have an elevated concentration of receptors, are known to be substantially more sensitive to insulin than are patients with Type I diabetes.

Insulin Excess

Insulin excess, produced in experimental animals by the cautious administration of increasing doses of insulin, is associated with a diminution in receptor concentrations and a reduction in sensitivity to administered insulin (37). In patients with insulin-secreting tumors (6), receptor concentrations decrease in proportion to the elevation in circulating insulin, which provides protection against the circulating insulin. In addition, there is frequently an increase in receptor affinity, which causes an increase in insulin binding to receptor and possibly correlates with unusual susceptibility to hypoglycemia.

Newborns with inappropriately elevated levels of circulating insulin frequently have hypoglycemia associated with an inappropriate elevation of receptors. Normally, it appears that insulin receptors are elevated in the fetus and progressively decrease with the approach of term and in the weeks to months following delivery. Thus, a fall in the concentration of insulin receptors appears to be a normal maturation process. The hypoglycemic infant, even when born at full term, frequently has many features of premature infants; we interpret the elevated concentrations of insulin receptors as possibly another manifestation of their immaturity.

Disorders of Receptor Design or Specificity Spillover

In another group of disorders characterized by hormone excess, one or more features of the clinical disorder are due to effects of the hormone interacting with receptors of a related hormone. The two examples presented here come from the insulin family, and other examples are cited in the references.

To provide perfect specificity, each hormone should have a single receptor (48). As noted above, two or more hormones can have receptors with some overlap in specificity. For example, insulin binds with high affinity to receptors that are coupled to metabolic events and also binds with lower affinity to the receptors for insulin-like growth factors; insulin at high concentrations stimulates growth-related processes (60). Similarly, insulin-like growth factors bind with high affinity to their own receptors, yielding growth-related events, and bind with lower affinity to the insulin receptor; at high concentrations insulin-like growth factors produce metabolic events. When hormone levels are elevated, the hormone can hyperstimulate through its own receptor and also trigger the receptor for a related hormone, which we term specific spillover.

INFANTS OF DIABETIC MOTHERS In an infant of a diabetic mother, high levels of glucose and amino acids from the maternal circulation traverse the placenta and hyperstimulate the infant's beta cells. The abnormally high level of circulating insulin acts on insulin receptors to produce an excess of metabolic events including excess deposition of fat and glycogen, and hypoglycemia during the first few post-natal hours. In addition, the insulin at high concentrations interacts with receptors for the insulin-like growth factors, thereby producing excess linear growth and macrosomia, which are characteristic of these infants as well as other infants with hyperinsulinemia.

HYPOGLYCEMIA WITH NON-ISLET CELL TUMORS   A related series of events may explain the hypoglycemia with some non-islet cell tumors; with these tumors, insulin levels are not elevated, but in about one third to one half of these patients elevated levels of insulin-like growth factors have been detected that appear to account for the hypoglycemia (32). Again, a disorder characterized by hypersecretion of one hormone has clinical manifestations due to features of the receptor—in this case, that one hormone may interact with receptors of another related hormone.

RECEPTORS FOR OTHER PEPTIDE HORMONES

Insulin-like Growth Factors

The insulin-like growth factors (IGF-I, IGF-II, somatomedin A, somatomedin C, and multiplication stimulating activity or MSA) are five closely related peptides that have a high degree of structural homology with proinsulin (60). The importance of IGF receptors in human disease states is still unclear. Progress in this area has been slow because of the difficulty in distinguishing among the wide variety of types of IGF receptors.

Despite these difficulties, studies have been made of IGF receptors in human diseases. Receptors for somatomedin C/IGF-I on circulating mononuclear cells of newborn infants are increased when compared with cells from normal adults (46). Hypopituitary children have increased numbers of somatomedin receptors, and administration of hGH (via an elevation in circulating levels of somatomedin) leads to a decrease in the number of somatomedin receptors (45). In cultured skin fibroblasts from a patient with leprechaunism, who had both insulin resistance and growth retardation, binding of both labeled insulin and labeled IGF-I were markedly depressed. The authors speculated that the same biochemical lesion may be responsible for both receptor defects (27).

Vasopressin Receptors

Specific receptors for anti-diuretic hormone and adenylate cyclase that is sensitive to vasopressin have been demonstrated in human monocytes (8).

Vasopressin deficiency was associated with a 50% increase in the level of hormone binding, which may account, at least in part, for the heightened sensitivity to low doses of vasopressin observed in these patients. In addition to the in vitro approach, methods for measurement of the vasopressin receptor compartment in the whole animal in vivo are probably applicable to humans as well (58).

Receptor for Growth Hormone and Prolactin

Receptors for hGH that have the pattern of specificity typical for human growth have been studied in cultured lymphocytes of the IM-9 line and in other lymphocyte cell lines. Despite the presence of hGH receptors on some human lymphocyte cell lines, other cells do not have detectable hGH receptors: circulating lymphocytes, other circulating human cells, immortalized (Epstein-Barr) B-lymphocytes, and cultured fibroblasts. The human adipocyte, which is sensitive to hGH, appears to be the most promising system in which to study the hGH receptor in human diseases (12, 23). For the growth receptors in humans, ^{125}I-hGH is competed for by hGH = bovid placental lactogens $\gg$ human placental lactogen $>$ human prolactin = nonprimate prolactins and growth hormones. For prolactin receptors, ^{125}I-hGH is competed for by hGH = prolactin = placental lactogen $\gg$ nonprimate GH (29). Prolactin receptors on human breast cancers are being studied (along with estrogen and progesterone receptors) to characterize their endocrine sensitivity and biological behavior.

Thyrotropin Receptors

There are two distinct types of sites that bind TSH. One receptor, present at high concentrations, is of relatively low affinity, binds best under unusual conditions (low pH and salt) and has a broad specificity; in addition to TSH it binds cholera toxin, gangliosides, normal gamma globulins, and thyroglobulin. Its role in thyroid function and disease is still speculative. The other receptor, present at much lower concentrations, has a much higher affinity, binds best under conditions that are closer to "physiological," and has a specificity that is limited to biologically active thyrotropins (40).

ANTI-RECEPTOR ANTIBODIES Patients with hyperthyroidism due to Graves' disease often have circulating immunoglobulins that mimic TSH in vivo and in vitro, including both early proximal events (e.g. activation of adenylate cyclase) and late distal effects (e.g. thyroid hormone release) (30, 51). These antibodies can cross the placenta and produce transient hyperthyroidism and goiter in newborns.

Immunoglobulins from these patients also inhibit the binding of labeled TSH to its receptor—in thyroid and extrathyroidal tissue (adipocyte and testis); from humans and nonprimates; membrane-bound and detergent-solubilized. These studies strongly suggest that the inhibition of TSH binding to its receptor is due to antibodies that bind directly to the TSH receptor. It is likely but not yet proven that the receptor-binding antibodies are also responsible for the TSH-like bioactivity (26, 42, 59).

The receptor binding antibodies are present in a majority of patients with Graves' disease, and the titer of antibody correlates with both hyperfunction (trapping of iodide or pertechnetate) and hyperplasia of the thyroid. About 10% of patients with Hashimoto's thyroiditis have similar antibodies.

The antibodies vary widely in their intrinsic activity and are often partial agonists. Some antibodies bind, do not activate, but do block TSH action —i.e. are antagonists. Blocking antibodies may be present in some patients with hypothyroidism (15).

THYROID TUMORS The receptor display on thyroid tumors may be altered (1, 10, 17). Undifferentiated cancers typically are devoid of TSH receptors; the adenylate cyclase is unresponsive to TSH, as expected, but is stimulated by PGE, indicating the basic integrity of the system. Some tumors develop "ectopic receptors"—i.e. they become responsive (activation of adenylate cyclase) to hormones that do not affect the normal gland (e.g. glucagon, ACTH, and GH). Ectopic receptors may account for apparent autonomous behavior of certain tumors (48).

Literature Cited

1. Abe, Y., Ichikawa, Y., Muraki, T., Homma, M. 1981. Thyrotropin (TSH) receptors and adenylate cyclase activity in human thyroid tumors: Absence of high affinity receptor and loss of TSH responsiveness in undifferentiated thyroid carcinoma. *J. Clin. Endocrinol. Metab.* 52:23-28
2. Archer, J. A., Gorden, P., Roth, J. 1975. Defect in insulin binding to receptors in obese man. Amelioration with calorie restriction. *J. Clin. Invest.* 55:166-74
3. Archer, J. A., Gorden, P., Gavin, J. R. III, Lesniak, M. A., Roth, J. 1973. Insulin receptors in human circulating lymphocytes: Application to the study of insulin resistance in man. *J. Clin. Endocrinol. Metab.* 36:627-33
4. Bar, R. S., Gorden, P., Roth, J., De Meyts, P., Kahn, C. R. 1976. Fluctuations in the affinity and concentration of insulin receptors on circulating monocytes of obese patients: Effects of starvation, refeeding, and dieting. *J. Clin. Invest.* 58:1123-35
5. Bar, R. S., Muggeo, M., Kahn, C. R., Gorden, P., Roth, J. 1980. Characterization of the insulin receptors in patients with the syndromes of insulin resistance and acanthosis nigricans. *Diabetologia* 18:209-16
6. Bar, R. S., Gorden, P., Roth, J., Siebert, C. W. 1977. Insulin receptors in patients with insulinomas: Changes in receptor affinity and concentration. *J. Clin. Endocrinol. Metab.* 44:1210-13
7. Beck-Nielsen, H., Pedersen, O., Swartz-Sorensen, N. 1978. Effects of diet on the cellular insulin binding and insulin sensitivity in young healthy subjects. *Diabetologia* 15:289-96
8. Block, L. H., Locher, R., Tenschert, W., Siegenthaler, W., Hofmann, T., Mettler, R., Vetter, W. 1981. ^{125}I-8-L-arginine vasopressin binding to human mononuclear phagocytes. *J. Clin. Invest.* 68:374-81

9. Bullock, L. P., Bardin, C. W. 1972. Androgen receptors in testicular feminization. *J. Clin. Endocrinol. Metab.* 35:935–37
10. Carayon, P., Thomas-Morvan, C., Castanas, E., Tubiana, M. 1980. Human thyroid cancer: Membrane thyrotropin binding and adenylate cyclase activity. *J. Clin. Endocrinol. Metab.* 51:915–20
11. D'Ercole, A. J., Underwood, L. E., Groelke, J., Plet, A. 1979. Leprechaunism: Studies of the relationship among hyperinsulinism, insulin resistance, and growth retardation. *J. Clin. Endocrinol. Metab.* 48:495–502
12. DiGirolamo, M., Eden, S., Isaksson, O., Smith, U. 1981. Specific binding of human growth hormone to human adipocytes. *Endocrinology* 108:101 (Abstr.)
13. Dons, R. F., Corash, L. M., Gorden, P. 1981. The insulin receptor is an age dependent integral component of the human erythrocyte membrane. *J. Biol. Chem.* 256:2983–87
14. Dons, R., Ryan, J. L., Gorden, P., Wachslicht-Rodbard, H. 1981. Erythrocyte and monocyte insulin binding in man: A comparative analysis in normal and disease states. *Diabetes* 30:896–902
15. Drexhage, H. A., Bottazzo, G. F., Bitensky, L., Chayen, J., Doniach, D. 1981. Thyroid growth-blocking antibodies in primary myxoedema. *Nature* 289:594–96
16. Fantus, I. G., Ryan, J. L., Hizuka, N., Gorden, P. 1981. The effect of glucocorticoids on the insulin receptor: An *in vivo* and *in vitro* study. *J. Clin. Endocrinol. Metab.* 52:953–60
17. Field, J. B., Bloom, G., Chou, M. C. Y., Kerins, M. E., Larsen, P. R., Kotani, M., Kariya, T., Dekker, A. 1978. Effects of thyroid-stimulating hormone on human thyroid carcinoma and adjacent normal tissue. *J. Clin. Endocrinol. Metab.* 47:1052–58
18. Flier, J. S., Bar, R. S., Muggeo, M., Kahn, C. R., Roth, J., Gorden, P. 1978. The evolving clinical course of patients with insulin receptor antibodies: Spontaneous remission or receptor proliferation with hypoglycemia. *J. Clin. Endocrinol. Metab.* 47:985–95
19. Flier, J. S., Kahn, C. R., Jarrett, D. B., Roth, J. 1976. Characterization of antibodies to the insulin receptor: A cause of insulin-resistant diabetes in man. *J. Clin. Invest.* 58:1442–49
20. Flier, J. S., Kahn, C. R., Jarrett, D. B., Roth, J. 1977. Autoantibodies to the insulin receptor: effect on the insulin-receptor interaction in IM-9 lymphocytes. *J. Clin. Invest.* 60:784–94
21. Freychet, P., Laudat, M. H., Laudat, P., Rosselin, G., Kahn, C. R., Gorden, P., Roth, J. 1972. Impairment of insulin binding to fat cell membrane in the obese hyperglycemic mouse. *FEBS Lett.* 25:339–42
22. Gambhir, K. K., Archer, J. A., Bradley, C. L. 1978. Characteristics of human erythrocyte insulin receptor. *Diabetes* 27:701–8
23. Gavin, J. R. III 1981. Growth hormone receptors in canine and human adipocytes: Probes for structural determinants of growth hormone binding and action. *Endocrinology* 108:164 (Abstr.)
24. Kahn, C. R., Baird, K. L., Flier, J. S., Jarrett, D. B. 1977. Effect of autoantibodies to the insulin receptor on isolated adipocytes. Studies of insulin binding and insulin action. *J. Clin. Invest.* 60:1094–1106
25. Kahn, C. R., Podskalny, J. M. 1980. Demonstration of a primary (?genetic) defect in insulin receptors in fibroblasts from a patient with the syndrome of insulin resistance and acanthosis nigricans Type A. *J. Clin. Endocrinol. Metab.* 50:1139–41
26. Kishihara, M., Nakao, Y., Baba, Y., Kobayashi, N., Matsukura, S., Kuma, K., Fujita, T. 1981. Interaction between (TSH) binding inhibitor immunoglobulins (TBII) and soluble TSH receptors in fat cells. *J. Clin. Endocrinol. Metab.* 52:665–70
27. Knight, A. B., Rechler, M. M., Romanus, J. A., Van Obberghen-Schilling, E. E., Nissley, S. P. 1981. Stimulation of glucose incorporation and amino acid transport by insulin and an insulin-like growth factor in fibroblasts with defective insulin receptors cultured from a patient with leprechaunism. *Proc. Natl. Acad. Sci. USA* 78:2554–58
28. Kobayashi, M., Olefsky, J. M., Elders, J., Mako, M. E., Given, B. D., Schwedie, H. K., Fiser, R. H., Hintz, R. L., Horner, J. A., Rubenstein, A. H. 1978. Insulin resistance due to a defect distal to the insulin receptor: Demonstration in a patient with leprechaunism. *Proc. Natl. Acad. Sci. USA* 75:3469–73
29. Lesniak, M. A., Gorden, P., Roth, J. 1977. Reactivity of non primate growth hormones and prolactins with human growth hormone receptors on cultured human lymphocytes. *J. Clin. Endocrinol. Metab.* 44:838–49
30. Levey, G., Pastan, I. H. 1970. Activation of thyroid adenyl cyclase by long-

acting thyroid stimulator. *Life Sci.* 9:67–73
31. Lippe, B. M., Kaplan, S. A., Golden, M. P., Hendricks, S. A., Scott, M. L. 1981. Carbohydrate tolerance and insulin receptor binding in children with hypopituitarism: Responses after acute and chronic human growth hormone administration. *J. Clin. Endocrinol. Metab.* 53:507–13
32. Megyesi, K., Kahn, C. R., Roth, J., Gorden, P. 1974. Hypoglycemia in association with extrapancreatic tumors: Demonstration of elevated plasma NSILA-s by radioreceptor assay. *J. Clin. Endocrinol. Metab.* 38:931–34
33. Moxley, R. T. III, Livingston, J. N., Lockwood, D. H., Griggs, R. C., Hill, R. L. 1981. Abnormal regulation of monocyte insulin-binding affinity after glucose ingestion in patients with myotonic dystrophy. *Proc. Natl. Acad. Sci. USA* 78:2567–71
34. Muggeo, M., Bar, R. S., Roth, J., Kahn, C. R., Gorden, P. 1979. The insulin resistance of acromegaly: Evidence for two alterations in the insulin receptor on circulating monocytes. *J. Clin. Endocrinol. Metab.* 48:17–25
35. Muggeo, M., Saviolakis, G. A., Wachslicht-Rodbard, H., Roth, J. 1982. Effects of chronic glucocorticoid excess in man on insulin binding to circulating cells: Differences between endogenous and exogenous hypercorticism. *J. Clin. Endocrinol. Metab.* In press
36. Olefsky, J. M. 1976. Decreased insulin binding to adipocytes and circulating monocytes from obese subjects. *J. Clin. Invest.* 57:1165–72
37. Olefsky, J. M. 1981. Insulin resistance and insulin action: An *in vivo* and *in vitro* perspective. *Diabetes* 30:148–62
38. Pedersen, O., Beck-Nielsen, H., Heding, L. 1980. Increased insulin receptors after exercise in patients with insulin-dependent diabetes mellitus. *N. Engl. J. Med.* 302:886–92
39. Pedersen, O., Hjollund, E., Beck-Nielsen, H., Kromann, H. 1981. Diabetes mellitus caused by insulin-receptor blockade and impaired sensitivity to insulin. *N. Engl. J. Med.* 304:1085–88
40. Pekonen, F., Weintraub, B. D. 1979. Thyrotropin receptors on bovine thyroid membranes: Two types with different affinities and specificities. *Endocrinology* 105:352–59
41. Perez-Corral, F., de la Vina, S., Carbo, M., Barrio, R., Yturriaga, R., Perez-Maceda, B., Alonso, M., Sorrano-Rios, M. 1980. Rabson syndrome: Model of insulin resistance due to decreased number and affinity of insulin receptors in erythrocytes. *Diabetologia* 19:306 (Abstr.)
42. Pinchera, A., Fenzi, G., Vitti, P., Macchia, E., Toccafondi, R., Baschieri, L. 1980. Relationship between TSH receptor and thyroid plasma membrane antigens for Graves' disease immunoglobulins. In *Endocrinology 1980,* ed. I. A. Cumming, J. W. Funder, F. A. O. Mendelsohn, pp. 126-29. NY:Elsevier/North Holland Biomedical Press
43. Podskalny, J. M., Klein, M. N., Kahn, C. R. 1981. A spectrum of alterations in insulin binding and action in fibroblasts from patients with insulin resistance. *Diabetes* 30:53 (Abstr.)
44. Prince, M. J., Olefsky, J. M. 1980. Direct *in vitro* effect of a sulfonylurea to increase human fibroblast insulin receptors. *J. Clin. Invest.* 66:608–11
45. Rosenfeld, R. G., Kemp, S. F., Gaspich, S., Hintz, R. L. 1981. *In vivo* modulation of somatomedin receptor sites: Effects of growth hormone treatment of hypopituitary children. *J. Clin. Endocrinol. Metab.* 52:759–64
46. Rosenfeld, R., Thorsson, A. V., Hintz, R. L. 1979. Increased somatomedin receptor sites in newborn circulating mononuclear cells. *J. Clin. Endocrinol. Metab.* 48:456–61
47. Roth, J. 1981. Insulin binding to its receptor: Is the receptor more important than hormone? *Diabetes Care* 4:27–32
48. Roth, J., Grunfeld, C. 1981. Endocrine systems: Mechanisms of disease, target cells and receptors. In *Textbook of Endocrinology,* ed. R. H. Williams, pp. 15–74. Philadelphia:W. B. Saunders. 1270 pp.
49. Roth, J., Lesniak, M. A., Bar, R. S., Muggeo, M., Megyesi, K., Harrison, L. C., Flier, J. S., Wachslicht-Rodbard, H., Gorden, P. 1979. An introduction to receptors and receptor disorders. *Proc. Soc. Exp. Biol. Med.* 162:3–12
50. Schilling, E. E., Rechler, M. M., Grunfeld, C., Rosenberg, A. 1979. Primary defect of insulin receptors in skin fibroblasts cultured from an infant with leprechaunism and insulin resistance. *Proc. Natl. Acad. Sci. USA* 76:5877–81
51. Smith, B. R. 1981. Thyrotropin receptor antibodies. In *Receptors and Recognition, Ser. B,* ed. R. J. Lefkowitz, 13:215–44. London: Chapman and Hall. 253 pp.
52. Soll, A. H., Kahn, C. R., Neville, D. M. Jr. 1975. The decrease in insulin binding

53. Taylor, S. I., Podskalny, J. M., Samuels, B., Roth, J., Brasel, D. E., Pokora, T., Engel, R. R. 1980. Leprechaunism: A congenital defect in the insulin receptor. *Clin. Res.* 28:408 (Abstr.)
54. Taylor, S. I., Roth, J., Blizzard, R. M., Elders, M. J. 1981. Qualitative abnormalities in insulin binding in a patient with extreme insulin resistance: Decreased sensitivity to alterations in temperature and pH. *Proc. Natl. Acad. Sci. USA* 78:7157–61
55. Tevaarwerk, G. J. M., Strickland, K. P., Lin, C. H., Hudson, A. J. 1979. Studies on insulin resistance and insulin receptor binding in myotonia dystrophica. *J. Clin. Endocrinol. Metab.* 49:216–22
56. Wachslicht-Rodbard, H., Muggeo, M., Kahn, C. R., Saviolakis, G. A., Harrison, L. C., Flier, J. S. 1981. Heterogeneity of the insulin-receptor interaction in lipoatrophic diabetes. *J. Clin. Endocrinol. Metab.* 52:416–25
57. Wachslicht-Rodbard, H., Gross, H. A., Rodbard, D., Ebert, M. H., Roth, J. 1979. Increased insulin binding to erythrocytes in anorexia nervosa: Restoration to normal with refeeding. *N. Engl. J. Med.* 300:882–87
58. Weitzman, R. E., Fisher, D. A. 1978. Arginine vasopressin metabolism in dogs. I. Evidence for a receptor-mediated mechanism. *Am. J. Physiol.* 235:E591–97
59. Zakarija, M., McKenzie, J. M. 1978. Zoological specificity of human thyroid-stimulating antibody. *J. Clin. Endocrinol. Metab.* 47:249–54
60. Zapf, J., Rinderknecht, E., Humbel, R. E., Froesch, E. R. 1978. Non-suppressible insulin-like activity (NSILA) from human serum: Recent accomplishments and their physiologic implications. *Metabolism* 27:1803–28
61. Zeleznik, A. J., Roth, J. 1978. Demonstration of the insulin receptor *in vivo* in rabbits and its possible role as a reservoir for the plasma hormone. *J. Clin. Invest.* 61:1363–74

Ann. Rev. Physiol. 1982. 44:653-66

POLYPEPTIDE AND AMINE HORMONE REGULATION OF ADENYLATE CYCLASE[1]

G. D. Aurbach

Metabolic Diseases Branch, National Institute of Arthritis, Diabetes, and Digestive and Kidney Diseases, National Institutes of Health, Bethesda, Maryland 20205

Introduction

The biological effects of many polypeptide and amine hormones are transmitted, upon interaction with specific receptors at the cell surface, through activation of the membrane-bound enzyme adenylate cyclase and generation thereby of cyclic 3',5'-AMP (cAMP), the intracellular regulator of enzyme and transport systems. The adenylate cyclase complex is composed of at least three distinct protein components—the receptor (R) itself; the catalytic (C) unit (cyclizes ATP to cAMP); and the guanine nucleotide regulatory unit (G unit). Functions of the G unit include interaction with the receptor, binding to guanine nucleotide, interaction with the catalytic unit, and possibly a GTPase activity. Work in several laboratories has elucidated further the function of the G unit, clarified its role in hormone and fluoride activation of adenylate cyclase, and isolated it in pure form.

I review here characteristics of hormone receptors coupled through this mechanism, particularly systems linked through the guanine nucleotide regulatory protein.

Glucagon Receptors

Rodbell et al (68, 69) showed that GTP enhanced glucagon activation of adenylate cyclase and inhibited glucagon binding to receptors on plasma membranes of the liver. They postulated that guanine nucleotides and hormones interacted with hepatic adenylate cyclase through an interdepen-

---

[1]The US Government has the right to retain a nonexclusive, royalty-free license in and to any copyright covering this paper.

dent system, probably in an allosteric fashion. Their results prompted the many subsequent investigations into guanine nucleotide effects on hormone action, including those of β-adrenergic agonists, parathyroid hormone, serotonin, dopamine, FSH, VIP, prostaglandins, and vasopressin (see the discussion of guanine nucleotide control of adenylate cyclase, below).

It is now believed that the guanine nucleotide regulatory protein (G unit) when bearing an active guanosine triphosphate nucleotide (GN) represents the "coupling factor" linking receptor function to activation of adenylate cyclase. Hence GN influences apparent receptor affinity as well as enzyme activity.

The structural requirements for glucagon binding to receptors and activation of adenylate cyclase have been examined (23) using chemically modified glucagon derivatives. Binding activity is more importantly a function of C-terminal regions of the molecule whereas cyclase activity is strictly dependent on structure near the amino terminus. One derivative, trinitrophenyl glucagon, is a fairly effective inhibitor in vitro (23).

There is evidence for two classes of glucagon receptors (73), and maximal activation of adenylate cyclase occurs at low receptor occupancy (9). It is possible that only the high-affinity receptors are coupled (i.e. directly associated with G units; see discussion of guanine nucleotide control of adenylate cyclase) and capable of activating adenylate cyclase. Low-affinity receptors might represent receptors lacking (due to prior activation and thus dissociation from) G units. Certain cells transformed in tissue culture lose glucagon responsiveness (67). Such cells are replete with low-affinity (? receptors dissociated from G units) but lack high-affinity receptors. Apparent high-affinity receptors (possibly bearing G units) can be transferred via fusion to pigeon erythrocyte membranes which thereby respond to glucagon (67).

β-Adrenergic Receptors

Work with β-adrenergic receptors has provided much of the information relating receptor binding characteristics to activation of adenylate cyclase and production of cAMP. Among hormonal substances, β-adrenergic agonists represent some of the simplest compounds, and an array of agonists and antagonists is available in chemically defined forms. A series of high-affinity radioactive ligands has been described and there are available many discrete cell systems—e.g. avian or amphibian red cells, mammalian reticulocytes, adipocytes, pinealocytes, fibroblasts and diverse cultured cell systems—for studies of receptor binding and biological responses to β-adrenergic agents. Table I is a list of radioactive ligands developed for this purpose. A covalent affinity ligand also has been described (66).

Table 1 Radioactive ligands for β-adrenergic receptors

Ligand	Comment	Reference
^{125}I-Hydroxybenzylpindolol	β-blocker	4
^{3}H-Alprenolol	β-blocker	43
^{125}I-Cyanopindolol	β-blocker	22
^{125}I-Pindolol	β-blocker	7
^{3}H-Hydroxybenzylisoproterenol	β-agonist	41

β-Adrenergic receptor systems have been categorized as $β_{-1}$ or $β_{-2}$ classes according to the order of potency of ligands interacting with receptors and affecting a biological response (40). The $β_1$ class of receptors characteristically shows the order of potency isoproterenol (I) > epinephrine (E) = norepinephrine (N); $β_2$ receptors characteristically show I > E > N. These properties are preserved in studies on solubilized receptors (11), and each type of receptor shows stereospecific selectivity for ligands. In order to be valid, all studies of receptor binding and activation of adenylate cyclase must show the appropriate stereospecificity and order of potency of agonists and antagonists interacting in these systems.

Regulation of β-Adrenergic Receptors

Many factors can control β-adrenergic receptors and adenylate cyclase responses. The light cycle controls the concentration of β-adrenergic receptors on pinealocytes (36, 70). Changes in thyroid status (37, 83), malignant transformation (54), denervation (5, 80), drugs (6, 17, 28, 29), viral transformation, and developmental changes (5, 14, 46, 49) are among the many factors (4) influencing receptor concentration. In addition, cells exposed for prolonged periods to β-adrenergic agonists show a decrease (down-regulation) in β-adrenergic receptor number. Hoffman & Lefkowitz have reviewed this and other aspects of desensitization of β-adrenergic receptors (33).

Studies with reticulocyte systems provided some of the original evidence that β-adrenergic receptors and adenylate cyclase represented separate components of the membrane. Maturation of the circulating erythrocyte is associated with loss of both β adrenergic receptors and adenylate cyclase activity (8). Adenylate cyclase activity, however, is lost more rapidly than β-adrenergic receptors from the cell membrane. Schramm provided further evidence that receptors were distinct proteins, separable from catalytic units (see discussion of transfer of cell receptors). Definitive studies now show that β-adrenergic receptors, a guanine nucleotide regulatory protein, and the catalytic moiety of adenylate cyclase itself are all separable proteins.

Caron et al have now purified β-adrenergic receptors from frog erythrocytes by affinity chromatography (11).

β-Adrenergic systems also have provided much of the evidence implicating the guanine nucleotide regulatory protein in coupling of receptor function to adenylate cyclase activation. Binding of β-adrenergic agonists to the receptor initially induces a high molecular weight form of the receptor, probably representing a complex containing hormone, receptor, and guanine nucleotide regulatory unit (47, 48, 81). Addition of activating guanosine triphosphate analogs causes dissociation of the complex with apparent release of free receptors. These experiments represent part of the evidence favoring the sequence of events in activation of adenylate cyclase discussed in the section below on Guanine Nucleotide Regulatory Protein.

Changes in membrane lipid and fluidity appear to influence hormone receptor control of adenylate cyclase activity. Addition of unsaturated fatty acids enhances the rate of adrenergic activation of adenylate cyclase in turkey erythrocyte membranes by enhancing the accessibility of active guanine nucleotides to the G unit (61). Hirata et al (32) reported that β-adrenergic agonists increase phospholipid methylation in rat reticulocyte ghosts. Increased synthesis of phosphatidyl choline causes an increase in membrane fluidity and apparent increase in hormone-responsive adenylate cyclase. They also showed that GTP increases the ability of isoproterenol to stimulate membrane transmethylation. Their experiments (32) appear to invoke a particular membrane orientation for transmethylation. It has not yet been clarified whether intact membrane geometry is required to demonstrate transmethylation. Totally intact geometry is clearly not required for activation of adenylate cyclase, which has been demonstrated exhaustively with broken cell (hence non-uniform membrane geometry) preparations. It will be important in the future to attempt to correlate changes in membrane transmethylation with adenylate cyclase activity in purified membrane preparations.

Parathyroid Hormone Receptors

Receptors for parathyroid hormone in renal membranes have been reported in a number of studies (10). Segre et al (77) and Nussbaum et al (60) have determined binding characteristics of agonists and antagonists to identify regions of the molecule that interact specifically with the receptor. Useful analogs developed by the latter workers include ^{125}I-NLE-8,18,Tyr-34 amide as radioactive ligand and NLE-8,18,Tyr amide 3–34, which is equipotent in receptor activity but is an inhibitor of PTH-stimulated adenylate cyclase in vitro (77). The partial bPTH sequence 1–27 shows an apparent dissociation constant of 2.2×10^{-5}M whereas PTH 10–34 yields an apparent constant of 2.7×10^{-6}M. Thus the carboxylterminal region clearly contributes to affinity and yet the aminoterminus including the first two amino acids is clearly required for biological activity. ACTH inhibits receptor binding of parathyroid hormone in certain systems; in some it even inhibits the biological effect of parathyroid hormone on adenylate cyclase activity. There are some amino acid sequence homologies between ACTH 1–11 and PTH 15–27, but the 15–27 region of parathyroid hormone has been excluded as a significant factor affecting affinity of PTH (60). Thus the effects of ACTH in inhibiting binding or an adenylate cyclase activity of PTH cannot currently be attributed to amino acid sequence homologies between the two molecules.

Regulation and Coupling of Parathyroid Hormone Receptors

Guanine nucleotides, particularly guanylylimidodiphosphate (Gpp(NH)p), potentiate activation of adenylate cyclase in the kidney membranes by parathyroid hormone (30). Recently Nissenson et al (58) have shown decided effects of guanylylimidodiphosphate on binding of PTH ligands to canine renal membranes. They found that at 10^{-4}M, Gpp(NH)p caused marked acceleration of release of iodinated bPTH 1–34 from canine but not chicken renal membranes. The finding with canine membranes implies that a PTH receptor high-affinity intermediate is formed analogous to that found with other receptor cyclase systems (see discussion of guanine nucleotide regulation of adenylate cyclase).

Regulation of parathyroid hormone receptors by parathyroid hormone in the circulation seems to occur in a fashion analogous with that observed in other hormone systems. Vitamin D deficiency with its attendant hypocalcemia and hypersecretion of parathyroid hormone appears to cause a decreased responsiveness in vivo and in vitro to the adenylate cyclase system upon exposure to exogenous or endogenous parathryoid hormone. This desensitization presumably reflects "down-regulation" of receptors by exposure to high concentrations of parathyroid hormone (12, 56, 82). These several observations support the concept that high circulating concentrations of parathyroid hormone lead to down-regulation of receptors; receptor concentrations under these circumstances, however, have yet to be measured.

Other Receptor–Adenylate Cyclase Systems Coupled Through Guanine Nucleotide Interactions

Guanine nucleotides decrease agonist affinity for receptors of many other adenylate cyclase systems including serotonin (65), dopamine (16), FSH (1), VIP (3), and prostaglandins (31). Enhanced agonist activation of adenylate cyclase with guanine nucleotides also has been found with a number of peptide and amine hormone systems (reviewed in 50).

Transfer of Receptors and Adenylate Cyclase Components to Foreign Membrane Systems

In 1976 Orly & Schramm (62) showed that turkey erythrocytes and Friend erythroleukemia cells could be fused and that the fusion product allowed coupling of the turkey erythrocyte's beta-adrenergic receptor to the adenylate cyclase of the Friend erythroleukemia cell. This seminal observation showed that hormone receptors may be mobile and capable of activating adenylate cyclase catalytic units by movement through the fluid medium of the plasma membrane. The study also confirmed the virtually established hypothesis that hormone receptors and catalytic (C) units are distinct molecules. Subsequent studies by the same group extended observations to show that cell-to-cell transfer may be possible with receptors in general (76), that membrane extracts containing receptors can be transferred to other cells (38), that receptors can be transferred to membrane preparations of foreign cells (21, 55, 75), and that the independent units of receptor (R), guanine nucleotide regulatory unit (G), and catalytic unit (C) can be combined in lipid extracts, reconstituted into phospholipid vesicles, and made to interact. The recent work of Citri & Schramm (15) emphasized anew the concept that R and G can move about, within the fluid phase of the membrane, to interact and that the activated G unit similarly can move about and interact with C units in the membrane. The result of this latter study also emphasized the likelihood that the order of reaction in activation of adenylate cyclase is as hypothesized in the section on guanine nucleotide regulation of adenylate cyclase. In the study of Citri & Schramm (15), it was proposed that hormones can activate an R-G complex after insertion of R and G separately into lipid micelles and that such reconstructed R-G complexes are devoid of catalytic units. This information, together with the discussion of guanine regulation of adenylate cyclase suggests that at least in the turkey erythrocyte membrane system the catalytic unit is an unnecessary component in the coupling process per se. This would be counter to the hypothesis proposed by Abramowitz et al (2). Neufeld et al (55) also showed that the β-adrenergic receptor in CYC^- (lacks G unit) membranes is capable of activating the G-C complex in Friend erythroleukemia cells. Trypsin inactivates the effective G unit (Gpp(NH)p-bound form) from turkey erythrocyte membranes, but upon incubation of extracts containing this inert G unit with CYC^- membranes the G unit is restored to an active one (55). This suggests that either there are enzymes capable of repairing the inactivated G unit or there are residual subunits of aberrant G units in CYC^- membranes that can combine with trypsin-inactivated G units from turkey membranes to produce an active system. Functional G units contain a 42-K subunit and 35-K subunit. Since CYC^- lacks the 42-K subunit (at least it is undetectable by either biological assay or ADP-ribosylation), it may well contain a normal 35-K subunit. The 35-K subunit, on the other hand, may be more susceptible to trypsin proteolysis than is the 42-K protein. Indeed, proteolytic inactivation experiments of Hudson et al (34) and Downs et al (in preparation) indicate that proteases release an ADP-ribosylated fragment from the 42-K subunit of hormone-activated G units. Thus it is intriguing to suggest that CYC⁻ membranes supply the 35-K subunit that is inactivated by the action of trypsin on the G unit of turkey erythrocyte membranes.

The β-adrenergic receptor of turkey erythrocyte membranes was inserted into human erythrocyte membranes, and a guanine nucleotide-induced shift in affinity was observed (35). These results suggested that the receptor obtained from the turkey erythrocyte membrane had interacted with the G unit in the human erythrocyte membrane and that the G unit indigenous to the turkey erythrocyte membrane had not been co-transferred with the receptor. Solubilized receptor also has been reconstituted into artificial membranes (27).

Guanine Nucleotide Regulatory Protein (G Unit)

This protein links hormone-receptor interaction to activation of adenylate cyclase. The G unit can interact with the receptor as well as the catalytic unit and probably shuttles from one to the other as determined by the hormone and the nature of the guanine nucleotide bound to the G unit. Probably the major effect of hormone interaction with the receptor is facilitation of guanine nucleotide exchange on the guanine nucleotide regulatory component (13, 18). In avian erythrocyte membranes and in Leydig cell membranes, incubation with hormone facilitates interaction with added guanine nucleotides detected as enhanced specific binding of added radioactive guanine nucleotide (13, 18, 20). The nature of guanine nucleotide bound to the regulatory unit governs the nature of adenylate cyclase activity obtained. In turkey erythrocyte membranes, the endogenously bound component under basal conditions is guanosine diphosphate (18), and in this form the G unit-adenylate cyclase complex is active when sodium fluoride is added. Release of endogenous GDP and substitution of it by a guanosine triphosphate analog allows direct activation of the catalytic unit by the G unit. The GTP bound to the regulatory unit is rapidly hydrolyzed to GDP, causing inactivation of the enzyme complex. Substitution of a "nonhydrolyzable" analog—e.g. guanylylimidodiphosphate (Gpp(NH)p)—causes persistent activation of adenylate cyclase. An intermediate state is formed with 5'-GMP bound to the regulatory unit; development of this state is dependent on hormone (isoproterenol) plus 5'-GMP (exchanges with bound endogenous guanine nucleotide). In this state fluoride activity is reduced;

addition of GDP (or GTP which is rapidly hydrolyzed to GDP) allows restoration of fluoride activity. Addition of Gpp(NH)p causes direct activation of the cyclase. Downs et al (18) have shown that a diphosphate form (GDP or GPCP) is necessary for activation by fluoride. Another analog, guanosine-2-thiodiphosphate (GDP-β-S), produces a complex with the regulatory unit that is inactive with or without addition of sodium fluoride (18). Thus the nature of the guanine nucleotide bound to the regulatory unit governs the type of activity obtained. Regardless of the nucleotide bound, addition of appropriate concentrations of a different guanine nucleotide plus hormone facilitates specific binding of the new nucleotide, and the activity obtained reflects the nature of the newly bound guanine nucleotide analog. Dufau et al (20) have provided evidence for specific LH-induced guanine nucleotide (Gpp(NH)p) binding to membranes prepared from Leydig cells. In the adrenal cortical system calcium appears to be required for ACTH-induced interaction of guanine nucleotide with G unit (52).

Several models for hormonal regulation of adenylate cyclase have been developed. One scheme represents a serial model wherein the hormone interacts initially to provide a H-R-G complex which in turn interacts with guanosine triphosphate yielding an activated G unit. The activated G unit then interacts with the catalytic unit to yield the active enzymatic state. The active state is deactivated by hydrolysis of bound GTP to bound GDP (GTPase activity), and the GDP-bound G unit is then available for recycling to form R-G. This scheme emphasizes guanine nucleotide interchange on the G unit as the major if not only effect of hormone receptor interaction. A separate view has been proposed by Abramowitz et al (2). They believe, on the basis of kinetic considerations and varying states of enzyme activity with different guanine nucleotides in certain systems, that the Cassel & Selinger (13) model, outlined above as the serial model, does not sufficiently account for hormone activation of adenylate cyclase. The two distinct models are illustrated in Figure 1. They (2) suggest that the enzyme exists in two conformations, inactive and active, and that the equilibrium between these conformations (whether active guanine nucleotides are bound or not) is influenced by the hormone. The major contention differentiating their model from others is their proposition that the hormone can influence equilibria of inactive enzyme towards active enzyme regardless of whether an activated guanine nucleotide is bound. Even in their model, however, the major transformation facilitated by the hormone is exchange of the guanine nucleotide bound. In many systems it may not be possible to test for the minute fraction of "active enzyme" bearing no GTP. On the other hand, major arguments can be brought forth to support the serial model: (*a*) Direct hormone-induced exchange of guanine nucleotides on guanine nucleotide regulatory protein has been established in turkey erythrocyte membranes, and it has been proven that addition of hormone facilitates this exchange (13, 18). (*b*) The H-R-G complex can be detected in gel filtration experiments (47, 48, 51); H-R-G complex can be detected before activation of C (39). (*c*) The G-C complex can be detected in density gradient centrifugation (64). (*d*) The G unit can be purified by affinity chromatography and eluted from the resin in an active (Gpp(NH)p-bound) or inactive (5'-GMP-bound) form. Trapping the G unit on the affinity resin is dependent only on hormone (and inclusion of 5'-GMP which exchanges for endogenous guanine nucleotides). The G Unit eluted from the affinity resin in its active form can activate adenylate cyclase directly in membranes lacking G units (19, 57, 63). (*e*) Hormones can activate (presumably through nucleotide exchange mechanism) G units directly even in preparations containing deleted or inactivated catalytic unit, i.e. enzyme (15). A critical question in deciding among these models is whether there is a natural endogenous hormone-independent rate of conversion of inactive adenylate cyclase to active enzyme bearing no bound effective guanine nucleotide. So far the existence of such an activated form has not been demonstrated by experiment.

G Unit and Receptor Affinity

In several systems, addition of guanine nucleotides changes apparent receptor affinity for agonists but not antagonists (42, 51, 81). In the absence of added guanine nucleotide, the apparent affinity for agonist is much higher (addition of the nucleotide causes a shift towards much reduced apparent affinity for agonist). This shift is not seen in certain mutant cells either lacking the guanine nucleotide regulatory component (CYC^- mutant cells) or those containing a G unit incapable of interacting with receptor (uncoupled mutant, UNC). Guanine nucleotide–induced receptor affinity shifts had not been observed with β-adrenergic receptor of turkey erythrocyte membranes. Recently, however, Lad et al (39) showed a small shift in agonist affinity in the turkey erythrocyte membrane system after prior incubation with isoproterenol and 5'-GMP. This maneuver generates a higher-affinity state of the receptor which probably represents a high-affinity complex between hormone receptor and G unit (81). Detection of guanine nucleotide–induced shift in agonist affinity is yet another means of identifying an interaction between the G unit and receptor.

Guanine Nucleotides in Nonlinked or Uncoupled AC Systems

Effects of guanine nucleotides on binding of hormones to receptors do not always reflect coupling of receptor activity to activation of adenylate cyclase. Uncoupled systems [e.g. UNC mutant of S49 cells (71)] show receptor binding but no change in affinity with guanine nucleotide. The UNC mutant

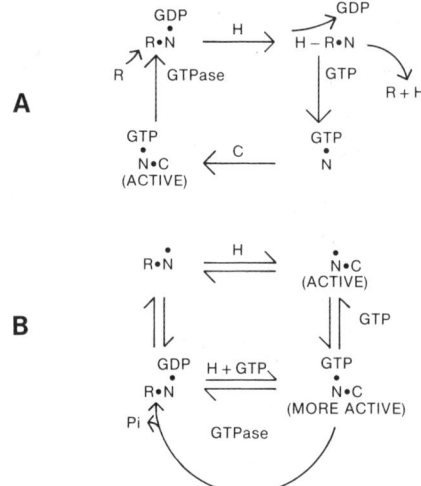

Figure 1 Proposed mechanisms for activation of adenylate cyclase. A. Serial model. Active state of enzyme dependent on guanosine triphosphate bound form of G unit which associates with catalytic unit (C). Hormone binding to receptor functions merely to facilitate exchange of guaninine nucleotide on G. unit. B. Equilibrium model. All forms of enzyme are in equilibrium (equilibrium constants not necessarily unity). Hormone influences conversion of inactive enzyme to active enzyme whether guanine nucleotide bound or not. The bound form nevertheless is postulated to be more active than the form not complexed with guanine nucleotide. [Modified from (2)]

G unit differs in charge from the normal protein (74). Other systems contain receptor complexes that are not necessarily coupled to activation of adenylate cyclase. For example, Amiranoff et al (3) found that guanine nucleotides shifted apparent binding affinity for both VIP and glucagon in isolated liver cells, but only the interaction of glucagon led to activation of adenylate cyclase. VIP in the intestine, however, does activate adenylate cyclase and such activation is controlled by guanine nucleotides. HTC cells show a similar phenomenon; they lack the catalytic unit of adenylate cyclase yet contain β-adrenergic receptors linked to guanine nucleotide regulatory units in that the characteristic agonist shift is still observed (71).

Another class of ligands, e.g. α-adrenergic, cholinergic, opiate, angiotensin cholecystokinin, and adenosine agonists that do not activate but rather inhibit (as shown specifically for α-adrenergic, cholinergic, and opiate agents) adenylate cyclase also show changes in receptor affinity effected by GTP [see review in (78)]. This information suggests that inhibitory effects on adenylate cyclase also may be mediated through a guanine nucleotide binding protein. Of interest in this regard is the finding that only the α_2 and not the α_1 subclass of α-adrenergic receptors are influenced by guanine nucleotides (33). It is specifically the α_2 class that inhibits adenylate cyclase.

An unsettled question is whether the same G unit may be capable of controlling the inhibitory as well as the stimulatory functions on adenylate cyclase.

Role of Guanine Nucleotide Regulatory Unit in Fluoride and Hormone Activation of Adenylate Cyclase

It is now clear the fluoride-stimulated activity also depends upon the G unit and the nature of the guanine nucleotide bound to it (see above). Affinity chromatography of the G unit can be achieved on GTP-linked sepharose after preactivating the unit with isoproterenol and 5'-GMP (57, 63 79). Adsorption to the affinity matrix is strictly dependent upon addition of hormone (isoproterenol), indicating that this regulatory component is specifically involved in hormonal activation of adenylate cyclase. The G unit can be eluted from the resin with a variety of guanine nucleotides—5'-GMP, GTP, GDP, GDP-β-S, or Gpp(NH)p (19). Assays of the eluates with AC⁻ membranes give activities dependent on the nature of the guanine nucleotide bound to the G unit (in the elution step). Fluoride stimulatable adenylate cyclase depends upon a guanosine diphosphate form bound to the G unit. Direct activation is achieved by adding the Gpp(NH)p bound form of the regulatory unit. Eluates prepared with 5'-GMP are inactive unless GDP + fluoride or Gpp(NH)p alone are added. These several observations as well as those of Northrup et al (59) indicate that the same regulatory protein coupling hormone-receptor to cyclase activity is also required for fluoride activation of the enzyme. Fluoride activation of adenylate cyclase can no longer be taken as a measure of catalytic unit activity.

Cholera Toxin and ADP-ribosylation of the Guanine Nucleotide Regulatory Unit

Cholera toxin is a multimer composed of subunits designated A and B [see (53) for review]. The B subunits serve in binding the toxin to gangliosides on the cell surface. Upon entering the cell the toxin is reduced and dissociates to yield the A subunit which is an enzyme, an ADP-ribosyltransferase that catalyzes transfer of ADP-ribose from NAD to a basic amino acid, usually arginine in protiens. The G unit is a substrate for cholera toxin-catalyzed ADP-ribosylation. ADP-ribosylation of the G unit enhances the sensitivity of the adenylate cyclase complex to GTP since the ADP-ribosylated form has reduced GTPase activity (thus decreasing the rate of deactivation of the active adenylate cyclase complex). In the intact cell this effect of cholera toxin is evidenced as enhanced sensitivity to hormone (72). One can utilize the ADP-ribosyltransferase activity of cholera toxin to label specifically the G unit with ^{32}P (^{32}P-labeled ADP-ribose). This forms a convenient marker for the G unit in fractionation procedures such as density gradient centrifugation, gel electrophoresis, and other purification procedures (19). Results in several laboratories show that the same protein required for hormone and fluoride activation of adenylate cyclase is also the substrate for ADP-ribosylation catalyzed by cholera toxin. Recent work on purification (see below) of the regulatory subunit from rabbit liver defines further the subunits of the regulatory protein and the type of subunit that is the substrate for ADP-ribosylation.

Purification of the Regulatory Component

Northup et al (59) have achieved virtually complete purification of the regulatory component from rabbit liver plasma membranes. The final preparation is represented by two major bands and one minor component on polyacrylamide gel electrophoresis and sodium dodecyl sulfate. The major bands represent protein subunits with molecular weights of 45,000 (45 K) and 35,000 (35 K). The minor component has a molecular weight of 52,000 (52 K). Only the 52-K and 45-K components are substrates for ADP-ribosylation (59). The 52-K component has not been found in either turkey or pigeon erythrocyte membranes, may be a precursor of the 45-K component, and is not absolutely required for activity. The 45-K and 35-K components have been resolved only on denaturing SDS electrophoretic gels and have not been successfully recombined to give an active protein. The purified holo-protein causes activation of adenylate cyclase in the presence of sodium fluoride with CYC⁻ (a mutant cell that lacks the G unit) membranes and isoproterenol-stimulated adenylate cyclase in UNC (uncoupled mutant) membranes (59).

Guanine Nucleotide Regulatory Protein in Disease

Studies on the genetic disorder pseudohypoparathyroidism (PHP) have revealed a deficiency of G units in a number of cases. Mature human erythrocytes contain G units (but no significant receptors or catalytic enzyme units) and represent a source to be tested for G unit activity. Many patients with classical Albright's hereditary osteodystrophy and PHP show a reduced complement of G units in their red cells as well as in platelets (24–26, 44, 45). It seems likely that this deficiency (apparently generalized in tissue distribution) accounts for the diminished hormonal responsiveness documented in this syndrome (44). Some patients with the classical syndrome show a normal complement of the regulatory protein and may represent a defect analogous to the UNC mutant of S49 cells or a defect not evident in the types of assays currently used (24–26, 44, 45). Patterns of inheritance of PHP have suggested a sex-linked defect in some families, an autosomal pattern in others. Tests on affected families for erythrocyte G unit activity are compatible with two or more different patterns of inheritance (26). Since the G unit contains at least two protein subunits, it is possible that each is controlled by a separate gene, one possibly sex-linked, the other possibly autosomal. Other hypotheses are also tenable.

ACKNOWLEDGMENT

I am indebted to Mrs. Lillian Perry for superb secretarial assistance.

Literature Cited

1. Abou-Issa, H., Reichert, L. E. Jr. 1979. *Endocrinology* 104:189–93
2. Abramowitz, J., Iyengar, R., Birnbaumer, L. 1980. *J. Biol. Chem.* 255:8259–65
3. Amiranoff, B., Laburthe, M., Rosselin, G. 1980. *Biochem. Biophys. Res. Commun.* 96:463–468
4. Aurbach, G. D., Brown, E. M. 1979. In *Contemporary Metabolism,* ed. N. Freinkel, pp. 193–233. NY: Plenum
5. Banerjee, S. P., Sharma, V. K., Kung, L. S. 1977. *Biochim. Biophys. Acta* 470:123–27
6. Banerjee, S. P., Sharma, V. K., Khanna, J. M. 1978. *Nature* 276:407–9
7. Barovsky, K. and Brooker, G. 1980. *J. Cycl. Nuc. Res.* 6:297–307
8. Bilezikian, J. P., Spiegel, A. M., Brown, E. M., Aurbach, G. D. 1977. *Mol. Pharmacol.* 13:775–85
9. Birnbaumer, L., Pohl, S. L. 1973. *J. Biol. Chem.* 248:2056–61
10. Brown, E. M., Aurbach, G. D. 1980. *Vit. Horm.* 38:205–56
11. Caron, M. G., Srinivasan, Y., Pitha, J., Kociolek, K., Lefkowitz, R. J. 1979. *J. Biol. Chem.* 254:2923–27
12. Carnes, D. L., Anast, C. S., Forte, L. R. 1978. *Endocrinology* 102:45–51
13. Cassel, D., Selinger, Z. 1977. *J. Cycl. Nuc. Res.* 3:11–22
14. Charlton, R. R., Venter, J. C. 1980. *Biochem. Biophys. Res. Commun.* 94:1221–26
15. Citri, Y., Schramm, M. 1980. *Nature* 287:297–300
16. Creese, I., Usdin, T. B., Snyder, S. H. 1979. *Mol. Pharmacol.* 16:67–76
17. Davies, A. O., Lefkowitz, R. J. 1980. *J. Clin. Endocrinol. Metab.* 51:599–605
18. Downs, R. W. Jr., Spiegel, A. M., Singer, M., Reen, S., Aurbach, G. D. 1980. *J. Biol. Chem.* 255:949–54
19. Downs, R. W. Jr., Reen, S. A., Levine, M. A., Aurbach, G. D., Spiegel, A. M. 1981. *Arch. Biochem.* 209:284–90
20. Dufau, M. L., Baukal, A. J., Catt, K. J. 1980. *Proc. Natl. Acad. Sci. USA* 77:5837–41
21. Eimerl, S., Neufeld, G., Korner, M., Schramm, M. 1980. *Proc. Natl. Acad. Sci. USA* 77:760–64
22. Engel, G. 1981. *Triangle* 19:69–76
23. Epand, R. M., Rosselin, G., Hoa, D. H., Cote, T. E., Laburthe, M. 1981. *J. Biol. Chem.* 256:1128–32
24. Farfel, Z., Bourne, H. R. 1980. *J. Clin. Endocrinol. Metab.* 51:1202–4
25. Farfel, Z., Brickman, A. S., Kaslow, H. R., Brothers, V. M., Bourne, H. H. 1980. *N. Engl. J. Med.* 303:237–42
26. Farfel, Z., Brothers, V. M., Brickman, A. S., Felix, C., Neer, R., Bourne, H. R. 1981. *Proc. Natl. Acad. Sci. USA* 78:3098–102
27. Fleming, J. W., Ross, E. M. 1980. *J. Cycl. Nucl. Res.* 6:407–19
28. Glaubiger, G., Lefkowitz, R. J. 1977. *Biochem Biophys. Res. Commun.* 78:720–25
29. Glaubiger, G., Tsai, B. S., Lefkowitz, R. J. Weiss, B., Johnson, E. M., Jr. 1978. *Nature* 273:240–42
30. Goltzman, D., Callahan, E. N., Tregear, G. W., Potts, J. T. Jr. 1978. *Endocrinology* 103:1352–1360
31. Heidenrich, K. A., Weiland, G. A., Molinoff, P. B. 1980. *J. Cycl. Nuc. Res.* 6:217–30
32. Hirata, F., Strittmatter, W. J., Axelrod, J. 1979. *Proc. Natl. Acad. Sci. USA* 76:368–72
33. Hoffman, B. B., Lefkowitz, R. J. 1980. *Ann. Rev. Pharmacol. Toxicol.* 20:581–608
34. Hudson, T. J., Roeber, J. F., Johnson, G. L. 1981. *J. Biol. Chem.* 256:1459–65
35. Jeffery, D. R., Charlton, R. R., Venter, J. C. 1980. *J. Biol. Chem.* 255:5015–18
36. Kebabian, J. W., Zatz, M., Romero, J. A., Axelrod, J. 1975. *Proc. Natl. Acad. Sci. USA* 72:3735–39
37. Kunos, G., Vermes-Kunos, I., Nickerson, M. 1974. *Nature* 250:779–81

38. Laburthe, M., Rosselin, G., Rousset, M., Zweibaum, A., Korner, M., Selinger, Z., Schramm, M. 1979. *FEBS Lett.* 98:41–43
39. Lad, P. M., Nielsen, T. B., Preston, M. S., Rodbell,M. 1980. *J. Biol. Chem.* 255:988–995
40. Lands, A. M., Arnold, A., McAuliff, J. P., Luduena, F. P., Brown, T. G. Jr. 1967. *Nature* 214:597–99
41. Lefkowitz, R. J., Williams, L. T. 1977. *Proc. Natl. Acad. Sci. USA* 74:515–19
42. Lefkowitz, R. J., Mullikin, D., Caron, M. G. 1976. *J. Biol. Chem.* 254:4686–92
43. Lefkowitz, R. J. 1979. *Ann. Int. Med.* 91:450–58
44. Levine, M. A., Downs, R. W. Jr., Marx, S. J., Lasker, R., Aurbach, G. D. 1981. In *Hormonal Control of Calcium Metabolism*, ed. D. V. Cohn, R. V. Talmage, J. L. Matthews, pp. 95–101. Amsterdam: Excerpta Medica
45. Levine, M. A., Downs, R. W. Jr., Singer, M., Marx, S. J., Aurbach, G. D., Spiegel, A. M. 1980. *Biochem. Biophys. Res. Commun.* 94:1319–24
46. Limas, C. J. 1979. *Biochim. Biophys. Acta* 588:174–78
47. Limbird, L. E., Lefkowitz, R. J. 1978. *Proc. Natl. Acad. Sci. USA* 75:228–32
48. Limbird, L. E., Gill, D. M., Lefkowitz, R. J. 1980. *Proc. Natl. Acad. Sci. USA* 77:775–79
49. Ludford, J. M., Talamo, B. R. 1980. *J. Biol. Chem.* 255:4619–27
50. Maguire, M. E., Ross, E. M., Gilman, A. G. 1977. *Adv. Cycl. Nuc. Res.* 8:1–83
51. Maguire, M. E., Van Arsdale, P. M., Gilman, A. G. 1976. *Mol. Pharmacol.* 12:335–39
52. Mahaffee, D. D., Ontjes, D. A. 1980. *J. Biol. Chem.* 255:1565–71
53. Moss, J., Vaughan, M. 1979. *Ann. Rev. Biochem.* 48:581–600
54. Mukherjee, A., Wong, T. M., Buja, L. M., Lefkowitz, R. J., Willerson, J. T. 1979. *J. Clin. Invest.* 64:1423–28
55. Neufeld, G., Schramm, M., Weinberg, N. 1980. *J. Biol. Chem.* 255:9268–9274
56. Nichols, G. A., Carnes, D. L., Anast, C. S., Forte, L. R. 1979. *Am. J. Physiol.* 236:401–9
57. Nielsen, T. B., Downs, R. W. Jr., Spiegel, A. M. 1980. *Biochem. J.* 190:439–43
58. Nissenson, R. A., Teitelbaum, A. P., Abbott, S. R., Pliam, N., Silve, C., Zitzner, L., Nyiredy, K., Arnaud, C. D. 1981. In *Hormonal Control of Calcium Metabolism*, ed. D. V. Kohn, R. V. Talmage, J. L. Matthews, pp. 44–54. Amsterdam: Excerpta Medica
59. Northup, J. K., Sternweis, P. C., Smigel, M. D., Schleifer, L. S., Ross, E. M., Gilman, A. G. 1980. *Proc. Natl. Acad. Sci. USA* 77:6516–20
60. Nussbaum, S. R., Rosenblatt, M., Potts, J. T. Jr. 1980. *J. Biol. Chem.* 255:10183–87
61. Orly, J., Schramm, M. 1975. *Proc. Natl. Acad. Sci. USA* 72:3433–37
62. Orly, J., Schramm, M. 1976. *Proc. Natl. Acad. Sci. USA* 73:4410–14
63. Pfeuffer, T. 1977. *J. Biol. Chem.* 252:7224–34
64. Pfeuffer, T. 1979. *FEBS Lett.* 101:85–89
65. Peroutka, S. J., Lebovitz, R. M., Snyder, S. H. 1979. *Mol. Pharmacol.* 16:700–708
66. Rashidbaigi, A., Ruoho, A. E. 1981. *Proc. Natl. Acad. Sci. USA* 78:1609–13
67. Reilly, T. M., Blecher, M. 1981. *Proc. Natl. Acad. Sci. USA* 78:182–86
68. Rodbell, M., Krans, M. J., Pohl, S. L., Birnbaumer, L. 1971. *J. Biol. Chem.* 46:1872–76
69. Rodbell, M., Lin, M. C., Salomon, Y. 1974. *J. Biol. Chem.* 249:59–65
70. Romero, J. A., Zatz, M., Kebabian, J. W., Axelrod, J. 1975. 258:435–36
71. Ross, E. M., Howlett, A. C., Ferguson, K. M., Gilman, A. G. 1978. *J. Biol. Chem.* 253:6401–12
72. Rudolph, S. A., Schafer, D. E., Greengard, P. 1977. *J. Biol. Chem.* 252:7132–39
73. Schaltz, L., Marinetti, G. V. 1972. *Science* 176:175–77
74. Schleifer, L. S., Garrison, J. C., Sternweis, P. C., Northup, J. K., Gilman, A. G. 1980. *J. Biol. Chem.* 10:2641–44
75. Schramm, M. 1979. *Proc. Natl. Acad. Sci. USA* 76:1174–78
76. Schramm, M., Orly, J., Eimerl, S., Korner, M. 1977. *Nature* 268:310–13
77. Segre, G. V., Rosenblatt, M., Reiner, B. L., Mahaffey, J. E., Potts, J. T. Jr. 1979. *J. Biol. Chem.* 254:6980–86
78. Spiegel, A. M., Downs, R. W. Jr. 1981. *Endocrinol. Rev.* 2:275–305
79. Spiegel, A. M., Downs, R. W. Jr., Aurbach, G. D. 1979. *J. Cycl. Nuc. Res.* 5:3–17
80. Sporn, J. R., Harden, T. K., Wolfe, B. B., Molinoff, P. B. 1976. *Science* 194:624–26
81. Stadel, J. M., DeLean, A., Lefkowitz, R. J. 1980. *J. Biol. Chem.* 255:1436–41
82. Tomlinson, S., Hendy, G. N., Pembertson, D. M., O'Riordan, J. L. H. 1976. *Clin. Sci. Med.* 51:59–69
83. Williams, L. T., Lefkowitz, R. J., Watanabe, A. M., Hathaway, D. P., Besch, H. R. 1977. *J. Biol. Chem.* 252:2787–89

CALMODULIN IN ENDOCRINE CELLS

Anthony R. Means and James G. Chafouleas

Department of Cell Biology, Baylor College of Medicine, Houston, Texas 77030

Introduction

In order to survive in a dynamic environment all living cells must be able to identify and respond to variation in specific extracellular signals. Peptide hormones comprise such signals in mammalian cells. Target cells recognize the hormone through specific receptors on the outer surface of the plasma membrane. The binding of hormone to receptor initiates a series of rapid events that eventually translates this external signal into a specific cellular response mediated by a selective alteration of the intracellular metabolism. The mechanism by which the extracellular event is transduced to an intracellular event is still not totally understood. The effect of β-adrenergic agents on cAMP metabolism led to the proposal that cAMP was the second messenger responsible for this transduction through the activation of a cAMP-dependent protein kinase. While this mechanism explains many hormonal events, it is not compatible with the studies of Hutson et al (45) and Cherrington (13), who demonstrated that neither cAMP metabolism nor activation of cAMP-dependent protein kinase were involved in the α-adrenergic activation of glycolysis and gluconeogenesis in rat liver. Subsequently Keppens et al (50) and Assimacopoulos-Jeanenett et al (3) presented data supporting the concept that the α-adrenergic response was mediated through Ca^{2+}. Indeed calcium is intimately involved in the regulation of secretion from endocrine cells. This ion enters cells by diffusion but is transported between organelles and out of the cell by an active process involving CA^+-Mg^{2+} ATPases. Changes in the intracellular concentration of this ion leads to alterations in contractility of the cells. Such changes are due to an activation of actomyosin ATPase and controled depolymerization of the cytoplasmic microtubule network.

Calmodulin as an Intracellular Receptor

It is now well accepted that Ca^{2+} plays a major role in the regulation of cellular activity. Berridge (5) and Rassmussen et al (79) have extensively reviewed the involvement of Ca^{2+} in the regulation of cell metbolism. The extent to which calcium is involved in cellular processes led Rassmussen (77) to propose that calcium be classed with the cyclic nucleotides as a second messenger to external stimuli. Indeed, the relationship between the cyclic nucleotides and calcium has been alluded to in several reviews on the subject (5, 77, 78, 80). Although the involvement of Ca^{2+} in the transduction of external stimuli into cellular responses has been well documented, the biochemical mechanisms mediating the Ca^{2+} effects have yet to be elucidated. Kretsinger (54, 55) has indicated the importance of the cytosolic Ca^{2+}-binding proteins and has suggested that the intracellular targets for Ca^{2+} would have to interact either with a great many different proteins, each with one specific function, or with one or several proteins, each with multiple functions. This latter group of proteins could be viewed as intracellular Ca^{2+} receptors.

The concept of a Ca^{2+} receptor mediating the Ca^{2+} regulation of key metabolic processes is attractive, since it would require that the cell had only to conserve the Ca^{2+} binding capability of one protein instead of many. Also, the use of the Ca^{2+}-protein receptor intermediate for regulation allows for greater diversity through changes in the tertiary structure of the protein. In recent years it has become apparent that one such Ca^{2+}-binding protein, calmodulin (CaM), may indeed be viewed as the eukaryotic Ca^{2+} receptor (5, 52, 64, 97). Several criteria must be satisfied before a protein may be classified as a Ca^{2+} receptor: (*a*) It must specifically bind Ca^{2+} with high affinity; (*b*) the binding of Ca^{2+} to the protein must be mandatory for its regulatory role; (*c*) it must regulate vital intracellular processes common to all eukaryotic cells; (*d*) It must be ubiquitous; and (*e*) because of its critical role in cell survival it must be highly conserved.

Calmodulin is a heat-stable 17,000 M_r multifunctional Ca^{2+} binding protein that meets all the criteria set forth for a Ca^{2+} receptor (23). It contains four equivalent Ca^{2+} binding sites with a K_d of 2.4×10^{-6}M which do not bind Mg^{2+} under physiological conditions (23). Ca^{2+} binding induces a more α-helical conformation of the protein, which preceeds activation of the calmodulin-dependent enzymes (23, 44, 51, 62). This protein has been shown to mediate the calcium regulation of many fundamental intracellular enzyme systems. To date 15 separate enzymes have been reported to require CaM as a regulatory component (51a). Calmodulin is ubiquitous in eukaryotes (11, 49, 88, 96). The highly conserved nature of this protein has been suggested by demonstrations that the amino acid sequence of CaM is invariant in structure in the cow, rabbit, and rat (22, 35, 36, 98) and in the sea pansy, *R. reniformis* (48), differing by no more than seven conservative amino acid substitutions. Moreover, each of these proteins contains four internally homologous calcium binding domains, and only one of the substitutions (in *R. reniformis*) occurs in these highly conserved regions. The extent to which this protein is conserved in eukaryotes was demonstrated by Chafouleas et al (11). Employing a radioimmunoassay for the protein they demonstrated that CaM from representative species of primative algae, slime molds, and coelenterates, to the more advanced plants and mammals, exhibited immunological identity. Taken together these data suggest that CaM is one of the most highly conserved as well as widely dispersed proteins studied.

Calmodulin regulates cyclic nucleotide metabolism in some systems by stimulation of both phosphodiesterase and adenylyl cyclase (15). These enzymes were first shown to be CaM-regulated in bovine brain. Cheung (14) first proposed the presence of a protein activator of phosphodiesterase, which proved to be CaM. Subsequently a variety of cells and tissues have been found to contain CaM-responsive cyclic nucleotide metabolizing systems. Wolff et al (101) have even shown that membranes from the prokaryote *Bordetella pertussis* contain a CaM-activatable adenylyl cyclase, and Iwasa et al (47) suggest that a CaM-like activity exists in *Escherichia coli*. It should be pointed out, however, that no protein exists in *E. coli* that is immunologically similar to CaM (11). Glycogen metabolism can also be affected by CaM. Phosphorylase kinase contains CaM as an integral subunit (16), and a CaM-dependent glycogen synthase kinase has recently been isolated by Payne et al (73).

Regulation of Calcium Flux

Calcium transport is also regulated by CaM. Whereas Ca^{2+} readily enters cells, its movement between organelles and removal from cells are active processes (21). The red blood cell plasma membrane was intially shown to contain a Ca^{2+}-Mg^{2+} ATPase regulated by a small Ca^{2+}-binding protein (31). This protein was subsequently shown to be CaM (33). Not only does CaM stimulate the ATPase activity of the Ca^{2+}-pump, but the binding of the protein to the membrane also results in Ca^{2+} extrusion (43). The data accumulated from the studies on the red blood cell suggest that energy-dependent Ca^{2+} removal is important in maintaining Ca^{2+} homeostasis and that CaM is intimately involved in regulation of this process. Calmodulin should be able to regulate the intracellular Ca^{2+} level in a rather specific manner. Normally free Ca^{2+} concentrations are approximately $10^{-7}M$ (55). At such concentrations, the Ca^{2+} binding sites on CaM would be unoccupied. When a stimulus increases this level to $5\mu M$, the CaM ion binding sites would become filled, thereby converting the protein to an active conformation. The Ca^{2+}-CaM complex would then associate with and activate the plasma membrane Ca^{2+} pump. The operational Ca^{2+}-Mg^{2+} ATPase would then actively transfer Ca^{2+} from the cytosol to the outside of the cell. When intracellular Ca^{2+} levels were reduced below μM the sites on CaM would empty, reversing its conformation and thus its association with the Ca^{2+} pump. Clearly this is an oversimplification of the process in the red cell. However, a closed system such as this should provide a simple kinetic mechanism for the maintenance of Ca^{2+} homeostasis. Moreover, these data suggest that an acute surge in Ca^{2+}, such as occurs in response to regulatory molecules that act on the cell surface, could be extruded from the cell by a similar mechanism. Two examples of this kind of effect can be found in adipocytes in response to epinephrine and pancreatic cells in response to glucose. Both of these cell types have been shown to contain CaM-dependent Ca^{2+} pumps associated with their plasma membranes (74, 75). Data suggest that the Ca^{2+}-pump associated with intracellular membranes such as cardiac muscle sarcoplasmic reticulum (59), plant microsomes (27), and synaptic plasma membranes isolated from mammalian brain also possess a CaM-regulatable Ca^{2+}-transport component (91).

Calmodulin and Protein Phosphorylation

The effects of calmodulin on cyclic nucleotide and glycogen metabolism as well as on Ca^{2+} transport may involve regulation of multiple enzymes and seem to occur in a wide variety of cell types. The same is true for the Ca^{2+} control of contractility, discussed in some detail below. The mechanism by which enzyme activity is affected by CaM has yet to be clearly established. The one possibility for a common action appears to be the activation of specific protein kinases. Phosphorylase kinase represents such a specific enzyme. Binding of Ca^{2+} to the δ-subunit (CaM) of this protein activates phosphorylase through phosphorylation. In addition, CaM promotes the inactivation of glycogen synthase by a similar mechanism. The possibility that calmodulin-mediated phosphorylation may be a primary mechanism for intracellular regulation is suggested by the work of Yamauchi & Fujisawa (104), who have demonstrated that most of the Ca^{2+}-dependent endogenous phosphorylation of rat brain cytosolic proteins requires calmodulin. DeLorenzo & Freedman (25) observed that Ca^{2+} stimulation of endogenous release of norepinephrine from isolated synaptic vesicles was associated with a rapid increase in phosphorylation of specific vesicle proteins. Schulman & Greengard (84) demonstrated that an endogenous heat-stable cytosolic protein was required for the calcium-dependent phosphorylation of synaptosomal membrane fractions from rat cerebral cortex. They also observed that authentic CaM could substitute for the endogenous protein. The authors therefore suggested that the calcium-dependent phosphorylation was mediated through CaM. Subsequently, DeLorenzo et al (26) showed that removal of an endogenous heat-stable protein from purified synaptic vesicles caused these vesicles to become refractory to Ca^{2+} stimulation. The Ca^{2+} stimulation of norepinephrine release and membrane protein phosphorylation could be restored to these depleted vesicles in a dose-response manner by the addition of authentic CaM. Schulman & Greengard (85) have now demonstrated that the Ca^{2+}-dependent phosphorylation of membrane proteins also occurs in many nonneuronal tissues. However, while these membrane phosphorylations all required Ca^{2+} and CaM, tissue-specific endogenous substrates for the kinases were observed.

The possibility of a coordinated regulation of brain membrane protein phosphorylation by both cAMP and Ca^{2+} is suggested by Sieghart et al (87). They observed that the phosphorylation of the brain-specific proteins Ia and Ib were regulated by both Ca^{2+} and cAMP. However, while both kinases phosphorylate these two proteins, they do so at different amino acid residues. Le Peuch et al (60) have demonstrated that this coordinated regulation is not limited to these two brain proteins. They have shown that the rate of Ca^{2+} uptake by cardiac sarcoplasmic reticulum is regulated by cAMP and Ca^{2+}-calmodulin-dependent phosphorylation of the membrane protein phospholamban. cAMP mediates its regulation through cAMP-dependent protein kinase, while the Ca^{2+}-dependent phosphorylation is mediated through a membrane-bound protein kinase, which requires both Ca^{2+} and CaM for activity. Both kinases phosphorylate different sites on the protein, and it is the Ca^{2+}-dependent phosphorylation that is mandatory for Ca^{2+} uptake. The cAMP-dependent phosphorylation, while incapable by itself of stimulating Ca^{2+} uptake, does amplify uptake when Ca^{2+}-dependent phosphorylation has taken place. The possibility that this membrane-bound CaM-dependent protein kinase may be similar to phosphorylase kinase, or in fact a form of phosphorylase kinase devoid of its δ-subunit (CaM), is suggested by the fact that it is capable of phosphorylating exogenous phosphorylase in the presence of CaM. Conversely, exogenous phosphorylase kinase is capable of phosphorylating phospholamban. These findings are intriguing when taken with the work of Drowning et al (9), who have shown that the preferential phosphorylation of a 40,000 Mr protein in synaptic plasma membranes following electrical stimulation can be mimicked by exogenous phosphorylase kinase but not by cAMP-dependent protein kinase. These studies suggest that the CaM regulation of protein phosphorylation mediated through phosphorylase kinase may be more extensive than previously thought. Of equal interest is the coordinated regulation of protein phosphorylation exhibited by cAMP and CaM, since CaM is in fact involved in both processes. While CaM plays a direct role in Ca^{2+}-dependent phosphorylation, it also plays an indirect role in cAMP-dependent phosphorylation by regulating cAMP metabolism through activation of both adenylyl cyclase and cyclic nucleotide phosphodiesterase.

Whether CaM mediates its regulatory function solely through protein phosphorylation or through other mechanisms, its ability to regulate is dependent on changes in the intracellular free Ca^{2+} concentration. Calmodulin-controlled systems would be constitutively turned either on or off in the presence of a constant intracellular level of Ca^{2+}. This, in fact, may be the mechanism by which external signals are transduced into intracellular responses in hormone action. Indeed the most characteristic aspect of hormone action is the occurrence of Ca^{2+} fluxes (5, 79).

Hormone-Mediated Calmodulin Redistribution

Although Ca^{2+}, fluxes are observed as a consequence of hormone action in target cells, it may not be a true increase or decrease in the intracellular levels of Ca^{2+}, but a mobilization and redistribution of membrane bound Ca^{2+} that is essential. This is suggested by studies employing chlorotetracycline to monitor the shifts in membrane-bound calcium. Le Burton et al (58) demonstrated that the ADP- or ionophore A23187-induced shape change in human platelets was temporarily associated with a decrease in membrane-bound calcium. A similar phenomenon was observed in rabbit neutrophils following stimulation with chemotactic factors (70, 71). In fact, the redistribution of Ca^{2+} is so rapid that the authors have suggested this mobilization may represent one of the initial molecular events following the binding of chemotactic factor to its membrane receptor.

The fact that CaM seems to regulate so many physiological processes suggests that alteration of the intracellular levels of this Ca^{2+} receptor could also be an important mechanism for control of cellular metabolism and motility. We have evaluated a variety of hormonally regulated systems to determine whether CaM is selectively elevated (12). The systems have included estrogen and progesterone stimulation of chick oviduct, androgen and rat prostate, corticosteroids in rat pituitary GH_1 cells, FSH and Sertoli cells, ACTH and Yl adrenal tumor cells, TSH and rat thyroid slices, prolaction and rat mammary gland, EGF and rat pituitary GH_3 cells, and GnRH in rat pituitary cells. In no instance is CaM selectively elevated as assessed by radioimmunoassay. This is true even in the chick oviduct, where estrogen results in a remarkable cytodifferentiation and the appearance of many new protein products. In the immature undifferentiated gland, during primary stimulation, withdrawal from hormone, or secondary stimulation the amount of CaM per cell remains unchanged.

Many peptide hormones and growth factors that interact with cell surface receptors promote the secretion of proteins, ions, or steroids (65). In the past few years it has become increasingly apparent that exocytosis of many cell proteins requires packaging of the secretory product in clathrin-coated vesicles. Linden et al (63) have revealed that CaM is a component of such vesicles isolated from brain and that the association is Ca^{2+}-dependent and of high affinity ($K_d = 10^{-9}$M). Since exocytosis is known to require Ca^{2+}, these studies raised the possibility that CaM might imply the Ca^{2+} sensitivity to the secretory process. Much indirect evidence exists that is compatible with this hypothesis. Anti-CaM compounds such as phenothiozines have been reported to inhibit glucose-mediated insulin release from pancreatic islet (53) and insulinoma cells (83), intestinal ion secretion (46), histamine release from mast cells (86), protein secretion from polymorphonuclear leukocytes (28, 69), and seretonin release from platelets (100). Although the mechanism for such responses remains unknown, it likely does not involve changes in the total cell content of CaM.

The use of pharmacological compounds such as the phenothiazines to identify cellular processes regulated by CaM deserves mention. Weiss and colleagues (99) revealed that phenothiazines bound to CaM in a calcium-dependent fashion and that the interaction of the drugs with CaM prevented the stimulation of phosphodiesterase activity. Further studies demonstrate that the binding of phenothiazines by CaM prevents the stimulation of a variety of enzymes by this protein. Why was the binding of drugs or enzymes calcium-dependent? It has now been revealed that the interaction of calcium with CaM exposes a highly lipophilic region of the molecule (56, 93). This lipophilic region binds the phenothiazines; this region of the protein also interacts with phosphodiesterase, myosin light-chain kinase, and Ca^{2+}-Mg^{2+} ATPase.

Van Eldik et al (95) have recently characterized the antigenic site on CaM that associates with a series of anti-rabbit antibodies. This site was shown to be restricted to an 18-amino-acid region in the C-terminal (IV domain) portion of the protein. Since these antibodies prevent the stimulation of enzyme and phenothiazine binding by CaM, it is likely the IVth Ca^{2+} binding domain of the protein is involved in these interactions. Another series of pharmacologic tools, the naphthalenesulfonamides or W-compounds, have recently been developed by Hidaka et al (41). These are considerably less hydrophobic but have similar affinities to the best phenothazines—i.e. K_ds for calmodulin of 10^{-6} to 10^{-7} M. The advantage of the W-compounds is that they more readily enter cells so that the incubation of cells with such drugs may lead to an indication of various reactions that require CaM as a regulatory component. It must be stressed, however, that the use of these drugs should not be taken as absolute proof of a CaM-dependent process. All of the drugs have many different actions within cells, including binding to cell surfaces and alteration of transport properties. Blackmore et al (6) have utilized two phenothiazines (chlorpromazine and trifluoperazine) to determine whether the response of isolated hepatocytes to α-adrenergic agonists involved CaM. These authors found that either drug completely inhibited phosphorylase activation and Ca^{2+} efflux in response to phenylephrine. Further experiments, however, revealed that the drugs prevented binding of the α-agonists to plasma membrane receptors. Therefore it was impossible to conclude whether or not CaM was involved. Until more specific inhibitors of CaM can be found, the results obtained from the use of pharmacological agents to probe CaM-mediated events in cells must be interpreted cautiously.

Since CaM is as component of virtually every intracellular compartment as well as of the plasma membrane, efforts have been made to determine whether cell surface–acting agents promote an alteration in its distribution. Distinct anatomical regions of the central nervous system, such as the corpus striatum, contain dopamine receptors that seem to be coupled to adenylyl cyclase (38). Calmodulin may mediate dopamine action, since phosphorylation of membrane proteins promotes the apparent release of CaM from membrane-bound to soluble form (31). Since a soluble CaM-dependent phosphodiesterase exists, it has been proposed that long-term stimulation of dopamine receptors is associated with an increase in the soluble CaM content, thereby activating PDE and decreasing receptor responsiveness. Similar data suggest interneuronal pathways exist where opiates increase soluble CaM via a release of dopamine and thus act as indirect dopamine agonists (39). Smoake & Solomon (89) have reported altered CaM distribution in liver cells from rats with streptozotocin-induced diabetes. These authors conclude that such changes might play a role in the alteration of cAMP metabolism known to exist in such pathological states.

The difficulties with interpretation of most CaM distribution studies is that the protein is assayed by its ability to stimulate a CaM-dependent enzyme. Since all such assays are Ca^{2+}-dependent and other CaM-binding proteins are likely to be present in each subcellular fraction, it is difficult to obtain quantitative values for CaM. This difficulty is circumvented when a radioimmunoassay is employed, since the assay can be performed in the presence of EGTA and is therefore Ca^{2+}-independent (11). The radioimmunoassay has been utilized to determine the quantity and subcellular distribution of CaM in the rat pituitary gonadotrope before and during GnRH-induced LH release (18). Indeed the distribution of CaM does change in response to GnRH. An initial rise in the percentage of CaM associated with the plasma membrane appears concomitantly with the depletion of cytoplasmic CaM. These changes occur temporarily in concert with secretion of LH. As the CaM begins to be cleared from the plasma membrane, its level increases first in the secretory granule and microsomal fractions before finally replenishing the cytoplasm. The magnitude of the changes between plasma membrane and cytoplasmic content of CaM are related to the dose of GnRH. Calmodulin redistribution is also hormone-specific, since analogs such as des[1] GnRH(2–10), which has no efficacy in promoting LH secretion, did not alter intracellular changes in CaM. Finally a budget of CaM content in all subcellular fractions revealed that GnRH did not increase total CaM, and more than 95% of the cellular CaM was recovered.

These data suggest that CaM may be important in the regulation of protein secretion but provide little information concerning the mechanism. At this juncture it is impossible to predict whether CaM redistribution is a cause or consequence of the secretory process. In the red blood cell (43), pancreatic islet (75), and adipocyte (74), CaM-activated ATPases are found in the plasma membrane and, at least in the adipocyte, the enzyme appears to be hormonally regulated. Plasma membranes from islet cells have also been reported to contain a CaM-stimulated adenylyl cyclase activity (94). Calmodulin is also a major component of postsynaptic membranes (34, 61, 102), has been proposed to mediate the Ca^{2+}-effects on synaptic transmission (31, 38, 39), and thus may play a role in neurotransmitter release (26). Finally, with the caveats discussed earlier, both of the anti-CaM drugs, trifluoperazine and naphthalenesulfonamides, also inhibit the receptor-mediated secretory process in a variety of systems.

Calmodulin and Receptor-Mediated Endocytosis

Receptor-mediated endocytosis is also a Ca^{2+}-dependent process and also involves clathrin-coated vesicles (32, 81). Although internalization of GnRH does not appear to be required for the LH release process, the gonadotrope response to this releasing hormone does include the pattern of patching, capping, and internalization observed for many cell surface–mediated ligand systems (19). This receptor redistribution pattern in the gonadotrope is mimicked by changes found in CaM associated with the plasma membrane when assessed by indirect immunofluorescence microscopy. Recruitment of clathrin-coated vesicles to the plasma membrane of human lymphoblastoid cells occurs following stimulation with multivalent anti-IgM antibodies (81). This recruitment is inhibited by the presence of anti-calmodulin drugs, and CaM is a component of such vesicles. Thus the appearance of CaM at the plasma membrane may be associated with the accumulation of coated pits involved in the receptor internalization process. Insulin, which also is internalized following cap formation, promotes the translocation of glucose transport activity from the microsomal or Golgi fractions to the plasma membrane (20, 92). Actin and myosin have also been reported to co-cap with several cell surface receptors (8, 29), and actin-containing matrixes have been isolated from *D. discoideum* (17), murine tumor cells (66), and lymphocyte plasma membranes (66) associated with various receptors. Thus the phenomenon of redistribution of new activities to the plasma membrane may be a generalized occurrence for plasma membrane receptor-mediated events. This redistribution suggests a mechanism by which CaM-regulated events could be affected without new protein synthesis. It is likely that CaM redistribution is secondary to alterations in the net flux or distribution of Ca^{2+} within the cell.

Calmodulin and Cell Motility

The structural basis of motility in smooth and nonmuscle cells involves the cellular cytoskeleton. The cytoskeleton is comprised of a few specific porteins. The structural elements can be regulated by a wide variety of molecules, one of which is Ca^{2+}. The components of the cytoskeleton most intimately involved with contractility and force generation required for endo- or exocytosis are the actin-based microfilaments (21). Many of the actin filaments (6 nm) are arranged in larger structures up to one μm in diameter (microfilament bundles) that contain other proteins such as myosin, tropomyosin, filamin and α-actinin. Actomyosin is present in all eukaryotes from amoeba to man. The myosin molecule of smooth and nonmuscle cells is composed of one pair of heavy chains (200,000 daltons) and 2 pairs of light chains (each consisting of a 17,000- and a 20,000-dalton component) (1). At one end of the heavy chain, the polypeptide exhibits a globular configuration; this portion contains both the actin binding site and an ATPase activity. The light chains are proposed to regulate both ATPase and actin binding activities in a manner that requires both Ca^{2+} and protein phosphorylation. Actin activation of the ATPase occurs only after phosphorylation of the light chains, and a direct relationship exists among calcium concentration, the degree of light-chain phosphorylation, and enzyme activity (90). Finally the degree of light-chain phosphorylation is positively correlated with tension development (57).

An enzyme responsible for phosphorylation of the 20,000 M_r light chain (LC_{20}) was found to be dependent upon Ca^{2+} for optimal activity and was termed myosin light-chain kinase (MLCK) (76). Subsequently, this enzyme was shown to be present in cardiac, smooth, and nonmuscle tissues and cells. The Ca^{2+} requirement was due to the fact that CaM was a regulatory component of the enzyme (103). Thus Ca^{2+} binds CaM, promoting the association with MLCK. This interaction activates the enzyme, resulting in phosphorylation of LC_{20}. Phosphorylation promotes a conformational change in the myosin that allows actin binding and resultant stimulation of the ATPase acivity. Hydrolysis of ATP provides the energy necessary for tension development and contractility. A phosphatase exists that specifically removes the phosphate from LC_{20}, which reverses the process (1). The degree of LC_{20} phosphorylation is directly proportional to tension development, and the degree of dephosphorylation is directly proportional to relaxation. This relationship is of physiological relevance and has been shown to occur in perfused tissue (42), isolated smooth muscle strips (10), skinned muscle fibers (4), and isolated cells (20a). It has been demonstrated in vitro that MLCK can be phosphorylated by the catalytic subunit of cAMP-dependent protein kinase (2). This phosphorylation inhibits the binding of CaM and has therefore been proposed as one mechanism for inhibiting the contractile response. However LC_{20} can also be phosphorylated in vitro by cAMP-dependent protein kinase (72). The phosphorylation occurs on the same serine residue as the reaction catalyzed by MLCK, prompting speculation that LC_{20} phosphorylated in this manner might also promote actin-activated ATPase activity. The physiological significance of the cAMP-dependent proten kinase phosphorylation of MLCK and LC_{20} remains to be determined. It is difficult to imagine a dual role for this enzyme that on the one hand would inactivate MLCK and on the other hand phosphorylate LC_{20}, thereby negating the previous phosphorylation.

The localization of CaM in interphase cells was shown to be primarily associated with the actin-containing microfilament bundles when assessed by indirect immunofluorescence microscopy (24). Recent studies from our laboratory have shown a similar localization of MLCK (37). Using the same procedure, others have shown that myosin (30), actin (40), and tropomyosin (27a) also are found associated with microfilament bundles in the cytoplasm. Collectively these studies reveal that a regulatory molecule (CaM), an enzyme (MLCK), a substrate (LC_{20}), and a second enzyme (myosin ATPase) all exist in a complex macromolecular structure, the microfilament bundle. Changes in Ca^{2+} concentration should be sufficient to trigger the entire series of reactions leading to contraction of nonmuscle cells. Thus the microfilament bundle can be considered analogous to the sarcomere in skeletal muscle. The Ca^{2+} receptor in nonmuscle cells is CaM, whereas the troponin system serves the same function in muscle. Evidence suggests that CaM, through activation of MLCK, may also play a role in the assembly of myosin into microfilament bundles in nonmuscle cells (82). This possibility points out a very important difference between the organelles involved in force generation in muscle and nonmuscle cells. The sarcomere is a stable structure, whereas microfilament bundles and microfilaments themselves are dynamic and can be stimulated to undergo rapid and dramatic reorganization under physiological conditions.

Conclusions

There is little question that CaM is important in endocrine cells. Since CaM appears to be constitutively expressed, the multiple actions of this Ca^{2+} receptor are probably secondary to changes in the net flux or distribution of Ca^{2+}. Preliminary evidence suggests that CaM can undergo intracellular redistribution in a Ca^{2+}-dependent manner. Such changes are probably mediated via such components of the cellular cytoskeleton as the microfilaments and microtubules. Elucidation of the mechanisms that underwrite the Ca^{2+}-dependence of endocrine cells will require development of new technology to allow alteration of the level of CaM in normal cells. We have recently isolated, cloned, and sequenced the peptide coding region of the CaM gene from the electroplax of the electric eel (66, 67). One way to affect changes in CaM content is to use vectors such as mutant Simian virus-40 to introduce multiple copies of the CaM gene into cells. One can then determine whether elevation of the CaM concentration produces changes in cell shape or function that are characteristic of hormone action.

Literature Cited

1. Adelstein, R. S. 1980. Phosphorylation of muscle contractile proteins. *Fed. Proc.* 39: 1544–73
2. Adelstein, R. S., Conti, M. A., Hathaway, D. R., 1978. Phosphorylation of smooth muscle myosin light chain kinase by the catalytic subunit of adenosine 3':5'-monophosphate-dependent protein kinase. *J. Biol. Chem.* 253: 8347–50
3. Assimacopoulos-Jeanenett, F. D., Blackmore, P. F., Exton, J. H. 1977. Studies on α-adrenergic activation of hepatic glucose output studies on role of calcium ion in α-adrenergic activation of phosphorylase. *J. Biol. Chem.* 252: 2662–69
4. Barron, J. T., Barany, M., Barany, K., 1979. Phosphorylation of the 20,000-dalton light chain of myosin of intact arterial smooth muscle in rest and in contraction. *J. Biol. Chem.* 254: 4954–56
5. Berridge, M. J. 1975. The interaction of cyclic nucleotides and calcium in the control of cellular activity. *Adv. Cyc. Nucl. Res.* 6: 1–98
6. Blackmore, P. F., El-Refai, M. F., Dehaye, J.-P. Strickland, W. G., Hughes, B. P., Exton, J. H. 1981. Blockade of hepatic α-adrenergic receptors and response by chlorpromazine and trifluoperazine. *FEBS Lett.* 123: 245–48
7. Bond, G. H., Clough, D. L., 1973. A soluble protein activator of (Mg^{2+}+Ca^{2+})-dependent ATPase in human red cell membranes. *Biochim. Biophys. Acta.* 323: 592–99
8. Bourguignon, L. Y..W., Tokuyasu, K. T., Singer, S. J., 1978. The capping of lymphocytes and other cells studied by an improved method of immunofluorescence staining of frozen sections. *J. Cell Physiol.* 95: 239–58
9. Browning, M., Bennett, W., Lynch, G., 1979. Phosphorylase kinase phosphorylates a brain protein which is influenced by repetitive synaptic activation. *Nature* 278: 273–75
10. Cassidy, P., Hoar, P. E., Kerrick, W. G. L. 1979. Irreversible thiophosphorylation and activation of tension in functionally skinned rabbit ileum strips by (^{35}S) ATP γ S. *J. Biol. Chem.* 254: 11148–53
11. Chafouleas, J. G., Dedman, J. R., Munjaal, R. P., Means, A. R., 1979. Calmodulin: development and application of a sensitive radioimmunoassay. *J. Biol. Chem.* 254: 10262–67
12. Chafouleas, J. G., Pardue, R. L., Brinkley, B. R., Dedman, J. R., Means, A. R., 1980. Effect of viral transformation on the intracellular regulation of calmodulin and tubulin. In *Calcium-Binding Proteins. Structure and Function,* ed. F. L. Siegel, E. Carafoli, R. H. Kretsinger,

D. H. MacLennan, R. H. Wasserman, pp. 189–96. Amsterdam: Elsevier
13. Cherrington, A. D., Assimacopoulos, F. D., Harper, S. C., Corbin, J. D., Park, C. R., Exton, J. H., 1976. Studies on the α-adrenergic activation of hepatic glucose output. *J. Biol. Chem.* 251: 5209–18
14. Cheung, W. Y. 1970. Cyclic 3', 5'-Nucleotide phosphodiesterase: demonstration of an activator. *Biochem. Biophys. Res. Commun.* 38: 533–38
15. Cheung, W. Y. 1980. Calmodulin plays a pivotal role in cellular regulation. *Science* 207: 19–27
16. Cohen, P., Burchell, A., Foulkes, J. G., Cohen, P. T. W., Nairn, A., Vanaman, T., 1978. Identification of the Ca^{2+}-dependent modulator protein as the fourth subunit of rabbit skeletal muscle phosphorylase kinase. *FEBS Lett.* 92: 287–93
17. Condeelis, J. S. 1979. Isolation of concanavalin A caps during various stages of formation and their association with actin and myosin. *J. Cell Biol.* 80: 751–58
18. Conn, P. M., Chafouleas, J. G., Rogers, D., Means, A. R., 1981. Gonadotropin releasing hormone stimulates calmodulin redistribution in the rat pituitary. *Nature* 292:264–65
19. Conn, P. M., Marian, J., McMillian, M., Rogers, D., 1980. Evidence for calcium mediation of gonadotropin releasing hormone action in the pituitary. *Cell Calcium* 1: 7–20
20. Cushman, S. W., Wardzala, L. J., 1980. Potential mechanism of insulin action on glucose transport in the isolated rat adipose cell. *J. Biol. Chem.* 255: 4758–62
20a. Daniel, J. L., Molish, I. R., Holmsen, H., Salganicoff, L., 1981. Phosphorylation of myosin light chain in intact platelets. In *Protein Phosphorylation*, ed. O. M. Rosen, E. G. Krebs, pp. 913–28. NY: Cold Spring Harbor.
21. Dedman, J. R., Brinkley, B. R., Means, A. R., 1979. Regulation of microfilaments and microtubules by calcium and cyclic AMP. *Adv. Cyc. Nucl. Res.* 11: 131–74
22. Dedman, J. R., Jackson, R. L., Schreiber, W. F., Means, A. R., 1978. Sequence homology of the Ca^{2+}-dependent regulator of cyclic nucleotide phosphodiesterase from rat testis with other Ca^{2+}-binding proteins. *J. Biol. Chem.* 253: 343–46
23. Dedman, J. R., Potter, J. D., Jackson, R. L., Johnson, J. D., Means, A. R., 1977. Physiocochemical properties of rat testis Ca^{2+}-dependent regulator protein of cyclic nucleotide phosphodiesterase: relationship of Ca^{2+}-binding, conformational changes and phosphodiesterase activity. *J. Biol. Chem.* 252: 8415–22
24. Dedman, J. R., Welsh, M. J., Means, A. R., 1978. Ca^{-2+}-dependent regulator: production and characterization of a monospecific antibody. *J. Biol. Chem.* 253: 7515–21
25. DeLorenzo, R. J., Freedman, S. D., 1978. Calcium-dependent neurotransmitter release and protein phosphorylation in synaptic vesicles. *Biochem. Biophys. Res. Commun.* 80: 183–91
26. DeLorenzo, R. J., Freedman, S. D., Yohe, W. B., Maurer, S. C., 1979. Stimulation of Ca^{2+}-dependent neurotransmitter release and presynaptic nerve terminal protein phosphorylation by calmodulin and a calmodulin like protein isolated from synaptic vesicles. *Proc. Natl. Acad. Sci. USA* 76: 1838–42
27. Dieter, P., Marme, D., 1980. Calmodulin activation of plant microsomal Ca^{2+} uptake. *Proc. Natl. Acad. Sci. USA* 12: 7311–14
27a. Drenckhahn, D., Groschel-Stewart, V. 1980. Localization of myosin, actin, and tropomyosin in rat intestinal epithelium: immunohistochemical studies at the light and electron microscope levels. *J. Cell Biol.* 86:475–83
28. Elferink, J. G. R., 1979. Chlorpromazine inhibits phagocytosis and exocytosis in rabbit polymorphonuclear leukocytes. *Biochem. Pharmacol.* 28: 965–68
29. Flanagan, J., Koch, G. L. E., 1978. Cross-linked surface Ig attaches to actin. *Nature* 273: 278–81
30. Fujiwara, K., Pollard, T. D., 1976. Fluorescent antibody localization of myosin in the cytoplasm, cleavage furrow, and the mitotic spindle of human cells. *J. Cell Biol.* 71: 848–75
31. Gnegy, M. E., Lau, Y. S., 1980. Effects of chronic and acute treatment of antipsychotic drugs on calmodulin release from rat striatal membranes. *Neuropharmacology* 19: 319–23
32. Goldstein, J. L., Anderson, R. G. W., Brown, M. S., 1979. Coated pits, coated vesicles, and receptor-mediated endocytosis. *Nature* 279: 679–85
33. Gopinath, R. M., Vincenzi, F. F., 1977. Phosphodiesterase protein activator of $(Ca^{2+} + Mg^{2+})$ ATPase. *Biochem. Biophys. Res. Commun.* 77: 1203–9
34. Grab, D. J., Berzins, K., Cohen, R. S., Siekevitz, P., 1979. Presence of cal-

modulin in postsynaptic densities isolated from canine cerebral cortex. *J. Biol. Chem.* 254: 8690–96
35. Grand, R. J. A., Perry, S. V., 1978. The amino acid sequence of the troponin C-like protein (modulator protein) from bovine uterus. *FEBS Lett.* 92: 137–42
36. Grand, R. J. A., Shenolikar, S., Cohen, P., 1981. The amino acid sequence of the δ-subunit (calmodulin) of rabbit skeletal muscle phosphorylase kinase. *Eur. J. Biochem.* 113: 359–67
37. Guerriero, V., Lagace, L., Dedman, J. R., Means, A. R., 1980. Myosin light chain kinase: Production and characterization of an antibody and localization in cells. *J. Cell Biol.* 87: 244a
38. Hanbauer, I., Gimble, J., Lovenberg, W., 1979. Changes in soluble calmodulin following activation of dopamine receptors in rat striatal slices. *Neuropharmacology* 18: 851–57
39. Hanbauer, I., Gimble, J., Sankaran, K., Sherard, R., 1979. Modulation of striatal cyclic nucleotide phosphodiesterase by calmodulin: Regulation by opiate and dopamine receptor activation. *Neuropharmacology* 18: 859–64
40. Herman, I. M., Pollard, T. D., 1979. Comparison of purified anti-actin and fluorescent heavy meromyosin staining patterns in dividing cells. *J. Cell Biol.* 80: 509–20
41. Hidaka, H., Yamaki, T., Totsuka, T., Asano, M., 1979. Selective inhibitors of Ca^{2+}-binding modulator of phosphodiesterase produce vascular relaxation and inhibit actin-myosin interaction. *Mol. Pharmacol.* 15: 49–59
42. High, C. W., Stull, J. T., 1980. Phosphorylation of myosin in perfused rabbit and rat hearts. *Am. J. Physiol.* 239: H756–64
43. Hinds, T. R., Larsen, F. L., Vincenzi, F. F., 1978. Plasma membrane Ca^{2+} transport: Stimulation by soluble proteins. *Biochem. Biophys. Res. Commun.* 81:455–61
44. Ho, H. C., Desai, R., Wang, J. H. 1975. Effect of Ca^{2+} on the stability of the protein activator of cyclic nucleotide phosphodiesterase. *FEBS Lett.* 50: 374–77
45. Hutson, N. G., Brumley, F. T., Assimacopoulos, F. D., Harper, S. C., Exton, J. H. 1976. Studies on the α-adrenergic activation of hepatic glucose output. I. Studies on the α-adrenergic activation of phosphorylase and gluconeogenesis and inactiation of glycogen synthase in isolated rat liver parenchymal cells. *J. Biol. Chem.* 251: 5200–8
46. Ilundain, A., Naftalin, R. J., 1979. Role of Ca^{2+}-dependent regulator protein in intestinal secretion. *Nature* 279: 446–48
47. Iwasa, Y., Yonemitsu, K., Matsui, K., Fukunaga, K., Miyamoto, E., 1981. Calmodulin-like activity in the soluble fraction of *Escherichia coli*. *Biochem. Biophys. Res. Commun.* 3: 656–60
48. Jamieson, G. A., Hayes, A., Blum, J. J., Vanaman, T. C., 1980. Structure and function relationships among calmodulins from divergent eukaryotic organisms. See Ref. 12, pp. 165–72
49. Kukiuchi, S., Yamazaki, R., Teshima, Y. L., Miyomoto, E., 1974. Multiple cyclic nucleotide phosphodiesterase activities from rat tissues and occurrences of a calcium-plus magnesium-ion-dependent phosphodiesterase and its protein activator. *Biochem. J.* 146: 109–20
50. Keppens, S., Varndenheede, J. R., DeWulf, H., 1977. On the role of calcium as second messenger in liver for the hormonally-induced activation of glycogen phosphorylase. *Biochem. Biophys. Acta* 496: 448–57
51. Klee, C. B., 1977. Conformational transition accompanying the binding of Ca^{2+} to the protein activator of 3', 5'-cyclic adenosine monophosphate phosphodiesterase. *Biochemistry* 16(5): 1017–24
52. Klee, C. B., Crouch, T. H., Richman, P. G., 1980. Calmodulin. *Ann. Rev. Biochem.* 49: 489–518
53. Krausz, Y., Willheim, C. B., Siegel, E., Sharp, G. W. G., 1980. Possible role for calmodulin in insulin release—studies with trifluoperazine in rat pancreatic islets. *J. Clin. Invest.* 66: 603–7
54. Kretsinger, R. H., 1976. Calcium-binding proteins. *Ann. Rev. Biochem.* 45: 239–66
55. Kretsinger, R. H., 1979. The informational role of calcium in the cytosol. *Adv. Cyc. Nucl. Res.* 11: 1–26
56. Laporte, D. C., Wierman, B. M., Storm, D. R., 1980. Calcium-induced exposure of a hydrophobic surface on calmodulin. *Biochemistry* 19: 3814–19
57. Lebowitz, E. A., Cooke, R., 1978. Contractile proteins of actomyosin from human blood platelets. *J. Biol. Chem.* 253: 5443–47
58. LeBurton, G. C., Dinerstein, R. J., Roth, L. J., Feenberg, H., 1976. Direct evidence for intracellular divalent cation redistribution associated with platelet shape change. *Biochem. Biophys. Res. Commun.* 71: 362–70

59. Lepeuch, C. J., Haiech, J., Demaille, J. G., 1979. Concerted regulation of cardiac sarcoplasmic reticulum calcium transport by cyclic adenosine monophosphate dependent and calcium-calmodulin-dependent phosphorylations. *Biochemistry* 18: 5150–57
61. Lin, C. T., Dedman, J. R., Brinkley, B. R., Means, A. R., 1980. Localization of calmodulin in rat cerebellum by immunoelectron microscopy. *J. Cell Biol.* 85: 473–80
62. Lin, Y. M., Liu, Y. P., Cheung, W. Y., 1974. Cyclic $3':5'$-nucleotide phosphodiesterase purification, characterization, and active form of the protein activator from bovine brain. *J. Biol. Chem.* 249: 4943–54
63. Linden, C. D., Dedman, J. R., Chafouleas, J. G., Means, A. R., Roth, T. F., 1981. Interactions of calmodulin with coated vesicles from brain. *Proc. Natl. Acad. Sci. USA* 78: 308–12
64. Means, A. R., Dedman, J. R., 1980. Calmodulin—an intracellular calcium receptor. *Nature* 285: 73–77
65. Means, A. R., Dedman, J. R., Tash, J. S., Tindall, D. J., Van Sickle, M., Welsh, M. J., 1980. Regulation of the testis Sertoli cell by follicle stimulating hormone. *Ann. Rev. Physiol.* 42: 59–70
65a. Means A. R., Tash, J. S., Chafouleas, J. G., 1981. Physiological implications of the occurrence, distribution and regulation of calmodulin in eukaryotic cells. *Physiol. Rev.* In press
66. Mescher, M. F., Jose, M. J. L., Balk, S. P., 1981. Actin-containing matrix assocated with the plasma membrane of murine tumor and lymphoid cells. *Nature* 289: 139–44
67. Munjaal, R. P., Chandra, T., Woo, S. L. C., Dedman, J. R., Means, A. R., 1981. A cloned calmodulin structural gene probe is complementary to DNA sequences from diverse species. *Proc. Natl. Acad. Sic. USA* 78: 2330–34
68. Munjaal, R. P., Dedman, J. R., Means, A. R., 1980. Isolation of the structural gene for calmodulin. *Ann. NY Acad. Sci.* 356: 110–18
69. Naccache, P. H., Molski, T. F. P., Alobaidi, T., Becker, E. L., Showell, H. J., Sha'Afi, R. I., 1980. Calmodulin inhibitors block neutrophil degranulation at a step distal from the mobilization of calcium. *Biochem. Biophys. Res. Commun.* 97(1): 62–68
70. Naccache, P. H., Showell, H. J., Becker, E. L., Sha'Afi, R. I., 1979. Involvement of membrane calcium in the response of rabbit neutrophils to chemotactic factors as evidenced by the fluorescence of chlorotetracycline. *J. Cell Biology* 83: 179–86
71. Naccache, P. H., Volpi, M., Showell, H. J., Becker, E. L., Sha'Afi, R. I., 1979. Chemotactic factor-induced release of membrane calcium in rabbit neutrophils. *Science* 203: 461–63
72. Noiman, E. S., 1980. Phosphorylation of smooth muscle myosin light chains by cAMP-dependent protein kinase. *J. Biol. Chem.* 255: 11067–70
73. Payne, M. E., Soderling, T. R., 1980. Calmodulin-dependent glycogen synthase kinase. *J. Biol. Chem.* 255(17): 8054–56
74. Pershadsingh, H. A., McDaniel, M. L., Landt, M., Bry, C. G., Lacy, P. E., Mcdonald, J. M., 1980. Ca^{2+}-activated ATPase and ATP-dependent calmodulin-stimulated Ca^{2+} transport in islet cell plasma membrane. *Nature* 288: 492–95
75. Pershadsingh, H. A., Landt, M., Mcdonald, J. M., 1980. Calmodulin-sensitive ATP-dependent Ca^{2+} transport across adipocyte plasma membranes. *J. Biol. Chem.* 255(19): 8983–84
76. Pires, E. M. V., Perry, S. V. 1977. Purification and properties of myosin light chain kinase from fast skeletal muscle. *Biochem. J.* 167: 137–46
77. Rasmussen, H., 1970. Cell communication, calcium ion, cyclic adenosine monophosphate. *Science* 170: 404–12
78. Rasmussen, H., Goodman, D. B. P., 1975. Calcium and cAMP as interrelated intracellular messengers. *Ann. NY Acad. Sci.* 253: 789–96
79. Rasmussen, H., Goodman, D. B. P., Friedman, N., Allen, J. E., Kurokawa, K., 1976. Ionic control of metabolism. In *Handbook of Physiology—Endocrinology*, 7: 225–64. Washington DC: Am. Physiol. Soc.
80. Rebhun, L. I., 1977. Cyclic nucleotides, calcium and cell division. *Int. Rev. Cytol.* 49: 1–54
81. Salisbury, J. L., Condeelis, J. S., Satir, P., 1980. Role of coated vesicles, microfilaments and calmodulin in receptor-mediated endocytosis by cultured B lymphoblastoid cells. *J. Cell Biol.* 87: 132–41
82. Scholey, J. M., Taylor, K. A., Kendrick-Jones, J., 1980. Regulation of nonmuscle myosin assembly by calmodulin-dependent light chain kinase. *Nature* 287: 233–35
83. Schubart, U. K., Fleischer, N., Erlichman, J., 1980. Ca^{2+}-dependent protein phosphorylation and insulin release in

intact hamster insulinoma cells. *J. Biol. Chem.* 255: 11063–66
84. Schulman, H., Greengard, P., 1977. Stimulation of brain membrane protein phosphorylation by calcium and an endogenous heat-stable protein. *Nature* 271: 478–79
85. Schulman, H., Greengard, P., 1978. Ca^{2+}-dependent protein phosphorylation system in membranes from various tissues, and its activation by "calcium-dependent regulator." *Proc. Natl. Acad. Sci. USA* 75. 5432–36
86. Sieghart, W., Theoharides, T. C., Alper, S. L., Douglas, W. W., Greengard, P., 1978. Calcium dependent protein phosphorylation during secretion by exocytosis in the mast cell. *Nature* 275: 329–31
87. Sieghart, W., Forn, J., Greengard, P., 1979. Ca^{2+} and cyclic AMP regulate phosphorylation of same two membrane-associated proteins specific to nerve tissue. *Proc. Natl. Acad. Sci. USA* 76: 2475–79
88. Smoake, J. A., Song, S. Y., Cheung, W. Y., 1974. Cyclic 3':5'-nucleotide phosphodiesterase distribution and developmental changes of the enzyme and its protein activator in mammalian tissues and cells. *Biochem. Biophys. Acta* 341: 402–11
89. Smoake, J. A., Solomon, S. S., 1980. Subcellular shifts in cyclic AMP phosphodiesterase and its calcium-dependent regulator in liver. Role of Diabetes. *Biochem. Biophys. Res. Commun.* 94: 424–30
90. Sobieszek, A., 1977. Calcium-linked phosphorylation of a light chain of vertebrate smooth muscle. *Eur. J. Biochem.* 73: 477–83
91. Sobue, K., Ichida, S., Yoshida, H., Yamazaki, R., Kakiuchi, S., 1979. Occurrence of a Ca^{2+} and modulator protein-activatable ATPase in the synaptic plasma membranes of brain. *FEBS Lett.* 99: 199–201
92. Suzuki, K., Kono, T., 1980. Evidence that insulin causes translocation of glucose transport activity to the plasma membrane from an intracellular storage site. *J. Biol. Chem.* 77: 2542–45
93. Tanaka, T., Hidaka, H., 1980. Hydrophobic regions function in calmodulin enzyme(s) interactions. *J. Biol. Chem.* 255: 11078–80
94. Valverde, I., Vandermeers, A., Anjaneyulu, R., Malaisse, W. J., 1979. Calmodulin activation of adenylate cyclase in pancreatic islets. *Science* 206: 225–27
95. Van Eldik, L. J., Watterson, D. M., 1981. Reproducible production of antiserum against vertebrate calmodulin and determination of the immunoreactive site. *J. Biol. Chem.* 256: 4205–10
96. Waisman, D., Stevens, F. C., Wang, J. H., 1975. The distribution of the Ca^{2+}-dependent protein activator of cyclic nucleotide phosphodiesterase in invertebrates. *Biochem. Biophys. Res. Commun.* 65: 975–82
97. Wang, J. H., Waisman, D. M., 1979. Calmodulin and its role in the second-messenger system. *Curr. Top. Cell. Regul.* 15: 47–107
98. Watterson, D. M., Sharief, F., Vanaman, T. C., 1980. The complete amino acid sequence of the Ca^{2+}-dependent modulator protein (calmodulin) of bovine brain. *J. Biol. Chem.* 255: 962–71
99. Weiss, R., Levin, R. M., 1978. Mechanism for selectively inhibiting the activation of cyclic nucleotide phosphodiesterase and adenylate cyclase by antipsychotic agents. *Adv. Cyc. Nucl. Res.* 9: 285–304
100. White, G. C. II, Raynor, S. T., 1980. The effects of trifluoperazine, an inhibitor of calmodulin on platelet function. *Throm. Res.* 18: 279–84
101. Wolff, J., Cook, G. H., Goldhammer, A. R., Berkowitz, S. A., 1980. Calmodulin activates prokaryotic adenylate cyclase. *Proc. Natl. Acad. Sci. USA* 77: 3841–44
102. Wood, J. G., Wallace, R. W., Whitaker, J. N., Cheung, W. Y., 1980. Immunocytochemical localization of calmodulin and a heat-labile calmodulin binding protein (CaM-BP$_{80}$) in basal ganglia of mouse brain. *J. Cell Biol.* 84: 66–76
103. Yagi, K., Yazawa, M., Kakiuchi, S., Oshima, M., Uenishi, K., 1978. Identification of an activator protein for myosin light chain kinase as a Ca^{2+}-dependent modulator protein. *J. Biol. Chem.* 253: 1338–40
104. Yamauchi, T., Fujisawa, H., 1979. Most of the Ca^{2+}-dependent endogenous phosphorylation of rat brain cytosol proteins requires Ca^{2+}-dependent regulator protein. *Biochem. Biophys. Res. Commun.* 90: 1172–78

AUTHOR INDEX

(Names appearing in capital letters indicate authors of chapters in this volume.)

A

Abbott, S. R., 657
Abe, H., 401, 402, 404-10
Abe, Y., 648
Abita, J. P., 468, 558
Åblad, B., 476
Abou-Issa, H., 657
Abramow, M., 161
Abramowitz, J., 658, 660, 662
Aceves, J., 156, 171, 195
Ackerly, J. A., 242, 245
Adam, A., 574, 578
Adams, P. A., 133
Adams, P. R., 319
Adams, R. N., 328
Adams, T. E., 585, 587
Adelstein, R. S., 676, 677
Ader, R., 577
Adler, J., 501-13, 553
Adouette, A., 519, 531
Adrian, R., 455
Adrian, R. H., 357-61, 363-65
Advis, J. P., 597-99
Afonson, S., 250
Agarwal, K. L., 633
Aggarwal, B. B., 599
Aghajanian, G., 270, 275
Agostini, B., 298, 300
Ahlquist, R. P., 476
Ahmed, Z., 431
Aiton, J. F., 393
Aiuchi, T., 521, 530
Ajabor, L., 583, 590
Akaike, N., 428, 531
Akasu, T., 393, 394, 398
AKERA, T., 375-88; 376-84, 446, 467
Akowitz, A. A., 305
Aksamit, R., 503
Alabaster, V. A., 227
Albers, R. W., 344, 402
Albert, J., 150
Alberts, A. W., 629
Albertson, B. D., 606, 609
Albright, C. D., 375
Albright, T., 597
Albuquerque, E. X., 465, 467
Alcantara, F., 545
Alemany, S., 542
Alerstam, T., 109, 111, 113, 114
Alexander, E. A., 170, 171, 195, 196, 205, 207, 209
Alexander, R. McN., 97, 99-101, 103, 133, 135-37
Alexander, R. W., 475, 479

Alfonzo, M., 338
Alken, R. G., 383
Allen, D., 456-59
Allen, D. G., 411
Allen, F., 161
Allen, G., 298-301, 309, 310
Allen, G. G., 194, 195
Allen, J. E., 668, 672
Allessie, M. A., 491, 492
Ally, A. I., 258, 262
Almers, W., 360-64, 367, 368, 429, 456
Alobaidi, T., 673
Alonso, C., 597
Alonso, G. L., 306
Alonso, M., 643
Alper, R. H., 276
Alper, S. L., 673
Alpern, R. J., 183
Alsen, C., 469
Alt, W., 543
Altamirano, M., 30, 31
Alter, I., 262
Altman, L. C., 288
Al-Ubaidi, F., 258, 269
Amer, S. M., 479
Amiranoff, B., 657, 662
Ammons, C., 243
Amsterdam, A., 329
Anand, M. B., 415, 416
Anast, C. S., 657
Andersen, J., 310
Andersen, J. P., 299
Anderson, D., 320
Anderson, D. J., 320
Anderson, F. L., 260
Anderson, G. J., 485, 486
Anderson, J. J., 628, 629
Anderson, J. P., 299
Anderson, M. J., 322-24
Anderson, M. W., 227, 229, 261, 262, 270, 271
Anderson, R. A., 502
Anderson, R. G., 329
Anderson, R. G. W., 326, 329, 675
Anderson, R. H., 490
Anderson, T. P., 285
Andersson, R., 413
Andreoli, T. E., 162, 183, 184, 186, 191, 192, 213
Andrews, P. M., 150
Andrews, W. W., 597-99
Angelo, L. S., 274
Anggard, E., 211, 257, 261
Angus, C. W., 328
Anholt, R., 319

Anjaneyulu, K., 632
Anjaneyulu, R., 675
Anner, B. M., 375, 378
Ansorge, S., 632
Antunes, J. L., 584, 586, 587, 598, 602
Antzelevitch, C., 488, 490, 491
Anuras, S., 31, 38
Aoyagi, M., 260
Appel, S. H., 325
Araki, S., 584, 587
Arato, J. S., 609
Archer, J. A., 641
Arfania, J., 587
Argibay, J. A., 364
Ariki, M., 306
Arimura, A., 585, 633
Armour, J. A., 486
Armstrong, C. M., 357
Armstrong, J. B., 502, 505
Armstrong, J. M., 262
Arnason, B. G. W., 328
Arnaud, C. D., 657
Arnett, T. R., 619
Arnold, A., 476, 655
Arnsdorf, M. F., 485
Arnstein, A., 574
Aronson, N. N. Jr., 633
Aronson, R. S., 486, 487
Arroyave, C. V., 289
Asai, H., 520
Asano, M., 673
Ash, J. F., 329
Ashcroft, F. M., 364, 428
Ashe, B. M., 629
Ashida, K., 469
Ashworth, J. M., 536
Asin, K. E., 578
Askari, A., 381
Ask-Upmark, E., 86
Assan, C. J., 248
Assimacopoulos-Jeanenett, F. D., 667
Assimacopoulos, F. D., 667
Aster, S. D., 264
Aswad, D., 505
Aswanikumar, S., 553, 555-57, 560
Ataka, M., 530
Aten, B., 630
Athanasoulis, C. A., 8
Atkinson, J. P., 150
Atkinson, L. E., 583, 587
Atrache, V., 259
Atsumi, S., 323
Atwood, H. L., 339
Aubert, M. L., 596-98, 601

683

AUTHOR INDEX

Audibert, A., 574, 578
Augustine, R., 347
AURBACH, G. D., 653-66; 655, 656, 659-61, 663, 664
Aures, D., 8, 9
Ausiello, D., 152, 173
Austen, K. F., 259, 284, 285, 288, 632
Avasthi, P. S., 152
Aves, E. W., 303
Axelrod, D., 322, 324, 328, 329
Axelrod, D. P., 322
Axelrod, J., 542, 560-62, 564, 565, 655, 656
Axelrod, R., 616, 623
Ayabe, T., 58, 62, 63
Azen, S., 189

B

Baba, Y., 648
Babior, B. M., 286
Bach, M. K., 259
Badger, T. M., 597
Baehler, R. W., 152
Baehner, R. L., 286
Baer, H., 464
Baez, S., 15, 23
Baghdassarian, A., 590
Bahl, C., 626
Bailey, J. C., 485
Bailey, W. H., 577
Bailie, M. D., 276
Bailin, G., 299, 310
Baines, A. D., 206
Baird, D. T., 590
Baird, K. L., 640
Bairey, N., 242, 245, 251
BAKER, M. A., 86-96; 86-90, 92
Baker, P. F., 359, 436, 439, 464, 465, 522
Baker, P. R., 58
Baker, R. M., 381
Baker, S. P., 475, 476, 480
Bakhle, Y. S., 227, 241, 258, 261, 264, 269, 270, 272, 273
Balciunas, J., 134
Balerna, M., 464, 466, 468, 470
Balinski, M., 523
Balk, S. P., 676, 678
Ball, C. B., 504
Ballenger, J. C., 579
Baltes, W., 541
Bandura, J. P., 485, 486, 490
Banerjee, S. P., 476, 655
Banerjee, U., 572
Bankir, L., 147, 203, 213-15
Bankowski, K., 577
Bar, F. W., 485
Bar, R. S., 639, 641-43, 645
Baranetsky, N. G., 602

Barany, K., 677
Barany, M., 677
Barchi, R. L., 359, 360, 364-66
Bardin, C. W., 583, 590, 639
Barfuss, D. W., 183, 184
Barger, A. C., 206
Barhanin, J., 466, 468, 470
Barkley, D. S., 535, 536, 546
Barnard, E. A., 322
Barnash, D. P., 116
Barnes, B. A., 242, 245
Barnes, G. E., 20, 58
Barnes, J. L., 152, 209
Barnes, L. D., 152
Barnes, P., 476
Barnett, D. B., 477
Baroody, R. A., 229, 261
Barovsky, K., 655
Barra, J., 544, 546
Barrand, P., 535, 544, 546
Barrantes, F. J., 319
Barratt, L. J., 194, 195, 209
Barrett, A. J., 633
Barrett, E. F., 361
Barrett, J. M., 158, 159, 203, 212
Barrett, J. N., 361
Barrio, R., 643
Barrnett, R. J., 227
Barron, J. T., 677
Barry, J., 584
Barry, W. H., 475, 479
Barth, L. G., 347
Barth, L. J., 347
Bartholomew, G. A., 133-35, 137, 138
Bartschat, D. K., 402, 415, 444
Baschieri, L., 648
Bashour, F. A., 65, 476
Baskin, R. J., 339, 344
Bassett, A. L., 485
Bassett, D. J. P., 270
Bassford, P. J. Jr., 628
Bassilian, S., 301
Bassingthwaighte, J. B., 430
Baste, C. A., 478-80, 633
Bastian, J., 358
Bastide, F., 300
Batbouta, J. C., 273
Battersby, C., 6, 75
Battilana, C. A., 210
Baudinette, R. V., 109, 114
Bauer, B., 426
Bauer, E., 633
Bauer, G. E., 632, 633
Bauer, H. C., 324
Bauer, R., 174
Bauer, U., 174
Bauereisen, E., 21
Baughman, W. L., 587
Baukal, A. J., 659, 660
Baum, S., 8
Baumann, H., 8, 9

Baxter, J. D., 320
Bayerdörffer, E., 191
Baylor, S., 451, 456
Bayne, E. K., 322, 323
Bazari, W. L., 541
Beamish, F. W. H., 123-25
Beaty, G. N., 357, 366
Beaudette, N. Y., 627
Beauge, L. A., 376, 377
Beaumont, A., 571, 573
Beaumont, W., 4
Beavo, J. A., 412
Bech, C., 93
Bechtel, P. J., 412
Beck, F., 174
Beck, W., 600
Becker, E., 412
Becker, E. L., 553-60, 563, 672, 673
Beck-Nielsen, H., 641, 642
Beckwith, J., 628
Beckwith, J. R., 628
Bedouelle, H., 628
Bedwani, J. R., 262
Beebe, D. P., 284
Beehler, R. W., 150
Beeler, G. W., 425-27
Beeler, G. W. Jr., 438, 442
Beeler, J., 454
Beeler, W., 459
Beeuwkes, R., 175
Beeuwkes, R. III, 147, 175
Begenisich, T., 357, 359
Behar, A. J., 4
Beidler, L. M., 527, 531
Beijer, H. J. M., 66
Beitins, I. Z., 597, 599, 606
Belamarich, F. A., 273
Beland, J., 265
Belchetz, P., 584
Belchetz, P. E., 584, 585, 588, 590, 598, 605, 606
Beleer, T. J., 337, 339
Bell, P. R. F., 6, 75
Belloc, F., 376
Bello-Reuss, E., 205
Belluzzi, J. D., 573, 579
Benedetti, E. L., 320
Bengele, H., 196
Bengele, H. H., 170, 171, 195, 205, 207
Benghezal, A., 578
Ben-Harari, R. R., 227, 273
Bennet-Clark, H. C., 135
Bennett, A., 271
Bennett, H. P. J., 631, 633, 634
Bennett, J. P., 337, 341
Bennett, J. W., 90
Bennett, L., 137
Bennett, W., 671
Benninger, C., 438, 440, 442
Benoit, R., 633
Bentfeld, M., 377

AUTHOR INDEX 685

Benuck, M., 249
Benzanilla, F., 456
Berbari, E. J., 492
Berchier, P., 574, 578
Bercovici, T., 320
Beress, L., 469
Berg, D. K., 322, 327
Berg, H. C., 501-3, 509, 511, 512, 543
Bergenstal, R. M., 631
Berger, J. D., 522
Berger, M., 109, 110, 113, 115
Berger, R. L., 406
Bergland, R. M., 621
Bergman, E. N., 60
Beringer, T., 347
Berkov, S., 250, 264
Berkowitz, S. A., 669
Berliner, R. W., 212
Berman, M. C., 304
Berman, P. W., 319, 328
Bern, H. A., 620
Bernard, P., 468
Berndt, T., 203-6
Bernstein, R. L., 536, 546
Bernstein, M. H., 93, 109-12
Berridge, M. J., 668, 672
BERRY, C. A., 181-201; 182-87
Berry, D. C., 88
Bers, D. M., 402, 438, 441, 443, 444
Bertaud, W. S., 347
Berti, F., 480
Bertler, A., 271
Bertrand, H. A., 302
Berzins, K., 675
Besch, H. R., 655
Besch, H. R. Jr., 376, 382, 402, 404, 413-15, 476, 480
Bessen, M., 523
Besser, G. M., 602
Best, P., 451, 452, 456
Best, P. M., 464
Betto, R., 348
Betz, H., 325-27
Betz, W., 323
Beubler, E., 52
Beug, H., 545
Bezanilla, F., 357, 358, 456
Bhalla, R., 476
Bhalla, R. C., 413
Bhathena, D., 152
Bhattacharya, A N., 583-87, 590
Bhattacharya, J., 210
Bhulla, R. C., 476, 480
Biagi, B., 184
Bianchi, C. P., 456
Biber, G., 74
Bidlack, J. M., 404, 405, 409
Biemesderfer, D., 168, 170, 171, 196

Bienvenue, A., 310
Biewener, A. A., Alexander, R. McN., Heglund, N. C., 103
Bilezikian, J. P., 477, 655
Bilezikjian, L. M., 408, 409
Billette, J., 490
Birchard, G. F., 93
Birdsall, N. J. M., 297
Birnbaum, D., 4
Birnbaum, M., 325
Birnbaumer, L., 653, 654, 658, 660, 662
Birnholz, J., 597
Biron, P., 271
Biryukova, T., 348
Bitensky, L., 648
Bito, L. Z., 229, 261
Bittsonen, R., 306
Black, R. A., 508
Blackburn, J. P., 271
Blackmore, P. F., 667, 674, 676
Blackwell, G. J., 259, 562
Blanck, J., 405
Blasie, J. K., 299, 311, 344
Blatt, C. M., 87
Blaustein, M. P., 431, 436, 439
Blecher, M., 654
Bleich, H. D., 556
Bligh, J., 88, 89
Blinks, J. R., 411, 451, 457-59
Blix, P. M., 631
Blizzard, R. M., 590, 643
Blobel, G., 320, 343, 556, 626, 627, 629
Bloch, B., 584
Bloch, R., 327
Bloch, R. J., 322
Block, E. R., 275, 276
Block, L. H., 646
Blondelet, M.-H., 544, 546
Bloom, F., 573, 574
BLOOM, F. E., 571-82; 227, 270, 274, 574-77
Bloom, G., 648
Bloom, S. R., 50, 263
Bloor, C., 90
Bloor, C. M., 289
Blosser, J. C., 325
Blostein, R., 377, 379
Blum, A. L., 76
Blum, J. J., 524, 669
Blundell, T. L., 622
Bobic, A., 475
Bogdanove, E. M., 587
Boggs, D. F., 93
Bogumil, J., 585
Bohlen, H. G., 15, 22, 30, 31, 46, 51, 72
Böhlen, P., 620, 633
Böhme, R., 541
Bohnet, H. G., 590
Bohus, B., 573, 574, 577, 578

Boileau, J. C., 271
Boineau, J. P., 492
Bokoch, G. M., 562
Boland, R., 297, 337-42, 344-46, 348
Bolis, L., 133
Boll, W., 304, 305
Bolt, D. J., 583
Bolte, H.-D., 394
Bolton, T., 456
Bond, J. H., 30, 32, 49, 75, 77
Bondi, A. Y., 368, 369
Bone, Q., 619
Bonewald, L. F., 634
Bonjour, J.-P., 204, 576
Bonke, F. I. M., 491, 492
Bonner, J. T., 535, 536, 539, 540, 546, 616
Bonta, I. L., 291
Bonting, S. L., 344, 378
Bonvalet, J. P., 203, 207
Bonventre, J. V., 213
Bookin, H. B., 578
Borg, J., 632
Borgeat, P., 259
Borjesson, B. W., 622
Bornet, E. P., 404, 405, 412, 459
Boss, B., 620
Bossen, E. H., 340
Bottazzo, G. F., 648
Boucher, R., 619
Boudoint, Y., 116
Boudry, J.-F., 204
Boughton-Smith, N. K., 9
Boulpaep, E. L., 154, 171, 194
Boura, A. L. A., 262
Bourdeau, J. E., 162, 191
Bourgeois, J.-P., 327
Bourguignon, L. Y. W., 676
Bourne, H. R., 664, 665
Bowen, J. C., 34, 79
Bowers, R. E., 261
Bowes, K. L., 39
Bowler, K., 85
Bowyer, D. W., 606, 609
Boyar, R., 590, 599, 600
Boyar, R. M., 590, 597, 600
BOYD, A., 501-17; 504, 506, 507, 512, 513
Boyd, H. M., 475
Boyd, J. A., 260
Boyden, N. T., 249, 264
Boyden, P. A., 487, 488
Boyer, P., 306
Boyer, P. D., 306, 337
Boyett, M. R., 430
Bozzaro, S., 540
Brachet, P., 535, 536, 540, 544-46
Bradley, C. L., 641
Braithwaite, A. W., 323
Branchaud, C. T., 609

AUTHOR INDEX

Brandenburg, D., 622
Brandt, N. R., 415
Brasel, D. E., 643
Brashler, J. R., 259
Braun, J., 329
Brawer, J. R., 576
Bray, D. F., 347
Bray, J. J., 325
Brazeau, P., 620, 633
Brazy, P. C., 182-85, 204, 206
Brdiczka, D., 347, 348
Brecker, G. A., 29, 31, 34
Brehm, P., 428, 519-24, 526, 530
Brendt, T. J., 193
Brennan, C., 133
Brennan, S. O., 631
Brenner, B. M., 152
Brenner, H. R., 327
Brenner, M., 540, 544
Brett, J. R., 124, 128
Breuker, E., 572
Brickman, A. S., 664, 665
Bridge, J. H. B., 440
Bridson, W. E., 586
Briggs, F. N., 310, 445
Brigham, K. L., 261, 274, 289
Brik, H., 325
Briley, M. S., 476
Brill, R. W., 124, 125, 128, 129
Brinkley, B. R., 669, 672, 675, 676
Brinkley, H. J., 583
Brinley, F. J. Jr., 440
Brittain, R. B., 476
Brobmann, G. F., 29, 31, 34
Brock, T., 413
Brodde, O.-E., 476
Brodie, B. B., 76
Brody, D. A., 490
BRODY, T. M., 375-88; 376, 377, 379-83, 446, 467
Brokaw, C. J., 133
Bronson, F. H., 583
Brooker, G., 655
Brosius, J., 622
Brostrom, M. A., 401
Brothers, V. M., 664, 665
Brouwers, H. A. A., 66
Brown, A. M., 428
Brown, D. A., 501-3, 509, 511
Brown, E. M., 655, 656
Brown, H., 426, 430, 431
Brown, J., 557
Brown, J. C., 633
Brown, L. C., 60, 63
Brown, M., 620
Brown, M. R., 263
Brown, M. S., 326, 329, 675
Brown, T. G., 476
Brown, T. G. Jr., 655
Browne, R., 49, 50
Browning, J. L., 522, 530

Browning, M., 671
Brownlee, G. G., 625
Brownstein, M., 620
Brownstein, M. J., 635
Brukner, L., 286
Brumley, F. T., 667
Brunette, M. G., 205
Brunner, H. R., 247, 248
Brunner, M., 536, 546
Bruns, F. J., 209
Brunsson, J., 51
Bry, C. G., 670, 675
Bryant, D. M., 109, 117
Bryant, S. H., 366
Buckalew, V. M. Jr., 383
Buggy, J., 576
Bugnon, C., 584
Buja, L. M., 476, 655
BULGER, R. E., 147-79; 152, 162, 167, 168, 175
Bulkey, G. B., 80
Bullen, B., 597
Bullen, B. A., 597
Buller, A. J., 338, 348
Bullock, L. P., 639
Bumpus, F. M., 244, 245, 249
Bunting, S., 260, 262
Buonassisi, V., 274
Burchell, A., 669
Burchell, A. R., 58
Burden, S., 327
Burden, S. J., 323, 328
Burg, M., 184, 189-91, 193, 196, 212
Burg, M. B., 162, 171, 183-86, 188, 191, 195, 196, 209
Burger, F. J., 85
Burgus, R., 620
Burke, J. A., 152
Burke, T. J., 152
Burke, W., 369
Burks, T. F., 15, 21, 23
Burls, R. F., 286
Burr, I. M., 590
Burrage, T. G., 322
Burres, S. A., 328
Bursztajn, S., 326, 329
Buschlen, S., 309
Butler, P. J., 109, 110, 113, 114
Butler, W. L., 536
Butler, W. R., 583-87
Butow, R. A., 300, 304
Butz, U., 536, 538, 539
By, A. W., 465
Bylund, D. B., 476
Byrne, B. C., 519
Byrne, B. J., 519

C

Cabanac, M., 87, 92
Cabeen, W. R. Jr., 440
Cabot, R. M., 39

Caflesch, C. R., 196
Cahalan, M., 357, 359
Cahalan, M. D., 428, 468
Caille, J., 358
Caldwell, P. R. B., 264
Callahan, E. N., 657
Camishion, R. C., 51
Campbell, D., 59
Campbell, D. T., 358-60
Campbell, K., 454
Campbell, K. P., 298, 337, 338, 342, 344
Campbell, W. B., 210
Campeau, L., 271
Canalis, E., 633
Cannon, J. K., 275, 276
Cannon, L. E., 344
Cano, R. P., 619
Cantini, M., 340
Cao, D. M., 542
Cappuccinelli, P., 536
Caputa, M., 87, 88, 90, 92
Caputo, C., 337, 338, 358, 455, 456
Carafoli, E., 337, 344, 402, 415, 416, 443, 444
Carayon, P., 648
Carbo, M., 643
Card, D. J., 326
Cardinal, J., 183
Cardinal, R., 488
Carette, B., 584
Carey, F. G., 126, 129, 130
Carey, R. J., 577
Carithers, R. W., 85
Carleton, J. S., 231
Carlquist, M., 633
Carlson, H. E., 602
Carlsson, B., 476
Carlsson, E., 476
Carlton, M. L., 243
Carmel, P. W., 584, 586-88, 598, 602
Carmeliet, E., 426, 429-31
Carmeliet, E. E., 430
Carneiro, J. J., 61, 62
Carnes, D. L., 657
Caron, M. G., 478, 655, 656, 661
Caroni, P., 402, 415, 416, 443, 444
Carp, H., 287
Carpenter, J-L., 329
Carr, D. H., 60
Carrell, R. W., 631
Carrette, J., 302
Carriere, S., 205
Carroll, R. J., 629, 633
Carson, M. P., 273
Carson, V., 475
Cartaud, J., 320
Carter, N. W., 193
Carvalho, M. G. C., 302

AUTHOR INDEX 687

Casals-Stenzel, J., 242, 245, 248
Casey, T. M., 133-35, 138
Caspar, R. F., 602
Cassel, D., 659-61
Cassidy, P., 677
Cassorla, F. G., 606, 609
Cassuto, J., 38
Castanas, E., 648
Castano, J. G., 542
Castellano, C., 573
Castillo, C. A., 250
Caswell, A. H., 415
Catravas, J. D., 263, 275, 276
Catt, K., 599, 606
Catt, K. J., 598, 599, 659, 660
Catterall, W. A., 465, 466, 468
Cavagna, G. A., 97, 99-101, 103
Cedgard, S., 38, 51
Celestian, J. F., 577
Cerreta, J. M., 251
Cerundolo, J., 633
Chabardes, D., 162, 205, 206, 212
Chabardès, G., 189, 192-94, 197
CHAFOULEAS, J. G., 667-82; 668, 669, 672-74
Chak, D. v., 304, 307
Chakraborty, P. K., 585
Chaloub, R. M., 302, 303
Chambers, D. A., 538
Champeil, P., 309-11
Chan, F. K., 546
Chan, S. J., 617, 625-30, 635
Chan, W., 263
Chandler, K., 451, 455, 456
Chandler, W. K., 357-59, 361, 363
Chandra, T., 678
Chang, A. C. K., 8
Chang, C. C., 376, 381
Chang, C. N., 556, 629
Chang, J. L., 586
Chang, K. J., 572
Chang, R. C. C., 633
Chang, R. J., 586, 602
Chang, R. K. C., 122
Changeaux, J-P., 320, 325, 326
Changeux, J.-P., 319, 320, 325-27
Chaplin, S. B., 112, 114-16
Chapman, D., 310
Chapman, J. B., 397
Chapman, L. W., 86, 88, 89
Chapman, R. A., 425, 436
Chappel, S. C., 588, 600, 603
Charbardes, D., 212, 213
Charbon, G. A., 34, 35, 66, 74
Charlton, R. R., 655, 659, 664
Charney, A. N., 45
Chassin, P., 98
Chaudhari, A., 262

Chayen, J., 648
Chelis, C., 302
Chelsky, D., 506, 508
Chen, F.-C. M., 475, 476
Cheng, J. B., 476
Chenieux-Guicherey, P., 475
Chenoweth, D. E., 284, 555
Chen-Yen, S.-H., 346
Chernoff, D., 284
Cherrington, A. D., 667
Chesnais, J. M., 431
Cheung, H. S., 244, 246-49, 264
Cheung, L. Y., 9, 10
Cheung, W. Y., 668, 669, 675
Chevallier, J., 300, 302, 304
Chiang, P. K., 561
Chiang, T. S., 249
CHIARANDINI, D. J., 357-72; 368, 369
Chicheportiche, R., 464-66, 468, 470
Chien, S. M., 31, 38
Chiesi, M., 303
Chigurupati, R., 378, 381
Chiu, A., 228, 230, 243, 264
Chiu, A. T., 229, 230, 242-44, 246
Chiu, T. H., 381
Chiu, V. C. K., 304
Choi, Y. R., 381
CHOU, C. C., 29-42; 19, 29-32, 34-39, 45
Chou, M. C. Y., 648
Chretien, M., 619, 633
Christadoss, P., 328
Christensen, G. C., 87
Christensen, J., 31, 38
Christian, C., 324
Christian, C. N., 323, 324
Chu, G., 405
Chu, L., 377
Chu, L. L. H., 632
Chuang, E. L., 211
Chubb, J., 381
Chung, A., 229, 242, 243, 245-47, 249
Chung, E. K., 486
Chung, J. M., 92
Church, N. S., 137
Churchill, S., 175
Chyn, T., 341
Chyn, T. L., 338, 343, 345
Ciani, S., 364
Ciaraldi, T., 473
Cieskowski, M., 8
Citri, Y., 658, 661
Citterio, G., 103
Civan, M. M., 346
Claessens, R., 574, 578
Clark, A. J., 435
Clark, D. A., 259
Clark, J., 99, 100

Clark, J. L., 633, 634
Clark, R. A., 80, 286, 287
Clark, T. L., 480
Clarke, M., 541
Clarke, S., 504, 505
Clausen, T., 381, 382
Claypool, W. D., 285
Clayton, R. N., 598
Clements, E., 230, 231
Clifton, D. K., 586
Clippinger, M. S., 414
Clique, A., 162, 212, 213
Close, R. I., 105
Clough, D. P., 89
Cloupeau, M., 134, 135, 137, 140
Coan, C., 300, 307, 310
Coan, C. R., 310
Coburn, R. F., 262
Coceani, F., 86
Coffey, A. K., 150
Coffino, P., 478
Cogan, M. G., 182-85
Cogen, P. H., 586
Coghlan, J. P., 242, 245
Cohen, H., 617
Cohen, I. S., 464
Cohen, J., 617
Cohen, M. H., 536, 545
Cohen, M. M., 65
Cohen, M. W., 322, 324
Cohen, P., 669
Cohen, P. T. W., 669
Cohen, R. S., 675
Cohen, S., 288
Cohen, S. A., 322-25
Cohn, D. V., 632, 633
Cohn, R., 59
Cohn, Z. A., 286
Colatsky, T. J., 464
Colley, D. P., 80
Collins, J. H., 541
Colucci, W. S., 247, 248
Colwell, R. K., 112, 115, 116
Comite, F., 585
Compere, S. J., 347
Condeelis, J. S., 329, 675, 676
Condon, M. E., 249
Conn, P. M., 674, 675
Conner, G., 453, 459
Connor, J. A., 431
Conte, F. A., 590, 597, 600, 602, 607, 609
Conti, M. A., 677
Conti-Tronconi, B., 328
Cook, G. H., 669
Cook, J. A., 260, 261
Cook, L., 572
Cooke, R., 676
Cooper, I., 620
Cooper, S., 538
Coquelin, A., 583

Coraboeuf, E., 431, 445, 465, 468
Corash, L. M., 641
Corbet, P. S., 133
Corbin, J. D., 667
Corcoran, B., 555, 556, 561, 564
Corcoran, B. A., 553, 555-57, 560, 561
Corey, E. J., 259
Corley, K. P., 585, 590, 597, 600, 605
Corman, B., 185, 205
Cornell, R., 341
Cornett, L. E., 476
Corrsin, S., 115
Costantin, L., 452
Costantin, L. L., 358
Costanzo, L. S., 193, 194
Cota, G., 366, 367
Cote, T. E., 654
Cotran, R. S., 150
Cottier, H., 287
Coukell, M. B., 545, 546
Coulombe, A., 465
Counter, S. A. Jr., 133
Couraud, F., 468, 469
Cox, J. L., 492
Cox, J. W., 150
Coy, D. H., 8, 571, 574, 575
Craddock, P. R., 289
Cramer, E. B., 553, 559, 564
Cranefield, P. F., 391, 395, 397, 485-87, 489
Crawford, R., 633, 635
Crawshaw, L. I., 92
Crayen, M., 167
Creese, I., 657
Creighton, T. E., 506
Crexells, C., 271
Crick, F., 622
Crompton, M., 337, 344
Cronau, L. H., 269, 275-77
Cronin, R. E., 152, 162
Cronk, M., 633, 634
Cronkite, D. L., 519
Cross, S. A. M., 227
Crouch, T. H., 416, 668
Crowder, M. E., 598
Crowe, L. M., 339
Crowley, T. J., 376
Crowley, W. F., 585, 588, 590
Crowley, W. F. Jr., 585, 605
Crumpton, C. W., 250
Crutchley, D. J., 227, 231, 251, 260
Cuatrecasas, P., 558, 572, 573
Cubeddu-Ximenez, L., 478
Cullen, M. J., 369
Culpepper, R. M., 162, 191, 192, 213
Cunningham, D. D., 630

Curfman, G., 247, 248
Curfman, G. D., 376
Curtis, M. B., 93
Cushman, D. W., 244, 246-49, 264
Cushman, S. W., 676
Cutler, G. B., 606, 609
Cutler, G. B. Jr., 585
Cuttelod, S., 248
Cutz, E., 263
Czech, M. P., 514

D

Dabney, J. M., 35
Dagher, G., 378
Dahlén, H. G., 590
Dahlén, S.-E., 259
Dahlöf, C., 476
Dahlof, C-G., 476
Dahlquist, F. W., 506, 508
Dailey, R. A., 587, 602
D'Alecy, L. G., 89, 90
D'Alnoncourt, N., 488
Daly, I. de Burgh, 435
Daly, J. W., 465, 467
Damato, A. N., 492
Damiani, E., 339, 348
D'Amore, P. A., 553
Dampney, R. A. C., 90
Daniel, E. E., 39, 40
Daniel, J. L., 677
Daniel, P. M., 86, 87
Daniels, C. J., 628, 629
Daniels, M. P., 323, 324
Danilo, P. Jr., 488
Danon, Y., 85
Darmon, M., 535, 536, 544, 546
Darnell, J. E., 622
Das, D. K., 275, 276
Das, M., 244, 246, 264
Data, J. L., 10
Date, T., 629
Dausse, J. P., 475
Daut, J., 396
David, C. N., 535, 536
David, C. S., 328
David, R., 600
Davidowitz, J., 368
Davidson, M. M. L., 413
Davies, A. O., 655
Davies, R. E., 103
Davies, T., 413
Davis, A. T., 501, 504, 513
Davis, D. D., 86
Davis, J., 557
Davis, J. M., 213, 214
Davis, L. D., 486, 487
Dawes, J. D. K., 86, 87
Dawkins, R., 616
Dawson, C. A., 270, 274, 275
Dawson, T. J., 98

Day, A. R., 229, 242, 243, 553, 555-57
Day, L. A., 628, 629
Dayhoff, M. O., 619
Deamer, D. W., 339, 344
Dean, W. L., 309
deAssis Leone, F., 305
DeBarry, J., 442, 467
de Bermudez, L., 193
DeBlas, A. L., 565
DeBoer, A., 487
De Champlain, J., 480
DeCherney, A., 583
Decker, J. L., 565
Dedman, J. R., 668, 669, 672-78
De Foor, P., 343
De Foor, P. H., 348
DeFranco, A. L., 504, 506, 508, 510
Deftos, L. J., 622
Degani, C., 337
DeGoede, J., 526
DeHaan, R. L., 426-28
Dehaye, J.-P., 674, 676
Deitmer, J. W., 431, 440, 441, 445, 521, 522, 529
De Jong, F. H., 590
De Johge, H. R., 416, 417
Dekker, A., 648
De Koning, J., 585
De Kort, C. A. D., 139
Delaney, J. P., 6, 77
de la Vina, S., 643
del Carmine, R., 565
Delcomyn, F., 134
De Lean, A., 477, 479, 656, 661
De Lean, A. L., 475, 477
de Leon, S., 343, 344
Délèze, J., 390
Dellman, H.-D., 576
DeLorenzo, R. J., 670, 671, 675
Del Vecchio, P., 233
Del Vecchio, P. J., 245, 251
Demaille, J. G., 346, 404, 405, 408-10, 412, 670
Dembinski, A., 8
deMeis, L., 298, 299, 302, 303, 306, 337, 407
Demers, J. M., 431
De Meyts, P., 641
Dennett, C. J., 492
Dennis, M. J., 320
Dennis, V. W., 182-85, 203-6
Denny, J. B., 241
DePamphilis, M. L., 502
dePetris, S., 329
de Peyer, J. D., 519-22, 524, 529
De Pont, J. J. H. H. M., 378
De Pover, A., 377

AUTHOR INDEX 689

D'Ercole, A. J., 643
Deroubaix, E., 465, 468
de Rouffignac, C., 147, 158, 159, 185, 193, 203-7, 212
Derstine, P. L., 543, 544
Desai, R., 668
Desai, U., 289
DeTorrente, A., 152
Devaux, P. F., 310
Devezeaux, D., 134, 135, 137, 140
Devillers, J. F., 134, 135, 137, 140
Devine, C. E., 347
Devreotes, P. N., 321, 326, 536, 543-45
Devynck, M-A., 378
De Wied, D., 571-79
De Witt, R. J. W., 539, 546
DeWulf, H., 667
Dey, R. D., 263
Dhalla, N. S., 415, 416
Dharmsathaphorn, K., 49
Diamond, L., 172
Diaz, M., 203, 213-15
Dickinson, S., 103
Dicou, E., 540, 545
Dicou, E. L., 544, 545
Diczfalusy, E., 599
Dierschke, D. J., 583-87, 590
Dieter, P., 541, 670
Dieterle, Y., 262
Diezi, J., 156, 171, 195
DiFrancesco, D., 426, 430, 431
DiGirolamo, M., 647
Dinauer, M. C., 544
Dinerstein, R. J., 672
Dingledine, R., 572
Dipolo, R., 358, 376, 377
DiRienzo, J., 629
DiRienzo, J. M., 626
Di Scala, V. A., 154, 158, 172
DIZON, A. E., 121-31; 122-24, 126-29
Dobberstein, B., 626, 627, 629
Dobbins, J. W., 49
Dobuler, K. J., 276
DOBYAN, D. C., 147-79; 152, 162
DOCHERTY, K., 625-38; 630
Dodge, F. A., 485, 489
Dodson, G. G., 622
Dogterom, J., 577
Dohrn, M, 600
Dollery, C. T., 476
Dolnik, V. R., 109, 112, 113
Donabedian, H., 558
Donald, D. E., 61, 62
Donaldson, V. H., 284
Doniach, D., 648
Donker, A. J., 248
Donley, J. R., 337
Donovan, B. T., 583

Donovan, P., 116
Donowitz, M., 45
Dons, R. F., 641
Dorer, F. E., 229, 242-44, 246
Dorge, A., 174
Dorner, A. J., 629
Doroshow, J. H., 286
Doucet, A., 171, 184, 190, 191, 196
Dougherty, J. P., 305
Doughty, M. J., 524
Douglas, W. W., 673
Dousa, T. P., 197
Dow, S., 172
Dowd, F., 413, 414
Dowd, F. J. Jr., 413, 414
Downey, H. F., 476
Downs, R. W. Jr., 659-64
Drabikowski, W., 414
Drachman, D. B., 326, 328, 329, 339
Drage, D. J., 536, 545
Draper, R., 504
Drazen, J. M., 259
Dreisin, R. B., 289
Drenckhahn, D., 677
Dresdner, K., 441
Drewes, J., 585
Drexhage, H. A., 648
Dreyer, A. C., 476
Dreyer, B. E., 251
Dribin, L. B., 361
Drobac, M., 276
Drummond, G. I., 402, 413-16, 478
Drummond, M., 402, 413-16
Dryl, S., 526, 530, 531
Duance, V. C., 300
DuBose, T. D., 193
DuBose, T. D. Jr., 208, 210
Duckett, G., 583
Dudel, J., 464
Duello, T. M., 598
Dufau, M. L., 599, 606, 659, 660
Duffus, W. P. H., 329
Duffy, P. A., 47
Dufy, B., 585
Dufy-Barbe, L., 585, 590
Duggan, P. F., 338
Duguid, J. R., 629
Dulhunty, A. F., 364
Dull, T. J., 622
Dunlap, K., 522, 523, 526, 528, 529
Dupont, Y., 299, 300, 304, 305, 309, 311
Durrer, D., 488, 490-92
Durston, A. J., 535
Dusting, G. J., 262
Duval, A., 360, 362, 363
Dyrenfurth, I., 585, 586
Dzau, V. J., 247, 248

E

Eades, C. G., 572
Earp, H. S., 412
Eaton, D. C., 361
Ebashi, S., 451
Ebert, M. H., 644
Echlin, P., 175
Eckert, R., 428, 519-24, 526, 530
Eckstein, F., 539, 546
Eden, S., 647
Edie, R., 485, 487, 488
Edvinsson, L., 274
Edwards, B. R., 212
Edwards, C., 349
Edwards, R. M., 197
Edwards, S., 504
Egan, R. W., 261
Eglow, R., 275, 276
Ehara, T., 426, 428
Ehara, Y., 585
Eichling, J. O., 274
Eimerl, S., 658
Einarson, B., 319
Einwachter, H. M., 438, 440, 442
Eipper, B. A., 630, 631, 633
Eisenach, J. C., 320
Eisenberg, M. M., 10
Eisenberg, R. S., 362, 364
Eisenman, G., 359
Eisner, D. A., 391, 392, 394-96, 398
Eitle, E., 543, 546
Ek, L., 476
Eklund, S., 34, 50, 51
Elashoff, J., 8, 9
Elders, J., 643
Elders, M. J., 643
Elferink, J. G. R., 673
Elftman, H., 99, 100
Elharrar, V., 485
Eling, T. E., 227, 229, 258, 260-62, 270, 271
Elizari, M. V., 485
Ellington, C. P., 135, 136
Ellinwood, W. E., 585, 586
Ellis, D., 431, 436, 438, 440, 441, 443, 445
El-Refai, M. F., 674, 676
El-Sherif, N., 485, 486, 488, 492
Elson, E. L., 322
Elson, J., 487
Emdin, S. O., 633-35
Endo, J., 457
Endo, M., 338, 451-53, 458
Engel, A., 319
Engel, A. G., 328
Engel, G., 655
Engel, R. R., 643

Engelman, D. M., 627
England, P., 458, 460
English, D. K., 286
Engstrom, P., 504, 506, 511, 513
Entman, M. L., 340, 402, 404, 405, 412, 459
Epand, R. M., 654
Epel, D., 347
Epstein, M., 306, 524
Epstein, W., 504, 513
Epting, R. J., 111, 114, 115, 133, 137
Erdmann, E., 376, 381, 394, 475
Erdos, E. G., 249, 260, 263, 264
Erlanger, J., 491
Erlichman, J., 673
Ernst, E. A., 59
Esch, F., 633
Esch, H., 133, 134, 140
Esmann, M., 377
Estensen, R. D., 632
Etlinger, J. D., 345
Euler, D. E., 487, 492
Evan, A., 170
Evan, A. P., 152
Evans, H. E., 87
Evans, J. M., 602
Evans, R. R., 284
Eveloff, J., 173, 191
Exton, J. H., 667, 674, 676
Ezerman, E. B., 338

F

Fabiato, A., 412, 425, 446, 451-53, 458
Fabiato, F., 412, 425, 446, 451-53, 458
Fahrenkrug, J., 50
Faiman, C., 598
Falcone, R. A., 492
Falk, G., 531
Falk, W., 555
Falkmer, S., 635
Fallon, J. T., 492
FAMBROUGH, D. M., 319-35; 319, 321-23, 325, 326, 349
Fanburg, B. L., 242, 246, 339
Fang, V. S., 599
FANTONE, J. C., 283-93; 286, 290, 291
Fantus, I. G., 642
Fara, J. W., 76
Farfel, Z., 664, 665
Farmer, G. J., 125
Farquhar, M. G., 150
Farre, J., 485
Farshidi, A., 492
Fasman, G. D., 627

Fassold, E., 307
Fasth, S., 34
Fawcett, A. A., 90
Fay, R. S., 523
Fearon, D. T., 285
Fedak, M. A., 98-101
Feenberg, H., 672
Fehr, J., 289
Fei, D. W., 242, 245
Feinsinger, P., 112, 114-16
Feldberg, W., 259
Feldman, D. A., 414, 416, 417
Feldman, S., 4
Felgentrager, J., 381
Felix, C., 664, 665
Fellmann, D., 584
Fellmann, P., 310
Fenn, W. O., 99, 100
Fenoglio, J. J., 486, 492
Fenoglio, J. J. Jr., 485
Fenzi, G., 648
Ferguson, H. C., 479
Ferguson, K. M., 661, 662
Ferin, J., 602
Ferin, M., 584-88, 598
Fernandez, H. N., 284, 555
Fernandez-Madrid, F., 273
Ferraz, C., 346
Ferreira, H. G., 531
Ferreira, S. H., 229, 241, 248, 249, 257, 261
Ferrier, G. R., 382, 487, 490, 491
Ferris, T. F., 150
Fessler, J. H., 625
Fessler, L. J., 625
Feuerstein, G., 261
Fevre, M., 597
Fiegel, V. D., 558
Field, J. B., 648
Figueredo, J., 154
Fillenz, M., 274
Finegan, J. -A. M., 402, 411
Fink, R., 361, 367, 429
Finkelstein, J., 590, 599
Finkelstein, J. W., 590, 600
Finney, J. O. Jr., 415
Fisch, C., 485, 486, 488
Fischbach, G. D., 319, 322-25, 327, 329
Fischbein, L. C., 560
Fischer, E. H., 346
Fischer, J. E., 50
Fischer, K., 469
Fiser, R. H., 643
Fisher, A. B., 262, 270, 275, 276
Fisher, D., 576
Fisher, D. A., 647
Fishman, A. P., 251, 269
Fishman, M. C., 383
Fitch, F. W., 328
Fitch, W., 59

Fitzpatrick, D. F., 413
Fitzsimons, J. T., 576
Flanagan, J., 676
Flandrois, R., 89
Flasch, H., 381
Flashner, M. S., 375-77
Flavin, M., 564
Fleckenstein, A., 426
Fleeger, J. L., 583, 590
Fleisch, H., 204
Fleischer, N., 673
Fleischer, S., 297, 299, 300, 343, 344, 348, 453, 459
Fleming, B. P., 430
Fleming, J. W., 402, 404, 413-15, 476, 659
Fletcher, D. J., 632, 633
Fletcher, M. P., 558
Flexner, C., 10
Flexner, L. L., 578
Flexner, T. B., 578
Flier, J. S., 383, 639, 640, 642, 643
Flink, J. R., 275-77
Flower, R. J., 259, 562
Fluragon, T. D., 288
Foidart, J. M., 150
Folkow, B., 13, 23, 25, 72
Fondacaro, J. D., 15, 19, 21, 30, 32, 34, 35
Forand, R., 633
Forbes, C. D., 284
Fordtran, J. S., 50
Forn, J., 671
Forrest, J. W., 325
Forrster, J. M., 80
Forslund, T., 248
Forte, L. R., 657
Forte, M., 519, 530, 531
Fortes, P. A. G., 381
Fortner, J. G., 58
Fosset, M., 442, 466-70
Foster, D. L., 590
Foster, P. R., 488
Foulkes, J. G., 669
Fouts, J. R., 271
Fowler, A. V., 628
Fox, A. A. L., 382-84
Fox, A. P., 428, 429
Fox, J. A., 211
Foxcroft, G. R., 583
Fozzard, H. A., 425, 427, 440, 441, 443, 444, 451, 459
Fraenkel, G. S., 525
Frank, E., 319, 322, 328
Frank, E. F., 322, 327
Frank, J., 457
Frank, J. S., 445
Frank, M., 467
Franke, J., 545
Frankenhaeuser, B., 365
Frantz, A. G., 586
Franzini, C., 349

AUTHOR INDEX 691

Franzini-Armstrong, C., 337-39
Fraser, H. M., 620
Frasier, S. D., 597, 599
Frazier, W. A., 538, 545
Freedman, M. H., 347
Freedman, S. D., 670, 671, 675
Freeman, J. A., 321
Freer, R. J., 553, 555-57
Frelin, C., 466, 468, 470
French, L. A., 4
Freychet, P., 639
Freygang, W. H., 357, 361, 363-65
Fridovich, I., 286
Friedman, K., 520, 521, 530
Friedman, N., 668, 672
Friedman, P. A., 191, 192
Friedman, P. L., 486
Friedman, W. F., 412
Friedrich, K., 632
Frindt, G., 209
Frisch, R. E., 597
Froehlich, J. P., 304-7, 406, 412
Froesch, E. R., 645, 646
Fromageot, P., 242
Fromm, H., 535
Frömter, E., 183-85, 187, 191, 192, 196
Frothingham, R., 558
Fry, C. H., 430, 431
Fry, F. E. J., 130
Fuchs, S., 329
Fudman, E. J., 557
Fuhrman, F. A., 85
Fujii, Y., 31, 36
Fujisawa, H., 670
Fujita, T., 648
Fujiwara, K., 677
Fujiwara, M., 469
Fukunaga, K., 669
Full, R. J., 134
Fulpius, B. W., 328
Funder, J. W., 480
Fung, L. K., 437, 438, 441
Furthmayr, H., 150, 627
Futrelle, R. P., 542
Fyhrquist, F., 248

G

Gabbay, K. H., 631
Gabe, M., 170
Gache, C., 305
Gaddum, J. H., 269
Gadsby, D. C., 391, 394, 395, 397, 464, 485
Gaffori, O., 576, 577
Gage, P. W., 362, 364
Gagnon, C., 561
Gagon-Brunette, M., 162
Gainer, H., 635
Galan Galan, F., 619

Galla, J. H., 152
Gallagher, J. J., 491
Gallagher, M., 572, 573
Gallavan, R. H., 29-31, 34, 36-39, 45
Gallavan, R. H., Jr., 30, 32
Gallien-Lartigue, O., 347
Gallin, E. K., 553, 559, 564
Gallin, J. I., 553, 554, 556, 558, 559, 563, 564
Gallis, J-L., 376
Gambhir, K. K., 641
Gamble, M., 630
Gammeltoft, S., 622
Gannon, B. J., 44, 73
Garan, H., 492
Garay, R. P., 378, 379, 476
Garcia, A. M., 307
Garcia, C.-R., 583
Garcia del Rio, C., 242, 245, 248
Gardner, J., 326
Gardner, J. M., 321-23, 326
Garfield, R. E., 322
Garg, D. K., 79
Garg, L., 184, 189-91, 193, 196
Gargus, J. J., 509
Garland, P. B., 322
Garofalo, R. S., 524
Garrahan, P. J., 306, 379
Garrison, J. C., 662
Garsky, V., 573
Garybobo, C. M., 310, 311
Gaspich, S., 646
Gauss-Miller, V., 553
Gautvik, K., 328
Gavin, J. R. III, 641, 647
Gavras, H., 247, 248
Gavras, I., 248
Gavrilov, V. M., 109, 112, 113
Gay, L. A., 361, 364
Gay, V. L., 583
Geddes, D. M., 271
Geelhoed, G. W., 606, 609
Geiger, B., 322
Geis, G. S., 92
Geisow, M., 329
Gelband, H., 485, 487
Gelbart, D. R., 210
Geller, D. M., 630, 631
Geller, J. S., 544
Gelman, S., 59
Genel, M., 605
Genest, J., 619
George, J. N., 413
Gerber, E., 413
Gergely, J., 348
GERISCH, G., 535-52; 535-46, 553
Gerkins, J. F., 10
Germak, J. A., 589, 590
Gershon, N., 324
Gerten, J., 246

Gertner, J. M., 605
Gertz, E. W., 415
Gessaman, J. A., 109
Gessner, K., 184, 185
Gettes, L. S., 427, 485
Geumei, A., 65
Gewitz, M., 276
Gewitz, M. H., 273
Ghatei, M., 263
Ghatei, M. A., 263
Ghysel-Burton, J., 381
Giachetti, A., 476, 480
Gibbons, W. R., 426, 430, 459
Gibo, D. L., 137
Giclas, P. C., 289
Giebisch, G., 156, 168, 170, 171, 184, 195, 196, 203
Giebisch, G. H., 194
Gigli, I., 284
Gil, M. G., 542
Gilai, A., 358
Gilbert, M., 491
Gilbert, P. E., 572
Gilbert, W., 626
Giles, W., 430
Giles, W. R., 426
Gilkey, J. C., 347
Gill, D. M., 479, 656, 661
Gillespie, J. I., 362
GILLIS, C. N., 269-81; 227, 262, 263, 269-71, 273-77
Gilly, W. F., 368, 369, 455
Gilman, A. G., 657, 661-64
Gilot, P., 574, 578
Gimble, J., 674, 675
Gingell, D., 530
Gingerich, R. L., 633
Gingold, M. P., 309
Ginsborg, B. L., 369
Gips, H., 586
Girgis, S. I., 619
Giroud, C. J. P., 609
Gitelman, S. E., 524
Gitler, C., 320
Given, B. D., 643
Glasel, J. A., 556
Glaser, B. M., 553
Glass, N. R., 124, 128
Glaubiger, G., 480, 655
Glazier, J. B., 242
Glenn, A. W., 242, 245
Glick, G., 58, 62, 63
Gliemann, J., 622
GLITSCH, H. G., 389-400; 390-96, 439
Gloor, P., 86
Glossmann, H., 475
Glover, T. D., 583
Gluckman, P. D., 601
Glynn, I. M., 375
Gnegy, M. E., 669, 674, 675
Gnodde, H. P., 585
Gocke, D. J., 246

Godfraind, T., 377, 381
Godynicki, S., 87
Goebelsmann, U., 586
Goetzl, E. J., 260, 287, 288, 557, 562, 563
Gold, P., 579
Goldberg, A. L., 326, 329
Goldberg, M., 205
Goldbeter, A., 544
Golden, M. P., 644
Goldenberg, R. L., 586
Goldfarb, S., 205
Goldfarb, V., 320
Goldfien, A., 476
Goldhammer, A. R., 669
Goldin, S. M., 375
Goldman, D. W., 563
Goldschmidt, M., 51
Goldsmith, P. C., 584
Goldspink, G., 112, 114, 115, 133
Goldstein, I., 288
Goldstein, I. M., 284
Goldstein, I. P., 597
Goldstein, J. L., 326, 329, 675
Goldstein, J. M., 286
Goldsworthy, G. J., 139
Goldyne, M., 257
Golterman, N. R., 633
Goltzman, D., 657
Gomez, C. M., 328
Gomoll, A. W., 479
Gomperts, B., 347
Gonzalez-Serratos, H., 358, 454, 459
Gooding, R. M., 123, 124, 126, 127, 129
Goodman, A. L., 589
Goodman, D. B. P., 668, 672
Goodman, R. L., 584, 590
Goodwin, D. W., 241, 242
Goodwin, F. K., 579
Goodyer, C. G., 609
Gopinath, R. M., 669
Gorden, P., 639, 641-43, 645-47
Gordon, J. L., 231
Gordon, S., 285
Gore, R. W., 30, 51, 72, 73
Gorenstein, C., 249
Gorman, A. L. F., 429, 522
Goth, A., 263
Gottschall, J., 150
Gottschling, D., 632
Gotz-Welbergen, A. V., 597
Gough, W. B., 488
Goy, M. F., 501, 504-7, 509, 510, 512
Goy, R. W., 586
Grab, D. J., 675
Graber, M. L., 196
Grabowski, W., 392, 394, 395
Graham, W. F., 242, 245

Grand, R. J. A., 669
Grandchamp, A., 154
Granelli-Piperno, A., 285
Granger, D. N., 18, 19, 21-23, 30, 33, 36, 37, 43, 46-49, 51, 52, 58-60, 78
Granger, H. J., 15-25, 58
Granstrom, E., 257
Grant, A. O., 487, 488
Grant, N., 573
Grantham, J., 212
Grantham, J. A., 184
Grantham, J. J., 171, 184, 188, 196, 197, 209
Grassi de Gende, A. O., 348
Grassmick, B., 19, 30-32, 36
Gray, A., 622
Gray, R. D., 309
Gray, R. R., 403, 404, 406, 408
Grayson, J., 74
Greeff, K., 382-84
Green, A. A., 535-38
Green, E., 565
Greén, K., 257
Green, N., 183-85, 190, 191
Green, N. M., 298-301, 309, 310
Greenbaum, L. M., 245, 251
Greenberg, I., 324
Greene, L. A., 634
Greene, L. L. J., 248
Greene, N. M., 276, 277
Greene, W. C., 347
Greenewalt, C. H., 109, 111-15
Greengard, P., 663, 670, 671, 673
Greenlee, R., 633
Greenlee, W. J., 264
Greenspan, K., 485, 486
Greenwald, R., 275, 276
Greenway, C. V., 57, 61, 62, 65, 76, 77
Greenway, D. C., 338, 343
Greger, R., 191, 192, 204
Greger, R. F., 205
Gregory, R. A., 633
Grenier, J. F., 31, 36, 37
Greven, H., 572, 573, 575
Greven, K., 573, 574
Gribic, M., 247, 248
Griffin, G. L., 285
Griffith, L. D., 167
Griggs, R. C., 642
Grim, E., 77
Grimm, D. J., 274
Grishin, E. V., 468
Grodin, J. M., 586
Grodzinska, L., 259
Groelke, J., 643
Gronhagen-Riska, C., 248
Gros, F., 326
Groschel-Stewart, V., 677
Gross, D. M., 264

Gross, E., 557
Gross, H. A., 644
Gross, J. B., 192, 193, 195
Gross, J. D., 545
Gross, K. B., 262
Gross, P. M., 90
Grossman, A., 602
Grossman, M. I., 7-10, 76
Grotendorst, G. R., 553
Gruber, K. A., 383
Gruener, R., 324
Gruenfeld, J.-P., 203, 213-15
GRUMBACH, M. M., 595-613; 590, 595-602, 604, 606, 607, 609
Grunfeld, C., 639, 640, 642, 643, 645, 648
Gryglewski, R. J., 259, 260, 262
Guerriero, V., 677
Guicheney, P., 476
Guillain, F., 309
Guillemin, R., 573, 576, 620, 633
Guilleux, J.-C., 412
Guimaraes-Motta, H., 302
Guimaraes-Mutta, H., 302, 303
Guisan, Y. J., 39
Gulik-Krzywichi, T., 470
Gunderson, C. B. Jr., 463
Gunn, D. L., 525
Gunnar, R. M., 487
Gunsalus, R. P., 628, 629
Guppy, M., 122
Gurd, F. N., 39
Guth, P. G., 72
GUTH, P. H., 3-12; 4-9, 72
Guy, M. N., 152
Guyda, H. J., 590

H

Ha, D. B., 338-40, 345
Haas, H. G., 390, 438, 440, 442
Haas, J. A., 203-6
Haase, W., 173, 191
Habal, M. B., 87
Habener, J. F., 625, 630
Haber, E., 242, 245, 476
Habersetzer-Rochat, C., 468
Habliston, D., 229-31
Habliston, D. L., 230
Haddy, F. S., 14
Haga, N., 522
Hagege, J., 170
Hageman, G. R., 486
Hagiwara, S., 361, 364, 426, 428, 429
Haglund, J., 38
Haiaml, L., 625
Haiech, J., 404, 405, 408-10, 670

# AUTHOR INDEX 693

Hails, C. J., 109, 117
Hainsworth, F. R., 109, 114-17
Haist, R. E., 630
Hakim, T. S., 274
Haldar, J., 577
Hales, C. N., 630
Hales, J. R. S., 89, 90
Haley, J., 633, 634
Hall, C. S. G., 609
Hall, D. A., 162, 213
Hall, E. M., 535, 536, 539, 546
Hall, M., 628
Hall, Z. W., 320, 323, 325, 327, 328
Hallback, D., 51
Halmi, K. A., 597
Halpin, R. A., 337-39
Halushka, P. V., 260, 261
Hamasaki, Y., 260, 265
Hamberg, M., 257, 260
Hamburger, R. J., 205
Hamilton, C. A., 476
Hamilton, J. W., 633
Hamilton, W. D., 616, 623
Hammarström, S., 257, 259, 260
Hammat, M. T., 246, 248
Hammerschmidt, D. E., 289
Hammon, K. J., 264
Hammond, G. L., 269, 275-77
Hamsen, H. S., 263
Hamwood, S. M., 606
Hanbauer, I., 674, 675
Hancock, A. A., 475, 477
Hanegan, J. L., 133, 138
Hanley, M. J., 209
Hannan, T. J., 573-75
Hansen, G. P., 168, 170
Hansen, J. L., 340
Hansen, J. W., 606
Hansen, O., 381, 382
Hansen Bay, C. M., 360
Hansma, H. G., 522, 525, 526, 530
Hansmann, M., 588, 590
Hanson, K., 36
Hanson, K. M., 14, 15, 23, 24, 57, 58, 62-66
Hara, K., 545
Hara, N., 260
Hara, R., 520
Harada, N., 467
Harayama, S., 504, 513
Hardebo, J. E., 273, 274
Harden, T. K., 475, 479, 655
Harken, A., 490
Harkin, M. M., 286
Harmanci, M. C., 154, 172
Harms, P. G., 583, 590
Harper, A. A., 8, 9
Harper, S. C., 667
Harris, A. J., 323
Harris, E., 264

Harris, J. B., 369
Harris, J. O., 275, 276
Harris, J. R., 346
Harris, L., 402
Harris, R. B., 244
Harris, R. H., 204
Harrison, L. C., 639, 642, 643
Harrison, S. C., 299, 311
Harrison, T. M., 625
Harrow, J. A. C., 415, 416
Hart, J. S., 109, 110, 113, 115
Hart, M. A., 230, 234, 235
Hartig, P. R., 505
Hartley, J. L., 244, 246
Hartman, B. K., 274
Hartman, P. E., 501, 504, 513
Hartzell, H. C., 322
Harvath, L., 555
Harvey, A. L., 319
Haslam, R. J., 413
Hasselbach, W., 298-302, 304, 306, 307, 311, 337
Hastings, D., 377
Hathaway, D. R., 655, 677
Hauger, R. L., 590
Haupert, G. T. Jr., 383
Hauser, D. C. R., 519, 531
Hausler, A., 585, 587, 590, 605
Hauswirth, O., 485
Havrankova, J., 620
Hawkins, H., 261
Hay, D., 152
Hayashi, H., 505, 506, 513
Hayashi, S., 6
Hayes, A., 524, 669
Haynes, D. H., 304
Hays, R. M., 158, 576, 577
Hayslett, J. P., 171, 194, 196
Hayward, J. N., 86-88
Hazelbauer, G. L., 501, 503, 504, 506, 511, 513
HEATH, J. E., 133-43; 133, 134, 138
HEATH, M. S., 133-43
Hebb, C. O., 269
Hebdon, G. M., 300, 301, 310
Heber, D., 585
Hebert, S. C., 162, 191, 192, 213
Hecht, H. H., 391
Hechtman, H. B., 261, 273, 277
Hedberg, A., 475, 476, 478
Hedblom, M. L., 504, 505
Heding, L., 641
Hedqvist, P., 259
Heger, J. J., 485
HEGLUND, N. C., 97-107; 97-101, 103-5
Hegstrand, L. R., 476, 477
Hegyvary, C., 378, 381
Heidenrich, K. A., 657
Heilmann, C., 347, 348

Hein, P. R., 586
Heinemann, S., 327, 328
Heinrich, B., 133-35, 137-40
Heinrikson, R. L., 629
Heinz, N., 381
Heistad, D. D., 90
Helle, K. B., 477
Heller, P. F., 305
Hellman, L., 590, 599, 600
Hellman, R. N., 376
Helman, S., 212
Helman, S. I., 195-97, 209
Helmond, F. A., 586
Henderson, E. J., 535, 536, 538, 545
Hendricks, S. A., 644
Hendy, G. N., 657
Hennessey, T., 520, 526, 530
Henrich, H., 21, 51, 72
Henry, D. P., 480
Henson, J., 289
Henson, P. M., 284, 288, 289
Hepburn, H. R., 134, 139
Heppner, F. H., 115
Heppner, R., 459
Heptinstall, R. H., 211
Herbert, E., 618
Herbette, L., 299, 311, 344
Herdey, A., 426, 427, 429
Herman, I. M., 677
Herman, R., 80
Hermann, H., 597
Herold, F. M., 509, 512
Herreid, C. F. II, 134
Herrmann, T. R., 298
Herron, M. J., 558
Hersch, M. I., 134, 139
Hershman, J. M., 602
Herz, A., 572
Herzmann, J., 606
Hesketh, T. R., 337, 341, 347
Hess, A., 368
Hess, B., 537, 544
Hess, D. L., 584, 586-88, 590, 602
Hess, M. W., 287
Heuser, J. E., 320, 324
Hicks, M. J., 405, 408-11
Hidaka, H., 469, 673
Hidalgo, C., 301, 310, 337, 341
Hierholzer, K., 188, 195, 196
Higashihara, E., 208, 210
Higdon, J. J. L., 115
High, C. W., 677
Hildebrand, E., 526
Hill, A. V., 97, 101, 103
Hill, H. R., 560
Hill, J. H., 285, 286
Hill, R. L., 642
Hille, B., 357-61, 426, 530
Hilmen, M., 501, 502, 504, 509
Hinds, T. R., 669, 675
Hino, N., 426, 428, 431

Hinshaw, L. B., 13
Hintermann, R., 544
Hintz, R. L., 643, 646
Hirabayashi, T., 524
Hirai, M., 381
Hiraoka, M., 391, 393
Hirata, F., 542, 561, 562, 564, 565, 656
Hirsch, J. G., 555
Hirsch, L. J., 58, 62, 63
Hirschmann, R., 264
Hiwatashi, K., 522
Hizuka, N., 642
Hjollund, E., 642
Ho, H. C., 668
Hoa, D. H., 654
Hoar, P. E., 677
Hobart, P., 633
Hobart, P. M., 635
Hobbs, A. S., 344, 402
Hobson, A. C., 508, 510
Hobson, W. C., 598
Hockel, G. M., 480
Hocking, B., 139
Hodess, A. B., 487
Hodgen, G. D., 589, 590
Hodgkin, A., 453
Hodgkin, A. L., 357-59, 361, 363, 364, 436, 439
Hodgkin, D. C., 622
Hoedemaeker, P. J., 248
HOFFMAN, B. B., 475-84; 476, 477, 479, 655, 662
Hoffman, B. F., 485-89
Hoffman, H. J., 606
Hoffman, J. F., 377
Hoffman, L., 329
Hoffmann, F., 589
Hoffmann, G. J., 328
Hoffmann, J.-Y., 377
Hoffmann, W., 310
Hoffsommer, R. D., 264
Hoffstein, S., 288
Hofler, J. G., 320, 321
Hofmann, T., 646
Hofnung, M., 628
Hogan, P. M., 486
Hogan, W. C., 376
Hogben, C. A. M., 76
Hogg, R. W., 509
Hohlweg, W., 600
Holcroft, S. W., 80
Holland, P. C., 337, 338, 340, 344, 345
Hollenberg, N. K., 247, 248
Holliger, C., 49, 50
Holmberg, C., 183, 185-87
Holm-Jensen, I., 134
Holmsen, H., 677
Holroyd, J. A., 299, 300
Holroyde, M. J., 458, 460
Hom, P. M., 437, 438, 441

Homcy, C. J., 476
Homma, M., 648
Hondeghem, L. M., 426, 428, 487, 488
Honerjäger, P., 466, 467
Hood, L. E., 319, 321, 633
Hooper, R. J. L., 597
Hoorntje, S. J., 248
Hoover, R. L., 152, 173
Hope, R. R., 485, 486, 492
Hopkinson, C. R. N., 583
Horackova, M., 426, 427, 436, 442, 466, 467
Hordof, A. J., 485, 487, 488
Horejsi, J., 606
Hori, H., 310
Horner, J. A., 643
Horowicz, P., 358, 360, 363, 364, 453, 456
Horowitz, L. N., 486, 487, 490, 492
Horres, C. R., 393
Horster, M., 173
Horwitz, A. F., 341
Hostetter, G., 577, 578
Hotchkiss, J., 585, 586
Hotta, Y., 467
Houglum, J., 259
Hourihan, J., 597
Houslay, M. D., 337, 341, 347
Houvenaghel, A., 34
Howard, B. D., 463
Howard, F. M., 328
Howlett, A. C., 661, 662
Hreno, A., 39
Hsieh, C. P., 35
Hsu, K. C., 264
Huang, C., 455
Huang, W., 381
Huang, W.-Y., 633
Hubbard, J. I., 325
Hubel, K. A., 51
Hudson, A. J., 642
Hudson, D. M., 93, 109, 110
Hudson, P., 633, 634
Hudson, T. J., 659
Huerta, M., 368
Huesgen, A., 535, 539, 542
Huet, C., 329
Hughes, B. P., 674, 676
Hughes, G. M., 133, 134
Hughes, J., 227, 270, 274, 571-73
Hughes, M., 468
Hughes, R. L., 59
Hugli, T. E., 284, 289, 555
Hugues, M., 468
Hui, C. S., 368, 369, 455
Hui, C.-W., 402, 413-16
Hülser, D., 536, 539, 542, 543
Hülsmann, W. C., 378

Hultberg, E., 476
Hulten, L., 34
Humbel, R. E., 645, 646
Humbert, F., 154, 168
Hume, J. R., 426
Hummel, D., 115
Humphrey, C. S., 50
Hung, L. T., 242
Hunkapiller, M. W., 319, 321, 633
Huppler, R. E., 572
Huseman, C. A., 602
Huser, J., 152, 170
Hussain, M. N., 276
Hussell, D. J. T., 109
Husson, H. P., 470
Hutchings, A., 231
Hutchison, J. S., 585, 587-90, 605
Hutson, N. G., 667
Hutter, O. F., 362, 364, 365
Huxtable, R. J., 275, 276
Hyman, A. L., 274
Hynes, R. A., 45

I

Ichida, S., 670
Ichikawa, I., 152
Ichikawa, Y., 648
Ierokomas, A., 299
Iino, T., 501, 502
Iino, Y., 195
Ikeda, Y., 32
Ikeler, T. J., 264
IKEMOTO, N., 297-317; 300, 301, 304-7, 310, 337, 341
Ikushima, S., 469
Ildefonse, M., 358, 360
Ilsley, S., 633
Ilundain, A., 673
Imae, Y., 501, 513
Imai, M., 183, 184, 186-89, 191-93, 195, 213
Imanishi, S., 487
Imbembo, A., 8
Imbert, M., 162, 205, 206, 212, 213
Imbert-Teboul, M., 162, 189, 192-94, 197, 212
Imura, H., 605
Ineichen, H., 347
Inesi, G., 298-300, 302, 303, 305, 307, 310, 337, 339, 451
Ingebretsen, W. R. Jr., 412
Inouye, M., 626, 629
Insel, P. A., 478
Intaglietta, M., 23
Inui, M., 402, 404-11
Irisawa, H., 390, 397, 425-31, 487

AUTHOR INDEX 695

Iriuchijima, J., 480
Irwin, R. I., 188
Isaksson, O., 647
Isenberg, G., 391, 395-97, 426-30
Ishi, K., 6
Ishiguro, K., 524
Ishikawa, H., 338, 349
Ito, K., 628
Ito, S., 391, 394
Iturizaga, M., 31, 36, 37
Iversen, L. L., 572
Iwasa, J., 467
Iwasa, Y., 669
Iwasawa, Y., 270, 275
Iyengar, R., 658, 660, 662
Izumi, T., 468, 469

J

Jacini, P., 103
Jack, D., 476
Jackson, B. A., 197
Jackson, R. C., 629
Jackson, R. L., 668, 669
Jacob, H. S., 289
Jacob, M., 322
Jacobson, A., 536
Jacobson, A. L., 415
JACOBSON, E. D., 71-82; 7, 8, 10, 15, 19, 21-23, 29-32, 34, 35, 71, 74, 76, 77, 79, 80
Jacobson, H. R., 182-88, 193
Jacobson, R. I., 605
Jacques, L., 323, 324
Jacques, Y., 466, 470
Jaffe, E. A., 260
Jaffe, L. F., 347
Jaffe, R. B., 586, 587
Jahn, T. L., 525
Jahr, H., 632
Jaimovich, E., 358, 360
Jakábová, M., 413
Jalife, J., 486, 488, 490, 491
Jamieson, G. A., 669
Jamieson, G. A. Jr., 524
Jamieson, J. D., 630
Jamison, K. R., 602
Jamison, R., 189, 210
Jamison, R. L., 203-5, 208, 210
Jandali, M. K., 578
Janoff, A., 287
Janse, M. J., 485, 488, 490
Jansson, G., 4, 78
Jarrett, D. B., 640, 642
Jastorff, B., 537, 545
Jay, E. W. K., 466
Jayes, A. S., 97, 99, 101
Jeffery, D. R., 659, 664

Jennings, H. S., 519, 520, 531
Jensen, R. A., 573-75
Jensen, T. F., 134
Jessell, T. M., 319, 324
Jessen, C., 88, 89
Jewelewicz, R., 585
Jewell, B. R., 430
Jewell, P. A., 86
Jilka, R. L., 337, 339, 341, 344
Jodal, M., 34, 50, 51
Johansen, K., 93
Johanson, A. J., 590
Johansson, B., 23, 24
Johansson, E. D. B., 586
Johansson, S., 413
John, M. J., 622
Johnson, A. K., 576
Johnson, A. R., 241, 249, 264
Johnson, B. J., 34
Johnson, D. K., 590
Johnson, E. A., 393, 397, 489
Johnson, E. M. Jr., 480, 655
Johnson, G. L., 478, 659
Johnson, J. D., 668
Johnson, K. J., 285, 286, 291
Johnson, L., 597
Johnson, L. M., 583
Johnson, M. M., 324
Johnson, P. C., 13-15, 23-25, 51, 57, 58, 72
Johnson, R. G., 46, 51
Johnson, V., 154
Johnston, C. I., 246, 248, 480
Johnston, P. A., 210
Jolesz, F., 347, 348
Jones, C. M., 262
Jones, D. J., 125
Jones, D. R., 109
Jones, G. S., 583, 590
Jones, J., 172
Jones, L. R., 402, 404, 412-15, 476
Jones, M. L., 285
Jordan, K., 47, 50, 52
Jorgensen, A. O., 338, 343, 459
Jorgensen, P. L., 378
Jornvall, H., 263, 633
Jorpes, J. W., 633
Jose, M. J. L., 676, 678
Josephson, M. E., 486, 490, 492
Josephson, R. K., 133
Joshua, H., 264
Juan, H., 52
Jubb, S. L., 577, 578
Jublz, W., 260
Judah, J. D., 630, 632
Judd, H., 590
Judd, H. L., 599
Judy, W. V., 480
Juliani, M. H., 538
Jundt, H., 445

Junod, A. F., 227, 262, 269, 270, 273, 274
Jutsum, A. R., 139

K

Kachadorian, W. A., 158, 172
Kachelhoffer, J., 31, 36, 37
Kadoma, M., 402, 404, 405, 407-9, 411, 412
Kadziela, W., 88, 90
Kahane, I., 558
Kahn, C. R., 639-44, 646
Kahn, J. R., 229, 243, 244, 246
Kaiser, G. A., 492
Kaissling, B., 147, 152, 158, 159, 162, 165, 167, 168, 170
Kaissling, K. E., 527
Kakebeeke, P. I. J., 536, 539, 546
Kakisawa, H., 466
Kakiuchi, S., 404, 405, 408, 409, 670, 676
Kalbitzer, H. R., 306
Kallenberg, C. G., 248
Kalnins, V., 343, 459
Kalnins, V. I., 338, 343
Kameyama, T., 345
Kammer, A. E., 133, 134, 137-40
Kamo, N., 531
Kampmann, W., 394
Kampp, M., 4
Kan, W. H., 479, 480
Kanazawa, T., 304-6
Kanazawa, Y., 631
Kaneko, H., 523
Kaneko, M., 99-101
Kang, A. H., 285, 553
Kang, K., 378
Kaniike, K., 404, 405, 412, 459
Kanwar, Y. S., 150
Kao, I., 328
Kao, W., 466
Kapen, S., 590, 599, 600
Kaplan, J., 286, 291
Kaplan, N. O., 412
Kaplan, S. A., 644
Kaplan, S. L., 590, 595-98, 600-2, 604, 607, 609
Kapp, B. S., 572, 573
Karagueuzian, H. S., 492
Karikas, G. A., 378
Karis, J., 486
Kariya, T., 648
Karli, J. N., 378
Karlin, A., 319
Karliner, J. S., 476
Karlish, S. J. D., 301, 307, 311, 375
Karnaky, K. J. Jr., 161

AUTHOR INDEX

Karnovsky, M. J., 152, 158, 160, 161, 173
Karsch, F. J., 584-86, 590
Kasai, M., 453
Kaschnitz, R., 625, 629
Kasell, J., 491
Käser-Glanzmann, R., 413
Kashgarian, M., 171, 196
Kaslow, H. R., 664
Kass, R. S., 426-28, 430, 431, 488
Kastin, A. J., 571, 574, 575
Kastor, J. A., 486, 492
Katongole, C. B., 583
Katz, A. I., 171, 184, 190, 191, 196
KATZ, A. M., 401-23; 401-6, 408-12, 415, 451, 460
Katz, B., 357, 361, 363
Katz, J., 590
Katz, L. N., 485
Katz, M. L., 60
Katz, S., 408, 409
Katzung, B. G., 426, 428, 487, 488
Kau, S. T., 154
Kauffman, G. L., 9
Kaufman, J., 626
Kaufmann, R., 426, 428
Kawakita, M., 301
Kawamura, S., 183, 184, 186-88
Kayman, S. C., 541
Kaziro, Y., 301
Kazlatchkine, M. D., 285
Kazumoto, F., 31, 36
Kebabian, J. W., 655
Keim, P., 629
Kelch, R. P., 585, 590, 595, 597, 600, 602, 604, 605
Kellaway, C. H., 259
Keller, H. U., 287
Kelley, P. R., 633
Kellogg, E. W., 286
Kelly, A. M., 338
Kemmler, W., 630, 632
Kemp, S. F., 646
Kemper, B., 625, 629
Kendeigh, S. C., 109, 112, 113
Kendrick-Jones, J., 677
Kennett, F. F., 402
Kent, G. N., 633
Kent, R. S., 479
Kenyon, J. L., 426, 430
Keogh, E., 584
Keogh, E. J., 584, 585, 587, 588, 602, 605
Keppens, S., 667
Ker, R. F., 97, 99, 101
Kerins, M. E., 648
Kern, R., 438, 440, 442
Kerr, J. C., 6, 8
Kerrick, G., 451, 452

Kerrick, W. G. L., 677
Kershaw, G. R., 248
Kessin, R. H., 545
Kessler, H. B., 286
Kessler, M., 440
Kessner, A., 402
Kettyle, W. M., 215
Kety, S. S., 75
Keye, W. R. Jr., 587
Khairallah, P. A., 245, 249
Khan, M. A., 139
Khan, S., 504, 508, 509
Khanna, J. M., 655
Khodorov, B. I., 466
Khodr, G. S., 620
Khosla, M. C., 244
Kidokoro, Y., 324
Kidwai, J. R., 632
Kihara, M., 501, 511
Kilgore, D. L., 93
Kim, D.-H., 376, 377, 379, 381
Kim, D. K., 58
Kimelberg, H. K., 378
Kimura, I., 413
Kimura, M., 413
Kimura, S., 633
Kindermann, W., 90
King, A. C., 545
King, M. G., 574, 575
King, T. E., 289
Kinne, D. W., 58
Kinne, R., 173, 191
Kinne, R. K. H., 191
Kinoshita, N., 401, 402, 406, 409
Kinsolving, C. R., 375
Kinter, L., 175
Kiode, S. S., 617
Kirchberger, M. A., 403-6, 408, 409, 411, 412, 460
Kirsch, G. E., 362
Kirschke, H., 632
Kirtland, W. H., 90
Kishihara, M., 648
Kitamura, S., 264
Kitazawa, T., 452, 459
Kitchell, J. F., 126
Klahr, S., 376
Klare, J., 391
Klebanoff, S. J., 286
Kleber, A. G., 485, 488
Klee, C. B., 416, 668
Kleene, S. J., 505, 510
Klein, C., 535, 537, 538, 544, 546
Klein, L. S., 262
Klein, M. N., 644
Klein, R. L., 346
Kleinman, H. K., 553
Kleinzeller, A., 172
Klibanski, A., 597
Klip, A., 298-300, 343

Klöckner, U., 426-28
Klose, R. M., 171
Kloss, S., 204
Kluger, M. J., 89, 90
Klymkowsky, M. W., 320, 324
Knepper, M., 184, 189-91, 193, 196
Knight, A. B., 646
KNOBIL, E., 583-93; 583-90, 597, 598, 602, 603, 605, 606
Knoblauch, M., 49, 50
Knoebel, S. B., 488
Knopf, K., 134
Knowles, A., 338
Knowles, A. F., 302
Knox, F. G., 193, 203-6
Kobatake, Y., 521, 530, 531
Kobayashi, M., 643
Kobayashi, N., 648
Kobayashi, S., 413, 467
Koch, G. L. E., 676
Kociolek, K., 655, 656
Koda, L. Y., 575
Koeppen, B. M., 196, 197
Koepsell, H., 158, 159, 212, 213
Koering, M. J., 586
Koerner, T., 242, 245
Kohatsu, S., 39
Kohlhardt, M., 426, 427, 429
Kohnert, K.-D., 632
Koide, Y., 412
Koiwai, O., 505, 506, 513
Koketsu, K., 393, 394, 398
Kokko, J. P., 162, 183-93, 195, 197, 204, 208-10, 212
Kokubun, S., 426-28, 430, 487
Kolassa, N., 302, 306
Kolin, A., 73
Koller, H., 435
Kolodej, A., 40
Komeda, Y., 501, 504
Kondoh, H., 504
Konijn, T. M., 535-37, 539, 540, 543, 545, 546
Kono, T., 676
Konrad, C., 475
Konrad, M., 626
Konturek, S. J., 8
Koo, A., 62
KOOB, G. F., 571-82; 574-77
Kootsey, J. M., 397, 489
Koppel, D. E., 322
Kopriwa, B. M., 576
Korbut, R., 259, 262
Korn, E. D., 541
Korner, H., 658
Korner, P., 475
Kort, E. M., 502, 512
Kosaha, K., 631
Koshland, D. E., 501, 503-10, 512

AUTHOR INDEX 697

Koshland, D. E. Jr., 554, 560, 562
Kosk-Kosicka, D., 339, 341, 342, 345
Kosterlitz, H. W., 572
Kot, P. A., 262
Kotake, H., 426-31
Kotani, M., 648
Kotchen, T. A., 152
Kountz, S., 59
Kovacs, G. L., 573, 574, 577
Kovács, L., 361, 455, 457
Kowalewski, K., 40
Kozima, T., 466
Kozlowski, G. P., 577, 578
Kozuka, M., 506
Krail, G., 622
Kramer, J. M., 635
Kranias, E. G., 406-10
Krans, M. J., 653
Krarup, N., 59
Krasne, S., 364
Krause, E.-G., 413, 416, 417
Krause, H., 426, 427, 429
Krausz, M. M., 261
Krausz, Y., 673
Kravietz, W., 475
Krebs, E. G., 401
Kredich, N. M., 561, 562
Kreil, G., 625, 626, 629
Kreisberg, J. I., 152, 172, 173
Krejs, G., 50
Krejs, G. J., 49, 50
Krens, F. A., 536, 540
Kretsinger, R. H., 668, 669
Kreutzer, D. L., 285, 289
Krey, L. C., 584-87
Krieger, D. T., 616, 620, 623, 635
Krieger, E. M., 248
Krikos, A., 504, 507, 513
Kriz, W., 147, 152, 158, 159, 162, 165, 167, 168, 170, 203, 212, 213
Kroeger, H., 347
Krogh, A., 15
Krol, R., 8
Kromann, H., 642
Kronenberg, R. S., 289
Krönert, H., 90
Ku, D. D., 382, 467
Kubler, M., 426, 427, 429
Kubota, T., 184
Kuczmarski, E. R., 541
Kuehl, F. A., 261
Kuffler, S. W., 368
Kugler, J. A., 602
Kuhlmann, J., 376
Kuhn, P., 90
Kuida, M., 260
Kukiuchi, S., 668
Kulczycki, A., 283
Kulin, H. E., 599, 602, 606

Kuma, K., 648
Kumagai, H., 524
Kumagai, S., 565
Kumamoto, K., 264
Kunau, R. T. Jr., 208, 211
KUNG, C., 519-34; 519-31
Kung, L. S., 655
Kunkel, R. G., 291
KUNKEL, S. L., 283-93; 291
Kunos, G., 479, 480, 655
Kunze, D. L., 394
Kuo, J. F., 478-80
Kupchan, S. M., 465
Kurihara, K., 521, 530, 531
Kurihara, S., 456-58
Kuriki, Y., 306
Kurobe, Y., 307
Kurokawa, K., 668, 672
Kurzmack, M., 300, 305, 307, 310
Kusano, M., 6
Kushmerick, M. J., 103
Kutty, M. N., 121
Kuzuya, T., 401, 404-10
Kvietys, P. R., 18, 19, 21, 23, 30, 32, 33, 36, 37, 43, 45, 48, 49, 51, 58-60, 78
Kwauk, S., 633
Kwok, S. C. M., 617, 635
Kyi, J. K. J., 39, 40

L

Labrecque, A. D., 629
Laburthe, M., 654, 657, 658, 662
Lacour, J. R., 89
Lacy, F., 189, 210
Lacy, F. B., 210
Lacy, P. E., 670, 675
Lad, P. M., 661
Lagace, L., 677
Laidlaw, Z., 15, 23
Laks, H., 276
Lambert, A. B., 109
Lambert, E. H., 328
Lameire, N. H., 203, 207
Lamers, J. M. J., 378, 404, 405, 413, 414, 416, 417
Lamprecht, J., 322
Lancaster, M. C., 85
Landas, S., 576
Lande, S., 578
Landgraf, W. C., 310
Landick, R., 628, 629
Landon, J., 631, 633, 634
Lands, A. M., 476, 655
Lands, W. E., 257
Landt, M., 670, 675
Lane, L. K., 375, 381, 404, 405, 412, 459
Lane, R. S., 22

Lang, F., 203-5
Lang, U., 597
Langendorf, R., 486
LANGER, G. A., 435-49; 379, 415, 431, 435, 437, 438, 440-42, 444-46
Langer, S. Z., 476
Lantz, B. M. T., 80
Lantz, C. H., 242, 245
Lanzillo, J. J., 246
Laporte, D. C., 673
Laragh, J. H., 246, 248, 250, 263, 264
LaRaia, P. J., 403, 404
LaRochelle, F. T. Jr., 212
Larochelle, J., 109-12
Larsen, F. L., 669, 675
Larsen, G. L., 289
Larsen, J. A., 59
Larsen, P. R., 648
Larsen, S. H., 502, 509
Larson, B., 111, 114
Larson, R. E., 103
Larsson, C., 211
Larsson, L.-I., 633
Lasker, R., 664
Lasserre, P., 376
Laszlo, D. J., 501, 512, 513
Lattimer, N., 262
Lau, Y. H., 477
Lau, Y. S., 669, 674, 675
Laudat, M. H., 639
Laudat, P., 639
Lautt, W. W., 58, 60, 61, 63
Lawrence, R. A., 286
Lawson, A. E., 62
LAZDUNSKI, M., 463-73; 305, 442, 464-70
Lazo, J. S., 275, 276
Lazure, C., 619
Lazzara, R., 485, 486, 492
Leary, W. P., 241-43, 248
LeBel, D., 298
Lebovitz, R. M., 657
Lebovitz, E. A., 676
LeBurton, G. C., 672
Lechene, C., 195, 196, 213
Lechene, C. P., 205
Lederer, W. J., 391, 392, 394-96, 398
Ledingham, J. G., 241-43, 248
Lee, A. G., 297, 530
Lee, B. C., 274
Lee, C. O., 440, 441
Lee, C. Y., 470
Lee, E. W., 426
Lee, J. B., 261
Lee, J. S., 48
Lee, K. S., 426
Lee, N., 629
Lee, T. L., 501, 504, 513
Leffert, H. L., 347

AUTHOR INDEX

LEFKOWITZ, R. J., 475-84; 475-80, 553, 556, 655, 656, 661, 662
Legant, P. M., 245, 251
Le Grimellec, C., 204
Legros, J. J., 574, 578
LeGrumellec, C., 174
Lehouelleur, J., 368
Leichtling, B. H., 478
Leigh, J. B., 309
Le Magnen, J., 527
LeMaire, M., 299, 309, 310
Le Moal, M., 574-77
Lenko, H. L., 597
Lennarz, W. J., 321
Lennon, V. A., 328
Lenoir, M-C., 442, 467
Lentz, K. E., 229, 243, 244, 246
Lentz, T. L., 322
Léoety, C., 360, 362, 363
Leonard, A. S., 4
Leonard, E. J., 555
Leone, R. M., 597
Léoty, C., 360, 362
Le Peuch, C. J., 346, 404, 405, 408-10, 412, 670
Le Peuch, D. A. M., 404, 405, 409
Lepow, I. H., 284, 288
Lequin, R. M., 590
Lernmark, Å., 630
Le Roith, D., 616, 617, 623
Lesk, A. M., 628
Lesniak, M. A., 616, 617, 620, 623, 639, 641, 642, 647
Leuenberger, P., 245, 251, 264
Leuenberger, P. J., 245, 251
Leung, N. L.-K., 415, 416
Levandowsky, M., 519, 531
Levchenko, T. S., 404, 413, 414
Leveille, G. A., 376, 378
Levey, G., 647
Levin, A. E., 524
Levin, R. M., 673
Levin, S. R., 602
Levine, E. A., 286
Levine, M., 229, 243, 244, 246
Levine, M. A., 661, 663, 664
Levine, S. D., 158
Levinsky, N. G., 209
Levinson, S. R., 360, 362
Levis, G. M., 378
Levitsky, D., 348
Levitsky, D. O., 404, 413, 414
Levitt, D. G., 75
Levitt, M. D., 30, 32, 49, 75, 77
Levitt, T. A., 327
Levy, L. R., 284
Levy-Marchal, C., 476
Lew, V. L., 531
Lewis, D. E., 300, 307, 310

Lewis, D. H., 271
Lewis, M., 616, 623
Lewis, R. A., 259
Lewis, R. V., 633
Lewis, T., 485
Lewis, U. J., 633, 634
Leyendecker, G., 586, 588, 590
Lezzi, M., 347
Li, C. H., 573
Li, H.-C., 409
Liang, D. C., 573
Liang, I. Y. S., 62
Liang, K. C., 573
Libby, P., 326, 329
Lichtenstein, L. M., 283
Lieberman, M., 393, 489
Liebman, J. E., 274
Lifschitz, M. D., 152, 203, 207
Lifson, N., 37, 47, 48
Lighthill, M. J., 109, 113, 115
Lilja, B., 271
Limas, C., 479
Limas, C. J., 412, 416, 479, 480, 655
Limbird, L. E., 479, 656, 661
Lin, B. J., 630
Lin, C. H., 642
Lin, C. T., 675
Lin, M. C., 653
Lin, M. H., 376, 378
Lin, Y. M., 668
Lind, K. E., 299
Lindall, A. W., 630
Lindberg, M., 599
Lindell, S. E., 271
Lindemann, J. P., 412
Linden, C. D., 673
Linden, D. C., 325
Linden, G., 468
Lindenmayer, G. E., 402, 415, 444
Lindsay, D. B., 61
Lindsay, D. G., 622
Lindsey, T. T., 633
Lindstrom, J., 319, 328
Linehan, J. H., 274, 275
Linford, R. H., 7, 8, 10, 76
Ling, K.-Y., 519, 522, 525, 527, 531
Ling, N., 573, 576, 620, 633
Lingham, R. B., 376
Link, D. P., 80
Linker, A., 150
Linzell, J. L., 61
Liotta, A. S., 616, 620, 623
Lioy, F., 250
Lippe, B. M., 644
Lipset, J. S., 245, 251, 263, 264
Lissaman, P. B. S., 115, 116
Lissitzky, S., 468, 469
Liu, S. C., 299, 311
Liu, S. K., 487, 488
Liu, Y. P., 668

Livingston, J. N., 642
Lloyd, B. L., 376, 382
Lo, M. M. S., 322
LoBuglio, A. F., 286
Locher, R., 646
Locker, G. Y., 286
Lockwood, D. H., 642
Lodish, H. F., 536
Loeb, H. S., 487
Loewenstain, W. R., 522
Lofstrom, J. B., 271
Loh, H. H., 573
Lohr, N. S., 264
Lombet, A., 464-66, 469, 470
Lomedico, P. T., 628
Lømo, T., 323, 325
Long, D., 4
Lonigro, A. J., 261
Lonnerholm, G., 171
Loo, D. D. F., 365
Loomis, W. F., 536
Looney, T. R., 104
Lopaschuk, G., 408, 409
Lord, J. A. H., 572
Loriaux, D. L., 585, 606, 609
Loring, R. H., 327
Losada, A., 543
Lough, J. W., 340
Louis, C. F., 299, 300, 405, 409
Louis, K. M., 586
Louw, G., 88
Louw, G. N., 133
Love, M. M., 286
Lovelace, D. E., 488
Lovenberg, W., 674, 675
Lowry, P. J., 631, 633, 634
Lowry, S. F., 9, 10
Lubbe, W. F., 485
Lucchesi, B. R., 382
Lucci, M. S., 193
Lucky, A. W., 599
Lüderitz, B., 394
Ludford, J. M., 655
Luduena, F. P., 476, 655
Ludwig, P., 341
Luff, A. R., 339, 348
Luft, F. C., 152
Lugo, J. E., 250
Lugtenburg, C., 588, 590
Lüllmann, H., 377
Lundgren, O., 4, 34, 38, 50, 51, 74, 75, 78
Lundholm, L., 413
Lüscher, E. F., 413
Lüttgau, H. C., 361, 390, 435
Lutz, J., 21
Lutz, M., 183
Lyman, C. P., 89
Lynch, C., 361, 363
Lynch, G., 671
Lynham, J. A., 413
Lynn, W. S., 288, 555

AUTHOR INDEX   699

M

Maack, T., 154
Macchia, E., 648
MacFerran, S. N., 49, 50, 52
MacGregor, R. R., 632, 633
Machemer, H., 519-24, 528, 529
Maciag, T., 633
MacIntyre, I., 619
Mackay, S. A., 542, 544
MacLachlan, T. L., 60, 63
MacLennan, D. H., 298-300, 337, 338, 340, 342-45, 459
MacLeod, D. P., 431
MacLeod, P., 527
Macnab, R. M., 501-4, 508, 509, 511, 554, 562
Macus, A. J., 260
MacVaugh, H., 487
Maddrell, S. H. P., 133
Madri, J. A., 150
Maeda, K., 501, 513
Maeda, T., 629
Maeda, Y., 539, 544
Maeno, H., 413
Magaldi, J. B., 204
Magilton, J. H., 87
Magnuson, J. J., 126, 129
Maguire, M. E., 657, 661
Mahaffee, D. D., 660, 664
Mahaffey, J. E., 656
Mahoney, D., 378, 381
Maihle, N. J., 524
MAILMAN, D., 43-55; 15, 21-23, 44, 45, 47-53, 78, 79
Main, I. H. M., 7-10
Mains, R. E., 630, 631, 633
Majima, T., 526
Majno, G., 274
Major, R., 578
Makino, S., 133
Makinose, M., 300, 302, 304-6
Mako, M. E., 631, 643
Malaisse, W. J., 675
Malchow, D., 535-46, 553
Malech, H. L., 553, 559, 564
Malesci, A., 634
Malik, A. B., 274
Malik, K. U., 262
Malm, J. R., 487
Malmstroem, K., 415, 416
Malnic, G., 171
Maluiy, G. M. O., 98
Malsten, C., 257
Malven, P. V., 583
Malvin, R. L., 576
Mandel, F., 304, 307, 406-10
Mandel, G., 507
Mandel, S., 269, 275-77
Manders, W. T., 476
Mandrino, M., 358

Manganiello, V., 565
Mangiapane, M. L., 576
Manning, M., 575-77
Manson, M. D., 509, 512
Manter, J. T., 99, 100
Maratos-Flier, E., 383
Marban, E., 426, 429, 441, 457
March, C. M., 586
Marchand, G., 205
Marchand, G. R., 203, 204
Marchesi, V. T., 227, 627
Marcus, M. L., 90
Marescaux, J., 31, 36, 37
Marfat, A., 259
Margaria, R., 99, 100, 103
Margreth, A., 338, 339, 348
Marian, J., 675
Marin, G. K., 377
Marín-Cao, D., 541
Marinetti, G. V., 475, 654
Markelonis, G. J., 324
Marks, D., 626
Marks, N., 249
Marlborough, D. I., 229, 242, 243
Marley, P. B., 262
Marmé, D., 541, 670
Maron, M. B., 270, 274
Marquardt, J., 299, 311, 344
Marsh, A. J., 376, 382
Marsh, D., 189
Marsh, J., 625-27, 629, 630
Marsh, J. D., 475, 479
Marshall, G., 585, 587, 588, 590, 597, 598, 605, 606
Marshall, G. R., 588
Marshall, J. C., 585, 590, 597, 600, 605
Marshall, L. M., 323
Marshall, M. W., 359, 360, 369
Marshall, P. E., 584
Martial, J. A., 320
Martin, G. R., 553
Martin, J. B., 623, 635
Martin, L., 229, 231
Martin, L. C., 243, 245-47
Martin, O. B., 303
Martin, W. R., 572
Martinez, J. L., 573
Martinez, J. L. Jr., 573-76
Martinson, J., 4, 5, 78
MARTONOSI, A., 337-55; 299, 337-42, 344-48, 454
Martonosi, A. N., 297, 300, 303, 305, 310, 338, 339, 343-45, 347
Martonosi, M. A., 341
Maruo, T., 617
Maruta, H., 541
Marver, D., 183, 196
Marx, S. J., 664
Mary-Rabine, L., 485, 487, 488
Masiar, E., 383

Masiar, P., 383
Mas-Oliver, J., 415
Mason, J., 174
Mason, J. W., 536, 546
Masoro, E. J., 302
Mathews, M. B., 625
Mathie, R. T., 59
Mato, J. M., 535-37, 540-43, 545
Matsubara, T., 487
Matsui, K., 669
Matsukuma, S., 535
Matsukura, S., 648
Matsumura, P., 501, 502, 504
Matsuzaki, Y., 260, 261, 265
Matt, H., 523
Matthey, R. A., 289
Maunsbach, A. B., 152, 154
Maurer, S. C., 671, 675
Maxfield, F. R., 329
Maxwell, L. C., 77
May, B., 93
May, M. L., 133, 134, 137, 138
May, W., 275, 276
Maycock, A. L., 264
Mayer, S. E., 411, 412
Maylie, J., 426, 431
Maynard, J. R., 251
McAllister, R. E., 425
McArthur, J., 599, 606
McArthur, J. W., 585, 588, 590, 597, 605
McAuliff, J. P., 476, 655
McCaffrey, T. V., 92
McCarthy, K., 289
McClellan, G., 454, 458-60
McClenagham, M. D., 413
McConnaughey, M. M., 402, 404, 413-15, 476
McConnell, E., 134
McCook, R. D., 92
McCord, J. M., 286
McCormack, J. T., 584, 586, 587, 590
McCormick, J. R., 286, 291
McCullough, T. E., 414
McDaniel, M. L., 670, 675
McDevitt, D. G., 61
Mcdonald, J. M., 670, 675
MCDONALD, T. F., 425-34; 426-28, 431
McDonald, T. J., 263
McDougal, J. G., 242, 245
McGaugh, J. L., 573-75
McGiff, J. C., 258, 261, 262
McGill, K. A., 337, 341
McGrath, P., 86
McGuigan, J. A. S., 426, 430
McGuire, R. A., 172
McIntosh, C. H. S., 633
McIntosh, D. B., 304
McIntyre, J. O., 299
McIssac, D., 383

McKee, W. G., 542
McKeen, C. R., 261
McKenzie, J. M., 648
McKeown, J., 182-85
McKeown, J. W., 204, 206
McKinley, D., 452, 453
McKinney, G. R., 479
McKinney, T. D., 196
McKinstry, D. N., 248
McLarnon, J. G., 365
McLaughlin, P. J., 619
Mc Lean, M. J., 469
McLeod, R., 185
McMahan, U. J., 323, 328
McMahon, T. A., 97, 104, 105
McMartin, C., 631, 633, 634
McMillian, M., 675
McNamara, E. R., 171, 195, 207
McNamara, J. J., 36
McNaughton, P. A., 430
McNeil, L. W., 585
McNeill, J. R., 65
McNeilly, A. S., 590
McNish, R., 381
Meade, R. C., 49
MEANS, A. R., 667-82; 668, 669, 672-78
Mecham, R. P., 285
.Medawar, P., 621
Meech, R. W., 361, 429
Meehan, J. P., 92
Meggs, L., 247, 248
Megyesi, K., 639, 642, 646
Mehra, R., 488
Meissner, A., 39
Meissner, G., 297, 300, 304, 342, 452, 453, 459
Mellander, S., 23, 24, 62
Mellins, R. B., 245, 251, 263, 264
Mello-Aires, M., 171
Melmed, S., 602
Melmon, K. L., 478
Melton, T., 501, 504, 513
Mendel, D. B., 212
Mendez, B., 320
Mendez, C., 382, 485, 487
Menguy, R., 7
Mercola, D. A., 622
Merlie, J., 327, 328
Merlie, J. P., 320, 321, 326
Merrill, S., 14
Merrill, W. W., 289
Merritt, C. R., 375
Mescher, M. F., 676, 678
Messing, R. B., 573-75
Metcalfe, J. C., 297, 337, 341, 347
Mettler, R., 646
Mevarech, M., 633
Meyer, D. I., 627
Meyer, P., 378, 475, 476

Michael, J., 260
Michael, L. H., 376, 381
Michalak, M., 338-42, 345, 414
Michalowski, J., 31, 32, 34, 36
Michelson, E. L., 486, 492
Michelson, J. D., 328
Michie, D. D., 234
Michler, A., 327
Michoud, P., 156, 171, 195
Midgley, A. R. Jr., 586
Migala, A., 298, 300
Miledi, R., 368
Mill, P. J., 133, 134
Millar, J. A., 246, 248
Millar, R. P., 633
Miller, J. B., 508, 512
Miller, M., 211, 577
Miller, M. E., 87
Miller, P. D., 152
Miller, R. J., 572, 573
Miller, T., 78
Miller, T. A., 35
Mills, C., 114
Mills, R. G., 325
Milstein, C., 625
Minneman, K. P., 475-78
Minnick, S., 576
Minoshia, S., 513
Miranda, F., 468
Misanko, B. S., 213
Mishell, D. R. Jr., 586
Miskin, R., 328
Mitsui, T., 310
Mittra, I., 633
Miyamoto, E., 669
Miyamoto, H., 453
Miyazaki, S., 364
Miyomoto, E., 668
Mizonishi, T., 32
Mobley, B., 453
Model, P., 556, 629
Moe, B., 486
Moe, G. K., 486, 487, 490, 491
Moeglen, J. M., 574, 578
Mohme-Lundholm, E., 413
Mojarad, M., 260
Moldow, C. F., 289
Moler, T., 7
Moler, T. L., 6, 9, 72
Molinoff, P. B., 475-78, 655, 657
Molish, I. R., 677
Moll, D., 339
Moller, J. V., 299
Molloy, C., 625
Molski, T. F. P., 673
Mommaerts, W. F. H. M., 338, 348
Moncada, S., 9, 258-60, 262
Monk, M., 545
Monroe, S. E., 586, 587
Montague, W., 325
Montal, M., 319

Montegut, M., 162, 212, 213
Moody, F. G., 73
Moolenaar, W. H., 526
MOORE, E. N., 485-97; 486, 487, 490, 492
Moore, R. C. Jr., 599
Moore, R. E., 470
Moossy, J., 584, 586
Mope, L., 458, 460
Morad, M., 411, 426, 431, 456
Morales, M. F., 310, 311
Morales-Aguilera, A., 366
Moran, J. M., 487
Morau, A., 184
Morcos, N. C., 402, 415, 416, 444
Morel, F., 162, 167, 171, 174, 184, 189-94, 197, 204-6, 212, 213
Moreno, A. H., 58
Moreno, F., 628
Morgan, J. F., 300, 301, 310
Morgan, T., 212
Morganroth, J., 486
Morgat, J. L., 242, 558
Moridera, K., 605
Morimoto, K., 428
Morimoto, T., 338, 343
Morita, H., 527
Morkin, E., 403, 404
Morley, J. E., 602
Morr, M., 537, 545
Morris, A. I., 51
Morris, H. R., 259, 262
Morrow, C. S., 465, 468
Morsink, H., 488
Mortara, M., 233, 234
Mortillaro, N. A., 15, 18, 19, 21-23, 30, 33, 36, 37, 43, 48, 49, 51, 58-60
Morton, J. J., 242, 245, 246, 248
Moses, A. M., 571, 576, 577
Moskowitz, M. A., 274
Moss, G. E., 598
Moss, J., 663
Mostov, K. E., 343
Moulopoulos, S. N., 378
Moult, P. J. A., 602
Moustafa, E., 476
Movva, R., 629
Moxley, R. T. III, 642
Mozhayeva, G. N., 468
Mroz, E., 196
Mucci, L., 479, 480
Muehlbauer, R. C., 204
Mueller, P. L., 601
Mueller, W. J., 486
Muggeo, M., 639, 642, 643
Mukherjee, A., 476, 655
Mukherjee, C., 288, 478
Mullens, I. A., 536, 537
Müller, E., 545

AUTHOR INDEX 701

Muller, W., 347, 348
Muller-Eberhard, H. J., 284
Mulligan, R. C., 625
Mullikin, D., 478, 479, 661
Mullikin-Kilpatrick, D., 479
Mullin, J. M., 172
Mullins, L. J., 396, 440, 441, 443, 445
Mumford, R. A., 629
Munck, A., 59
Munjaal, R. P., 668, 669, 674, 678
Muraki, T., 648
Muramatsu, I., 469
Murayama, Y., 204
Murofushi, H., 524
Murphy, A., 300, 301, 310
Murphy, A. J., 300-2, 310
Murphy, R. C., 259
Murphy, R. D., 262
Murphy, W. R., 480
Murthy, V. S., 76, 77
Muskavitch, M. A., 512
Musso, E., 430, 486
Mustafa, S. J., 22
Mutt, V., 260, 263, 633
Myerburg, R. J., 485
Myers, C. E., 286
Myers, C. H., 167
Myers, J. L., 203, 205
Myrick, P., 116

N

Naccache, P. H., 559, 563, 672, 673
Nachshen, D. A., 431
Nachtigall, W., 135
Nadel, E. R., 92
Naegel, G. P., 289
Nafrawi, A. F., 65
Naftalin, R. J., 673
Naftolin, F., 583, 590
Nagai, M., 90
Nagao, Y., 32
Nagasaka, T., 89
Nägele, B., 545, 546
Nahorski, S. R., 477
Nairn, A., 669
Naitoh, Y., 520-25, 527, 530
Nakai, Y., 584-88, 602, 605
Nakajima, S., 358, 362
Nakajima, Y., 451-53
Nakamura, H., 299, 310, 337, 339, 341
Nakamura, K., 6, 626, 629
Nakamura, R. M., 586
Nakanishi, K., 466
Nakano, J., 261
Nakao, Y., 648
Nalbandov, A. V., 583
Nanjundiah, V., 536, 539, 542-45

Nankin, H. R., 583
Narahashi, T., 463-66
Narebski, J., 88, 90
Nath, J., 564
Nathan, C. F., 286
Nathan, D., 485
Nathan, R. D., 426-28
Nathanson, M., 86, 88
Nathanson, N. M., 320
Naumov, A. P., 468
Nawrath, H., 426, 428, 431
Nayler, W. G., 415
Nealon, T. F., 58
Needham, A. D., 90
Needleman, P., 260
Neer, E. J., 475, 479
Neer, R., 664, 665
Negus, V., 87
Neher, E., 357, 360
Neill, J. D., 587, 602
Neill, W. H., 122-24, 126, 127, 129, 130
Nelson, C. A., 565
Nelson, D. L., 519, 520, 522, 524, 526, 530, 531
Nelson, E., 274
Nelson, J. A., 75, 81
Nelson, P. G., 323-25
Nelson, R. D., 558
Nesbitt, K., 271
Nett, T. M., 598
Neufeld, G., 658
Neville, D. M. Jr., 639
New, W., 438, 442
Newell, P. C., 535-38
Newman, B. G., 137
Newman, L. J., 288
Newton, H. J., 583, 590
Ng, K. K. F., 261
Ng, R. C. K., 162, 191
Niall, H., 633, 634
NIALL, H. D., 615-24; 617, 621, 622
Niarchos, A. P., 246, 248, 250, 263, 264
Nicholas, T. E., 274
Nicholls, M. R., 630
Nichols, G., 657
Nichols, R. A., 362
Nickel, E., 320, 347, 348
Nickerson, M., 655
Nicolaou, K., 261
Nicola Siri, L., 362, 366-68
Nicoll, C. S., 620
Nicolson, S. W., 133
Niedel, J., 557, 558
Niedel, J. E., 558
Niedergerke, R., 435, 458
Nielsen, T. B., 661, 663
Niemeyer, R. S., 241, 242
Nies, A. S., 10, 61
Nikolaev, N. A., 134
Nillius, S. J., 586

Nilsson, G., 263
Nilsson, K., 413
Nimura, Y., 405
Nirenberg, M., 565
Nishida, E., 524
Nishikori, K., 413
Nishimoto, A. Y., 402, 438, 443, 444
Nishimura, M., 426-28, 430, 487
Nishino, N., 629
Nissen, M. C., 575
Nissenson, R. A., 657
Nissley, S. P., 646
Niswender, G. D., 586
Noack, E., 381
Noble, D., 396, 425, 446
Noble, P. H., 51
Noble, S., 426, 430
Nocke, W., 586
Noda, K., 396
Noe, B. D., 631-33
Noiman, E. S., 677
Noma, A., 390, 397, 425-31, 487, 531
Nonner, W., 360
Norberg, R. A., 112, 116, 136, 137
Norberg, U. M., 109, 111-13, 115, 116
Nordgren, S., 34
Norman, R. L., 586-88, 600, 603
Norris, C. P., 20-22, 25, 58
Northup, J. K., 662-64
Norton, T. R., 468, 469
Nosyreva, E. D., 468
Noyes, B. A., 633
Nozawa, Y., 531
Nudd, L. M., 437, 438, 445
Nussbaum, S. R., 656, 657
Nyhof, R., 45
Nyiredy, K., 657

O

Oates, J. A., 10
O'Brodovich, H. M., 263
Ocetkiewics, A., 259
Ochi, R., 426, 428, 431
O'Day, D. H., 535
O'Dea, R. F., 560, 561
Odell, W. D., 586, 607, 609
O'Donnell, S. R., 476, 477
O'Donoghue, J. K., 245
O'Dorisio, T., 150
O'Dorisio, T. M., 50
Ody, C., 262, 270, 273
Oehme, M., 440
Oertel, D., 521, 522, 524, 528, 529
Oetliker, H., 451, 456
Offermeier, J., 476

O'Flaherty, J. T., 289
Ogawa, Y., 451-53
Ogura, A., 521, 522, 524, 528, 529
Oh, T. H., 324
Ohlsson, J. T., 244
Ohman, U., 31, 36-38
Ohmori, F., 401, 402, 405-11
Ohnoki, S., 338
Ohta, Y., 393, 394, 398
Ojeda, S. R., 597-99
Oka, T., 381
O'Keefe, G. III, 536, 546
Okita, G. T., 380
Olambiwonnu, N. O., 599
Olden, K., 321, 324, 327, 329
Olefsky, J. M., 639, 641, 643
Ole-Moi Yoi, O., 632
Olgaard, M. K., 381
Oliver, J., 167
Oliver, J. R., 475
Oliveros, J. C., 578
O'Loughlin, K. M., 597
Olson, R. D., 571, 574
Olsson, I., 285, 286
Olsson, R. A., 403, 404
Omoto, C. K., 523
Ondetti, M. A., 244, 247-49, 264
Ondeyka, D. L., 264
O'Neil, R. G., 171, 195, 197
Ong, S. M., 412
Ontjes, D. A., 660, 664
Oparil, S., 242, 245, 251
Opie, L. H., 485
Orange, R. P., 259
Orci, L., 154, 161, 168, 329
Ordal, G. W., 504, 508
Ordonez, N. G., 170
Orida, N., 324
Orida, N. K., 322, 324
O'Riordan, J. L. H., 657
Orkin, L. R., 15, 23
Orloff, J., 171, 197, 212
Orlogff, J., 183-85
Orlow, S. J., 545
Orly, J., 656, 658
Orning, L., 259
Ornston, M. K., 502
Orr, F. W., 285
Orr, W., 289
Ortiz, O., 396
Ortmann, R., 478
Orton, T. C., 270, 271
Osame, M., 319
Osborne, M., 323
Osgood, R. W., 150, 152, 171, 195, 207-11
Oshima, M., 676
Oshiro, G., 61, 65
Osman, H., 89

Osman, M. M., 251
Osswald, H., 203, 204
Osterrieder, W., 426
Oswald, R. E., 321
Otani, A., 284
Otoguro, M., 487
Otte, S. C., 633
Owen, J. E., 572
Owen, P. J., 274
Owens, K., 402
Owens, R. E., 583, 590
Owman, C., 273, 274
Oxender, D. L., 628, 629
Oyamada, M., 260
Oyer, P. E., 631

P

Padula, R. T., 51
Pagano, M., 597
Page, E., 392, 394
Page, I. H., 245, 249
Page, R. B., 621
Page, S., 458
Page, S. G., 368
Pai, J.-K., 259
Palade, G. E., 274, 630
Palade, P. T., 363-68, 429
Palfi, F. J., 376, 382
Pallett, M. J., 137
Pallotta, J. A., 383
Palmiter, R. D., 347
Pan, P., 535, 536, 539, 540, 546
Panczenko, B., 259
Panet, R., 301
Pang, D. C., 310
Pang, L. M., 263
Panke, W. F., 58
Panksepp, J., 574
Papahadjopoulos, D., 530
Papavassiliou, F., 195-97
Papkoff, H., 599
Pappano, A. J., 430, 467
Pappone, P. A., 360-62
Paraskevova, J., 383
Pardue, R. L., 672
Parish, R. W., 544
Park, C. M., 619
Park, C. R., 667
Park, Y. C., 345, 346
Parker, C. M., 347
Parker, C. W., 347, 562
Parker, D. C., 583, 599
Parker, I., 451, 457
Parker, J. C., 48, 74, 75
Parker, L. N., 607, 609
Parker, R. E., 22, 33
Parker, S. R., 510
Parkinson, J. S., 501, 504, 505, 508, 510, 511
Parma, R., 209
Parnham, M. J., 291

Parvizi, N., 583
Passonneau, J. V., 305
Pastan, I. H., 329, 647
Patchett, A. A., 248, 264
Patel, G. K., 49
Patlak, C. S., 183-86, 188
Patrick, J., 319, 328
Patton, J. M., 587, 602
Patzelt, C., 625-27, 629, 633
Paule, W. J., 86
Paunier, L., 597
Pavasuthipaisit, K., 587
Pawlik, W., 15, 21-23, 34, 35
Pawlik, W. W., 15, 19, 21, 34, 35
Payet, M. D., 426, 430, 431
Payne, L. G., 264
Payne, M. E., 669
Peacey, M. J., 545
Peach, M. J., 242, 244, 245
Peachey, L. D., 358
Pearse, A. G. E., 263
Pearson, J. D., 231
Pedersen, D. H., 488
Pedersen, O., 641, 642
Pedley, T. J., 133
Pekonen, F., 647
Pellegrino, C., 349
Peltz, G., 541
Pelzer, D., 426, 428
Pembertson, D. M., 657
Pena, G., 245, 247
Penman, S., 324
Pennell, J., 189
Penny, R., 590, 597, 599
Pennycuick, C. J., 109-16, 127
Pensky, J., 284
Peon, J., 490, 491
Peper, K., 464, 466, 467
Pereira da Silva, L., 535
Perez, H. D., 284, 286
Perez, R., 588
Perez, T. C., 30, 31
Perez-Corral, F., 643
Perez-Maceda, B., 643
Perhuch, J. L. Jr., 479
Periyasamy, S. M., 381
Perkins, J. L., 585
Perkins, J. P., 479
Perlman, D., 583
Pernollet, M-G., 378
Peroutka, S. J., 274, 657
Perrelet, A., 154, 168, 329
Perrin, G., 87
Perry, M. A., 74, 75
Perry, S. V., 669, 670
Pershadsingh, H. A., 670, 675
Persson, H., 476
Pert, C. B., 553, 555-57, 616, 623
Peter, E. T., 4
Peter, K., 162, 167

Peter, S., 158, 162, 203, 212
Peters, P. D., 175
Peters, T., 377
Peterson, E. R., 264
Peterson, J. D., 630, 632
Peterson, S., 273
Petrone, W. F., 286
Petrucci, D. A., 301
Pette, D., 347, 348
Pfeifer, W. D., 578
Pfeuffer, T., 661, 663
Philipp, G., 381
Philipson, K. D., 402, 438, 443, 444
Phillips, D., 289
Phillips, D. K., 476
Phillips, M. I., 576
Phoenix, C. H., 586
Piascik, M. T., 408, 409
Pick, A., 485, 486
Pick, U., 300, 301, 307, 311
Pickett, R. D., 270, 271
Pickett, W. C., 557, 562
Pictet, R., 633
Pictet, R. L., 635
Pierau, Fr.-K., 90
Pierzchala, P. A., 629
Pietra, G. G., 269
Pik, K., 480
Pike, M. C., 553, 556, 561, 562
Pikula, S., 342
Pilar, G., 368
Pilarska, M., 299, 338, 339, 341, 342
Pinchera, A., 648
Pinkus, G. S., 632
Pinshow, B., 98
Piper, E. L., 585
Piper, P. J., 229, 231, 259, 262
Piperno, G., 524
Pires, E. M. V., 676
Pirtle, I., 629
Pirtle, R., 629
Pitha, J., 655, 656
Pitlick, F. A., 251
Pitt, B., 276
PITT, B. R., 269-81; 275-77
Pitts, A. M., 172
Pitts, B. J. R., 402, 413, 414, 443
Plant, T. D., 364
Plant, T. M., 583-88, 590, 598, 602, 603, 605, 606
Plattner, H., 523
Pleschka, K., 90
Plet, A., 643
Pliam, N., 657
Ploegh, H. L., 344
Plotz, E. J., 586, 588, 590
Podleski, T. R., 322, 324
Podolsky, R., 452
Podskalny, J. M., 643, 644

Poe, S. L., 402, 415
POHL, C. R., 583-93; 588-90
Pohl, S. L., 653, 654
Pokora, T., 643
Polak, J. M., 263
Pollard, T. D., 677
Polsky-Cynkin, R., 246
Pomerantz, D. K., 583
Ponce-Hornos, J. E., 442
Pongratz, H., 88
Ponzio, G., 470
Poo, M.-M., 322, 324
Poole-Wilson, P. A., 431
Poppert, D., 475
Portius, H. J., 394
Posner, B. I., 576
Post, C., 271
Post, J. A., 63-65
Post, R. L., 300, 375, 383
Post, R. M., 579
Postlethwaite, A. E., 285, 553
Potel, M. J., 542
Potreau, D., 368
Potter, E. L., 167
Potter, J. D., 408, 409, 458, 460, 668
Potter, L. T., 320, 475, 476, 480
Potts, J. T. Jr., 622, 625, 630, 656, 657
Poujeol, P., 193, 204, 205
Pousse, A., 31, 36, 37
Pouyssegur, J., 466, 470
Powell, J. R., 478-80
Powers, J. C., 629
Pradayrol, L., 633
Prager, R., 302, 306
Prasad, A. V., 598
Prawel, D. A., 134
Preece, M. A., 597
Preiksaitis, H., 479, 480
Prentiss, R. A., 30, 32
Presl, J., 606
Pressler, V., 36
Preston, M. S., 661
Prestronk, D., 326
Pretlow, T. G. II, 172
Pricam, C., 154, 168
Prichard, M. M. L., 86, 87
Priest, R. E., 478-80
Prince, M. J., 641
Pringle, J. W. S., 133
Prives, J., 324
Prives, J. M., 321, 327, 329
Profera, C., 359, 464
Proppe, D., 377, 469
Prosser, C. L., 527
Proverbio, F., 377
Pryor, J. L., 171
Prystowsky, E. N., 485
Pszolla, M., 158, 167, 168
Pucacco, L. R., 193

Pucell, A. G., 337
Pudimat, P. A., 324
Pugh, L. G. C. E., 101
Pullen, R. A., 622
PUMPLIN, D. W., 319-35; 324, 329
Punzengruber, C., 302, 306
Purcell, E. M., 543
Puri, R. B., 632
Puri, V., 599
Pusch, H., 392, 394, 395
Pytkowsk, B., 31, 32, 34, 36

Q

Quagliata, F., 291
Qualizza, P. B., 188
Quigley, J. P., 632, 633
Quigley, M. E., 602
Quinn, P., 476
Quinn, P. S., 625-27, 629, 630, 632

R

Raab, J. L., 98, 127
Rabin, D., 585
Racker, E., 300, 302, 306, 307, 337, 338
Radding, J. A., 556
Rader, R. D., 90
Radzyner, M., 49, 50
Raff, M. C., 329, 347
Raffo, A., 409
Raftery, M. A., 319-21
Rahe, C. H., 583, 590
Rahmsdorf, H. J., 540, 541
Raichle, M. E., 274
Rainey, R. C., 133
Raisman, R., 476
Raizada, M. K., 576
Rakoff, J. S., 586
Rakowski, R., 451, 455
Rall, T. W., 401
Ramanathan, R., 519, 531
Ramanathan, S., 476, 480
Ramon, J., 431
Ramwell, P. W., 262
Randall, D. J., 125
Randall, W. C., 486, 487
Rang, H. P., 391
Rankin, L. I., 152
Rao, G. M. M., 125
Rao, V. K. M., 632
Rapoport, S. I., 390
Rappaport, A. M., 62
Rashidbaigi, A., 654
Raskin, P., 49, 50
Rasmussen, H., 668, 672
Rastegar, A., 171, 196
Ratcliffe, J. G., 631, 633, 634
Ratnoff, O. D., 284

Rau, W., 196
Rauch, B., 304
Rauh, J., 524
Rautenberg, W., 93
Ravazzola, M., 619
Ravdin, P., 322, 324
Ravdin, P. M., 322
Rawlins, J. E., 133
Ray, K., 458, 460
Ray, R., 465, 468
Raymond, G., 368
RAYNER, J. M. V., 109-19; 109-16, 136, 137
Raynor, S. T., 673
Rayns, D. G., 347
Reader, R. W., 501, 502, 504, 512
Rebar, R., 583, 585, 586, 590
Rebar, R. W., 599
Rebhun, L. I., 564, 668
Rechler, M. M., 643, 646
Rector, F. C., 183
Rector, F. C. Jr., 182-87, 209
Redding, T. W., 633
Reder, R. F., 485, 487, 488
Redfern, P., 360
Redfors, S., 38, 51
Redors, S., 38
Reed, J. D., 4, 8, 9
Reed, P. W., 562
Reed, R. B., 597
Reeder, S., 440
Reen, S., 659-61
Reen, S. A., 661, 663, 664
Rees, L. H., 602, 631, 633, 634
Reeves, J. J., 585
Reeves, J. P., 415, 416, 442, 443
Rega, A. F., 306
Rehfeld, J. F., 633-35
Reich, E., 285, 328, 632
Reichel, K., 523
Reichert, L. E. Jr., 657
Reid, J., 249
Reilly, T. M., 654
Reimann, E. M., 401
Reinberg, T., 469
Reineck, H. J., 152, 171, 195, 207-11
Reiner, B. L., 656
Reiness, C. G., 320, 323, 325, 327, 328
Reinhard, J. F., 274
Reinlib, L., 443
Reis, M. A., 325
REITER, E. O., 595-613; 590, 599, 600, 602, 606
Reiter, M., 467
Reithmeier, R. A. F., 298-300, 343, 344
Reitz, M. E., 229, 261
Remtulla, M. A., 408

Remy, R., 578
RENAUD, J. F., 463-73; 465, 466, 468-70
Renkin, E. M., 16
Rennels, M. L., 274
Rennke, H. G., 152
Repaske, D. R., 511
Repke, D. I., 405, 408, 409, 411, 460
Repke, K., 435
Repke, K. R. H., 394
Requena, J., 445
Requena, M., 30, 31
Rerkins, J. P., 478
Resko, J. A., 585, 586, 588, 600, 603
Reuter, H., 401, 411, 415, 417, 425-27, 429, 431, 435, 438, 439, 442, 445, 459, 460, 464, 467
Revel, J.-P., 320
Revelle, R., 597
Revello, P. T., 505
Revenko, S. V., 466
Reyes, E., 297, 338-41, 345, 346, 348
Reyes, F. I., 598
Reynolds, D. G., 6, 8, 45, 73
Reynolds, G. T., 347
Reynolds, H., 289
Reynolds, R. C., 264
Rhee, H. M., 381
Ribet, A., 633
Ricchiuti, N. V., 437, 438
Rice, V. P., 633
Rich, A., 625
Rich, B. H., 599
Richards, A. G., 134
Richards, G. E., 607, 609
Richards, P. D. I., 78
Richards, S. A., 93
Richardson, D. W., 588-90
Richardson, J., 265
RICHARDSON, P. D. I., 18, 19, 21, 23, 33, 57-69; 58-66
Richet, G., 170
Richman, D. P., 328
Richman, P. G., 416, 668
Richter, B., 408, 409
Rick, R., 174
Rickaby, D. A., 270, 274, 275
Ridderstrale, Y., 171
Ridgway, E. B., 347
Ridgway, H., 501, 504
Ridgway, H. F., 504, 509
Riedel, V., 545
Riesen-Willi, U., 347
Rietbrock, N., 376
Rigand, J. L., 310, 311
Rigter, H., 572-76, 578
Rikkers, L. F., 75, 81

Riley, A. L., 209
Riley, W., 175
Rinderknecht, E., 645, 646
Ring, E. J., 8
Rink, T., 457
Rink, T. J., 441, 457
Rinkenberger, R. L., 485
Rios, E., 455, 457
Rister, M., 286
Ritchie, A. C., 476
Ritchie, J. M., 360, 391, 463-65
Rivier, C., 620
Rivier, J., 575-77, 585, 620
Robbins, A. R., 381
Robbins, D. C., 631
Robbins, E. G., 572
Robblee, L. S., 273
Roberts, A. J., 246, 248, 250, 263, 264
Roberts, J. L., 125
Roberts, J. M., 476
Robertson, A., 536, 545
Robertson, B., 479, 480
Robertson, D. M., 599
Robinson, A., 586
Robinson, F. M., 264
Robinson, J. D., 375-77, 381
Robinson, N. L., 139
Robinson, R. B., 477
Robinson, R. R., 154, 168, 170, 182, 183, 188, 205, 206
Robinson-White, A., 273
Roblero, J., 241-44, 248-51
Rocha, A. S., 190, 191, 197, 204, 212
Rochat, H., 468, 469
Roche-Bender, N., 599
Roch-Ramel, F., 183, 189
Rockson, S. G., 476
Rockstein, M., 133
Rodbard, D., 644
Rodbell, M., 515, 653, 661
Rodriguez, H. J., 376
Rodriguez, R. J., 286
Rodriguez-Sierra, J. F., 586
Roeber, J. F., 659
Roesinger, B., 154
Roeske, W. R., 475, 476, 480
Roffwarg, H., 590, 599, 600
Rogart, R. B., 360, 463-65
Rogers, D., 674, 675
Rogers, P. A. W., 73
Rogers, R. M., 619
Rogowski, R. S., 310
Roigaard-Peterson, H., 633
Roinel, N., 174, 193
Rolett, E. L., 411
Roll, F. J., 150
Rollag, M. D., 597
Rolland, R., 590
Rolls, B. J., 576
Romanus, J. A., 646

AUTHOR INDEX

Romero, J. A., 655
Romey, G., 464, 466, 468-70
Rommerts, F. F. G., 620
Romsos, D. R., 376, 378
Roos, B., 287
Roos, W., 536, 538, 539, 542, 544
Root, A., 583
Root, A. W., 599
Rose, B., 522
Rose, G. D., 628
Rose, J. C., 262
Rosen, A. M., 631
Rosen, M. R., 477, 485, 487, 488
Rosenberg, A., 643
Rosenblatt, H., 587, 602
Rosenblatt, M., 627, 656, 657
Rosenfeld, R., 646
Rosenfeld, R. G., 646
Rosenfield, R. L., 599
Rosenthal, J., 325
Rosenthal, N. P., 361
Rosenzweig, J. L., 620
Ross, E. M., 657, 659, 661-64
Ross, G., 4, 64
Ross, G. T., 586, 606
Ross, J. Jr., 412
Rosselin, G., 639, 654, 657, 658, 662
Rossi, B., 305
Rossier, C., 536, 543, 546
Rossier, J., 574, 633
Rossman, L. G., 599
ROTH, J., 639-51; 616, 617, 620, 623, 639-48
Roth, J. A., 269-71, 273
Roth, J. C., 595, 604
Roth, L. J., 672
Roth, R. A., 270, 273, 275, 276
Roth, S. I., 339
Roth, T. F., 673
Rothrock, J. W., 264
Rotundo, R. L., 321-23
Roufa, D., 297, 338-41, 345-48
Rouffignac, C. de, 159
Rougier, O., 358
Rouse, D., 162, 191
Rousset, M., 658
Rowe, G. G., 250
Rowe, P. H., 583
Roy, C., 152, 153
Roy, G., 358, 360
Rubenstein, A. H., 630, 631, 643
Rubin, A. H., 347
Rubin, B., 247, 248, 264
Rubin, L. L., 319
Rubinow, D. R., 579
Rubinson, K. A., 359, 464, 465
Ruddy, S., 284
Rüdel, R., 396, 451, 457, 464

Rudolph, S. A., 663
Ruesta, V. J., 490
Ruf, W., 36
Rugg, E. L., 477
Ruiz-Ceretti, E., 426, 430, 431
Rumrich, G., 204
Ruoho, A. E., 654
Rush, B. D., 61
Ruskin, J. N., 492
Russ, D., 583
Russell, J. H., 630
Russell, J. T., 454, 635
Rutili, G., 48
Rutter, W. J., 633, 635
Ruyle, W. V., 264
Ryan, J. L., 641, 642
RYAN, J. W., 241-55; 225-31, 233-35, 237, 241-51, 264
Ryan, K. J., 585
Ryan, T. E., 300
RYAN, U. S., 223-39; 225-35, 237, 241-43, 245, 246, 248-51, 264
Ryder, S. W., 634
Ryerson, G., 275, 276

S

Sabatini, D. D., 338, 343
Sabo, E. F., 244, 247-49, 264
Sachs, G., 172
Sacktor, B., 412
Safer, B., 459
Sahig, M. K., 632
Saibene, F. P., 99, 100
SAID, S. I., 50, 257-68; 260, 261, 263-65
SAIMI, Y., 519-34; 521, 526, 528, 529
Saint Legar, D., 302
Saito, A., 299, 344
Saito, T., 487
Sakakibara, J., 467
Sakio, H., 260, 261
Sakmann, B., 327, 360
Salama, G., 456
Saldeen, T., 271
Salgado, H. C., 248
Salganicoff, L., 677
Salisbury, J. L., 329, 675
Salmon, J. A., 259
Salmons, S., 348
Salomon, D., 562
Salomon, Y., 653
Salpeter, M. M., 324, 327
Salpeter, S. R., 320
Saltzman, H. A., 286
Salvato, P. D., 79
Salvatori, S., 348
Salviati, G., 339, 348
Salzberg, B., 456
Samhoun, M. N., 262

Sampieri, F., 468
Samuels, B., 643
Samuelsson, B., 257, 259-61
Sánchez, J. A., 357, 362, 366-68
Sancho, J. M., 383
Sanders, D. J., 4
Sanders, M., 90
Sandman, C. A., 574
Sandoval, I., 93
Sanes, J. R., 323
Sankaran, K., 674, 675
Sano, T., 393, 490
Santen, R. J., 583, 587, 590
Santner, S. J., 599
Sanui, H., 347
Sapp, K., 583, 590
Sarantakis, D., 573
Sargent, P. B., 323, 328
Sarna, S. K., 39
Sartore, S., 340
Sarzala, M. G., 310, 338, 339, 341, 342, 345, 414
Sasaki, S., 213
Satir, B. H., 524
Satir, P., 329, 523, 675
Sato, M., 531
Satow, Y., 521, 522, 524, 526-30
Sattler, R. W., 469
Saubermann, A., 175
Saubermann, A. J., 175
Saunders, G. F., 628
Saunders, J. H., 382, 487
Saunders, M. J., 299, 300
Sauviat, M. P., 431, 465, 470
Savage, S. B., 137
Saviolakis, G. A., 642, 643
Sawanobori, T., 393
Sawers, R. S., 590
Sawyer, W. H., 571, 575-77
Scales, D., 298, 299, 339
Scanlon, S., 538
Scarpa, A., 299, 311, 344, 346, 416, 459
Schackelford, J. S., 52
Schafer, D. E., 663
Schafer, J. A., 183, 184, 186
Schaffalitzky de Muckadell, O. B., 50
Schaffer, S., 459
Schallenberger, E., 590
Schally, A. V., 8, 571, 574, 575, 585, 633
Schalow, G., 368
Schaltz, L., 654
Schanne, O. F., 426, 430, 431
Schapiro, H., 66
Schatzmann, H.-J., 393
Scheidegger, C., 544
Schein, S. J., 521, 522, 524, 528, 529

AUTHOR INDEX

Schellekens, L. A., 590
Scherlag, B. J., 485, 486, 492
Scherzer, H., 290
Scheuer, P. J., 470
Schiaffino, S., 338, 340
Schibeci, A., 345, 346
Schiebinger, R. J., 606, 609
SCHIFFMANN, E., 287, 553-68; 553-57, 560-65
Schiller, A., 154, 158
Schilling, E. E., 643
Schilling, W. P., 402, 415
Schirpke, B., 413, 415-17
Schleifer, L. S., 662-64
Schlessinger, J., 322
Schlosberg, A. J., 274
Schmidt, C. F., 75
Schmidt, H., 368, 466
Schmidt, J. S., 632
Schmidt, S., 632
Schmidt, W., 536
Schmidt-Nielsen, K., 93, 98, 109, 111, 112, 114, 127, 133
Schmitt, F. O., 491
Schmitt, H., 475
Schmitt, O., 466
Schneider, H. P. G., 590
Schneider, I. G., 602
Schneider, J. A., 401, 417, 431
Schneider, M., 451, 455, 457
Schneider, M. F., 338, 361
Schneiderman, J. H., 62
Schnermann, J., 158, 159, 212-14
Schocken, D. D., 477
Schoemaker, J., 588, 590
Schoen, F. J., 275, 276
Scholey, J. M., 677
Scholtholt, J., 59, 61, 64, 65
Scholtz, H., 475, 476
Scholz, H., 381, 417, 426, 427, 439
Schopman, F. J. G., 491, 492
Schor, N., 152
Schouella, D., 137
Schrager, P., 358, 360
Schramm, K., 656, 658, 661
Schrauwen, E., 34
Schreiber, W. E., 669
Schreiner, G. F., 329
Schrier, R. W., 152
Schubart, U. K., 673
Schubiger, K., 536, 540
Schuchmann, K. L., 109, 115
Schuetze, S. M., 319, 322, 327
Schug, G., 633
Schuiling, G. A., 585
Schulkes, A. A., 50
Schulman, H., 347, 670, 671
Schultz, D. R., 233, 243, 251
Schultz, J. E., 522, 524
Schultz, W., 264

Schulz, R., 572
Schumann, H. J., 476
Schuurkes, J. A. J., 34, 35, 66
Schuurmans Stekhoven, F., 344
Schwaiger, M., 30, 32, 34, 35
Schwartz, A., 304, 307, 348, 375, 376, 381, 404-10, 412-14, 459
Schwartz, G. J., 195
Schwartz, J. M., 196
Schwartz, L., 276
Schwartz, M., 14, 628
Schwartz, M. J., 183, 196
Schwartz, M. M., 158-61
Schwartz, T. W., 633
Schwarz, H., 545, 546
Schwarz, J. R., 464
Schwarz, W., 357, 359
Schwedie, H. K., 643
Schweitz, H., 466, 468, 470
Scoggins, B. A., 242, 245
Scollan, E. L., 574, 575
Scott, A. P., 631, 633, 634
Scott, M. L., 644
Scott, T., 307
Seagrave, R. C., 85
Sealey, J. E., 248
Sealy, W. C., 491
Seavy, B. K., 633, 634
Sebbane, R., 320, 321
Seegal, B. C., 264
Seeherman, H. J., 98
Seeley, J., 183, 185
Seeley, T., 133
Segal, D., 573
Segal, D. C., 574
Segal, S. J., 617
Segel, L. A., 544
Segre, G. V., 656
Seidah, N. G., 619, 633
Seides, S. F., 492
Seiler, M. W., 150
Seitz, N., 435, 438, 439
Seldin, D. C., 632
Seldin, D. W., 183, 184, 186-88, 209
Seligmann, B. E., 559
Selinger, Z., 301, 658-61
Semba, T., 31, 32, 36
Semple, P. F., 245, 246
Sen, A. K., 376
Senior, R. M., 285, 286
Seppa, H., 553
Seppa, H. E. J., 553
Seppa, S., 553
Seraydarian, K., 300, 301, 338, 348
Seriguchi, D. G., 468, 469
Seron, X., 574, 578
Server, A. C., 634
Setchell, B. P., 61
Severson, D. L., 478
Seyama, I., 467

Seyler, L. E., 605
Sha'Afi, R. I., 329, 559, 563, 672, 673
Shaffer, B. M., 536, 544, 545
Shainberg, A., 325, 345
Shamoo, A., 454
Shamoo, A. E., 298, 300, 404, 405, 409
Shand, D. G., 10
Shank, R. D., 630
Shannon, W. A. Jr., 263
Shapiro, R. I., 545
Shapro, D., 273
Share, L., 571, 576, 577
Sharegi, G. R., 193, 194
Sharief, F., 669
Sharma, H., 150
Sharma, R. V., 476, 480
Sharma, V. K., 476, 655
Sharman, D. F., 271
Sharp, G. W. G., 673
Sharp, P. J., 583
Sharpe, L. G., 274
Sharpe, R. M., 590, 620
Shaw, J., 289
Shaw, J. O., 289
Shear, M., 478
Sheehan, K. L., 602
Shelley, T., 6
Shen, L-P., 635
Shenolikar, S., 669
Shenton, J. C., 583
SHEPHERD, A. P., 13-27; 15-25, 45, 74, 75, 77
Shepro, D., 261, 273, 277
Sher, L., 633
Sherard, R., 674, 675
Sheridan, M. A., 264
Sherlag, B. J., 492
Sherman, B. M., 597
Sherwin, R. S., 49
Sherwood, L. M., 246
Sheth, N. A., 583
Shibata, S., 468, 469
Shibolet, S., 85
Shichi, H., 514
Shieh, I.-S., 382
Shields, B., 629
Shields, R., 58
Shigekawa, M., 305, 402, 405, 408-11
Shigenobu, K., 469
Shiloach, J., 616, 617, 623
Shimoni, Y., 426, 430
Shiner, J. S., 445
Shing-Sheu, S., 441, 443, 444
Shirai, N., 467
Shiraishi, T., 59, 64, 65
Shire, J., 633, 634
Shofstall, R. E., 633
Shollenberger, C. A., 115, 116
Shooter, E. M., 634

AUTHOR INDEX 707

Shore, P. A., 76
Short, R. V., 583
Shoshan, V., 298
Shotzberger, G. S., 467
Showell, H., 289
Showell, H. J., 289, 553, 555-59, 563, 672, 673
Shrager, P., 359, 464
Shuman, H., 346, 454, 459
Shuster, S., 575
Shusterman, C. L., 526
Siebert, C. W., 645
Siegel, E., 673
Siegel, R. E., 324
Siegelbaum, S., 427, 428, 430
Siegelbaum, S. A., 426-30
Siegenthaler, W., 646
Siegfried, W. R., 116
Sieghart, W., 671, 673
Siekevitz, P., 675
Sigulem, D., 154
Sigurdsson, S. B., 23, 24
Sigworth, F. J., 359, 360, 464
Sih, C. J., 259
Silbert, J. E., 285
Siler, T. M., 587
Siler-Khodr, T. M., 620
Silhavy, T. J., 628
Silman, R. E., 597, 609
Silva, J. L., 309
Silve, C., 657
Silver, A., 269
Silverman, A. J., 584
Silverman, M., 501, 502, 504, 505, 509
Silverstein, S. C., 286
Simchowitz, L., 560
Simmons, R. L., 558
SIMON, M., 501-17; 501, 502, 504-7, 509, 512, 513
Simon, M. I., 501, 502, 504
Simon, W., 440
Simons, A. H. M., 588, 590
Simons, P. A., 586
Simonton, C. A., 154, 182, 183, 188, 206
Simpson, F. O., 347
Simpson, J. R., 576
Simson, M. B., 492
Singer, D. H., 487
Singer, M., 659-61, 664
Singer, S. J., 329, 676
Singh, B. N., 485
Singh, D., 413
Singh, R. N. P., 633, 634
Sit, S. P., 30, 32, 45
Sitar, D. S., 65
Sivarajah, K., 262
Sivaramakrishnan, S., 414
Sizonenko, P. C., 590, 597
Sjodin, R. A., 396
Sjoqvist, A., 34, 50
Sjovall, H., 51

Skeggs, L. T., 229, 243, 244, 246
Skehel, J. J., 627, 628
Skews, B. W., 134, 139
Sklar, C. A., 597, 601, 607, 609
Skou, J. C., 375, 377
Slater, C. R., 323
Sloane, B., 346
Smallberg, S. A., 579
Smeby, R. R., 245, 249
Smellie, W. S. A., 242, 245, 248
Smettan, G., 405
Smigel, M. D., 663, 664
Smilowitz, H., 322
Smirnov, V. N., 404, 413, 414
Smith, B. R., 640, 647
Smith, D. R., 259
Smith, D. S., 234
Smith, E., 4-8
Smith, E. E., 20, 58
Smith, G. A., 337, 341, 347
Smith, J., 261
Smith, M. A., 585
Smith, R. A., 492, 504
Smith, T. W., 376, 475, 479
Smith, U., 227, 229, 230, 234, 241-43, 251, 647
Smoake, J. A., 668, 674
Smy, J. R., 8, 9
Smyth, D. G., 634
Snyder, S. H., 249, 274, 476, 635, 657
Snyderman, R., 285, 553, 556, 561, 562
Sobel, A., 320
Sobieszek, A., 676
Sobue, K., 670
Soderling, T. R., 669
Soergel, K. H., 49
Soffer, R. L., 229, 244, 246, 247, 250, 264
Sohtell, M., 184
Sok, D.-E., 259
Sokol, H. W., 215
Solaro, R. J., 445, 458, 460
Solcia, E., 263
Sole, M. J., 276
Solez, K., 211
Soll, A. H., 10, 639
Soll, D. R., 539
Solomon, S. S., 674
Soltoff, S., 262
Somers, R. L., 514
Somlyo, A. P., 322, 346, 454, 459
Somlyo, A. V., 346, 454, 459
Sommerschild, H., 328
Song, S. Y., 668
Sonnenberg, A., 76
Sonnenberg, H., 208
Sonnenblick, E. H., 415
Sorensen, R. L., 630

Sorrano-Rios, M., 643
Soumireu-Mourat, B., 573
Spalding, B. C., 359, 360, 464
Spargo, B. H., 170
SPEAR, J. F., 485-97; 486, 487, 490, 492
Speihler, V., 573
Sperber, I., 149
Sperelakis, N., 401, 417, 431, 467, 469, 471, 485
Spiegel, A. M., 655, 659-64
Spiehler, V. R., 573
Spielman, S. R., 486
Spielman, W. S., 203, 204
Spies, H. G., 585-88, 600, 603
Spigelman, L., 630
Spilberg, I., 560
Spitzer, N. C., 346
Spitznagel, S. K., 284
Sporn, J. R., 655
Spotnitz, A., 485, 487, 488
Spragg, J., 632
Springer, M. S., 501, 504-7, 509, 510, 512
Springer, W. R., 505, 560
Spudich, J. A., 541
Sreter, F. A., 347, 348
Srinivasan, Y., 655, 656
Stadel, J. M., 656, 661
Stahl, D., 80
Stalcup, S. A., 245, 251, 263, 264, 276
Stalder, H., 262
Stam, A. C. Jr., 415
Stamford, B. A., 92
Stamler, J. F., 576
Standen, N. B., 361, 363-65
Stanfield, P. R., 357, 361-67, 428
Stanley, E. F., 328
Stanton, B., 171
Stanton, B. A., 168, 170, 171
Stark, R. D., 57, 61
Starling, E. H., 269
Staubus, A. E., 75, 81
Staum, B. B., 205
Stead, W. W., 203, 205
Steadman, J. H., 630
Steck, T. L., 536, 543, 544
STEFANI, E., 357-72; 357, 362, 366-69
Stehlik, D., 306
Stein, J. H., 150, 152, 171, 195, 203, 207-11
Stein, L., 573, 579
Stein, R., 633
Stein, S., 633
Steinbach, A. B., 368
Steinbach, J. H., 323, 327, 360
Steinberg, A. D., 565
Steinberg, H., 270, 275, 276
Steiner, A. L., 412

STEINER, D. F., 625-38; 617, 625-35
Steiner, R. A., 586
Steinhardt, R. A., 436, 439
Steitz, T. A., 627
Stendahl, O. I., 564
Stene-Larsen, G., 477
Stenson, W., 560
Stenson, W. F., 562
Stephenson, E., 451, 453, 454, 457
Stern, A. S., 633
Stern, P., 212
Sternfeld, J., 535, 536
Sternweis, P. C., 662-64
Stetson, D. L., 168, 170, 196
Stevens, C. F., 357
STEVENS, E. D., 121-31; 124, 126, 129, 130, 133
Stevens, F. C., 668
Stewart, D. J., 376
Stewart, J., 159, 212
Stewart, J. F., 486
Stewart, J. M., 241-44, 246, 248-51
Stewart, P. S., 298, 300
Stewart, T. A., 264
Stimler, N. P., 289
Stinis, H. T., 404, 405, 413, 414
St. Louis, P. J., 402, 404, 413-17
Stock, J. R., 505, 510
Stocking, R. A., 92
Stokes, J. B., 195, 196, 210, 211
Stolwijk, J. A. J., 92
Stone, C. A., 264
Stoner, L. C., 183, 189, 193, 194
Storm, D. R., 673
Storm, S. R., 392, 394
Story, B. D., 476
Story, H. E., 86
Stossel, T. P., 560, 564
Strader, C. B. D., 320
Strader, C. D., 319-21
Strampp, A., 617
Strand, J. C., 261
Stratis, J. P., 501, 504, 513
Straus, E., 634
Strauss, A. W., 629
Strawbridge, R. A., 476
Strichartz, C. R., 360
Strichartz, G., 464
Strickland, K. P., 642
Strickland, W. G., 674, 676
Strittmatter, W. J., 542, 656
Strominger, J. L., 344
Strother, J. T., 61
Stroud, R. M., 320, 324
Stroufova, A., 606
Struck, C. J., 475

Strum, J. M., 227, 270, 274
Struve, T., 588, 590
Stuart, A. E., 361
Stull, J. T., 411, 677
Stya, M., 324
Styne, D. M., 597-99, 602, 604, 606, 607
Su, Y.-F., 478, 479
Suarez-Kurtz, G., 451, 457
Suehiro, A., 36
Suehiro, G. T., 36
Sugano, Y., 89
Sugiyama, H., 323, 324
Suki, W. N., 162, 191
Suko, J., 302, 306
Sulakhe, P. V., 402, 404, 413-17, 476
Sullivan, J. V., 406
Sullivan, S. J., 554, 558
Sullivan, T. J., 283
Sumida, M., 304, 306, 307
Summers, R., 579
Sun, A. M., 630
Sun, F. F., 262
Surawicz, B., 391, 394, 487
Sures, I., 622
Surve, A. H., 586
Sutherland, C. L., 347
Sutherland, E. W., 401
Sutko, J. L., 415, 416, 442, 443
Suzuki, F., 490
Suzuki, K., 676
Suzuki, Y., 524
Svanvik, J., 15, 17, 22, 25, 74
Svenson, R. H., 491
Swamy, B. V., 65
Swan, A. A. B., 269
Swan, K. G., 10, 74
Swan, K. G. R., 73
Swanobori, T., 489
Swartz, M. H., 415
Swartz-Sorensen, N., 641
Sweadner, K. J., 375, 383
Sweet, C. S., 264
Swerdloff, R. S., 585, 586
Swift, C. S., 87
Szentivanyi, A., 413
Szidon, P., 242, 245, 251
Szot, S., 287

T

Tabler, M., 546
TADA, M., 401-23; 298, 302, 303, 337, 401-12, 415, 451, 460
Tager, H. S., 625-27, 629-31, 633, 634
Tague, L. L., 22, 35
Tai, H.-H., 258-61
Tainer, J. A., 555
Takahashi, K., 522
Takahashi, M., 521, 522

Takakuwa, Y., 305
Takeda, K., 382, 383
Takeya, K., 467
Takigawa, S., 490
Takisawa, H., 298, 302, 305, 337, 409, 410
Talamo, B. R., 655
Talano, J. V., 487
Taleb, L., 205
Talley, J. V., 559
Talmadge, K., 626
Talner, N., 276
Tamborlane, W. V., 605
Tan, A., 492
Tanabe, H., 521, 530
Tanaka, M., 451-53
Tanaka, T., 673
Tanford, C., 299
Tangstrand, B., 476
Tanner, J. M., 596, 597
Tarnavsky, G. K., 585
Tarr, M., 184
Tarrab-Hazdai, R., 320, 329
Tash, J. S., 673
Tasler, J., 8
Taub, D., 264
Taugner, R., 154, 158
Taylor, A. E., 15, 19, 21, 22, 30, 33, 36, 37, 43, 46-49, 51, 52
Taylor, B. L., 501, 509, 512, 513
TAYLOR, C. R., 97-107; 87-89, 97-101, 103-5, 127, 128
Taylor, E., 306, 307
Taylor, E. W., 297, 306, 307
Taylor, F. B., Jr., 285
Taylor, G. W., 259, 262
Taylor, K. A., 677
Taylor, P., 138, 140
Taylor, R., 357
Taylor, R. B., 329
Taylor, R. R., 376, 382
TAYLOR, S. I., 639-51; 643
Taylor, S. R., 358, 451, 452, 457
Taymor, M. L., 587
Tazieff-Depierre, F., 468
Tedesco, P., 509, 512
Teitelbaum, A. P., 657
Temma, K., 376, 377, 379, 381-83
ten Broeke, J., 264
Ten Eick, R. E., 426, 428, 487
Tenschert, W., 646
Tenu, J. P., 302
TEPPERMAN, B. L., 71-82; 35
Tepperman, K., 326
Terasaki, M., 347
Terasawa, E., 586
Terborgh, J., 112, 115, 116

AUTHOR INDEX 709

ter Haar, M. B., 583
Terragno, N. A., 261
Terreros, D. A., 184
Teshima, Y. L., 668
Tevaarwerk, G. J. M., 642
Texter, E. C. Jr., 14
The, T. H., 248
Theoharides, T. C., 673
Thesleff, S., 360
Thibault, G., 619
Thiede, E. W., 526
Thiele, J., 522
Thielen, J., 392, 394, 395
Thilo, L., 538
Thody, A. J., 575
Thoenes, W., 167
Thomas, D. D., 310, 337, 341
Thomas, G. J. Jr., 628, 629
Thomas, M. V., 429
Thomas, R., 185
Thomas, R. C., 395, 396
Thomas, S. P., 109, 111, 112
Thomas-Morvan, C., 648
Thompson, G. A., 531
Thompson, I. E., 587
Thompson, J. A., 572
Thomson, K., 536
Thor, P., 8
Thorley-Lawson, D. A., 298, 300, 301, 310
Thorpe, V., 4
Thorsett, E. D., 264
Thorsson, A. V., 646
Thrall, R. S., 291
Thurau, K., 174
Thys, H., 97, 99, 100
Tickle, I. J., 622
Tierney, D. F., 250
Tifft, C. P., 248
Tikkanen, I., 248
Till, G., 285
Tillack, T. W., 297, 338-42, 344-46, 348
Tilley, L. P., 487, 488
Tillisch, J. H., 437, 438, 441
Tillotson, D., 364, 428, 521-23
Timko, P. L., 111, 116
Timsit-Berthier, M., 578
Tindall, D. J., 673
Tindall, G. T., 587, 602
Tippins, J. R., 259, 262
Tirri, R., 85
Tisher, C. C., 152, 154, 161, 167, 168, 170, 182, 183, 188 193, 195, 197, 206, 209
Titchen, D. A., 60
Toccafondi, R., 648
Toews, M. L., 505, 510
Tokuyasu, K. T., 676
Tomchik, K. J., 545
Tomitz, M., 627
Tomlinson, S., 657

Tong, S. W., 298, 299
Tonomura, Y., 298, 302-6, 310, 311, 337, 378, 402, 405-10
Toon, P. A., 297
Toorchen, D., 545
Torre-Bueno, J. R., 109-12
Totsuka, T., 673
Touabi, M., 204
Touboul, P., 490
Tourneur, Y., 470
Touvay, C., 193, 205
Track, N. S., 263
Tracy, H. J., 633
Traill, T., 271
Traut, J., 542
Trautwein, W., 391, 396, 397, 426-28, 431, 438, 442, 464, 466, 467
Tree, M., 242, 245, 248
Tregear, G. W., 242, 245, 657
Trethewie, E. R., 259
Trevan, D. J., 545
Trinh-Trang-Tan, M.-M., 203, 213-15
Trinnaman, B. J., 298-301, 309
Tripathi, O., 426, 427
Tristram, E. W., 264
Troehler, U., 204
Troen, P., 583
Trosko, J. E., 376, 381
Troup, P. J., 488
Troutman, S., 183
Troutman, S. L., 183, 184, 186
Troy, J. L., 152
Trumble, W. R., 415, 416
Trump, B. F., 167
Tsagaris, T. J., 260
Tsai, B. S., 244, 480, 655
Tsai, C. C., 583, 585, 586, 590
Tsang, A. S., 545, 546
Tsartos, A. S., 328
Tse, J., 478-80
Tseng, L. F., 573
Tsien, R. W., 401, 411, 412, 425-31, 441, 457, 486, 488
Tsien, R. Y., 441, 457
Tso, W. W., 501, 502, 504, 512
Tsou, R. C., 587, 602
Tsuchii, A., 530
Tsuda, Y., 428
Tubiana, M., 648
Tucker, V. A., 109-11, 115, 134, 136
Tugwell, D. R., 585
Tume, R. K., 303
Tune, B., 183-85
Tune, B. M., 183-85, 188
Tunstall, J., 436
Turini, G. A., 247, 248
Turino, G. M., 245, 251
Turnberg, L. A., 51
Turner, R. J., 184
Turner, S. R., 555

Tutwiler, G. F., 633
Tyan, G.-G., 536, 540
Tyczynski, M., 88, 90
Tyler, G., 80
Tyllstrom, J., 15, 17, 22, 25

U

Uchitel, O. D., 369
Udenfriend, S., 633
Ueda, T., 530
Uenishi, K., 676
Uhm, D. Y., 441
Ulbricht, W., 357, 464, 466
Ullrich, A., 622
Ullrich, K. J., 195-97, 204
Ulm, E. H., 264
Unanue, E. R., 329
Underwood, L. E., 643
Unkeless, J., 285
U'Prichard, D. C., 476
Urban, I., 573
Usdin, T. B., 657
Uther, J. B., 492
Utsunomiya, T., 261

V

Vagne, M., 263
Vahouny, G. V., 402
Vaitukaitis, J. L., 586
Vale, W., 585, 620
Vale, W. W., 585
Valentich, J. D., 172
Valenzuela, P., 320
Valk, T. W., 585, 590, 597, 600, 605
Valleau, J. D., 19, 21, 22, 46
VALTIN, H., 203-19; 172, 203, 211, 212, 214, 215
Valverde, I., 675
Van Alstyne, E., 402, 415
Vanaman, T., 669
Vanaman, T. C., 524, 669
Van Arsdale, P. M., 661
Van Capelle, F. J. L., 488, 490-92
van de Meene, G. C. J., 535
Vandenberg, G., 583, 585-87, 590
Van der Drift, C., 509, 512
Vanderkooi, J. M., 299, 339
Vanderlaan, W. P., 633
van der Mark, F., 74
Vandermeers, A., 675
Vanderoalm, T. J., 8
Van Derstrappen, G., 14
Vander Tuig, J. G., 376
Van der Werf, P., 505
vander Zee, H., 274
VandeWide, R. L., 602
Vande Wiele, R. L., 585, 587, 588

Vandlen, R. L., 320
van Driel, R., 536, 539, 543, 546
Vane, J. R., 9, 227, 229, 231, 241, 257-62, 264, 269, 272
Van Eeden, A. V. P., 378
van Eldik, L. J., 524, 673
van Haastert, P. J. M., 536, 540
Van Hee, R. H., 6
Van Houten, J., 519-21, 525, 530, 531
van Houten, M., 576
Vanhoutte, P. M., 6
van Kessel, H., 588, 590
Van Obberghen-Schilling, E. E., 646
van Osnabrugge, G. J. C., 588, 590
Van Ree, J. M., 573, 574, 578, 579
van Rees, G. P., 585
Van Riezen, H., 572, 573, 575, 578
Van Sickle, M., 673
Van Wimersma Greidanus, T. B., 577
Van Wimersma Greidanus, T. J. B., 577, 578
Van Winkle, W. B., 402, 404, 405, 412, 459
Varani, J., 285
Vargo, T., 583
Varndenheede, J. R., 667
Varner, A. A., 602
Varney, D., 162
Varney, D. M., 213
Varnum, B., 539
Vasquez, B. J., 573-75
Vassalle, M., 390, 397, 430, 485, 486
Vassalli, J. D., 285, 632
Vassas, J. M., 431
Vassil, T. C., 264
Vassort, G., 426, 427, 436, 442, 466, 467
Vatner, S. F., 476
Vaughan, E. D. Jr., 242, 245
Vaughan, H., 565, 663
Vaughan-Jones, R. D., 392, 394
Vaughan-Neil, E. F., 276
Vaughan Williams, E. M., 368
Vaughn, P. C., 365
Vaught, W. G., 585
Veliquette, J., 573-75
Venkatachalam, M. A., 150, 158-61
Venkatasubramanian, K., 287, 557, 561, 562
Venosa, H., 358
Venosa, R. A., 358, 360
Venter, J. C., 274, 412, 655, 659, 664

Verbeke, N., 377
Vereecke, J., 426, 429, 431
Vergara, C., 301
Vergara, J., 357, 456
Verge, P., 285, 286
Vergera, J., 456
Verhave, T., 572
Verjovski-Almeida, S., 302, 303, 305, 309
Vermes-Kunos, I., 655
Verney, E. D., 269
Vernon, A., 99-101, 103
Versteeg, D. H., 578
Verveen, A. A., 526
Vestweber, D., 546
Vetter, W., 646
Vianna, A. L., 298, 299, 302, 304, 306, 337, 407
Vigne, P., 470
Vilas, U., 625
Villiger, A., 49, 50
Vincent, A., 328
Vincent, J. D., 585
Vincent, J. P., 466, 468, 470
Vincenzi, F. F., 669, 675
Virji, M. A. G., 632
Visintainer, J., 556
Vitadello, M., 340
Vitkauskas, G., 558
Vitti, P., 648
Viveros, O. H., 560
Vleugels, A., 431
Vogel, G., 538
Vogel, Z., 323, 324
Volpe, P., 339, 348
Volpi, M., 559, 672
von Wilbrandt, W., 435
Vukovich, R. A., 248

W

Wachslicht-Rodbard, H., 639, 641-44
Wade, J. B., 154, 158, 168, 170, 171, 196
Waeber, B., 247, 248
Waechter, C. J., 321
Wagenaar, E. B., 347
Wagner, H. H., 464
Wahl, S. M., 555, 557
Waisman, D., 668
Waisman, D. M., 668
Wakshull, E., 322, 323
Waku, K., 342
Waldman, H. M., 262
Waldo, A. L., 492
WALKER, L. A., 203-19
Wallace, A. G., 491
Wallace, K. B., 276
Wallace, L. J., 538
Wallace, R. A., 625
Wallace, R. W., 675
Wallach, E., 583

Wallentin, I., 15, 17, 22, 25
Wallick, E. T., 375, 376, 381
Wallin, J. D., 209
Walsh, D. A., 401, 414
Walsh, J. H., 10, 50
Walsh, J. R., 622
Walsh, M. P., 346
Walter, M. F., 523, 524
Walter, P., 627
Walter, R., 578, 579
Waltman, A. C., 8
Walus, K. M., 29, 30, 34, 35
Wand, I., 262
Wang, C.-T., 299, 344
Wang, E. A., 504, 505, 507, 510
Wang, F. L., 244
Wang, J. H., 668
Wang, T., 304, 307, 348, 406-10
Wangensteen, O. H., 4
Wanstall, J. C., 476, 477
Warabi, H., 565
Ward, A., 540
Ward, M. R., 369
WARD, P. A., 283-93; 284-86, 288-91, 554
Ward, P. E., 264
Wardlaw, S., 586
Wardzala, L. J., 676
Warner, A. E., 364, 365
Warner, F. D., 523
Warnock, D. G., 182-88
Warnock, R., 262
Warren, G. B., 297, 337, 341, 347
Warren, M., 588
Warren, M. P., 597
Warrick, H. M., 509, 512
Watanabe, A. M., 376, 382, 402, 404, 412-15, 476, 480, 655
Watanabe, Y., 426-28, 430, 524
Waterfield, A. A., 572
Watkins, M. L., 184, 186
Watterson, D. M., 524, 669, 673
Wayland, H., 72
Webb, P. W., 121, 128
Webb, R. C., 413
Webb, W. W., 322
Webster, R. O., 284, 289
Wedner, H. J., 540, 560
Weening, J. J., 248
Weglicki, W. B., 402, 415
Wei, E., 573
Wei, J.-W., 476
Weibel, J., 172
Weick, R. F., 585-88
Weidley, E., 572
Weidmann, S., 459, 485
Weigele, J. B., 359, 360
Weiland, G. A., 657

AUTHOR INDEX 711

Weinberg, C. B., 320, 323, 328
Weinberg, N., 658
Weinberger, S. B., 574
Weiner, H., 590
Weingart, R., 431, 488
Weingartner, H., 579
Weinhold, P. A., 414, 416, 417
Weininger, D., 129
Weinrib, A. B., 75, 81
Weintraub, B. D., 647
Weis-Fogh, T., 133-39
Weiss, B., 480, 655
Weiss, G., 585-87, 590
Weiss, J., 626
Weiss, J. M., 577
Weiss, M. B., 492
Weiss, R., 673
Weiss, S. J., 286
Weissman, G., 288
Weitzman, E., 590, 599
Weitzman, E. D., 590, 600
Weitzman, R., 576, 602
Weitzman, R. E., 647
Weksler, B. B., 260
Weldon, P. R., 324
Wellens, H. J. J., 485
Welling, D. J., 184
Welling, L. W., 184
Wells, H. S., 46, 51
Wellsmith, N. V., 402, 415
Welsh, M. J., 673, 677
Wendler, G., 134
Wendt, I. R., 437, 438
Wennemark, J. R., 486, 490
Wennogle, L. P., 319
Wentz, A. C., 583, 590
Wessels, M. R., 478, 479
West, N. H., 109
Westenskow, D. R., 75, 81
Westertep, K. R., 109, 117
Westgaard, R. H., 325
Whalen, G. E., 49
Wharton, J., 263
Whitaker, C., 228, 230, 233, 234, 243, 264
Whitaker, J. N., 675
White, F., 90
White, G. C. II, 673
White, M. K., 277
White, N., 578
White, R. R., 286
White, S. S., 597-99
White, T., 626
Whitlock, J. P. Jr., 347
Whittle, B. J. R., 7 10
Wick, U., 535, 536, 539-44
Wicklein, D., 45
Wickner, W., 628, 629
Wide, L., 586
Wier, W., 457, 458
Wierman, B. M., 673
Wiersma, D. A., 270, 273
Wiesner, K., 466

Wiest, S. A., 381
Wiggins, J. R., 391
Wight, A., 341
Wikberg, J., 413
Wilborn, W. H., 43, 49, 51
Wildt, L., 585-88, 590, 597, 598, 605, 606
Wiley, D. C., 627, 628
Wiley, H. S., 625
Wilke, B., 632
Wilkin, P. J., 134, 136, 137, 140
Wilkins, R. H., 586
Wilkinson, G. R., 61
Wilkinson, S., 558
Will, H., 404, 405, 413-17
Willerson, J. T., 476, 655
Willett, L. B., 583
Willheim, C. B., 673
Williams, A. J., 415
Williams, B. A., 134
Williams, G. H., 247, 248
Williams, L. T., 475, 553, 556, 655
Williams, R. F., 590
Williams, R. S., 476
Williams, T. L., 364
Williamson, J., 459
Williamson, J. R., 416
Willingham, M. C., 329
Wilms-Schopman, F., 488
Wilson, I. A., 627, 628
Wilson, I. B., 244
Wilson, S. C., 583
Windhager, E. E., 193, 194
WINEGRAD, S., 451-62; 346, 445, 457-60
Winek, R., 476
Wingert, T. D., 602
Winker, F., 302, 306
Winne, G., 45, 48, 53
Winter, J. S. D., 598
Wise, C. D., 573
Wise, R. M., 402, 445
Wise, W. C., 260, 261
Wissig, S. L., 167
Wit, A. L., 485-88, 492
Withers, P. C., 111, 115, 116
WITHRINGTON, P. G., 57-69; 58-66
Witman, G. B., 523, 524
Witschi, J., 597
Woakes, A. J., 110, 114
Woan, J. M., 245, 251, 264
Woelders, H., 540
Wolf, L. L., 109, 114-17
Wolf, S., 4
Wolfe, B. B., 475, 655
Wolfe, J. H. N., 261
Wolfe, P. B., 536, 546
Wolff, J., 631, 669
Wolgast, M., 74
Wollemann, M., 480

Wollenberger, A., 404, 405, 413-17
Wollmer, A., 622
Wong, D., 406, 411
Wong, D. T.-M., 137
Wong, K., 286
Wong, P. T. S., 338
Wong, P. Y.-K., 262
Wong, T. M., 476, 655
Woo, S. L. C., 678
Wood, C. L., 477
Wood, D. C., 520
Wood, E., 459
Wood, J. G., 675
Wood, S. P., 622
Woodcock, E., 480
Woodcock, E. A., 480
Woodhall, C., 182, 183, 188
Woodhall, P. B., 154, 183, 193, 206
Woodrow, M. L., 416
Worthington, C. R., 299, 311
Wray, H. L., 403, 404, 406, 408
Wright, D. G., 563
Wright, F. S., 193, 194
Wu, F. S., 345, 346
Wu, M. T., 264
Wu, R. H. K., 590
Wu, T. Y.-T., 133
Wu, W. C., 49
Wu, W. C.-S., 320
Wuhrmann, P., 347
Wunderer, G., 468
Wunsch, E., 8
Wurster, B., 536, 538-40, 544-46
Wurster, R. D., 90, 92
Wurtman, R. J., 597
Wuster, M., 572
Wuttke, W., 590, 600
Wyllie, H. J., 229
Wyvratt, M. J., 264

Y

Yablonski, M. D., 47, 48
Yaffe, D., 345
Yagi, K., 676
Yagil, G., 345
Yalow, R. S., 634
Yamada, M., 401, 402, 404-11
Yamada, S., 300, 301, 304-6, 310, 476, 480
Yamaguchi, M., 378
Yamaji, T., 584-87, 590
Yamaki, T., 673
Yamamoto, S., 376, 377, 379, 381, 382
Yamamoto, T., 298, 302-6, 310, 337, 402, 405-10
Yamamura, H. I., 475, 476, 480

Yamane, S., 133
Yamauchi, T., 670
Yamazaki, R., 668, 670
Yanagihara, K., 425, 427, 487
Yanai, M., 466
Yang, H. Y. T., 249
Yanofsky, C., 628, 629
Yarger, W. E., 197, 209
Yasumasu, I., 525
Yasuoka, K., 301
Yate, D. W., 300
Yazawa, M., 676
Yee, V. J., 182, 183, 186-88
Yeh, R. P., 546
Yen, S.-S., 346
Yen, S. S. C., 583, 585-87, 590, 599, 602
Ying, S-Y., 620
Yingst, D. R., 377
Yip, C. C., 632
Yohe, W. B., 671, 675
Yonemitsu, K., 669
Yoshida, H., 670
Yoshida, T., 264

Yoshimoto, Y., 605
Youdim, M. B. H., 227, 270
Young, B. A., 88
Young, D., 167
Young, J. R., 586
Youngberg, S. P., 203, 204
Yturriaga, R., 643
Yu, B. P., 302
Yu, S.-D., 626
Yuen, H. S. H., 124, 128, 130

Z

Zabin, I., 628
Zach, R., 116
Zacharias, L., 597
Zahn, H., 622
Zajac, S., 40
Zakarian, S., 634
Zakarija, M., 648
Zalin, R. J., 325
Zamboni, A., 97, 99, 100
Zamudio, R., 597
Zapf, J., 645, 646

Zatz, M., 655
Zeiler, R., 488
Zeithin, I. J., 34
Zeleznik, A. J., 641
Zettner, B., 381
Zierler, K. L., 75
Zigmond, S. H., 554-58
Zimmerman, E., 586
Zimmerman, E. A., 584, 588
Zimmerman, M., 629
Zimniak, A., 338
Zimniak, P., 338, 342
Zinner, M. J., 6, 8, 45
Zipes, D. P., 485, 487, 488
Ziskind, L., 320
Zitzner, L., 657
Zubrzycka, E., 338-43, 345
Zühlke, H., 630, 632
Zukin, R. S., 505, 512
Zurawski, G., 628, 629
Zürcher, A. F., 585
Zurier, R. B., 291
Zweibaum, A., 658
Zwizinski, C., 629

SUBJECT INDEX

A

Absorption, intestinal
  see Intestinal absorption
Acanthosis nigricans
  insulin resistance associated with, 642
Acetylcholine
  effect on slow inward calcium current in heart, 431
  intestinal blood flow and motility effects of, 33, 35
Acetylcholine receptor, 319
  accelerated degradation in myasthenia gravis, 328
  antibody-induced accelerated degradation of, 328–29
  arrangement in clusters in mature myotubes, 322–324
  biosynthesis of, 325
    assembly, processing and intracellular transport, 321–322
    factors affecting, 323–325
    incorporation into plasma membrane in, 322–323
    translation of messenger RNA in, 320–321
  degradation of, 325–329
    acceleration by cross-linking, 328–329
    mechanism of turnover, 325–326
    turnover rates, 326–328
  glycosylation role in stabilization of, 326–327
  molecular heterogeneity of, 320
  muscle activity effects on, 325
  structure of, 319–320
  turnover in skeletal muscle, 319–329
  unknown factors capable of increasing amount of, 323–325
Acid secretion in stomach
  see Gastric acid secretion
Acid-base balance
  collecting duct role in, 171–172
Acidification of urine
  collecting tubule role in, 196–197
  heterogeneity in proximal tubule, 185
Acidosis
  influence on liver circulation, 59

Aconitine
  action on fast sodium channel in heart, 465–467
Acromegaly
  insulin receptors in, 641–642
ACTH
  processing of prohormone of, 633
Actin-myosin interaction
  calcium role in skeletal muscle, 297, 676–677
  calmodulin role in non-muscle cell contraction, 676–677
Action potential
  calcium channels involvement in, 366–368
    in *Paramecium*, 522–523
  chloride channels involvement in, 364–366
  conduction velocity related to rate of depolarization of, 489
  early computations in muscle, 358
  electrogenic sodium pumping effect on, 396–397
  generation in *Paramecium* from receptor potential, 521–522
  ionic movements in generation of, 357
    in myocardium, 373–374
  potassium channels involvement in, 361–364
  sodium channels involvement in, 358–361
  sodium pump in generation of, 389
Active transport
  of calcium
    see ATPase, calcium-activated
  of sodium, 389
    see Sodium pump
Adenine nucleotides
  presence in pulmonary vascular endothelium, 227, 231, 233
  caveolae intracellulares in metabolism of, 235
  See also Adenylate cyclase; ATPase; Cyclic AMP
Adenosine
  role in autoregulation of intestinal circulation, 22
Adenosine diphosphate
  See ADP

Adenosine monophosphate
  see Cyclic AMP
Adenosine triphosphate
  see ATP
Adenylate cyclase
  ADP-ribosylation of guanine nucleotide regulatory unit effect on, 663–664
  beta-adrenergic receptor activation of, 655–656, 658–659
  calmodulin regulation of, 669
  catecholamine activation of, 478–480
  fluoride-stimulated activity of, 663
  glucagon receptor activation of, 653–654
  guanine nucleotides in uncoupled systems, 661–663
  parathyroid hormone receptor activation of, 657
  periodic activity changes in chemotaxis of *Dictyostelium*, 543–545
  polypeptide and amine hormone regulation of, 653–665
  transfer of components and receptors to foreign membrane systems, 658–659
  See also Cyclic AMP
ADH
  see Vasopressin
ADP
  metabolism by pulmonary vascular endothelium, 231–233
ADP-ribosylation
  of guanine nucleotide regulatory unit, 663–664
Adrenarche, 595
  control in puberty, 607–609
  as independent process from gonadarche, 607
Adrenergic agents, beta
  cyclic AMP as second messenger in actions of, 667
  effect on liver circulation, 62–63
Adrenergic receptors
  beta
    desensitization of catecholamines to, 478–481

713

hypersensitivity related to depletion of catecholamines, 480–481
propranolol withdrawal syndrome related to, 480–481
radioactive ligand agonists for study of, 654–655
in regulation of activity of adenylate cyclase, 655–656, 658–659
subtypes in heart, 476–478
quantitative determination of, 477–478
catecholamine regulation of, 478–481
in heart, 475–481
direct demonstration of, 475–476
See also Catecholamines; Sympathetic nervous system
Aequorin
in calcium measurements within sarcoplasmic reticulum, 457–458
Aerotaxis, 511
Afterdepolarizations, 487
Aldosterone
effect on connecting tubule transport actions, 193–194
lungs role in increased secretion in heart failure, 247
See also Mineralocorticoids
Alkalosis
influence on liver circulation, 59
Alloxan pulmonary edema prostaglandins role in, 260
Alveolar macrophages
chemotactic factors produced in, 288
Amenorrhea
gonadotropin pulses in, 590
Amines, biogenic
effect on stomach blood flow, 6–8
metabolism in lungs, 269–278
disposition limited by intrapulmonary metabolism, 270
disposition limited by transport, 269–270
magnitude of clearance, 271–272
pathophysiology of clearance, 275–277
physiologic significance of clearance of, 271–274
unaffected by passage through lungs, 271

Amino acids
functionally important sites of calcium ATPase, 300–302
4-Aminopyridine
effect on potassium channels in muscle, 362
Aminopyrine clearance, 76
AMP
see Cyclic AMP
Anaphylatoxins, 284
Anaphylaxis, 283
eosinophil chemotactic factors of, 288
Androgens
influence on gonadal function, 609
Angiography of splanchnic circulation, 80
Angiotensin
drinking behavior induced by, 576
effect on intestinal blood flow and motility, 35
effects on lungs, 263–264
hepatic vascular actions of, 65–66
metabolism in lungs, 242–247, 264
in vitro enzyme kinetics of, 244–245
in vivo enzyme kinetics of, 245–247
relation to hypertension, 250–251
sites of hydrolysis of, 243
Angiotensin converting enzyme
hypertension therapy with inhibitors of, 247–251
hypoxia effect on, 251
inhibition associated with lung disease, 248
inhibition related to changes in bradykinin metabolism, 248–249
inhibitors of, 264
localization of
at caveolae intracellulares of pulmonary vascular endothelium, 235
at nonpulmonary sites, 264
in pulmonary vascular endothelium, 228–231
in lungs
ability to discriminate among homologs of angiotensin, 244–245
kinetics of, 246–247
multiple substrates of, 264
role in regulating blood pressure, 222
See also Angiotensin; Bradykinin

Anorexia nervosa
FSH and LH release in, 605
insulin receptors in, 644
Antidiuretic hormone
see Vasopressin
Arachidonic acid
biosynthesis of, 257–259
chemotactic factors derived from, 287–288
derivations as leukocyte chemoattractant, 557, 562–563
inhibition of synthesis of, 259–260
metabolism in lungs, 257–263
release in lungs, 260
related to lung injury, 260–261
Arcuate nucleus
in hypothalamic control of pituitary-ovarian axis, 588–589
increased sensitivity to inhibitory effects of sex steroids in childhood, 601
in pulsatile gonadotropin secretion, 584–585, 595–596
See also Hypothalamus
Arrhythmias
abnormal automaticity, 485–487
ectopic impulses induced by injury current, 488
electrogenic sodium pumping effects against, 397–398
mechanisms of, 485–492
reentry, 491–492
slow conduction, 489–491
triggered activity, 487–488
Aspirin
effect on gastric acid secretion, 10
ATP
effect on myocardial sodium pump, 377
effect on stoichiometry of sodium-calcium exchange in heart, 445
role in muscular efficiency during locomotion, 103
ATPase
calcium-activated, 337
in cilia action in Paramecium, 524
conformational changes related to steps coupled with enzyme reaction, 309–311
coupled with calcium pump, 302–309

SUBJECT INDEX 715

disposition of segments
  within sarcoplasmic
  membrane, 299
functionally important
  amino acid residues
  in, 300–302
molecular motion studies
  of, 310–311
muscle activity related to
  concentration of,
  347–348
muscle content during
  sarcoplasmic
  reticulum
  development, 338–339
in myocardial sarcolemma,
  402
in myocardial sarcoplasmic
  reticulum, 402–11
organization of
  polypeptides in,
  299–300
phosphoenzyme
  intermediates involved
  with, 304–306
phosphorylation site of,
  300
structure of, 298–302
subunit-subunit interaction
  and kinetics of,
  307–309
X-ray diffraction studies
  of, 311
sodium and potassium, 375
binding sites for cardiac
  glycosides on,
  383–384
concentrations in various
  myocardial cells, 376
inhibition of, 380–384
role in leukocyte
  chemotaxis, 559
turnover rates in
  myocardial cells,
  376–378
Autoimmune processes
in insulin resistance, 642–643
in myasthenia gravis, 328
in thyroid disease, 647–648
Automaticity
abnormalities of, 485–487
slow response, 487
Autonomic nervous system
in control of stomach
  circulation, 4–6
See also Central nervous
  system
Autoregulation
of hepatic circulation, 57–58
Autoregulatory escape from
  vasoconstrictor influence, 4
Aversive conditioning
endorphins effect on, 573

B

vasopressin effect on,
  577–578
Avoiding reaction in
  Paramecium, 524–525

B

Bacteria
behavior of, 615–616
chemotaxis of, 501–515
  methylation of transducers
    in, 505–510
  motility and, 501–502
  responses to changes in
    protonmotive force,
    512
  sensing system in, 502–505
  thermotactic response, 513
  transducer proteins in,
    505–507
motility of, 501–502
physiological aspects of
  behavior of, 501–503
sensing system of
  central components,
    504–505
  chemoreceptors, 503–504
  genetics of, 503–505
  sensing system of, 502–503
  transducers, 504–507
  weak acid repellents and pH
    taxis, 511–512
Baroreceptors
influence on liver blood flow,
  62
Basement membrane, 180–182
Batrachotoxin
action on fast sodium
  channel in heart,
  465–467
Behavior
neuropeptides effect on,
  571–579
approaches in study, 572
endorphins, 572–576
vasopressin, 576–579
of unicellular organisms,
  615–616
See also Chemotaxis
Bernoulli effect, 135
Beta-adrenergic receptors
see Adrenergic receptors,
  beta
Bicarbonate
reabsorption in collecting
  tubule, 196–197
reabsorption in proximal
  tubule, 183–185
ion mobility ratio with
  chloride and, 187
Bigeminy, 486
Biogenic amines
see Amines, biogenic

Birds
brain cooling in, 93
flight energetics in, 109–117
See also Flight energy
Bleomycin
effect on pulmonary amine
  clearance, 275
Blood pressure
influence of lung metabolism
  of polypeptides on,
  250–251
See also Hypertension
Bombesin-like peptide in lungs,
  263
Bradykinin
caveolae intracellulares in
  pulmonary vascular
  endothelium metabolism
  of, 235
effects of, 263
  on liver circulation, 64
  on lung vasculature, 250
metabolism in lungs,
  242–244, 264
hypoxia effect on, 251
in vitro kinetics of, 244
inhibition of angiotensin
  converting enzyme
  related to 248–249
by pulmonary vascular
  endothelium, 226–227,
  229–230
relation to hypertension, 250
release of prostaglandins
  induced by, 260
sites of hydrolysis of, 243
synthesis in lungs, 263
Brain cooling
in birds, 93
corneal evaporation and, 93
in exercise, 89–90
in humans, 92
importance in heat tolerance,
  85
in mammals, 85–92
with panting, 88
in warm environment, 88–89
Brain temperature
body temperature
  relationship to, 87–90
cerebral vasculature
  temperature and, 87–92
control of
  in birds, 93
  in mammals, 85–92
during exercise, 89
factors determining, 85–86
temperature in cerebral
  circulation as factor, 86
upper respiratory evaporation
  related to, 88–93
venous drainage to cavernous
  sinus related to, 87

SUBJECT INDEX

in warm-bodied fish during swimming, 130
See also Brain cooling
Breathing
see Panting

C

C5a
  role as leukocyte chemoattractant, 555–556
Calcitonin
  effect on urinary excretion, 192–193
  evolution of function of, 620
  studies on gene for, 619
Calcium
  active transport of
    see ATPase, calcium-activated; Calcium pump
  binding proteins synthesized in sarcoplasmic reticulum, 342–343
  in cilia action, 523–524
  coordinated regulation of phosphorylation with cyclic AMP, 671–672
  effect on acetylcholine receptor, 325
  effect on myocardial sodium pump, 377–378
  influence on calcium-ATPase synthesis in muscle, 345–347
  influx into *Paramecium*, 522–523
  ionic channels in heart, 425–432
    blockers of, 426
    inactivation of, 427–429
    in myocardial sarcolemma, 417
    selectivity and saturation of, 426
  ionic channels in muscle, 366–368
    selectivity of, 367–368
    site of localization of, 367
    in slow muscle fibers, 369
  ionic channels in *Paramecium*, 527–528
    role in action potential generation in *Paramecium*, 521–522
  in leukocyte chemotaxis, 559–560, 563
  in myocardial contraction, 374
  Paramecium reaction to, 527, 530
  receptor for, 668–669

regulation of accumulation in platelets, 413
regulation of muscle cell concentration of, 344–349
release from cardiac sarcoplasmic reticulum, 451–460
  calcium effect on, 452
  and cardic cycle, 459–460
  changes in intracellular membranes and, 455–457
  chloride effect on, 452–454
  measurements of intracellular calcium concentrations during, 457–459
  mechanisms triggering, 451–454
role in muscle contraction and relaxation, 297
role in regulation of cell metabolism, 668–669
as second messenger, 667–669
slow inward current in heart, 425–432
  activation of, 426
  blockers of, 426
  calcium-activated potassium conductance, 429–430
  calcium-induced inactivation of, 427–429
  physiological and pharmacological interventions, 430–431
  properties of, 425–429
  steady state, 427–428
sodium exchange in heart with, 435–447
  membrane carrier role in, 437–438
  role in contractile control, 445–446
  sodium binding in sarcolemma membrane related to, 436–438
  stoichiometry of, 444–445
  in transport across sarcolemmal membrane, 439–444
transport of
  calmodulin in regulation of, 669–670
  in connecting tubule, 194
  in sarcoplasmic membrane, 300, 304, 337
  in thick ascending limb of Henle's loop, 191

See also ATPase, calcium-activated; Calmodulin
Calcium pump
  ATPase reaction coupled with, 302–309
  sequential steps in, 304–309
  calmodulin involvement with, 669–670
  mechanism involved in transport in muscle sarcoplasmic reticulum, 297–298
  in myocardial cells, 402–417
  phosphoenzyme intermediates involved with, 304–306
  phospholamban phosphorylation effect on, 405–406, 409–411
  structure and function of protein of, 297–312
  three configurations of protein of, 302, 303
  See also ATPase, calcium-activated
Calmodulin
  adenylate cyclase regulation by, 669
  calcium transport regulated by, 669–670
  and cell motility, 676–677
  in cilia action, 524
  dopamine action mediated by, 674
  in GnRH-induced LH release, 674–675
  hormone-mediated redistribution of, 672–675
  in myosin heavy chain dephosphorylation, 541
  phenothiazines in study of, 673–674
  in phosphorylation of cardiac sarcoplasmic reticulum, 408–409
  properties of, 668–669
  and protein phosphorylation, 670–672
  as receptor, 668–669
  and receptor-mediated endocytosis, 675–676
  role in endocrine cells, 668–678
Calsequestrin, 338, 402
  cell-free synthesis of, 343–344
  in heavy fraction of sarcoplasmic reticulum, 453

SUBJECT INDEX 717

synthesis during sarcoplasmic reticulum development, 342–343
Capillary density and intestinal blood flow, 16
Capillary filtration coefficient, 78
Capillary recruitment, 15–16
Captopril
  complications associated with, 248
  therapeutic uses of, 247–248
Carbon dioxide
  influence on intestinal circulation, 21–22
  liver circulation related to tension of, 59
Carbonic anhydrase activity in collecting ducts, 171–172
Carboxypeptidase
  effect on chemotactic factors, 284
Cardiac glycosides
  effect on sodium, potassium ATPase, 380–382
  endogenous ligands of receptor of, 383–384
  myocardial toxicity of, 381–382
  receptor for therapeutic action of, 381
Cardiotoxins
  aconitine, 465–467
  action on cardiac plasma membranes, 463–471
  batrachotoxin, 465–467
  ervatamine, 470
  grayanotoxins, 465–467
  palytoxin, 470
  saxitoxin, 463–465
  scorpion and sea anemone toxins, 468–469
  snake venom, 470
  on sodium ionic channels, 463–470
  in study of differentiation of cardiac sodium channels, 469–470
  tetrodotoxin, 463–465
  veratridine, 465–467
Catecholamines
  actions on heart, 475
  adenylate cyclase activation by, 478–480
  desensitization of beta-adrenergic receptors to, 478–481
  effects on liver circulation, 62–63
  phospholamban phosphorylation in cardiac sarcoplasmic reticulum related to

inotropic action of, 411–412
  receptors in heart, 475–481
  regulation of adrenergic receptors by, 478–481
  See also Adrenergic receptors; Epinephrine; Norepinephrine
Cathepsin B
  in prohormone conversion to hormone, 632–633
Caveolae intracellulares, 234–235
Cavernous sinus, 86–87
CCK
  intestinal blood flow and motility effects of, 34–35
Cell coupling
  conduction velocity related to, 489–490
Cell membrane
  of myocardial cell, 373
Cellular slime moulds
  see Dictyostelium discoideum
Central nervous system
  in control of liver circulation, 61–62
  in control of onset of puberty, 595–609
  in control of ovarian function, 583–590
    arcuate nucleus and pulsatile gonadotropin secretion, 583–585
  in control of stomach blood flow, 3–6
  hypothalamic control of pituitary ovarian axis, 588–590
  influence on intestinal absorption, 51
  ovarian control of gonadotropin secretion, 585–588
Cerebral circulation
  brain temperature related to temperature in, 87–92
  breathing rate related to, 90–92
  circle of Willis, 86
  importance of temperature as factor affecting brain temperature, 86
  rate of blood flow related to heat stress in, 90–92
  venous drainage to cavernous sinus, 86–87
Chemoattractants
  arachidonic acid derivatives, 557
  C5a, 555–556
  cell surface binding sites for, 537–539

cyclic AMP as, 535–539, 543–546
folic acid, 536
formylated peptides, 556–557
of leukocytes, 555–557
membrane potential changes stimulated by, 559
oligopeptides, 536
oxygen, 536–537
regulation of binding to receptor in leukocytes, 557–559
regulation of destroying enzyme, 545–546
in various Dictyostelium species, 536–537
Chemoreceptors
  in bacterial chemotaxis, 503–504
Chemotactic mediators
  arachidonic acid metabolites in generation of, 287–288
  cell-derived, 285–289
  complement-derived, 284–285
  experimental studies on role in lung injury, 289–291
  and immune complex-associated lung injury, 290–291
  inhibitors in lung, 284–285
  intrapulmonary effects on lungs, 289–290
  intravascular effects on lungs, 289
  involved in lung injury, 283–291
    experimental studies on, 289–291
    origin and regulation of, 283–289
  lymphokines, 288
  mast cell-derived, 288
  oxygen metabolites in generation of, 286–287
  produced by alveolar macrophages, 288
Chemotaxis
  bacterial, 501–515
    adaptation and de-adaptation in, 505–506
    attractants, 501
    central components of sensing system in, 504–505
    chemoreceptors in, 503–504
    comparator function of transducers in, 507
    control of methylation in, 507–510

718 SUBJECT INDEX

evolution of, 514
genetics of, 503–505, 508–510
motility and, 501–502
phosphotransferase system in, 510–511, 513
repellents, 501
responses to changes in protonmotive force, 512
sensory strategy and, 502–503
thermotactic response, 513
transducers in, 504–507
weak acid repellents and pH taxis, 511–512
in *Dictyostelium*, 535–536
cell behavior during, 542
chemoattractants in various species, 536–537
cyclic AMP as chemoattractant in 535–539, 543–546
functions of, 535–536
gradient detection in, 543
periodic synthesis and release of cyclic AMP in, 543–545
reactions probably involved in, 539–542
of leukocytes, 553–565
attractant binding to receptor in, 557–559
biochemical characteristics of, 555–564
chemoattractants involved in, 555–557
cyclic nucleotide changes in, 560
general characteristics of, 553–555
ionic events involved in, 559–560
in locomotion, 563–564
measurement of, 555
membrane potential changes in, 559
methyl-accepting proteins of, 505–510
phospholipid metabolism in, 561–563
protein methylation metabolism in, 560–561
methyl-accepting proteins of, 505–510
receptor-mediated events in, 554–555, 559–563
in *Paramecium*
action potential generation in, 521–522

avoiding reaction, 525–526
calcium influx in, 522–523
ciliary reversal in, 523–524
hyperpolarization and speed control of cilia, 524–525
prospects in study of, 531–532
receptor potential generation in, 520–521
responses to chemicals, 527, 580
in study of cell behavioral response, 499–500
universality in biological systems, 499
variety of cells seen in, 553
See also Chemoattractants
Chick embryo heart cells
sodium pump in, 393
Chloride
effect on calcium release from cardiac sarcoplasmic reticulum, 452–454
ionic channels in muscle, 364–366
in slow fibers, 368
reabsorption of
cotransport process with sodium, 191–192
ion mobility ratio with bicarbonate and, 187
ion mobility ratio with sodium and, 186–188
in proximal tubule, 183, 185–188
in thick ascending limb of Henle's loop, 190–192
Cholecystokinin
and gastric blood flow, 8
influence on liver circulation, 64
prohormone conversion to, 633, 634
Cholera toxin
in ADP-ribosylation, 663–664
effects of toxin on intestinal absorption, 50–51, 53
Chronic granulomatous disease
abnormal membrane potential changes in chemotactic response, 559
Cilia
beat cycle in *Paramecium*, 523
calcium role in action of, 523–524
hyperpolarization and speed control of, 524
repellents to, 530

Circadian rhythm of
gonadotropin release in puberty, 599–600
Circulation
See Cerebral circulation; Gastrointestinal blood flow measurement; Intestinal circulation; Liver circulation; Stomach circulation
Clearance methods for gastrointestinal blood flow, 76
Clearance of tritiated water, 78–79
Colchicine
effect on leukocyte subcellular structure, 554
Collecting ducts, 167–172
in acid-base balance, 171–172
antidiuretic hormone effect on, 172
heterogeneity of sodium reabsorption in, 207–209
intercalated cells of, 168–171
phosphate reabsorption in, 204–205
in potassium excretion, 170–171
subdivisions of, 194
transport processes in, 194–197
Comparative biology
bird flight energetics, 109–117
brain cooling studied by, 85–93
energetics of locomotion in endothermic insects, 133–140
in evolution of hormones, 615–623
link between energetics and mechanics of locomotion studied by, 97–98
locomotion in fish, 121–130
Complement
chemotactic factors derived from, 284–285
inhibitors of, 284–285
Conductance, intestinal during absorption, 48–49
Conduction system in heart
see Arrhythmias
Conduction velocity in heart, 489–491
action potential depolarization role in, 489
cell coupling role in, 489–490
cell excitability influence on, 490–491

SUBJECT INDEX 719

Congestive heart failure
  lungs role in secondary hyperaldosteronism of, 247
Connecting tubule, 165–167
  classification of, 167
  transport processes in, 193–194
  two cell types in, 165
Corticosteroids
  see Glucocorticoids
Countercurrent exchange system
  in intestinal absorption, 53
  medullary recycling of urea in, 212
  nephron heterogeneity role in, 211–216
  role in brain cooling during exercise, 89–90
  vasopressin enhancement of sodium transport and, 214
Coupling ratio
  and electrogenic mechanism of sodium pump, 395
Cyclic AMP
  in active calcium transport in cardiac sarcoplasmic reticulum, 405–406
  attraction by *Dictyostelium*, 535–539, 543–546
  coordinated regulation of phosphorylation with calcium, 671–672
  destruction in *Dictyostelium*, 545–546
  in *Dictyostelium* locomotion, 542
  in leukocyte chemotaxis, 560
  in mediating inotropic effect of catecholamines, 412
  as mediator in *Dictyostelium* chemotaxis, 616
  periodic synthesis and release in *Dictyostelium*, 543–545
  receptors for, 537–538
  as regulator of acetylcholine receptor levels, 325
  as second messenger, 667
  short-term responses as mediators of chemotaxis, 539
  stimulation of own production, 535–536
  See also Adenylate cyclase
Cyclic GMP
  in leukocyte chemotaxis, 560
  role in chemotaxis in *Dictyostelium*, 540–541
  See also Guanine nucleotide regulatory protein

Cyclooxygenase, 287
  in prostaglandin synthesis, 257–258
Cysteinyl residues
  in calcium ATPase polypeptides, 300–301
Cytochalasin B
  effect on leukocyte subcellular structures, 554

D

Desensitization of
  beta-adrenergic receptors, 478–481
  to isoproterenol, 478–480
  relevance to hypertension, 479–481
Diabetes mellitus
  insulin receptors in, 641, 645–46
Dictyostelium discoideum
  cell behavior during chemotaxis in, 542
  chemoattractants in various species of, 536–537
  chemotaxis in, 535–546
  cyclic AMP as chemoattractant in, 535–539, 543–546, 616
  folic acid as chemoattractant in, 536, 538–539
  gradient detection in, 543
  reactions involved in chemotactic response of, 539–542
  cyclic GMP regulation, 540–541
  myosin heavy chain dephosphorylation, 541
  phospholipid methylation, 541–542
  short-term responses to cyclic AMP, 539
  regulation of enzymes in, 545–546
Digitalis
  See Cardiac glycosides
Distal tubule, 161–164
  cortical ascending thick limb of loop of Henle, 162
  distal convoluted tubule, 162
  hormone sensitivities in, 167
  divisions of, 161, 189–190
  electrolyte concentrations in nuclei of, 175
  medullary ascending thick limb of loop of Henle, 161–162

  transport processes in, 192–193
Dopamine
  calmodulin role in action of, 674
  effect of lung passage on, 271
  effect on liver circulation, 63

E

Ectopic heart rhythm, 486
  induced by injury current, 488
Elastase
  chemotactic factors activated by, 285–287
  inhibitor of, 249
Elastic strain energy, 99
Electrogenic sodium pump, 389–398
Electroneutral sodium pump, 389, 394–395
Endocytosis
  calmodulin and receptor-mediated, 675–676
Endorphins
  and appetitively motivated learning, 574–576
  and aversively motivated behaviors, 572–574
  clearance from blood, 575
  effect on memory, 573–574
  processing of prohormone of, 633
  vasopressin secretion stimulated by, 576
Endothelium
  See Pulmonary vascular endothelium
Endothermic insects
  energetics of locomotion in, 133–140
Energy
  conversion to work in locomotion, 97–106
  elastic strain, 99
  during flight
    see Flight energy
  mechanical during locomotion, 99–101
  metabolic cost of generating force during locomotion, 103–105
  storage and recovery in elastic elements during locomotion, 103
Energy consumption
  comparative approach to locomotion, 97–98
  efficiency during locomotion, 101–103

in fish locomotion, 121–130
  caloric input related to, 126–128
  in circulating blood, 125
  efficiency during tuna locomotion, 123–124
  in ion-osmoregulation, 125–126
  in irrigating gills, 125
  thermoconservation and, 129–130
  thrust/drag relationships and, 128–129
in flying bird, 109–117
  See also Flight energy
as function of speed and body size, 98–101
in insect locomotion, 133–140
  flight, 135–140
  jumping, 135
  walking and running, 134–135
measurement in flying birds, 109–110
rate during locomotion, 104–105
Enkephalin
  processing of prohormone of, 633
Enkephalinase inhibitor, 249
Enzymes
  induction effect on liver circulation
  kinetics of metabolic reaction in lungs, 244–247
  localization in pulmonary vascular endothelium, 223–237
  See also Angiotensin converting enzyme; ATPase
Eosinophil chemotactic factors of anaphylaxis, 288
Epidermal growth factor
  prohormone processing of, 634
Epinephrine
  affinity for beta-receptor subtypes, 476
  effect on intestinal smooth muscle, 34
  effect on liver circulation, 63
  effect on slow inward calcium current in heart, 430, 431
  effect on stomach blood flow, 6–7
  lack of effect of passage through lungs on, 271
Ervatamine
  as cardiotoxin, 470

Escherichia coli
  effect of toxin on intestinal absorption, 50–51
  flagella and motility of, 502
Estradiol
  in control of gonadotropin secretion, 585–588
  maturation of positive feedback on gonadotropins, 606–607
  negative feedback on gonadotropins, 602–603
Estrogen
  positive feedback effect on gonadotropins, 597
  role in adrenarche, 608–609
  See also Estradiol
Evolution
  of peptide hormones, 615–623
  conservation of structure in, 622
  gene duplication in, 617–619
  gene repression in, 619–620
  single hormone functioning in new role in, 620–621
Exercise
  brain cooling in, 89–90
  lungs role in physiologic hypertension of, 247
Exocytosis
  in hormone-receptor mediated activity, 673
Extracellular space
  nephron heterogeneity in maintaining fluid balance with expansion of, 206–211
  prostaglandins role in maintenance of, 210–211

F

Fatty acids
  effect on myocardial sodium pump, 378
Fibronectin
  in glomerulus, 150
Fick principle, 74, 76
Fish
  energetics of locomotion in, 121–130
  thermoconservation and energetics in, 129–130
Flagella
  energization of motion of, 502
  motility in *Escherichia coli* related to, 502

regulation of rotation in bacterial chemotaxis, 508–509
Flight energy
  in birds, 109–117
    aerodynamic powers involved in, 110
    during boundary flight, 115
    efficiency of, 112
    during foraging, 116
    during forward flight, 110–112
    during gliding and soaring, 114
    during group flight, 115–116
    during hovering, 114–115
    measurement of, 109–110
    metabolic power curves in, 111–112
    power consumption and, 110–112
    power-velocity curves in, 111, 112
    size as variable, 112–114
    strategies for improving performance, 114–117
    theoretical models of, 111–112
  in endothermic insects, 135–140
    aerodynamic forces involved in, 135–137
    gliding and soaring, 137
    heat transfer, 137–138
    indirect measures of power in, 138–139
    oxygen consumption, 138
    power measurements during, 138–139
    power requirements in, 135–138
    utilization of flight fuels, 139
    during warm-up, 139–140
Flow meter
  in measurement of gastrointestinal blood flow, 73–74
Fluorescence microscopy
  in evaluation of gastrointestinal circulation, 72
Fluorescent excitation analysis, 81
Fluoride
  stimulation of adenylate cyclase activity by, 663
Folic acid
  as chemoattractant, 536

SUBJECT INDEX 721

inactivation in *Dictyostelium*, 546
receptors for, 538
Follicle stimulating hormone (FSH)
  change in ratio to LH during puberty, 604–605
  difference from LH in inhibitory effect of GnRH on, 585
  discordance with LH release during maturation, 604–605
  levels from fetal life through puberty, 598–599
  pulsatile release in puberty, 599
  pulsatile secretion of, 583–585
  See also Gonadotropins
Formylated peptides
  as leukocyte chemoattractant, 556–557
  regulation of binding to receptor, 557–559
Friction
  effect on muscle work in locomotion, 101
FSH
  See Follicle stimulating hormone

G

Gastric acid secretion
  clearance methods in evaluation of, 76
  nervous system control of, 3–6
  stomach blood flow and, 9–10
Gastrin
  acid secretion and, 9–10
  effect on stomach blood flow, 8
  prohormone conversion to, 633
Gastrointestinal blood flow measurement, 71–78
  angiography in, 80
  capillary filtration coefficient in, 78
  clearance methods, 76
  clearance of tritiated water in, 78–79
  clinical techniques for, 79–81
  criteria for selection of technique, 71
  direct visualization techniques, 71–73
  fluorescent excitation analysis in, 81

fractionation of isotopes and microspheres in, 76–77
functional techniques for, 77–79
in vitro vascular segments in, 79
indicator dilution methods in, 74–75
intraperitoneal xenon washout method for, 80–81
morphologic studies in, 73
noncannulating flow meters in, 73–74
oxygen tension in, 79
venous outflow method for, 73
video dilution technique for, 80
washout techniques in, 75–76, 80–81
See also Intestinal circulation; Liver circulation
Genes
  calcium role in expression in muscle, 345–347
  duplication in peptide hormone evolution, 617–619
  nucleotide sequence in genes coding for hormones, 622
  repression in evolution of hormones, 619–620
Genetics
  of bacterial chemotaxis, 503–505, 508–510
  of hormone secretion, 616–617
Gliding energetics, 114
  in endothermic insects, 137
Glomerular basement membrane
  morphology of, 150–152
Glomerular filtration rate
  vasopressin and heterogeneity of, 213–215
Glomerulus
  endothelium in acute renal failure, 152
  glycoproteins in, 150
  glycosaminoglycans in, 150
  podocyte in, 150–152
  types of cells cloned from epithelium of, 152
Glucagon
  activation of adenylate cyclase in uncoupled system, 662

effect on intestinal absorption, 49
and gastric blood flow, 8
hepatic vascular actions of, 64
proglucagon conversion to, 633
regulation of bound receptor activation of adenylate cyclase, 653–654
Glucocorticoids
  inhibition of liberation of arachidonic acid, 259
leukocyte chemotaxis effects of, 562
Glucose
  effect on intestinal circulation, 19
  reabsorption in proximal tubule, 183–185
  See also Insulin
Glycine
  reabsorption in proximal tubule, 183, 184
Glycosaminoglycans in glomerulus, 150
GMP
  see Cyclic GMP
GnRH
  see Gonadotropin releasing hormone
Golgi apparatus
  in sarcoplasmic reticulum synthesis, 343
  in synthesis of acetylcholine receptor, 321–322
Gonadarche, 595
Gonadotropin releasing hormone (GnRH)
  calmodulin role in LH release induced by, 674–675
  continuous infusion inhibition of gonadotropin secretion, 584–585, 598
  difference in inhibitory effect on FSH and LH, 585
  functional inefficiency prepubertally, 604
  inhibition of release during childhood, 600–603
  rate of release and ovarian function, 589–590
  rhythmic discharges of, 583–585
  synthetic hormone interaction with pituitary gonadotropin secretion, 603–605
  therapeutic use in amenorrhea, 588, 590

Gonadotropins
  arcuate nucleus and pulsatile
      secretion of, 583–585
  continuous infusion of
      GnRH inhibitory effect
      on, 584–585, 598
  disorders characterized by
      infrequency of, 590
  episodic release in puberty,
      599–600
  estradiol negative feedback
      on, 602–603
  in gonadal dysgenesis, 602
  levels from fetal life through
      puberty, 598–599
  maturation of estradiol
      positive feedback effect
      on, 606–607
  ovarian control of secretion
      of, 585–588
    estradiol role, 585–588
    hypothalamus versus
        pituitary as site of,
        586–588
    progesterone role, 585–586,
        588, 590
  patterns of secretion of,
      597–598
  prepubertal restraint of
      release of
    intrinsic CNS inhibitory
        mechanism, 602
    negative feedback
        mechanism, 600–603
  release following GnRH
      administration, 604–605
  synthetic GnRH interaction
      with pituitary secretion
      of, 603–605
  See also Follicular
      stimulating hormone;
      Leuteinizing hormone
Grayanotoxins
  action on fast sodium
      channel in heart,
      465–467
Growth hormone
  insulin receptors in deficiency
      of, 644
  insulin resistance associated
      with excess of, 641–642
  receptor of, 647
GTP
  role in hormone-receptor
      interactions, 659–660
  See also Cyclic GMP;
      Guanine nucleotide
      regulatory protein
Guanine nucleotide regulatory
    protein, 653
  and beta-adrenergic receptor
      interaction, 656

cholera toxin and
    ADP-ribosylation of,
    663–664
  in disease, 664–665
  in fluoride-stimulated
      adenylate cyclase
      activity, 663
  and glucagon receptor
      interaction, 653–654
  mechanism of, 659–662
  and parathyroid hormone,
      657
  purification of, 664
  and receptor affinity, 661
  receptor complex with, 658
  in uncoupled adenylate
      cyclase systems, 661–663
  variety of receptor-adenylate
      cyclase systems coupled
      through, 657

H

Heart
  adrenergic receptors in,
      475–481
  arrhythmia mechanisms,
      485–492
  cardiotoxins effect on,
      463–471
  injury currents in, 488
  slow inward calcium current
      in, 425–432
Heat exchange
  nasal cavity properties for,
      87
  See also Brain cooling
Heat transfer
  by insects in flight, 137–138
Henle, loop of
  in classifying nephrons,
      148–149
  cloning of cells of, 173
  species variation in, 149
  thick ascending limb,
      161–162
    cortical, 161
    heterogeneity of sodium
        reabsorption in, 213
    medullary, 161–162
    morphologic subdivisions
        of, 190
    in providing maximal
        medullary
        hypertonicity, 191
    role in heterogeneity of
        sodium reabsorption,
        210
    transport processes in,
        190–192
  thin limb segments, 157–161

ascending limb of
    long-looped nephrons,
    161
  descending short limb, 158
  four regions of, 157
  lower descending limb of
      long-looped nephrons,
      160–161
  role in heterogeneity of
      sodium reabsorption,
      210
  role in urine osmolality
      changes, 189
  transport processes in,
      188–189
  upper descending limb of
      long-looped nephron,
      159–160
  urea exchange in, 159, 212
Hepatic artery flow
  see Liver circulation
Histamine
  acid secretion and, 9–10
  effect on liver circulation, 65
  effect on stomach blood flow,
      7–8
  intestinal blood flow and
      motility effects of, 35
  lack of effect of passage
      through lungs, 271
  receptor role in gastric
      circulation, 7–8
Histidine
  as functionally important site
      of calcium ATPase, 302
Histones
  in regulation of sarcoplasmic
      reticulum protein
      synthesis, 347
Hormones
  biosynthesis from larger
      polypeptides, 625–635
  conformational changes in,
      628–629
  loop model of, 626–628
  pore model of, 626
  preprotein cleavage
      mechanisms in, 629
  prohormone processing in,
      630–634
  proteases involved in,
      631–634
  trigger hypothesis of,
      628–629
  calmodulin role in
      receptor-mediated
      activity, 672–675
  evolution of
    conservation of structure
        in, 622
    first appearance in,
        615–617

SUBJECT INDEX 723

gene duplication in,
617–619
gene repression in,
619–620
patterns of, 617–622
single hormone functioning
in new role in,
620–621
gastrointestinal
effect on stomach blood
flow, 8
influence on liver
circulation, 63–64
lack of value of blood levels
of locally active
peptides, 621
lung clearance of, 241–252
receptors for, 639–640
as activator of target cell,
640
alterations in human
diseases, 639–648
fundamental functions of,
639
growth hormone, 647
guanine nucleotide
regulatory protein in
interaction of,
659–662
insulin, 641–646
insulin-like growth factors,
646
methods of characterizing,
640–641
prolactin, 647
regulation of interactions
after binding, 653–665
second messenger initiated
by binding to,
639–640
thyrotropin, 647–648
vasopressin, 646–647
sequence variability in
species, 622
in unicellular organism,
616–617
Hovering flight energetics,
114–115, 138–139
Hydrogen ion
transport in collecting
tubules, 196–197
See also Acidification of
urine
5-Hydroxytryptamine (5HT)
lung metabolism of
effect on lung circulation,
273 274
limited by transport,
269–270
magnitude of, 271–273
oxygen lung toxicity
mediated through
change in, 276

platelet aggregation
modified by, 273
as marker of pulmonary
endothelial cell damage,
275
possible role in pulmonary
hypertension, 276–277
Hyperthermia, intestinal
motility effects on, 33
postprandial, 18–19
Hypertension
angiotensin converting
enzyme inhibitors in
treatment of, 247–251
desensitization of
beta-adrenergic receptors
relevance to, 479–481
lungs role in
exercise-induced, 247
Hypoglycemia
insulin levels with non-islet
cell tumors, 646
See also Insulin
Hypogonadotropic
hypogonadism
response to GnRH in, 605
Hypothalamo-pituitary ovarian
axis, 588–590
ontogeny of, 608
Hypothalamus
arcuate nucleus role in
gonadotropin secretion,
584–585, 595–596
control of pituitary-ovarian
axis by, 588–590
in estrogen-induced negative
feedback on
gonadotropins, 602–603
medial basal, 584
role in estrogen and
progesterone feedback
actions, 586–588
role in gonadotropin release,
584–585
temperature of, 92
See also Brain cooling;
Brain temperature
Hypoxia
effect on angiotensin
converting enzyme, 264
effect on lung metabolic
functions, 251

I

Immune complexes
and neutrophil-dependent
lung injury, 290–291
Indicator dilution methods
for gastrointestinal blood
flow, 74–75
Injury currents inducing
ectopic impulses, 488

Insects
energetics of locomotion in,
133–140
flight, 135–140
terrestrial, 134–135
Insulin
antibodies against, 642–643
deficiency of, 644
excess of, 645
in newborns, 645
proinsulin conversion to,
631–632
receptors of
in acromegaly, 641–642
in diabetes mellitus, 641
disorders of specificity of,
645–646
in hypoglycemia with
non-islet cell tumors,
646
in obesity, 641
post-binding defects of,
644
in primary target cell
disorders, 643–644
resistance to
extreme, 642–644
moderate, 641–642
structure-function
characteristics of, 622
supersensitivity to, 644
Insulin-like growth factors
receptors of, 646
Intestinal absorption
blood flow related to, 43–54
transport mechanisms in,
44–45
washout effect, 52–53
conductance changes during,
48–49
countercurrent exchange role
in, 53
effective mucosal blood flow
and, 44
hemodynamics interaction
with, 43–54
liver blood flow after, 60–61
oxygen delivery during,
45–46
regulation of, 49–52
bacterial toxins in, 50–51
glucagon in, 49
neural, 51
opiates and, 50
prostaglandins in, 52
somatostatin in, 49–50
vasoactive intestinal
peptide in, 50
Starling forces influence on,
46–48
Intestinal circulation
absorption related to, 53–54

SUBJECT INDEX

arterial hypoxia effect on, 17–18
capillary density and, 16
chemicals and neurotransmitters effect on, 33–36
colloid osmotic pressure influences on, 46–48
in different tissue layers, 30, 32–33, 35–37
effective mucosal blood flow, 44
glucose effects on, 19
local control of, 13–25
during absorption, 45–46
glucose role in, 19
interaction between metabolic and myogenic mechanisms in, 23–25
interstitial hypoxia role in, 22–23
metabolic feedback signal controlling, 21–23
metabolic model of, 14–19
myogenic theory of, 14, 23–25
pressure-flow relationships in, 13–14, 19–21
reactive hyperemia and, 18
luminal distension effect on, 36–38
microspheres in measurement of, 77
motility related to, 29–40
blood flow effect on motility, 38–40
motility effect on blood flow, 29–38
postprandial hyperemia, 18–19
precapillary sphincter role in, 14–15
problems in measurement of, 1
regulation of, 1–2
See also Intestinal circulation, local control of
transport mechanisms and, 44–45
various chemicals effect on, 33–36
villous microvasculature and, 43–44
See also Gastrointestinal blood flow measurement; Liver circulation; Stomach circulation
Intestinal distension
effect on blood flow, 36–38
Intestinal motility
blood flow effect on, 38–40

effect on intestinal blood flow, 29–38
chemical and neurotransmitter studies of, 33–36
luminal distension and, 36–38
mechanisms of, 29–31
rhythmic and tonic contractions, 31–33
effect on lymphatics, 37–38
ischemia effects on, 38–40
various chemicals effect on, 33–36
Ion mobility ratio
and proximal tubule reabsorption, 186–188
Ionic channels
in heart
calcium, 417
slow-inward calcium current, 425–432
sodium pump, 375–384, 389–398
sodium-calcium exchange, 435–447
in *Paramecium* locomotion, 526–529
in skeletal muscle, 357–369
calcium, 366–368
chloride, 364–366
in slow (tonic) fibers, 368–369
sodium, 358–361
Ion-osmoregulation
in fish locomotion, 125–126
Isoproterenol
desensitization of heart to, 478–479
effect on acid secretion, 9
effect on stomach blood flow, 6–7
Isotopes
fractional distribution in measuring gastrointestinal blood flow, 76–77

J

Jumping
energetics in insects, 135
Juxtamedullary nephrons
arteriole patterns of, 148
comparison of proximal tubule reabsorption with superficial nephrons, 185–186
connecting tubule in, 165
efferent arterioles of, 181
loops of Henle in, 149
proximal segments of, 182
proximal tubule in, 154, 182

role in heterogeneity of sodium excretion, 209–211
role in phosphate reabsorption, 204–206
selective response of filtration rate to vasopressin, 214–215
thick ascending limb of Henle's loop in, 190
thin loops of Henle in, 189
See also Nephrons, heterogeneity of

K

Kallikrein
proinsulin converting activity of, 632
Kidney
heterogeneity of tubular transport function in, 181–197
see Nephrons, heterogeneity of
morphology of
basic principles underlying study of, 147
cloning of renal cell populations in study of, 172–173
collecting ducts, 167–172
connecting tubule, 165–167
distal tubule, 161–164
glomerular basement membrane, 150–152
in vitro culture of nephron cell populations in, 173–174
new techniques in, 172–173
proximal tubule, 152, 154–157
renal epithelial cell lines in study of, 172
thin limb segments, 157–161
three types of renal corpuscles in, 147–148
X-ray microanalysis in study of, 174–175
See also Nephrons
Kininase
see Angiotensin converting enzyme

L

Laminin in glomerulus, 150
Learning
endorphins effect on appetitively motivated, 574–576

SUBJECT INDEX 725

aversively motivated, 572–574
vasopressin effect on, 577–578
appetitively motivated, 578
aversively motivated, 577–578
Leukocytes
chemoattractants of, 555–557
chemotactic factors derived from, 285–289
chemotaxis of, 553–565
biochemical characteristics of, 555–564
cyclic nucleotide changes in, 560
general characteristics of, 553–555
ionic events involved in, 559–560
in locomotion, 563–564
measurement of, 555
membrane phospholipid metabolism in, 561–563
membrane potential changes in, 559
protein methylester metabolism in, 560–561
receptor-mediated events in, 554–555, 559–563
regulation of attractant-receptor binding in, 557–559
sensing system in, 554
role in response to lung injury, 283
Leukotrienes
biosynthesis in lungs, 259
chemotactic properties of, 287–288
effects caused by, 259
inactivation of, 262
LH
see Luteinizing hormone
Linoleic acid
chemotactic activity of, 288
Lipids
importance in sarcoplasmic membrane development, 341–342
influence on beta-adrenergic receptor regulation of adenylate cyclase, 656
Lipomodulin, 565
Lipoxygenase, 287
Liver circulation
arterial blood composition changes affecting, 59
blood gas tensions influence on, 59–60

enzyme induction effect on, 61
extrinsic regulation of, 61–66
autacoids in, 64–65
catecholamines in, 62–63
gastrointestinal hormones in, 63–64
nervous system in, 61–62
vasoconstrictor peptides in, 65–66
intrinsic vasoregulation of, 57–59
hepatic arterial-portal interactions in, 58–59
pressure-flow autoregulation, 57–58
venous pressure elevation and, 58
liver function related to, 60–61
metabolic stimulation effect on, 61
osmolarity influence on, 60
portal venous blood composition changes affecting, 59–60
postprandial effects on, 60–61
principle determinants of, 57
See also Portal vascular system
Locomotion
calmodulin role in cell motility, 676–677
comparative biology approach to, 97–106
in *Dictyostelium*, 542
differences on land and in water, 122
energy consumption and speed and body size in, 98–101
in fish, 121–130
circulating the blood in, 125
efficiency of, 123–124
energy cost of, 122–129
ion-osmoregulation in, 125–126
irrigating gills in, 125
metabolic costs based on caloric inputs, 126–128
oxygen uptake measures in study of, 123–124
speed related to mass, 126–128
thermoconversion and, 129–130
thrust/drag relationships and, 128–129
frictional losses in, 101
in insects
flight, 135–140

terrestrial, 134–135
in leukocytes
chemotaxis-mediated effects, 563–564
general characteristics of, 553–555
link between energetics and mechanics of, 97–106
metabolic cost of generating force during, 103–105
muscular efficiency during, 101–103
in *Paramecium* adaptation, 526
attraction and repulsion mechanisms, 525
basic schema of, 520–524
hyperpolarization and speed control in, 524
ionic channels role in, 526–529
speed of, 524–525
rate of muscle performance of work in, 99–101
speed and body size determinants of, 99–101
storage and recovery of energy in elastic elements during, 103
See also Chemotaxis; Flight energy
Loop model of hormone biosynthesis, 626–628
Loop of Henle
see Henle, loop of
Lungs
acute inflammatory reactions in, 283
angiotensin metabolism in, 264
inhibition in disease, 248
biogenic amine clearance by in control of lung circulation, 273–274
extrapulmonary significance, 272–273
and fluid balance, 274
intrapulmonary significance, 273–274
magnitude of, 271–272
oxygen toxicity mediated through changes in, 276
pathophysiology of, 275–277
pulmonary hypertension mediated through, 276–277
bradykinin metabolism in, 264
chemotactic factors involved in injury of, 283–291
cell-derived, 285–289

complement-derived, 283–285
experimental studies implicating, 289–291
hormone clearance by, 241–252
injury related to prostaglandin release, 260–261
metabolic functions of, 222
amines, 269–278
on angiotensin converting enzyme, 229–231, 247–251, 264
broad range of chemical compounds handled by, 226
hypoxia effect on, 251
involving adenine nucleotides, 227, 231, 233
on prostaglandins and related lipids, 229, 231, 257–263
selectivity of hormones and drugs handled by, 226
structural bases for, 223–237
on vasoactive peptides, 263–265
physiologic hypertension mediated by, 247
prostaglandin metabolism in biosynthesis, 257–259
physiologic and pathologic influences on, 262–263
release of prostaglandins in, 260–261
sites of, 260
uptake and inactivation, 261–262
thrombogenesis modified by action on biogenic amines, 273
two cooperating systems of, 226
See also Pulmonary vascular endothelium
Luteinizing hormone (LH)
calmodulin role in GnRH-induced release of, 674–675
change in ratio to FSH during puberty, 604–605
difference from FSH in inhibitory effect of GnRH on, 585
discordance with FSH release during maturation, 604–605

levels from fetal life through puberty, 598–599
midcycle surge of, 606
periodicity of release of, 597–598
pulsatile release in puberty, 599–600
pulsatile secretion of, 583–585
sleep production of, 599–600
See also Gonadotropins
Lymph flow
changes during intestinal absorption, 47–48
intestinal motility effects on, 37–38
Lymphocytes
chemotactic factors derived from, 288
Lysine
as functionally important site of calcium ATPase, 301–302
Lysosomes
degradation of acetylcholine receptors in, 325–326

M

Macula densa, 161, 162
Macrophages, alveolar
chemotactic factors produced in, 288
Magnesium
effect on myocardial sodium pump, 377
role in calcium pump in heart, 415–416
in skeletal muscle, 302–309
Mammals
control of brain temperature, 85–92
Masculinization
insulin resistance associated with, 642
Medial basal hypothalamus
see Hypothalamus
Membranes
potential changes induced by chemoattractants, 559
See also Sarcoplasmic reticulum
Memory
endorphins effect on, 573–574
opiates effect on, 573
vasopressin effect on, 577–578
Menstrual cycle
neuroendocrine model of, 588–590

Mesenteric artery blood flow measurement, 80
Mescaline
disposition in lungs, 270
Messenger, second
see Second messenger
Messenger RNA
in biosynthesis of acetylcholine receptor, 320–321
calcium effect on processing in muscle, 345–347
Metabolic model for regulation of intestinal flow, 14–19, 23–25
Methionine-enkephalin, 633
effects on intestinal motility, 35
Methylation
of chemotaxis proteins, 505–507
control in bacterial, 507–510
GTP influence on beta-adrenergic receptor reactions, 656
of phospholipids in chemotaxis
in *Dictyostelium,* 541–542
in leukocytes, 561–562
of proteins in leukocyte chemotaxis, 560–561
Microspheres
use in measuring gastrointestinal blood flow, 77
Mineralocorticoids
effect on tubular reabsorption in collecting tubules, 195–196
in connecting tubules, 193–194
effect on urinary acidification, 197
See also Aldosterone
Monoamine oxidase
action in lungs, 270
Morphine
effect on intestinal absorption, 50
intestinal blood flow and motility effects of, 33–36
See also Endorphins
Motility, intestinal
see Intestinal motility
Muscle
contraction correlated with sarcoplasmic reticulum development, 347–348
ionic channels in, 357–369
calcium, 366–368
chloride, 364–366
potassium, 361–364

SUBJECT INDEX 727

in slow fibers, 368–369
sodium, 358–361
regulation of calcium
concentration in,
344–349
regulation of contraction and
relaxation of, 297
turnover of acetylcholine
receptors in, 319–329
See also ATPase,
calcium-activated;
Sarcoplasmic reticulum
Muscle work
efficiency during locomotion,
101–103
energy cost of force
generation during
locomotion, 103–105
friction effect on, 101
metabolic cost during
locomotion, 103–105
rate of cross-bridge cycling
and energy consumption
during, 105–106
rate of performance of,
99–101
two main variables
determining, 97–98
See also Energy; Energy
consumption
Myasthenia gravis
accelerated degeneration of
acetylcholine receptor
in, 328
Myocardial contraction
changes in intracellular
membranes during,
455–457
physiological function of
sarcoplasmic reticulum
during, 459–460
sodium-calcium exchange in
control of, 445–446
Myocardial membrane, 373
electrogenic mechanism of
sodium pump in,
389–398
regulation and function of
sodium pump in,
375–384
Myogenic theory of
autoregulation, 23–25
Myosin heavy chain
dephosphorylation, 541

N

Naloxone
memory effects of, 573
in study of opiate activity,
572

Nephrons
classification of
based on cortical position,
147–148
based on short- vs
long-looped, 148–149
collecting ducts, 167–172
connecting tubule, 165–167
distal tubule, 161–164
heterogeneity of
axial, 203
biological importance of,
203–216
in maintaining fluid
balance with
expansion of
extracellular space,
206–211
in maintaining phosphate
balance, 203–206
in medullary recycling of
urea, 212
prostaglandins role in,
210–211
reduction in nephrogenic
diabetes insipidus, 215
of transport function,
181–197
in vasopressin action in
water conservation,
212–215
vasopressin influence on,
213–215
in water conservation,
211–216
See also Juxtamedullary
nephron
in vitro culture of cell
populations of, 173–174
intra- versus inter-, 181–182
proximal tubule, 152,
154–157
subdivision of, 181
superficial versus
juxtamedullary, 181–182
thin limb segments of,
157–161
Nerve growth factor
prohormone processing of,
634
Nervous system
see Brain cooling; Central
nervous system;
Hypothalamus
Neuromuscular junction
see Acetylcholine receptor
Neuropeptides
behavioral effects of, 571–579
endorphins, 572–576
vasopressin, 576–579
Neurotoxins
see Cardiotoxins

Newborns
insulin receptors in, 645
Norepinephrine
affinity for beta receptor
subtypes, 476
effect on calcium uptake into
sarcoplasmic reticulum,
374
effect on liver circulation, 63
effect on stomach blood flow,
6–7
possible role in pulmonary
hypertension, 276–277
pulmonary clearance of
limited by transport,
269–270
magnitude of, 271–272
Nucleotides
see Adenylate cyclase; Cyclic
AMP; Guanine
nucleotide regulatory
protein
Nutrition
influence on onset of puberty,
597

O

Obesity
insulin receptors in, 641
Opiates
effect on pain perception, 572
memory deficits due to, 573
naloxone in studying activity
of, 572
See also Endorphins
Orthokinesis, 525
Osmolarity
liver circulation influenced
by, 60
thin limbs of Henle's loop
role in urine changes of,
189
Osmotic pressure
and intestinal absorption,
46–48
Ovary
central nervous system
control of, 583–590
control of gonadotropin
secretion by, 585–588
hypothalamic control of
pituitary-ovarian axis,
588–590
in menstrual cycle, 588–590
See also Estradiol
Ovulation
infrequency during puberty,
606–607
Oxygen
as chemoattractant, 536–537

effects of high concentration
  on pulmonary
  endothelial cell amine
  metabolism, 276
metabolites related to
  neutrophil-mediated
  tissue injury, 286–287
prostaglandin metabolism in
  lungs influenced by,
  262–263
Oxygen consumption
  of flying insects, 138
  during locomotion, 98–99
  See also Energy consumption
Oxygen tension
  in measurement of
    gastrointestinal blood
    flow, 79
Oxygen uptake
  in estimating total cost of
    tuna locomotion,
    123–124
Oxygenation
  in autoregulation of intestinal
    circulation, 14–23
  intestinal absorption and,
    45–46
  intestinal luminal nutrients
    effect on, 45–46
  liver circulation related to,
    59–60
  thermoconservation in fish
    related to, 129–130

P

PAH
  secretion into proximal
    tubule, 188
Pain perception
  opiates effect on, 572
Palytoxin, 470
Panting
  brain cooling with, 88
  coordinated control of upper
    respiratory evaporation
    and blood flow during,
    90–92
Paramecium
  adaptation of locomotive
    behavior, 526
  avoiding reaction of, 519,
    525–526
  basic schema of behavior of,
    520–524
  action potential to calcium
    influx, 522–523
  calcium influx to ciliary
    reversal, 523–524
  receptor potential to action
    potential, 521–522

stimulus to receptor
  potential, 520–521
Parasympathetic nervous
  system
  stomach blood flow
    influenced by, 5
Parathyroid hormone (PTH)
  effect on connecting tubule,
    194
  effect on distal tubule, 162
  proparathyroid hormone
    conversion to, 632–633
  receptor of
    binding to, 656–657
    regulation and coupling of,
    657
Pentagastrin
  effect on gastric blood flow, 8
  influence on liver circulation,
    63
  intestinal blood flow and
    motility effects of, 34–35
Peptidases
  in lungs, 241–252
Peptide hormones
  evolution of, 615–623
  See also Hormones
Peptides
  see Neuropeptides;
    Polypeptide hormones
pH
  influence on liver circulation,
    59–60
Phenothiazines
  in study of calmodulin,
    673–674
Phenylethylamine
  disposition in lungs, 270
Phosphate
  nephron heterogeneity in
    maintenance of, 183,
    185, 203–206
Phosphatidylglycerol
  chemotactic activity of, 288
Phosphodiesterase
  calmodulin regulation of, 669
  in chemotaxis response of
    Dictyostelium, 544–546
Phosphoenzyme in calcium
  pump
  isomerization of
    intermediates, 304–306
Phospholamban
  in cardiac sarcoplasmic
    reticulum, 402
  phosphorylation in cardiac
    sarcoplasmic reticulum,
    403–411
  calmodulin-dependent,
    408–409
  elementary steps of
    ATPase in, 406–408

inotropic effect of
  catecholamines
  mediated by, 411–412
molecular model of,
  409–411
physiological relevance in,
  411–413
structural characteristics of,
  405
Phospholipids
  in development of
    sarcoplasmic reticulum,
    341–342
  metabolism in leukocyte
    chemotaxis, 561–563
    arachidonate metabolism,
    562–563
  modulation of
    phospholipase A2, 562
  phospholipid turnover,
    561–562
  methylation involvement in
    Dictyostelium
    chemotaxis, 541–542
Phosphorylation
  calmodulin in, 670–672
  of cardiac sarcolemma, 404,
    413–417
  of cardiac sarcoplasmic
    reticulum, 402–413
  coordinated regulation by
    calcium and cyclic AMP
Phosphotransferase system
  in bacterial chemotaxis,
    510–511
  response to sugars, 513
Pituitary
  adrenal androgen stimulating
    hormone of, 609
  fetal synthesis of
    gonadotropins, 598
  interaction of synthetic
    GnRH with, 603–605
  role in estrogen-induced
    positive feedback on
    gonadotropins, 606
  as site of estrogen-feedback
    action, 586–587
Plasma membrane
  cardiotoxin effect on,
    463–471
  hormone receptors activation
    of adenylate cyclase by
    movement through,
    658–659
  incorporation of
    acetylcholine receptors
    in, 322–323, 326
  See also Sarcoplasmic
    reticulum
Plasminogen activator
  of chemotactic factors, 285

SUBJECT INDEX 729

proinsulin converting enzyme
  activity of, 632
Platelets
  phosphorylation reaction in
    regulation of calcium
    accumulation, 413
  pulmonary metabolism of
    biogenic amines effect
    on, 273
  thromboxane A2 synthesis
    in, 260
Podocytes in glomerulus,
  150–152
Polymorphonuclear leukocytes
  see Leukocytes
Polypeptide hormones
  extrinsic labeling of, 242
  hormone biosynthesis from
    precursors, 625–635
  lung metabolism of, 241–252
  See also Angiotensin;
    Bradykinin
Pore model of hormone
  biosynthesis, 626
Portal hypertension
  vasopressin in alleviation of,
    66
Portal vascular system
  blood gas tensions influence
    on, 59–60
  catecholamine effects on, 63
  interaction with hepatic
    artery, 58–59
  neural influences on, 61–62
  osmolarity effects of, 60
  vasopressin effects on, 66
  See also Liver circulation
Postprandial hyperemia, 18–19
Potassium
  calcium-activated ionic
    currents in heart,
    429–430
  collecting duct role in
    urinary excretion of,
    170–171
  effect on cardiac glycoside
    action on sodium pump,
    381
  ionic channels in muscle,
    361–364
    delayed rectifier, 361–362,
      369
    inward rectifier, 363–364
    slow channels, 363, 369
    in slow muscle fibers, 369
  ionic channels in
    Paramecium, 527–528
  ionic movement in muscle
    action potential
    generation, 357, 521–522
  Paramecium reaction to,
    527, 530

reabsorption in connecting
  tubule, 195–196
  See also ATPase, sodium and
    potassium
Power
  in bird flight
    consumption of, 110–112
    induced, 110
    parasite, 110
    profile, 110
  in insect flight
    measurement of, 138–139
    requirements of, 135–138
Precapillary sphincter, 14–15
Prehormones
  see Preproteins
Preproteins, 625
  cleavage mechanisms of, 629
  in hormone biosynthesis
    from larger polypeptides,
    625–629
  loop model for, 626–628
  pore model for, 626
  trigger hypothesis for,
    628–629
ProACTH/endorphin, 633
Progastrin, 633
Progesterone
  effect on gonadotropin
    secretion, 585–586, 588,
    590
Proglucagon, 633
Prohormones, 630–634
Proinsulin
  conversion to insulin,
    631–632
Prolactin
  evolution of function of, 620
  receptor of, 647
Proline
  as insect flight fuel, 139
Propancreatic polypeptide, 633
Propanolol
  withdrawal effects, 480–481
Proproteins, 625, 630
  processing in hormone
    biosynthesis, 630–634
  significance of 'pro' sequence
    in, 630
Prosomatostatin, 633
Prostacyclin
  biosynthesis in lungs,
    257–258
  inhibition of, 259–260
  sites of, 260
  and gastric blood flow, 9
  lung injury mediated by, 261
  minimal degradation during
    circulation through
    lungs, 262

Prostaglandins
  effect on acid secretion, 10
  effect on collecting tubule
    transport functions, 195
  effect on gastric blood flow, 9
  effect on intestinal motility,
    35
  effect on water transport in
    collecting tubule, 197
  hepatic vascular actions of,
    65
  influence on intestinal
    absorption, 52
  intestinal secretion increased
    by, 52
  metabolism in lungs, 257–263
    biosynthesis, 257–260
    inhibition of synthesis,
      259–260
    pathologic influences on,
      262–263
    pregnancy effects on, 262
    by pulmonary vascular
      endothelium, 229, 231
    release into circulation,
      260
    uptake and inactivation,
      261–262
  pathologic states associated
    with release of, 260–261
  role in heterogeneity of
    volume expansion
    affecting juxtamedullary
    nephrons, 210–211
Proteases
  involved in prehormone
    conversion to hormone,
    631
  in proinsulin conversion to
    insulin, 632
Proteins
  conformational changes in,
    628–629
  synthesis of
    see Hormones, biosynthesis
      from larger
      polypeptides
Proteolysis, post translational
  in polypeptide hormone
    biosynthesis, 625–635
  See also Hormones,
    biosynthesis from larger
    polypeptides
Protonmotive force
  in bacterial chemotaxis, 512
Proximal tubule, 152, 154–157
  comparison of reabsorption
    in convoluted and
    straight segments,
    184–185
  electrolyte concentrations in
    nuclei of, 175

heterogeneity of reabsorption by, 183–188
ion permeabilities and reabsorption by, 186–188
morphologic subdivisions of, 182
paracellular pathway role in, 152, 154
phosphate reabsorption in, 204–206
role in plasma turnover of low molecular weight proteins, 154, 157
secretion into, 188
sodium reabsorption in, 209–210
transport processes in, 182–188
Pseudohypoparathyroidism guanine nucleotide regulatory protein in, 664–665
Puberty
circadian rhythm of gonadotropin release in, 599–600
control of adrenarche in, 607–609
epidemic release of gonadotropins in, 599–600
events characterizing, 595
gonadal levels from fetal life through, 598–599
gradual acceleration of arcuate oscillator during, 590
influences on age, 596–597
neuroendocrine control mechanisms and onset of, 595–609
restraint of onset of intrinsic CNS inhibitory mechanism, 602
negative feedback mechanism, 600–603
timing of onset of, 596–597
trend of earlier age on, 596–597
Pulmonary edema, alloxan-induced prostaglandin role in, 260
Pulmonary hypertension altered pulmonary amine metabolism role in, 276–277
Pulmonary vascular endothelium
angiotensin metabolism in, 243–247
angiotensin-converting enzyme localized in, 228–231

biogenic amine metabolism in and amine disposition, 270
receptor binding and, 274
related to lung injury, 275–277
bradykinin metabolism by, 226–227, 229–230, 243–244
caveolae intracellulares of, 234–235
cell cultures in study of, 233
cellular lining of, 226
electron microscope cytochemistry in localizing adenine nucleotide in, 227
hemostatic potential of, 233
metabolic functions of, 221–222
anatomic bases for, 223–230
cellular bases for, 230–234
hypoxia effect on, 251
kinetics of, 244–247
prostaglandin metabolism, 229, 231
subcellular bases for, 234–235
paraendocrine function of, 233
varied properties of, 221
See also Lungs
Purkinje fibers
effect of blocking slow inward calcium current in, 427–428
sodium pump contribution to cardiac membrane potential in, 390–392

R

Reabsorption, tubular
in collecting segments, 194–197
in connecting tubule, 193–194
in distal tubule, 192–193
in maintaining fluid balance with expansion, 206–211
of phosphate, 203–206
in proximal tubule, 182–188
heterogeneity of, 182–188
reflection coefficient and, 188
relative ion permeability and, 185–188
secretion, 188
in thick limb of Henle's loop, 190–192
in thin limb of Henle's loop, 188–189

in water conservation, 211–216
Reactive hyperemia in intestine, 18
Receptor potential
generation in *Paramecium*, 520–521
in generation of action potential, 521–522
Receptors
for acetylcholine in skeletal muscle, 319–329
See also Acetylcholine receptors
adrenergic, 475–481, 655–656
See also Adrenergic receptors
affinity effects of guanine nucleotides, 661
alterations in human diseases, 639–648
antibodies against, 642–643, 647–648
for biogenic amines in pulmonary vascular endothelium, 274
in breast cancer, 647
for C5a chemoattractant, 555–556
calcium, 668
See also Calmodulin
calmodulin as, 668–669
and endocytosis mediated by, 675–676
for chemoattractants in *Dictyostelium*, 537–538
for cyclic AMP, 537–538
for folic acid, 538–539
for formylated peptides, 556–557
gene duplication theory for evolution of, 618
glucagon, 653–654
growth hormone, 647
guanine nucleotide regulatory protein role in hormone interaction with, 658–662
histamine, 7–8
insulin
in disorders of receptor specificity, 645–646
in extreme insulin resistance, 642–644
in hypoglycemia with non-islet cell tumors, 646
in insulin deficiency, 644
in insulin excess, 645
in moderate insulin resistance, 641–642
in newborns, 645

SUBJECT INDEX 731

in supersensitivity to
   insulin, 64
for insulin-like growth
   factors, 646
methods of characterizing,
   640–641
parathyroid hormone
   binding characteristics for,
      656–657
   regulation and coupling of,
      657
of peptide hormones,
   639–640
   alterations in human
      disease, 639–648
   fundamental functions of,
      639–640
   in hormone biosynthesis
      from larger
      polypeptides, 627–628
post-binding defects in, 644
prolactin, 647
qualitative abnormalities in,
   643–644
quantitative defects in, 643
regulation of chemoattractant
   binding to, 557–559
regulation of interactions
   after binding to,
   653–665
second messenger initiation
   by binding to, 639–640
in thyroid tumors, 648
thyrotropin, 647
transfer with adenylate
   cyclase components to
   foreign membrane
   systems, 658–659
translocator, 627–628
vasopressin, 646–647
Reentry, 491–492
Reflection coefficients and ion
   reabsorption, 188
Renal corpuscles
   three types based on cortical
      position, 147–148
   See also Kidney; Nephrons
Renal failure, acute
   glomerular endothelium in,
      152
Renal physiology
   see Kidney; Nephrons
Reproductive physiology
   central nervous system
      control of ovarian
      function, 583–590
   neuroendocrine mechanisms
      and onset of puberty,
      595–609
   neuroendocrine model of
      menstrual cycle,
      588–590

regulatory systems
   controlling, 595–596
Respiratory evaporation
   brain temperature related to,
      88–93
Respiratory system
   see Lungs; Pulmonary
      vascular endothelium
Resting potential, 396
RNA
   see Messenger RNA
Running
   energetics in insects,
      134–135
   See also Locomotion

S

Salt
   see Sodium
Sarcolemma in myocardium
   calcium channel in, 417
   calcium movements across,
      415–416
   functional significance of
      phosphorylation of,
      416–417
   membrane of, 402
   protein kinase-catalyzed
      phosphorylation of,
      413–415
   sodium-calcium exchange in
      relationships in ionic
      binding, 436–438
   role in contractile control,
      445–446
   stoichiometry of, 444–445
   in transport across
      membrane, 439–444
Sarcoplasmic reticulum
   in heart
      calcium release from,
         451–460
      calcium-ATPase regulation
         in, 406–411
      changes in properties
         during cardiac cycle,
         455–460
      cyclic AMP-mediated
         control of calcium
         transport in, 405–406
      membrane of, 402
      phospholamban in, 405
      phosphorylation of,
         402–413
      physiological relevance of
         phospholamban in,
         411–413
   in skeletal muscle, 337
   ATPase reaction coupled
      with calcium pump in,
      302–309

calcium ATPase in,
   298–309
calcium pump protein of,
   297–312
calcium transport site in,
   300, 304
development of, 337–349
function of, 337
role in contraction, 297
X-ray diffraction analysis
   of, 311
Saxitoxin
   action on fast sodium
      channel in heart,
      463–465
Sciatin
   effect on increasing
      acetylcholine receptor of
      myotube, 324
Scorpion toxin
   action on fast sodium
      channels in heart,
      468–469
Sea anemone toxin
   action on fast sodium
      channels in heart,
      468–469
Second messenger, 639
   calcium as, 667–669
   cyclic AMP as, 667
Secretin
   effect on gastric blood flow,
      8
   effect on intestinal motility,
      34
   influence on liver circulation,
      64
Secretion into proximal tubule,
   188
Serotonin
   effect on liver circulation,
      65
   effect on lung fluid balance,
      274
Signal hypothesis, 626
Sinoatrial node cells
   sodium pump contribution to
      cardiac membrane
      potential in, 390
Sleep
   gonadotropin release during,
      599–600
Slime moulds
   see *Dictyostelium discoideum*
Slow-reacting substance of
   anaphylaxis (SRS-A)
   biosynthesis in lungs, 259
   role in lung disease, 259
Smoking
   chemotactic factors generated
      in lung associated with,
      286, 287

Smooth muscle
  chemicals and
    neurotransmitters effects
    on, 33–36
Snake venom cardiotoxin, 470
Soaring
  in endothermic insects, 137
Sodium
  active transport of
    see Sodium pump
  calcium exchange in heart
    with, 435–447
    calcium binding in
      sarcolemmal
      membrane and,
      436–438
    membrane carrier role in,
      437–438
    role in contractile control,
      445–446
    stoichiometry of, 444–445
    in transport across
      sarcolemmal
      membrane, 439–444
  intracellular concentrations
    in renal cells, 175
  ionic channels in heart
    biochemical properties of,
      465
    cardiotoxins specific for,
      463–470
  ionic channels in muscle,
    358–361
    chemical studies of,
      359–360
    effect of denervation on,
      360–361
    ionic selectivity of, 359
    kinetics of, 358–359
    in mammals, 360
    single channel
      conductance, 360
    in slow fibers, 368–369
  ionic current in *Paramecium*,
    526–528
  ionic movement in muscle
    action potential
    generation, 357
    Paramecium reaction to,
      527, 530
  permeability in thin limb
    segments, 158, 160–161
  reabsorption of
    in collecting tubule,
      195–196
    in connecting tubule, 193
    cotransport process with
      chloride, 191–192
    in proximal tubule,
      182–188
    in thick ascending limb of
      Henle's loop, 190–192
    vasopressin and, 212–213

slow inward current in heart
  and abnormal
  automaticity, 486–487
urinary excretion of
  blood flow distribution
    among nephrons and,
    206–207
  heterogeneity of tubular
    function and, 207–211
  See also ATPase, sodium and
    potassium
Sodium pump
  in action potential of cardiac
    cells, 389
  ATPase involved in, 375
  See also ATPase, sodium
    and potassium
  contribution to cardiac
    membrane potential,
    390–393
    in atrial and ventricular
      preparations, 392–393
    in chick embryo heart
      cells, 393
    in Purkinje fibers, 390–392
    in sinoatrial cells, 390
  control in intact cells,
    375–380
  electrogenic, 379, 389–398
    antiarrhythmic effects of,
      397–398
    contribution to cardiac
      membrane potential,
      390–392
    coupling ratio and, 395
    effect on action potential,
      396–397
    effect on cardiac automatic
      activity, 397
    effect on cardiac resting
      potential, 396
    electroneutral versus,
      394–395
    experimental data on,
      393–394
    inhibition in myocardial
      membrane, 380–384
      cardiac glycosides effect of,
      380–382
    relation to positive
      inotropic effects,
      381–382
    measurement of currents
      generated by, 398
    regulation in myocardial
      membrane, 375–384
    enzyme concentrations in,
      376
    enzyme turnover rates in,
      376–378
    sodium/potassium
      counter-transport ratio
      in, 378–380

vanadate in, 382–383
reserve capacity in cardiac
  muscle, 380
Somatomedin receptors, 646
Somatostatin
  effect on gastric blood flow, 8
  effect on intestinal
    absorption, 49–50
  as local regulatory agent in
    several sites, 621
  prosomatostatin conversion
    to, 633
Splanchnic circulation
  see Gastrointestinal blood
    flow measurement;
    Intestinal circulation;
    Liver circulation;
    Stomach circulation
Starling forces
  and intestinal absorption,
    46–48
Stomach
  ischemia effect on motility
    of, 40
  temperature increase after
    meal in warm-bodied
    fish, 130
Stomach circulation
  acid secretion related to,
    9–10
  clearance methods in
    measurement of, 76
  factors controlling, 3–9
    biogenic amines, 6–8
    hormones, 8
    neural, 3–6
    prostaglandins, 9
  See also Gastrointestinal
    blood flow measurement
Substance P
  in lungs, 263–265
  physiologic role of, 265
Swimming
  energetics in warm-bodied
    fish, 121–130
Sympathetic nervous system
  influence on intestinal
    absorption, 51
  role in liver circulation,
    61–62
  stomach blood flow
    influenced by, 4–5
  See also Adrenergic
    receptors;
    Catecholamines; Central
    nervous system

T

Tachycardia, ventricular
  reentry mechanism in, 492

SUBJECT INDEX 733

Tachyphylaxis
 of beta-adrenergic receptors, 478–481
Temperature
 cerebral circulation effects of, 90–92
 endothermic insects and energetics of locomotion, 133–140
 thermoconservation and locomotion energetics in warm-bodied fish, 129–130
 warm-bodied fish and energetics of locomotion, 121–130
 See also Brain cooling; Brain temperature
Tetracaine
 effect on potassium channels in muscle, 362
Tetraethylammonium
 effect on potassium channels in muscle, 362, 364
Tetrodotoxin
 action on fast sodium channel in heart, 463–465
 effect on acetylcholine receptor, 325
Thallium
 effect on potassium channels in muscle, 364
Thermotaxis, 511, 513
Thrombogenesis
 pulmonary metabolism of biogenic amines effect on, 273
Thromboxanes
 biosynthesis in lungs, 257–258
 sites of, 260
 inhibition of, 259
Thyroid stimulating hormone (TSH)
 autoantibodies against, 647–648
 receptors for, 647
Thyroid tumors
 receptors in, 648
Torpedo electroplax
 acetylcholine receptor in, 319–321
Transducers
 as chemoreceptors, 505
 methylation of, 505–507
 comparator function of, 507
 nature of signal to flagella, 508
 in Paramecium behavior, 520
 role in bacterial chemotaxis, 504–507

Translocators in hormone biosynthesis, 627–628
Transport functions in kidney tubules
 see Reabsorption, tubular
Transverse membrane system (T-system)
 in muscle contraction, 297
Trigeminy, 486
Trigger hypothesis of hormone biosynthesis, 628–629
Triggered activity in heart, 487–488
Tryptophan
 fluorescence changes related to conformational changes in calcium ATPase, 309
Tuna
 energetics of locomotion in, 121–130
 in giant bluefin tuna, 127–128
 in small tropical tuna, 126–127
Tympanic membrane temperature, 92

U

Urea
 medullary recycling of, 212
 secretion into proximal tubule, 188
 thin limb segment exchange of, 159
 transport in collecting tubule, 197
Urinary concentration
 nephron heterogeneity role in, 211–216
 See also Reabsorption, tubular

V

Vagal stimulation
 acid production secondary to, 9–10
 effect on stomach circulation, 5–6
 liver blood flow effects of, 62
 vagotomy effect on, 6
Vanadate
 as regulator of myocardial sodium pump, 382–383
Vasoactive intestinal peptide (VIP)
 effect on intestinal absorption, 50
 and gastric blood flow, 8

guanine nucleotides in activation of adenylate cyclase, 662
 in lungs, 263
 physiological role of, 265
Vasoactive peptides
 effect on liver circulation, 65–66
 metabolism in lungs, 263–265
 See also Angiotensin; Bradykinin
Vasopressin
 and appetitively motivated learning, 578
 and aversively motivated behavior, 577–578
 collecting duct as site of action of, 172
 effect on collecting tubule, 197
 effect on distal tubule, 162
 effect on gastric acid flow, 8
 evolution of function of, 620
 hepatic vascular actions of, 66
 and heterogeneity of filtration rates in single nephrons, 213–215
 influence on countercurrent system, 213
 lack of effect on connecting tubule, 193
 memory effects of, 577–578
 opiate and endorphin stimulation of release of, 576
 primary action of, 576–577
 receptors of, 646–647
 role in angiotensin-induced drinking behavior, 576
 and salt transport in thick ascending Henle's loop, 212–213
 vasoconstriction effect of, 577
Venous pressure
 effect of elevation on hepatic circulation, 58
Ventricular tachycardia
 reentry mechanism in, 492
Verapamil
 as blocker of slow inward calcium current in heart, 426
Veratridine
 action on fast sodium channel in heart, 465–467
Villi, intestinal
 microvasculature of, 43–44
 neural regulation of circulation and absorption in, 51

W

Walking
  energetics in insects, 134–135
Washout techniques
  in measuring gastrointestinal blood flow, 75–76, 80–81
Water
  nephron heterogeneity in conservation of, 211–216
  medullary recylcing or urea and, 212
  vasopressin and, 212–215
  reabsorption of
    in collecting tubule, 197
    in proximal tubule, 186–188
Wolff-Parkinson-White syndrome
  reentry in, 491, 492

X

Xenon
  in fluorescent excitation analysis, 81
  in washout studies, 75–76, 80–81
X-ray diffraction analysis
  of sarcoplasmic reticulum membranes, 311
X-ray microanalysis
  kidney morphology studied by, 174–175

Z

Zinc
  effect on potassium ionic channels, 362

CUMULATIVE INDEXES

CONTRIBUTING AUTHORS, VOLUMES 40–44

A

Agus, Z. S., 43:583–95
Akera, T., 44:375–88
Alpert, N. R., 41:521–38
Armstrong, D. T., 42:71–82
Arnolds, D. E. A. T., 40:185–216
Arruda, J. A. L., 40:43–66
Aukland, K., 42:543–55
Aurback, G. D., 44:653–66

B

Baer, P. G., 42:589–601
Baker, M. A., 44:85–96
Bárány, K., 42:275–92
Bárány, M., 42:275–92
Bardin, C. W., 43:189–98
Basbaum, A. I., 40:217–48
Beaulieu, M., 41:5-5–69
Becker, R., 43:189–98
Beeuwkes, R. III, 42:531–42
Behrman, H. R., 41:685–700
Bell, P. D., 42:557–71
Belloni, F. L., 40:67–92
Bennett, A. F., 40:447–69
Bergofsky, E. H. 42:221–33
Berry, C., 44:181–201
Blantz, R. C., 42:573–88
Bloom, F. E., 44:571–82
Bollenbacher, W. E., 42:493–510
Borgeat, P., 41:555–69
Bowers, C. W., 43:673–87
Boyd, A., 44:501–17
Brody, M. J., 42:441–53
Brody, T. M., 44:375–88
Bulger, R., 44:147–79
Bundgaard, M., 42:325–36
Butler, J., 42:187–98

C

Calder, W. A. III, 43:301–22
Castellini, M. A., 43:343–56
Chabardès, D., 43:569–81
Chafouleas, J. G., 44:667–82

Chapman, J. B., 41:507–19
Cheng, S.-L., 43:189–98
Chiarandini, D. J., 44:357–72
Chou, C. C., 44:29–42
Coceani, F., 41:669–84
Cohen, B., 40:527–52
Cohen, M. L., 43:91–104
Cokelet, G. R., 42:311–24
Cole, K. S., 41:1–24
Coleridge, H. M., 42:413–27
Coleridge, J. C. G., 42:413–27
Conger, J., 42:603–14
Cranefield, P. F., 41:459–72
Crawshaw, L. I., 42:473–91
Crim, L. W., 41:323–35
Culver, B. H., 42:187–98

D

Darian-Smith, I., 41:141–57
Davis, B., 41:369–81
Davis, R. W., 43:343–56
Dedman, J. R., 42:59–70
DeLuca, H. F., 43:199–209
Dennis, V. W., 41:257–71
Dizon, A. E., 44:121–31
Dobbs, R. E., 40:307–43
Dobyan, D., 44:147–79
Docherty, K., 44:625–38
Dockray, G. J., 41:83–95
Donald, D. E., 42:429–39
Downing, S. E., 42:199–210
Drouin, J., 41:555–69
Duling, B. R., 42:373–82

E

Eisenberg, E., 42:293–309
Elde, R., 41:587–602
Eldrige, F., 43:121–35
Eskin, A., 40:501–26

F

Fabiato, A., 41:473–84
Fabiato, F., 41:473–84
Fambrough, D., 44:319–35
Fanestil, D. D., 43:637–49
Fantone, J., 44:283–93

Farner, D. S., 42:457–72
Ferland, L., 41:555–69
Fields, H. L., 40:217–48
Fink, G., 41:571–85
Fisher, K. A., 42:261–73
Fishman, A. P., 42:211–20
Flaim, S., 43:455–76
Forte, J. G., 42:111–26
Friesen, H. G., 42:83–96

G

Ganong, W. F., 40:377–410
Gardner, J. D., 41:55–66
Gerisch, G., 44:535–52
Gibbs, C. L., 41:507–19
Gil, J., 42:177–86
Gilbert, L. I., 42:493–510
Gillis, C. N., 44:269–81
Gilmer, P. J., 41:653–68
Gilula, N. B., 43:479–91
Glitsch, H. G., 44:389–400
Goldfarb, S., 43:583–95
Goldman, R. D., 41:703–22
Goodwin, A. W., 41:141–57
Gore, R. W., 42:337–57
Gospodarowicz, D., 43:251–63
Gottschalk, C. W., 41:229–40
Granger, N. A., 42:493–510; 43:409–18
Grantham, J. J., 40:249–77
Greene, L. E., 42:293–309
Grossman, M. I., 41:27–33
Grumbach, M., 44:595–613
Gunn, R. B., 42:249–59
Gunsalus, G., 43:189–98
Guth, P. H., 44:3–12

H

Habener, J., 43:211–23
Hall, D. A., 40:249–77
Hall, V. E., 43:1–5
Hamrell, B. B., 41:521–38
Handler, J., 43:611–24
Harris, R. H., 41:653–68
Harris, W., 43:689–710
Hartshorne, D. J., 43:519–30
Haywood, J. R., 42:441–53

735

Heath, J. E., 44:133–43
Heath, M. S., 44:133–43
Heglund, N., 44:97–107
Heindel, J. J., 42:37–57
Heller, H. C., 41:305–21
Hertzberg, E. L., 43:479–91
Heymann, M. A., 43:371–83
Hilton, S. M., 42:399–411
Hökfelt, T., 41:587–602
Hoffman, B., 44:475–84
Holmes, K. C., 43:553–65
Homsher, E., 40:93–131
Houk, J. C., 41:99–114
Hudson, J. W., 41:287–303

I

Ikemoto, N., 44:297–317
Imbert-Teboul, M., 43:569–81
Imura, H., 43:265–78
Insel, P., 43:625–36
Irish, J. M. III, 40:249–77
Iwamoto, H. S., 43:371–83

J

Jacobson, E. D., 44:71–81
Jaffe, R. B., 43:141–62
Johansson, B., 43:359–70
Johnson, O., 41:141–57
Jolesz, F., 43:531–52
Jones, R. S., 41:67–82

K

Kalia, M. P., 43:105–20
Kaplan, J. G., 40:19–41
Katz, A., 44:401–23
Katzenellenbogen, B. S., 42:17–35
Kean, C. J., 40:93–131
Keeton, W. T., 41:353–66
Klitzman, B., 42:373–82
Knobil, E., 44:583–93
Koob, G. F., 44:571–82
Kontos, H. A., 43:397–407
Kooyman, G. L., 43:343–56
Kostyuk, P. G., 41:115–26
Kotite, N., 43:189–98
Kream, B. E., 43:225–38
Kregenow, F., 43:493–505
Krulich, L., 41:603–15
Kung, C., 44:519–34
Kunkel, S., 44:283–93
Kupfermann, I., 42:629–41
Kurtzman, N. A., 40:43–66
Kvietys, P. R., 43:409–18

L

Labrie, F., 41:555–69
Lagace, L., 41:555–69

Lands, W. E. M., 41:633–52
Lane, L. K., 41:397–411
Langer, G. A., 44:435–49
Larrea, F., 43:189–98
Lawrence, T. S., 43:479–91
Lazdunski, M., 44:463–73
Le Douarin, N. M., 43:653–71
Lee, J. C., 42:199–210
Lefkowitz, R., 44:475–84
Le Lièvre, C. S., 43:653–71
Leung, P. C. K., 42:71–82
Levy, M. N., 43:443–53
Licht, P., 41:337–51
Liedtke, A. J., 43:455–76
Lopes da Silva, F. H., 40:185–216

M

MacDonald, P. C., 43:163–88
Machen, T. E., 42:111–26
Macklem, P. T., 40:157–84
Mailman, D., 44:43–55
Malik, K. U., 43:597–609
Marston, S. B., 41:723–36
Martin, P. J., 43:443–53
Martonosi, A., 44:337–55
McDonagh, P. F., 42:337–57
McDonald, T., 44:425–34
McEwen, B. S., 42:97–110
McGiff, J. G., 42:589–601
Means, A. R., 42:59–70; 44:667–82
Menaker, M., 40:501–26
Millhorn, D. E., 43:121–35
Milsted, A., 41:703–22
Moore, E. N., 44:485–97
Morel, F., 43:569–81
Morris, B. J., 40:377–410
Moss, R. L., 41:617–31
Mulieri, L. A., 41:521–37
Murphy, R. A., 41:737–48
Musto, N., 43:189–98
Myers, J. L., 41:257–71
Myers, W. C., 41:67–82

N

Nadel, J. A., 41:369–81
Nakai, Y., 43:265–78
Nasjletti, A., 43:597–609
Nathanielsz, P. W., 40:411–45
Navar, L. G., 42:557–71
Neely, J., 43:419–30
Nellis, S., 43:455–76
Nelson, D. O., 43:281–300
Niall, H. D., 44:615–24

O

Öbrink, K. J., 42:111–26
Olsson, R. A., 43:385–95
Orci, L., 40:307–43
Orloff, J., 43:611–24

P

Pack, A. I., 43:73–90
Page, E., 43:431–41
Park, C. S., 43:637–49
Pastan, I., 43:239–50
Peck, E. J. Jr., 42:615–27
Peter, R. E., 41:323–35
Peterson, B. W., 41:127–40
Phillipson, E. A., 40:133–56
Phipps, R. J., 41:369–81
Pitt, B. R., 44:269–81
Ploth, D. W., 42:557–71
Pohl, C. R., 44:583–93
Prosser, C. L., 43:281–300
Pumplin, D., 44:319–35

R

Rabinowitz, M., 41:539–52
Radda, G. K., 41:749–69
Ramwell, P. W., 41:653–68
Raisz, L. G., 43:225–38
Raphan, T., 40:527–52
Raymond, V., 41:555–69
Rayner, J. M. V., 44:109–19
Rector, F. C., 41:197–210
Reibel, D. K., 43:419–30
Reid, I. A., 40:377–410
Reiter, E. O., 44:595–613
Renaud, J. F., 44:463–73
Reuter, H., 41:413–24
Richardson, P., 44:57–69
Riddiford, L. M., 42:511–28
Robinson, A. G., 40:345–76
Robison, G. A., 42:37–57
Rogart, R., 43:711–25
Rose, R. C., 42:157–71
Rosell, S., 42:259–71
Roth, J., 44:639–51
Rudolph, A. M., 41:383–95; 43:371–83
Ryan, J. W., 44:241–55
Ryan, U. S., 44:223–39

S

Said, S. I., 44:257–68
Saimi, Y., 44:519–34
Sanborn, B. M., 42:37–57
Saz, H. J., 43:323–41
Schafter, J. A., 41:211–27
Schiffmann, E., 44:553–68
Schloss, J. A., 41:703–22
Schneider, M. E., 43:507–17
Scholander, P. F., 40:1–17
Schrier, R. W., 42:603–14
Schulz, I., 42:127–56
Schwartz, A., 41:397–411
Seeley, P. J. 41:749–69
Seif, S. M., 40:345–76
Serön-Ferré, M. 43:141–62
Shepherd, A. P., 44:13–27
Shepherd, J. T., 42:429–39

CONTRIBUTING AUTHORS 737

Shibata, Y., 43:431–41
Shiu, R. P. C., 42:83–96
Siemankowski, R. F., 43:519–30
Simon, M., 44:501–17
Simpson, E. R., 43:163–88
Smith, J., 43:653–71
Snavely, M. D., 43:625–36
Soll, A. H., 41:35–53
Somjen, G. G., 41:159–77
Sonnenschein, R. R., 43:1–5
Sparks, H. V. Jr., 40:67–92
Spear, J. F., 44:485–97
Sperelakis, N., 41:441–57
Spyer, K. M., 42:399–41
Sreter, F. A., 43:531–52
Stanton, B., 41:241–56
Starger, J., 41:703–22
Stead, W. W., 41:257–71
Stefani, E., 44:357–72
Steiner, D., 44:625–38
Stevens, C. F., 42:643–52
Stevens, E. D., 44:121–31
Stolze, H. H., 42:127–56
Stuesse, S. L., 43:443–53
Szurszewski, J. H., 43:53–68

T

Tada, M., 44:401–23
Takahashi, J. S., 40:501–26
Tash, J. S., 42:59–70
Taylor, C. R., 44:97–107
Taylor, S., 44:639–51
Tepperman, B., 44:71–82
Tindall, D. J., 42:59–70
Touw, K. B., 42:441–54
Tregear, R. T., 41:723–36

U

Ullrich, K. J., 41:181–95
Unger, R. H., 40:307–43
Ussing, H. H., 42:1–16

V

Valtin, H., 44:203–19
van Sickle, M., 42:59–70
Vary, T. C., 43:419–30
Vasilenko, D. A., 41:115–26
Vassalle, M., 41:425–40

W

Wagner, P. D., 42:235–47
Walker, L., 44:203–19
Wallick, E. T., 41:397–411
Walsh, J. H., 41:35–53
Wang, L. C. H., 41:287–303
Ward, P., 44:283–93
Warnock, D. G., 41:197–210
Wasserstein, A., 43:583–95
Weems, W., 43:9–19
Weisbrodt, N., 43:21–31
Welsh, M. J., 42:59–70
Westfall, T. C., 42:383–97
White, F. N., 40:471–99
Williamson, J. R., 41:485–506
Willingham, M., 43:239–50
Willis, J. S., 41:275–86
Wilson, J. D., 40:279–306
Winegrad, S., 44:451–62
Wingfield, J. C., 42:457–72
Wit, A. L., 41:459–72
Withrington, P., 44:57–69
Wolfe, L. S., 41:669–84
Wood, J. D., 43:33–51
Wray, J. S. 43:553–65

Y

Yerna, M.-J., 41:703–22

Z

Zak, R., 41:539–52
Zelis, R., 43:455–76
Zigmond, R., 43:673–87

CHAPTER TITLES, VOLUMES 40–44

CELL AND MEMBRANE PHYSIOLOGY

Membrane Cation Transport and the Control of Proliferation of Mammalian Cells	J. G. Kaplan	40:19–41
Recent Studies on Cellular Metabolism by Nuclear Magnetic Resonance	G. K. Radda, P. J. Seeley	41:749–69
Co- and Counter-Transport Mechanisms in Cell Membranes	R. B. Gunn	42:249–59
Split Membrane Analysis	K. A. Fisher	42:261–73
Gap Junctional Communication	E. L. Hertzberg, T. S. Lawrence, N. B. Gilula	43:479–91
Osmoregulatory Salt Transporting Mechanisms: Control of Cell Volume in Anisotonic Media	F. M. Kregenow	43:493–505
Membrane Charge Movement and Depolarization-Contraction Coupling	M. F. Schneider	43:507–17
Regulation of Smooth Muscle Actomyosin	D. J. Hartshorne and R. F. Siemankowski	43:519–30
Development, Innervation, and Activity-Pattern Induced Changes in Skeletal Muscle	F. Jolesz, F. A. Sreter	43:531–52
X-Ray Diffraction Studies of Muscle	J. S. Wray, K. C. Holmes	43:553–65
Structure and Function of the Calcium Pump Protein of Sarcoplasmic Reticulum,	N. Ikemoto	44:297–317
Turnover of Acetylcholine Receptors in Skeletal Muscle	D. W. Pumplin, D. M. Fambrough	44:319–35
The Development of Sarcoplasmic Reticulum Membranes	A. Martonosi	44:337–55
Ionic Channels in Skeletal Muscle	E. Stefani, D. J. Chiarandini	44:357–72

CELLULAR NEUROBIOLOGY

Receptors for Amino Acids	E. J. Peck Jr.	42:615–27
Role of Cyclic Nucleotides in Excitable Cells	I. Kupfermann	42:629–41
Biophysical Analyses of the Function of Receptors	C. F. Stevens	42:643–52

CHEMOTAXIS

Bacterial Chemotaxis,	A. Boyd, M. Simon	44:501–17
The Physiological Basis of Taxes in Paramecium	Ching Kung, Y. Saimi	44:519–34
Chemotaxis in Dictyostelium,	G. Gerisch	44:535–52
Leukocyte Chemotaxis	E. Shiffmann	44:553–68

CIRCULATION: SYSTEMIC

The Peripheral Circulation: Local Regulation	H. V. Sparks, F. L. Belloni	40:67–92
Rheology and Hemodynamics	G. R. Cokelet	42:311–24
Transport Pathways in Capillaries—In Search of Pores	M. Bundgaard	42:325–36
Fluid Exchange Across Single Capillaries	R. W. Gore, P. F. McDonagh	42:337–57
Neuronal Control of Microvessels	S. Rosell	49:359–71
Local Control of Microvascular Function: Role in Tissue Oxygen Supply	B. R. Duling, B. Klitzman	42:373–82
Neuroeffector Mechanisms	T. C. Westfall	42:383–97

738

Central Nervous Regulation of Vascular Resistance	S. M. Hilton, K. M. Spyer	42:399–411
Cardiovascular Afferents Involved in Regulation of Peripheral Vessels	H. M. Coleridge, J. C. G. Coleridge	42:413–27
Autonomic Regulation of the Peripheral Circulation	D. E. Donald, J. T. Shepherd	42:429–39
Neural Mechanisms in Hypertension	M. J. Brody, J. R. Haywood, K. B. Touw	42:441–53
Vascular Smooth Muscle Reactivity	B. Johansson	43:359–70
Factors Affecting Changes in the Neonatal Systemic Circulation	M. A. Heymann, H. S. Iwamoto, A. M. Rudolph	43:371–83
Local Factors Regulating Cardiac and Skeletal Muscle Blood Flow	R. A. Olsson	43:385–95
Regulation of the Cerebral Circulation	H. A. Kontos	43:397–407
The Splanchnic Circulation: Intrinsic Regulation	D. N. Granger, P. R. Kvietys	43:409–18

CNS ACTIVITIES

Brainstem Control of Spinal Pain-Transmission Neurons	H. L. Fields, A. I. Basbaum	40:217–48
Brainstem Mechanisms for Rapid and Slow Eye Movements	T. Raphan, B. Cohen	40:527–52
Regulation of Stiffness by Skeletomotor Reflexes	J. C. Houk	41:99–114
Spinal Interneurons	P. G. Kostyuk, D. Vasilenko	41:115–26
Reticulospinal Projections to Spinal Motor Nuclei	B. W. Peterson	41:127–40
Posterior Parietal Cortex: Relations of Unit Activity of Sensorimotor Function	I. Darian-Smith, K. Johnson, A. Goodwin	41:141–57
Extracellular Potassium in the Mammalian Central Nervous System	G. G. Somjen	41:159–77
From the Neural Crest to the Ganglia of the Peripheral Nervous System	N. M. Le Douarin, J. Smith, C. S. Le Lièvre	43:653–71
Influence of Nerve Activity on the Macromolecular Content of Neurons and Their Effector Organs	R. E. Zigmond and C. W. Bowers	43:673–87
Neural Activity and Development	W. A. Harris	43:689–710
Sodium Channels in Nerve and Muscle Membrane	R. Rogart	43:711–25

CNS HIGHER FUNCTIONS

Physiology of the Hippocampus and Related Structures	F. H. Lopes da Silva, D. E. A. T. Arnolds	40:185–216

COMPARATIVE PHYSIOLOGY
See end of Index

DIGESTIVE SYSTEM

Neural and Hormonal Regulation of Gastrointestinal Function: An Overview	M. I. Grossman	41:27–33
Regulation of Gastric Acid Secretion	A. Soll, J. H. Walsh	41:35–53
Regulation of Pancreatic Exocrine Function In Vitro: Initial Steps in the Actions of Secretagogues	J. Gardner	41:55–66
Regulation of Hepatic Biliary Secretion	R. S. Jones, W. C. Myers	41:67–82
Comparative Biochemistry and Physiology of Gut Hormones	G. Dockray	41:83–95
Mechanisms of Gastric H^+ and Cl^- Transport	J. G. Forte, T. E. Machen, K. J. Öbrink	42:111–26

Title	Authors	Citation
The Exocrine Pancreas: The Role of Secretagogues, Cyclic Nucleotides, and Calcium in Enzyme Secretion	I. Schulz, H. H. Stolze	42:127–56
Water-Soluble Vitamin Absorption in Intestine	R. C. Rose	42:157–71
The Intestine as a Fluid Propelling System	W. A. Weems	43:9–19
Patterns of Intestinal Motility	N. W. Weisbrodt	43:21–31
Intrinsic Neural Control of Intestinal Motility	J. D. Wood	43:33–51
Physiology of Mammalian Prevertebral Ganglia	J. H. Szurszewski	43:53–68
Stomach Blood Flow and Acid Secretion	P. H. Guth	44:3–12
Local Control of Intestinal Oxygenation and Blood Flow	A. P. Shepherd	44:13–27
Relationship Between Intestinal Blood Flow and Motility	C. C. Chou	44:29–42
Relationships Between Intestinal Absorption and Hemodynamics	D. Mailman	44:43–55
Physiological Regulation of the Hepatic Circulation	P. D. I. Richardson, P. G. Withrington	44:57–69
Measurement of Gastrointestinal Blood Flow	B. L. Tepperman, E. D. Jacobson	44:71–82

ENDOCRINES

Title	Authors	Citation
Sexual Differentiation	J. D. Wilson	40:279–306
Localization and Release of Neurophysins	S. M. Seif, A. G. Robinson	40:345–76
The Renin-Angiotensin System	I. A. Reid, B. J. Morris, W. F. Ganong	40:377–410
Endocrine Mechanisms of Parturition	P. W. Nathanielsz	40:411–45
Mechanism of Action of Hypothalamic Hormones in the Adenohypophysis	F. Labrie, P. Borgeat, J. Drouin, M. Beaulieu, L. Lagacé, L. Ferland, V. Raymond	41:555–69
Feedback Actions of Target Hormones on Hypothalamus and Pituitary, with Special Reference to Gonadal Steroids	G. Fink	41:571–85
Localization of Hypophysiotropic Peptides and Other Biologically Active Peptides Within the Brain	R. Elde, T. Hökfelt	41:587–602
Central Neurotransmitters and the Secretion of Prolactin, GH, LH, and TSH	L. Krulich	41:603–15
Actions of Hypothalamic-Hypophysiotropic Hormones on the Brain	R. L. Moss	41:617–31
The Biosynthesis and Metabolism of Prostaglandins	W. E. M. Lands	41:633–52
Cellular Mechanisms of Prostaglandin Action	R. H. Harris, P. W. Ramwell, P. J. Gilmer	41:653–68
The Role of Prostaglandins in the Central Nervous System	S. Wolfe, F. Coceani	41:669–84
Prostaglandins in Hypothalamo-Pituitary and Ovarian Function	H. R. Behrman	41:685–700
Dynamics of Steroid Hormone Receptor Action	B. S. Katzenellenbogen	42:17–35
The Role of Cyclic Nucleotides in Reproductive Processes	B. M. Sanborn, J. J. Heindel, A. G. Robison	42:37–57
Regulation of the Testis Sertoli Cell by Follicle Stimulating Hormone	A. R. Means, J. R. Dedman, J. S. Tash, D. J. Tindall, M. van Sickle, M. J. Welsh	42:59–70
Interactions of Steroids and Gonadotropins in the Control of Steroidogenesis in the Ovarian Follicle	P. C. K. Leung, D. T. Armstrong	42:71–82

CHAPTER TITLES 741

Mechanism of Action of Prolactin in the Control of Mammary Gland Function	R. P. C. Shiu, H. G. Friesen	42:83–96
Binding and Metabolism of Sex Steroids by the Hypothalamic-Pituitary Unit: Physiological Implications	B. S. McEwen	42:97–110
The Fetal Adrenal Gland	M. Serón-Ferré, R. Jaffe	43:141–62
Endocrine Physiology of the Placenta	E. R. Simpson, P. C. MacDonald	43:163–88
Extracellular Androgen Binding Proteins	C. W. Bardin, N. Musto, G. Gunsalus, N. Kotite, S.-L. Cheng, F. Larrea, R. Becker	43:189–98
Recent Advances in the Metabolism of Vitamin D.	H. F. DeLuca	43:199–209
Regulation of Parathyroid Hormone Secretion and Biosynthesis	J. F. Habener	43:211–23
Hormonal Control of Skeletal Growth	L. G. Raisz and B. E. Kream	43:225–38
Receptor-Mediated Endocytosis of Hormones in Cultured Cells	Ira H. Pastan, M. C. Willingham	43:239–50
Epidermal and Nerve Growth Factors in Mammalian Development	D. Gospodarowicz	43:251–63
"Endorphins" in Pituitary and Other Tissues	H. Imura, Y. Nakai	43:265–78
Behavioral Effects of Neuropeptides: Endorphins and Vasopressin	G. F. Koob, F. E. Bloom	44:571–82
The Role of the Central Nervous System in the Control of Ovarian Function in Higher Primates	C. R. Pohl, E. Knobil	44:583–93
Neuroendocrine Control Mechanisms and the Onset of Puberty	E. O. Reiter, M. M. Grumbach	44:595–613
The Evolution of Peptide Hormones	H. D. Niall	44:615–24
Post-Translational Proteolysis in Polypeptide Hormone Biosynthesis	K. Docherty, D. F. Steiner	44:625–38
Receptors for Peptide Hormones: Alterations in Diseases of Humans	J. Roth, S. I. Taylor	44:639–51
Polypeptide and Amine Hormone Regulation of Adenylate Cyclase	G. D. Aurbach	44:653–66
Calmodulin in Endocrine Cells	A. R. Means, J. G. Chafouleas	44:667–82

HEART

Biochemical Mechanism of the Sodium Pump	E. T. Wallick, L. K. Lane, A. Schwartz	41:397–411
Properties of Two Inward Membrane Currents in the Heart	H. Reuter	41:413–24
Electrogenesis of the Plateau and Pacemaker Potential	M. Vassalle	41:425–40
Propagation Mechanisms in Heart	N. Sperelakis	41:441–57
Cardiac Arrhythmias	P. F. Cranefield, A. L. Wit	41:459–72
Calcium and Cardiac Excitation-Contraction Coupling	A. Fabiato, F. Fabiato	41:473–84
Mitochondrial Function in the Heart	J. R. Williamson	41:485–506
Cardiac Heat Production	C. L. Gibbs, J. B. Chapman	41:507–19
Heart Muscle Mechanics	N. R. Alpert, D. D. Hamrell, L. A. Mulieri	41:521–38
Molecular Aspects of Cardiac Hypertrophy	R. Zak, M. Rabinowitz	41:539–52
Control of Energy Metabolism of Heart Muscle	T. C. Vary, D. K. Reibel, J. R. Neely	43:419–30
Permeable Junctions Between Cardiac Cells	E. Page and Y. Shibata	43:431–41
Neural Regulation of the Heart Beat	M. N. Levy, P. J. Martin, S. L. Stuesse	43:443–53
Cardiocirculatory Dynamics in the Normal and Failing Heart	R. Zelis, S. F. Flaim, A. J. Liedtke, S. H. Nellis	43:455–76

Myocardial Membranes: Regulation and Function of the Sodium Pump	T. Akera, T. M. Brody	44:375–88
Electrogenic Na Pumping in the Heart	H. G. Glitsch	44:389–400
Phosphorylation of the Sarcoplasmic Reticulum and Sarcolemma	M. Tada, A. M. Katz	44:401–23
The Slow Inward Calcium Current in the Heart	T. F. McDonald	44:425–34
Sodium-Calcium Exchange in the Heart	G. A. Langer	44:435–49
Calcium Release from Cardiac Sarcoplasmic Reticulum	S. Winegrad	44:451–62
The Action of Cardiotoxins on Cardiac Plasma Membranes	M. Lazdunski, J. F. Renaud	44:463–73
Adrenergic Receptors in the Heart	B. B. Hoffman, R. J. Lefkowitz	44:475–84
Mechanisms of Cardiac Arrhythmias	J. F. Spear, E. N. Moore	44:485–97

KIDNEY, WATER, AND ELECTROLYTE METABOLISM

Relationship of Renal Sodium and Water Transport to Hydrogen Ion Secretion	J. A. L. Arruda, N. A. Kurtzman	40:43–66
Studies of Isolated Renal Tubules in Vitro	J. J. Grantham, J. M. Irish III, D. A. Hall	40:249–77
Sugar, Amino Acid, and Na^+ Cotransport in the Proximal Tubule	K. J. Ullrich	41:181–95
Proton Secretion by the Kidney	D. Warnock, F. Rector	41:197–210
Rheogenic and Passive Na^+ Absorption by the Proximal Nephron	J. A. Schafer, T. E. Andreoli	41:211–27
Renal Nerves and Sodium Excretion	C. W. Gottschalk	41:229–40
Potassium Transport in the Nephron	G. Giebisch, B. Stanton	41:241–56
Renal Handling of Phosphate and Calcium	V. W. Dennis, W. W. Stead, J. L. Myers	41:257–71
The Vascular Organization of the Kidney	R. Beeuwkes III	42:531–42
Methods for Measuring Renal Blood Flow: Total Flow and Regional Distribution	K. Aukland	42:543–55
Distal Tubular Feedback Control of Renal Hemodynamics and Autoregulation	L. G. Navar, D. W. Ploth, P. D. Bell	42:557–71
Segmental Renal Vascular Resistance: Single Nephron	R. C. Blantz	42:573–88
Hormonal Systems and Renal Hemodynamics	P. G. Bauer, J. C. McGiff	42:589–601
Renal Hemodynamics in Acute Renal Failure	R. W. Schreier, J. Conger	42:603–14
Distribution of Hormone-Dependent Adenylate Cyclase in the Nephron and Its Physiological Significance	F. Morel, M. Imbert-Teboul, D. Chabardès	43:569–81
PTH, Calcitonin, Cyclic Nucleotides and the Kidney	Z. S. Agus, A. Wasserstein, S. Goldfarb	43:583–95
The Renal Kallikrein-Kinin and Prostaglandin Systems Interaction	A. Nasjletti, K. U. Malik	43:597–609
Antidiuretic Hormone	J. S. Handler, J. Orloff	43:611–24
Catecholamines and the Kidney: Receptors and Renal Function	P. A. Insel, M. D. Snavely	43:625–36
Steroid Hormones and the Kidney	D. D. Fanestil, C. S. Park	43:637–49
Recent Advances in Renal Morphology	R. E. Bulger, D. C. Dobyan	44:147–79
Heterogeneity of Tubular Transport Processes in the Nephron	C. A. Berry	44:181–201
Biological Importance of Nephron Heterogeneity	L. A. Walker, H. Valtin	44:203–19

MUSCLE

Skeletal Muscle Energetics and Metabolism	E. Homsher, C. J. Kean	40:93–131

CHAPTER TITLES 743

Cytoplasmic Fibers in Mammalian Cells: Cytoskeletal and Contractile Elements	R. D. Goldman, A. Milsted, J. A. Schloss, J. Starger, M.-J. Yerna	41:703–22
The Crossbridge Theory	R. T. Tregear, S. B. Marston	41:723–36
Filament Organization and Contractile Function in Vertebrate Smooth Muscle	R. A. Murphy	41:737–48
Phosphorylation of the Myofibrillar Proteins	M. Barany, K. Barany	42:275–92
The Relation of Muscle Biochemistry to Muscle Physiology	E. Eisenberg, L. E. Greene	42:293–309

NERVOUS SYSTEM
See CNS; COMPARATIVE PHYSIOLOGY

PANCREAS: ENDOCRINE FUNCTION

Insulin, Glucagon, and Somatostatin Secretion in the Regulation of Metabolism	R. H. Unger, R. E. Dobbs, L. Orci	40:307–43

PREFATORY CHAPTERS

Rhapsody in Science	P. F. Scholander	40:1–17
Mostly Membranes	K. S. Cole	41:1–24
Life With Tracers	H. H. Ussing	42:1–16
The Annual Review of Physiology: Past and Present	V. E. Hall, R. R. Sonnenschein	43:1–5

RESPIRATORY SYSTEM

Respiratory Adaptations in Sleep	E. A. Phillipson	40:133–56
Respiratory Mechanics	P. T. Macklem	40:157–84
Control of Mucus Secretion and Ion Transport in Airways	J. A. Nadel, B. Davis, R. J. Phipps	41:369–81
Fetal and Neonatal Pulmonary Circulation	A. M. Rudolph	41:383–95
Organization of Microcirculation in the Lung	J. Gil	42:177–86
Mechanical Influences on the Pulmonary Microcirculation	B. H. Culver, J. Butler	42:187–98
Nervous Control of the Pulmonary Circulation	S. E. Downing, J. C. Lee	42:199–210
Vasomotor Regulation of the Pulmonary Circulation	A. P. Fishman	42:211–20
Humoral Control of the Pulmonary Circulation	E. H. Bergofsky	42:221–33
Ventilation-Perfusion Relationships	P. D. Wagner	42:235–47
Sensory Inputs to the Medulla	A. I. Pack	43:73–90
Central Determinants of Respiratory Rhythm	M. I. Cohen	43:91–104
Anatomical Organizations of Central Respiratory Neurons	M. P. Kalia	43:105–20
Central Regulation of Respiration by Endogenous Neurotransmitters and Neuromodulators	F. L. Eldridge, D. E. Millhorn	43:121–35
Structural Bases for Metabolic Activity	U. S. Ryan	44:223–39
Processing of Endogenous Polypeptides by the Lungs	J. W. Ryan	44:241–55
Pulmonary Metabolism of Prostaglandins and Vasoactive Peptides	S. I. Said	44:257–68
The Fate of Circulating Amines within the Pulmonary Circulation	C. N. Gillis, B. R. Pitt	44:269–81
Chemotactic Mediators in Neutrophil-Dependent Lung Injury	J. C. Fantone, S. L. Kunkel, P. A. Ward	44:283–93

COMPARATIVE PHYSIOLOGY:
BODY TEMPERATURE

Hibernation: Cellular Aspects	J. S. Willis	41:275–86
Hibernation: Endocrinologic Aspects	J. W. Hudson, L. C. H. Wang	41:287–303

Hibernation: Neural Aspects	H. C. Heller	41:305–21
Temperature Regulation in Vertebrates	L. I. Crawshaw	42:473–91
The Role of Nervous Systems in the Temperature Adaptation of Poikilotherms	C. L. Prosser, D. O. Nelson	43:281–300
Brain Cooling in Endotherms in Heat and Exercise	M. A. Baker	44:85–96

ENDOCRINOLOGY

Insect Endocrinology: Regulation of Endocrine Glands, Hormone Titer, and Hormone Metabolism	L. I. Gilbert, W. E. Bollenbacher, N. A. Granger	42:493–510

ENERGETICS

Energetics and Mechanics of Terrestrial Locomotion	C. R. Taylor, N. C. Heglund	44:97–107
Avian Flight Energetics	J. M. V. Rayner	44:109–19
Energetics of Locomotion in Warm-Bodied Fish	E. D. Stevens, A. E. Dizon	44:121–31
Energetics of Locomotion in Endothermic Insects	J. E. Heath, M. S. Heath	44:133–43
Insect Endocrinology: Action of Hormones at the Cellular Level	L. M. Riddiford	42:511

ORIENTATION

Avian Orientation and Navigation	W. T. Keeton	41:353–66

RESPIRATION AND METABOLISM

Activity Metabolism of Lower Vertebrates	A. F. Bennett	40:447–69
Comparative Aspects of Vertebrate Cardiorespiratory Physiology	F. N. White	40:471–99
The Physiology of Circadian Pacemakers	M. Menaker, J. S. Takahashi, A. Eskin	40:501–26
Scaling of Physiological Processes in Homeothermic Animals	W. A. Calder III	43:301–22
Energy Metabolisms of Parasitic Helminths: Adaptations to Parasitism	H. J. Saz	43:323–41
Physiology of Diving in Marine Mammals	G. L. Kooyman, M. A. Castellini, R. W. Davis	43:343–56

REPRODUCTION

Reproductive Endocrinology of Fishes: Gonadal Cycles and Gonadotropin in Teleosts	R. E. Peter, L. W. Crim	41:323–35
Reproductive Endocrinology of Reptiles and Amphibians: Gonadotropins	P. Licht	41:337–51
Reproductive Endocrinology of Birds	D. S. Farner, J. C. Wingfield	42:457–72

ORDER FORM   ANNUAL REVIEWS INC.

Please list the volumes you wish to order. If you wish a standing order (the latest volume sent to you automatically each year), indicate volume number to begin order. Volumes not yet published will be shipped in month and year indicated. Prices subject to change without notice.

ANNUAL REVIEW SERIES — Prices Postpaid, per volume	Regular Order Please send:	Standing Order Begin with:
Annual Review of ANTHROPOLOGY Vols. 1–8 (1972–79): $17.00 USA; $17.50 elsewhere Vols. 9–10 (1980–81): $20.00 USA; $21.00 elsewhere Vol. 11 (avail. Oct. 1982): $22.00 USA; $25.00 elsewhere	Vol(s). _____	Vol. _____
Annual Review of ASTRONOMY AND ASTROPHYSICS Vols. 1–17 (1963–79): $17.00 USA; $17.50 elsewhere Vols. 18–19 (1980–81): $20.00 USA; $21.00 elsewhere Vol. 20 (avail. Sept. 1982): $22.00 USA; $25.00 elsewhere	Vol(s). _____	Vol. _____
Annual Review of BIOCHEMISTRY Vols. 28–48 (1959–79): $18.00 USA; $18.50 elsewhere Vols. 49–50 (1980–81): $21.00 USA; $22.00 elsewhere Vol. 51 (avail. July 1982): $23.00 USA; $26.00 elsewhere	Vol(s). _____	Vol. _____
Annual Review of BIOPHYSICS AND BIOENGINEERING Vols. 1–9 (1972–80): $17.00 USA; $17.50 elsewhere Vol. 10 (1981): $20.00 USA; $21.00 elsewhere Vol. 11 (avail. June 1982): $22.00 USA; $25.00 elsewhere	Vol(s). _____	Vol. _____
Annual Review of EARTH AND PLANETARY SCIENCES Vols. 1–8 (1973–80): $17.00 USA; $17.50 elsewhere Vol. 9 (1981): $20.00 USA; $21.00 elsewhere Vol. 10 (avail. May 1982): $22.00 USA; $25.00 elsewhere	Vol(s). _____	Vol. _____
Annual Review of ECOLOGY AND SYSTEMATICS Vols. 1–10 (1970–79): $17.00 USA; $17.50 elsewhere Vols. 11–12 (1980–81): $20.00 USA; $21.00 elsewhere Vol. 13 (avail. Nov. 1982): $22.00 USA; $25.00 elsewhere	Vol(s). _____	Vol. _____
Annual Review of ENERGY Vols. 1–4 (1976–79): $17.00 USA; $17.50 elsewhere Vols. 5–6 (1980–81): $20.00 USA; $21.00 elsewhere Vol. 7 (avail. Oct. 1982): $22.00 USA; $25.00 elsewhere	Vol(s). _____	Vol. _____
Annual Review of ENTOMOLOGY Vols. 7–25 (1962–80): $17.00 USA; $17.50 elsewhere Vol. 26 (1981): $20.00 USA; $21.00 elsewhere Vol. 27 (avail. Jan. 1982): $22.00 USA; $25.00 elsewhere	Vol(s). _____	Vol. _____
Annual Review of FLUID MECHANICS Vols. 1–12 (1969–80): $17.00 USA; $17.50 elsewhere Vol. 13 (1981): $20.00 USA; $21.00 elsewhere Vol. 14 (avail. Jan 1982): $22.00 USA; $25.00 elsewhere	Vol(s). _____	Vol. _____
Annual Review of GENETICS Vols. 1–13 (1967–79): $17.00 USA; $17.50 elsewhere Vols. 14–15 (1980–81): $20.00 USA; $21.00 elsewhere Vol. 16 (avail. Dec. 1982): $22.00 USA; $25.00 elsewhere	Vol(s). _____	Vol. _____
Annual Review of MATERIALS SCIENCE Vols. 1–9 (1971–79): $17.00 USA; $17.50 elsewhere Vols. 10–11 (1980–81): $20.00 USA; $21.00 elsewhere Vol. 12 (avail. Aug. 1982): $22.00 USA; $25.00 elsewhere	Vol(s). _____	Vol. _____
Annual Review of MEDICINE: Selected Topics in the Clinical Sciences Vols. 1–3, 5–15, 17–31 (1950–52, 1954–64, 1966–80): $17.00 USA; $17.50 elsewhere Vol. 32 (1981): $20.00 USA; $21.00 elsewhere Vol. 33 (avail. Apr. 1982): $22.00 USA; $25.00 elsewhere	Vol(s). _____	Vol. _____
Annual Review of MICROBIOLOGY Vols. 15–33 (1961–79): $17.00 USA; $17.50 elsewhere Vols. 34–35 (1980–81): $20.00 USA; $21.00 elsewhere Vol. 36 (avail. Oct. 1982): $22.00 USA; $25.00 elsewhere	Vol(s). _____	Vol. _____
Annual Review of NEUROSCIENCE Vols. 1–3 (1978–80): $17.00 USA; $17.50 elsewhere Vol. 4 (1981): $20.00 USA; $21.00 elsewhere Vol. 5 (avail. Mar. 1982): $22.00 USA; $25.00 elsewhere	Vol(s). _____	Vol. _____
Annual Review of NUCLEAR AND PARTICLE SCIENCE Vols. 9–29 (1959–79): $19.50 USA; $20.00 elsewhere Vols. 30–31 (1980–81): $22.50 USA; $23.50 elsewhere Vol. 32 (avail. Dec. 1982): $25.00 USA; $28.00 elsewhere	Vol(s). _____	Vol. _____
Annual Review of NUTRITION Vol. 1 (1981): $20.00 USA; $21.00 elsewhere Vol. 2 (avail. July 1982): $22.00 USA; $25.00 elsewhere	Vol(s). _____	Vol. _____

(continued on reverse)

Annual Review of PHARMACOLOGY AND TOXICOLOGY
  Vols. 1–3, 5–20 (1961–63, 1965–80): $17.00 USA; $17.50 elsewhere
  Vol. 21 (1981): $20.00 USA; $21.00 elsewhere
  Vol. 22 (avail. Apr. 1982): $22.00 USA; $25.00 elsewhere   Vol(s). _____ Vol. _____

Annual Review of PHYSICAL CHEMISTRY
  Vols. 10–21, 23–30 (1959–70, 1972–79): $17.00 USA; $17.50 elsewhere
  Vols. 31–32 (1980–81): $20.00 USA; $21.00 elsewhere
  Vol. 33 (avail. Nov. 1982): $22.00 USA; $25.00 elsewhere   Vol(s). _____ Vol. _____

Annual Review of PHYSIOLOGY
  Vols. 18–42 (1956–80): $17.00 USA; $17.50 elsewhere
  Vol. 43 (1981): $20.00 USA; $21.00 elsewhere
  Vol. 44 (avail. Mar. 1982): $22.00 USA; $25.00 elsewhere   Vol(s). _____ Vol. _____

Annual Review of PHYTOPATHOLOGY
  Vols. 1–17 (1963–79): $17.00 USA; $17.50 elsewhere
  Vols. 18–19 (1980–81): $20.00 USA; $21.00 elsewhere
  Vol. 20 (avail. Sept. 1982): $22.00 USA; $25.00 elsewhere   Vol(s). _____ Vol. _____

Annual Review of PLANT PHYSIOLOGY
  Vols. 10–31 (1959–80): $17.00 USA; $17.50 elsewhere
  Vol. 32 (1981): $20.00 USA; $21.00 elsewhere
  Vol. 33 (avail. June 1982): $22.00 USA; $25.00 elsewhere   Vol(s). _____ Vol. _____

Annual Review of PSYCHOLOGY
  Vols. 4, 5, 8, 10–31 (1953, 1957, 1959–80): $17.00 USA; $17.50 elsewhere
  Vol. 32 (1981): $20.00 USA; $21.00 elsewhere
  Vol. 33 (avail. Feb. 1982): $22.00 USA; $25.00 elsewhere   Vol(s). _____ Vol. _____

Annual Review of PUBLIC HEALTH
  Vol. 1 (1980): $17.00 USA; $17.50 elsewhere
  Vol. 2 (1981): $20.00 USA; $21.00 elsewhere
  Vol. 3 (avail. May 1982): $22.00 USA; $25.00 elsewhere   Vol(s). _____ Vol. _____

Annual Review of SOCIOLOGY
  Vols. 1–5 (1975–79): $17.00 USA; $17.50 elsewhere
  Vols. 6–7 (1980–81): $20.00 USA; $21.00 elsewhere
  Vol. 8 (avail. Aug. 1982): $22.00 USA; $25.00 elsewhere   Vol(s). _____ Vol. _____

SPECIAL PUBLICATIONS
Prices Postpaid, per volume   Regular Order Please send:

Annual Reviews Reprints: Cell Membranes, 1975–1977
  (published 1978)   Soft cover: $12.00 USA; $12.50 elsewhere   _____ copy(ies)

Annual Reviews Reprints: Cell Membranes, 1978–1980
  (published 1981)   Hardcover $28.00 USA; $29.00 elsewhere   _____ copy(ies)

Annual Reviews Reprints: Immunology, 1977–1979
  (published 1980)   Softcover $12.00 USA; $12.50 elsewhere   _____ copy(ies)

History of Entomology
  (published 1973)   Clothbound $10.00 USA; $10.50 elsewhere   _____ copy(ies)

Intelligence & Affectivity: Their Relationship During Child Development, by Jean Piaget
  (published 1981)   Hardcover $8.00 USA; $9.00 elsewhere   _____ copy(ies)

Telescopes for the 1980s
  (avail. Aug. 1981) Hardcover $27.00 USA; $28.00 elsewhere   _____ copy(ies)

The Excitement & Fascination of Science, Volume 1
  (published 1965)   Clothbound $6.50 USA; $7.00 elsewhere   _____ copy(ies)

The Excitement & Fascination of Science, Volume 2
  (published 1978)   Hardcover $12.00 USA; $12.50 elsewhere   _____ copy(ies)
                     Soft cover $10.00 USA; $10.50 elsewhere   _____ copy(ies)

To: ANNUAL REVIEWS INC, 4139 El Camino Way, Palo Alto, CA 94306 USA   (Tel. 415-493-4400)
Please enter my order for the publications checked above.

Amount of remittance enclosed $ _____   California residents, please add applicable sales tax.
Please bill me ☐   Prices subject to change without notice.
Institutional purchase order # _____

Name _____

Address _____

_____ Zip Code _____

Signed _____ Date _____

☐ Please send free copy of the current *Prospectus* each year.
☐ Send free brochure listing contents of recent back volumes for Annual Review(s) of _____

191464

DATE DUE